Thieme

Color Doppler Sonography in Gynecology and Obstetrics

Werner O. Schmidt, M.D.
Professor and Chairman
Department of Gynecology and Obstetrics
The Saarland University Hospital
Homburg, Germany

Asim Kurjak, M.D., Ph.D.
Professor
Department of Gynecology and Obstetrics
Sveti Duh Hospital
University of Zagreb
Zagreb, Croatia

With contributions by:

F. Bahlmann
G. Bernaschek
W. Braendle
B. Breyer
J. Deutinger
A. K. Ertan
M. Friedrich
A. Funk
U. Gembruch
T. Golaszewski
D. Grab
H.-J. Hendrik
G. Hetzel
T. Hitschold
M. Holländer
B. Hüneke
A. Kleinkauf-Houcken
J. König
R. Kubale
S. Kupesic
A. Kurjak
A. Lindinger
C. Lindner
H. Madjar
D. Mink
K. R. Reitnauer
H.-D. Rott
W. Schmidt
K. Sterzik
I. Tossounidis
K. Vetter
C. Villena-Heinsen
H.-J. Voigt
E. Weiss
T. Zodan

579 illustrations
90 tables

Thieme
Stuttgart · New York

Library of Congress Cataloging-in-Publication Data

Farbdopplersonographie in Gynäkologie und Geburtshilfe. English.
Color Doppler Sonography in gynecology and obstetrics / [edited by] Werner O. Schmidt, Asim Kurjak ; with contributions by F. Bahlmann ... [et al.] ; [translator, Terry Telger].

p. ; cm.
Includes index.
ISBN 3-13-130131-7 (GTV : alk. paper) -- ISBN 1-58890-179-3 (TNY : alk. paper)

1. Color Doppler ultrasonography. 2. Duplex ultrasonography. 3. Generative organs, Female--Ultrasonic imaging. 4. Breast--Cancer--Ultrasonic imaging. 5. Fetus--Ultrasonic imaging. 6. Ultrasonics in obstetrics.

[DNLM: 1. Genital Diseases, Female--ultrasonography--Atlases. 2. Pregnancy Complications--ultrasonography--Atlases. 3. Ultrasonography, Doppler, Color--methods--Atlases. 4. Ultrasonography, Prenatal--methods--Atlases. WP 17 F219 2005a] I. Schmidt, Werner, Prof. Dr. II. Kurjak, Asim. III. Bahlmann, F. IV. Title.

RG107.5.U4F3713 2005
618'.047543--dc22

2004022003

This book is an authorized and updated translation of the German edition published and copyrighted 2000 by Georg Thieme Verlag, Stuttgart, Germany. Title of the German edition: Farbdopplersonographie in Gynäkologie und Geburtshilfe.

Translator: Terry Telger, Fort Worth, Texas, USA

Illustrator: Angelika Kramer, Stuttgart, Germany

Rüdigerstraße 14
70469 Stuttgart, Germany
http://www.thieme.de
Thieme New York, 333 Seventh Avenue,
New York, NY 10001 USA
http://www.thieme.com

Cover design: Angelika Kramer, Stuttgart, Germany
Typesetting by Druckhaus Götz GmbH, D-71636 Ludwigsburg, gesetzt auf CCS Textline (Linotronic 630)
Printed in Germany by Grammlich, Pliezhausen

ISBN 3-13-130131-7 (GTV)
ISBN 1-58890-179-3 (TNY) 1 2 3 4 5

Important note: Medicine is an ever-changing science undergoing continual development. Research and clinical experience are continually expanding our knowledge, in particular our knowledge of proper treatment and drug therapy. Insofar as this book mentions any dosage or application, readers may rest assured that the authors, editors, and publishers have made every effort to ensure that such references are in accordance with **the state of knowledge at the time of production of the book.** Nevertheless, this does not involve, imply, or express any guarantee or responsibility on the part of the publishers in respect to any dosage instructions and forms of applications stated in the book. **Every user is requested to examine carefully** the manufacturers' leaflets accompanying each drug and to check, if necessary in consultation with a physician or specialist, whether the dosage schedules mentioned therein or the contraindications stated by the manufacturers differ from the statements made in the present book. Such examination is particularly important with drugs that are either rarely used or have been newly released on the market. Every dosage schedule or every form of application used is entirely at the user's own risk and responsibility. The authors and publishers request every user to report to the publishers any discrepancies or inaccuracies noticed. If errors in this work are found after publication, errata will be posted at www.thieme.com on the product description page.

Preface

Color Doppler sonography has opened new scopes for our specialty not just in perinatology but also in gynecology and oncology. This new technology has enabled the precise visualization and detailed examination of highly complex vascular structures.

The present book, which contains over 40 contributions from leading experts, is intended to present relevant information in an easy-to-read format that will be an authoritative guide in both routine and specialized diagnostic situations. Practical tips and guidelines will help the reader to use color Doppler ultrasound appropriately and effectively. Reference curves and documentation guidelines are integrated into the various chapters.

This book is being published at a time when the clinical relevance of color Doppler sonography has become an extremely topical issue. By providing a comprehensive review of the literature and a detailed subject index, the editors and contributing authors have tried to bring newcomers up to date on the research aspects and clinical applications of color Doppler imaging.

This book can also provide a valuable reference for interested colleagues who are preparing for national registry examinations.

The editors wish to thank their coauthors for their eager willingness to collaborate on this project. We also acknowledge the consistently prudent and constructive help of Dr. A. K. Ertan from our center, who coordinated the contributions, and of the staff of Thieme Medical Publishers. Their efforts helped to speed the book to its completion.

It was the explicit goal of the editors and contributors to provide the interested reader and ultrasound user with a reference that would be of substantial value for everyday clinical work.

Homburg, Saar and
Zagreb, Kroatien

Werner Schmidt
Asim Kurjak

Contributors

Franz Bahlmann, M.D.
Supervising Physician
Gynecological Clinic
University of Mainz
Mainz, Germany

Gerhard Bernaschek, M.D.
Professor
University Hospital of Vienna
Department of Prenatal Diagnosis
and Therapy
Vienna, Austria

Wilhelm Braendle, M.D.
Professor
University Hospital Eppendorf
Department of Gynecological
Endocrinology and Reproductive Sciences
Hamburg, Germany

Branko Breyer, M.D.
Professor
Zagreb, Croatia

Josef Deutinger, M.D.
Professor
Department of Prenatal
Diagnosis and Therapy
Vienna, Austria

A. Kubilay Ertan, M.D.
Supervising Physician
The Saarland University Hospital
Homburg, Germany

Michael Friedrich, M.D.
Supervising Physician
The Saarland University Hospital
Homburg, Germany

Andreas Funk, M.D.
Associate Professor
Director Department of Obstetrics
and Gynecology
District Hospital Nürtingen
Nürtingen, Germany

Ulrich Gembruch, M.D.
Professor
Department of Obstetrics
Medical Center of Obstetrics
Bonn, Germany

Thomas Golaszewski, M.D.
Specialist for Gynecology and Obstetrics
Amstetten, Austria

Dieter Grab, M.D.
Associate Professor and
Supervising Physician
Department of Gynecology
University of Ulm
Ulm, Germany

Hans-Joachim Hendrik, M.D.
Supervising Physician
The Saarland University Hospital
Department of Prenatal Diagnosis
Homburg, Germany

Gert Hetzel
Siemens AG
Sales Dept. Sonography
Erlangen, Germany

Thomas Hitschold, M.D.
Professor and Director
Gynecological Clinic
Municipal Hospital Worms
Worms, Germany

Martin Holländer, M.D.
Department of Gynecology
The Saarland University Hospital
Homburg, Germany

Bernd Hüneke, M.D.
Professor and Head
Department of Prenatal Medicine
Gynecological Clinic and Polyclinic
University Hospital Eppendorf
Hamburg, Germany

A. Kleinkauf-Houcken, M.D.
Specialist for Gynecology and Obstetrics
Hamburg, Germany

Jochem König, M.D.
The Saarland University Hospital
Institute for Medical Informatics
Homburg, Germany

Reinhard Kubale, M.D.
Associate Professor
Institute for Radiology, Sonography,
and Nuclear Medicine
Pirmasens, Germany

Professor Sanja Kupesic, M.D., Ph.D.
Sveti Duh Hospital
Department of Obstetrics
Zagreb, Croatia

Professor Asim Kurjak, M.D., Ph.D.
Sveti Duh Hospital
Department of Obstetrics
Zagreb, Croatia

Angelika Lindinger, M.D.
Professor
The Saarland University Hospital
Department of Pediatric Cardiology
Homburg, Germany

Christoph Lindner, M.D.
Professor
Gynecological Hospital
Elim-Hamburg
Hamburg, Germany

Helmut Madjar
Associate Professor
Specialist Gynecology
German Clinic for Diagnostics
Wiesbaden, Germany

Dieter Mink, M.D.
Associate Professor
The Saarland University Hospital
Department of Gynecology
Homburg, Germany

Karin Reitnauer, M.D.
The Saarland University Hospital
Institute of Pathology
Homburg, Germany

Hans-Dieter Rott, M.D.
Professor
University of Erlangen-Nuremberg
Institute of Human Genetics
Erlangen, Germany

Werner O. Schmidt, M.D.
Professor and Chairman
Department of Gynecology and Obstetrics
The Saarland University Hospital
Homburg, Germany

Karl Sterzik, M.D.
Professor
Praxisklinik Frauenstrasse
Ulm, Germany

Isaak Tossounidis, M.D.
Department of Gynecology
The Saarland University Hospital
Homburg, Germany

Klaus Vetter, M.D.
Professor
Department of Obstetrics
Neukölln Hospital
Berlin, Germany

Carlos Villena-Heinsen, M.D.
Associate Professor
Municipal Hospital Baden-Baden
Gynecological Clinic
Baden-Baden, Germany

Hans-Joachim Voigt, M.D.
Professor and Director
Gynecological Clinic
Westpfalz Klinikum GmbH
Kaiserslautern, Germany

Erich Weiss, M.D.
Associate Professor
District Hospital Böblingen
Gynecological Clinic
Böblingen, Germany

Tina Zodan, M.D.
Gynecological Clinic
Universitätsspital Zürich
Zurich, Switzerland

Contents

Physical and Technical Principles

1 Physical and Technical Principles of Color Doppler Sonography

R. Kubale and G. Hetzel

2 Safety Aspects of Doppler and Color Doppler Sonography

H.-D. Rott

Infertility Evaluation and Assisted Reproduction

3 Uterine Blood Flow in Fertile and Infertile Women

S. Kupesic and A. Kurjak

4 Uterine Causes of Infertility

S. Kupesic and A. Kurjak

5 Changes in Uterine and Ovarian Perfusion with the Onset of Menopause

A. Kurjak and S. Kupesic

6 Color Doppler Sonography for the Optimization of Assisted Reproduction

D. Grab and K. Sterzik

7 Pulsed Doppler and Color Duplex Sonography in the Assessment of Tubal Patency

B. Hüneke, A. Kleinkauf-Houcken, C. Lindner, and W. Braendle

8 Abnormalities of Corpus luteum Function

S. Kupesic, A. Kurjak, and T. Zodan

9 Interventional Ultrasound in Reproductive Medicine

S. Kupesic, A. Kurjak, and A. K. Ertan

Obstetric Ultrasound

10 Doppler Ultrasound Examinations from Folliculogenesis to Early Pregnancy

S. Kupesic, A. Kurjak, and W. Schmidt

11 Evaluation of Early Placentation and the Embryonic Circulation with Doppler Ultrasound

A. Kurjak and S. Kupesic

12 Color Doppler Sonography in Ectopic Pregnancy

H. J. Voigt

13 Physiology of Doppler Flow in Maternal Vessels during Pregnancy

T. Golaszewski, J. Deutinger, and G. Bernaschek

18 Chronic Placental Insufficiency

H. J. Hendrik, A. K. Ertan, and W. Schmidt

19 Severely Abnormal Doppler Findings and Perinatal Abnormalities

A. K. Ertan, H. J. Hendrik, and W. Schmidt

20 Fetal Doppler Findings in Late Pregnancy

K. Vetter

21 Intrapartum Fetal Heart Rate Changes and Doppler Sonography

E. Weiss

22 Color Doppler Ultrasound in Fetal Echocardiography

A. Lindinger

23 Use of Color Doppler in Echocardiography

U. Gembruch

24 Structure of the Human Placenta and Pathomorphological Changes in Placental Insufficiency

K. R. Reitnauer

25 Morphological and Morphometric Studies of the Placenta with Doppler Abnormalities of the Fetal Umbilical Arteries*

T. Hitschold

Gynecological Ultrasound

26 Sonographic and Doppler Sonographic Examination of Uterine Anomalies

S. Kupesic and A. Kurjak

27 Doppler Examination of the Normal Endometrium and Benign Endometrial Changes

S. Kupesic and A. Kurjak

28 Transvaginal Color Doppler and Pulsed Doppler Diagnosis of Benign Changes in the Uterine Myometrium

A. Kurjak and S. Kupesic

29 Use of Color Doppler in the Evaluation of Suspicious Endometrial Findings

M. Holländer, C. Villena-Heinsen, A. K. Ertan, and W. Schmidt

30 Malignant Uterine Tumors

A. Kurjak and S. Kupesic

31 Blood Flow Changes in Cervical Carcinoma Treated by Primary Chemotherapy

C. Villena-Heinsen, A. K. Ertan, M. Holländer, and W. Schmidt

32 Color Doppler Imaging of Benign Adnexal Masses—A Spectrum of Findings

A. Kurjak, S. Kupesic, and A. K. Ertan

33 Malignant Adnexal Tumors

A. Kurjak, S. Kupesic, and A. K. Ertan

34 Three-Dimensional Power Doppler Sonography in Evaluating the Angiogenesis of Ovarian Tumors

A. Kurjak, S. Kupesic, and B. Breyer

35 Ovarian Cancer Screening

A. Kurjak, W. Schmidt, A. K. Ertan, and S. Kupesic

36 General Aspects of the Ultrasound Investigation of Blood Flow in Breast Tumors

C. Villena-Heinsen, M. Holländer, and W. Schmidt

37 Color Doppler Sonography in the Diagnosis of Breast Cancer

H. Madjar

38 Usefulness of the Minimum Resistance Index in the Benign–Malignant Discrimination of Breast Tumors

C. Villena-Heinsen, M. Friedrich, J. König, and W. Schmidt

39 Effect of Menopausal Status and Hormone Replacement on Therapy Resistance Indices and Blood Flow Velocities in Breast Tumors

C. Villena-Heinsen, I. Tossounidis, and W. Schmidt

40 Benign–Malignant Tumor Discrimination and Prognostic Evaluation of Breast Tumors with Color Doppler Sonography

C. Villena-Heinsen, A. K. Ertan, D. Mink, and W. Schmidt

Abbreviations

5-HETE	hydroxyeicosatetraenoic acid
ABCD profile	amniotic fluid index, biometry, kinetochardiography, Doppler examination
AEDF	absent end-diastolic flow
AFI	amniotic fluid index
AGA	appropriate for gestational age
AR	autoregression
ARED	absent or reverse end-dialostic flow
ASD	atrial septal defect
AV	atrioventricular
BI	balance index
bpm	beats per minute
CATCH 22	cardiac defects, abnormal facies, thymic hypoplasia, cleft palate, hypocalcemia, and chromosome 22
CDS	color duplex sonography
CFM	color-flow mapping
CGRP	calcitonin gene-releated peptide
CIN	intraepithelial neoplasia of the cervix
CRL	crown-rump length
CT	computed tomography
CTG	cardiotocography
CVI	color velocity imaging
CW	continuous wave
DA	ductus arteriosus
EDRF	endothelial-derived relaxing factor
EEG	electroencephalogram
EFI	extended amniotic fluid index
EFSUMB	European Federation of Societies for Ultrasound in Medicine and Biology
EP	ectopic pregnancy
ERF	endothelial relaxing factor
ES	extrasystole
EST	erythrocyte sedimentation rate
ET	embryo transfer
FFT	fast Fourier transfer
FHR	fetal heart rate
FIGO	International Federation of Gynecology and Obstetrics
FIP	frequency index profile
FNA	fine-needle aspiration
FSH	follicle-stimulating hormone
GIFT	gamete intrafallopian transfer
GnRH	gonadotropin-releasing hormone
GRIT	Growth Restriction Intervention Trial
H&E	hematoxylin and eosin
hCG	human chorionic gonadotrophin
hMG	human menopausal gonadotrophin
HPL	human placental lactogen
HR	heart rate
HRT	hormone replacement therapy
HSG	hysterosalpingography
Hy-Co-Sy	hysterosalpingo-contrast-sonography
ICA	internal carotid artrey
ICSI	intracytoplasmic sperm injection
ICU	intensive care unit
I_{SPTA}	spatial peak time average intensity
IUD	intrauterine device
IUGR	intrauterine growth retardation
IUI	intrauterine insemination
IVF	in-vitro fertilization
IVF-ET	in-vitro fertilization and embryo transfer
KCTG	kinetocardiotocography
LH	luteinizing hormone
LMP	last menstrual period
LPD	luteal phase defect/deficiency
LUF syndrome	luteinized but unruptured follicle syndrome
MAGS	microscopic angiogenesis grading system
MCA	middle cerebral artery
MEM	maximum entropy method
MI	mechanical index
MIP	maximum intensity projection
MRI	magnetic resonance imaging
Mv_{max}	peak systolic velocity
NC	nuchal cord
NS	not significant
OAPR	overall predictive ratio
ODS	Output Display Standard
PAS	periodic acid-Schiff
PCO	polycystic ovary (syndrome)
$PGF_{2\alpha}$	prostaglandin $F_{2\alpha}$
PI	pulsatility index
PID	Pelvic inflammatory disease (PID)
PIH	pregnancy-induced hypertension
PIV	pulsatility index for veins
Po_2, Pco_2	oxygen [carbon dioxide] partial pressure
PPO	postmenopausal palpable ovaries
PRF	pulse repetition frequency
PROST	pronucleus-stage transfer
PVIV	peak velocity index for veins
PW	pulsed wave
PZD	partial zona dissection
RI	resistance index
ROC	receiver operating characteristic
ROI	region of interest
SGA	small-for-gestational-age
SUZI	subzonal insemination
TASAV	temporal average of spatial average velocities
TEM	transparent energy mode
TET	intratubal embryo transfer
TGC	time gain compensation
THI	tissue harmonic imaging
TI	thermal index
TKD	tocodynamometry
TORCH	toxoplasmosis, rubella, cytomegalovirus, and herpes
TVCD	transvaginal color Doppler sonography
UA	umbilical artery
VSD	ventricular septal defect
ZIFT	zygote intrafallopian transfer

Physical and Technical Principles

1 Physical and Technical Principles of Color Doppler Sonography

R. Kubale and G. Hetzel

Historical Development

Duplex sonography and its offshoot, color duplex sonography, are relatively new methods that combine the pulsed echo technique of sectional image formation with the Doppler evaluation of blood flow. It is based on B-mode ultrasound, which permits an accurate morphological description of gynecological and obstetric findings as a result of increasingly higher resolution and faster frame rates. The first attempts to measure blood flow date back to Satomura[12]. It was not until the early 1970s that Pourcelot[11] was able to use continuous wave (CW) Doppler to investigate cerebrovascular diseases. Since 1977, authors have described Doppler ultrasound applications in the abdomen and especially in obstetrics and gynecology[3, 13, 16]. During the past 15 years, these applications have become widely adopted and established through the use of color-flow techniques and ongoing refinements in instrumentation[10].

Below we shall explore the physical and technical principles that are essential for optimally utilizing and interpreting B-mode and color Doppler sonography and avoiding misinterpretations.

B-Mode Sonography

Physical Principles of Echo Production

The ultrasound wave is a density and pressure wave that propagates through a medium, its behavior essentially conforming to the laws of sound and optics. Sound attenuation, sound reflection, and resonance phenomena can be utilized as means of identifying defects in the insonated medium. The pulse-echo principle is most commonly used in medicine and will be discussed below in greater detail.

Acoustic impedance. The echoes that are displayed in an ultrasound image are based on reflections that occur when the sound wave travels through media of different acoustic impedance. Acoustic impedance (Z) is a material-dependent (i.e. tissue-dependent) quantity that is the product of the density of the material (ϱ) and the velocity of sound in it (c). For the time being, we shall assume that this velocity is a material- and tissue-independent constant (equation 1) (see New Technical Processes and Approaches, p. 25).

$$Z = c \times \varrho \qquad (1)$$

Because the sound wave consists of alternating high and low pressure peaks, the tissue that transmits the wave is transiently compressed, and the sound velocity changes: the "peaks" at a higher pressure move faster than the "valleys" at a low pressure. This phase-dependent change in sound velocity distorts and steepens the original sine wave, analogously to a water wave breaking on the shore. As a result, the reflected wave acquires a component that is a multiple of the emitted basic frequency (harmonic energy). These frequencies, which create unwanted signals in the conventional B-mode image (see Principle of Echo Detection and Scanning Techniques, p. 3), can be utilized for specific "harmonic" imaging. This technique, called tissue harmonic imaging (THI), makes it possible to examine even patients who are technically difficult to scan (see New Techniques of Signal Acquisition and Processing, p. 25).

The acoustic impedance of tissues has a value similar to that of water. The exact value varies with the composition of the tissue (Table 1.1). Note that the acoustic impedances of soft tissues are markedly different from the impedances of air and bone.

Table 1.1 Impedance values and degree of reflection in biological tissues

Medium	Sound velocity c (m/s)	Specific density ϱ (g/cm^3)	Impedance Z (g cm^{-2} s^{-1})	Reflection factor R (relative to H_2O)
Water	1496	0.997	1.491	0.000
Fat	1476	0.928	1.370	0.042
Muscle	1568	1.058	1.660	0.054
Liver tissue	1570	1.055	1.660	0.054
Bone	3360	1.850	6.200	0.614
Air	331	0.0012	0.3972	0.999

Impedance (Z) is defined as the product of the specific density (ϱ) of the tissue and the sound velocity (c). An incident ultrasound beam is partially reflected at interfaces between tissues of different impedance. The degree of the reflection is measured by the reflection factor R in relation to water.

Reflection. Reflection occurs at every interface between two media that have different acoustic impedances. The reflected portion of the ultrasound wave (the echo) increases with the magnitude of the acoustic impedance difference. The amount of reflection that occurs at the interface between two different tissues with impedances Z_{tissue1} and Z_{tissue2} is described by the reflection factor R (Table 1.**1**) as given by equation (2):

$$R = \frac{Z_{\text{tissue 1}} - Z_{\text{tissue 2}}}{Z_{\text{tissue 1}} + Z_{\text{tissue 2}}} \qquad (2)$$

If the impedance mismatch between the tissues is small, the transmitted ultrasound wave will retain enough energy to produce additional echoes in deeper tissue layers. But when a large impedance mismatch exists, such as tissue–air or tissue–bone interfaces, almost all of the incident sound energy is reflected. Objects located behind such an interface cannot be visualized with ultrasound.

Refraction. Reflection is strongly angle-dependent and is subject to the same laws as optical refraction. If the sound strikes the interface between different tissues at a perpendicular angle, most of the reflected wave will return to the transducer. But if the sound strikes the interface at an oblique angle, only part of the reflected wave will reach the transducer, and the wave will deviate from its original direction as it continues on through the second medium (refraction).

Scattering. If the sound beam encounters a rough surface or small reflectors whose diameters are significantly smaller than the wavelength of the sound, the beam will be reflected or "scattered" in various directions. This occurs not just at interfaces but everywhere in the tissue. Because of scattering, homogeneous soft tissue is depicted as an interference pattern with a coarse or fine texture, depending on the transducer frequency (Speckle).

Absorption and attenuation. In addition to losses due to reflection and scattering, some of the ultrasound energy is absorbed in a frequency-dependent fashion as it overcomes cohesion and relaxation forces in the medium. The amplitude of the sound pressure declines exponentially with the thickness of the medium traversed—i.e., it is attenuated. Attenuation refers to the relationship of the initial sound pressure to its pressure after the sound has propagated for a given distance. It depends on the distance traveled by the sound pulse, the frequency of the transducer, and a material-specific constant. The following simplified formula (equation 3) is valid for soft tissues and for frequencies from 0.2 MHz to 100 MHz:

$$\text{Attenuation (dB)} = \text{frequency (MHz)} \times \text{distance traveled (cm)} \qquad (3)$$

Principles of Ultrasound Instrumentation

Principle of Echo Detection and Scanning Techniques

Ultrasound pulses are generated and processed by means of piezoelectric elements that are assembled into a transducer. In principle, a transducer is both a transmitter and a receiver of sound. An electric pulse (pulse length 0.5–2 wave trains of 1 μs duration) excites the active elements of the transducer. The resulting mechanical vibration propagates through the tissue. The mechanical vibration is reflected from a target object, and the returning sound wave induces the transducer elements to generate an electric signal. The time (t) between the emission of the pulse and the reception of the echo is a measure of the distance (z) of the transducer elements from the reflecting object (pulse-echo principle) (equation 4):

$$z = {}^{1}/_{2}\, ct \qquad (4)$$

where c is the velocity of sound and t is the time from pulse transmission to echo reception.

B-mode image. A two-dimensional sectional image is produced when a number of adjacent ultrasound beams are transmitted and received in the same plane. In the B-mode technique (for "brightness mode"), the echoes are displayed at a brightness that is proportional to their amplitude. The brightness-modulated signals that are received (ultrasound lines) are temporarily stored in a matrix. The contents of this matrix are transferred to a monitor at correlative sites and assembled into a geometrically correct ultrasound sectional image (Fig. 1.**1**).

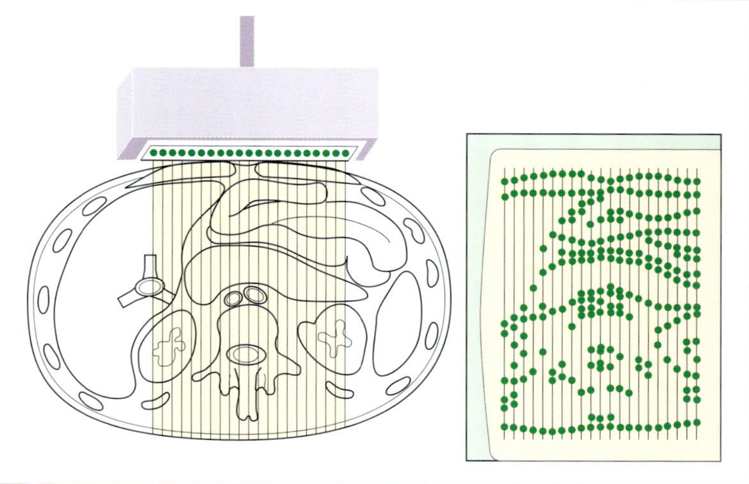

Fig. 1.**1** Production of an ultrasound image, illustrated for a linear array transducer. Transverse abdominal scan through the liver, kidneys, pancreas, and spine. When a pulse is transmitted, portions of the ultrasound energy are reflected from the organ surfaces and from interfaces within the organ parenchyma. These echoes are assigned to specific locations based on the transit time for a known sound velocity. The sum of the brightness-modulated points yields a sectional image in which the liver, aorta, pancreas, and kidneys can be evaluated.

The first ultrasound imagers were based on the manual movement of a single transducer over the patient (compound scan). These were initially replaced by mechanical transducers in which an element was rotated to produce a fan-shaped beam pattern.

Transducer arrays. Most scanners today employ transducer arrays consisting of individual elements arranged in a closely spaced row. Each array has a certain number of elements, in some cases more than 196. The elements are fired (excited) in groups, each of which transmits and receives an ultrasound line. Starting on one side of the array, the edge line is acquired first. Adding an element on one side of the group and turning off an element on the other side, the active group is shifted across the array. As the next group transmits and receives, the tissue is scanned in a sequential fashion. As this process is continued, a sectional image is produced. If the elements in the array are arranged in a straight row, the transducer is classified as a linear array. If the elements are arranged along a curve, the unit is called a curved or convex array (Fig. 1.**2**).

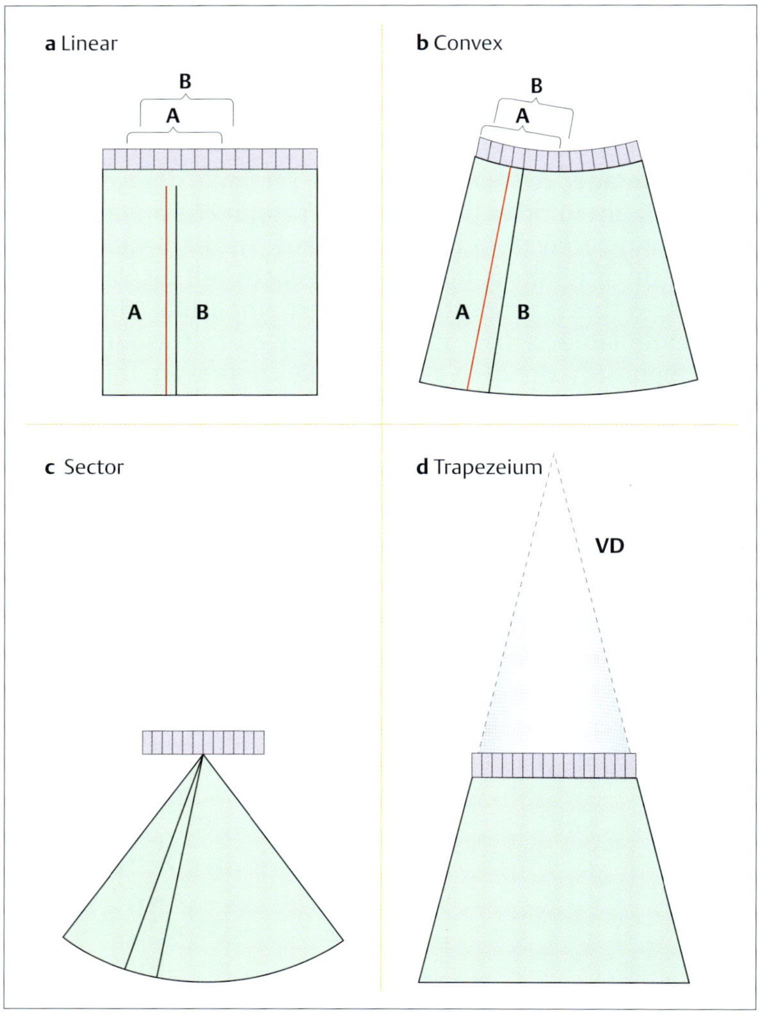

Fig. 1.**2** Functional principles of different transducer types.
a In a linear array, the elements are fired in offset groups at times A and B, producing a rectangular beam pattern.
b In a convex array, the transducer has a curved surface resulting in a fan-shaped beam.
c Sector scan with a phased-array transducer. All the elements are active.
d Trapezoid scan. This unit is operated like a phased array with a "virtual pivot point" (VD) located behind the transducer.

Electronic beam steering. A phased array scanner can scan the tissue in a sector- or trapezoid-shaped pattern. All of the elements in the array are active (group width m equals the number of elements n). A sector-shaped scan is produced by electronic beam steering (Fig. 1.**3**). The wavefront is steered by imposing time delays on the excitation pulses that are applied to the elements in the array. In this process the individual narrow elements (of width approximately 0.5λ; λ = wavelength) behave like "Huygen point emitters," meaning that each element is the source of a spherical wave in relation to the scan plane. All of the spherical waves add together to form a wavefront that is angled away from the array as determined by the time delay (Fig. 1.**3**). To accomplish electronic steering in an array, the elements must be spaced sufficiently close together, i.e., there must be a sufficiently fine element pitch or high element density. The individual elements must be so narrow that their directional characteristic is broad enough to allow adequate lateral signal transmission. The returning echoes must not be extinguished by the width of an individual element.

Recent technical developments include transducers with a trapezoidal beam pattern. The "trapezoid scan" is produced by sweeping a linear-array image field and represents the special case of a sector scan with a virtual pivot point located behind the transducer surface (Fig. 1.**2**). Only linear arrays that have a high element density can produce a trapezoid scan (see New Developments in Transducer Technology, p. 23).

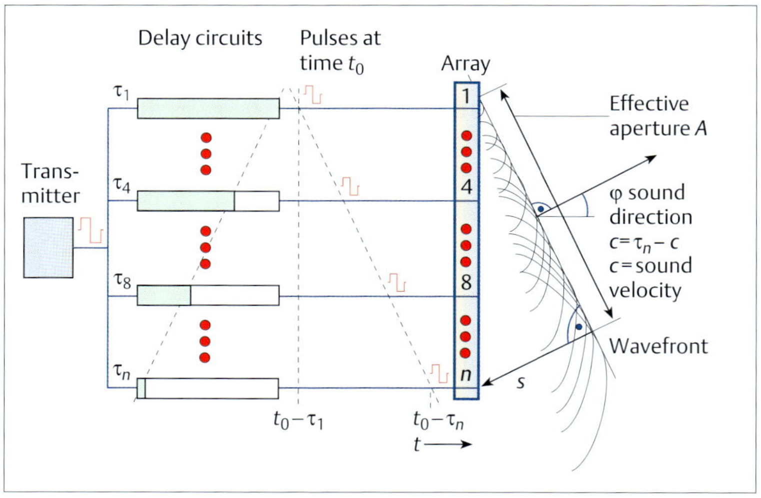

Fig. 1.**3** Phased array with electronic beam steering. The phased array transducer uses a variable electronic time delay (τ_i) to change the direction of the beam. If the time delay τ_i is greater than τ_n, the excitation pulses in the transducer will reach element n of the array first, before reaching element 1. The superimposed wave segments will combined to form a wavefront that is steered away from the transducer axis by the angle φ, depending on the time delay.

Transducer Selection: Penetration Depth, Beam Pattern, and Resolution

The selection of a transducer or scanning technique (see Principle of Echo Detection and Scanning Techniques, p. 4) is guided by the nature of the desired information and by anatomical considerations. Important decision-making criteria are the size of the region under study, the necessary scanning depth, and the necessary resolution.

Field-of-view width and penetration depth. Since the linear array produces a wide near field, this type of array is preferred for scanning superficial regions. The convex array has a broader far field with good near-field resolution, making it an ideal compromise for abdominal scanning. The phased array has a sector format that is particularly useful for difficult imaging conditions, as in echocardiography for scanning deeper structures. The trapezoid scan has advantages similar to the convex scan in terms of good near-field resolution and far-field width. An additional advantage, as in the phased array, is a small, straight transducer face that is easier to couple to the body surface for numerous applications.

Penetration depth and transmission frequency. The penetration depth is the maximum distance between the sound transmitter (transducer) and the deepest site in the tissue from which a reflection (echo) can still be acquired. Owing to the frequency-dependent absorption of ultrasound in tissue, the penetration depth is not only tissue-dependent but also frequency-dependent. For this reason, the selected transmission frequency is a tradeoff between penetration depth and desired resolution and thus depends on the nature of the study. High frequencies are well suited for the high-resolution imaging of superficial structures, while low frequencies are better for imaging deeper structures. The maximum penetration depth is inversely proportional to the transmission frequency.

Resolution. Resolution defines the smallest distance between two point reflectors that can still be recognized as separate points. Axial resolution denotes resolution along the axis of the beam. Lateral resolution describes resolution at right angles to the beam axis (Fig. 1.**4**).

Axial resolution. Axial resolution is determined chiefly by the spatial length of the emitted ultrasound pulse. Short pulse lengths combined with a broadband transducer (see below) provide the best axial resolution. Typical values for axial resolution are 2λ (λ = wavelength). Consequently, the axial resolution is usually better than the lateral resolution and increases with the ultrasound frequency.

Lateral resolution and beam pattern. Lateral resolution is one of the basic parameters for describing the quality of image formation and affects the diagnostic performance of the system as a whole. Generally it is in the range of 3–5 wavelengths and depends on the ultrasound beam pattern.

Figure 1.**5** shows a beam pattern in which all of the intensities are normalized to the maximum sound pressure. It is easier to analyze resolution in this type of diagram. Owing to interference effects, even a flat transducer produces a con-

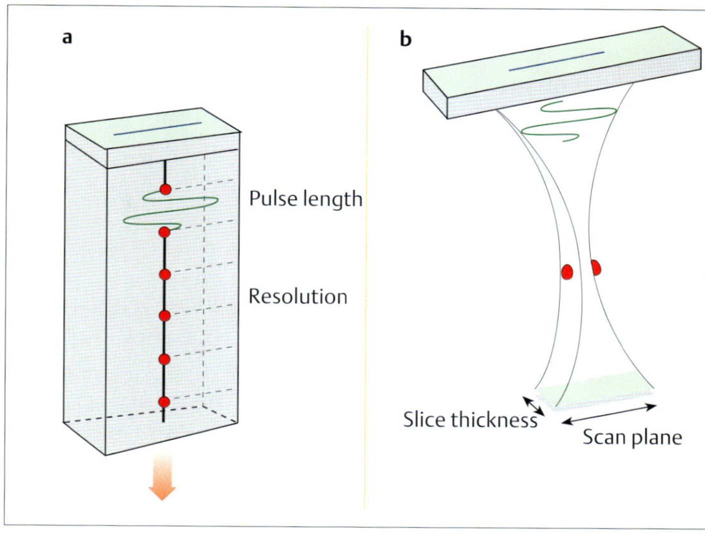

Fig. 1.**4** Definition of axial and lateral resolution.
a Two interfaces or point reflectors located at different depths along the beam axis can still be recognized as separate structures if the distance between them is greater than one-half the spatial length of the emitted pulse. Despite depth-dependent and frequency-dependent tissue absorption with an increasing shift of the echoes toward lower frequencies, the axial resolution of the image remains fairly constant.
b Lateral resolution depends on the varying width of the ultrasound beam with increasing depth. Two adjacent point reflectors in the beam can still be identified as separate points if the distance between them is greater than the "6-dB beam width" at the depth of interest. This parameter is useful for describing the spatial resolution of two points in the transducer plane and for estimating the effect of point reflectors or interfaces that are directly adjacent to the imaged surface in the longitudinal direction (scan plane–thickness phenomenon).

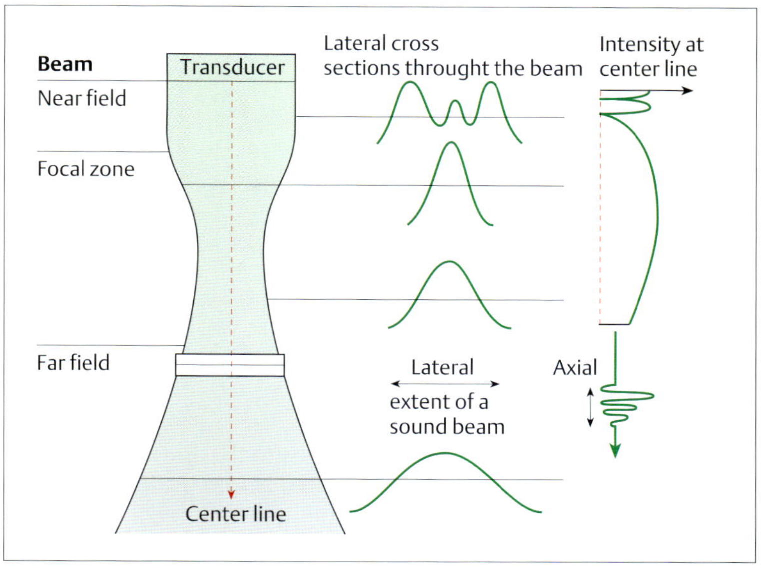

Fig. 1.5 Near field, far field, and focal zone of an ultrasound beam (the area insonated by an emitted ultrasound pulse). The near field has an inhomogeneous interference structure that greatly limits the image quality. Resolution is best in the focal zone, where the beam is constricted. As the beam diverges again in the far field, there is an increasing deterioration of lateral resolution.

stricted beam pattern with a natural focus. The lateral resolution is best within the focal zone of the transducer. The depth of the focal zone is always frequency-dependent. The focal zone Z_f of a flat transducer (with a natural focus) is defined by equation (5):

$$Z_f = k A^2 / \lambda \quad (5)$$

where A is the aperture, or the active transducer length used for transmission and reception, and λ is the frequency-dependent wavelength.

The boundary between the focal zone of the transducer and the near field, which is marked by inhomogeneities (extinction phenomena), is defined as twice the value of the lateral resolution at the focus. This near-field portion of the image is difficult or impossible to interpret because of artifacts. The far field begins at twice the value of the lateral resolution at the focus. The focus can be moved toward the transducer, thereby improving resolution, by the use of acoustic lenses, such as ones made of silicone rubber, by curving the transducer face (mechanical focusing), or by electronic focusing. For a given frequency, it is advantageous to use the largest possible aperture with suitable focusing. The advantages of a large aperture are greater penetration depth and better lateral resolution. The aperture can be reduced during reception by modifying the number of active elements (dynamic aperture) to minimize interference in the near field.

Resolution in the *z* axis. A third, frequently neglected, parameter is the resolution in the plane perpendicular to the transducer (*z*-axis resolution). This parameter is described by the slice thickness from which reflectors and scatterers located outside the transducer plane can generate an echo in the transducer plane (see Artifacts and Pitfalls, p. 8). The slice thickness focus of a given transducer is dependent on the transmission frequency.

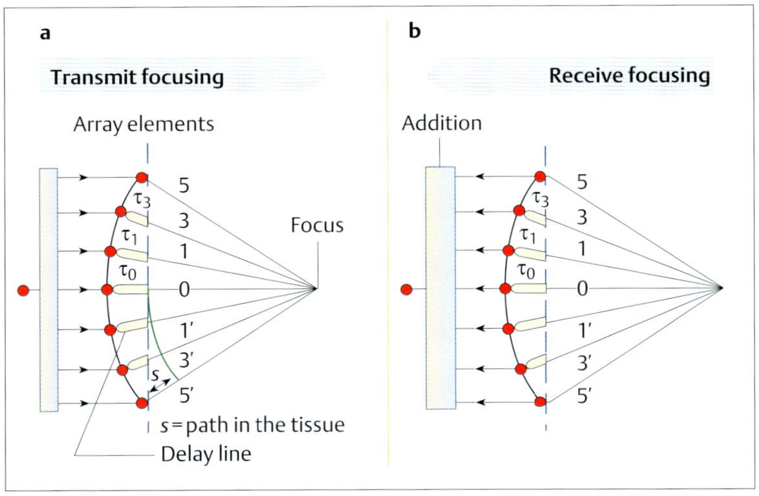

Fig. 1.**6** Transmit and receive focusing.
a In transmit focusing, the excitation pulses to the array elements are delayed such that the outer crystals are fired before the inner ones, allowing the beam to be focused at different depths.
b In receive focusing, echoes returning from the point designated as the focus arrive at element 0 first, where the return path is the shortest. If the echo-induced signals from that element are delayed by time τ_0, they can be time- and phase-matched to the signals from elements 5 and 5′. If all of the echo-induced signals are delayed according to the differences in their return times, all the elements will yield coherent signal components, so that adding the delayed signals will give an optimum echo display for the focal point.

Signal Processing: Focusing, Preprocessing, and Postprocessing

Focusing. Every type of transducer that is to be connected to an ultrasound imaging system requires special front-end circuitry that contains a transducer-specific interface. This circuitry includes all the components that link the transducer elements to the processing channels. Most of the front-end electronics is for transmit and receive focusing. As noted above, the beam of a flat transducer converges to a natural focus. While this focus can be modified by acoustic lenses, true variable adjustment of the depth of the focal zone can be achieved only by electronic means. In all types of arrays, focusing in the scan plane is accomplishing by means of variable electronic delay circuitry. Focusing can be done in both the transmit and receive modes (Fig. 1.**6**).

Receive focusing. In receive focusing, the echo signals that return to the array first are delayed by the difference in arrival times from the more peripheral signals. The goal is to ensure a coherent display of all signals that are received from the variable focal point.

Transmit focusing. The transmitted beam can be focused by delaying the excitation pulses in the array. Multizone focusing can be achieved only by scanning sequentially at different focal depths, since the focus of the wavefront can no longer be changed after the pulse has left the array. While transmit focusing allows the user to define the number and location of the focal zones, multizone focusing results in a slower frame rate. For example, three focal zones triple the time required to generate an image, and so the frame rate is reduced by a factor of 3.

Dynamic focusing. Other focusing possibilities are available in the receive mode. By varying the delay times during reception of the echo signals, it is possible to keep the focus at the depth from which echoes are momentarily being received. This process, called dynamic focusing, does not reduce the frame rate and operates automatically throughout echo reception, without user input, according to an optimized algorithm. The electronic focusing described above is a feature of all electronic transducers.

Time gain compensation. Preprocessing refers to the part of the signal processing circuit in which scan data are manipulated before they reach the scan converter. It begins with the removal of unwanted effects caused by the sound attenuation and center frequency changes that occur with increasing depth. Echoes from deeper objects are attenuated more than shallower echoes on their way to the transducer. Time gain compensation (TGC) is a preprocessing technique that compensates for this attenuation by amplifying the echoes from more distant reflectors as a function of their depth. Otherwise the echo signals from deeper sources would appear increasingly dark. Generally the TGC is set manually, because "homogeneous image appearance" and "region of interest" are not quantities that can be preprogrammed. One way to optimize the setting is to take a subcostal transverse scan of the liver, for example, and then adjust the TGC curve until the brightness of the liver appears homogeneous. If scanning must be done through fluid (e.g., ascites) or through the uterus, the TGC should be reduced accordingly to avoid tissue effects of overwhelming brightness.

Dynamic frequency filtering. Another preprocessing step is dynamic frequency filtering. The frequency-dependent attenuation of the sound, with a depth-dependent spectral shift toward lower frequencies, makes this a necessary step. Concurrent dynamic frequency filtering compensates for this shift effect and improves the signal-to-noise ratio in deeper regions.

Multifrequency and broadband technique. For these techniques to provide useful image improvement, the transducer must have the capacity for broadband excitation and broadband signal processing. Other advantages of broadband transducers are a shortened pulse length for better axial resolution (see Fig. 1.**4**) and the capacity for multifrequency excitation. In this way a single transducer can be used to image superficial structures at high resolution as well as structures located at greater depths. The multifrequency technique and broadband technique significantly expand the performance range of transducers and simplify the examination.

Edge enhancement. Other steps are demodulation and low-pass filtering and the ability to modify the shape of the echo signal in preprocessing. Edge enhancement, for example, can be used to steepen the slope of a change in signal level, thus providing better differentiation of echoes across an interface. (Edge enhancement is accomplished by high-pass filtering of the demodulated echo signals and is applied only to line-of-sight data.) Tissues appear to have a more fine-grained texture, and interfaces perpendicular to the beam direction (e.g., vessel wall boundaries) are more clearly defined. The effect, then, is an apparent improvement of axial resolution.

Dynamic range. The dynamic range makes it possible to match the contrast level of the ultrasound image to diagnostic requirements. From the large range of echo intensities that can be handled by modern systems, the user can select the dynamic range from the smallest to the brightest visible B-mode echoes that appears best for the current application. A smaller dynamic range is useful for defining contours, measuring the cranial biparietal diameter, and vascular imaging. A higher dynamic range is useful for evaluating the structure of tissues like the placenta, as this will define even the faintest tissue echoes.

Scan converter. The function of the scan converter is to convert the data acquired by various scanning techniques (e.g., in polar coordinates) into a correct spatial display pattern (e.g., in cartesian coordinates) on the viewing screen. Because data readout from the scan converter follows a television format (video) or computer format (e.g., SVGA), the ultrasound information requires intermediate (digital) storage on or before reaching the scan converter.

Correlation of the B-mode image. "Correlation" of the B-mode image is an averaging process that is carried out in the image memory. The time-averaging process can be structured in such a way that when new information is entered into memory, a portion of the stored (old) information from the previous image is retained in memory, and the new value is added to it. The effect is similar to "afterglow" (persistence) on the monitor screen. Correlation can be useful in slowly changing image sequences (e.g., in the abdomen, peripheral vessels, etc.), where it improves the signal-to-noise ratio. With fast-moving processes (e.g., fetal cardiac imaging), setting the correlation too high can cause a smearing of structures with loss of image quality.

Postprocessing. In postprocessing, the gray-scale depiction of the image information on the monitor can be manipulated by altering the data transfer characteristics between the scan converter and the monitor. This may be done, for example, to enhance or suppress low-level echoes in the final image.

Analysis of B-Mode Information and Artifacts

Optimizing the Examination and Analyzing the B-Mode Information

Homogeneous display of echo texture. After selecting the transducer that is appropriate for the desired information and the region under study (see Transducer Selection: Penetration Depth, Beam Pattern, and Resolution, p. 5), the user adjusts the transmission power output, focal depth, receiver gain, and TGC curve to obtain a homogeneous display of echo texture within the area of interest. It is good practice to adjust these parameters in, say, a subcostal oblique scan of the liver. If fluid is present in the near field (e.g., ascites), the TGC should be reduced to avoid effects of too much brightness.

The principal criteria for evaluating organs such as the liver, spleen, kidneys, uterus, and ovaries are size, shape, delineation, and changes in the basic echo pattern. The change may be hypoechoic, hyperechoic, or isoechoic. These pattern changes are helpful in detecting lesions such as hepatic, uterine and ovarian tumors or morphological changes in the placenta.

Artifacts and Pitfalls

Ultrasound image formation is based on certain physical assumptions, some idealized, such as those of constant sound velocity and attenuation or the assumption of straight-line sound propagation in the body. Deviations from these assumptions lead to artifacts. Some of these phenomena are diagnostically useful and furnish additional information on tissue properties, while others can lead to misinterpretation.

"Echo enhancement." Acoustic enhancement behind cysts is a diagnostically useful artifact. With the assumption of constant attenuation, the TGC curve produces a depth-dependent signal amplification in order to achieve uniform echo amplitudes from different depths (see Signal Processing: Focusing, Preprocessing, and Postprocessing, p. 6). However, the low degree of absorption and reflection in fluids causes too much signal gain, resulting in enhancement of echo brightness behind the fluid (Fig. 1.7).

"Shadowing by a strong reflector." Another potentially useful artifact is the distal shadowing that occurs behind strong reflectors. It may be caused by substances such as bone, stones, and air, which have an acoustic impedance very different from that of adjacent soft tissues (Table 1.1). Because of this large impedance mismatch, much of the ultrasound energy is reflected back toward the transducer. If the "reflector" is insonated at an approximate right angle, it produces a very bright echo (Fig. 1.7). The associated distal shadow reinforces the impression of a renal stone or similar lesion. However, if the reflector is insonated at a very oblique angle relative to the scan plane, most of the energy is reflected away. Shadowing occurs, but the reflector is not represented by a bright echo in the B-mode image.

Other shadowing effects are seen at the edge of circular objects. If the beam intercepts an oblique interface at a tangential angle, deflection of the beam may occur, producing a distal acoustic shadow. The interfaces are only partially defined.

Reverberations. Reverberations are artifacts caused by multiple reflections between two highly reflective interfaces. The prolonged transit time leads to faulty spatial encoding, creating a cascade of parallel echo bands spaced at uniform intervals. Reverberation artifacts can be eliminated by insonating the reflecting surface at an oblique angle.

Comet-tail artifact. A comet-tail or resonance artifact is also caused by multiple reflections. When reflective interfaces are located very close together with weakly attenuating tissue between them, they will generate many closely spaced echoes that form a long, bright stripe located distal to the actual reflector. This can occur with minute air bubbles, for example, or when structures with a high sound velocity, such as cholesterol crystals or metallic foreign bodies, are encountered by the beam, setting up multiple internal reflections.

Mirror-image artifacts. Another special case is the mirror-image artifact. When an object is located in front of a strong reflector, it will be "insonated" by that reflective interface. The returning echo is reflected back again and, because of its prolonged transit time, forms a second, virtual image of the object behind the strong reflector (Fig. 1.7).

Slice-thickness artifacts. Slice-thickness artifacts appear as areas situated at right angles to the scan plane that appear to fill in the echo-free space within small, fluid-filled objects. The artifact can be eliminated by changing the beam direction or using a different probe.

Side-lobe artifacts. Signals from reflectors located outside the scan plane can produce "ghost echoes" due to side-lobe effects. Side lobes are sound pressure peaks that are located outside the main beam and are directed obliquely outward from it. These artifacts have become rare, however, as a result of refinements in equipment.

Refraction and diffraction artifacts. Refraction and diffraction artifacts occur at interfaces between media that have very different ultrasound velocities. An example is the interface between bone (higher sound velocity) and silicone or other plastics (lower sound velocity).

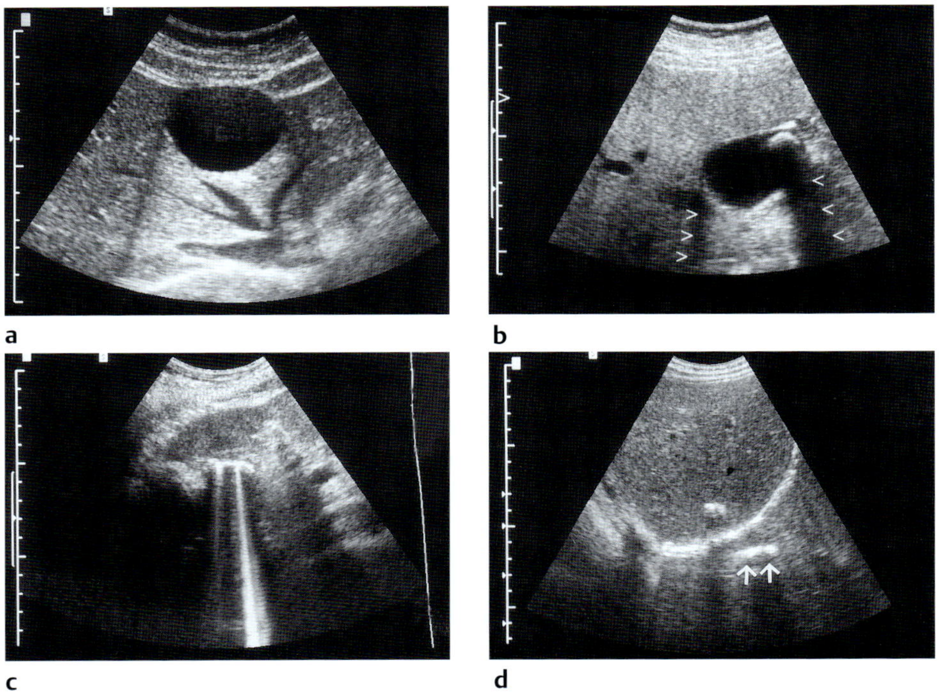

Fig. 1.7 B-mode artifacts: examples from abdominal imaging.
a Hepatic cyst with posterior acoustic enhancement due to overamplification of the ultrasound signals.
b Gallbladder with a stone (bright echo with a posterior acoustic shadow) and edge shadows (arrows).
c Comet-tail artifact caused by multiple reflections from air.
d Hepatic calcification with a mirror-image artifact projected on the opposite side of the diaphragm.

Duplex and Color Doppler Sonography

Physical Principles of Motion Detection

Doppler effect. The oldest approach to motion detection is the Doppler technique. It is based on the Doppler effect, first described mathematically by the physicist Christian Johann Doppler (1803–1853). He observed that the light emitted from stars that were moving toward the earth was shifted toward the blue end of the spectrum, i.e., toward higher frequencies and shorter wavelengths ("blue shift"). Conversely, a red shift was observed when the earth and star were moving apart. The same principle occurs in acoustics. The sound of an automobile engine changes its pitch when it moves relative to an observer. The change is most noticeable at the moment when the vehicle is passing the observer.

Frequency shift. When the Doppler effect is applied in medical diagnostics, the moving objects are red blood cells that are moving toward or away from the transducer at various velocities (Fig. 1.**8**). The echo signals reflected back to the transducer from the moving blood cells show a frequency shift of Δf relative to the original transmitted frequency f. This shift depends on the magnitude and direction of the blood flow velocity and is described mathematically by the equation (6):

$$\Delta f = (2fv \cos \Theta)/c \qquad (6)$$

where Δf represents the detectable Doppler shift between the transmitted frequency (f) and the frequency of the reflected signal. It is proportional to the velocity v of the blood cells and to the cosine of the angle Θ between the Doppler sampling beam and the vessel axis (Fig. 1.**8**). It is inversely proportional to the sound velocity in the medium being scanned (approximately 1540 m/s) and is usually within the audible range of 500–20,000 Hz in medical examinations. The factor 2 in the equation relates to the fact that the Doppler effect is operative twice in the pulse-echo technique: once when the pulse travels from the transducer to the moving blood cells, and again when the reflected signal returns to the receiver.

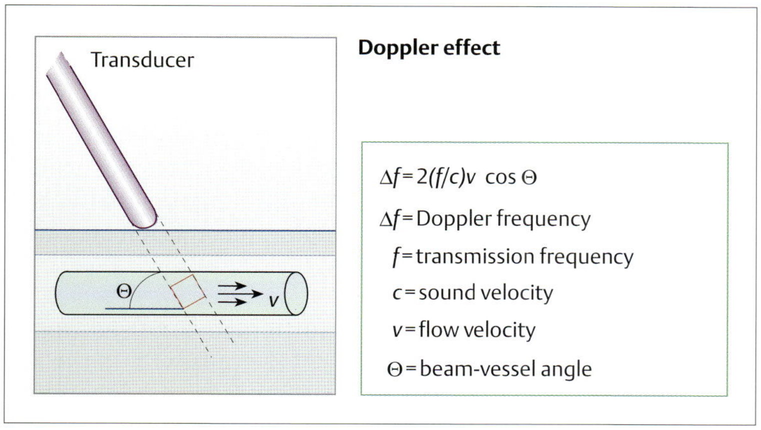

Fig. 1.**8** Principle of Doppler sonography.

The techniques of Doppler shift acquisition are classified as one-dimensional or two-dimensional.

One-dimensional or spectral techniques. In the one-dimensional techniques, the vessel is scanned by one beam and the flow velocities are analyzed along the path of that beam. The information acquired is displayed as a time-dependent Doppler spectrum. The spectral sampling region in one-dimensional techniques may cover the full depth of the beam (continuous-wave or CW Doppler) or only a specific, selected site along the path of the beam (pulsed-wave or PW Doppler) (Fig. 1.**9**).

Two-dimensional techniques. Two-dimensional techniques yield a spatial distribution of Doppler information that is acquired from multiple sampling sites. Multiple beams are transmitted, the received signals are selectively analyzed according to their depth, and the result is superimposed on the B-mode image in color-encoded form. The size of the analyzed area may cover the entire B-mode image, or it may be confined to a window called the "color box."

The image may depict the mean velocity in the volume element and the flow direction (color duplex sonography, CDS), or it may represent the sum of the squares of the amplitudes as a measure of the total number of moving red cells or the intensity of the blood flow (power Doppler imaging). Other techniques are based on the velocity-related position change in characteristic echoes (time-domain correlation or CVI)[1] or on the digital amplitude subtraction of two pulses emitted at different times (bi-flow technique)[18]. The latter technique yields a display with high spatial resolution, similar to digital subtraction angiography. At present, both of these techniques can be used only for superficial scanning under favorable insonation conditions. They have little or no practical importance in gynecology, so they need not be discussed here.

Technical Principles and Equipment Settings

One-Dimensional Techniques of Motion Detection (CW and PW Doppler)

CW Doppler. Continuous wave (CW) Doppler employs two separate crystals, one of which continuously transmits sound waves while the other continuously receives the reflected echoes (Fig. 1.**9a**). Because of its continuous operation, CW Doppler cannot provide depth discrimination or spatial localization of an echo. The advantage of the technique is that even very high flow velocities can be definitively analyzed (see below).

PW Doppler. Pulsed wave (PW) Doppler makes it possible to sample flow velocities of selected areas. The same crystals are used for transmitting and receiving, and short pulses are transmitted into the body as in B-mode imaging. A designated time interval is allowed for the pulse to travel to the desired Doppler sampling site and return to the transducer, at which point the gate for echo reception is briefly opened (Fig. 1.**9b**). The ex-

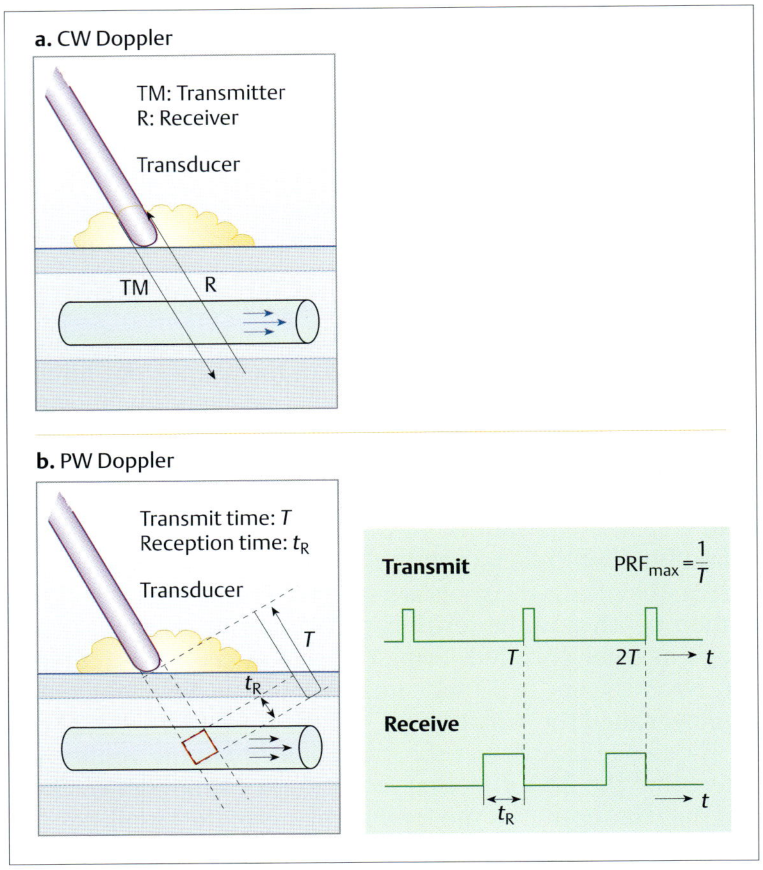

Fig. 1.9 One-dimensional techniques: principles of CW and PW Doppler.
a CW Doppler. The transducer contains two crystals, each of which transmits and receives continuously. This instrument can detect even high Doppler frequency shifts with no aliasing or other artifacts. The disadvantage is that all the vessels insonated by the beam yield one mixed signal, making it impossible to assign the signal to a specific vessel.
b PW Doppler. The same piezoelectric crystals are used to transmit and receive, and are driven in pulses. The echoes are received at a selected time interval (range gate), which depends on the transit time T that the signal takes to reach the transducer from the desired depth. The reception time t_R determines the size of the sample volume that will be analyzed.

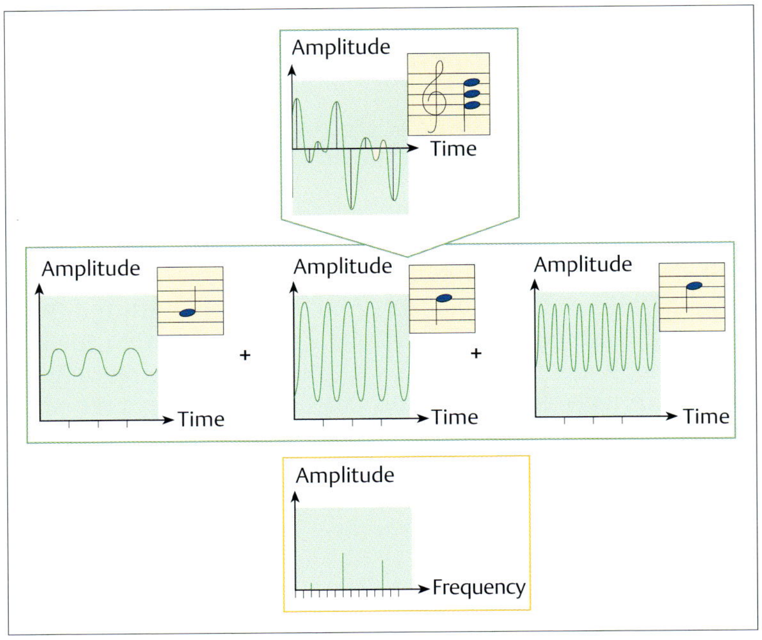

Fig. 1.**10** Principle of the fast Fourier transform (FFT). Just as the human ear is able to break down a musical chord into its separate notes (e.g., G, B, and D), the FFT is designed to resolve a frequency spectrum into its constituent frequencies. The result is displayed as an amplitude–frequency distribution (bottom graph).

aminer selects the size and depth of the range gate on the basis of visual guidance in the B-mode or color Doppler image.

Demodulation. The first step in obtaining velocity information from the echo signal is demodulation, which separates the Doppler signal (Doppler frequency Δf) from the reference or transmitted frequency. Demodulation is carried out in a phase detector by combining the echo signal with a reference signal, followed by rectification and low-pass filtering.

Spectral analysis. The echo returning from the blood to the transducer contains a mixture of frequencies that depend on the velocity distribution of the individual blood cells. In the next step, therefore, the frequencies contained in the Doppler signal are analyzed according to their distribution and amplitudes. This spectral analysis is accomplished by a mathematical algorithm called the fast Fourier transform (FFT). Just as the human ear can resolve a musical chord into its constituent notes (Fig. 1.**10**), the FFT processor can break down the demodulated signal into its underlying (harmonic) Doppler components and their amplitudes.

The frequency resolution and temporal resolution of the spectrum depend on the number of flow samples (generally between 64 and 256) that are used in computing the FFT algorithm. The greater the number of samples that go into the computation, the higher the frequency resolution. With 128-point FFT, the information from 128 consecutive Doppler signals (transmitted pulses) is broken down into the underlying frequencies. With the next transmitted pulse, the calculation results in a consecutive complete spectrum with 128 frequencies.

Doppler spectrum. The result of the analysis is displayed either as an amplitude–frequency curve for a given point in time (Fig. 1.**10**) or as a time-varying frequency spectrum (Doppler spectrum) that covers one or more cardiac cycles. In the latter display, which is commonly used in medicine, the amplitude of the Doppler frequencies at each point in time is represented by the brightness of the pixels (Fig. 1.**11**).

The Doppler spectrum forms the basis for a more detailed analysis, making it possible to determine and quantitate flow velocities and volumes and also to calculate resistance values that can demonstrate prestenotic or poststenotic changes in the vascular bed (see Analysis of Doppler Information and Artifacts, p. 19).

Pulse repetition frequency. One disadvantage of pulsed Doppler flowmetry is that the transit time of the pulses (T) in the abdomen ranges from approximately 0.03 to 0.26 ms, depending on the scanning depth (e.g. 3–20 cm). This limits the shortest possible time interval between two successive pulse transmis-

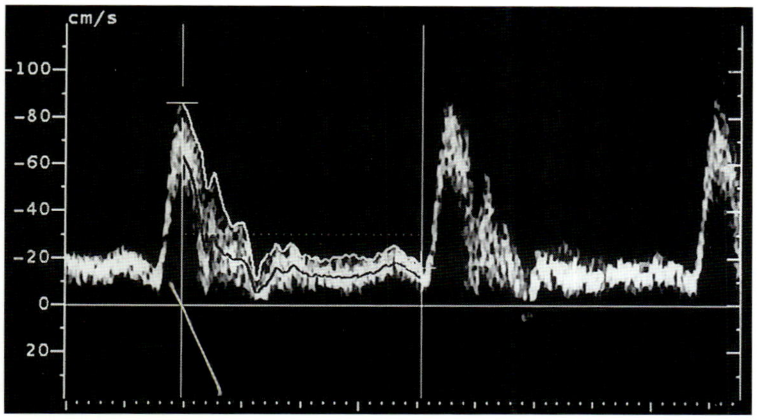

Fig. 1.11 Time-varying frequency spectrum (Doppler spectrum). The spectrum was sampled from an artery over two cardiac cycles and plotted by FFT analysis. The horizontal axis is time, and the vertical axis is the Doppler frequency shift or, after angle correction, the flow velocity. The brightness of the image points in the spectrum is proportional to the amplitude of the particular Doppler shift. In the first cardiac cycle the peak velocities (bright line at the top of the trace) and the mean velocities (dark line at the center) are shown as curves overtime course.

Relevant instrument settings:

Doppler frequency:	5.5 MHz
Pulse repetition frequency:	520 Hz
Wall filter:	50 Hz
Angle correction:	63°

sions. Hence, the rate at which the separate pulses are transmitted (pulse repetition frequency, PRF, equation 7) cannot exceed $1/T$ (e.g., 25.6 or 3.8 kHz) without causing errors of depth discrimination.

$$\mathrm{PRF} = \frac{1}{T} = \frac{c}{2s} \qquad (7)$$

In equation (7), s is the depth (m), PRF is the pulse repetition frequency (Hz), and c is the sound velocity (1540 m/s).

Two-Dimensional Techniques of Motion Detection (Color and Power Doppler)

Color Duplex Sonography

Whereas pulsed spectral Doppler displays the velocity distribution over time at a designated sampling site, color duplex scanning analyzes the flow velocities at a number of sites that are distributed over all or part of the sectional image. Color-flow imaging, unlike spectral Doppler, uses autocorrelation rather than Fourier analysis.

Autocorrelation. Recall that in FFT analysis each Doppler signal is sampled from the gate up to 256 times for each time interval in the spectrum, and the corresponding number of frequencies are calculated for that location. By contrast, autocorrelation in color Doppler imaging compares (correlates) all the demodulated Doppler signals along a color scan line that are sampled at a given time interval after transmission with those of the subsequent scan at the same location (Fig. 1.**12**). The result is a correlation vector that is characterized by its amplitude and phase difference. The phase difference at each image point in the color line supplies the mean frequency shift, and this value is directly proportional to the mean flow velocity. The variance can also be determined and displayed to indicate turbulence.

Kasai and Nameka first used the mathematical process of autocorrelation to calculate blood flow in 1982. An autocorrelation $A(t)$ of the function $g(t)$ is described by equation (8):

$$A(t) = \int_{-\infty}^{+\infty} g'(\tau)\, g(\tau + t)\, dt \qquad (8)$$

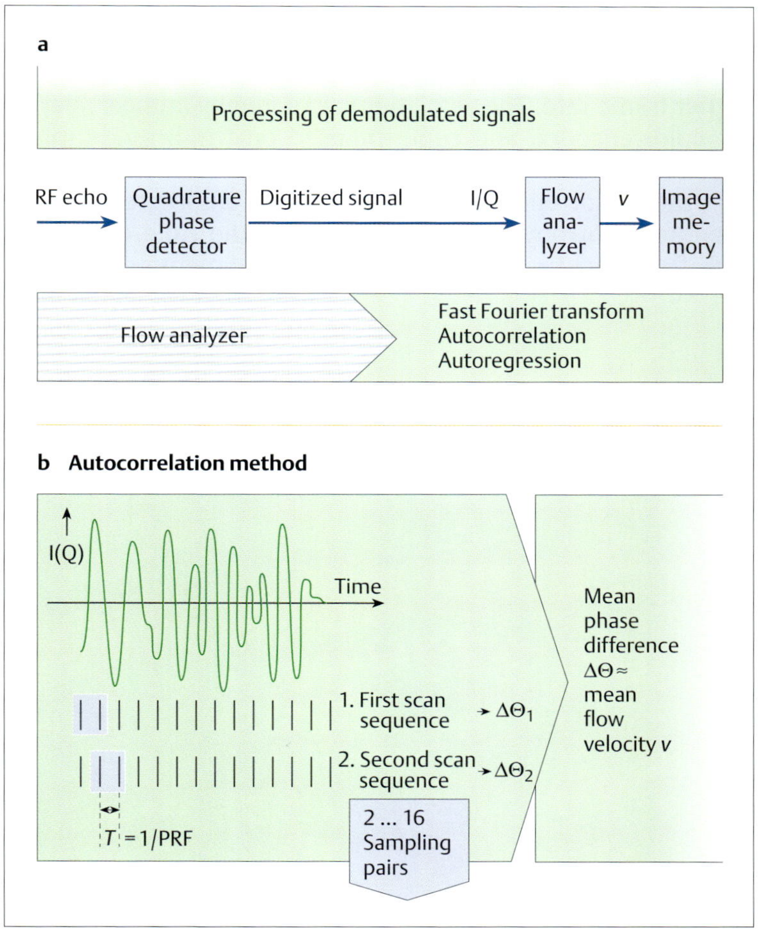

Fig. 1.**12** Two-dimensional Doppler techniques. Principle of signal analysis (illustrated for autocorrelation).
a Common principle of signal processing in all autocorrelation, autoregression, and FFT methods. The high-frequency signal is demodulated in the quadrature phase detector. This demodulation is done twice, with a phase shift of 90° (I, Q). After it has been filtered by suitable wall filters, the signal (which is digitized in modern systems) is fed to the actual analyzer. A fast Fourier transform, autocorrelation, or autoregression is carried out, depending on the method used. The result is usually stored as color information in image memory.
b In the most commonly used autocorrelation method, pairs of successive samplings are tested for their phase difference in a flow analyzer. This difference is a measure of the flow velocity v. Time averaging the phase differences in the selected scan sequence yields the mean velocity v, which is color-encoded and superimposed on the B-mode image. (After Haerten 1993.[5])

The function $g'(t)$ is the complex conjugate of $g(t)$. Both functions are superimposed and multiplied by each other, and their product is integrated from $-\infty$ to $+\infty$.

This mathematical folding process yields the value of the autocorrelation at time t. One of the first practical applications of autocorrelation was in electrical engineering to filter out very weak signals from noise and measure them. In ultrasound likewise, autocorrelation can determine the mean frequency shifts of extremely weak blood-flow echoes, whose amplitudes are up to 40 dB fainter than B-mode echoes, against a background of noise. To increase sensitivity at low signal-to-noise ratios, the measurement is repeated several times per scan line. The resulting correlation vectors are added to give the amplitude-weighted temporal mean value of the phase difference, from which the mean velocity can be computed. As a result, the time needed to generate a color image line is several times that needed to produce a gray-scale line. When there is a reasonably high signal-to-noise ratio as in cardiological studies, the number of samplings can be reduced to achieve a higher frame rate. The velocities calculated by autocorrelation are color-encoded and superimposed on the B-mode image (Fig. 1.**13**).

Other color-encoding techniques are autoregression, the maximum entropy method (MEM), and the time-domain or high-frequency cross-correlation technique mentioned earlier.

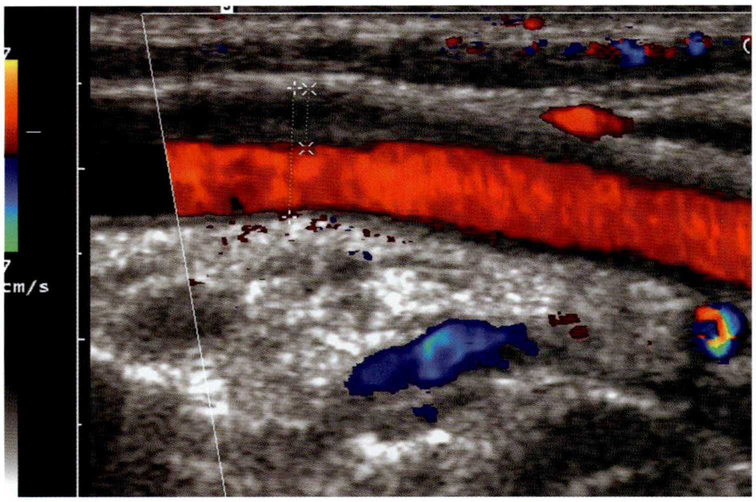

Fig. 1.**13** Longitudinal color duplex scan of the common carotid artery. The white frame outlines the color window in which the mean flow velocity is analyzed for each pixel and superimposed on the B-mode image in color-encoded form. The color scale on the left indicates the direction of color encoding (red = flow toward the transducer, blue = flow away from the transducer). Areas devoid of flow are shown in gray scale. The advantage of color encoding is evident when compared with the pure B-mode image just to the left of the color window: The hypoechoic wall thickening, which is clearly visible in the color duplex image, contrasts poorly with the clear lumen in the B-mode image.

Relevant system settings:

B-mode center frequency:	7.2 MHz
Doppler frequency:	5.14 MHz
Pulse repetition frequency:	868 Hz
Wall filter:	F5 (corresponds to 100 Hz)

Autoregression. Autoregression (AR) techniques are a spin-off of radar technology. Unlike autocorrelation, which analyzes successive pulses at designated time intervals, the time intervals are either varied or, in the case of MEM, randomly selected[7]. In practice, the theoretically higher accuracy and resolution of autoregression techniques have been offset by an unfavorable signal-to-noise ratio.

Cross-correlation. The cross-correlation technique compares the high-frequency signal of a pair of echo wavetrains to determine the velocity-related position change of characteristic echoes. An advantage of cross-correlation is its ability to detect even high velocities[8]. The need to have a signal-to-noise ratio higher than 6 dB has limited the routine abdominal use of this method, which can be used only for superficial vessels and in thin patients. Haerten[5] and Liu[9] have summarized the advantages and disadvantages of the techniques.

Color encoding. It is customary to use a red/blue color encoding format in which the color value indicates flow direction and the color brightness indicates flow velocity. There are color encoding maps that are particularly well suited for analyzing slow flows and providing good color fill. In some color-flow maps, green pixels are added to signify variance in flow velocities, indicating turbulence.

Motion detector. In some systems the scan data pass through a motion detector prior to autocorrelation in order to separate the essentially stationary tissue echoes from vascular echoes in the B-mode and color images and to eliminate motion artifacts. This filter can accurately discriminate between color-flow and gray-scale information in the presence of slow flows.

Time averaging (ensemble size). The accuracy of flow-data sampling in color Doppler can be increased by placing a number of transmitted pulses (4–20) at the same site for each color line and performing autocorrelation on all of the sequentially acquired data. Usually the sonographer can set the "time average" to determine the number of samples that are acquired.

Time averaging can increase color sensitivity (improved signal-to-noise ratio for color Doppler), but the higher number of transmitted pulses per color line prolongs the sampling time and reduces the frame rate.

Power Doppler Imaging (or Energy Mode)

Whereas color Doppler involves the direction-dependent detection and display of blood flow, power Doppler supplies information on the overall quantity of blood flow. The quantities of the flow components calculated by the correlator are detected without regard for direction. They are squared, integrated over time, and displayed in a color-encoded format. Integrating all the signals significantly improves sensitivity, making it possible to image not just the flow in major vascular trunks but also tissue perfusion (Fig. 1.**14**).

Power Doppler is also called transparent energy mode (TEM). This term indicates that power information is added to the B-mode image in such a way that, as in color duplex sonography, it does not mask the B-mode information in areas devoid of flow and leaves that information unchanged. This allows for better anatomical orientation.

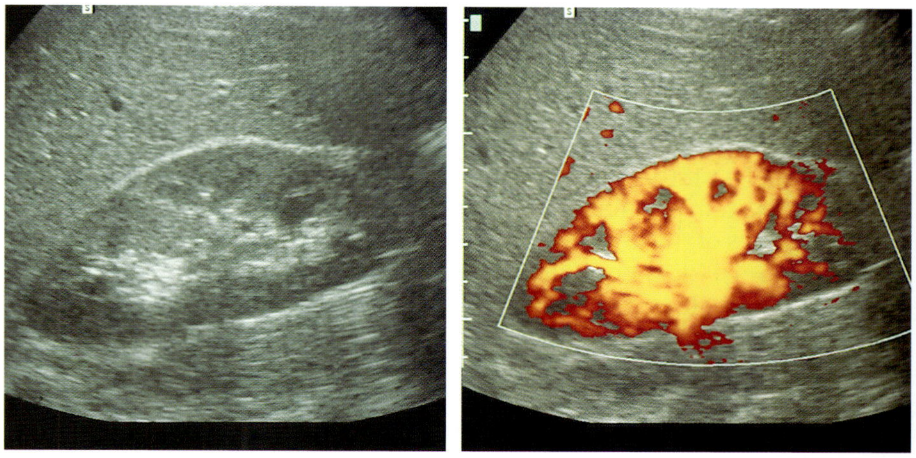

a b

Fig. 1.**14** Comparison of B-mode and power Doppler, illustrated for renal perfusion.
a Lateral scan demonstrates the liver and a normal-size kidney in longitudinal section. There is normal differentiation of the parenchyma and central echo complex.
b When color duplex (not shown here) was turned on, the branch points of the segmental and interlobar arteries could be identified, but only power Doppler can demonstrate parenchymal perfusion. There is no evidence of a complete or segmental infarction.

Selection of Methods and Physical Limitations

The advantages of the older one-dimensional techniques are their relatively low cost and availability. On the negative side, their use is time-consuming, especially in abdominal studies, and it is not possible to evaluate the perfusion of tissues and tumors.

Angle dependence. Because CW and PW Doppler and color duplex sonography are Doppler techniques, the color information is angle-dependent, much like the information in spectral Doppler analysis. For a constant flow velocity in a vessel, the color shading and intensity can vary depending on the angle between the Doppler beam and the vessel axis. For example, when a straight, uniformly perfused vascular segment is scanned with a convex transducer array, the varying angles at which the beams intercept the vessel lead to a color reversal accompanied by a (narrow) black area at the site where the beam is perpendicular to the vessel axis (Fig. 1.**15 a**).

Although the power mode is also a Doppler technique and therefore must be angle-dependent in principle, a signal can still be acquired even when the beam–vessel angle is 90° (Fig. 1.**15 b**). This is because the power spectrum has sufficient bandwidth to detect and integrate all nonzero spectral components. There are always transducer segments that insonate the pixel volume at an angle other than 90° and can thus contribute to the power Doppler image (see Fig. 1.**19**). On the other hand, color Doppler information represents only the mean frequency, and so a 90° incident beam does not yield a color-flow signal. As a result, power Doppler is largely independent of the beam–vessel angle and is not affected by aliasing, since the direction of the velocity does not enter into the calculation.

Aliasing. Aliasing is an effect familiar to motion picture viewers, when the spokes in the wheels of a passing stagecoach appear to rotate backward. Pulsed Doppler, with its pulse repetition frequency, is analogous to the frames of a movie camera: With a time delay of *T* between frames, the viewer cannot tell

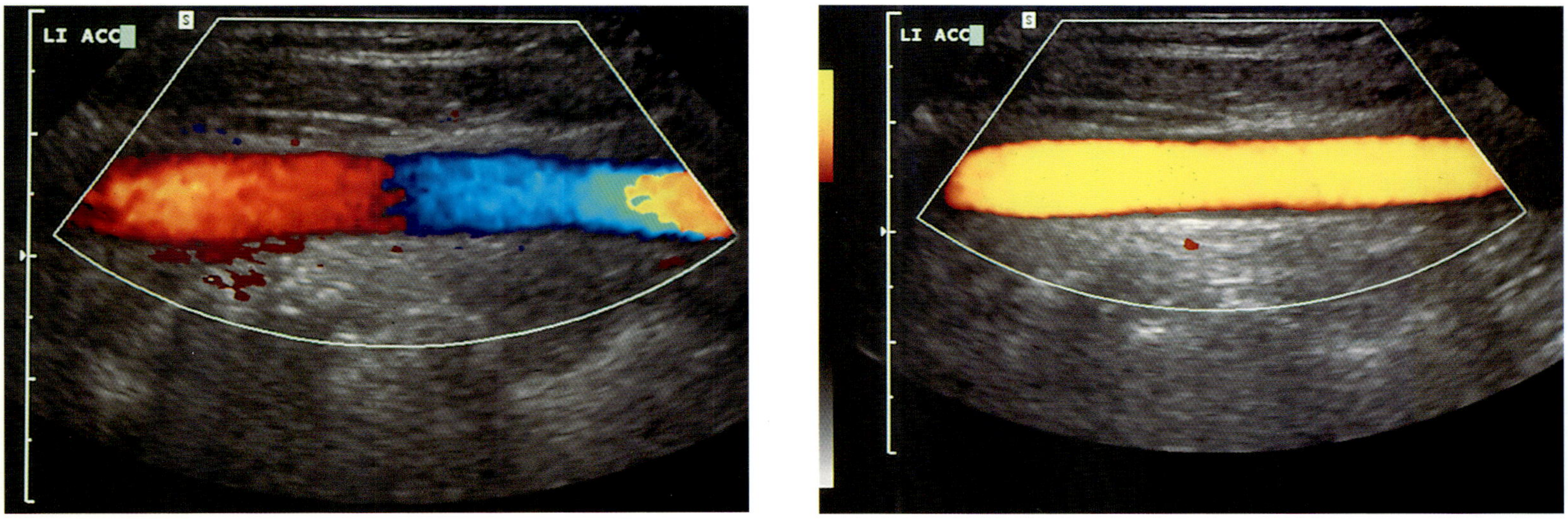

a b

Fig. 1.**15** Comparison of color duplex sonography and power Doppler: angle dependence.
a Longitudinal scan of a vessel with a trapezoidal beam. The standard color Doppler mode shows typical angle dependence of color encoding: blood on the left side is flowing away from the transducer and is encoded red, while blood on the right side is flowing toward the transducer and is encoded in the "opposite" color, blue. The blood on the far right appears light blue because the vessel is scanned at a small Doppler angle relative to the transducer ("pseudojet"). Since the beam-vessel angle is 90° at the center of the vessel, no flow is detectable and that area appears black ("pseudo-occlusion").
b When power Doppler is used at the same beam–vessel angle, the entire vessel is uniformly filled with color.

whether the wheel has turned, say, 225° forward or −135° backward (Fig. 1.16). In any sequence of images, the human brain always interprets the motion as minimal. Thus it will interpret rotation as occurring in reverse if the amount of rotation between two images is greater than one-half the period of the rotation, i.e., 180°. Similarly, when the FFT processor in an ultrasound system calculates the Doppler frequencies from the sampled values of the demodulated Doppler signals at the beat frequency $T = 1/\text{PRF}$, it does so under the assumption of minimal frequency values. For example, if the actual frequency is $^3/_2$ PRF, it will be interpreted and displayed as an aliasing frequency of $^1/_2$ PRF.

Nyquist limit. The highest frequency that can be detected without aliasing, called the Nyquist limit, is equal to one-half the PRF in both flow directions in directional Doppler (equation 9).

$$\Delta f_{max} \leq {}^1/_2\ \text{PRF} \quad \text{(Nyquist limit)} \qquad (9)$$

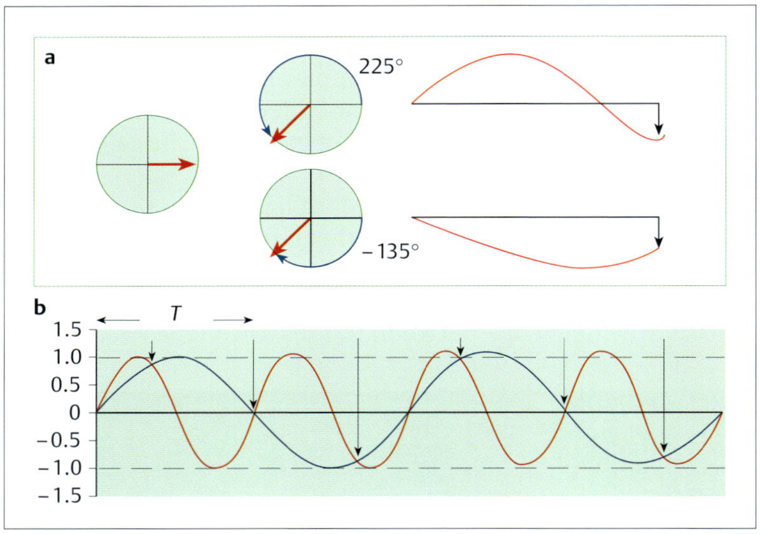

Fig. 1.16 Principle of aliasing.
a PW Doppler and the individual images in color duplex sonography are analogous to the individual frames in a movie camera. If the sampling time (*T*) is too long, the viewer cannot tell whether the wheel has rotated 245° forward or 135° backward. Since the brain always interprets motion in image sequences as being minimal, more than 180° of rotation (between two frames) will appear as reverse motion on the film.
b Aliasing also occurs in the recording of Doppler spectra. When the frequency to be measured (red curve) is sampled (arrows) at a rate of less than one-half the pulse repetition frequency, the underlying frequency cannot be unambiguously identified. The sampled points trace out a curve of lower frequency (blue curve). This causes a “wrapping around” of the spectral trace in the Doppler spectrum, as high positive frequencies that cannot be displayed above the baseline are truncated and appear as negative frequencies at the bottom of the spectrum. Aliasing in color Doppler causes the flow velocities to be displayed in the opposite color. When the sampling rate is equal to at least one-half the pulse repetition frequency (the Nyquist criterion), the curve can be plotted at the correct frequencies.

Baseline shift. When aliasing occurs in a Doppler spectrum, positive frequencies above the Nyquist limit are displayed as negative frequencies at the bottom of the spectral trace (Fig. 1.17). In color Doppler, this “wrap-around” effect appears as a zone of bright color reversal. This can be corrected in superficial vessels by increasing the PRF (Fig. 1.17, bottom) and in deeper vessels by shifting the baseline downward or upward (Fig. 1.18). This can double the range of detectable velocities both in the Doppler spectrum and in the color-flow image, provided the flow direction is not a concern. The maximum measurable velocity can be determined from the considerations on one-dimensional techniques and from the Doppler formula. The Doppler formula yields information on the maximum flow velocity that can be detected without aliasing. This limiting frequency is equal to the pulse repetition frequency PRF. The

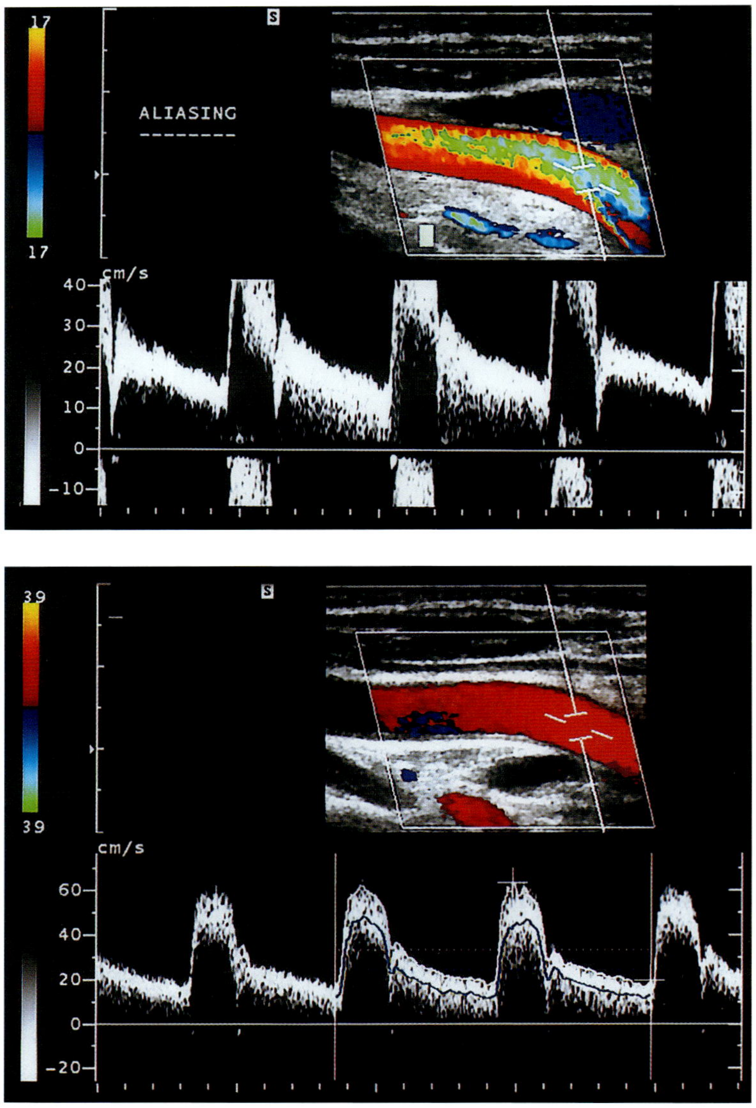

Fig. 1.17 Aliasing in color and spectral Doppler, before and after adjustment of the pulse repetition frequency (PRF).
a With a PRF of 2256 Hz, aliasing appears as a color reversal in the color-flow image and as a wrap-around of the truncated spectral peaks in the Doppler spectrum.
b When the PRF is increased to more than 5000 Hz for color Doppler or 4000 Hz for spectral Doppler, a normal, artifact-free color and spectral display of flow information is obtained.

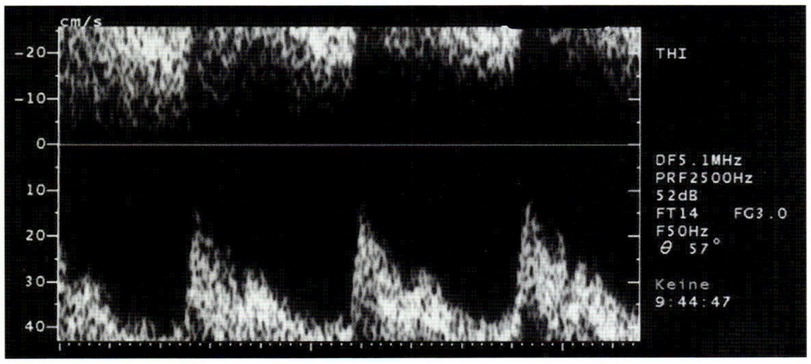

a

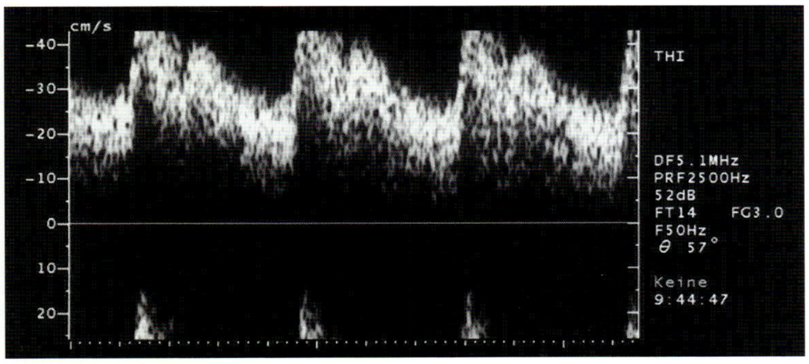

b

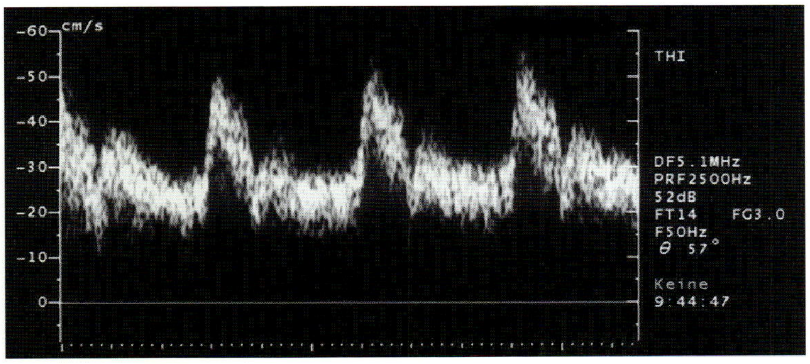

c

Fig. 1.**18** Aliasing due to incorrect placement of the spectral baseline.
a Severe aliasing caused by setting the baseline too high. Almost half of the spectrum has been truncated and displayed below the baseline.
b After an initial baseline shift, most of the spectrum is already displayed normally.
c When the baseline is lowered further, almost the entire PRF is available for sampling, and the Doppler spectrum is free of artifacts. Reverse flow components cannot be identified, however.

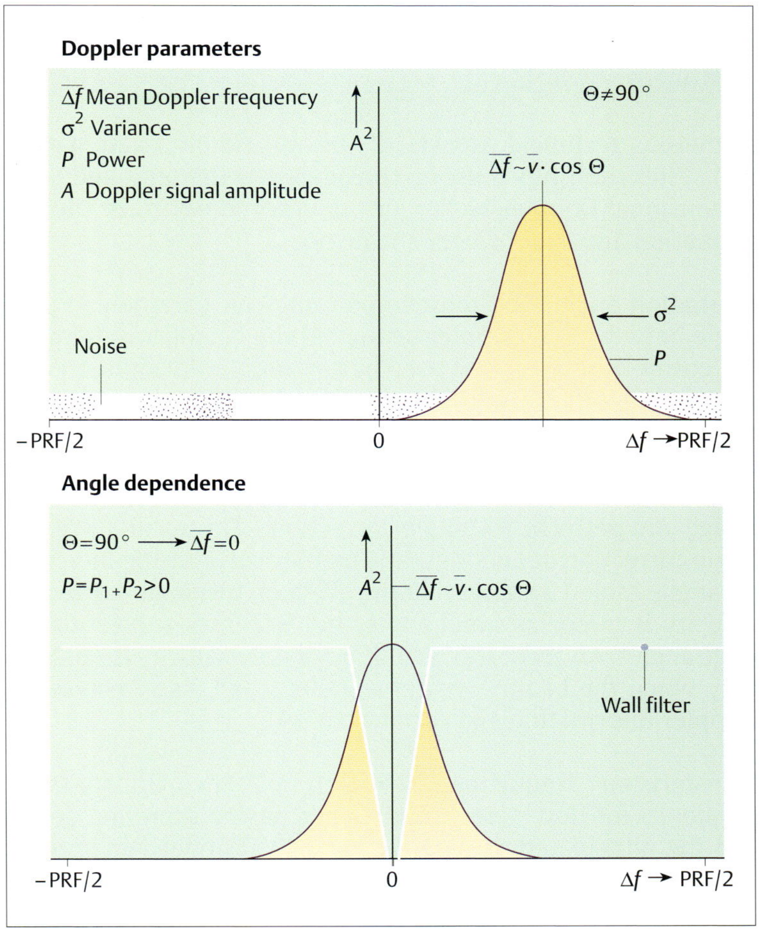

Fig. 1.**19** Doppler parameters in the amplitude–frequency curve. In both diagrams, the squares of the amplitudes of the flow components are plotted over the frequency axis.
a Doppler parameters in context. Color Doppler presents the mean Doppler frequency ($\overline{\Delta f}$). The variance (σ^2) is a measure of turbulence, corresponding to the bandwidth of the spectrum. Power Doppler is based on the area under the curve. Spectral Doppler displays all frequency components over the time axis.
b Special case of a 90° insonation angle. In color Doppler, a mean flow velocity of zero is calculated in both directions from the point where the beam is perpendicular. In power Doppler, on the other hand, the various directional components are added together, still allowing flow detection to occur.

maximum detectable velocity, then, decreases with increasing depth, and it increases as the transmission frequency is reduced.

Because all of the flow components are integrated over a time interval, power Doppler is very sensitive in detecting perfusion but is also much more susceptible to relative movements of the organs or transducer and therefore requires considerable user skill.

Comparison of spectral Doppler, color Doppler, and power Doppler. The relationship among spectral Doppler, color Doppler, and power Doppler is reviewed in Fig. 1.**19**. The figure shows the amplitudes (squared) of all the Doppler frequencies that can be measured for each sample volume and time interval:

- Spectral Doppler displays all of the amplitudes in the range gate in a time-varying, brightness-modulated spectral trace.
- In color Doppler, only the mean frequency shift $\Delta f = kx \cos \Theta$ at each point in the image is color-encoded. The variance is a measure of the bandwidth of the Doppler spectrum, i.e., how diverse the flow velocities are at the sampling site.
- Power Doppler encodes each pixel with the sum of all amplitudes, i.e., the area under the curve (amplitudes squared) averaged over time.

Technical Aspects of the Examination and Equipment Settings

Numerous technical aspects of the ultrasound examination and system settings must be considered in the one- and two-dimensional Doppler techniques in order to optimize the examination and avoid diagnostic errors.

Insonation angle. The most important scan parameter is the angle between the Doppler beam and the vessel axis. Because the cosine of this angle is used in the Doppler formula, the insonation angle should be kept well below 90° during the examination. When the angle is perpendicular, a Doppler signal is not recorded. To determine the velocity v from the Doppler frequency, it is necessary to measure the angle in the B-mode image and perform an "angle correction." The accuracy of the angle correction depends greatly on the size of the beam – vessel angle. Table 1.**2** shows that the effect on the angle-corrected velocity is minor at small angles but becomes substantial at large angles. An accurate calculation of flow velocity is possible only when the beam – vessel angle is small (see Analysis of Doppler Information and Artifacts, p. 19).

Transmission frequency. Selection of the transmission frequency for flow detection depends on the sampling depth and the anticipated velocity. As described under Penetration Depth (p. 5), high ultrasound frequencies are strongly attenuated with depth. This is particularly true of the Doppler signal, which is considerably weaker than the intensity of the echoes reflected from vessel walls and organs. At the same time, there are conditional upper and lower limits to the detectability of Doppler frequencies for technical reasons. Because the Doppler shift frequency is directly proportional to the velocity to be measured, it is advantageous to use a low transmission frequency for the measurement of high velocities. By the same token, a high transmission frequency should be used for low velocities.

Power output and receiver gain (Doppler or color gain). These settings should be adjusted so that color completely fills the vessel lumen with no extraluminal "blooming" and the displayed Doppler spectrum is free of noise and superimposed signals.

Table 1.**2** Effect of beam-vessel angle on angle correction errors

Beam-vessel angle	Correction factor (1/cos Θ)	Correction error
30°	1.15	± 3%
45°	1.41	± 6%
60°	2.00	± 9%
72°	3.24	± 15%
75°	3.86	± 21%
80°	5.76	± 30%

Angle correction errors are calculated by assumption of a detection accuracy of ± 3%

Velocity scale. In both spectral and color Doppler imaging, the scan frequency or velocity scale should be set for the anticipated range of velocities. Setting the scale too high makes it more difficult to detect and analyze the signals; venous signals in particular may go undetected. If the flow velocity in the vessel exceeds the maximum value set on the velocity scale (PRF) in one flow direction, this higher velocity will be misinterpreted as a high velocity in the opposite direction. One possible remedy is to increase the PRF. If this is no longer possible, the beam–vessel angle can be deliberately increased, or the scanning frequency may be decreased as described above. When aliasing occurs in the color-flow image, the color changes are always in the range of higher velocities (e.g., white or yellow). If the flow direction is actually changed owing to turbulence or flow reversal, the color changes run through black (Fig. 1.**20**). Power Doppler can mask these phenomena, which are often helpful diagnostically. On the other hand, the angiogram-like appearance of the vessels in power mode can aid in the detection of vascular lesions (Fig. 1.**20**).

Wall filter and motion-artifact filter. These filters suppress troublesome pulsations and flash artifacts, but they can also filter out diagnostically relevant low frequencies causing slow flows to be missed (Fig. 1.**21**). The same applies to color-write priority (color balance). If the priority is set too high in the B-mode image, flow may not be detected (see Artifacts and Pitfalls, p. 20).

Time averaging, color box, line density. Parameters such as time averaging, which increases sensitivity, and the line density reduce the frame rate. To keep the color frame rate in an acceptable range, it is good practice to limit the color-flow information to a color box, which should be of minimal size in its lateral extent and depth. In some examinations, such as fetal heart imaging, it is also helpful to reduce the color line density (at the cost of decreased color spatial resolution) in order to ensure adequate temporal resolution for cardiac imaging.

Another way to increase the color frame rate is to use a system that offers parallel processing of the echo signals from a transmitted pulse (see New Technical Processes and Approaches, p. 24). With this type of system, the information from multiple adjacent scan lines can be processed simultaneously.

Recommendations on key equipment settings for CW, PW, and color Doppler examinations are summarized in Table 1.**3**.

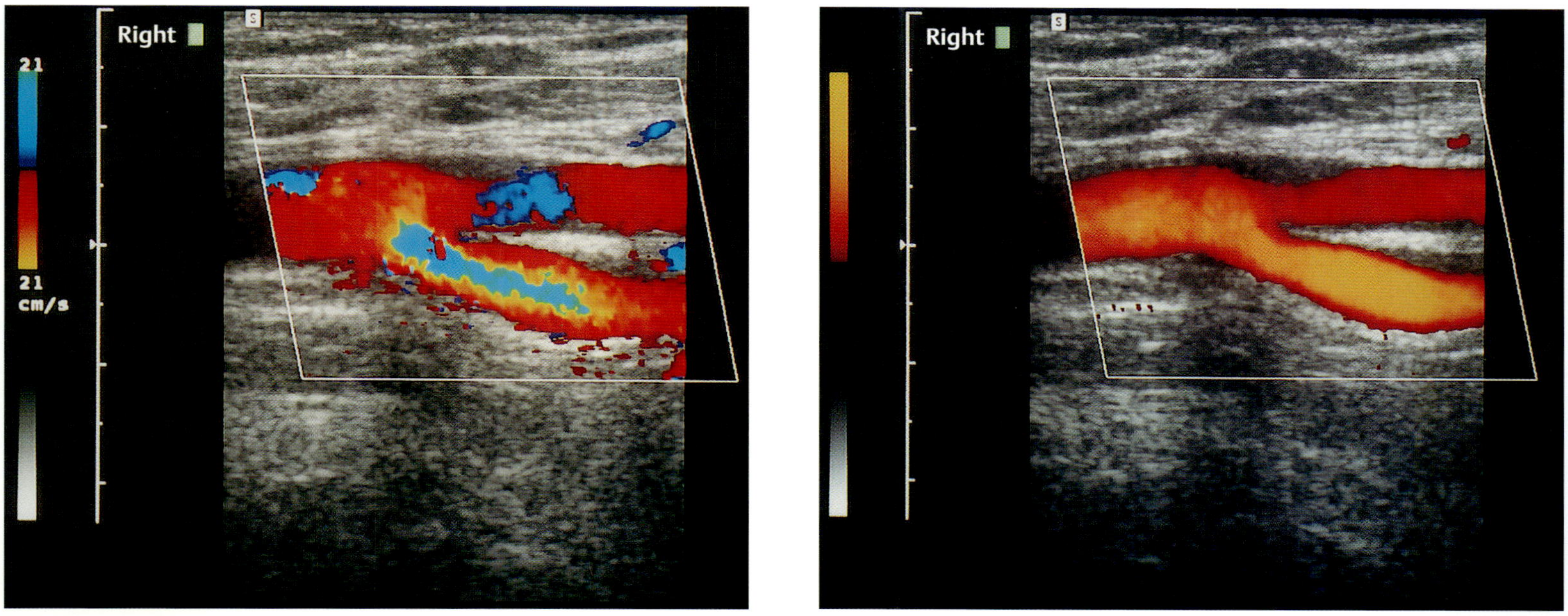

a b

Fig. 1.**20** Comparison of turbulence and aliasing in color duplex and power Doppler imaging.
a Color duplex image of a vascular bifurcation shows local wall thickening and zones of turbulence. Because the lower vessel branches off at an angle from the main vessel, the change of angle between the color box and vessel axis causes apparent flow acceleration in the lower vessel. Color aliasing is recognized by noting that the color change is in the range of the brighter colors on the velocity scale (e.g., white or yellow). With a true change in flow direction due to turbulence or flow reversal, the color would by running through black.
b Power Doppler shows a homogeneous vessel lumen in which local wall thickening at the origin of the upper vessel appears dark. Turbulence and aliasing are not visualized, but the local wall lesion is clearly appreciated in the angiogram-like view.

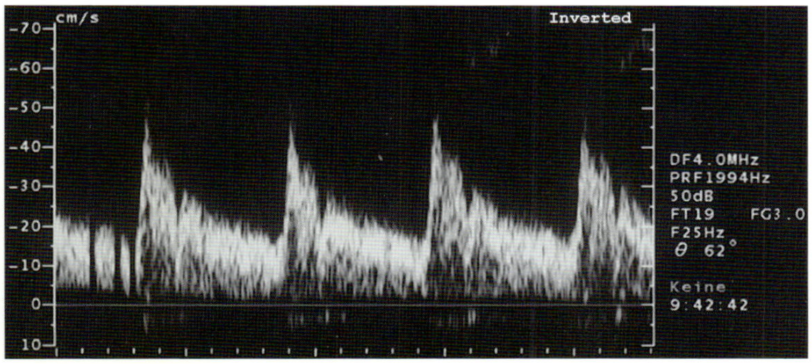

a

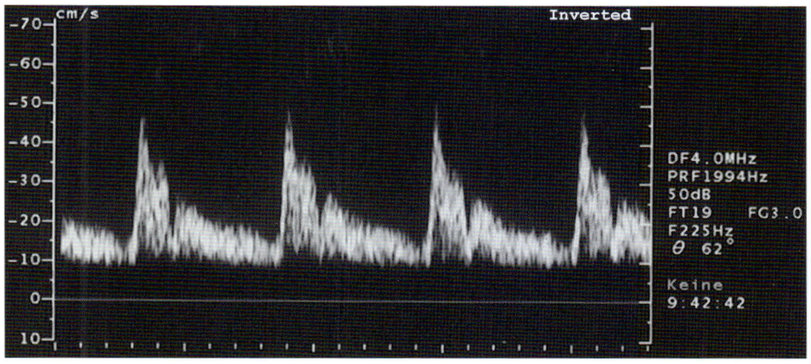

b

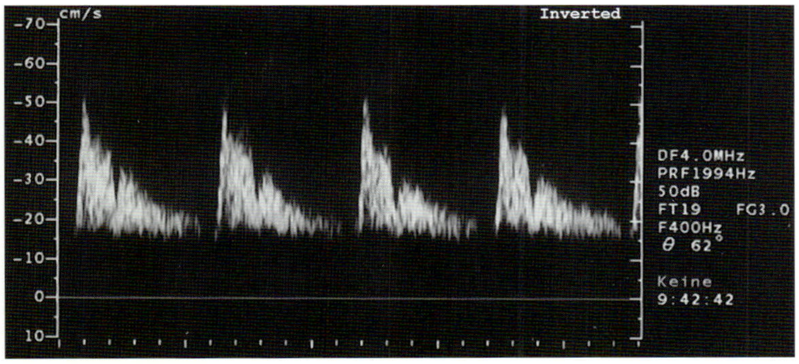

c

Fig. 1.**21** Effect of wall filter setting on frequency detection in spectral Doppler.
a Normal spectrum of an artery with high residual diastolic flow and low resistance in the distal vascular bed. Wall filter is set at 25 Hz.
b When the wall filter is increased to 225 Hz, velocities in the range from 0 to 10 cm/s are not displayed (angle correction from the B-mode image is constant for all settings in **a**–**c**).
c When the wall filter is further increased to 400 Hz, velocities in the range from 0 to 20 cm/s are lost, creating an apparent spectral pattern of absent end-diastolic flow. Tumor and tissue perfusion and slow venous flows may go undetected.

Table 1.**3** Recommended system settings for CW, PW and color Doppler examinations

Settings	CW	PW	Color Doppler	Settings, problems, and optimization
Transmit power, transmit intensity	+	+	+	The necessary transmit power depends on the examination conditions (sound absorption) and on depth. The power setting for obstetric examinations should be as low as possible.
Receiver gain	+	+	+	Set the gain to produce a spectrum with minimal noise and no superimposed signals or a color-flow image in which a normal vascular segment is completely filled with color and there is no color blooming or superimposed color noise.
Pulse repetition frequency (PRF), velocity scale	+	+	+	Set the PRF to fully utilize the range of measurable frequencies. A PRF of 1000 Hz is recommended for veins, 1500–3000 Hz for arteries. May have to increase the PRF if aliasing occurs, or may have to selectively increase the Doppler angle for deep abdominal scanning. In some cases it may also be necessary to reduce the scanning frequency (see equation 6).
Baseline setting	+	+	+	Adjust so that the range of measurable frequencies is optimally utilized. Shifting the baseline is helpful if aliasing occurs.
Wall filter and flash-artifact filter	+	+	+	Troublesome wall pulsations and motion artifacts are reduced, but low velocities may not be detected. Filter setting of 50–200 Hz is recommended for arteries, <100 Hz for veins. Use of other filters such as motion and flash-artifact filters can reduce sensitivity, especially for the detection of slow flows.
Color priority	–	–	+	Postprocessing parameter that controls the balance between gray-scale echoes and color pixels. If the color priority is set too low, it can suppress flow signals in the B-mode image. To search for thrombosis or reduce noise, it should be set for slight gray-scale priority.
Line density for B-mode/color flow (synonym: spatial resolution)	–	–	+	Improves the lateral resolution of the B-mode or color image at the cost of a reduced frame rate. If fast motion (e.g., fetal heart) or high velocities are to be detected, the line density should be reduced.
Temporal resolution	–	–	+	Sets the number of transmitted pulses used to compute an ultrasound color line or the number of acoustic lines used in autocorrelation. Increasing temporal resolution improves the signal-to-noise ratio and sensitivity. The disadvantage is a reduced frame rate.
Correlation (persistence)	–	–	+	Affects the composition of color information by the weighted addition of old and new information. The higher the number, the more old information is retained. Time averaging the color pixel values helps to image all flow in the vessel lumen but slows the color presentation. The disadvantage of a high setting is less appreciation of temporal dynamics.
Spatial averaging	–	–	+	One- or two-dimensional filtering of the color image based on the surrounding color pixels or color information. Reduces color noise without affecting the frame rate, but can lead to nonvisualization of very small vessels.

Analysis of Doppler Information and Artifacts

Optimizing the Examination and Analyzing the Doppler Information

Settings before the start of the examination. After selecting the proper transducer for the desired information and the proper frequency for Doppler interrogation, the examiner should adjust and optimize the settings listed in Table 1.**3** to obtain a well-defined spectrum that is free of noise and aliasing (Fig. 1.**22**). For examinations of the fetal aorta and maternal arteries, the power and gain settings should be reduced to obtain a clear spectral window below the trace. For two-dimensional color flow imaging, the PRF, gain, and any filter settings should first be adjusted in a normal vascular segment so that color uniformly fills the vessel lumen. We recommend the following technique. After identifying the vessel and angling the color box in proper relation to the vessel axis, adjust the power setting, PRF, and gain until a noisy color image is obtained. Then lower the gain setting until the color noise is gone.

Information. Information on flow obstructions at the sampling site itself and on conditions in the proximal and distal vascular beds can be derived from the signal frequencies, which are a measure of flow velocities; from the amplitudes, which are

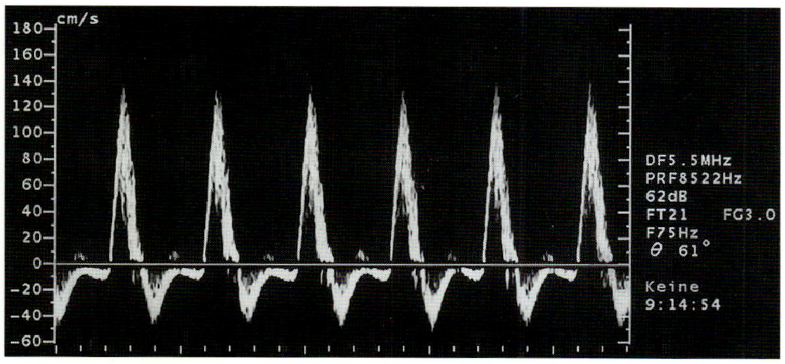

a

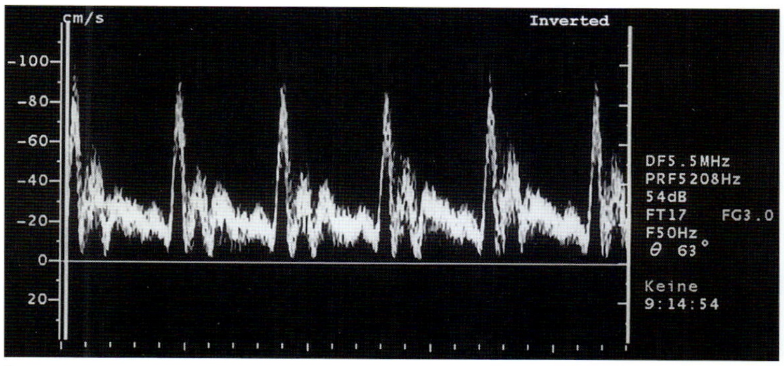

b

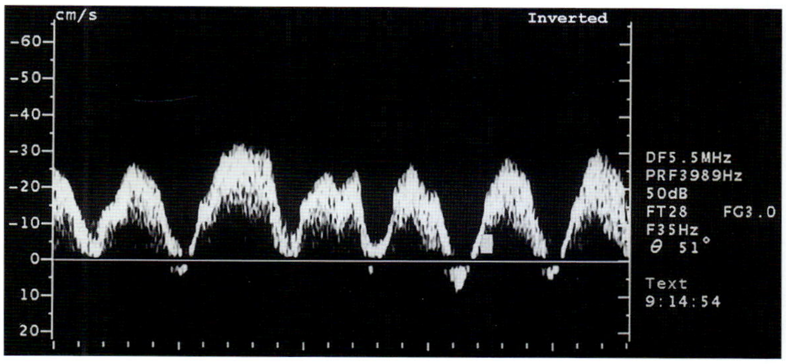

c

Fig. 1.**22** Dependence of the Doppler spectrum on resistance in the distal vascular bed.
a Spectrum of an artery with high resistance in the distal vascular bed. This resistance leads to a rapid decline of flow after systole, with a reverse flow component appearing in diastole. All the red blood cells move at approximately the same velocity during systole, creating a narrow frequency band with a clear "sonic window."
b Spectrum of an artery with low resistance in the distal vascular bed, with persistent residual diastolic flow.
c Venous signal from the inferior vena cava. Unlike the bandlike spectrum of peripheral veins or the portal vein, the superior and inferior vena cava show cardiac modulation of their flow, which can easily be confused with an arterial flow pattern.

proportional to the number of moving blood cells; and from the time changes over each cardiac cycle. Color Doppler imaging and spectral analysis are considered to be complementary in the evaluation and interpretation of Doppler information. When applied together, both color and spectral Doppler can be a valuable source of qualitative, semiquantitative, and quantitative information.

Analysis of the Doppler Spectrum

Waveform. The Doppler spectrum can be described in terms of the spectral bandwidth and the shape of the waveform (envelope curve), which is basically characterized by its steepness, peak systolic velocity, and end-diastolic velocity. Moving left to right, the trace begins its upstroke at the start of systole and concludes at end diastole. One or two cardiac cycles are always evaluated for spectral analysis. The shape of the spectrum depends on local findings and on the organ that is supplied by the sampled vessel (Figs. 1.**22**, 1.**23**).

Spectral bandwidth. If all of the blood cells are moving at the same velocity, the spectrum displays a narrow frequency band with a typical clear "sonic window" below the trace (Fig. 1.**22**). This is a normal finding in early systole for all larger vessels. Vessel wall irregularities produce turbulence, causing initial spectral widening or filling-in of the spectral window. Starting at about 50% stenosis, a jet develops owing to local flow acceleration. With proper angle correction, this acceleration can be identified in the Doppler spectrum.

Pulsatility index (PI). Several techniques have been devised for analyzing the spectral Doppler waveform. Gosling[4] was the first to describe an index for analyzing pulsatility. This is a

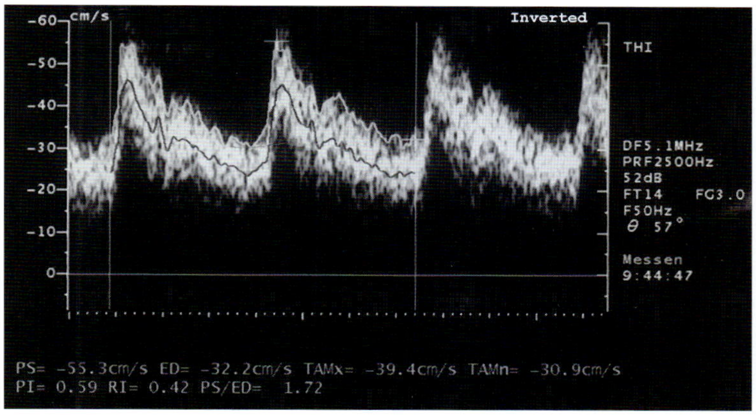

Fig. 1.**23** Doppler spectrum with parameters for spectral analysis (S/D, RI, PI). This spectrum is from a vessel with a low-resistance distal bed. After the operator manually marks the starting and end points of two cardiac cycles, most scanners will automatically calculate the key flow parameters and Doppler indices. While the velocity measurements are angle-dependent and require corresponding angle correction in the B-mode image, all of the indices listed below are independent of the beam angle because in each case the correction factor cancels out of the equation.

S: peak systolic velocity	(here: 55.3 cm/s)
D: end-diastolic velocity	(here: 32.3 cm/s)
Tav: time-average peak velocity	(here: 39.4 cm/s)
Pulsatility index PI = (S–D)/TAV	(here: 0.59)
Resistance index RI = (S–D)/S	(here: 0.42)
Stuart index S/D	(here: 1.72)
Maulik D/TAV	

measure of the difference in flow between systole and diastole during the cycle; it was first determined by calculation from the Fourier analysis. A simplified version is the Pulsatility index (PI). This is calculated by dividing the difference between the peak systolic and end-diastolic frequencies (D) by the time average of the maximum frequency shift (TAV) (Fig. 1.**23**).

Resistance index (RI). Pourcelot introduced the resistance index, an angle-independent measure of pulsatility in which S describes the peak systolic frequency and D the end-diastolic frequency. Many sonographers use simplified indices like that of Stuart and Drumm[15], which is the ratio of the peak systolic frequency S and end-diastolic frequency D. Maulik proposed the ratio D/TAV, in which the maximum end-diastolic frequency is normalized to the mean Doppler shift frequency of the outer envelope curve.

***S/D* ratio, *A/B* ratio.** Other parameters for waveform analysis were introduced in 1983 by Campbell et al.[2] and Thompson et al.[17]. These were based on curve fitting and on descriptions of steepness, maximum frequency, and other defined slopes. But from the many indices that have been devised for semiquantitative analysis, only the S/D ratio (also called also the A/B ratio), the resistance index (RI), and the pulsatility index (PI) have become widely used. The formula $RI = 1 - 1/(S/D)$ is used to convert between the RI and S/D ratio.

Quantitative information. Measurements of the Doppler shift frequencies with angle correction supply quantitative information on flow velocity and also on volume flow when the cross-sectional area is considered. Owing to errors in angle determination, the effect of filters, inaccuracies in estimating the true center frequency, and problems in determining the exact vascular cross section used in calculations, the importance of volume measurements has dwindled in favor of angle-independent semiquantitative parameters (PI, RI, S/D).

Analysis of Color Information

While the Doppler spectrum portrays the time course of flow at a selected sampling site, the color information supplied by two-dimensional Doppler techniques describes the flow conditions that prevail within a volume element at a given point in time. Color changes, like spectral shape, are influenced by physiological and pathophysiological states in the sampled vessel and in the proximal and distal vascular beds. Besides qualitative information (flow: yes/no), color image analysis focuses on the time course of color saturation and the color distribution:

Reverse flow components. Physiological changes in flow and flow direction occur at vessel branch points such as the carotid bifurcation, consisting of flow reversal or a reverse flow component in high-resistance vessels such as the aorta and the iliac arteries (Fig. 1.**24**). Low-resistance vessels like the middle cerebral artery and umbilical artery (in the second and third trimesters) show unidirectional flow that persists even in diastole, appearing as slightly darker residual flow. The presence of a reverse flow component in these vessels is a pathological sign and can be confirmed and described semiquantitatively by analyzing the Doppler spectrum. Apparent changes are caused by changes of insonation angles within a vessel (Figs. 1.**15**, 1.**24**) and by the occurrence of aliasing when the PRF is set too low. Other pathological changes are seen in association with malformations, stenoses, and occlusions.

Stenoses. The oldest indication for Doppler scanning was the detection of stenoses. These sites are recognized in color duplex sonography by typical local flow acceleration, which is usually manifested by a bright color jet or aliasing (Fig. 1.**25**). Measurements of maximum flow velocity help to confirm and quantitate the stenosis. Velocities of 140–180 cm/s are suspicious, while higher values confirm that a stenosis is present. Additional criteria include eddy flow located within and just past the stenosis and poststenotic dilatation, which is often visible in the B-mode image. Other, indirect criteria are the presence of distal poststenotic damping and increased prestenotic resistance, which causes a “flickering” of the color pixels. Poststenotic damping causes a loss of pulsatility in the color image and decreased brightness modulation in the spectrum.

Whereas the diagnosis of an intra-abdominal stenosis is generally based purely on hemodynamic criteria in the color-flow image, the color duplex scanning of superficial vessels with a high-resolution transducer can additionally define the morphology of the stenotic lesions (Fig. 1.**25**).

Thromboses and arteriovenous (AV) fistulae. Color Doppler may also demonstrate voids in the color-filled lumen signifying thrombosis or a coarse, mosaic-like vibration artifact (“confetti sign”). This may be an accompanying feature of a high-grade stenosis, or it may signify an AV fistula (Fig. 1.**26**). The lesions are differentiated by performing a spectral analysis and checking for secondary fistula signs such as increased flow in the feeding artery and/or an “arterialized” spectrum of the draining vein.

Tumors. The detection of blood flow aids in identifying tumors, and the specific blood flow pattern can advance the differential diagnosis. Newer approaches, such as calculating the inflow rate of ultrasound contrast medium in tumors or in the kidney, can furnish quantitative information on tissue perfusion and renal function.

Artifacts and Pitfalls

Artifacts and possible diagnostic errors can result from improper settings or from unexpected physical or equipment limitations. Some artifacts cause loss of sensitivity, while others give a false-positive indication of flow.

Absent (False-Negative) Flow Detection

Besides the power and gain settings, the detection of flow depends on the wall-filter and motion artifact-filter settings (Fig. 1.**21**) and on the selected PRF. Other system-specific parameters are often underestimated in their significance. These include the color priority, which controls the representation of color in the B-mode image. Setting the color priority too low

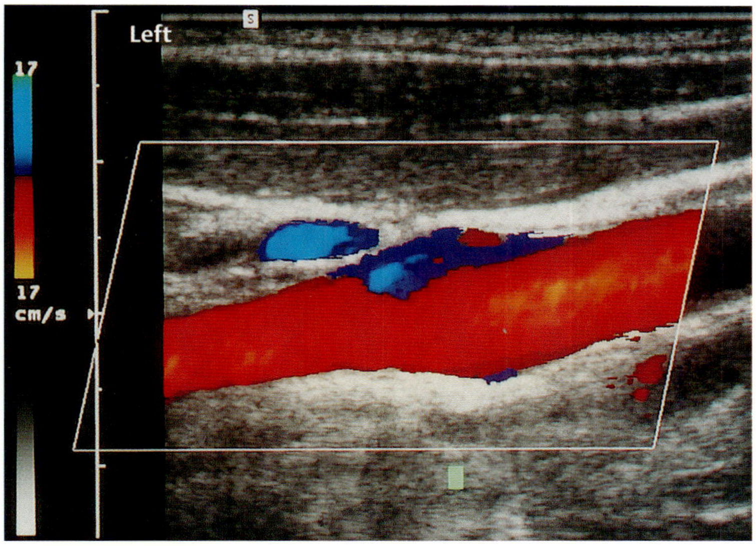

a

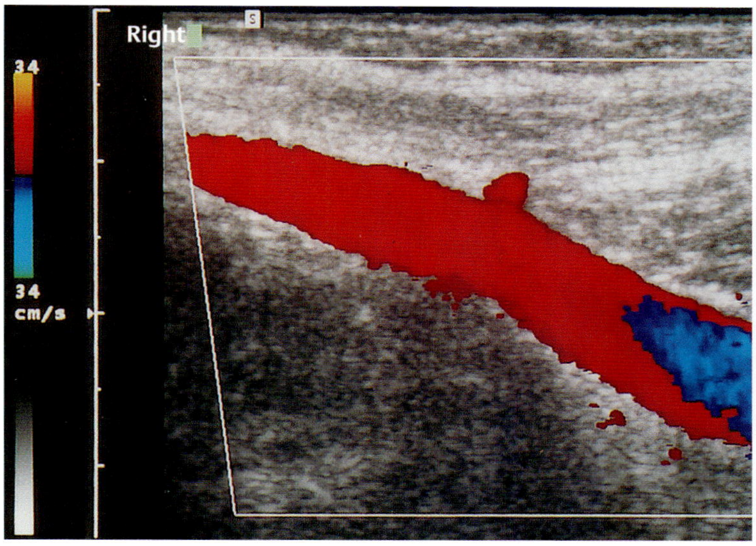

b

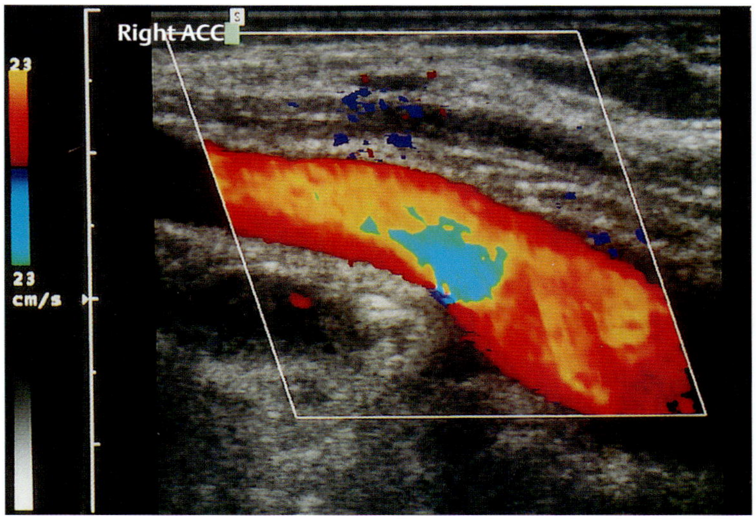

c

Fig. 1.**24** Difference between physiological flow reversal, reverse flow component, and aliasing.
a Physiological flow reversal in the carotid sinus. Zones of eddy flow at the carotid bifurcation are encoded in blue.
b Reverse flow component in a vessel with high diastolic resistance. The cranially directed flow is encoded in red. The reverse flow component, encoded in blue, is bordered by a dark zone representing the zero baseline (compare with spectrum in Fig. 1.**22 a**).
c Aliasing due to apparent flow acceleration ("pseudojet"). The cause is the smaller Doppler angle that results from the curvature of the vessel. When angle correction was applied, a normal flow velocity was detected in all segments.

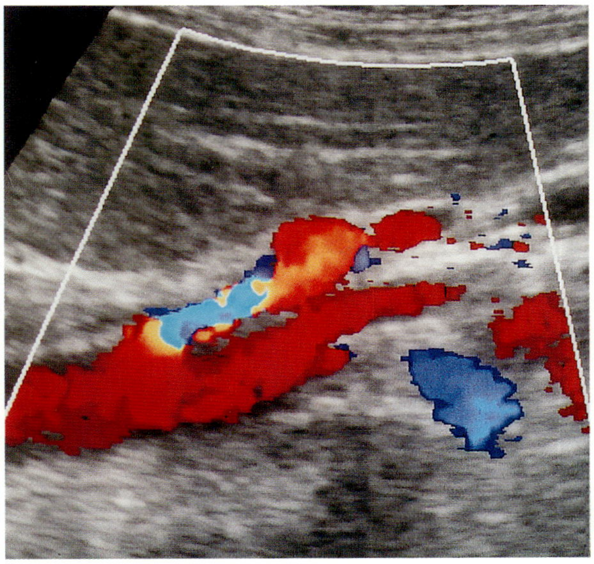

a

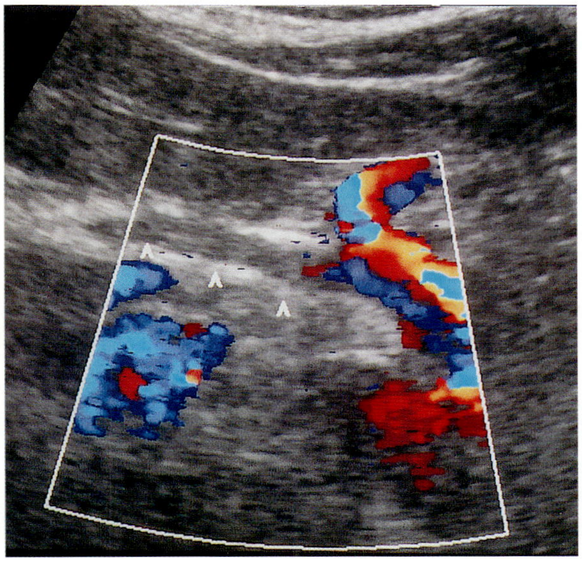

b

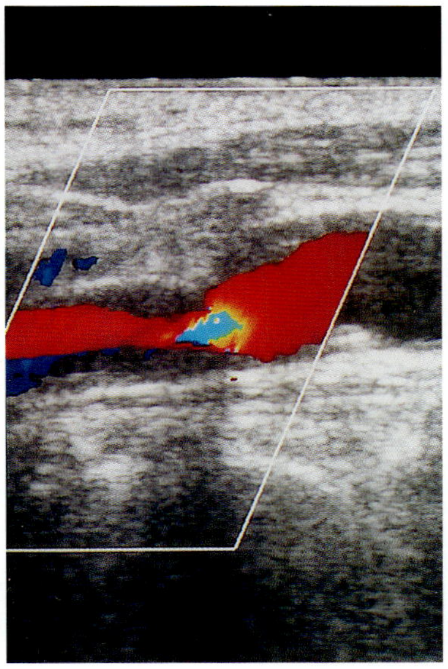

c

Fig. 1.**25** Color duplex images of stenoses in the neck and in abdominal arteries.
a Longitudinal scan through the liver and aorta shows a high-grade stenosis of the celiac trunk with aliasing and moderate poststenotic dilatation. The aorta and the origin of the superior mesenteric artery appear normal.
b Transverse scan through the vena cava (encoded blue), aorta, brachiocephalic trunk, splenic artery, and occluded common hepatic artery. The insonation angle is large, but scans at all angles confirmed the absence of flow signals.
c Seventy-percent stenosis at the origin of the internal carotid artery, with a short jet phenomenon. Angle-corrected Doppler measurements showed a rise in the peak frequency to more than 400 cm/s. The cause is a tubular wall thickening 1.5 cm long with a proximal filiform constriction.

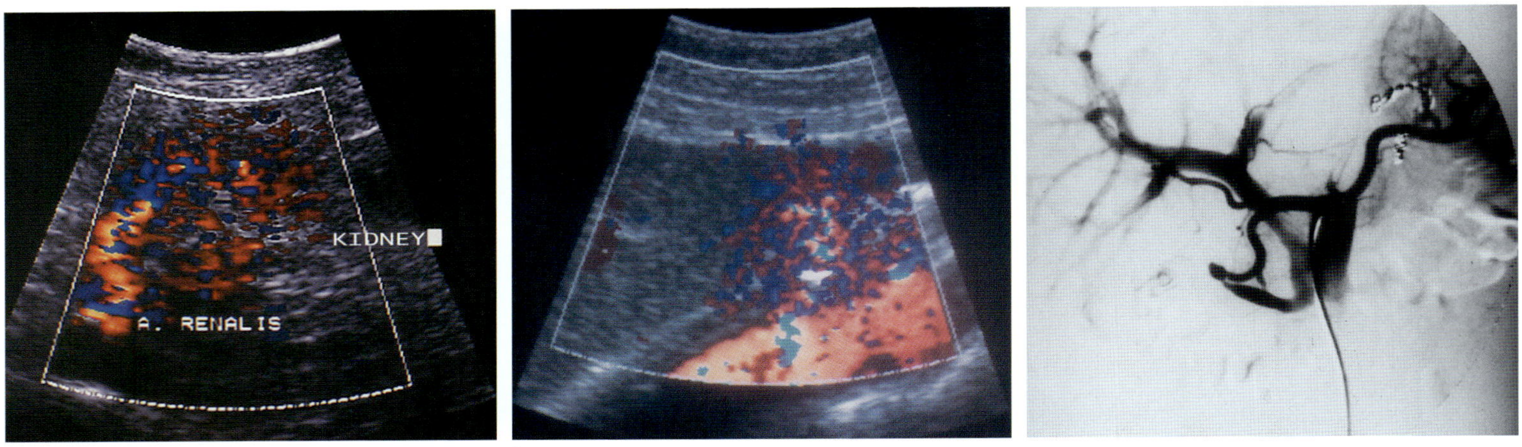

a b c

Fig. 1.**26** Vibration artifacts associated with high-grade stenosis and AV fistula.
a Patient 1: Vibration artifact in the lesser pelvis due to a high-grade stenosis (angiography demonstrated a 90% transplant renal artery stenosis).
b Patient 2: Coarse, continuous vibration artifact located to the right of the midline, anterior to the aorta.
c Patient 2: Angiogram demonstrates an AV fistula between the gastroduodenal artery and a branch of the superior mesenteric vein, causing immediate opacification of the portal vein.

can lead to spotty color flow (Fig. 1.**27**) or can mimic thrombosis. Technical problems such as too large a Doppler angle or calcifications in the scanned region can be recognized in the B-mode image and avoided. It should be noted that acoustic shadowing varies with the insonation angle and that angling the color box can produce an oblique shadow (Fig. 1. **28**).

False-Positive Flow

Image clutter, pulsation artifacts, or gross motion artifacts (especially in power Doppler) can mimic the presence of intraluminal flow or tissue perfusion.

Pulsation and motion artifacts. Image clutter (flash artifact) may be caused by setting the gain too high or the PRF too low. Soft-tissue pulsations or motion artifacts can produce color signals that are misinterpreted as flow. Transmitted cardiac pulsations can produce an apparent flow pattern, especially in cysts located near the diaphragm. Confetti-like artifacts associated with high-grade stenoses and AV fistulae are caused by the vibration of perivascular soft tissues and structures in response to rapid jet currents.

Slice-thickness artifacts and "phantom vessels." Slice-thickness artifacts are particularly common in older scanners. They can cause diagnostic errors in color duplex scanning, as flow information from neighboring vessels may be projected into an occluded or thrombosed vessel. "Ghost vessels" or "phantom vessels" can appear behind strong reflectors such as the diaphragm or pleura (Fig. 1.**28**), mimicking a second vessel that not only contains color signals but can even yield a Doppler spectrum.

a
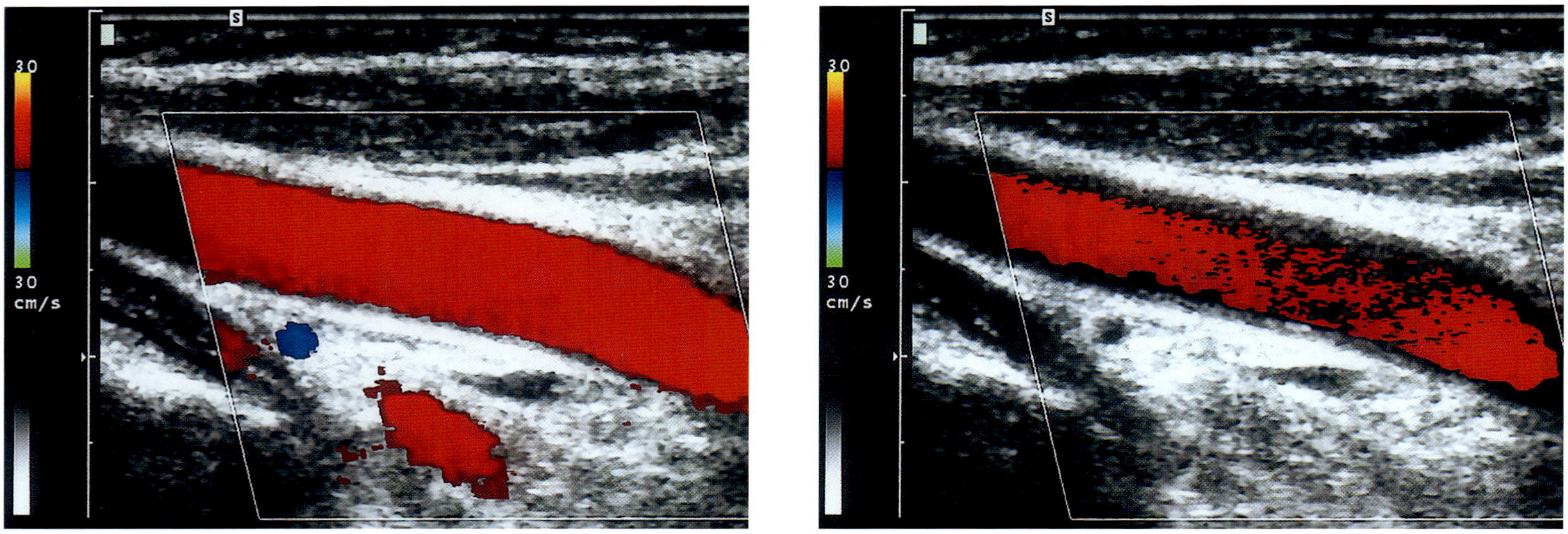

b

Fig. 1.**27** Color Doppler artifacts and pitfalls. I: Effect of color priority.
a Longitudinal scan of an artery with optimum settings (Table 1.**3**) shows complete filling of the vessel lumen.
b When color priority is decreased, the intraluminal color signal becomes spotty.

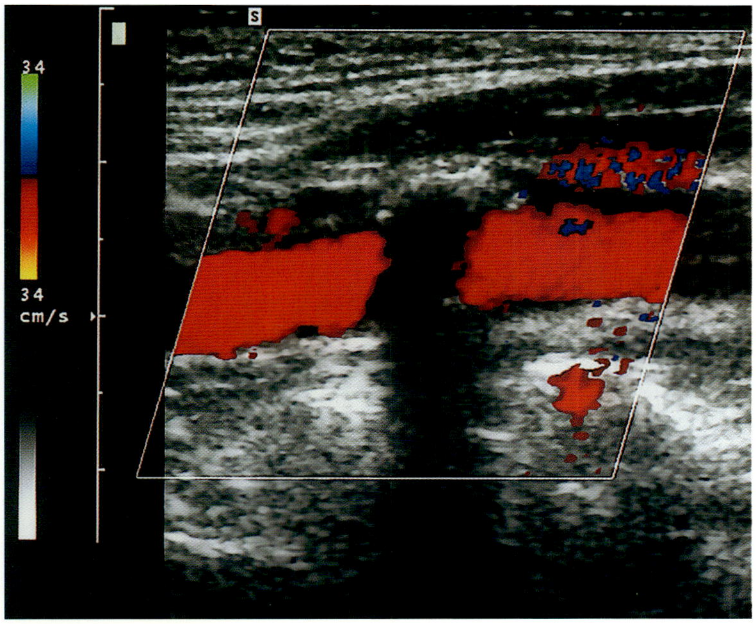

a

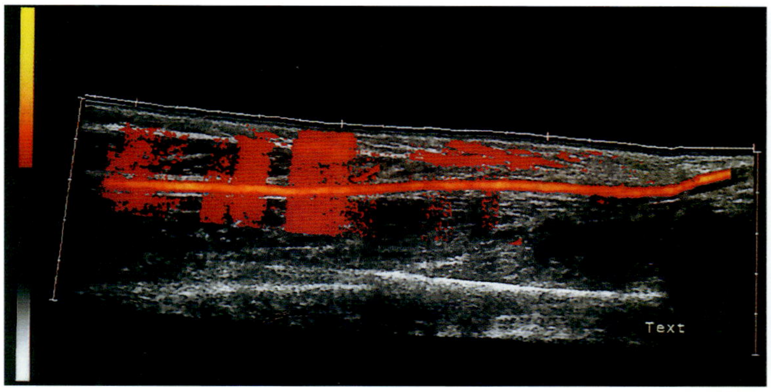

b

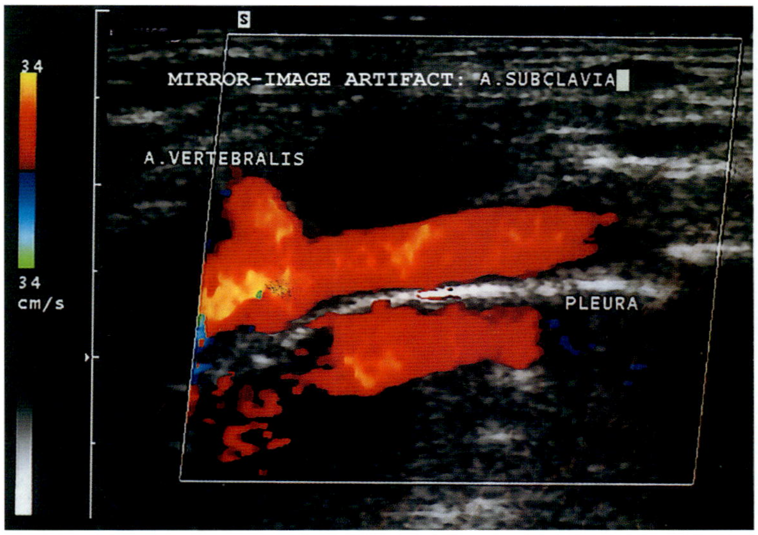

c

Fig. 1.**28** Color Doppler artifacts. II: Shadowing, motion artifacts, and mirror-image artifact.
a Calcific plaque with a heavy acoustic shadow. The shadow runs perpendicularly in the B-mode image, but the shadow in the color-flow image is oblique owing to the angle of the color box.
b Longitudinal scan of the posterior tibial artery, with motion artifacts (flash artifacts) induced by movement of the transducer.
c Mirror-image artifact along the pleural apex. At the top of the image is the subclavian artery with the origin of the vertebral artery. Below the subclavian is a "phantom vessel" caused by the reflection of color Doppler signals from the pleura, which is oblique to the beam.

New Technical Processes and Approaches

Technical advances in ultrasonography have been felt most strongly in the areas of transducer technology and manufacture; computer hardware with the capabilities of digital high-speed electronics; and new approaches to signal acquisition, filtering, and processing.

New Developments in Transducer Technology

Transducers with an increasing density of piezoelectric elements are offering higher resolution along with new, flexible scanning formats based on the steering capabilities of arrays (see Principle of Echo Detection and Scanning Techniques, p. 4).

Trapezoid scan. When a linear transducer is operated as a phased array, a sector field can be added to each side to expand the linear scan into a trapezoid scan. This provides a broader field of view and makes it possible to evaluate even hard-to-scan vascular segments such as the origin of the vertebral artery or common carotid artery from the brachiocephalic trunk.

2 D and multi-D arrays. Other innovations in transducer technology are two-dimensional (2 D) and multidimensional (multi-D) arrays. Unlike conventional array designs (one-line arrays), the 1.5 D array consists of multiple parallel rows of elements ($n \times m$ matrix). The rows are driven in the transmit and receive modes such that the slice thickness (elevation plane) can be kept optimally thin in both the near and far fields (Fig. 1.**29**).

The smallest elevation aperture (i.e., the center row of elements) is used for imaging in the near field, while the largest elevation aperture (i.e., all rows) is used for deeper scanning to help achieve good penetration depth and slice-thickness resolution. The greatest resolution increase occurs in the plane perpendicular to the transducer (Fig. 1.**29**).

Further subdivision and miniaturization of the array lines has led to the design of 2 D arrays that can scan along as well as perpendicular to the conventional scan plane. This makes it possible to acquire data from a pyramid-shaped volume and generate sectional images with any desired orientation.

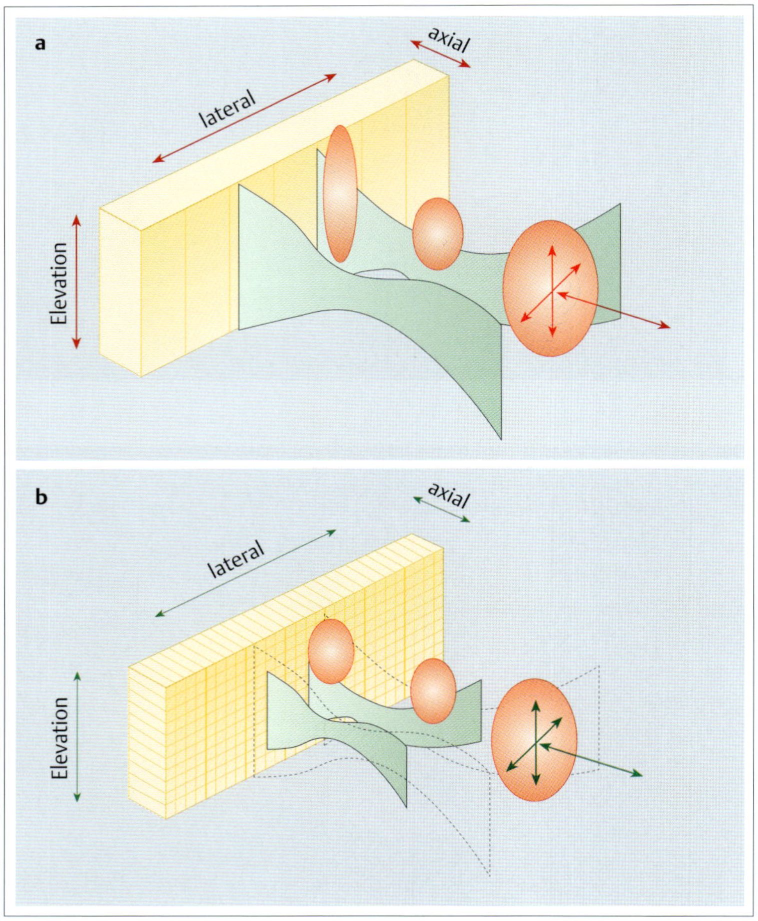

Fig. 1.**29** Improved resolution with a multi-dimensional transducer array. Slice-thickness focusing of a conventional transducer with a single row of elements (**a**) and a multi-dimensional array (**b**). While the axial resolution remains unchanged, the resolution perpendicular to the transducer plane (elevation plane) can be significantly improved by focusing.

New Techniques of Signal Acquisition and Processing

High-speed digital technology. The latest generation of ultrasound scanners can fully digitize the received echo signals while the signals are being acquired. They are also distinguished by their high computer capacity. The high-speed digital technology that is used in these systems combines the advantages of optimum signal scanning and digital technology (precision, signal fidelity, stability, signal dynamics) with a very high temporal resolution for transit time and phase correction. These optimal focusing conditions provide excellent lateral and axial resolution with very high signal dynamics, making it easier to appreciate diagnostic details.

Synthetic aperture. Other improvements result from enlarging the active surface area of the transducer (the aperture). This requires accurate, digital interim storage of the high-frequency echo signals from the individual array elements in unchanged form. This makes it possible to enlarge the aperture beyond the conventional theoretical maximum. In the synthetic aperture (e.g., SynAps) of the Sonoline Elegra system, echoes from portions of the array's various transmitted pulses are temporarily stored in digital form, digitally focused along with other adjacent subapertures, and utilized for image formation. As an example, Fig. 1.**30** shows the principle and effect of increasing the aperture by a factor of 2. Other factors can also be achieved. The more the aperture is enlarged, the better the lateral resolution and the greater the penetration depth and the ability to recognize details. The practical resolution limit is reached when all elements in the array are used to compute the information for every point in the ultrasound image.

Parallel processing. Another advantage of high-speed digital electronics is the ability to parallel-process multiple ultrasound lines from the information of one transmitted pulse. The practical advantages of parallel processing are a faster frame rate, especially in the two-dimensional Doppler techniques, and an improved ability to analyze rapid, dynamic processes and appreciate details.

Special filtering techniques. When special filtering techniques are applied, nonlinear tissue properties can be utilized to form the image. One cause of nonlinear tissue properties is the effect of ultrasound pressure on sound propagation velocity in the tissue, especially when high sound pressures are used. Because of this mechanism and the inherent differences in sound propagation velocity in different tissues, leading to different transit times, the pulse emitted by the transducer undergoes distortion in the tissue. If we analyze the frequency content of the received echoes, we find that they contain certain harmonic multiples, or overtones, of the basic frequency (Fig. 1.**31**).

Tissue harmonic imaging. Second-harmonic waves, or wave components at twice the transmitted frequency, are especially

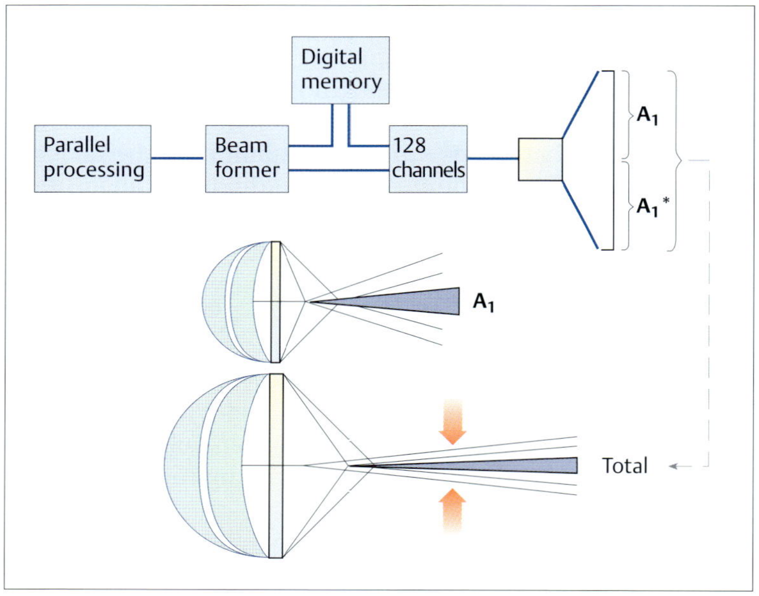

Fig. 1.**30** Synthetic aperture scanning: technical principle and effect. Each aperture consists of 128 elements, for example. With a digital interim memory, it is possible to combine the imaging information from multiple subapertures (two shown here: A_1, A_1*) into a large total aperture of 256 elements. The advantages are greater penetration depth and improved lateral resolution.

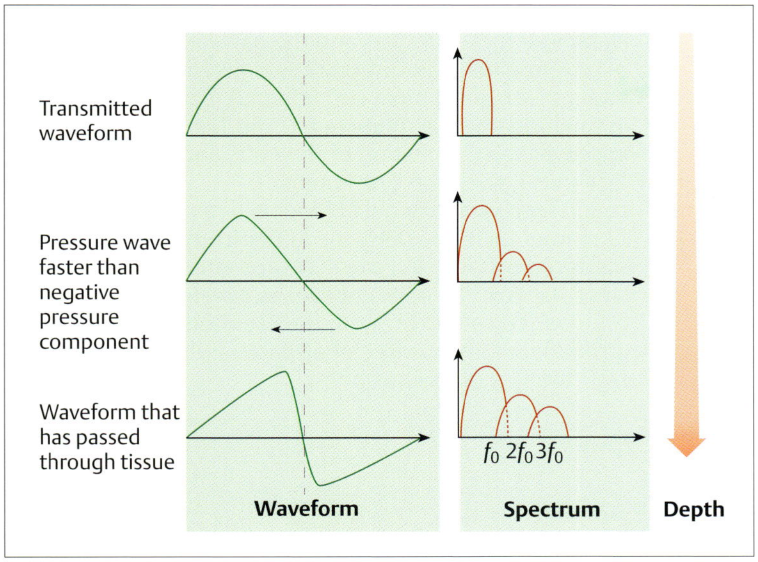

Fig. 1.**31** Effect of nonlinear sound propagation in tissue. A sound wave typically begins as a sine wave. The tissue is compressed, however, by the positive component of the pressure wave, and so the sound velocity is temporarily increased. This leads to a more rapid propagation of the positive part of the pressure wave (right arrow). During the rarefaction phase of the pressure wave, the tissue can expand again and the sound velocity decreases. As a result, the negative pressure component of the sound wave is conducted more slowly. This phase-dependent change in sound velocity distorts and steepens the original sine wave, analogously to a wave breaking on the shore. The effect increases with increasing depth. The result is that the reflected wave acquires a harmonic multiple of the basic transmitted frequency ("harmonic energy"). These frequencies, which are troublesome in a conventional B-mode image, can be specifically utilized for "tissue harmonic imaging" (THI). This technique can be used successfully even in patients who are technically difficult to scan.

prominent in the returned spectrum. Tissue harmonic imaging (THI) involves isolating these components and suppressing the basic frequency components. This technique has already been used to obtain "clearer," artifact-free images in patients who are technically difficult to scan.

Phase-inversion or pulse-inversion techniques. The phase-inversion or pulse-inversion techniques can be used to suppress the fundamental transmission spectrum and use only the harmonic frequency components over a wide frequency band to form the image (wideband tissue harmonic imaging, ensemble tissue harmonic imaging). Instead of a single pulse, two identical pulses are transmitted into the tissue per image line. The first is a standard ultrasound pulse, and the second is an exact copy with its phases inverted. The reflected and backscattered pulse pairs are subtracted (Fig. 1.**32**). All of the unchanged, "linear" signals cancel out. In this way all the signals from the tissue that are changed owing to nonlinear effects are selected out and can be used for image formation, independent of filtering techniques. The effect is cumulative and increases with penetration depth. The THI technique is most effective at moderate and greater depths. Further penetration past a depth of 12–15 cm is limited by absorption. This technique provides significantly greater image improvement than traditional filtering techniques for extracting the nonlinear (harmonic) components of the returned signal. The phase-inversion technique

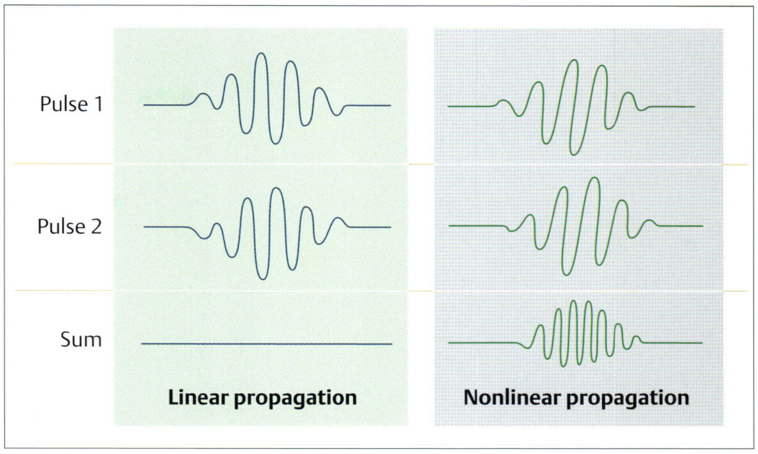

Fig. 1.**32** Principle of the phase-inversion or pulse-inversion technique. Two identical pulses whose phases are inverted (shifted 180°) are transmitted into the body (top and center rows: pulse 1 and pulse 2). When the returning signals are added, the linear (unchanged) echoes cancel out (lower left). The signals that have been distorted due to nonlinear effects (right) add together and can be used for image formation. This process suppresses the fundamental (basic) frequencies and the odd harmonic signal components while amplifying the even harmonic components, especially the second harmonics.

has already proved successful clinically in enhancing the visualization of smaller lesions (Fig. 1.**33**). When harmonic imaging is combined with ultrasound contrast agents, even the smallest metastatic lesions can be identified (Fig. 1.**34**).

Programmable image processor. By incorporating a special programmable image processor into the ultrasound system, it is possible to raise the computer capacity for image processing (several billion operations per second) to a level that was previously reserved for external workstations. In this way, various image processing tasks that formerly had to be done on separate systems (offline) can now be performed directly (online) and in real time. An example of this is adaptive contrast enhancement. The result is ultrasound images that are better suited to the physiology of human vision, making it possible to conduct the examination even in a bright room.

Panoramic imaging. Another application based on the programmable image processor is panoramic imaging, first introduced as the SieScape technology in the Sonoline Elegra system. In panoramic imaging, an extended field of view is obtained by manually scanning a standard transducer over the region of interest without a position sensor. This information is first displayed on the monitor in standard fashion. Based on the level of agreement between the individual real-time images, correlational algorithms are used to determine the transducer position on the body. This positional information is then used to seamlessly add the new image to the previous image. Thus when the transducer is moved along one plane or around an area, a real-time panoramic image is obtained. In this way the operator can freely define an image format that is optimum for the specific anatomy (Fig. 1.**35**). Panoramic imaging has been combined with power Doppler flow imaging to produce a color-encoded panoramic imaging process marketed as Color SieScape.

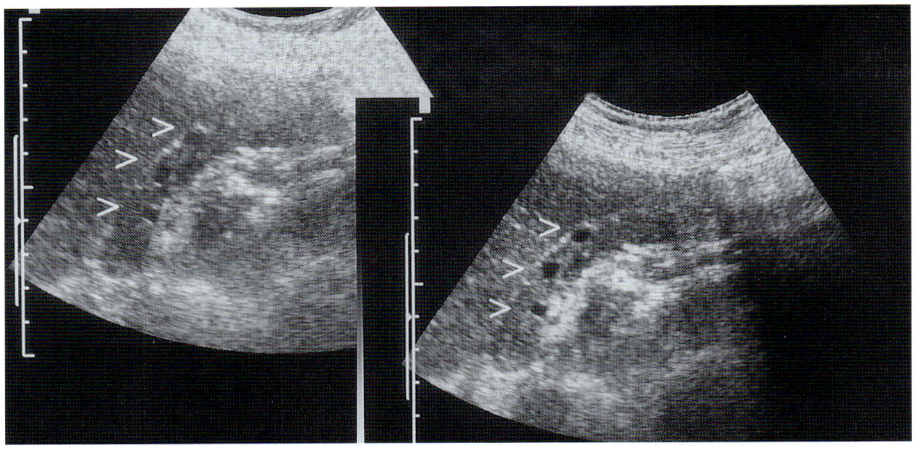

a

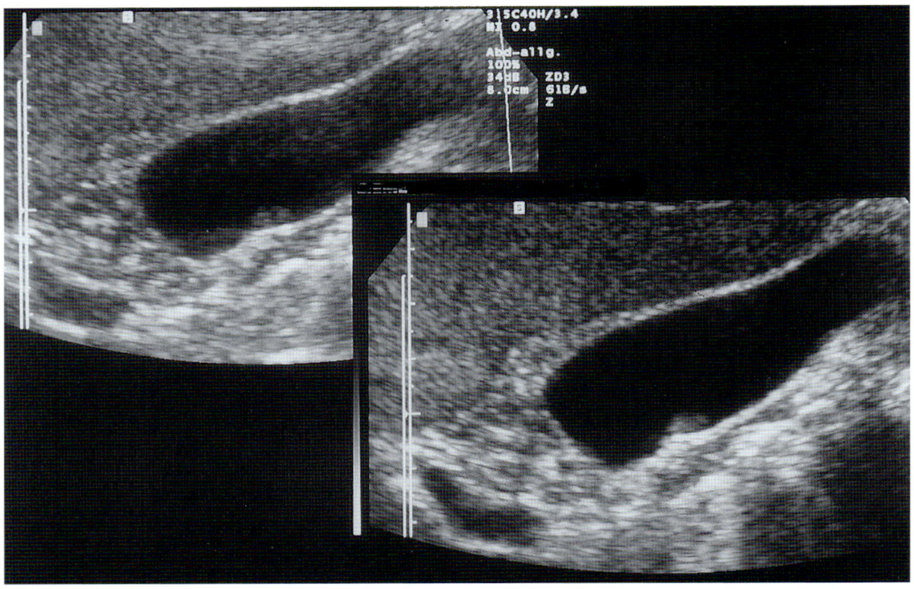

b

Fig. 1.**33** Effect of tissue harmonic imaging (THI), illustrated for renal cysts and the gallbladder.
a Longitudinal scan through the right kidney of a patient with small renal cysts that are barely detectable in the fundamental mode (left). The THI mode (right) improves the delineation of the cysts and gives an artifact-free view of the cyst lumen.
b Enlarged view of a gallbladder with a polyp. In the fundamental mode (left) the lumen contains spurious scattered echoes, and there are accentuated semicircular artifacts in the adjacent liver as a result of reverberations with a curved array. In the THI mode (right), the gallbladder lumen is free of artifacts and the other artifacts are less pronounced.

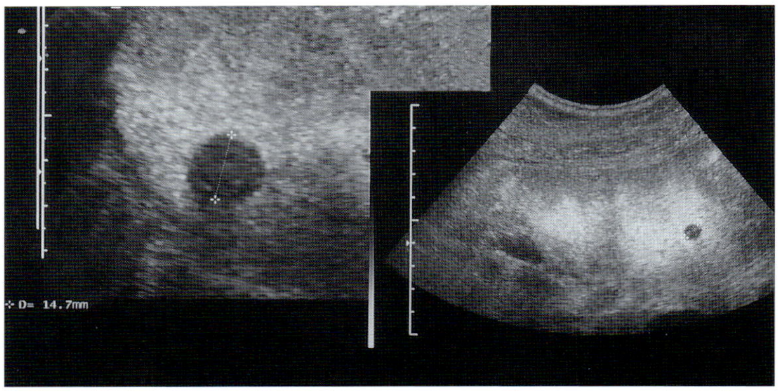

Fig. 1.**34** Effect of contrast harmonic imaging in the detection of small hepatic metastases following the injection of 300 mg/ml Levovist. After 2.5 minutes the contrast agent has become concentrated in the reticuloendothelial system (RES). The transmit power is increased, causing the contrast microbubbles to vibrate and burst. This “acoustic emission” produces hyperechoic enhancement of the liver parenchyma.
Foci that lack a RES do not enhance, and remain hypoechoic. As a result, even small metastases that are difficult or impossible to detect on plain images can be identified by their sharp contrast with the surrounding hepatic tissue.

2 Safety Aspects of Doppler and Color Doppler Sonography

H.-D. Rott

Doppler sonography, with its ability to define and evaluate the fetal blood supply, has been an important addition to diagnostic ultrasound in obstetrics. Because these techniques require power outputs and ultrasound intensities that are considerably higher than in B-mode imaging, the essential safety that has been established for diagnostic ultrasound in general cannot be assumed for all applications. Consequently, the sonographer should be familiar with potential bioeffects and their physical mechanisms in order to avoid any risks that may be associated with the examination.

Mechanisms of Tissue Effects

The biological effects of ultrasound are based largely on heating and cavitation. These effects are dependent on different sound field parameters and tissue properties and can therefore occur separately from each other. Other primary physical effects have no bearing on the clinical safety of ultrasound.

Heating

Ultrasound energy that is transmitted into tissue is partially reflected, partially scattered, and partially absorbed and converted to heat. The degree of heating depends on different properties of the ultrasound field and the exposed tissue. A key factor is the spatial peak time average intensity (I_{SPTA}). In pulsed Doppler flowmetry, however, the power output of the device is a more important factor than field intensity[4, 23]. Ultrasound frequency is also a factor, since higher frequencies are absorbed more strongly, and therefore the applied energy is concentrated over a smaller volume.

Absorption coefficient. Sound absorption in biological tissues increases with the protein content of the tissue[19]. Water and body fluids such as urine and amniotic fluid absorb almost no sound energy, and their absorption coefficient α is between 0.002 and 0.003 dB cm^{-1} MHz. Soft tissues such as brain, liver, kidney, and muscle have approximately the same absorption ($\alpha = 0.4–0.6$ dB cm^{-1} MHz). Bone is by far the strongest absorber ($\alpha = 5–10$ dB cm^{-1} MHz), placing it at greatest thermal risk. The secondary heating of soft tissues in close proximity to bone should also be considered when evaluating risk[21, 25]. Acoustically, the human embryo is equivalent to soft tissue. By the end of the first trimester, however, absorption increases owing to early mineralization of the bone. Thereafter it continues to increase during the rest of the fetal period.

Heat conduction and perfusion. Heat conduction and blood flow contribute to the elimination of tissue heat. They are the major factors that determine the final temperature that is reached, especially on prolonged exposure. Whereas heat conduction can be calculated to a good approximation, the perfusion in different tissues and organs varies substantially, and so the cooling component in different tissues can only be roughly estimated. With brief ultrasound exposure, perfusion contributes little to tissue cooling[5].

Hyperthermia. An increase in temperature may be a risk, since cell division is inhibited above 30 °C, and temperatures higher than 41 °C can cause cell death with sufficiently prolonged exposure. A key factor besides the amount and duration of tissue heating is the sensitivity of the affected tissue[3]. The embryo-fetal brain is extremely sensitive to temperature increases. Although brain tissue has a low absorption coefficient, secondary heating from the more strongly absorbing cranial bone must also be taken into account[4]. Animal experiments have shown that whole-body hyperthermia above 41 °C can be teratogenic, depending on the duration of exposure[11, 24].

Cavitation

Noninertial and inertial cavitation. Cavitation refers to the sound-induced formation and dynamic behavior of cavities and gas bubbles, which can produce a variety of physical, chemical, and biological effects in the medium[25]. Two types of cavitation occur: noninertial cavitation, which describes the prolonged resonant oscillations of preexisting gas microbubbles that do not collapse; and inertial cavitation, in which preexisting microbubbles ("cavitation nuclei") rapidly expand in the rarefaction phase of the ultrasound wave and collapse again in the subsequent compression phase. This process can generate local pressure amplitudes greater than 1000 MPa and local temperature peaks in excess of 1000 °C. This may be associated with electrical arcing ("sonoluminescence") and the formation of free radicals (OH^-, H^+, H_2O_2, etc.) with mutagenic properties that are indistinguishable from the radicals produced by ionizing radiation[15, 25]. A continuum exists between stable and transient cavitation.

Sound-pressure amplitudes. Cavitation is a threshold effect. The critical field parameter that determines whether cavitation occur is the negative sound-pressure amplitude p^- (by consensus, this amplitude is stated as a positive value in MPa). Under certain physical conditions, however, the signs of the pressure values may change when an ultrasound wave is reflected, so that the reflected wave acquires a negative pressure amplitude equal to the positive pressure amplitude of the incident wave. As a result of this, the positive pressure amplitude may also be a relevant quantity. The cavitation threshold increases at higher frequencies.

Cavitation nuclei and cavitation threshold. Human tissue is fairly resistant to cavitation because generally it does not contain cavitation nuclei. Such nuclei are presumed to exist, however, following infusions, administration of echo contrast agent, gas gangrene infections, and open injuries. The negative sound-pressure amplitude necessary for cavitation to develop is unknown. In shock-wave lithotripsy, cavitation is known to occur at negative pressures that exceed 10 MPa, but modern sonographic techniques employ fields whose negative peak pressure does not exceed 5 MPa (Fig. 2.**1a**). Cavitation on a large scale would seriously affect imaging because the induced bubbles would cause extreme increases in scattering and absorption. So far these effects have not been observed in ultrasound examinations. Cavitation, then, is unlikely to be a risk factor in the ultrasound techniques in current use. It should be added, however, that echo contrast agents greatly lower the threshold for cavitation (see Ultrasound Contrast Agents, p. 31).

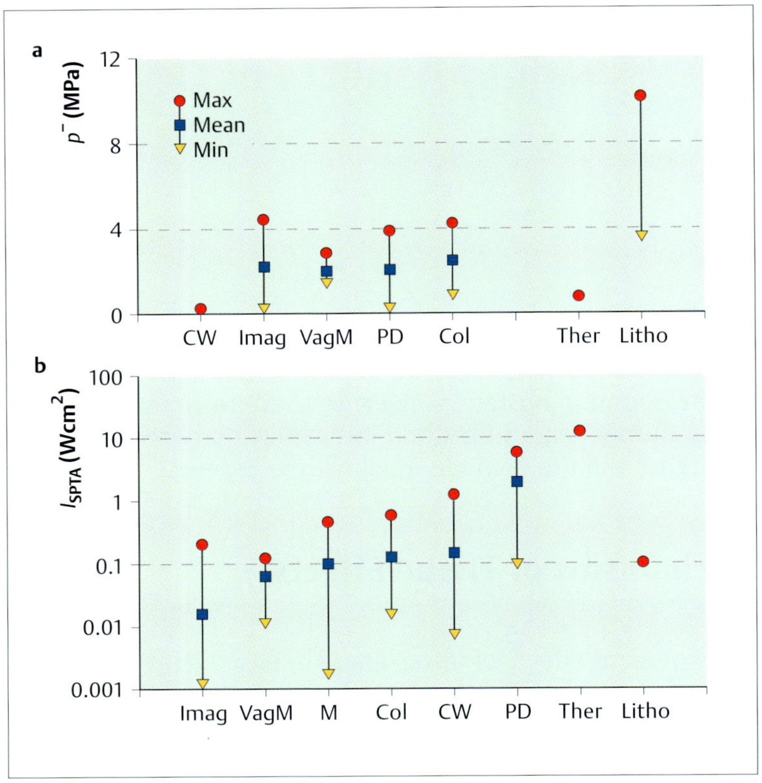

Fig. 2.**1** Minimum (min), maximum (max) and mean values of the negative peak pressure p^- (**a**) and I_{SPTA} intensity (**b**) of various European ultrasound scanners in various modes. The corresponding values of therapeutic and lithotripsy devices are also shown for comparison. Imag: B-mode imaging; VagM: M-mode with a vaginal probe; M: conventional M-mode; Col: color Doppler; CW: continuous-wave Doppler; PD: pulsed Doppler; Ther: physiotherapy; Litho: lithotripsy. (After reference 6.)

Risk Assessment of Various Ultrasound Techniques

The risks associated with various ultrasound techniques depend chiefly on the likelihood of a biologically significant temperature rise and the potential for inducing cavitation. The field parameters of current diagnostic instruments that have a bearing on these effects are reviewed in Fig. 2.**1**. Recent measurements have essentially confirmed these values while also showing that newer systems tend to have higher field intensities but do not have higher negative peak pressures[14].

Duplex Sonography

Duplex sonography combines B-mode imaging with pulsed Doppler flowmetry. B-mode provides the sectional image for locating the site of interest and positioning the Doppler sample volume, while flowmetry yields the desired information on blood flow velocity.

Heating. Biological risks have been ruled out for B-mode imaging. Flowmetry requires the additional use of a Doppler beam of greater pulse length and higher intensity (I_{SPTA}). The intensity of this stationary Doppler beam may fall within the upper range of therapeutic intensities. Animal experiments have shown that Doppler pulses can produce a biologically significant temperature rise in soft tissues, and that this effect can no longer be considered harmless[4,23] (Fig. 2.**2**). The heating effect can become particularly hazardous when bone located below the Doppler sample volume is exposed to the ultrasound beam, since bone has a high absorption capacity and undergoes rapid heating. It was found, for example, that when fetal guinea-pig brain was insonated within the intact skull, the temperature rise was greater near the bone on the opposite side of the skull than at the center of the brain[4]. Based on these findings, it is conceivable that instruments with a high power output and prolonged scan time (>30 seconds) can induce temperature rises during flowmetry that cannot be considered harmless.

Cavitation. Because the negative peak pressure is no greater in Doppler ultrasound than in B-mode and does not exceed 5 MPa, cavitation should not occur.

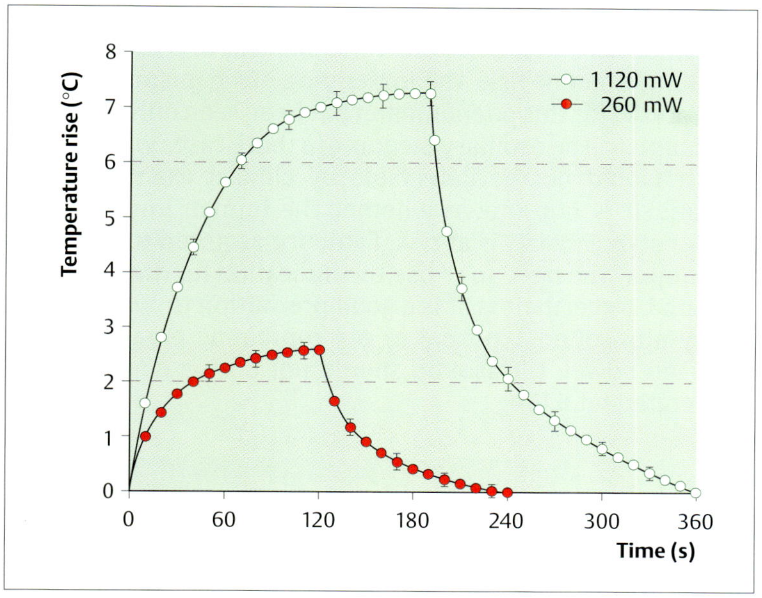

Fig. 2.2 Temperature rise over exposure time at the center of isolated guinea-pig brain.
Ultrasound field parameters: frequency 3.3 MHz, pulse duration 6.25 µs, PRF 4 kHz.
Upper curve: total power output 1120 mW, I_{SPTA} 2.5 W/cm.
Lower curve: total power output 260 mW, I_{SPTA} 2.9 W/cm
(After reference 4.)

Color Doppler

In color Doppler systems, blood flow velocities are determined not just in a single sample volume but over a larger "region of interest" as in B-mode imaging. These velocities are color-encoded according to their direction and magnitude and are superimposed over the B-mode image.

Heating. Because the Doppler beam is not stationary as in duplex sonography but is swept across the region of interest, the absorbed energy is distributed over a larger tissue volume, resulting in a substantially lower time average intensity. A useful rule of thumb is that the I_{SPTA} intensities in color Doppler are 10 times higher than in conventional B-mode imaging, while the intensities in duplex scanning are 100 times higher (Fig. 2.**1b**). There are considerable differences, however, among different scanners and modes of operation. For example, as the region of interest is increasingly narrowed in its lateral dimension, the thermal effect is increased. When the region of interest is narrower than 2 mm, the thermal conditions are practically the same as in duplex scanning. As a general rule, however, the thermal risks associated with ordinary clinical use are small.

Cavitation. As in duplex scanning, cavitation should not occur in color Doppler imaging.

Power Doppler

The ultrasound exposure in power Doppler imaging is the same as in color Doppler. The processing mode is different, however, as power Doppler displays signal amplitudes rather than blood cell velocities. This type of signal processing can detect even slow flow velocities, providing a more accurate display of vessel volumes. The thermal effect corresponds to that of color Doppler.

Color Velocity Imaging (CVI)

While this technique does not employ the Doppler principle, it does bear some similarities to color Doppler. The velocity and direction of blood flow are determined by the computer comparison of successive B-mode echo images. The principle is that B-mode can detect acoustic inhomogeneities ("streaks") in flowing blood and can track their motion. CVI requires a large computer capacity but uses low ultrasound intensities and is thermally harmless.

Transvaginal Scanning

A frequent concern in transvaginal scans is excessive exposure of the embryo owing to the proximity of the endovaginal probe. The opposite should be true, however, since the short range leads to less absorption, and therefore a lower power setting can safely be used. However, manufacturers have offset this advantage by using higher frequencies, which, while improving image resolution, also cause greater sound absorption that necessitates higher intensities. As a result, the average intensities in transvaginal B-mode scanning are somewhat higher than in abdominal scanning, but they are slightly lower in transvaginal pulsed Doppler[7]. The thermal exposure to the embryo or fetus is approximately the same as in transabdominal sonography.

Ultrasound Contrast Agents

Gas-containing ultrasound contrast agents are being used with increasing frequency. They appear to be immunologically safe, and microbubble sizes on the order of 10 µm do not affect capillary blood flow.

Cavitation. At the same time, ultrasound contrast agents considerably lower the threshold for cavitation. It is unknown whether administration of echo contrast agent is sufficient to induce cavitation by diagnostic ultrasound. Stable cavitation would increase the echogenicity of the contrast agent, but local hemolysis would also be expected to occur. So far, however, there have been no clinical reports of hemolytic effects[18]. Inertial cavitation would also be associated with free radical formation. This would contrast with the similar effect of ionizing radiation in the distribution pattern in the tissue: presumably the radicals would form chiefly in the intercellular compartment owing to the lower viscosity and would form intravascularly only at sites where the contrast agent is excited by ultrasound. It is unclear at present whether this is apt to produce mutagenic effects. Since no experimental data have yet been published on this issue, a definitive risk assessment cannot be made. On the whole, the risk does not appear to be particularly high[13]. There are no concrete objections to using ultrasound contrast agents during pregnancy, provided they are limited to the visualization of maternal tissues.

Exposure of Gas-Containing Tissues

Recent studies in various laboratory mammals such as mice, rabbits, pigs, and monkeys have shown that diagnostic insonation of the lung can induce local extravasation of capillary red cells if the negative pressure amplitudes is above 1 MPa. Greater negative pressures are needed to produce this effect in larger animal species than in small animals. The same effect has been observed at slightly higher thresholds (≤ 2 MPa) in gas-filled bowel (review in reference 20). Apparently this effect is nonthermal and requires air-filled cavities, and therefore the fetal lung is not at risk. The underlying mechanism is still unclear, and so far this phenomenon has been seen only in laboratory animals. The capillary bleeding in these cases was so slight that it would not be detectable by clinical examination in humans. It is not known whether the human lung, with its greater alveolar size, is at risk. Evidence acquired to date does not support the need to revise the indication for diagnostic ultrasound. Nevertheless, it is a good precaution to avoid unnecessary ultrasound exposure to gas-containing organs like the lung and bowel. Ultrasound should not pose a danger to the nonaerated fetal lung.

Measures to Limit Risk

Ultrasound field parameters. Users should know the ultrasound field parameters in the various modes that are relevant to clinical safety. International Standard 1157 of the IEC[16] requires that manufacturers provide users with necessary information on equipment exposure parameters. The German Union of Fund-Affiliated Physicians approves for clinical use only equipment that conforms to this standard[17].

Some ultrasound equipment displays output indices on the monitor screen in accordance with the requirements of the Output Display Standard (ODS)[2]. However, the user must be given appropriate information in advance in order to interpret the indices that are displayed (thermal index TI and mechanical index MI)[9].

Training. A final essential aspect of risk abatement is the qualified training of examiners. This applies both to the risk-free use of diagnostic ultrasound equipment and to the avoidance of diagnostic errors.

Recommendations

Various national and international institutions have published recommendations on the prevention of health risks from diagnostic ultrasound[1,22,24,25]. In all of these recommendations, it is assumed that the ultrasound energy transmitted into the body should be kept low as a precautionary measure, even if bioeffects are not expected to occur. The recommendations are summarized below.

General Recommendations

1. The power output of ultrasound equipment should be as low as possible, and the gain should be set as high as possible.
2. The operator should know whether the equipment continues to emit ultrasound in the freeze-frame mode and, if necessary, should break transducer–skin contact.
3. Ultrasound examinations should be medically indicated. There are no safety objections to the routine B-mode ultrasound screening of all pregnancies.
4. Air-containing organs such as the lung, stomach, and bowel should not be exposed unnecessarily to ultrasound.
5. The use of ultrasound contrast agents requires a definite indication. Their use should be preceded by an individual risk–benefit analysis for the patient.

Pulsed Doppler

6. Pulsed Doppler for flowmetry should be activated only after the vessel of interest has been identified with color Doppler and the sample volume (range gate) has been defined. The sample volume should be kept small, because some instruments compensate for attenuation with time gain compensation (TGC) even in the pulsed Doppler mode.
7. In fetal and neonatal examinations, exposure of bone below the region of interest should be minimized or avoided.
8. The Doppler scan time should be kept short to avoid possible excessive heating and should not exceed 30 seconds if possible. Repeat sampling, if necessary, should be done after a delay of 60 seconds.
9. Fever potentiates the thermal risk, and therefore a shorter scan time should be used in febrile patients.
10. Routine screening of the fetal and placental circulation with Doppler ultrasound cannot be recommended for every pregnancy owing to continued uncertainties about risks.

Appendix: Statements on the Biological Safety of Diagnostic Ultrasound Fields

EFSUMB Statement on the Clinical Safety of Diagnostic Ultrasound[10]

Diagnostic ultrasound has been widely used in clinical medicine for many years with no proven deleterious effects. However, as the use of ultrasound increases, with the introduction of new techniques, with a broadening of the medical indications for ultrasound examinations, and with increased exposure, continuous vigilance is essential to ensure its continued safe use.

A broad range of ultrasound exposure is used in the different diagnostic modalities currently available. Doppler imaging and measurement techniques may use higher exposures than those used in B- and M-modes, with pulsed Doppler techniques having the potential for the highest levels.

The recommendations contained in this statement assume that commercial ultrasound equipment conforming to international safety standards is being used, and that it is used prudently, by competent personnel who are trained in safety matters.

B- and M-modes. Based on scientific evidence of ultrasound-induced biological effects to date, there is no reason to withhold B- or M-mode scanning for any clinical application, including the routine clinical scanning of every woman during pregnancy. Some techniques, such as tissue harmonic imaging and coded excitation, may use higher exposures than conventional imaging. The user is advised to monitor the TI and MI values on the screen and to avoid unnecessarily high values and prolonged exposure times. Scanning for 3 D imaging does not introduce any additional safety considerations.

Doppler modes (color flow, power Doppler imaging and spectral pulsed Doppler). Exposures used in these Doppler modes are commonly higher than for B- and M-modes. The highest powers, and therefore the greatest potential for thermal effects, occur in spectral pulsed Doppler mode at high settings of the output power and scale controls and in Doppler imaging modes when using narrow or deep color boxes.

The informed use of Doppler ultrasound is not contraindicated. However, at maximum machine output settings, significant thermal effects at bone surfaces cannot be excluded. The user is advised to monitor the TI and MI (mechanical index) values on the screen, and to act prudently to limit these, and exposure times, when scanning sensitive structures and regions of bone and gas.

Doppler for fetal heart monitoring (CTG). The power levels used for fetal heart monitoring (CTG) are sufficiently low that the use of this modality is not contraindicated on safety grounds, even when it is to be used for extended periods.

Gas-filled contrast agents. Particular safety considerations are associated with the use of gas-filled contrast agents. High values of mechanical index should be used only when required for a particular clinical study.

Ultrasound exposure during pregnancy. The embryonic period is known to be particularly sensitive to any external influences. Until further scientific information is available, investigations should be carried out with careful control of output levels and exposure times.

With increasing mineralization of the fetal bone as the fetus develops, the possibility of heating fetal bone increases. The user should prudently limit exposure of critical structures such as the fetal skull or spine, particularly during Doppler studies.

Safety considerations for other organs. Particular care should be taken to reduce the risk of heating during investigations of the eye. Extra care is also appropriate when carrying out neonatal cardiac and cranial investigations.

(1994, revised 1998, 2002, and 2003)

WFUMB Statement on Thermal Effects in Clinical Applications[25]

B-mode imaging. Known diagnostic ultrasound equipment, as used today for simple B-mode imaging, operates at acoustic outputs that are not capable of producing harmful temperature rises. Its use in medicine is, therefore, not contraindicated on thermal grounds. This includes endoscopic, transvaginal and transcutaneous applications.

Doppler. It has been demonstrated in experiments with unperfused tissue that some Doppler diagnostic equipment has the potential to produce biologically significant temperature rises, specifically at bone–soft tissue interfaces. The effects of elevated temperatures may be minimized by keeping the time for which the beam passes through any one point in tissue as short as possible. Where output power can be controlled, the lowest available power level consistent with obtaining the desired diagnostic information should be used. Although the data on humans are sparse, it is clear from animal studies that exposures resulting in temperatures less than 38.5 °C can be used without reservation on thermal grounds. This includes obstetric applications.

Transducer heating. A substantial source of heating may be the transducer itself. Tissue heating from this source is localized to the volume in contact with the transducer.

(1991, revised 1996)

References

1 American Institute of Ultrasound in Medicine (AIUM): Bioeffects Considerations for the Safety of Diagnostic Ultrasound. AIUM, Bethesda Md. 1988

2 American Institute of Ultrasound in Medicine and National Electrical Manufacturers Association (AIUM/NEMA): Standard for Real Time Display of Thermal and Mechanical Acoustic Output Indices on Diagnostic Ultrasound Equipment. AIUM, Rockville Md. 1992

3 Barnett SB, Rott H-D, Ter Haar G, Ziskin MC, Maeda K, Nyborg W: The sensitivity of biological tissue to ultrasound. Ultrasound Med. Biol. 23 (1997) 805–812

4 Bosward KL, Barnett SB, Wood AKW, Edwards MJ, Kossoff G: Heating of guinea-pig fetal brain during exposure to pulsed ultrasound. Ultrasound Med. Biol. 19 (1993) 415–424

5 Carstensen EL, Child SZ, Norton S, Nyborg WL: Ultrasonic heating of the skull. J. Acoust. Soc. Am. 87 (1990) 1310–1317

6 Duck FA, Martin K: Trends in diagnostic ultrasound exposure. Phys. Med. Biol. 36 (1991) 1423–1432

7 European Committee for Ultrasound Radiation Safety—The Watchdogs: Tutorial Paper: Transvaginal ultrasonography—Safety aspects. Eur. J. Ultrasound 1 (1994) 355–357

8 European Federation of Societies for Ultrasound in Medicine and Biology (EFSUMB): Guidelines for the safe use of Doppler ultrasound for clinical applications. Eur. J. Ultrasound 2 (1995) 167–168

9 European Federation of Societies for Ultrasound in Medicine and Biology (EFSUMB): Tutorial Article: Thermal and mechanical Indices. Eur. J. Ultrasound 4 (1996) 145–150

10 European Federation of Societies for Ultrasound in Medicine and Biology (EFSUMB): Clinical Safety Statement for Diagnostic Ultrasound 2003. EFSUMB Newsletter 17(1) (2003) 15

11 European Federation of Societies for Ultrasound in Medicine and Biology (EFSUMB): Tutorial: Thermal teratology. Eur. J. Ultrasound 9 (1999) 281–283

12 European Federation of Societies for Ultrasound in Medicine and Biology (EFSUMB): Tutorial: Acoustic cavitation and lung haemorrhage. Eur. J. Ultrasound 9 (1999) 277–280

13 European Federation of Societies for Ultrasound in Medicine and Biology (EFSUMB): Tutorial: Safety of ultrasonic contrast agents. Eur. J. Ultrasound 9 (1999) 195–197

14 Henderson J, Willson K, Jago JR, Whittigham TA: A survey of the acoustic outputs of diagnostic ultrasound equipment in current clinical use. Ultrasound Med. Biol. 21 (1995) 699–705

15 Heusinger H: Comparison of the reactions induced by ultrasound and gamma rays in aqueous lactose solutions. Ultrasonics 28 (1990) 30–36

16 International Electrotechnical Commission (IEC): International Standard 1157: Requirements for the Declaration of the Acoustic Output of Medical Diagnostic Ultrasonic Equipment. Genf 1992

17 Kassenärztliche Bundesvereinbarung (KBV): Qualitätsvoraussetzungen gemäß § 135 Abs. 2 SGB V zur Durchführung von Untersuchungen in der Ultraschalldiagnostik (Ultraschallvereinbarung) vom 10.02.1993. Dtsch. Arztebl. 90 (1993) B390–403

18 Nanda NC: Echocontrast enhancers – how safe are they? Advances in Echo Contrast 2 (1993) 97–110

19 National Council on Radiation Protection and Measurements (NCRP): Exposure Criteria for Medical Diagnostic Ultrasound: I. Criteria based on Thermal Mechanisms. NCRP Report No 113, Bethesda Md. 1992

20 Rott H-D: Capillary lung bleeding from exposure to diagnostic ultrasound – A literature review. BMUS Bullentin 5 (1997) 20–21

21 Rott H-D: Ultraschalldiagnostik: Neuere Bewertung der biologischen Sicherheit. Dtsch. Arztebl. 93 (1996) A1533–1537

22 Strahlenschutzkommission (SSK) des Bundesministeriums für Umwelt, Naturschutz und Reaktorsicherheit: Empfehlungen zur Patientensicherheit bei Anwendungen der Ultraschalldiagnostik in der Medizin. Empfehlungen der Strahlenschutzkommission (Heft 14). Empfehlungen und Dokumentationsteil. Fischer, Stuttgart 1998

23 Ter Haar GR, Duck FA, Starritt HC, Daniels S: Biophysical characterisation of diagnostic ultrasound equipment – preliminary results. Phys. Med. Biol. 34 (1989) 1533–1542

24 World Federation for Ultrasound in Medicine and Biology (WFUMB): Symposium on Safety and Standardisation in Medical Ultrasound: Issues and Recommendations Regarding Thermal Mechanisms for Biological Effects of Ultrasound. Barnett SB, Kossoff G (eds.): Ultrasound Med. Biol. 18/9 (Special Issue) 1992

25 Barnett SB, Ter Haar GR, Ziskin MC, Rott HD, Duck FA, Maeda K: International recommendations and guidelines for the safe use of diagnostic ultrasound in medicine. Ultrasound Med. Biol. 26 (2000) 355–66

Infertility Evaluation and Assisted Reproduction

3 Uterine Blood Flow in Fertile and Infertile Women

S. Kupesic and A. Kurjak

Uterine Blood Supply

Anatomy. The uterus derives most of its blood supply from the uterine arteries, with a small amount supplied by the ovarian arteries. The uterine arteries branch into the arcuate arteries (Fig. 3.**1**), which project circumferentially into the outer third of the myometrium. The arcuate arteries divide further into the radial arteries, which then branch into the basal arteries and spiral arteries after crossing the myometrial–endometrial boundary. The relatively short basal arteries terminate in a capillary bed that supplies the stratum basale of the endometrium. The spiral arteries project farther into the endometrium and open into a broad capillary bed that supplies the stratum functionale of the endometrium.

Menstrual cycle. Interestingly, only the spiral arteries show demonstrable anatomical changes during the menstrual cycle[24]. These arteries contract when menstruation occurs, possibly as a result of falling estrogen and progesterone levels. This vasoconstriction induces local hypoxia, ischemia, and cell death in the stratum functionale. The distal portion of the arteriolar vascular bed and the capillary bed are then sloughed along with the stratum functionale. The basal arteries do not react to falling estrogen and progesterone levels and thus serve to maintain the integrity of the stratum basale during the menstrual cycle[12]. A 3-fold to 5-fold increase in endometrial thickness occurs during the next menstrual cycle. To ensure this, a completely new capillary bed arises from the rapidly growing proximal portions of the spiral arteries. The process begins with the growth of new capillaries from the stratum basale[12]. A new intima and muscular coat develops in the capillary walls[20].

Implantation. The angiogenesis that occurs during the proliferative and secretory phases of the menstrual cycle serves to prepare the endometrium for the implantation of a fertilized ovum[13]. After fertilization has occurred, implantation begins with the attachment of the blastocyst to the endometrial surface. As the trophoblastic cells invade the endometrium, maternal capillaries are breached, causing profound physiological and structural changes to occur in the uterine vascular bed[21]. One of the first changes is increased vascular permeability at the implantation site[3]. This is followed by a metabolic activation of the endometrium in preparation for angiogenesis[3]. Subsequent placentation is facilitated by the changes in the maternal vascular bed. Further details on this process can be found in the section on the Doppler examination of early placentation and embryonic blood flow (p. 103).

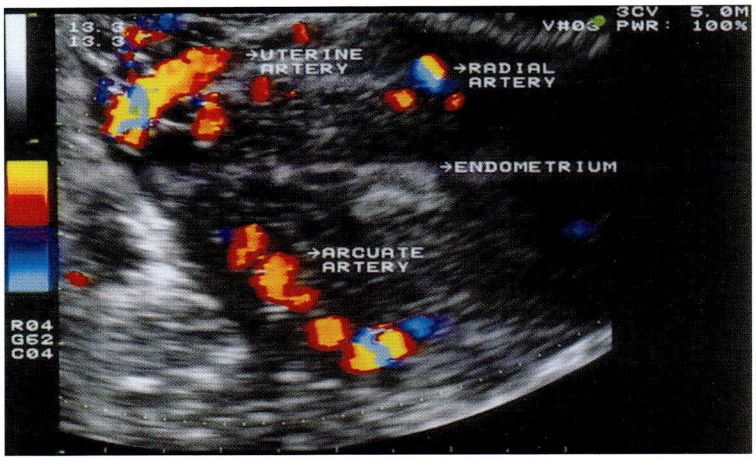

Fig. 3.**1** Transvaginal image of an arcuate arterial network.

Changes in Uterine Blood Flow during the Menstrual Cycle

More than any other available examination technique, transvaginal color Doppler sonography requires an accurate knowledge of the location of the uterus, myometrium, endometrium, and the vessels that supply them[5, 11, 17, 18].

It is known that uterine perfusion depends largely: on the age of the patient, the phase of the menstrual cycle, and other special factors such as pregnancy or neoplasia[19]. These factors will be analyzed below in detail.

Blood Flow Parameters in the Uterine Arteries

Complex relationships exist between ovarian hormone levels in the peripheral venous blood and the blood flow parameters in the uterine arteries[2, 8, 9, 23]. During the proliferative phase, a small amount of end-diastolic blood flow can be detected in the uterine arteries in the majority of women (Fig. 3.**2**). The resistance index (RI) until day 13 of a 28-day cycle is in the range 0.88 ± 0.04. Steer et al.[23] reported that diastolic flow was no

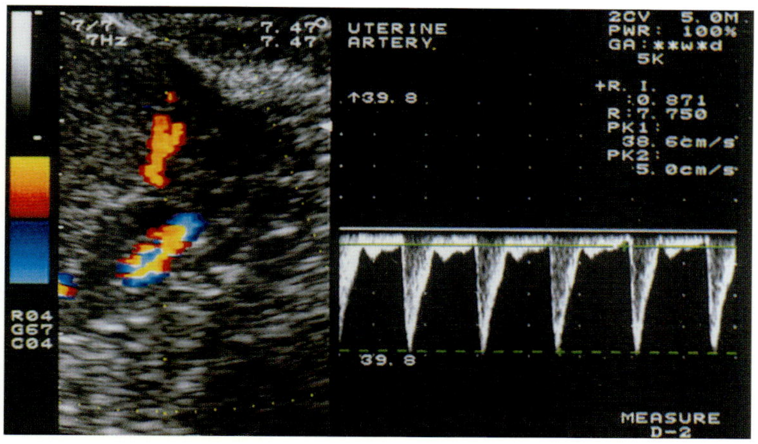

Fig. 3.**2** The uterine artery appears to the left of the cervix at the junction of the cervix and uterine corpus. The blood flow velocity in the uterine artery during the proliferative phase is characterized by low end-diastolic blood flow and a high vascular resistance (RI = 0.87).

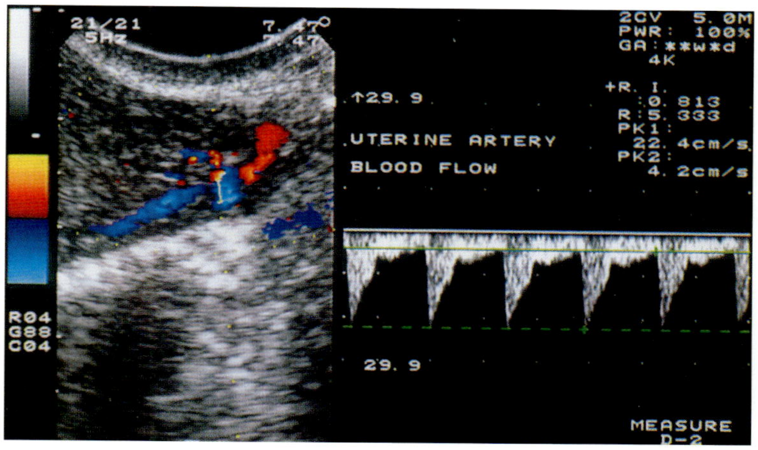

Fig. 3.**3** The blood flow velocity in the uterine artery during the secretory phase is characterized by a higher velocity and lower resistance index (RI = 0.81).

longer detectable in the uterine arteries on the day of ovulation. Goswamy and Steptoe[8] found an increasing resistance index and systole/diastole ratio during the postovulatory fall in serum estradiol levels. Increased resistance in the uterine arteries was measured three days after the surge in luteinizing hormone (LH), and Scholtes et al.[22] found that the pulsatility index (PI) in the uterine arteries was highest on day 16 of the menstrual cycle.

These results could be due to increased uterine contractility[10] and increased compression of the vessels traversing the uterine wall, thus reducing the vessel diameters and causing greater flow resistance. During the normal menstrual cycle, a sharp rise of end-diastolic blood flow velocities can be demonstrated between the proliferative and secretory phases[18]. It is particularly noteworthy that the lowest vascular resistance coincides with the period of maximal corpus luteum function (RI = 0.84 ± 0.04), when implantation most commonly occurs (Fig. 3.**3**). It seems logical that the blood flow to the uterus is highest in the luteal phase, and this increase has been reported by Kurjak et al.[18], Goswamy et al.[8,9], Steer et al.[23], and Bataglia et al.[1]. The persistence of a low RI in the luteal phase suggests that the relaxation effect on the uterine arteries continues until the start of menstruation. Zaidi et al.[27] described a circadian rhythm in uterine artery blood flow, independent of fluctuating hormone levels.

Blood Flow Parameters in the Radial and Spiral Arteries

Blood flow fluctuations similar to those in the main trunk of the uterine artery were observed in the radial and spiral arteries following the introduction of transvaginal color Doppler and pulsed Doppler sonography[14] (Fig. 3. **4**). Our results show that the postovulatory fall of serum estradiol levels coincides with a postovulatory rise of the RI in the myometrial vessels (Fig. 3.**5**).

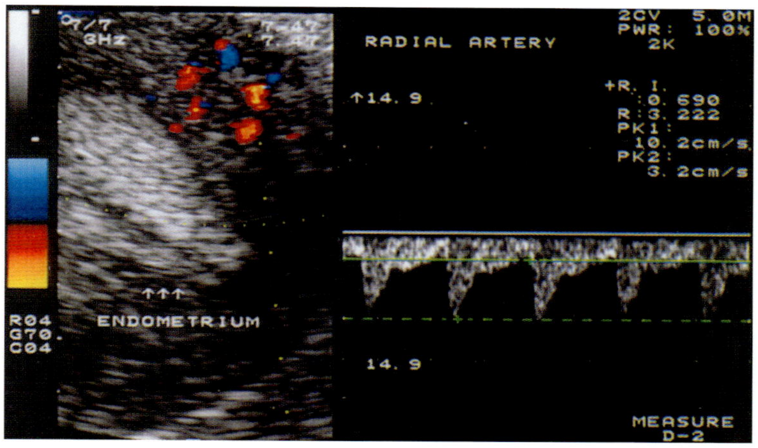

Fig. 3.**4** Doppler trace of the right radial arteries. Note the position of the cursor in the left half (in the myometrium) for sampling the Doppler spectrum. Lower velocities and a lower resistance (RI = 0.69) are measured in the examined vessels.

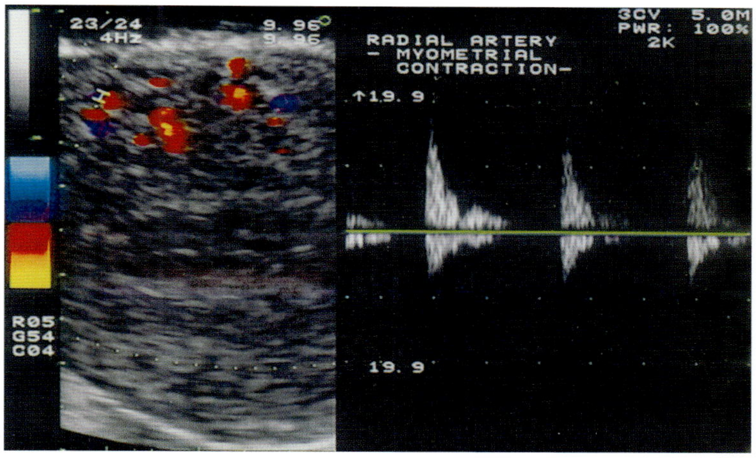

Fig. 3.**5** Blood flow velocities in the radial arteries during the periovulatory period. The suspension of end-diastolic blood flow (right) is a transient effect based on myometrial contractions.

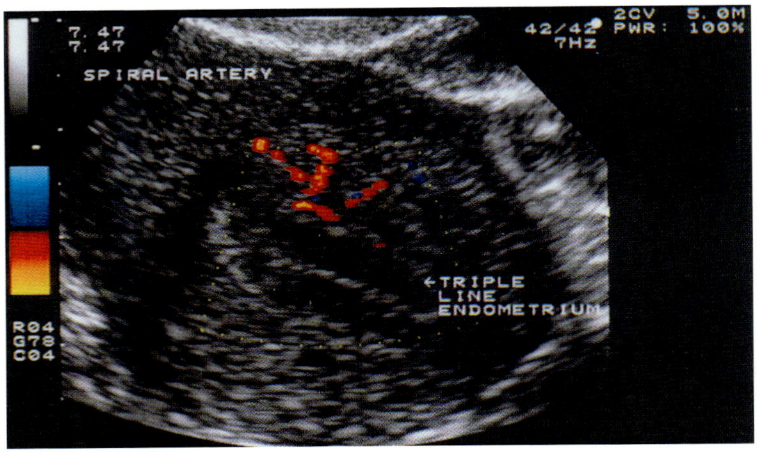

Fig. 3.6 Color signals from the spiral arteries in the periphery of the three-line endometrium.

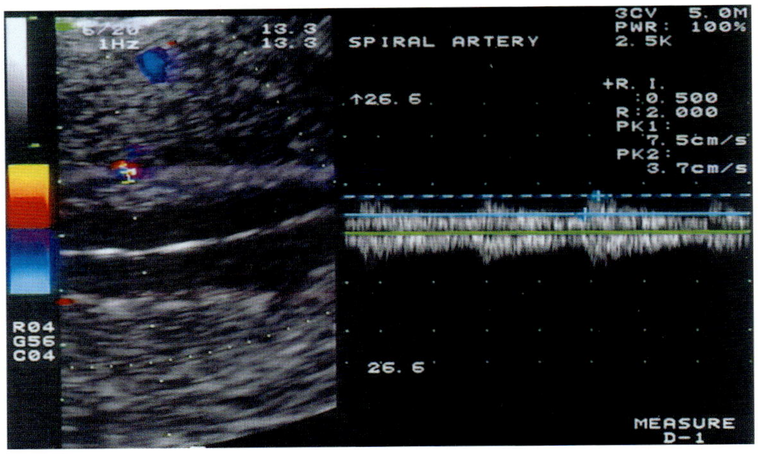

Fig. 3.7 Blood flow velocities in the spiral arteries during the periovulatory period. A scan on the day of ovulation shows a decreased resistance index (RI = 0.50) and increased blood flow velocity.

It has been shown that increased uterine contractility is associated with a decrease in endometrial blood flow[10]. It is well known that the endometrium undergoes marked structural and functional changes during the menstrual cycle. The histological changes include a remarkable degree of new blood vessel formation (Fig. 3.**6**). The spiral arteries show increased development during the menstrual cycle. The increase in endometrial blood flow depends strongly on the blood flow in the uterine artery, arcuate arteries, and radial arteries. When the spiral arteries in a normal cycle are compared with the uterine arteries, which have larger diameters, the spiral arteries are found to exhibit lower blood flow velocities ($p < 0.05$) and a lower vascular resistance ($p < 0.05$) (Fig. 3.**7**). It appears that **the data obtained by analyzing endometrial blood flow permit a more accurate assessment of the success rate of implantation and of unexplained fertility problems than the evaluation of uterine artery blood flow alone.**

Uterine Blood Flow in Infertile Women

As more experience is gained with transvaginal color and pulsed Doppler ultrasound, this examination has assumed an important role in the management of infertile women. A steadily increasing RI has been measured in the uterine arteries during anovulatory cycles (Fig. 3.**8**). Moreover, end-diastolic blood flow cannot be detected in some infertile women[9] (Fig. 3.**9**). The data are still insufficient, however, to establish whether absent end-diastolic blood flow is associated with infertility and low pregnancy rates.

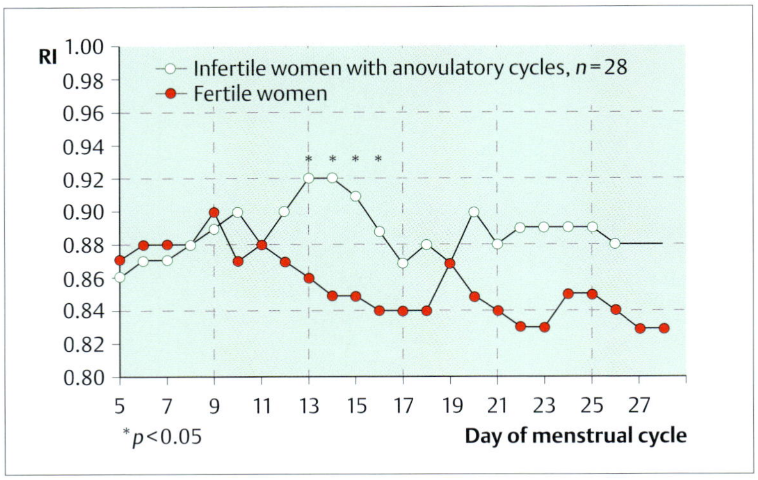

Fig. 3.8 Uterine artery blood flow changes in fertile and infertile women during the menstrual cycle.

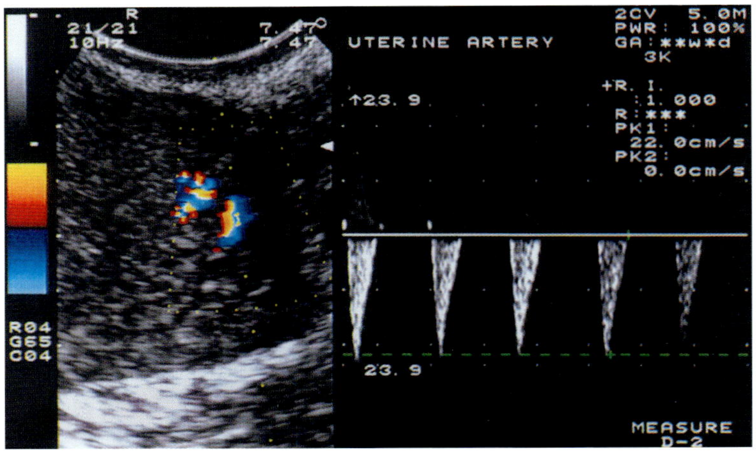

Fig. 3.9 Absence of end-diastolic blood flow in both uterine arteries in a patient with primary infertility.

Uterine Blood Flow and Fertilization Rate

The assessment of uterine artery blood flow could be used as a means of detecting unfavorable uterine blood flow before a planned embryo transfer.

Steer et al.[23] calculated the likelihood of pregnancy on the day of embryo transfer based on pulsatility values in the uterine arteries. According to their findings, the likelihood of pregnancy was greatest when medium PI values were measured in the uterine artery. In 35% of cases, pregnancy did not occur when the mean PI value before the embryo transfer was greater than 3.0.

Tsai et al.[25] investigated the prognostic value of uterine perfusion on the day of human chorionic gonadotropin (hCG) administration in patients who were undergoing intrauterine insemination. They calculated the PI of the ascending branch of the uterine artery on the day of hCG administration and compared the vascular resistance in the uterine artery with the outcome of intrauterine insemination. No pregnancy occurred when the PI was greater than 3. The fertilization rate was 18% when the PI was less than 2 and 19.8% when the PI was between 2 and 3. These data indicate that the measurement of uterine perfusion on the day of hCG administration may be of value in predicting the success of intrauterine insemination.

Zaidi et al.[28] investigated whether the assessment of uterine artery blood flow by transvaginal color Doppler scanning on the day of hCG administration in patients undergoing in-vitro fertilization (IVF) could predict pregnancy and implantation rates. One hundred thirty-five patients undergoing 139 IVF cycles were analyzed. The results of the study suggest that the measurement of the uterine artery PI on the day of hCG administration can predict the likelihood of successful implantation, since the highest pregnancy rate (34.7%) was associated with a uterine artery PI between 2 and 3. If possible, hCG should be administered when the uterine artery PI is less than 3 in order to achieve a high implantation rate.

Further studies are needed, however, to investigate the precise relationship between uterine artery blood flow and the likelihood of pregnancy. In embryo-transfer patients who are found to have unfavorable uterine blood flow indices, it might be prudent to freeze the embryos and reschedule the transfer for a later date. It may be possible to improve uterine blood flow by administering estradiol[6] or progesterone[4].

Endometrial Imaging

One of the greatest problems in the current practice of in vitro fertilization is the transfer of multiple embryos in an effort to increase the pregnancy rate. This leads to an increased number of multiple pregnancies, which are associated with poorer perinatal outcomes and higher obstetric risks than singleton pregnancies. It is known that the likelihood of pregnancy depends strongly on the embryo itself and on favorable uterine blood flow. Instead of performing an endometrial biopsy, which can lead to injury and bleeding at the implantation site, color Doppler sonography can be used as a means of detecting favorable uterine blood flow [16].

Zaidi et al.[26] examined 96 patients undergoing IVF treatment on the day of hCG administration by transvaginal color Doppler sonography. They assessed endometrial thickness, endometrial morphology, the presence or absence of subendometrial or intraendometrial blood flow, and intraendometrial vascular penetration on the day of hCG administration and related the results to pregnancy rates. The overall pregnancy rate was 32.3%. There were no significant differences between the pregnant and nonpregnant groups with regard to endometrial thickness, subendometrial peak systolic blood flow velocity, or subendometrial pulsatility index. However, the absence of subendometrial blood flow was always associated with a failure of implantation.

Conclusions

Transvaginal color and pulsed Doppler sonography is a highly reproducible study that is fast and easy to perform. It may predict the likelihood of successful implantation, thereby reducing the incidence of multiple gestation. The measurement of uterine blood flow could provide a noninvasive tool for investigating the uterine milieu and could help us to learn more about the pathophysiology of infertility, especially in cases with an undetermined cause.

Uterine Blood Flow in the Normal Cycle and during Ovarian Stimulation with Confirmed Ovulation

Kupesic and Kurjak[27] measured blood flow velocities in the uterine, radial, and spiral arteries during the periovulatory period of normal cycles and in stimulated cycles with confirmed ovulation. Seventy-eight patients with a male infertility factor were examined daily in an IVF clinic. In normal cycles, the uterine artery PI was 3.16 two days before ovulation and decreased to 2.22 on the day before ovulation. This difference was not observed in stimulated cycles. The mean PI was 3.06 and remained unchanged during the periovulatory period. Distinct waveforms could be recorded from the endometrium and myometrium during this period. The PI in the radial and spiral arteries showed higher values in stimulated cycles than in normal cycles.

Clomiphene citrate. It is known that clomiphene citrate occupies the estrogen receptors in estrogen-sensitive tissues and in this way influences the growth of the endometrium[7]. A strong correlation was found between endometrial thickness and blood flow velocities. This did not apply to the patient group that had received clomiphene citrate/human menopausal gonadotropin (hMG) stimulation and exhibited normal endometrial development. The authors found absent end-diastolic blood flow in the spiral arteries in 55.6% of these cases (Fig. 3.**10**). No differences in endometrial thickness or perfusion were seen between the hMG-stimulated patients and the nonstimulated patients with normal cycles.

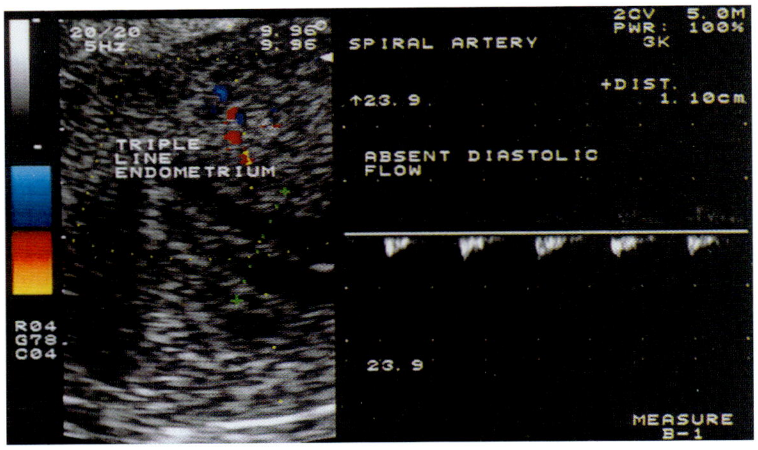

Fig. 3.**10** Blood flow velocities in the spiral arteries of a triple-line endometrium (left). The absence of end-diastolic blood flow is a sign of poor endometrial perfusion and correlates with poor endometrial receptivity.

Conclusion. Blood flow changes in the spiral arteries can be considered an accurate predictor of implantation success rates in patients undergoing IVF and embryo transfer. In patients who are found to have unfavorable uterine blood flow in the current treated menstrual cycle, the embryos should be frozen and the transfer postponed for a normal cycle or a cycle with favorable endometrial blood flow in response to therapy.

References

1 Battaglia C, Larocca E, Lanzani A, Valentini M, Genanzzani AR: Doppler ultrasound studies of the uterine arteries in spontaneous and IVF stimulated ovarian cycles. Gynecol. Endocrinol. 4 (1990) 245–250
2 Bourne TH, Jurkovic D, Waterstone J, Campbell S, Collins WP: Intrafollicular blood flow during human ovulation. Ultrasound Obstet. Gynecol. 1 (1991) 53–59
3 Christofferson R, Nilsson BO: Morphology of the endometrial microvasculature during early placentation in the rat. Cell Tissue Res. 253 (1988) 209–220
4 de Ziegler D, Bessis R, Frydman R: Vascular resistance of uterine arteries: physiological effects of estradiol and progesterone. Fertil. Steril. 55 (1991) 775–779
5 DuBose TJ, Hill LW, Henningan JW Jr: Sonography of arcuate uterine blood vessels. J. Ultrasound Med. 4 (1985) 229–233
6 Ford SP, Reynolds RP, Farley DB: Interaction of ovarian steroids and periarterial alpha-1-adrenergic receptors in altering uterine blood flow during the estrous cycle of gilts. Am. J. Obstet. Gynecol. 150 (1984) 480–484
7 Glissant A, de Mouzon J, Frydman R: Ultrasound study of the endometrium during in vitro fertilization cycles. Fertil. Steril. 44 (1985) 786–790
8 Goswamy RK, Steptoe PC: Doppler ultrasound studies of the uterine artery in spontaneous ovarian cycles. Hum. Reprod. 3 (1988) 721–726
9 Goswamy RK, Williams G, Steptoe PC. Decreased uterine perfusion a cause of infertility. Hum. Reprod. 3 (1988) 955–959
10 Hauksson A, Akerlund M, Melin P: Uterine blood flow and myometrial activity at menstruation, and the action of vasopressin and a synthetic antagonist. Br. J. Obstet. Gynaecol. 95 (1988) 898–904
11 Jurkovic D, Jauniaux E, Kurjak A, Cambell S: Transvaginal color Doppler assessment of the uteroplacental circulation in early pregnancy. Obstet. Gynecol. 77 (1991) 365–369
12 Kaiserman-Abramof IR, Padykula HA: Angiogenesis in the postovulatory primate endometrium: The coiled arteriolar system. Anat. Rec. 224 (1989) 479–489
13 Khong TY, De Wolf F, Robertson WB, Brosens I: Inadequate maternal vascular response to placentation in pregnancies complicated by preeclampsia and by small-for-gestational age infants. Br. J. Obstet. Gynaecol. 93 (1986) 1049–1059
14 Kupesic S, Kurjak A: Uterine and ovarian perfusion during the periovulatory period assessed by transvaginal color Doppler. Fertil. Steril. 60 (1993) 439–443
15 Kupesic S, Kurjak A, Stilinovic K: The assessment of female infertility. In Kurjak, A (ed.): An Atlas of Transvaginal Color Doppler. (Parthenon Publishing), Carnforth 1994, 171–199
16 Kupesic S, Kurjak A, Vujisic S, Petrovic Z: Lutheal phase defect: comparison between Doppler velocimetry, histological and hormonal markers. Ultrasound Obstet. Gynecol. 9 (1997) 105–112
17 Kurjak A, Kupesic-Urek S: Normal and abnormal uterine perfusion. In Jaffe R, Warsof LS (eds.): Color Doppler Imaging in Obstetrics and Gynecology. McGraw Hill, New York 1992, 255–263
18 Kurjak A, Kupesic-Urek S, Schulman H, Zalud I: Transvaginal color flow Doppler in the assessment of ovarian and uterine blood flow in infertile women. Fertil. Steril. 56 (1991) 870–873
19 Long MG, Boultbee JE, Hanson ME, Begent JHR: Doppler time velocity waveform studies of the uterine artery and uterus. Br. J. Obstet. Gynaecol. 96 (1989) 588–593
20 Ramsey EM, Donner ME: Placental Vasculature and Circulation. Sounders, Philadelphia 1980, 1–52
21 Ramsey EM, Donner MV: Placental vasculature and circulation in primates. In Kaufmann P, Miller RK (eds.) Trophoblast Research, Vol. 3; Placental Vascularization and Blood flow. Plenum Press, New York 1988, 217–233
22 Scholtes MCW, Wladimiroff JW, van Rijen HJM, Hop WCJ: Uterine and ovarian flow velocity waveforms in the normal menstrual cycle: a transvaginal study. Fertil. Steril. 52 (1989) 981–985
23 Steer CV, Mills CV, Campbell S: Vaginal color Doppler assessment on the day of embryo transfer (ET) accurately predicts patients in an *in vitro* fertilization programme with suboptimal uterine perfusion who fail to become pregnant. Ultrasound Obstet. Gynecol. 1 (1991) 79–82
24 Torry RJ, Rongish BJ: Angiogenesis in the uterus: potential regulation and relation to tumor angiogenesis. Am. J. Reprod. Immunol. 27 (1992) 171–179
25 Tsai YC, Chang JC, Tai MJ, Kung FT, Yang LC, Chang SY: Relationship of uterine perfusion to outcome of intrauterine insemination. J. Ultrasound Med. 15 (1996) 633–636
26 Zaidi J, Campbell S, Pitroff R, Tan SL: Endometrial thickness, morphology, vascular penetration and velocimetry in predicting implantation in an in vitro fertilization program. Ultrasound Obstet. Gynecol. 6(3) (1995) 191–198
27 Zaidi J, Jurkovic D, Campbell S, Pitroff R, McGregor A, Tan SL: Description of circadian rhythm in uterine artery blood flow during the periovulatory period. Hum. Reprod. 10(7) (1995) 1642–1646
28 Zaidi J, Pitroff R, Shaker A, Kyei-Mensah A, Campbell S, Tan SL: Assessment of uterine artery blood flow on the day of human chorionic gonadotropin administration by transvaginal color Doppler ultrasound in an in vitro fertilization program. Fertil. Steril. 5(2) (1996) 377–381

4 Uterine Causes of Infertility

S. Kupesic and A. Kurjak

For fertilization to occur, the uterine cavity must create an optimum environment for the successful transport of spermatozoa from the cervix to the fallopian tubes. A normal structure of the uterine mucosa, with normal glandular secretions and blood flow, is also essential for implantation and placentation. Uterine abnormalities such as polyps, leiomyomas, malignant tumors, infections, and intrauterine adhesions and synechiae can contribute to fertility problems. Accordingly, numerous studies have been done to look for a correlation between sonographic parameters (endometrial thickness and echogenicity) and uterine receptivity to conception. This chapter explores the role of transvaginal and color Doppler sonography in evaluating endometrial changes during the menstrual cycle and in the detection of uterine abnormalities.

Ultrasound Detection of Uterine Abnormalities

Congenital Anomalies

Congenital anomalies of the uterus are diagnosed in 38–55% of women with habitual abortion[33,47].

Examination during the secretory phase. The ultrasound diagnosis of uterine anomalies such as a septate uterus, bicornuate uterus, or uterus didelphys is most sensitive and specific during the secretory phase of the menstrual cycle. The accuracy of the ultrasound diagnosis depends on the severity of the anomaly[23]. The endometrium is most clearly defined during the secretory phase, when separate echogenic lines demarcate the endometrium from the surrounding hypoechoic myometrium, making it easier to detect contour abnormalities (Fig. 4.**1**). Accordingly, a careful transvaginal ultrasound examination in a patient with a unicornuate uterus can demonstrate a single uterine cornu with an atypical endometrial echo[24].

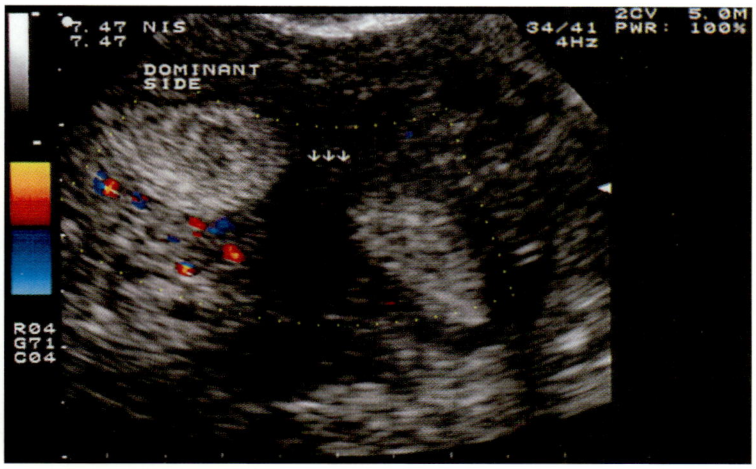

Fig. 4.**1** Uterine duplication anomaly demonstrated by transvaginal color Doppler sonography. Two separate endometrial images can be seen during the secretory phase of the menstrual cycle.

Large intracavitary fluid collections should raise suspicion of hymenal or even vaginal atresia.

Sonohysterography. Sonohysterography refers to the continuous sonographic visualization of the uterine cavity during and after the instillation of contrast medium into the uterus[11]. This method can accurately demonstrate uterine septa, for example, and can define their extent. In a study by Randolph et al.[41], the results of sonohysterography agreed with hysteroscopic findings in 53 of 54 patients, corresponding to a sensitivity of 98% and a specificity of 100%.

The interested reader can find more information on the ultrasound and color Doppler evaluation of uterine anomalies in Chapter 26 (p. 259 – 265).

Endometrial Polyps

Endometrial polyps are known to have causal significance in habitual abortion and infertility. The polyps appear sonographically as diffuse or focal sites of endometrial thickening. On sonohysterography, an intracavitary polyp appears surrounded by echo-free fluid, and its site of attachment can generally be identified[11] (Fig. 4.**2**). If the examination is performed in the follicular phase of the cycle, it is unnecessary to instill fluid in order to detect increased endometrial thickness (Fig. 4.**3**). Generally, however, polyps can be detected more easily in the periovulatory and secretory phases of the cycle by fluid instillation.

The use of transvaginal color Doppler sonography makes it possible to identify small arteries that supply the endometrial polyps (Fig. 4.**4**). Three-dimensional ultrasound permits an accurate examination of the uterine cavity in which the polyps can be clearly visualized and delineated from the uterine wall (Fig. 4.**5**).

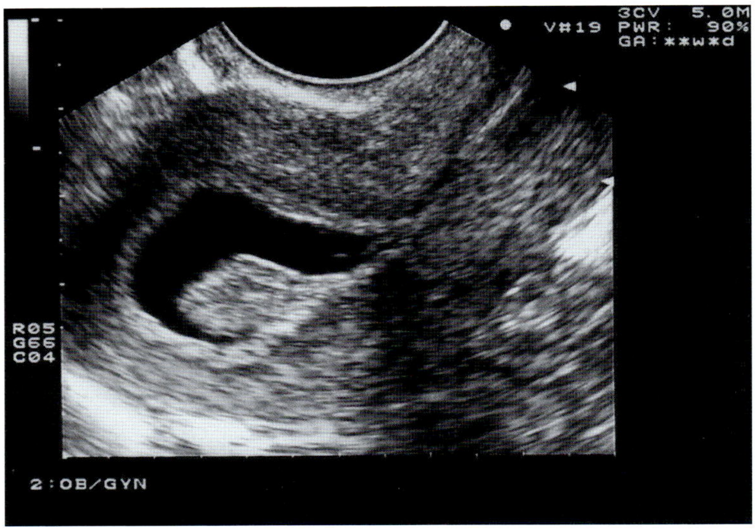

Fig. 4.2 Transvaginal ultrasound examination of the uterus after the instillation of isotonic saline solution in an infertile patient with acyclic bleeding episodes. The area of increased endometrial thickness is a polyp.

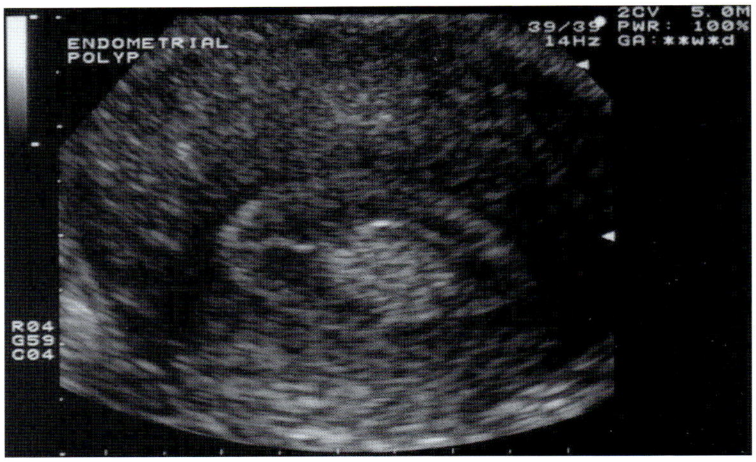

Fig. 4.3 Transvaginal ultrasound examination of the uterus shows a focal area of increased echogenicity: an endometrial polyp.

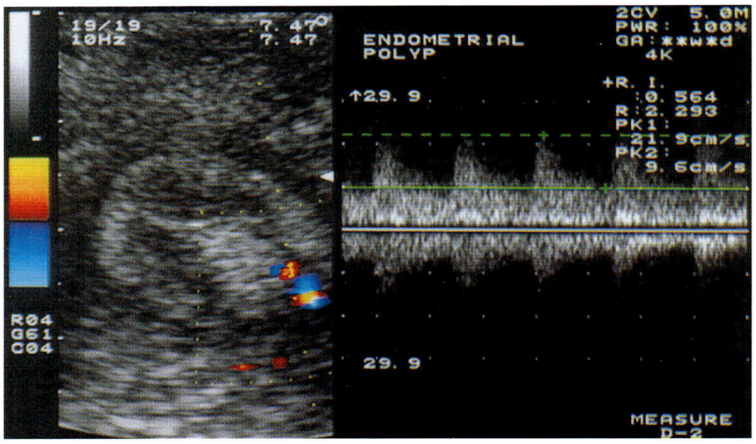

Fig. 4.4 Transvaginal color Doppler examination shows a focal area of increased echogenicity bounded by an area with a normal vascular distribution. This pattern is typical of an endometrial polyp. The moderate to slightly increased resistance index in pulsed color Doppler (RI = 0.69) is consistent with a benign uterine mass.

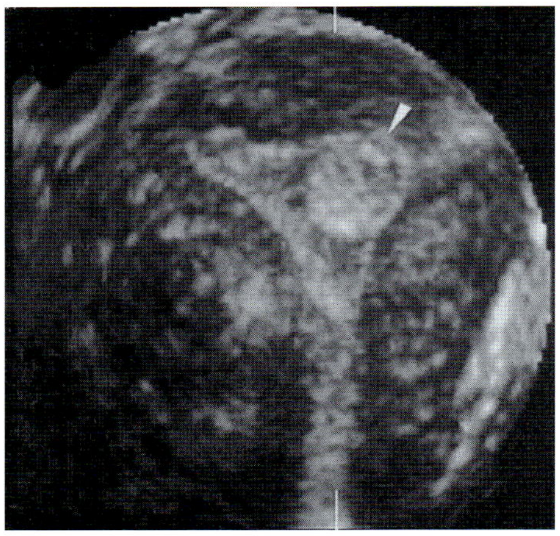

Fig. 4.5 Endometrial polyp demonstrated by three-dimensional ultrasound. The polyp appears as a localized area of endometrial thickening that does not completely fill the uterine cavity.

Submucous Leiomyomas

Sonographic appearance. The diagnosis of a submucous leiomyoma (fibroid) is based on a distortion of the uterine contour, uterine enlargement, and a change in echo pattern (Fig. 4.**6**).

Because leiomyomas contain variable amounts of smooth muscle and connective tissue, these benign tumors vary in their sonographic features[41]. They range from hypoechoic to hyperechoic, depending on the proportions of smooth muscle and connective tissue. Central ischemia can occur as the tumor grows and outstrips its blood supply, generally followed by varying stages of degeneration. The most frequent source of calcifications in the uterus is the degenerative calcification of a leiomyoma. Cystic, myxomatous, and hyaline forms of degeneration can also occur. Given the large range of variation in tumor appearance, it is not uncommon to confuse sub-

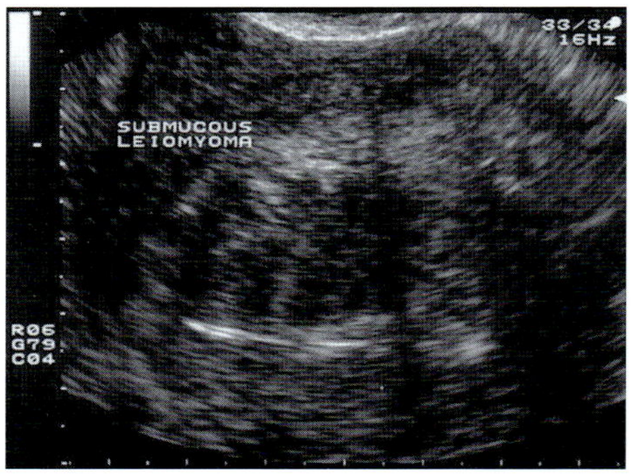

Fig. 4.6 Transvaginal scan in a patient with an isoechoic submucous leiomyoma.

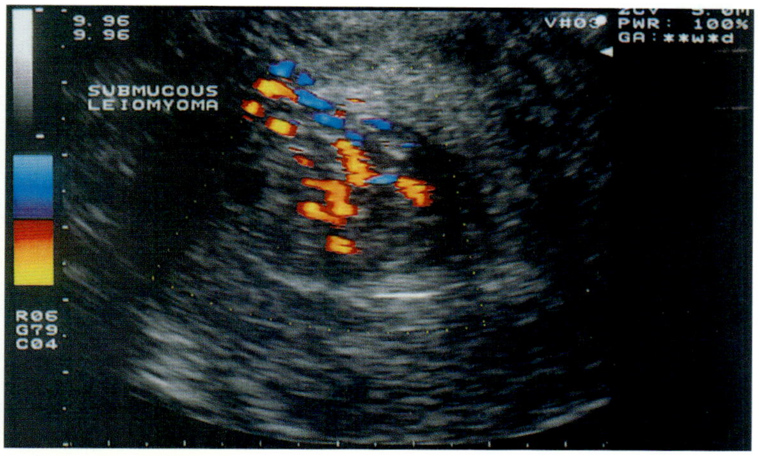

Fig. 4.7 Same patient as in Fig. 4.6. Color Doppler demonstrates a large vessel in proximity to the leiomyoma.

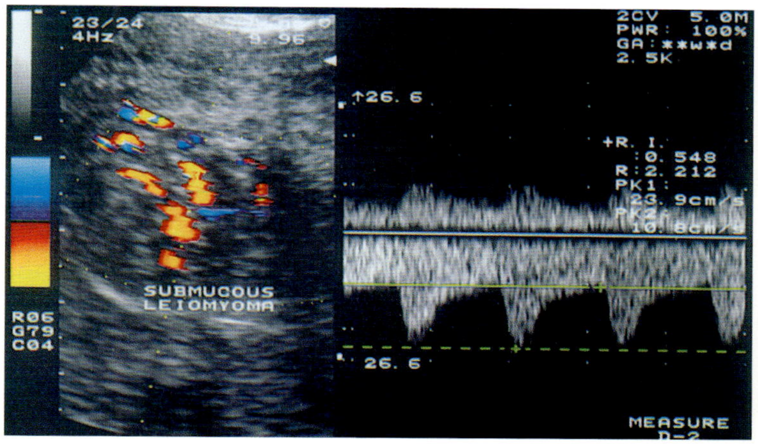

Fig. 4.8 Same patient as in Fig. 4.7. The color signals on pulsed Doppler examination (right) indicate moderate vascular resistance (RI = 0.55).

mucous leiomyomas with endometrial polyps, endometrial carcinoma, blood, or even mucus.

Fedele et al.[20] investigated the accuracy of transvaginal sonography in the detection of small submucous leiomyomas in patients who underwent transvaginal ultrasound scanning and hysteroscopy as a prelude to hysterectomy. The sensitivity/specificity of transvaginal sonography in this study was comparable to that of hysteroscopy.

Effects on the endometrium. The uterine environment does not appear to be conductive to implantation of the fertilized ovum in patients with submucous leiomyomas. The blood supply may also be inadequate for this purpose[30].

Deligdish and Loewenthal[13] conducted a histological study of the endometrium in patients with submucous leiomyomas. Atrophic changes were found in the endometrial glands and stroma in areas of endometrium that were located over or opposite to the leiomyoma, while the glands at the tumor margins tended to be hyperplastic. Increased vascularity and elevated estrogen levels were also observed.

Farrer-Brown et al.[19] were able to detect arterial obstruction and venous dilatation within the endometrium overlying a leiomyoma. This suggests that submucous leiomyomas cause a reduction in blood flow, which suppresses the release of hormones that are necessary for normal endometrial development. Ultimately these changes can lead to endometrial atrophy and inadequate placentation. Additionally, submucous leiomyomas can hamper both the growth of the fetus and the normal enlargement of the uterus[50].

Blood supply. Leiomyomas grow centripetally by the proliferation of smooth muscle cells and connective tissue, and they form a pseudocapsule consisting of compressed muscle fibers. As a result, most blood vessels are found at the periphery of the mass on color Doppler examination (Fig. 4.7). Blood vessels located at the center of leiomyomas are usually associated with necrosis, degenerative changes, or inflammatory processes. These vessels show a lower resistance index (RI) than peripherally located vessels and are occasionally misinterpreted as malignant tumor angiogenesis[32]. The flow resistance in the vessels supplying leiomyomas depends not only on their size but also on their location in the uterus. The blood flow characteristics of leiomyoma-feeding vessels show significant differences between the vessels of subserous, intramural, and submucous leiomyomas. The low resistance found in subserous leiomyomas can be attributed to the fact that blood vessels are distributed to these tumors over a very small area (Fig. 4.8). These vessels are surrounded by loose connective tissue and are therefore dilated with very little resistance to flow. This contrasts with the higher-resistance vessels that supply submucous and intramural leiomyomas. The high basal tonus of the myometrium, which surrounds intramural and submucous leiomyomas, could account for the difference in hemodynamic parameters.

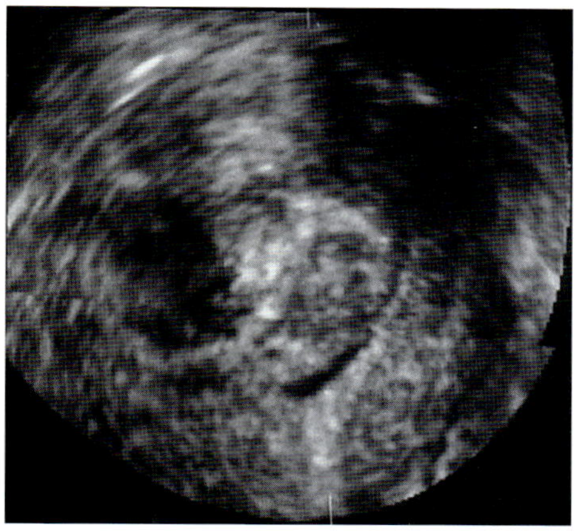

Fig. 4.9 A submucous leiomyoma completely fills the uterine cavity. Visualized by 3 D ultrasound.

Three-dimensional ultrasound is most accurate for defining the spatial relationship between submucous leiomyomas and the uterine cavity (Fig. 4.9).

Kurjak et al.[32] performed transvaginal color-flow Doppler examinations in 101 patients with palpable uterine leiomyomas and in 60 healthy women. The mean RI measured at the periphery of the leiomyomas was 0.54, and the mean PI was 0.89. Histopathological examination confirmed benign uterine tumors in all cases, even when the RI was very low. Low RI values were found in cases with necrotic, degenerative, or inflammatory changes within the leiomyoma. Increased blood flow velocity and a decreased RI (mean RI = 0.74) were found in both uterine arteries of the leiomyoma patients.

Adenomyosis

Adenomyosis, characterized by the ingrowth of endometrial tissue into the myometrium, is usually asymptomatic but may be associated with uterine bleeding, pain, and infertility. At ultrasound, a diffusely enlarged uterus with no signs of leiomyomas and an intact endometrium should suggest the possibility of adenomyosis[9]. Occasionally the middle layers of the myometrium may show altered echogenicity in severe cases. Numerous small cysts have also been described within the myometrium as an expression of adenomyosis[47].

The sensitivity and specificity of transvaginal sonography in the detection of this benign condition are 86% and 50%, respectively[9]. Color Doppler imaging can demonstrate increased vascularity, which is characterized chiefly by a moderate vascular resistance (Fig. 4.**10**).

Endometritis

Chronic endometritis is characterized by increased echogenicity, thickness, and vascularity of the endometrium[30]. The most frequent cause of chronic endometrial infection is tuberculosis. The active phase of the infection is associated with increased rates of ectopic pregnancy and abortion. Transvaginal sonography can demonstrate calcified pelvic lymph nodes or even small, irregular calcifications in the adnexal region as well as suggestive deformities of the uterine cavity with no prior history of abortion or curettage that would indicate intrauterine adhesions. In the acute stage of endometritis, the vascular resistance at the periphery of the endometrium is low to moderate. Once irreversible tissue destruction has occurred, no blood flow can be observed. Transvaginal scanning can demonstrate the abnormal morphology of the endometrium, making it necessary to take bacterial cultures and institute broad-spectrum antibiotic treatment. A 1- to 2-month course of conjugated estrogens should be administered for the prevention of intrauterine adhesions after endomyometritis. This treatment promotes regeneration of the endometrium, and pulsed Doppler flowmetry can confirm this response by showing a marked rise of end-diastolic blood flow velocity in the spiral arteries.

Asherman Syndrome

In 1948, Asherman described intrauterine adhesions in eight patients[4]. The adhesions can result from endometrial damage leading to the development of fibrous bands and synechiae in the uterine cavity. The endometrial damage may be caused by overvigorous curettage of the uterine cavity following a miscarriage but more often results from curettage in an advanced pregnancy. Tuberculosis can also cause intrauterine synechiae in rare cases. These can produce cordlike adhesions of varying thickness, resulting in partial or complete obliteration of the uterine cavity. The pattern of menstrual bleeding is usually hyporrheic to amenorrheic.

In some patients with endometrial adhesions in Asherman syndrome, ultrasound shows areas with no detectable endometrium next to completely normal-appearing areas. The adhesions appear as irregularities of endometrial texture or hyperechoic bridges within the uterine cavity (Fig. 4.**11**).

> Schlaff and Hurst[42] examined seven women with amenorrhea due to severe Asherman syndrome. Transvaginal sonography showed a well-developed endometrial stripe in three of the seven women, while three others had no demonstrable endometrium. All the women with a well-developed endometrial structure had adhesions occluding the lower uterine segment and had a resumption of normal menses and normalization of the uterine cavity after hysteroscopy. By contrast, the women with minimal endometrial structure and no identifiable uterine cavity did not benefit from surgical treatment. This study suggests that the endometrial pattern seen with transvaginal sonography is highly predictive of the surgical and clinical outcome in patients with severe Asherman syndrome characterized by complete obstruction of the uterine cavity at hysterosalpingography.

Doppler flow imaging of the uterine cavity (Fig. 4.**11**) does not show increased vascularity associated with intrauterine synechiae. These lesions are more clearly identified during men-

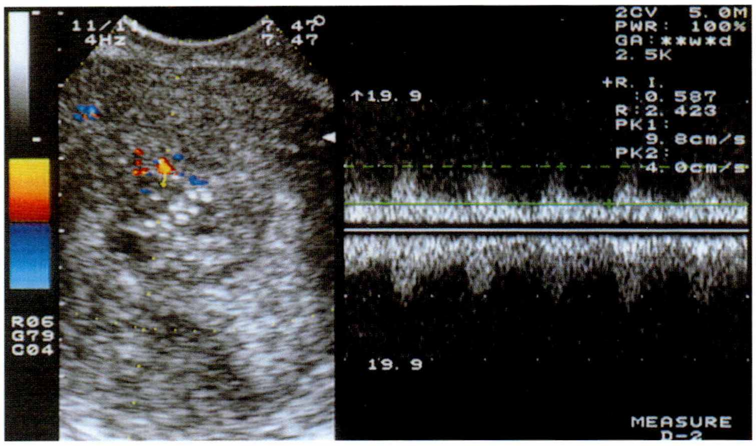

Fig. 4.**10** Diffusely enlarged uterus with a thickened "Swiss cheese" endometrium shows increased vascularity (left). Doppler flow analysis (right) indicates a low blood flow velocity and moderate flow resistance (RI = 0.59).

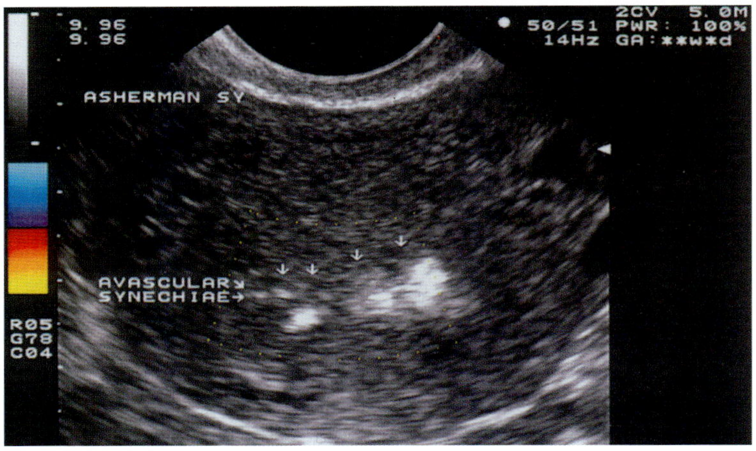

Fig. 4.**11** Transvaginal ultrasound scan in an infertile patient with intrauterine synechiae. Note the hyperechoic bridges within the uterine cavity. Color Doppler shows no signs of increased vascularity.

struation or when outlined by intracavitary fluid in sonohysterography. Three-dimensional ultrasound examination in Asherman syndrome shows a significant reduction of endometrial volume in all affected areas of the uterine cavity (Fig. 4.**12**).

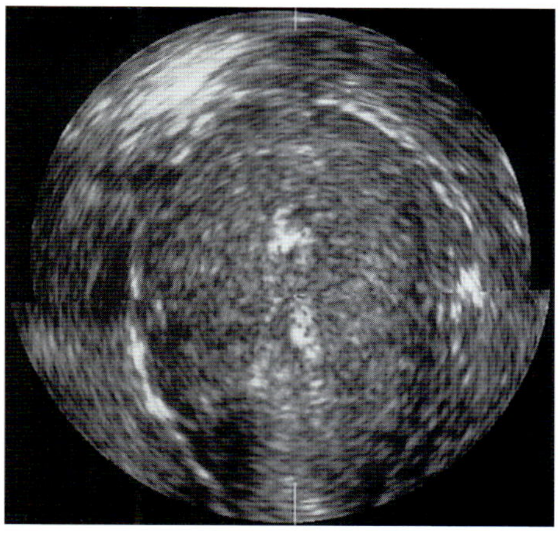

Fig. 4.**12** Three-dimensional ultrasound image shows an irregular ▷ uterine cavity with a significantly reduced endometrial volume.

Ultrasound Detection of Endometrial Causes of Infertility

Effect of Endometrial Thickness and Morphology on Fertility

Ultrasonography is a noninvasive modality that permits an accurate and reproducible assessment of endometrial texture and thickness.

Cycle-dependent changes. With its responsiveness to estradiol and progesterone, the endometrium exhibits changes in echo pattern during the various phases of the menstrual cycle. The endometrium appears as a thin, echogenic stripe in the postmenstrual phase, and in the proliferative phase it becomes isoechoic to the myometrium. As ovulation approaches, the endometrium shows increased echogenicity owing to the development of endometrial glands and increased glandular secretions[23]. A hypoechoic halo sign probably corresponds to the inner layer of the myometrium, while the hypoechoic area within the endometrium results from edema of the pars compacta. During the secretory phase, a progressive rise of acoustic signal is seen in response to progesterone. These endometrial changes are a result of increased mucus secretion and the development of a spiral phenomenon with a progression of the changes from the base of the endometrium toward the surface.

Endometrial structure and implantation. There is considerable variation in the results that have been reported on endometrial assessment with ultrasound.

Gonen and Caspar[26] described three different types of endometrial pattern at the time of follicular aspiration for subsequent in-vitro fertilization. In their opinion, the three-layered endometrium is more favorable for successful implantation than the homogeneous hyperechoic endometrium or the intermediate, isoechoic type of endometrium. Also, endometrial thickness was greater in the group of patients who achieved pregnancy (8.7 ± 0.4 mm) than in the group who did not (7.5 ± 0.2 mm).

Other authors[15,25,43] report that the implantation of embryos is drastically reduced when the endometrial thickness is less than 6 mm, while the most favorable thickness for implantation is in the range of 9–10 mm. In the opinion of Smith et al.[44], both the thickness and the echo pattern of the endometrium are important factors in assisted reproduction programs. Other studies[40,48,49] have also shown statistically significant correlations between endometrial thickness, endometrial texture, and pregnancy rates.

Kepic et al.[27] state that the endometrial thickness and structure as well as follicular size and estradiol level are key parameters in determining pregnancy rates.

On the other hand, Fleischer et al.[21,22] found no correlation between endometrial thickness and implantation in their studies.

These contradictory results can be explained by the variable ultrasound appearance of the endometrium at different times (day of human chorionic gonadotropin (hCG) administration, day of follicular aspiration, day of embryo transfer). An ultrasound examination on the day of hCG administration appears to yield the most reliable data, since progesterone production in this phase of the cycle has not yet altered the endometrial structure.

Li et al.[34] studied the prevalence of retarded endometrial development in the luteal phase of infertile women ($n = 142$) and fertile women ($n = 68$). The prevalence of retarded endometrial development was significantly higher in the infertile group than in the control group (14 % versus 4.4 %). When the authors divided the infertile subjects into four subgroups of different etiology, they found that women with endometriosis had a significantly higher prevalence of abnormal endometrial development (29 %), while the prevalence in women with a tubal or male infertility factor was not significantly higher than in the fertile controls. Twenty-one percent of the women with idiopathic infertility had acyclic endometrial development. The results of the study are summarized in Table 4.**1**.

Endometrial blood flow. An important technical advance has been the combination of transvaginal pulsed color Doppler ultrasound with real-time imaging. With this method, it is possible to evaluate uterine receptivity by examining the perfusion of the uterine arteries. Moreover, the resistance indices of the uterine arteries represent an important prognostic factor for pregnancy and delivery[5,45,46]. The endometrium derives its

Table 4.**1** Comparison of age, duration of infertility, length of follicular phase, length of luteal phase, and prevalence of retarded endometrial development in four groups of infertility patients and one group of women with normal fertility

	Group 1 (tubal cause) (*n* = 34)	**Group 2 (male cause) (*n* = 21)**	**Group 3 (endometriosis) (*n* = 48)**	**Group 4 (idiopathic) (*n* = 48)**	**Group 5 (normal) (*n* = 68)**
Age (years)	32.5 ± 4.0 (NS)	30.8 ± 4.0 (NS)	34.0 ± 2.9 (NS)	32.7 ± 4.4 (NS)	33.4 ± 4.0 (NS)
Duration of infertility (years)	6.1 ± 3.3 (NS)	6.8 ± 2.7 (NS)	6.9 ± 2.9 (NS)	6.0 ± 3.3 (NS)	–
Length of follicular phase (days)	14.3 ± 3.2 (NS)	13.7 ± 2.1 (NS)	13.9 ± 2.1 (NS)	14.5 ± 2.4 (NS)	13.6 ± 1.8
Length of luteal phase (days)	13.2 ± 1.0 (NS)	13.1 ± 1.6 (NS)	11.9 ± 1.5 (NS)	12.7 ± 1.8 (NS)	12.9 ± 1.5
Prevalence of retarded endometrial development (histological dating with traditional criteria), *n*/*n* (%)	1/34 (2.9) (NS)	3/39 (7.7) (NS)	6/21 (29) ($p < 0.01$)	10/48 (21) ($p < 0.01$)	3/68 (4.4)

With kind permission of Li et al.[34]. The results shown (except for the prevalence of retarded endometrial development) are mean values ± standard deviation (SD). The results in the four groups of infertility patients and one group of fertile controls were individually compared with fertile subjects using a two-sample *t*-test or 2 × 2 contingency table analysis. NS = not significant.

4

blood supply from the branches of the uterine arteries. The radial arteries pass through the myometrium and form two types of terminal branches: straight and spiral. The straight branches, called the basal arteries, supply the basal layers of the endometrium. The spiral branches, called the spiral arteries, pass through the endometrium and supply the stratum functionale[3]. The spiral arteries, unlike the basal arteries, are very sensitive to hormonal influences during the menstrual cycle.

Kupesic and Kurjak[28] were the first to describe the perfusion of the spiral arteries during the periovulatory period in spontaneous and stimulated cycles with both sonographically and hormonally confirmed ovulation. The spiral arteries in spontaneous cycles had a PI of 1.13 on the day before ovulation, compared with a PI of 2.32 in stimulated cycles, and showed a declining vascular resistance. In patients with three or more clomiphene-stimulated cycles, the endometrial thickness was significantly lower than in patients with spontaneous cycles or first-time clomiphene stimulation (Table 4.**2**). A significantly greater endometrial thickness was found throughout the follicular phase in patients after human menopausal gonadotropin (hMG) stimulation compared with patients who had been stimulated with clomiphene/hMG. Distinct flow velocity waveforms were obtained from the endometrium in 80% of the women who received clomiphene stimulation for the first time. By contrast, clear spiral artery waveforms were obtained in only 16.7% of the women who had received clomiphene stimulation for three or more cycles. When spiral artery blood flow is analyzed in relation to the type of stimulation applied, a significant difference ($p < 0.001$) is found between clomiphene/hMG stimulation and the other forms. Clomiphene citrate induces a downregulation of estrogen receptors in estrogen-sensitive tissue, affecting both the growth and morphology of the endometrium[2, 10, 51, 52].

Kupesic and Kurjak[28] noted a strong correlation between endometrial thickness and blood flow velocity. This did not apply to patients stimulated with clomiphene/hMG, however: while these patients displayed normal endometrial growth, 55.6% of them had no detectable diastolic flow in the endometrium.

Zaidi et al.[53] assessed endometrial thickness, endometrial morphology, and the presence or absence of subendometrial or intraendometrial color Doppler flow in 96 women undergoing IVF treatment for infertility. The results of this study, done on the day of

Table 4.**2** Endometrial thickness in 27 spontaneous cycles and 51 stimulated cycles

	Days before (–) and after (+) ovulation				
	–3	**–2**	**–1**	**0**	**+1**
Spontaneous cycles (*n* = 27)	8 ± 1.1*	10 ± 1.2	12 ± 1.4	12 ± 1.5	13 ± 1.2
Cycles stimulated with CC (*n* = 15)	7 ± 1.5	9 ± 1.4	11 ± 1.4	12 ± 1.2	13 ± 1.6
Three or more cycles stimulated with CC (*n* = 12)	4 ± 1.5	6 ± 2.0	7 ± 2.0	7 ± 1.8	7 ± 2.0
CC/hMG (*n* = 16)	5 ± 1.5	6 ± 2.0	8 ± 2.0	9 ± 2.5	9 ± 2.0
hMG (*n* = 8)	6 ± 1.8	8 ± 2.0	11 ± 1.8	12 ± 1.8	12 ± 1.8

With kind permission of Kupesic and Kurjak (28). * Values are stated in mm as mean ± standard deviation. CC = clomiphene citrate. hMG =

hCG administration, were correlated with pregnancy rates. The overall pregnancy rate was 32.3%. No significant differences were found between the pregnant and nonpregnant groups with regard to endometrial thickness. Also, the pregnancy rates based on endometrial morphology were not significantly different ($p > 0.05$). However, the absence of detectable endometrial blood flow was always associated with failure of implantation ($p < 0.05$). There was no significant difference in pregnancy rates related to different depths of vascular penetration (subendometrial zone, outer hyperechoic zone, or inner hyperechoic zone).

In both of these studies on the evaluation of endometrial blood flow, the use of pulsed Doppler ultrasound was recommended for assessing uterine receptivity and investigating cases of unexplained infertility.

Luteal phase defect. Another important clinical problem is luteal phase defect (luteal phase deficiency, LPD), which is defined as a delay in the histological development of the endometrium of more than two days compared with normal endometrial development on a given day of the cycle[12, 37]. In the past, various methods have been used to evaluate endometrial function: histology, electron microscopy, histochemistry, immunohistochemistry, hysteroscopy, and the measurement of endometrial proteins in the plasma or in endometrial washings. All of these tests are invasive and uncomfortable and might conceivably interfere with successful implantation. This led Doherty to use transvaginal sonography as a noninvasive means of evaluating the endometrium in the luteal phase and identifying patients with LPD[17].

A special Doppler study was performed to investigate the relationship between the color Doppler appearance of segmental uterine and ovarian blood flow and the histological results of endometrial biopsies[29]. The spiral arteries of the subjects in the control group had an RI of 0.53 ± 0.04 in the periovulatory phase, 0.50 ± 0.02 in the midluteal phase, and 0.51 ± 0.04 in the late luteal phase. Resistance measurements in the spiral arteries in the patients with LPD indicated higher values during the periovulatory phase (RI = 0.70 ± 0.06, $p < 0.001$), midluteal phase (RI = 0.72 ± 0.6, $p < 0.001$), and late luteal phase (RI = 0.72 ± 0.04, $p < 0.001$). Resistance measurements in superficial and deep ovarian vessels showed a significant difference between the normal control group and the patients with LPD.

Thus, examination of the corpus luteum and small endometrial vessels with pulsed Doppler ultrasound can be helpful in the assessment of luteal phase adequacy.

Effect of Age on Endometrial Function

Navot[36] and Edwards[18] cited advancing age as the principal cause of the decline in female fertility. Both studies showed that amenorrheic patients and women over 40 years of age who received donor oocytes achieved higher implantation and pregnancy rates than women over age 40 who had regular menstrual cycles and received their own oocytes. Thus, oocyte quality appears to be a more important determinant of pregnancy rates in this age group than endometrial receptivity. The best way to compensate for a fertility decline is by implanting a donor oocyte in an artificially induced cycle. Batista et al.[6] observed normal secretory endometrial function and normal endometrial maturation during the luteal phase in women over age 40 with normal menstrual cycles. Their results clearly indicate that the failure of implantation due to endometrial causes is not the major factor in the decline of fertility in this population.

Kurjak and Kupesic[31] performed numerous ultrasound examinations throughout the menstrual cycle in 120 healthy fertile women, 85 postmenopausal women, and 45 postmenopausal women receiving hormone replacement therapy. They found significant changes in flow velocity waveforms recorded from the ovarian, uterine, radial and spiral arteries, correlating with the age of the patients. The fact that the RI of the uterine artery does not change significantly during the first years of menopause supports the thesis that the aging process initially affects the uterus less than the ovaries. Accordingly, the uterine environment can be manipulated more easily during the menopausal years through proper hormonal stimulation.

Endometrial Peristalsis

Birnholtz[8] was the first to report on movements of the endometrium as a reflection of myometrial activity. These contractions usually start during the follicular phase and become more frequent around the time of ovulation. At that time the contractions are directed toward the uterine fundus and assist sperm transport to promote fertilization. The earliest studies on these contractions employed transabdominal ultrasound and thus do not permit a quantification of the movements.

Oike et al.[39] used transvaginal sonography to observe endometrial movements during the proliferative phase of the menstrual cycle. They were unable to detect contractions during the secretory phase. Abramowitz and Archer[1] and De Vries et al.[14] used transvaginal ultrasound in 1990 as a means of classifying endometrial movements by their intensity and frequency. Videotape is an ideal method of observing endometrial peristalsis; the tape is played back at a higher speed to analyze endometrial movements. De Vries et al.[14] performed 46 endovaginal examinations in 42 women. They found that contractions occurred in all phases of the menstrual cycle except during menstruation. Lyons et al.[35] published comparable results in 1991.

When Oike et al.[38] correlated endometrial activity with endocrine parameters, they found a strong correlation between endometrial peristalsis and the serum estradiol level. A rise in the progesterone level appears to decrease the frequency of endometrial and myometrial movements. Preliminary results reported by Abramowitz and Archer suggest that a disturbance of contractile peristalsis may have causal significance in some cases of idiopathic infertility.

Cervical Factor

The cervical mucus plays a key role in the fertilization process[7]. It enables sperm to survive for up to 48 hours in the acidic vaginal milieu. The consistency of the cervical mucus is subject to predominantly hormonal influences and thus to changes during the menstrual cycle. Especially when infection is present, the mucus contains markedly increased amounts of leukocytes and other phagocytes, which adversely affect sperm survival. Additionally, antibodies to sperm can be demonstrated in up to 10% of women of childbearing age[7]. Although these antibodies are not cytotoxic, they can have a negative impact on sperm motility.

The female cervix can be evaluated with transabdominal and transvaginal ultrasound, or it can be directly examined with a speculum during the gynecological examination. The length of the cervical neck between the external and internal os should be accurately measured by transvaginal imaging.

The width of the cervical os, like cervical gland secretions, depends on the estrogen level. The production of thin cervical mucus during the periovulatory phase at midcycle is accompanied by a high blood flow velocity in both uterine arteries, which can be verified by transvaginal color Doppler imaging.

References

1 Abramowitz JS, Archer DF: Uterine endometrial peristalsis—a transvaginal ultrasound study. Fertil. Steril. 54 (1990) 451–454

2 Aksel S, Saracoglu OF, Yeoman RR, Wiebe RH: Effects of the clomiphene citrate on cytosolic estradiol and progesterone receptor concentrations in secretory endometrium. Amer. J. Obstet. Gynecol. 155 (1986) 1219–1223

3 Applebaum M: The menstrual cycle, menopause, ovulation induction, and in vitro fertilization. In Copel JA, Reed KL (eds.): Doppler ultrasound in obstetrics and gynecology. Raven Press, New York 1995, 71–86

4 Asherman JG: Amenorrhea traumatica (atretica). J. Obstet. Gynaecol. Br. Emp. 55 (1948) 23

5 Bassil S, Magritte JP, Roth J, Nisolle M, Donnez J, Gordts S: Uterine vascularity during stimulation and its correlation with implantation in in-vitro fertilization. Hum. Reprod. 8 (1995) 1497–1501

6 Batista M, Cartledge TP, Zellmer AM et al.: Effects of aging on menstrual cycle hormones and endometrial maturation. Fertil. Steril. 64 (1995) 492–499

7 Beck VW: The cervical factor. In Garcia CR, Mastroianni L, Amelar RD, Dubin L (eds.): Current Therapy of Infertility 3. Decker, Toronto 1988, 118–121

8 Birnholtz J: Ultrasonographic visualization of endometrial movements. Fertil. Steril. 41 (1984) 157–158

9 Brosens JJ, de Souza NM, Barker FG, Paraschos T, Winston RM: Endovaginal ultrasonography in the diagnosis of adenomyosis uteri: identifying the predictive characteristics. Br. J. Obstet. Gynaecol. 102(6) (1995) 471

10 Check JH, Dieterich C, Lurie D: The effect of consecutive cycles of clomiphene citrate therapy on endometrial thickness and echo pattern. Obstet. Gynecol. 86 (1995) 341–345

11 Cullinan JA, Fleischer AC, Kepple DM, Aenoco AL: Sonohysterography: a technique for endometrial evaluation. Radiographics 15 (1995) 501–514

12 Dawood YM: Corpus luteum insufficiency. Curr. Opin in Obstet. Gynecol. 6 (1994) 121–127

13 Deligdish L, Loewenthal M: Endometrial changes associated with myomata of the uterus. J. Clin. Pathol. 23 (1970) 676

14 De Vries K, Lyons FA, Ballard G, Levi CS, Lindsay DJ: Contractions of the inner third of the myometrium. Amer. J. Obstet. Gynecol. 162 (1990) 679–682

15 Dickey RP, Olar TT, Curole DN, Taylor SN, Ryne PH: Endometrial pattern of thickness associated with pregnancy outcome after assisted reproduction technologies. Hum. Reprod. 7 (1992) 418–421

16 Dodson M: The endometrium. In Dodson M (ed.): Transvaginal ultrasound. Churchill Livingstone, New York 1995, 73–103

17 Doherty CM, Silver B, Binor Z, Wood Molo M, Radwanska E: Transvaginal ultrasonography and the assessment of luteal phase endometrium. Amer. J. Obstet. Gynecol. 168 (1993) 1702–1709

18 Edwards RG, Morcos S, MacNamee M, Balamaceda JP, Walters DE, Asch R: High fecundity of amenorrhoeic women in embryo-transfer programmes. Lancet 338 (1991) 292–294

19 Farrer-Brown G, Beilby JO, Tarbit MH: Venous changes in the endometrium of myomatous uteri. Obstet. Gynecol. 38 (1971) 743

20 Fedele L, Bianchi S, Dorta M et al.: Transvaginal ultrasonography versus hysteroscopy in the diagnosis of uterine submucous myomas. Obstet. Gynecol. 77 (1991) 745–748

21 Fleischer AC, Herbert CM, Sacks GA, Wentz AC, Entman SS: Nonconception cycles of IVF-ET. Fertil Steril. 46 (1986) 442–446

22 Fleischer AC, Herbert CM, Sacks GA, Wentz AC, Entman SS, James AE Jr: Sonography of the endometrium during conception and nonconception cycles of in vitro fertilization and embryo transfer. Fertil Steril. 46 (1986) 442–447

23 Fleischer AC, Keppe DM: Benign conditions of the uterus, cervix and endometrium. In Nyberg A, Hill LM, Bohm-Velez M, Mendelson EB (eds.): Transvaginal ultrasound. Mosby Year Book, St. Louis 1992, 21–43

24 Funk A, Fendel H: Sonography diagnosis of congenital uterine abnormalities. Z. Geburtshilfe Perinatol. 192 (1988) 77–88

25 Glissant A, de Mouzon J, Frydman R: Ultrasound study of the endometrium during IVF cycles. Fertil. Steril. 44 (1985) 786–790

26 Gonen Y, Casper RF: Prediction of implantation by the sonographic appearance of the endometrium during controlled ovarian stimulation for in vitro fertilization. J. In Vitro Fert. Embryo. Transf. 146 (1990) 146–152

27 Kepic T, Applebaum M, Valle J: Preovulatory follicular size, endometrial appearance, and estradiol levels in both conception and nonconception cycles: A retrospective study. 40th Annual Clinical Meeting of the American College of Obstetricans and Gynecologists, April 1992, 20 (abstract)

28 Kupesic S, Kurjak A: Uterine and ovarian perfusion during the periovulatory period assessed by transvaginal color Doppler. Fertil. Steril. 60 (1993) 439–443

29 Kupesic S, Kurjak A, Vujisic S, Petrovic Z: Luteal phase defect: comparison between Doppler velocimetry, histological and hormonal markers. Ultrasound Obstet. Gynecol. 9 (1997) 105–112

30 Kurjak A, Kupesic S: Benign uterine conditions. In: Kurjak A (ed.): An Atlas of Transvaginal color Doppler. Parthenon Publishing, Carnforth 1994, 247–317

31 Kurjak A, Kupesic S: Ovarian senescence and its significance on uterine and ovarian perfusion. Fertil. Steril. 64 (1995) 532–537

32 Kurjak A, Kupesic S, Miric D: The assessment of benign uterine tumor vascularization by transvaginal color Doppler. Ultrasound Med. Biol. 18 (1992) 645–649

33 Kutteh W, Carr B: Recurrent pregnancy loss. Textbook of reproductive medicine. Appleton and Lange, Norwalk 1993, 559–570

34 Li TC, Dockery P, Cooke ID: Endometrial development in the luteal phase of women with various types of infertility: comparison with women of normal fertility. Hum. Reprod. 6 (1991) 325–330

35 Lyons EA, Ballard G, Taylor PH, Levi CS, Zhieng XH, Kredentser JV: Characterization of subendometrial myometrial contractions throughout the menstrual cycle in normal fertile women. Fertil. Steril. 55 (1991) 771–774

36 Navot D, Bergh PA, Williams MA et al.: Poor oocyte quality rather than implantation failure as a cause of age-related decline in female fertility. Lancet 337 (1991) 1375–1377

37 Noyes RW, Hertig AT, Rock J: Dating the endometrial biopsy. Fertil. Steril. 1 (1950) 3–25

38 Oike K, Ishihara K, Kikuchi S: A study of the endometrial movement and serum hormonal level in connection with uterine contraction. Acta Obstet Gynaecol Jpn 42 (1990) 86–92

39 Oike K, Obata S, Tagaki K, Matsuo K, Ishihan K, Kikuchi S.: Observation of endometrial movements with transvaginal sonography. J. Ultrasound Med. 7 (1988) 899

40 Rabinowitz R, Laufer N, Lewin A et al.: The value of ultrasonographic endometrial measurement in the prediction of pregnancy following in vitro fertilization. Fertil. Steril. 45 (1986) 824–828

41 Randolph JR, Ying YK, Maier DB, Schmidt CL, Riddick DH: Comparison of real-time ultrasonography, hysterosalpingography and laparotomy/hysteroscopy in the evaluation of the uterine abnormalities and tubar patency. Fertil. Steril. 46 (1986) 828–832

42 Schlaff WD, Hurst BS: Preoperative sonographic measurement of endometrial pattern predicts outcome of surgical repair in patients with severe Asherman's syndrome. Fertil. Steril. 63 (1995) 410–413

43 Sher G, Herbert C, Massarani G, Jacobs MH: Assessment of the undergoing IVF-ET. Hum. Reprod. 6 (1991) 232–237

44 Smith B, Porter R, Ahuja K, Craft I: Ultrasonic assessment of endometrial changes in stimulated cycles in an in vitro fertilization and embryo transfer program. J. In Vitro Fert. Embryo. Trans. 1 (1984) 233–238

45 Steer CF, Campbell S, Tan S et al.: The use of transvaginal color flow imaging after in vitro fertilization to identify optimum uterine conditions before embryo transfer. Fertil Steril. 57 (1992) 372–376

46 Sterzik K, Grab D, Sasse V, Hutter W, Rosenbusch B, Terinde R: Doppler sonographic findings and their correlation with implantation in an in vitro fertilization program. Fertil. Steril. 52 (1989) 825–828

47 Stray-Pedersen B, Stray-Pedersen S: Etiological factors and subsequent reproductive performance in 195 couples with a prior history of habitual abortion. Amer. J. Obstet. Gynecol. 148 (1984) 140–146

48 Thickman D, Arger P, Turek R, Biasco L. Mintz M, Coleman B: Sonographic assessment of the endometrium in patients undergoing in vitro fertilization. J. Ultrasound Med. 5 (1986) 197–210

49 Welker BG, Dembruch U, Diedrich K, Al-Hasani S, Krebs D: TVS of the endometrium during oocyte pick-up in stimulated cycles for IVF. J. Ultrasound. Med. 1 (1989) 233–238

50 Winkel AC: Diagnosis and treatment of uterine pathology. In Carr BR, Blackwell RE (eds.): Textbook of reproductive medicine. Appleton and Lange, Norwalk 1993, pp. 481–505

51 Transabdominal ultrasonographic evaluation of endometrial thickness in clomiphene citrate-stimulated cycles in relation to conception. J. Clin. Ultrasound 22 (1994) 109–112

52 Yagel S, Ben-Chetrit A, Anteby E, Zacut D, Hochner-Celnikier D, Ron M: The effect of ethynil estradiol on endometrial thickness and uterine volume during ovulation induction by clomiphene citrate. Fertil. Steril. 57 (1992) 33–36

53 Zaidi J, Campbell S, Pittrof R, Tan SL: Endometrial thickness, morphology, vascular penetration and velocimetry in predicting implantation in an in vitro fertilization program. Ultrasound Obstet. Gynecol. 6 (1995) 191–198

5 Changes in Uterine and Ovarian Perfusion with the Onset of Menopause

A. Kurjak and S. Kupesic

There has been a growing trend in recent decades for women to postpone childbearing until 30 to 40 years of age. Many of these women are faced with infertility problems, however. The fact that natural fertility is very low at 45 years of age and older presents a challenge to the various specialties that are involved in the treatment of infertility. These include ultrasonography, especially transvaginal color Doppler and pulsed Doppler imaging. This innovative technique provides a unique, noninvasive method of evaluating the normal and the abnormal female pelvis.

Decline of Fertility in the Perimenopausal Period

The perimenopausal years, which begin at 40 years of age, represent the transition between the reproductive period and the postmenopausal period. The decline in the fertility of couples with advancing age has been amply documented. It is believed that the age-related decline in pregnancy rates is caused by functional inadequacy of the ovaries and of the endometrium. Two major problems are the reduced ability of the zygote to implant and the aging of the oocytes. Numerous studies have addressed the question of which factor poses the greater problem.

Uterine Receptivity

Some authors attribute the decline in fertility with aging to a decline in uterine receptivity. Ezra and Schenker[12] described an increased abortion rate of genetically normal embryos in the population of older women, which may be attributable to uterine dysfunction. Highly sensitive tests for human chorionic gonadotropin (hCG) indicate that up to 30% of pregnancies are lost between implantation and the 6th week of gestation[45].

Oocyte Quality

The frequency of both euploid and aneuploid abortions increases with maternal age. Experience with the oocytes of younger women who were donors for older women shows that aging oocytes, rather than endometrial factors, are the principal cause of decreased fertility. Some specialists believe that the significantly higher pregnancy rate with donor oocytes is attributable to their better quality.

Navot et al.[31] studied 35 infertile women over age 40 who had failed at attempts to conceive with their own oocytes. Oocytes were donated by 29 younger women (mean age 33.4 ± 0.7 years) undergoing in-vitro fertilization (IVF). The rate of implantation per embryo transferred was higher with donated oocytes (14.7%) than with self-oocytes (3.3%) in the women over 40 years of age ($p < 0.01$).

To further study the effect of aging on reproductive outcome, pregnancy results were compared between the young donors and older recipients. The clinical pregnancy and delivery rates for the donors (33% and 23%) and recipients (40% and 30%) were not significantly different. These data suggest that the age-related decline in female fertility is related to oocyte quality and is correctible by oocyte donation.

Drews et al.[11] found that similar pregnancy and livebirth rates were achieved when donor oocytes from the same women were given to one woman over age 40 and another under age 40. In a second report[39], uterine receptivity as determined by clinical pregnancy rates was similar for oocyte recipients over age 40 and young IVF surrogates when both groups received oocytes from young donors. The pregnancy rate declined when surrogates received oocytes from women over 40 years of age.

In another study, Navot et al.[32] evaluated 38 ovum donors throughout 102 oocyte donations. They documented 51 cycles in younger recipients (35.8 ± 3.1 years) and 51 cycles in older recipients (44.0 ± 3.1 years). They found that the capacity to conceive and to carry a pregnancy to term when oocyte quality is controlled appears to be independent of uterine aging in the fifth decade of life.

Borini et al.[3] attempted to determine the potential of the aging uterus in terms of implantation, pregnancy, abortion, and obstetric complications in postmenopausal women over age 50 who received donated oocytes. They found that women from 50 to 62 years of age can become pregnant with donated oocytes when they receive adequate hormone replacement therapy.

It is also important, however, to consider the effect of pregnancy on preexisting maternal diseases and the rising risks of preeclampsia, hypertension, and diabetes mellitus. Patients 40 years of age or older who wish to undergo IVF therapy with their own oocytes should be thoroughly counseled by their doctor. The following risks should be mentioned[42]:

- A 30–50% reduction in pregnancy potential
- A rising risk of chromosome abnormalities
- Abortion and stillbirth

When these risks are known, an additional test of ovarian capacity can help to identify the women for whom IVF, other forms of assisted reproduction, or surgical intervention are the most appropriate forms of treatment.

Ovarian Function

Follicle-stimulating hormone (FSH). Toner et al.[43] report that when age, infertility etiology, and semen quality are taken into account, FSH is the best predictive parameter of ovarian function. The combination of age and basal FSH in treated patients increases the accuracy of the prognosis and can provide an index for the functional ovarian reserve ("ovarian age").

Women over age 40 with a favorable hormonal profile respond well to assisted reproduction, whereas women of any age with a basal FSH level > 20 IU/l respond poorly to ovarian stimulation. Moreover, it is rare to find FSH levels above 25 IU/l in an existing pregnancy. Pregnancy is most likely to occur when the FSH level is between 10 and 20 IU/l.

E_2. Another parameter that is useful in the prediction of "ovarian age" is the basal E_2, with values higher than 50 pg/ml indicating a poor ovarian reserve. For this reason, it is best to consider age, FSH, luteinizing hormone (LH), and E_2 for the optimum prediction of ovarian response. Provocative tests of ovarian function are probably superior to static tests, but they are difficult to perform and are not widely practiced.

Age. Fitzgerald et al.[13] studied the effect of age on follicular growth and endometrial thickness. Ultrasound examinations were done to confirm ovulation and to measure follicular and endometrial growth. Ovulation occurred later in older women, with an increase in the mean follicular phase length from 13.9 days (20–25 age group) to 15.9 days (37–45 age group; $p < 0.05$). The mean maximum follicular diameter before ovulation was significantly smaller in older women: 16.7 mm (37–45 years), 21.3 mm (32–36 years), and 19.6 mm (21–25 years). The maximum endometrial thickness during the luteal phase was greatest in older women: 15.9 mm (37–45 years), 12.1 mm (21–25 years; $p < 0.001$). While the levels of ovarian steroids showed no differences, the serum gonadotropin levels during menstruation were higher in older women.

These data point to significant age-related differences in the pituitary–ovarian axis and endometrial thickness that influence the management of older women in medically assisted reproduction programs.

Meldrum[29] emphasizes in his review that decreased implantation with aging is associated with a high incidence of delayed or absent secretory transformation of the endometrium. In patients who were treated with physiological amounts of progesterone replacement, a marked regression of implantation was found with increasing age. Treatment with high doses of progesterone significantly improved the implantation rates. Thus, oocytes donated by young women and an elevated progesterone level can correct the age-related deficits in older women.

Effects of Estradiol and Progesterone on Vascular Resistance

Rhythmic changes in uterine blood flow during the menstrual cycle are sometimes related to the ratio of progesterone and estrogen in the blood[15,22,47]. The higher the ratio of estrogen to progesterone, the greater the blood flow in the uterine vascular bed[10,16,24]. Progesterone antagonizes the vasodilator effect of estrogen on the uterus[6,39], the magnitude of this inhibition depending upon the ratio of the two steroids[6].

Sympathetic Innervation of the Uterus

The periarterial sympathetic vasoconstrictor nerves of the uterus are recognized as important factors in the regulation of uterine blood flow. Exposure to progesterone increases the vasoconstrictor effect of these nerves, while exposure to estrogen decreases it[18,19]. The results of Ford et al.[17] indicate that ovarian steroids affect the function of uterine periarterial sympathetic nerves by altering the number of alpha-adrenergic receptors. This may contribute to the marked changes in uterine blood flow observed during the estrous cycle of pigs.

Estrogen Effect

To determine whether ovarian hormones at physiological levels affect uterine vascular resistance, De Ziegler et al.[9] studied young women with loss of ovarian function who received physiological amounts of exogenous estradiol and progesterone. Their results show that in the absence of endogenous estrogen production by the ovaries, the uterine arteries have a high vascular resistance as indicated by low systolic Doppler flow and high PI (pulsatility index) values.

Goswamy and Steptoe[21] theorize that persistent diastolic flow during the early follicular phase is a more common phenomenon in multiparous women than in nulliparae. They observed a profound alteration of the Doppler flow pattern reflecting a marked decline of vascular resistance following the transdermal administration of estradiol (0.1–0.4 mg/day). This observation is consistent with the hypothesis that the periovulatory decline in vascular resistance is mediated by estradiol[34]. Since estrogen receptors have been identified in the wall of the uterine arteries, it is reasonable to suppose that estradiol has a direct effect on uterine Doppler flow. It is postulated that the effect of estrogen on uterine arterial vascular resistance is directly related to the plasma level of biologically active estrogen and that a direct dose–response relationship exists[37].

Other possible mechanisms include the modulation of the production and/or secretion of various vasoactive substances such as prostaglandins[40], CGRP (calcitonin gene-related peptide[29]), and ERF (endothelial relaxing factor [30]) by estradiol.

There is no question that transvaginal Doppler flow studies of the uterine arteries are a valuable tool for assessing the biological efficacy of various estrogen treatments. This is particularly important in evaluating the effects of postmenopausal hormone replacement therapy in women with increased hepatic estrogen metabolism resulting in lower plasma estradiol levels.

Progesterone Effect

So far, the effect of progesterone on the pulsatility index (PI) in humans has not been definitively evaluated. Progesterone acts as a vasoconstrictor in lower mammalian species[23,35]. This has raised concern that a sequential or continuous progesterone regimen administered for endometrial protection[23] may cause a partial or complete reversal of the vasodilator action of estrogen.

Hillard et al.[23] used transvaginal color and pulsed Doppler to measure the PI in the uterine arteries during and after estrogen-only and combined estrogen–progesterone therapy. In all patients, the addition of norethindrone acetate (0.7 mg/day) or medroxyprogesterone acetate (10 mg/day) partially antagonized the response to transdermal estradiol. The PI was approximately 34% lower than the pretreatment value during the combined estrogen–progesterone phase of treatment, but it was 13% higher compared with the estradiol-only phase. It is unclear whether the effects of progesterone on uterine arterial PI are related to a decrease in protection against arterial disease on estrogen replacement therapy when progesterone is added.

Effect of Age on Ovarian and Uterine Perfusion

Authors' Study

Kurjak and Kupesic analyzed the relationship between age ("ovarian aging") and uterine and ovarian perfusion[26]. One hundred and ninety women were studied: 91 fertile women with normal menstrual cycles, 65 postmenopausal women, and 34 postmenopausal women receiving hormone replacement therapy.

Timing of examinations. Doppler signals from the ovarian artery were characterized by low flow velocity and high resistance. The women in the control group were examined between days 5 and 10 of their cycle and every day thereafter until clinical ovulation occurred. The values after ovulation were measured at least twice during the luteal phase. The ovary that contained the follicle or corpus luteum was classified as "dominant." In the postmenopausal group of women, the examinations were performed at randomized intervals.

Vessels examined. The uterine artery was scanned lateral to the cervix at the level of the corpus–cervix junction. The average value for both uterine arteries was taken. Radial artery waveforms were recorded within the myometrium, and the spiral arteries were sampled at the level of the myometrial–endometrial junction. During each examination, the resistance index (RI) was automatically computed from the Doppler waveform using the formula: (Systole–diastole) : systole. The ovarian artery was scanned lateral to the ovary. If a clear signal was not obtained, the sample volume was moved across the ligament until an arterial signal was identified.

Ovarian Artery

Women with normal cycles. Longitudinal studies in which the ovarian artery was examined in women with normal cycles demonstrated narrow systolic waveforms with an RI of 0.92 ± 0.08 (mean ± SD) in the early postmenstrual phase. In the periovulatory phase (days 12–14 of the cycle), the Doppler traces recorded from the ovarian artery on the dominant side showed marked broadening with continuous diastolic flow, indicating a marked decrease in vascular resistance (RI = 0.86 ± 0.04) (Fig. 5.**1**). Doppler measurements repeated in the luteal phase demonstrated a further rise of end-diastolic flow velocity, which was most apparent on day 21 of the cycle. This finding was reflected in a significant decline of the RI to 0.83 ± 0.04. The nondominant ovarian artery did not show the cyclic variations described for the dominant side.

Postmenopausal women. In postmenopausal women who had had their last menstrual period 1–5 years earlier, Doppler measurements of the ovarian artery showed no significant changes compared with the nondominant side in the healthy, fertile controls. When the values in the postmenopausal group were compared with ovarian vascular resistance in the early follicular and luteal phase of women with normal cycles, significant differences were observed ($p < 0.01$, $p < 0.001$). The ovarian artery in the postmenopausal group showed a marked interruption of diastolic flow signals in 55% of the women. Absence of diastolic flow in the ovarian artery was detected in

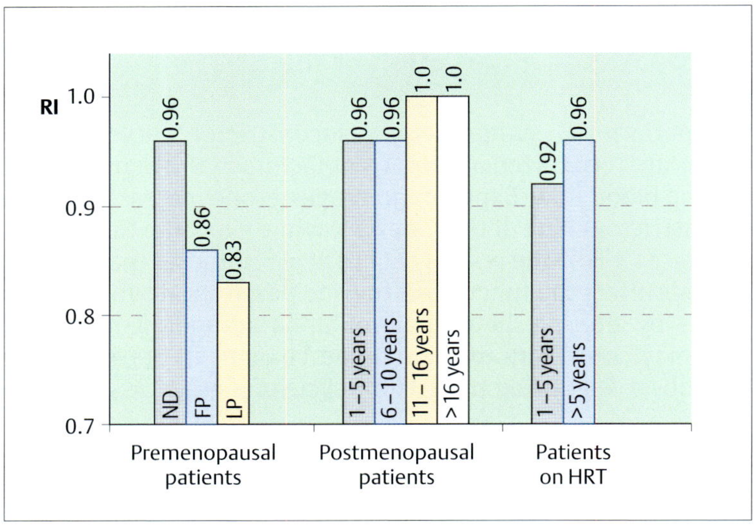

Fig. 5.**1** Ovarian artery blood flow in premenopausal and postmenopausal women (with and without HRT). ND = nondominant ovarian artery; FP = follicular phase; LP = luteal phase, years = years after menopause; HRT = hormone replacement therapy.

most women who had been menopausal for 6–10 years (84.2%), while interrupted diastolic flow and an RI of 1.0 were consistently found in women who had been menopausal for more than 11 years.

Patients on hormone replacement therapy (HRT). We found no significant differences when we compared the ovarian artery RI of patients who had been on HRT for less than five years with patients who had received the therapy for more than five years. We also found no difference in ovarian artery resistance between patients receiving HRT and untreated patients. We were unable to record Doppler flow signals from the ovarian parenchyma of our postmenopausal subjects, regardless of whether or not they were receiving HRT.

Uterine Artery

Women with normal cycles. The RI values of the right and left uterine arteries correlated strongly in every evaluation, and so we were able to use both values for statistical analysis. The mean RI of the uterine artery during the proliferative phase was 0.88 ± 0.04. Doppler measurements of the uterine artery during the luteal phase showed a decreased RI value (0.84 ± 0.004) (Fig. 5.**2**).

Postmenopausal women. A significant correlation was found between the baseline RI and time after menopause in the postmenopausal group of patients. The highest impedance values were found in the women who had been menopausal for the longest time. Absence of uterine artery diastolic flow was seen in 15% of the women who had been in menopause for 1–5 years (Fig. 5.**3**). A marked interruption of uterine artery diastolic flow was found in 31.6% of the women who had been menopausal for 6–10 years and in 54.5% of the women who had been menopausal for 11–15 years. Finally, 79.2% of the women who had been menopausal for more than 16 years showed an absence of diastolic flow, which signifies high vascular impedance.

Patients on HRT. The uterine artery RI was significantly lower in patients who were receiving HRT. Increased diastolic flow and hence a lower RI were particularly evident in patients who had been taking hormones for more than six years.

Radial Arteries

The change in the flow velocity patterns of the radial arteries in premenopausal and postmenopausal women (with or without HRT) was similar to the flow patterns observed in the uterine arteries. The mean RI of the radial arteries decreased from 0.74 in the proliferative phase to 0.68 in the luteal phase (Fig. 5.**4**). The RI values showed a tendency to increase after menopause. Patients who were receiving HRT showed a broadening of the systolic waveform and continuous flow signals during diastole, attributable to decreased resistance in the radial arteries.

Spiral Arteries

A significant difference in spiral artery resistance was found between the proliferative phase (RI = 0.64) and the luteal phase

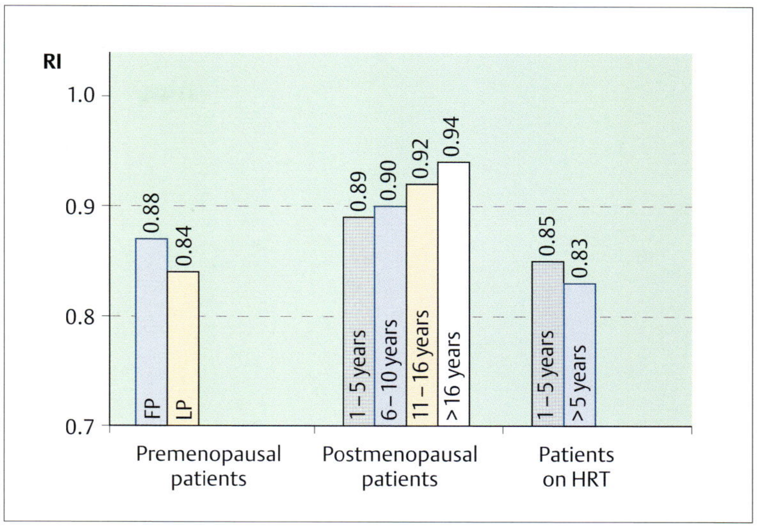

Fig. 5.**2** Uterine artery blood flow in premenopausal and postmenopausal women (with and without HRT). FP = follicular phase; LP = luteal phase; years = years after menopause; HRT = hormone replacement therapy.

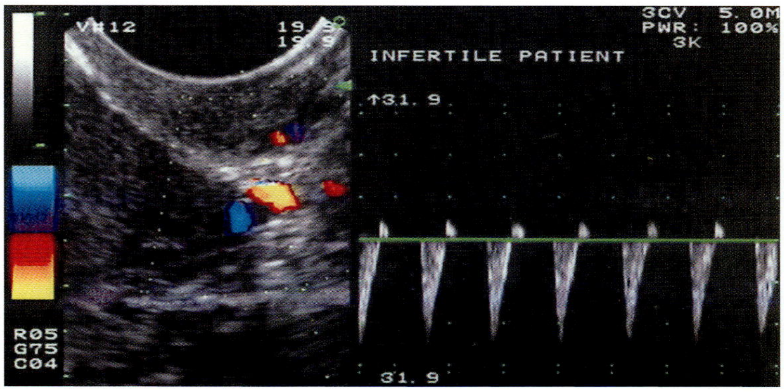

Fig. 5.**3** Absence of diastolic flow and reverse flow in the uterine artery of a postmenopausal patient.

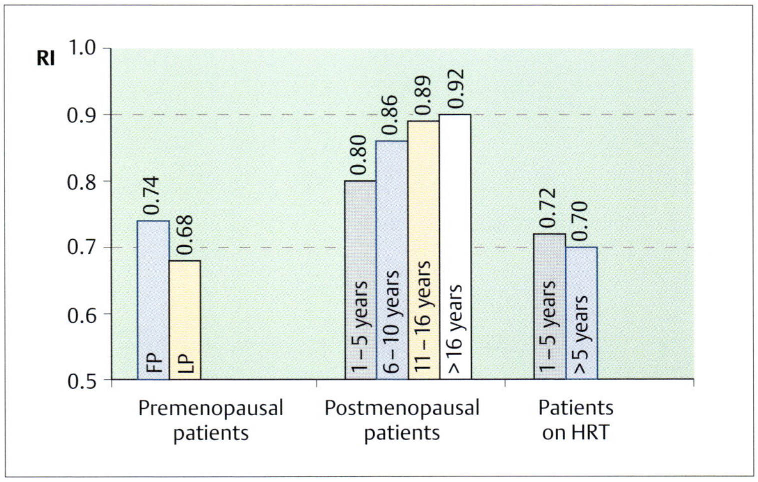

Fig. 5.**4** Radial artery blood flow in premenopausal and postmenopausal women (with and without HRT). FP = follicular phase; LP = luteal phase; years = years after menopause; HRT = hormone replacement therapy.

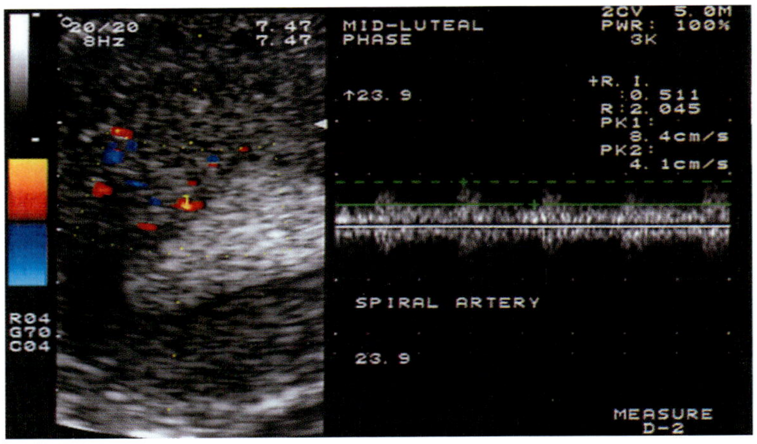

Fig. 5.**5** Hyperechoic endometrial layers demonstrated by transvaginal sonography. Analysis of the Doppler waveform of the spiral arteries (right) shows decreased resistance in the midluteal phase of the menstrual cycle (RI = 0.51).

(RI = 0.50) in the control group of healthy fertile women (Fig. 5.**5**). Clear Doppler signals could be recorded from the spiral arteries in only 30% of the women who had been menopausal for 1–5 years (Table 5.**1**). This postmenopausal group also showed significantly higher impedance in the spiral arteries than the control group. In women who had been menopausal for more than six years, we were unable to record flow signals from the periphery of the endometrium. The spiral artery detection rates were higher ($p < 0.001$) in women who were receiving HRT (Table 5.**2**). The flow velocity waveforms of the spiral arteries in these women showed a decreased RI ($p < 0.01$) compared with postmenopausal women not receiving HRT.

Interpretation of the Results

The results of our Doppler study agree with other clinical studies that have documented a sudden change in ovarian function after 40 years of age[7, 11, 29, 30, 31, 38, 44, 46].

Ovarian blood flow during the menstrual cycle. The use of color Doppler has simplified the analysis of sequential changes in intraovarian blood flow. The highest resistance is observed on the first day of the menstrual cycle, the lowest on the day of the LH peak. With transvaginal color Doppler, it is also possible to observe perfused areas at the periphery of the follicle. The RI is 0.54 shortly before ovulation, starts to decline two days before ovulation, and reaches its low point of 0.44 at the time of ovulation. The mature corpus luteum normally has a diameter of 1–3 cm and shows low impedance values (mean RI 0.43). On just the 23rd day after menstruation, the corpus luteum begins to undergo regressive changes. Decreased blood flow velocities and an increased RI (mean value 0.49) are the typical signals of these changes[1, 5, 8, 25, 28, 48].

Ovarian blood flow after menopause. Absent diastolic flow in the ovarian artery was a common finding in the early postmenopausal period and was consistently present in women who had been in menopause for more than 11 years. We could detect no intraovarian blood flow in the group of postmenopausal women. This may be due to a progressive increase in fibroblasts and connective tissue accompanied by a decline in circulating estrogen levels. For this reason, any signals that are detected with color Doppler ultrasound in the postmenopausal ovary should be considered highly suspicious for abnormal neovascularization and should prompt a detailed investigation with pulsed Doppler.

Table 5.**1** Detection rates of the uterine, spiral, and radial arteries in postmenopausal patients

Number of patients	Duration of menopause (years)	Detection rate of uterine artery (%)	Detection rate of radial artery (%)	Detection rate of spiral artery (%)
15	1–5	100	100	30
14	6–10	100	89.5	0
17	11–15	100	76.2	0
19	> 15	100	33.3	0
Total 65		100	76.2	7.14

With permission from Kurjak and Kupesic[27].

Table 5.**2** Detection rates of the uterine, spiral, and radial arteries in postmenopausal patients receiving HRT

Number of patients	Duration of menopause (years)	Detection rate of uterine artery (%)	Detection rate of radial artery (%)	Detection rate of spiral artery (%)
21	1–5	100	100	35.7
17	>5	100	94.1	17.6
Total 38		100	97.7	28.8

With permission from Kurjak and Kupesic[27].

Uterine blood flow after menopause. We found continuous diastolic flow in the uterine arteries of all healthy, fertile control subjects. The uterine arteries of postmenopausal women showed increasing vascular impedance that was manifested by a narrow systolic waveform and a high RI. Bonilla Musoles et al.[2] found increased vascular resistance in the uterine arteries after the onset of menopause.

It should be noted, however, that the change in vascular impedance and absence of diastolic flow in the ovarian arteries are the most conspicuous signs. The fact that the uterine artery RI does not change significantly during the initial postmenopausal years supports the thesis that the aging process affects the uterus less than was previously supposed[27].

Hormone replacement therapy. The profound decline of vascular resistance in our patients who were receiving HRT was manifested by changes in the Doppler signal patterns of the radial arteries and uterine artery. Other, earlier Doppler studies[9,23,49] documented a marked decline of uterine artery resistance following treatment with physiological amounts of transdermally administered E_2. This implies that it is relatively easy to manipulate the uterine system with controlled hormone regimens. The authors discovered a rapid and profound response of uterine blood flow to hormone replacement therapy. The cardiovascular benefits are apparent and are seen even in women who do not start hormone replacement therapy until late in menopause.

At the same time, there are other variables that can influence uterine perfusion after menopause. A possible persistence of residual ovarian activity and the use of vasoactive drugs can decrease the values of uterine vascular resistance. Even dietetic and psychosocial factors can affect the local uterine blood flow in postmenopausal women.

Conclusions. These data confirm and support the observation that oocytes donated by young women, together with appropriate hormonal support of the endometrium, can help to overcome the problems of deficient uterine receptivity and high abortion rates that occur in postmenopausal women[3].

There is no doubt that transvaginal color and pulsed Doppler ultrasound provide a noninvasive method that will help us to understand the possible effects of aging on female fertility.

References

1 Baber RJ, McSweeney MB, Gill RW: Transvaginal pulsed Doppler ultrasound assessment of blood flow to the corpus luteum in IVF patients following ET. Br. J. Obstet. Gynaecol. 95 (1988) 1226–1230

2 Bonilla Musoles F, Marti MC, Ballester MJ: Normal uterine arterial blood flow in postmenopausal women assessed by transvaginal color Doppler sonography: the effect of hormone replacement therapy. J. Ultrasound Med. 14 (1995) 497 –501

3 Borini A, Bafaro G, Violini F, Bianchi L, Casadio V, Flamigni C: Pregnancies in postmenopausal women over 50 years old in an oocyte donation program. Fertil. Steril. 63 (1995) 261

4 Bourne TH, Hillard TC, Whitehead MI, Crook D, Campbell S: Oestrogens, arterial status, and postmenopausal women. Lancet 335 (1990) 1471–1472

5 Bourne T, Jurkovic D, Waterstone J, Campbell S, Collins WP: Intrafollicular blood flow during human ovulation. Ultrasound Obstet. Gynecol. 1 (1991) 53–59

6 Caton D, Abrams RM, Clapp JF, Barron D: The effect of exogenous progesterone on the rate of blood flow of the uterus of ovariectomized sheep. Q. J. Exp. Physiol. 59 (1974) 231

7 Check JH, Lurie D, Callan C, Baker A, Benfer K: Comparison of the cumulative probability of pregnancy after in vitro fertilization-embryo transfer by infertility factor and age. Fertil. Steril. 61 (1994) 257–261

8 Collins WP, Jurkovic D, Bourne TH, Kurjak A, Campbell S.: Ovarian morphology, endocrine function and intrafollicular blood flow during peri-ovulatory period. Hum. Reprod. 6 (1991) 319–324

9 De Ziegler D, Bessis R, Frydman R: Vascular resistance of uterine arteries: physiological effects of estradiol and progesterone. Fertil. Steril. 55 (1991) 775–779

10 Dickson WM, Bosc MJ, Locatelli A: Effect of estrogen and progesterone on uterine blood flow in castrate sows. Am. J. Physiol. 217 (1969) 1431–1437

11 Drews M, Bergh P, Williams M, Grunfeld L, Garrisi G, Navot D: Age related decline in female infertility is not due to diminished capacity of the uterus to sustain embryo implantation. 48th annual meeting of the American Fertility Society. New Orleans 1992, Abstract 0-086, S39

12 Ezra Y, Schenker JG: Appraisal of in vitro fertilization. Eur. J. Obstet. Gynecol. Reprod. Biol. 48 (1993) 127–133

13 Fitzgerald CT, Seif MW, Killick SR, Elstein M: Age related changes in the female reproductive cycle. Br. J. Obstet. Gynaecol. 101(3) (1994) 229–233

14 Ford SP: Control of uterine and ovarian blood flow throughout the estrous cycle and pregnancy of the ewe, sow and cow. J. Anim. Sci. 55, (Suppl. 2) (1973) 32–34

15 Ford SP, Chenault JR, Echterncamp SE: Uterine blood flow of cows during the estrous cycle and early pregnancy: effect of the conceptus on the uterine blood supply. J. Reprod. Fertil. 56 (1979) 53–58

16 Ford SP, Reynolds LP: role of adrenergic receptors in mediating estradiol-17 beta stimulated increases in uterine blood flow of cows. J. Anim. Sci. 57 (1983) 665–669

17 Ford SP, Reynolds LP, Farley DB, Bhatnagar RK, van Orden DE: Interaction of ovarian steroids and periarterial alpha-1-adrenergic receptors in altering uterine blood flow during the estrous cycle of gilts. Amer. J. Obstet. Gynecol. 5 (1984) 480–484

18 Ford SP, Weber LJ, Stormshak F: In vitro response of ovine and bovine uterine arteries to prostaglandin $F_{2\alpha}$ and periarterial sympathetic nerve stimulation. Biol. Reprod. 15 (1976) 58–64

19 Ford SP, Weber LJ, Stormshak F: Role of estradiol 17 beta and progesterone in regulating constriction of ovine arteries. Biol. Reprod. 17 (1977) 480–486

20 Furchgott RF, Zawadzki JW: The obligatory role of endothelial cells in the relaxation of arterial muscle by acetyl choline. Nature (London) 288 (1980) 373–376

21 Goswamy RK, Steptoe PC: Doppler ultrasound studies of the uterine artery in spontaneous ovarian cycles. Hum. Reprod. 3 (1988) 721–725

22 Henricks DM, Guthlie HD, Handlin DL: Plasma estrogen, progesterone and luteinizing hormone levels during the estrous cycle in pigs. Biol. Reprod. 6 (1975) 210–216

23 Hillard TC, Bourne T, Whitehead MI, Cryford TB, Collins WP, Campbell S: Differential effects of transdermal estradiol and sequential progestogens on impedance to flow within the uterine arteries of postmenopausal women. Fertil. Steril. 58 (1992) 959–963

24 Killam AP, Rosenfeld C, Battaglia FC, Makowski EL, Meschia G: Effect of estrogens on the uterine blood flow in oophorectemized ewes. Amer. J. Obstet. Gynecol. 115 (1973) 1045–1050

25 Kupesic S, Kurjak A: Uterine and ovarian perfusion during the periovulatory period assessed by transvaginal color Doppler. Fertil. Steril. 60 (1993) 439–443

26 Kupesic S, Kurjak A, Babic MM: Uterine and ovarian perfusion changes from reproductive maturity to menopause. ACDS 12 (1996) 79–87

27 Kurjak A, Kupesic S: Ovarian senescence and its significance on uterine and ovarian perfusion. Fertil. Steril. 64 (1995) 532–538

28 Kurjak A, Kupesic-Urek S, Schulman H, Zalud I: Transvaginal color flow Doppler in the assessment of ovarian and uterine blood flow in infertile women. Fertil. Steril. 56 (1991) 870–873

29 Meldrum DR: Female reproductive aging-ovarian and uterine factors. Fertil. Steril. 59 (1993) 1–5
30 Menken J, Trussel J, Larsel U: Age and infertility. Science 233 (1986) 1389–1394
31 Navot D, Bergh PA, Wiliams MA et al.: Poor oocyte quality rather than implantation failure as a cause of age-related decline in female fertility. Lancet 337 (1991) 175–177
32 Navot D, Drews MR, Bergh PA et al.: Age-related decline in female fertility is not due to diminished capacity of the uterus to sustain embryo implantation. Fertil. Steril. 61 (1994) 97–101
33 Padilla, SL, Garcia JE: Effect of maternal age and number of in vitro fertilization procedures on pregnancy outcome. Fertil. Steril. 52 (1989) 270–273
34 Perrot-Applanat M, Groyer-Picart MT, Garcia E, Lorenzo F, Milgrom E: Immunocytochemical demonstration of estrogen and progesterone receptors in muscle cells of uterine arteries in rabbits and humans. Endocrinology 123 (1988) 1511 –1514
35 Resnik R: The effect of progesterone on estrogen induced uterine blood flow. Gynecol. Invest. 128 (1976) 251–254
36 Resnik R, Brink GW, Plumer MH: The effect of progesterone on estrogen-induced uterine blood flow. Amer. J. Obstet. Gynecol. 128 (1977) 251–255
37 Sarrel PM: Ovarian hormones and circulation. Maturitas 590 (1990) 297–298
38 Sauer MV, Paulson RJ, Lobo RA: Pregnancy after 50: application of oocyte donation to women after natural menopause. Lancet 341 (1993) 321–323
39 Serafini P, Tran C, Tan T, Norbryhn G, Batzofin J: Oocyte aging is the main factor responsible for the decline in fertility with chronological advancement. Evidence from the IVF surrogacy and egg donation programmes. 48th Annual meeting of the American Fertility Society. New Orleans 1992, Abstract 0–011, 5
40 Steinleitner A, Stancyzk FZ, Levin JN, d'Ablaing G, Vijod MA, Shabhagian VL: Decreased in vitro production of 6-keto-prostaglandin $F_{1\alpha}$ by uterine arteries from postmenopausal women. Amer. J. Obstet. Gynecol. 161 (1989) 1677–1681
41 Stevenson JC, McDonald DWR, Waren RC, Booker MW, Whitehead MI: Increased concentration of circulating calcitonin gene related peptide during normal human pregnancy. Br. Med. J. 29 (1986) 1329–1330
42 Toner JP, Flood JT: Fertility after the age of 40. Obstetrics and Gynecology Clinics of North America 20 (1993) 261–272
43 Toner JP, Philput CB, Jones GS, Muasher SJ: Basal follicle-stimulating hormone level is a better predictor of in vitro fertilization performance than age. Fertil. Steril. 55(4) (1991) 784–791
44 Van Noord-Zaadstra BY, Looman CWN, Alsbach H, Rabbena JDF, Te Velde ER, Karbaat J: Delaying child bearing: effect of age on fecundity and outcome of pregnancy. Br. Med. J. 302 (1991) 1361–1365
45 Wilcox AJ, Weibereg CR, O'Connor JF et al.: Incidence of early loss of pregnancy. N Engl. J. Med. 319 (1988) 189–194
46 Wiro MS, Schewchuk AB: Pregnancy outcome in 242 conceptions after arteficial insemination with donor sperm and effects of maternal age on the prognosis for successful pregnancy. Amer. J. Obstet. Gynecol. 148 (1984) 518–524
47 Yuthasastrakosol P, Palmer WM, Howland BE: Luteinizing hormone, oestrogen and progesterone levels in peripheral serum of anoestrous and cyclic ewes as determined by radioimmunoassay. J. Reprod. Fertil. 43 (1975) 57–62
48 Zalud I, Kurjak A: The assessment of luteal blood flow in pregnant and non-pregnant women by transvaginal color Doppler. J. Perinat. Med. 18 (1990) 215–221

6 Color Doppler Sonography for the Optimization of Assisted Reproduction

D. Grab and K. Sterzik

The inability to conceive is a problem that affects some 10–15% of all couples. The causes of infertility are multifactorial in many cases. In addition to primary ovarian, hypothalamic-pituitary, extragenital endocrine, psychogenic, immunological, genetic, tubal, uterine, cervical, and vaginal causes of infertility, there are growing numbers of cases that result from a male infertility factor or have an unexplained cause ("idiopathic infertility"). In 30% of cases, the causes of infertility relate to both partners[22].

Functional Evaluation of the Endometrium

For all therapeutic approaches that are applied in assisted reproduction (Table 6.**1**), it must be considered that the functional status of the endometrium is a major determinant of successful implantation. The outcome of any procedure in reproductive medicine depends critically upon whether the embryo encounters an endometrium that is receptive to implantation. Theoretical considerations led Paulson et al.[34] to conclude that a lack of endometrial receptivity in hormone-induced cycles is responsible for two-thirds of all treatment failures.

Histological studies. Histological studies have shown that different stimulation protocols lead to different distributions of endometrial abnormalities, with clomiphene showing a particularly high association with significant endometrial dysfunction ranging to complete atrophy[8,42,51]. Based on these findings, Dallenbach et al.[8] state that endometrial biopsy histology should be an essential study in the investigation and treatment of infertility. The disadvantage of histological examination is its invasiveness.

Sonomorphological studies. The value of sonomorphological studies of the endometrium in monitoring the cycles of infertile patients is discussed controversial. While it is true that modern high-frequency endovaginal probes can accurately define the sonoanatomy of the endometrium[7], research to date has been unable to establish whether ultrasound can supply a clinically useful prognosis with regard to endometrial receptivity[21,44]. Our own studies indicate that histological findings are not consistently reflected in either the thickness or echo pattern of the endometrium.

Table 6.**1** Methods in reproductive medicine

Method	Abbreviation
Intrauterine insemination	IUI
In-vitro fertilization and embryo transfer	IVF-ET
Gamete intrafallopian transfer	GIFT
Pronucleus stage transfer	PROST
Zygote intrafallopian transfer	ZIFT
Intratubal embryo transfer	TET
Intracytoplasmic sperm injection	ICSI
Partial zona dissection	PZD
Subzonal insemination	SUZI

Pulsed Doppler ultrasound. Hemodynamic measurements of uterine and ovarian blood flow with Doppler ultrasound represent a new approach in the diagnosis of infertility. These measurements are based upon studies by Taylor et al.[45]. These authors recorded waveforms from the uterine and ovarian arteries by transabdominal pulsed Doppler scanning and distinguished them from other blood vessels in the lesser pelvis by their location and distinctive flow patterns.

Another advance was the development of Doppler-capable endovaginal transducers. Feichtinger et al.[12] were able to obtain nonsuperimposed scans of the uterine and ovarian blood vessels by the transvaginal route and investigated uterine and ovarian blood flow under physiological and pathological conditions.

Color Doppler sonography. The use of color Doppler sonography can provide an anatomically precise view of the vascular supply in the lesser pelvis[3,13]. Uterine and ovarian blood vessels can be rapidly identified and evaluated with this technique (Figs. 6.**1** and 6.**2**).

Goswamy et al.[16] presumed that a relationship existed between hemodynamic parameters and implantation rates on the basis of their Doppler investigations of uterine blood flow. Shortly thereafter, we were able to confirm this observation in our own patients[18,43]. We later studied a larger population to determine whether Doppler measurements of uterine blood flow could be useful in predicting the outcome of implantation[17].

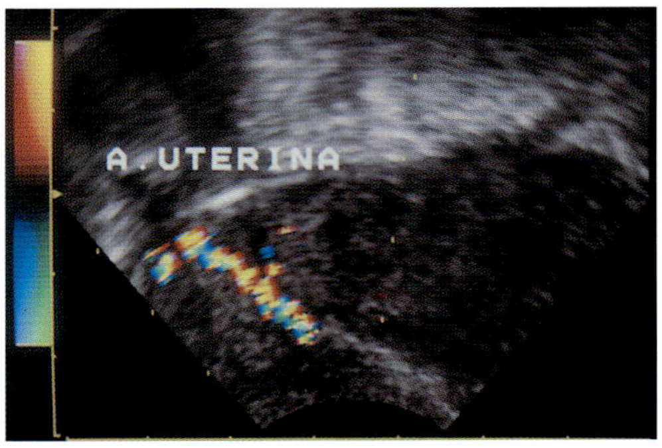

Fig. 6.1

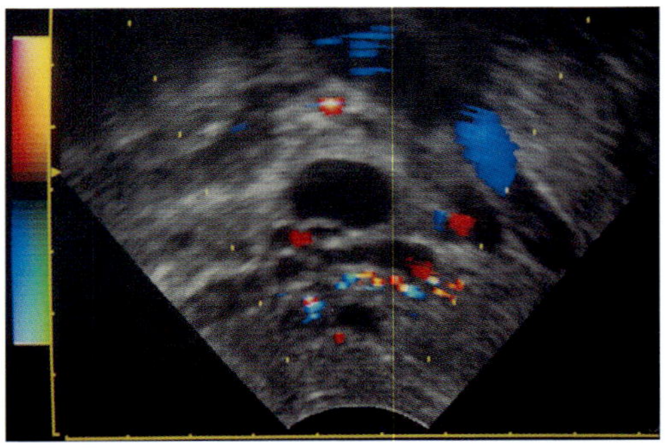

Fig. 6.2

Fig. 6.1 Color Doppler image of the ascending main branch of the uterine artery. Parasagittal scan at the level of the uterine cervix.

Fig. 6.2 Color Doppler image of ovarian blood flow in a spontaneous cycle.

Authors' Studies

Patients and Methods

Patients. Our study population consisted of 124 patients with a tubal or male infertility factor who were undergoing in-vitro fertilization (IVF) after a standard hormone stimulation protocol with human menopausal gonadotropin (hMG) and human chorionic gonadotropin (hCG)[42]. Sonographic and Doppler sonographic examinations were performed in a total of 68 spontaneous cycles and 161 treated cycles.

Methods of assisted reproduction and controls. Ultrasound-guided transvaginal follicular aspiration was performed in 105 patients. The harvested oocytes were inseminated with washed spermatozoa from the male partner within 12 hours after collection. Embryo transfer was performed at the 4-cell or 8-cell stage.

Nineteen patients underwent pelviscopic follicular aspiration with subsequent gamete intrafallopian transfer (GIFT).

In both the IVF-ET and GIFT groups, the endometrium was endosonographically evaluated and uterine blood flow was measured just prior to follicular aspiration.

Sonographic and Doppler sonographic examinations were performed in 36 patients during the proliferative phase of a spontaneous cycle, at midcycle, in the secretory phase of the cycle, and between days 8 and 10 of the next hormone-stimulated cycle.

Comparison group. Consistent with the definition of our study population, no biopsies were performed in the stimulated cycles. For comparison, we formed a group of 53 women with normal cycles in whom oocyte fertilization did not occur after hormonal stimulation with gonadotropins during the study period. These patients underwent an endometrial biopsy on the day of the planned embryo transfer.

Results. Pregnancy was confirmed sonographically in 34 of 124 patients (27%) following successful fertilization of the oocytes and embryo transfer ($n = 105$) or GIFT ($n = 19$) (29 singleton pregnancies, 3 twin pregnancies, and 2 triplet pregnancies). A total of six abortions and one tubal pregnancy were recorded. Twenty-seven pregnancies culminated in a live birth (22%).

Examination Procedures

Endometrial Sonography

Ultrasound examinations of the endometrium were performed with a 5 MHz endovaginal probe. The uterus was scanned in sagittal section, depending on the position of the anterior fornix (anteflexed uterus) or posterior fornix (retroflexed uterus).

Endometrial thickness was determined as twice the thickness of the endometrial layer, disregarding the periendometrial vascular zone[7] (Fig. 6.**3**).

The endometrium was also classified by its echo pattern using the grading system proposed by Smith et al.[38] (Figs. 6.**4**– 6.**7**).

The endosonographic findings were related to the phase of the cycle (spontaneous cycles), the result of the histological examination, and the clinical course (implantation rate).

Histology

A uterine fundus biopsy was performed in 70 patients during the secretory phase of the spontaneous cycle ($n = 17$) or stimulated cycle ($n = 53$). The biopsy material was immediately fixed in formalin and embedded in paraffin. Sections were prepared

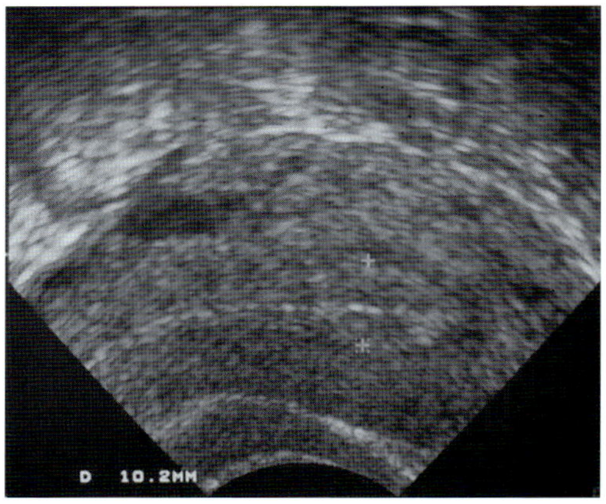

Fig. 6.3 Transvaginal scan of the anteflexed uterus from the anterior fornix. The boundaries of the endometrium are marked with cursors.

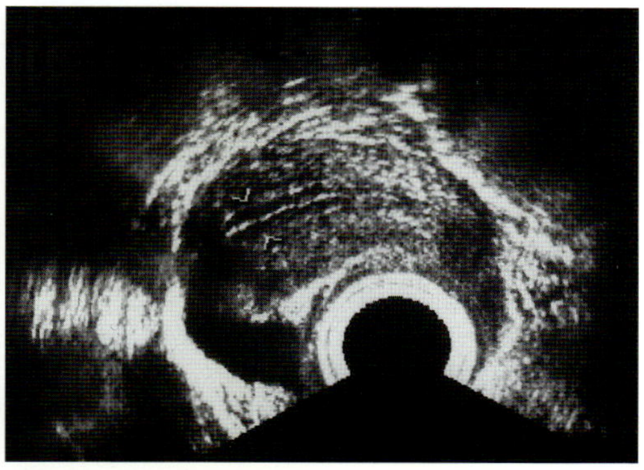

Fig. 6.4 The endometrium is echo-free except for a prominent central echo (grade D pattern of Smith et al. 1984[38]).

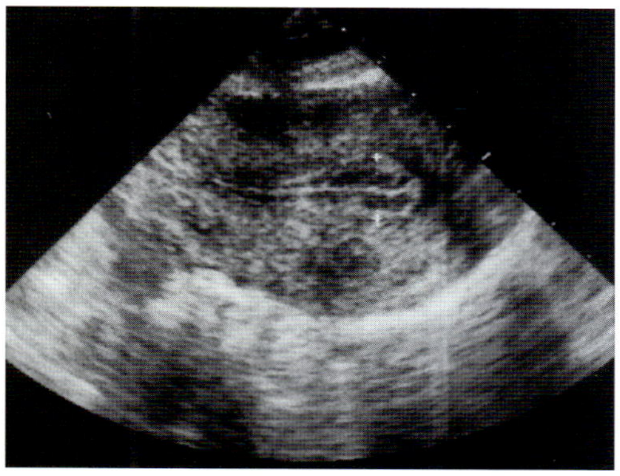

Fig. 6.5 Increasing echogenicity of the endometrium. The endometrium is always less echogenic than the surrounding myometrium, however (grade C pattern of Smith et al. 1984[38]).

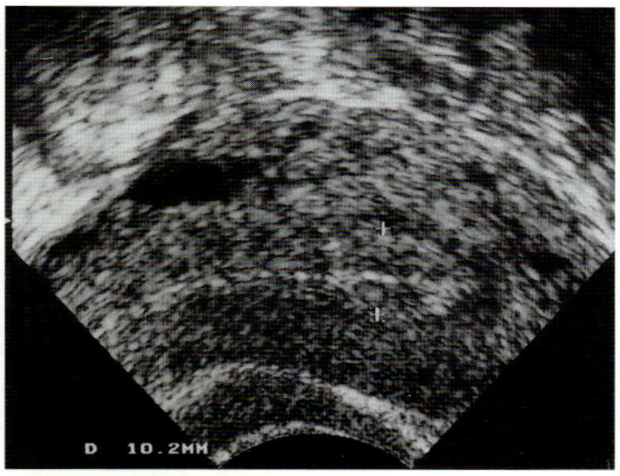

Fig. 6.6 Endometrial echogenicity is comparable to that of the myometrium. A hypoechoic zone (periendometrial vascular zone) can be identified between the myometrium and endometrium (grade B pattern of Smith et al. 1984[38]).

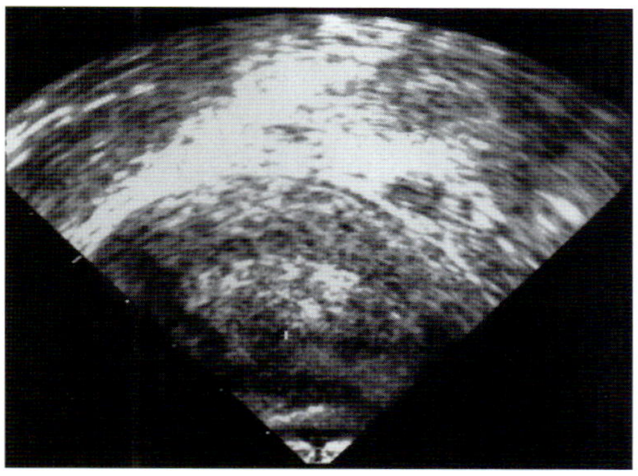

Fig. 6.7 The endometrium is more echogenic than the myometrium (grade A pattern of Smith et al. 1984[38]).

and stained with hematoxylin–eosin or van Gieson stain, and the specimens were evaluated according to the criteria of Noyes et al.[28] All histological examinations were performed at the Prof. Dallenbach Institute in Mannheim.

Scanning electron microscopic examinations were additionally performed in a subgroup of patients (Figs. 6.**8**– 6.**10**).

Doppler Sonography

Equipment. The Doppler studies of uterine blood flow were performed with an endovaginal probe using duplex technique. The probe was a mechanical 5 MHz transducer (B-mode) combined with an integrated 4.5 MHz Doppler transducer (Combison 320–5 with 300 Doppler module, Kretztechnik, Zipf, Austria). The color Doppler scanner was equipped with a 5 MHz endovaginal probe (Toshiba 270 A, Neuss, Germany).

Parameters. The probe was positioned in the vaginal fornix, and the uterine artery was scanned in parasagittal section at the level of the cervix. The Doppler beam was then directed toward the uterine vascular bundle at an angle that produced a maximum frequency shift. The waveforms were subjected to semiquantitative analysis using the resistance index[35] and pulsatility index[14]. These values reflect the degree of vascularity and functional state of the distal arteriolar and capillary bed. High indices signify a low degree of vascularity with a correspondingly high vascular resistance, while low indices reflect a low vascular resistance and high degree of vascularity in the distal bed.

Semiquantitative measurements. Semiquantitative measurements were performed on the day of follicular aspiration in all 124 patients and in the previous spontaneous cycle of 39 patients. Twenty-three other patients were examined in the spontaneous cycle but not in the subsequent, stimulated cycle. Thus the semiquantitative parameters determined for spontaneous cycles are based on a population of 62 patients.

Quantitative measurements. Quantitative measurements are possible only if the angle of the Doppler beam is known. This requires a precise longitudinal image of the vessel to be examined. Color Doppler can provide an extended view of the ascending main branch of the uterine artery (Fig. 6.**11**). A total of 272 quantitative measurements of systolic and mean uterine blood flow velocities were performed in 32 untreated cycles and in 33 hormone-stimulated cycles.

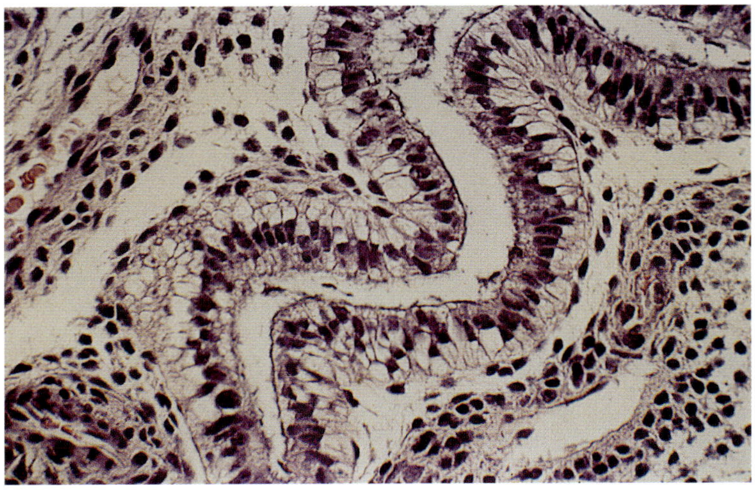

Fig. 6.**8** Normal histological endometrial structure in the secretory phase of the cycle.
a Light micrograph (×240).

b Scanning electron micrograph (×1250).

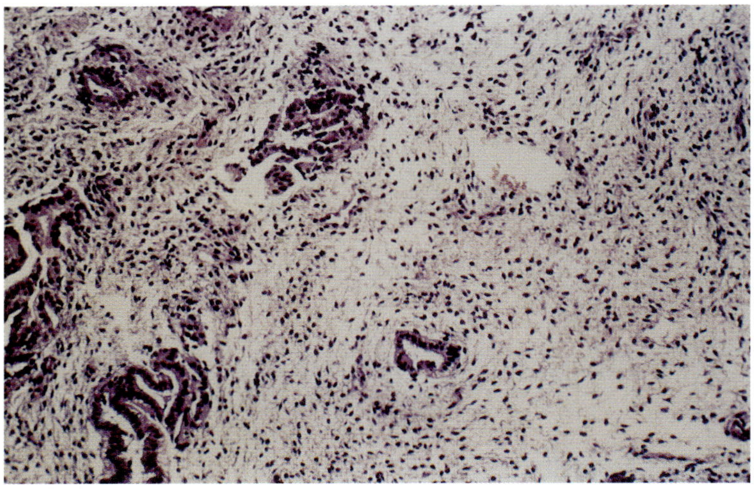

Fig. 6.**9** Deficient secretion.
a Light micrograph (×100).

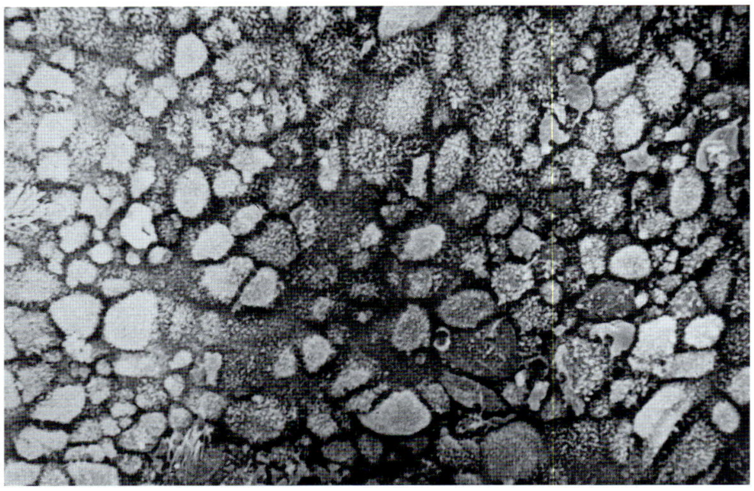

b Scanning electron micrograph (×1250).

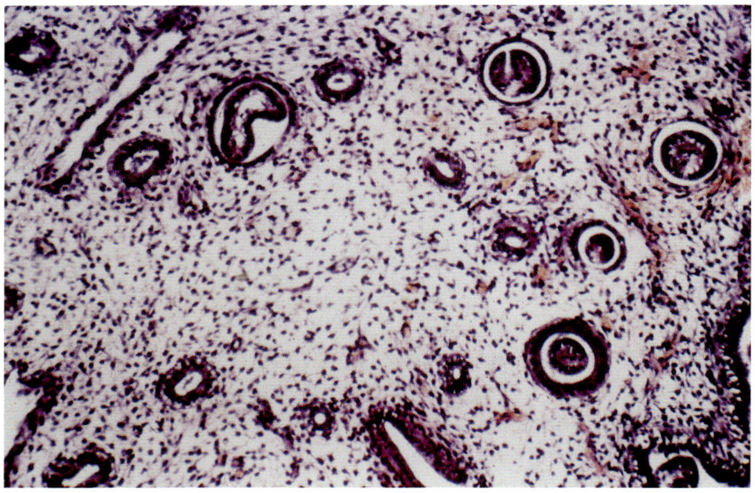

Fig. 6.**10** Abortive secretion.
a Light micrograph (×100).

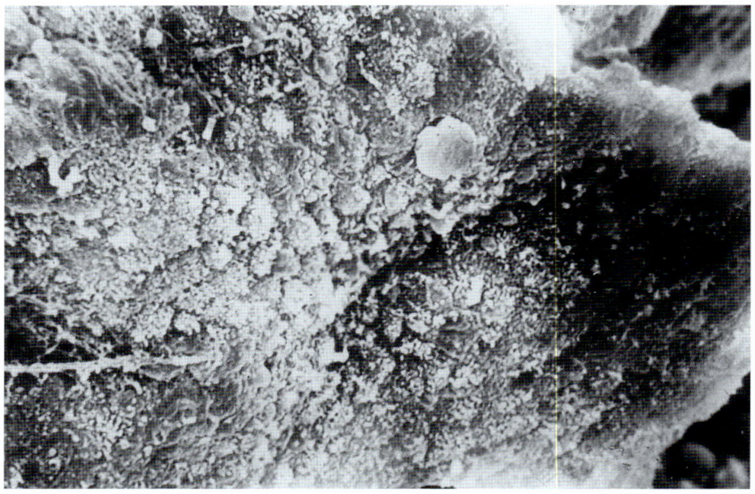

b Scanning electron micrograph (×1250).

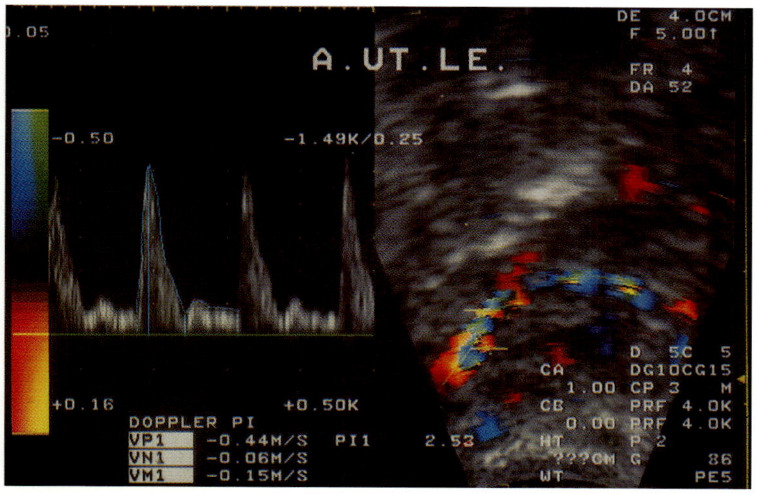

Fig. 6.**11** Color Doppler image of the uterine artery in parasagittal section. The sample volume has been placed over the vessel lumen at an insonation angle of 52°. At left the Doppler spectrum is recorded from that area.

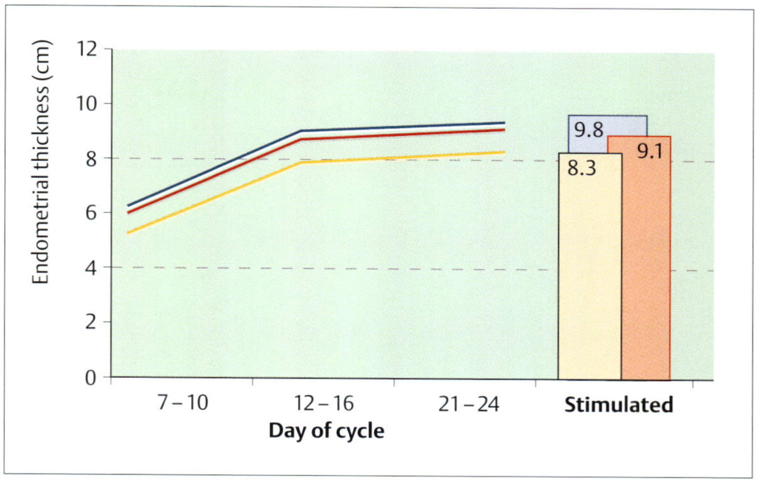

Fig. 6.**12** Endometrial thickness measured endosonographically in spontaneous cycles and after hormonal stimulation. Mean values ± standard deviation.

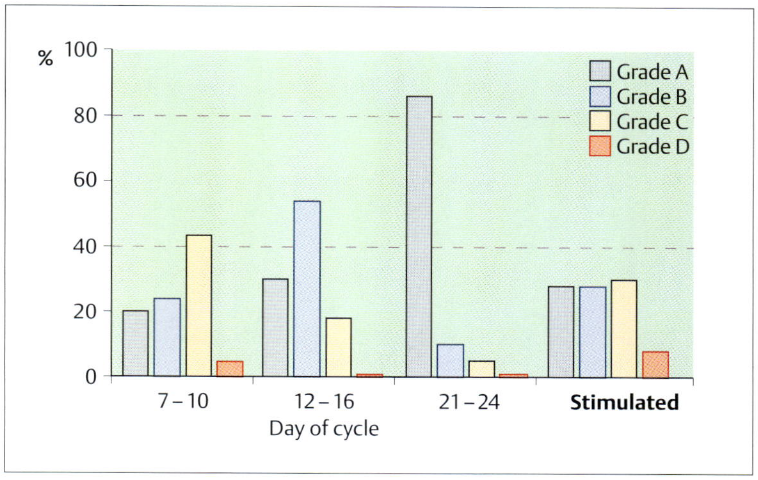

Fig. 6.**13** Endometrial echo patterns (grades A–D after Smith et al. 1984[38]) in spontaneous cycles and after hormonal stimulation.

Evaluation. Both the semiquantitative and quantitative measurements were performed bilaterally. For statistical analysis, the mean value of the measurements taken on the right and left sides was used for each parameter.

The hemodynamic parameters were correlated with the sonomorphological, hormonal, histological, and clinical parameters (implantation rate). The dependence of hemodynamic parameters on the phase of the cycle and on hormonal stimulation was also evaluated.

Results

Endometrial Sonography

Endometrial thickness. Figure 6.**12** shows the sonographically determined endometrial thickness in spontaneous cycles and after hormonal stimulation (mean values ± standard deviation). Analysis of variance indicates a highly significant increase of endometrial thickness over the course of the cycle ($p < 0.001$). Intergroup testing (*t*-test) shows a highly significant difference in endometrial thickness between the midproliferative phase and midcycle ($p < 0.001$). The values measured at midcycle and in the mid-secretory phase do not differ significantly ($p = 0.2$). The endometrial thickness in these phases corresponds to the values measured in the stimulated cycles ($p = 0.5$).

Echo pattern. A characteristic change of echo pattern was observed in the spontaneous cycles (Fig. 6.**13**). While a grade C pattern[38] was noted most frequently during the proliferative phase, grades A and B were more prevalent at midcycle. The echo pattern in the secretory phase was almost exclusively grade A. The echo pattern in the stimulated cycles most closely resembled the pattern observed in the proliferative phase.

Implantation rate. Neither the endometrial thickness nor the echo pattern showed a definite correlation with implantation rate. No significant difference was found in endometrial thickness following successful or unsuccessful implantation (9.5 mm versus 9.1 mm; $p = 0.71$). Figure 6.**14** shows the distribution of echo patterns by successful or failed implantation. Clinical pregnancies occurred in association with all four endometrial grades as defined by Smith et al.[38].

Histology

Sonographic findings and histological diagnosis. Most endometria in the secretory phase were assigned to grade A as defined by Smith et al.[38], regardless of the histological diagnosis. Two endometria that were grade B by sonographic criteria were either healthy or showed deficient transformation by histological evaluation. Also, the sonographic endometrial thickness did not consistently match the histological thickness: the widths measured sonographically were sometimes larger with nonreceptive histology than in the receptive cases, and the

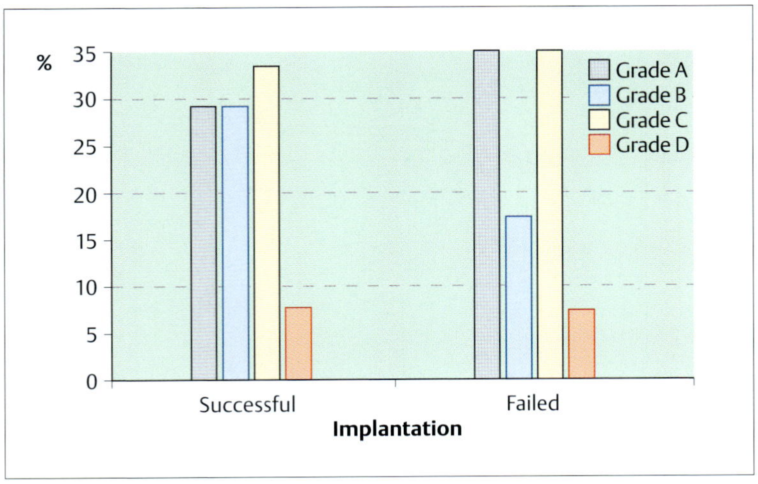

Fig. 6.**14** Distribution of endometrial echo patterns (grades A–D after Smith et al. 1984[38]) in stimulated cycles, shown separately for successful and failed implantation.

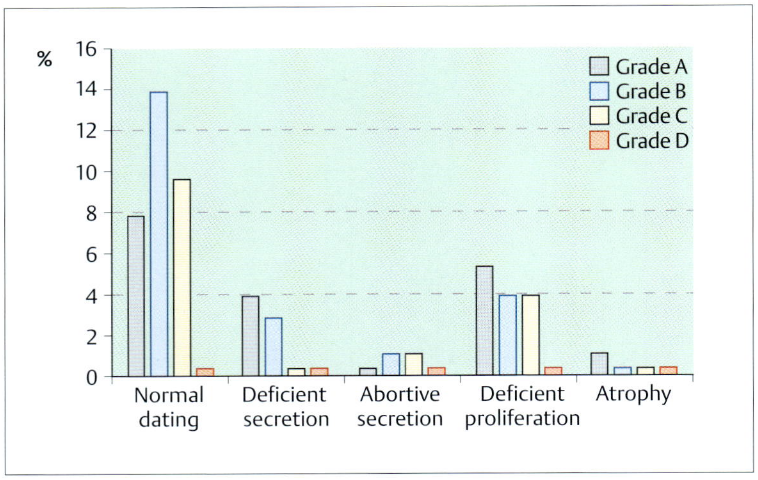

Fig. 6.**15** Endometrial echo patterns (grades A–D after Smith et al. 1984[38]) following hormonal stimulation related to the histological evaluation of endometrial biopsy specimens taken on the day scheduled for embryo transfer.

greatest endometrial thicknesses (19 and 14 mm) were measured in the two cases that showed deficient proliferation.

In a comparison group of 53 patients who were biopsied on the scheduled day of embryo transfer after failure of oocyte fertilization, the endometrium was histologically healthy in 57 % of the cases. Comparing the sonographically classified endometrial patterns with the histological diagnoses (Fig. 6.**15**), we find that in cases with normal histology the distribution of sonographic findings most closely matched the conditions seen in the periovulatory phase of the spontaneous cycle. However, because all sonographic grades occur in association with nearly all the histological diagnoses, we must conclude that the echogenicity of the endometrium is not a reliable predictor of the histological diagnosis.

Findings after stimulation. In a subgroup of patients, light microscopy was supplemented by scanning electron microscopy of the endometrium after hormonal stimulation with hMG/hCG and the findings were compared with biopsies taken from spontaneous cycles.

In cycles stimulated with gonadotropins, 9 of 16 biopsies that had been taken on the day of the planned embryo transfer showed normal dating, while the other 7 were deficient. Compared with nonstimulated cycles, the stimulated endometria showed a greater number of ciliary cells with longer and more prominent cilia.

Degenerative developmental changes usually did not affect the whole endometrium but were confined to a localized area. Secretion was rarely impaired. Both apocrine and droplet secretion could be demonstrated even in areas that appeared morphologically “deficient.“

Doppler Sonography

Figures 6.**16**– 6.**18** show the values measured in spontaneous cycles and the subsequent cycles stimulated with gonadotropins. The median resistance values in the spontaneous cycles were lower and the blood flow velocities higher than in the previous nonstimulated cycles. The differences were small, however, and only some were statistically significant.

Resistance index. The median resistance index (RI) of the uterine artery in the stimulated cycle was less than in the spontaneous cycle (Fig. 6.**16**), but the difference was not statistically significant ($p = 0.14$). A comparison of the different cycle phases and the stimulated cycles using regression analysis showed a correlation coefficient of 0.21 for the proliferative phase. The correlation coefficient was 0.28 in the periovulatory phase and 0.35 in the secretory phase.

Pulsatility index. The pulsatility index (PI) of the uterine artery was significantly lower than in the periovulatory phase ($p < 0.001$) and secretory phase ($p < 0.01$) of the corresponding spontaneous cycles (Fig. 6.**17**). The PI in the proliferative phase correlated weakly with the corresponding values measured in the stimulated cycle (correlation coeffi-

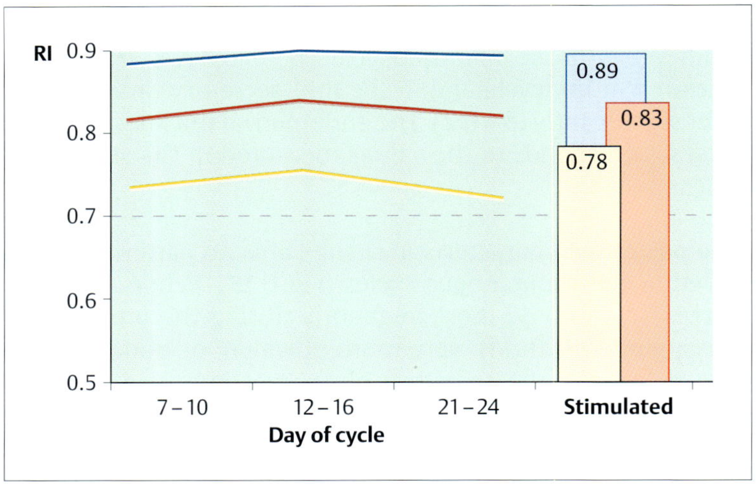

Fig. 6.**16** Resistance index (RI) of the uterine artery in spontaneous and stimulated cycles (median values plus upper and lower quartiles).

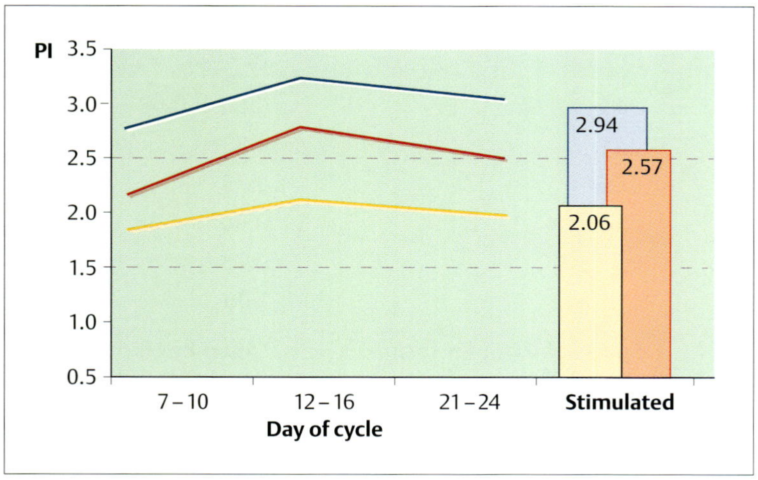

Fig. 6.**17** Pulsatility index (PI) of the uterine artery in spontaneous and stimulated cycles (median values plus upper and lower quartiles).

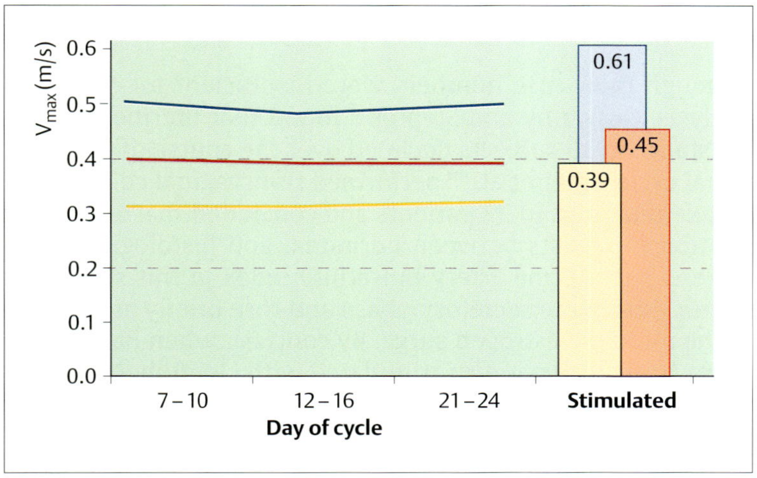

Fig. 6.**18** Peak systolic velocity (V_{max}) of the uterine artery in spontaneous and stimulated cycles (median values plus upper and lower quartiles).

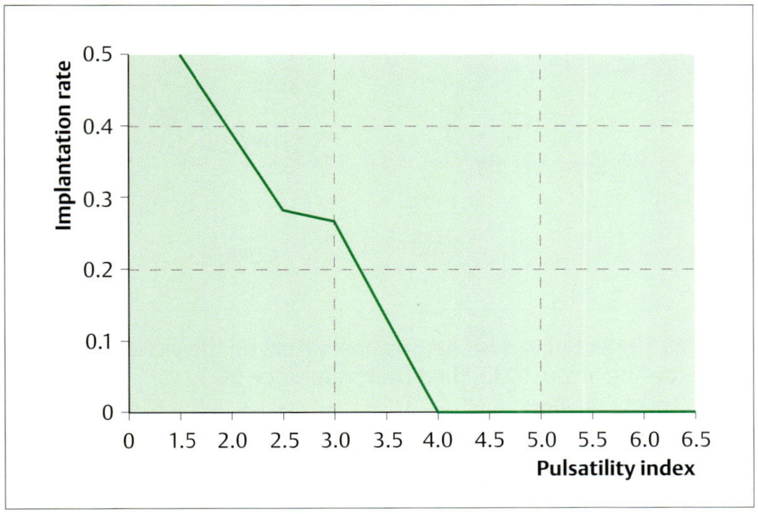

Fig. 6.**19** Implantation rate as a function of pulsatility index (PI).

cient = 0.15). The correlation coefficient was 0.39 at midcycle and 0.34 in the secretory phase.

Blood flow velocities. The peak systolic flow velocities and the mean intensity-weighted uterine flow velocities were both higher in the stimulated cycles than in the previous untreated cycles.

The peak blood flow velocity (Fig. 6.**18**) after hormonal stimulation was significantly higher than the values measured in the periovulatory and secretory phases ($p < 0.05$, Wilcoxon test). The correlation between spontaneous and stimulated cycles was higher in the proliferative phase (correlation coefficient = 0.54) and secretory phase (correlation coefficient = 0.55) than at midcycle (correlation coefficient = 0.30).

Similar patterns were observed for the intensity-weighted mean blood flow velocity. This velocity, too, was higher than the values measured in spontaneous cycles, but the difference was statistically significant only in the periovulatory phase ($p < 0.01$, Wilcoxon test). It slightly exceeded the 0.05 significance level during the secretory phase ($p = 0.07$, Wilcoxon test). The correlation coefficient between stimulated and spontaneous cycles was 0.53 in the proliferative phase, 0.47 in the secretory phase, and 0.17 in the periovulatory phase.

Correlation of Doppler Measurements with Implantation

Semiquantitative studies of uterine blood flow were performed in all patients on the day of follicular aspiration. When the measured values are related to clinical course, we find that cases with successful implantation show significantly lower resistance values than cases in which pregnancy was not achieved. The Wilcoxon test indicated a 0.045 probability of error for the resistance index and 0.03 for the pulsatility index. Table 6.**2** shows the mean values, standard deviations, median values, 25th and 75th percentiles (upper and lower quartiles), and the extreme values for the resistance index and pulsatility index for patients with successful and failed implantations. Both groups showed the same age distribution (successful implantation: 32.8 ± 3.8 years; failed implantation: 31.1 ± 3.2 years, NS), hormone parameters (successful implantation: estradiol 1190 ± 76 pg/ml, progesterone 0.92 ± 0.07 ng/ml; failed implantation: estradiol 1262 ± 65 pg/ml, progesterone 0.89 ± 0.06 ng/ml, NS), and number of transferred embryos (successful implantation: 2.4 ± 0.6; failed implantation: 2.5 ± 0.04, NS).

The implantation rate declines with increasing pulsatility index (Fig. 6.**19**). No pregnancies were achieved above a pulsatility index of 3.5 or above a resistance index of 0.95. A pulsatility index over 3.5 or a resistance index over 0.95 signified a nonreceptive endometrium with a specificity of 100% and a sensitivity of 14%. The positive predictive value for these cutoff limits is 100%.

Table 6.**2** Resistance index (RI) and pulsatility index (PI) of uterine vessels in successful and failed implantations. The difference between the two groups is statistically significant ($p < 0.05$, Wilcoxon test)

RI	**Successful implantation ($n = 27$)**	**Failed implantation ($n = 90$)**
Mean value ± SD	0.81 ± 0.06	0.84 ± 0.08
Median	0.78	0.83
Lower quartile	0.76	0.78
Upper quartile	0.88	0.90
Minimum	0.67	0.73
Maximum	0.93	1.00
PI	**Successful implantation ($n = 27$)**	**Failed implantation ($n = 90$)**
Mean value ± SD	2.25 ± 0.61	2.82 ± 1.16
Median	2.23	2.59
Lower quartile	1.91	2.09
Upper quartile	2.61	3.05
Minimum	0.88	1.30
Maximum	3.41	7.05

Discussion of the Role of Doppler Examinations

The ovary and uterus are the only organs in which significant neoangiogenesis takes place under physiological conditions in adults. The flow velocity waveforms of the uterine arteries reflect the architecture of the terminal vascular branches. They also relate to the functional status of the downstream arterioles, as vasoconstriction can produce characteristic changes in the Doppler waveforms[27].

Effect of Sex Hormones on Uterine Blood Flow

Neuroanatomical studies show that uterine blood flow is controlled by a complex substrate that is sensitive to sex steroids[31, 32, 50]. Besides the classic neurotransmitters norepinephrine and acetylcholine, researchers have identified a number of neuropeptides that act on the vessel wall of uterine arteries (Fig. 6.**20**)[20, 30, 49]. With Doppler ultrasound, it is possible to study uterine and ovarian blood flow under the influence of endogenous and exogenously administered sex hormones[9]. This type of study can advance our understanding of how vascular processes are involved in the pathophysiology of reproductive biology[3, 4, 5]. It also provides a simple, noninvasive tool for measuring the effect of pharmacological treatment regimens on uterine and ovarian hemodynamics.

Blood Flow Changes during the Menstrual Cycle

There is considerable debate in the literature over whether uterine and ovarian blood flow follow a cyclical pattern. The data published to date suggest that uterine and ovarian perfusion increase at the middle of the cycle[12, 15, 16, 23, 24, 40]. Feichtinger et al.[12] performed repeated transvaginal ultrasound measurements of uterine and ovarian blood flow in spontaneous cycles and concluded that the resistance index is higher at the beginning and end of the cycle than at midcycle, although their case numbers were insufficient for statistical analysis. Goswamy and Steptoe[15] found that uterine vascular resistance in 16 subjects declined over the course of the menstrual cycle. Steer et al. [40] performed transvaginal color Doppler examinations in 23 patients and concluded that a complex relationship exists between hormonal and histological parameters. The uterine artery pulsatility index in this study was lowest during the secretory phase and rose briefly at the time of the midcycle estrogen surge. By contrast, when Battaglia et al.[2] examined 19 patients stimulated with clomiphene/hCG or hMG/hCH, they found that the pulsatility index fell during the

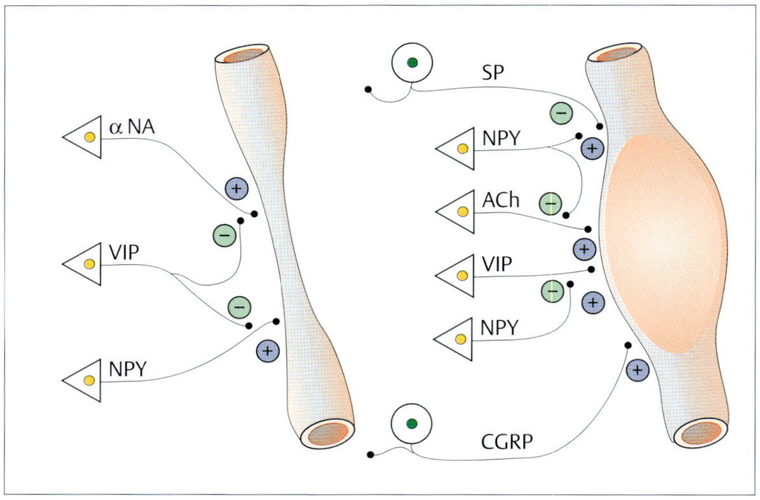

Fig. 6.**20** Effect of various neurotransmitters on the vessel wall tonus of the uterine artery. (Modified from reference 50.)

ACh Acetylcholine
αNA Norepinephrine
VIP Vasoactive intestinal polypeptide
NPY Neuropeptide Y
SP Substance P
CGRP Calcitonin gene-related polypeptide

proliferative phase and rose during the luteal phase. Meanwhile, Scholtes et al.[37] found in the transvaginal examination of 16 women with regular cycles that the uterine artery pulsatility index was involved only marginally in the cyclic variations observed in ovarian blood flow patterns. Long et al.[25] could find no significant differences in the pulsatility index of uterine artery waveforms before and after ovulation.

In our own study population, we observed a slight but statistically significant rise of resistance indices in the periovulatory period. Quantitative measurements showed a largely constant median peak systolic velocity of approximately 40 cm/s, regardless of the phase of the cycle[19].

In interpreting the data, it should be considered that uterine blood flow in spontaneous cycles is apparently subject to a circadian rhythm. When Zaidi et al.[53] measured uterine artery blood flow during the proliferative phase, they found a significantly higher pulsatility index in the morning and a significantly lower blood flow velocity than in the evening.

Blood Flow in Stimulated Cycles

So far there have been no reports of hemodynamic measurements performed in spontaneous and stimulated ovarian cycles in the same population. Battaglia et al.[2] performed comparative Doppler examinations of uterine blood flow in 10 spontaneously ovulating controls and in 19 patients receiving clomiphene/hCG or hMG/hCG stimulation in an IVF program. Except for two cases in the stimulated group that had strongly elevated RI values and showed a premature fall of estradiol levels, the proliferative phase was characterized by a fall of the pulsatility index in both the treated and untreated groups. The index values were higher in the stimulated patients than in the control group. Between days 7 and 13 of the cycle, the PI in the treated group fell from 5.2 ± 0.8 to 3.4 ± 0.8 while the PI in the control group declined from 4.4 ± 0.6 to 2.3 ± 0.3.

In our own population, we found that the resistance values after gonadotropin stimulation were lower and the blood flow velocities higher than in the previous, nonstimulated cycles. The pulsatility index proved to be a more statistically rewarding parameter than the resistance index. The difference between the two indices is related to methodology: given the quantities that enter into the pulsatility index, waveforms with absent or greatly decreased diastolic flow can be differentiated better with this parameter than with the resistance index[26]. The increased blood flow velocities and decreased pulsatility index after gonadotropin treatment can be explained by the significantly higher estradiol levels in the treated cycles[9]. The higher pulsatility index that Battaglia et al.[2] described in stimulated cycles may result from the study design. They obtained their measurements in two contrasting groups rather than in a uniform population, so there may have been significant, uncontrolled differences between the control group and treated group. Moreover, they employed two different stimulation regimens, one with clomiphene/hCG and one with hMG/hCG. Our own results indicate a lower vascular resistance in the uterine artery after hormonal stimulation than in spontaneous cycles.

Tekay et al.[46] found no change in uterine or ovarian blood flow in nine patients with ovarian hyperstimulation syndrome compared with 21 stimulated control patients. After the symptoms subsided, however, the uterine artery pulsatility index was significantly lower in the five patients who had become pregnant than in the pregnant controls.

Oyesnaya et al.[33] found a positive correlation between oocyte yield and the vascular index of the follicles in an IVF-ET population. This index was defined as the ratio of follicles with a demonstrable pulsatile pattern to the total number of follicles. The timing of hCG administration could be optimized with the aid of these measurements. Another study showed that blood flow measurements in the stroma of suppressed ovaries can be used to evaluate ovarian responsiveness to stimulation[10].

Uterine Blood Flow and Implantation

An even more important clinical issue is the correlation between Doppler parameters and implantation. Goswamy et al.[16] found a significantly lower pregnancy rate in a selected clinical population with four or more unsuccessful IVF attempts than in patients with fewer attempts. Doppler ultrasound studies of uterine blood flow showed decreased uterine perfusion in 48% of the patients who had previous unsuccessful IVF attempts. Drawing on a previous study in 16 subjects[15], the authors classified the flow velocity waveforms by qualitative criteria based on the presence and pattern of diastolic flow. A complete or partial absence of diastolic flow was classified as decreased uterine perfusion, while continuous diastolic flow was classified as normal uterine perfusion. Goswamy and Steptoe found improvement of uterine perfusion in 31 of 38 patients who had been pretreated with estradiol valerate, and 15 of these patients subsequently conceived with in-vitro fertilization. The authors performed transabdominal scans with a full-bladder technique, noting that excessive bladder filling could alter the uterine waveforms and lead to an erroneous classification[15].

Our own group of authors scanned the uterine artery of IVF patients by the transvaginal route. They found that the vascular resistance was lower in patients with successful implantation than in patients who failed to conceive[18]. Uterine waveforms sampled on the day of follicular aspiration were significantly lower in the successful patients than in those who did not conceive[43]. The results of more recent studies have confirmed this finding (Table 6.**3**). Zaidi et al. [52] achieved similar results in color Doppler examinations of the endometrium, finding that nonvisualization of subendometrial and intraendometrial vessels indicated poor endometrial receptivity. Salle et al.[36] integrated uterine blood flow and sonomorphological criteria into a score for predicting uterine receptivity.

The data indicate that the implantation rate and pulsatility index are inversely correlated with each other. In all studies published to date, no pregnancies occurred above a critical cutoff level of approximately 3 to 3.5. Although values below this cutoff have little prognostic value owing to the broad overlap between the pregnant and nonpregnant populations, the results indicate that severely decreased uterine perfusion can be an important cause of infertility.

Table 6.3 Association between implantation and uterine vascular resistance. Review of published studies

Authors	*n*	Pregnancy rate (%)	RI		*p*	PI		*p*	No implantation when:
			Pregnant	Not pregnant		Pregnant	Not pregnant		
Strohmer et al. (1991)	105	12		NI	–	2.32 ± 0.84	2.84 ± 1.29	< 0.05	NI
Steer et al. (1992)[41]	82	34		NI	–	2.08 ± 0.43	2.62 ± 0.85	< 0.05	PI > 3
Spernol et al. (1993)[39]	117	17		NI	–	1.86 ± 0.48	2.11 ± 0.55	< 0.05	PI > 3.09
Favre et al. (1993)[11]	185	21		NI	–	2.80 ± 0.48	2.90 ± 0.55	NS	PI > 3.55
Cacciatore et al. (1996)[6]	200	35	0.85 ± 0.04	0.87 ± 0.04	< 0.05	2.45 ± 0.54	2.66 ± 0.39	< 0.05	RI > 0.95 or PI > 3.3
Zaidi et al. (1996)[54]	139	25		NI	–	2.52 ± 0.50	2.64 ± 0.80	NS	NI
Tekay et al. (1996)[48]	32*	13		NI	–	3.33 (2.04–3.91)	3.02 (2.13–6.72)	NS	
	25	36				2.47 (1.52–3.77)	2.38 (1.79–4.87)	NS	PI > 4
Our data	124	22	0.81 ± 0.06	0.84 ± 0.08	< 0.05	2.25 ± 0.61	2.82 ± 1.16	< 0.05	RI > 0.95 or PI > 3.5

* = Frozen embryo transfer.
NI = not indicated.
NS = not significant.

Conclusions for the Clinical Management of Assisted Reproduction

Study results indicate that Doppler measurements of uterine blood flow can be incorporated into the clinical management of assisted reproduction. Very high resistance values in the uterine artery Doppler waveforms reflect poor endometrial receptivity and imply a poor outcome of assisted reproductive procedures. In our own study, uterine vascular resistance above the 90th percentile signified a nonreceptive endometrium with 100% specificity and a positive predictive value of 100%. Doppler measurements of uterine blood flow can be integrated into the transvaginal ultrasound examination of the lesser pelvis with only a slight increase in scanning time and no additional patient discomfort. Low vascular resistance values indicate normal endometrial receptivity. Higher-order multiple pregnancies can be avoided by using stimulation sparingly and by limiting the number of transferred embryos, although when uterine artery vascular resistance is increased the capabilities of assisted reproduction should be fully utilized owing to the markedly decreased implantation rate in these cases. When uterine vascular resistance is greatly increased, it should be concluded that the endometrium is nonreceptive. These patients should not be subjected to the physical and emotional ordeal of in-vitro fertilization without appropriate prior hormonal treatment.

Summary

Hormonal influences. Color Doppler sonography provides a simple, noninvasive means of investigating uterine and ovarian blood flow under the influence of endogenous and exogenous sex hormones. Our own studies in 68 spontaneous cycles and 161 cycles treated with gonadotropins show that uterine vascular resistance rises slightly in response to endogenous progestins, while the systolic blood flow velocities (V_{max}) remain constant at approximately 40 cm/s.

Significantly lower uterine vascular resistance and significantly higher blood flow velocities were measured following hormonal stimulation with gonadotropins.

Endometrial receptivity. Doppler measurements of uterine blood flow can be used in the assessment of endometrial receptivity. High uterine vascular resistance implies a markedly reduced chance of fertilization. Our own measurements and the data published by other groups of authors indicate that decreased uterine perfusion can be a significant obstacle to implantation in infertile patients. Very high vascular resistance in the uterine arteries reflects poor endometrial receptivity and suggests that assisted reproductive techniques will be unsuccessful. Although the sensitivity of uterine artery impedance measurements is only 14% owing to the numerous factors that affect receptivity, resistance values above the 90th percentile predict a nonreceptive endometrium with 100% specificity and a positive predictive value of 100%. This disorder cannot be detected by the use of classic noninvasive and invasive diagnostic tests. With a severe decrease in uterine perfusion, the likelihood of implantation is so low that assisted reproduction therapies should be discontinued without appropriate preliminary treatment.

Comments. The data indicate that Doppler ultrasound studies of uterine blood flow should be incorporated into the clinical management of assisted reproduction. The Doppler measurements can be added to the transvaginal ultrasound evaluation of the lesser pelvis with no additional patient discomfort and can significantly advance the differential diagnosis and treatment of infertility.

References

1 Aytoz A, Ubaldi F, Tornaye H, Nagy ZP, Van Steirteghem A, Devroey P: The predictive value of uterine artery blood flow measurements for uterine receptivity in an intracytoplasmatic sperm injection program. Fertil. Steril. 68 (1997) 935–937

2 Battaglia C, Larocca E, Lanzani A, Valentini M, Ganazzani AR: Doppler ultrasound studies of the uterine arteries in spontaneous and IVF stimulated cycles. Gynecol. Endocrinol. 4 (1990) 245–250

3 Bourne TH: Transvaginal color Doppler in gynecology. Ultrasound Obstet. Gynecol. 1 (1991) 309–312

4 Bourne TH, Hagström H-G, Hahlin M et al.: Ultrasound studies of vascular and morphologic changes in the human corpus luteum during the menstrual cycle. Fertil. Steril. 65 (1996) 753–758

5 Bourne TH, Reynolds K, Waterstone J et al.: Paracetamol-associated luteinized unruptured follicle syndrome: effect on intrafollicular blood flow. Ultrasound Obstet. Gynecol. 1 (1991) 420–425

6 Cacciatore B, Simberg N, Fusaro P, Tiitinen A: Transvaginal Doppler study of uterine artery blood flow in in vitro fertilisation-embryo transfer cycles. Fertil. Steril. 66 (1996) 130–134

7 Cornet D, Giacomini P, Pereira-Coelho A, Salat-Baroux J: Etude échographique de l'épaisseur de l'endomètre au cours des cycles spontanés et stimulés: applications à la fécondation in vitro. J. Gynecol. Obstet. Biol. Reprod. 15 (1986) 621–625

8 Dallenbach C, Sterzik K, Dallenbach-Hellweg G: Histologische Endometriumbefunde bei Patientinnen am Tage eines geplanten Embryo-Transfers. Geburtshilfe Frauenheilkd. 47 (1987) 623–629

9 De Ziegler D, Bessis R, Frydman R: Vascular resistance of uterine arteries: physiological effects of estradiol and progesterone. Fertil. Steril. 55 (1991) 775–779

10 Engmann L, Sladkevicius P, Agrawal R, Bekir JS, Campbell S, Tan SL: Value of ovarian stromal blood flow velocity measurement after pituitary suppression in the prediction of ovarian responsiveness and outcome of in vitro fertilization treatment. Fertil. Steril. 71 (1999) 22–28

11 Favre R, Bettahar K, Grange G et al.: Predictive value of transvaginal uterine Doppler assessment in an in vitro fertilization program. Ultrasound Obstet. Gynecol. 3 (1993) 350–353

12 Feichtinger W, Putz M, Kemeter P: Transvaginale Doppler-Sonographie zur Blutflußmessung im kleinen Becken. Ultraschall 9 (1988) 30–36

13 Fleischer AC: Ultrasound imaging – 2000: assessment of utero-ovarian blood flow with transvaginal color Doppler sonography, potenzial clinical applications in infertility. Fertil. Steril. 55 (1991) 684–691

14 Gosling RG, King DH: Ultrasonic angiology. In: Harcus A, Adamson L (eds.): Arteries and veins. Churchill, Livingstone 1975, 61–98

15 Goswamy RK, Steptoe PC: Doppler ultrasound studies of the uterine artery in spontaneous ovarian cycles. Hum. Reprod. 3 (1988) 721–726

16 Goswamy RK, Williams G, Steptoe PC: Decreased uterine perfusion—a cause of infertility. Hum. Reprod. 3 (1988) 955–959

17 Grab D: Zur Rezeptivität des Endometriums: Stellenwert der Sonographie und Dopplersonographie im unbehandelten Menstruationszyklus und nach hormoneller Stimulation. Habilitationsschrift, Ulm 1994

18 Grab D, Sasse V, Terinde R, Sterzik K: Der vaskuläre Widerstand der Aa. uterinae als prognostisches Kriterium für eine erfolgreiche Implantation? Fertilität 5 (1989) 61–64

19 Grab D, Stoijc S, Königbauer U, Sterzik K: Uterine Blutströmungsprofile im Spontanzyklus von Sterilitätspatientinnen. Fertilität 8 (1992) 1–4

20 Hansen V, Schifter S, Allen J, Maigaard S, Forman A: Effects of human calcitonin gene-related peptides and substance P on human intracervical arteries. Gynecol. Obstet. Invest. 25 (1988) 258–261

21 Isaacs JD, Wells DS, Williams DB, Odem RR, Gast MJ, Strickler RC: Endometrial thickness is a valid monitoring parameter in cycles of ovulation induction with menotropins alone. Fertil. Steril. 65 (1996) 262–266

22 Knörr K, Knörr-Gärtner H, Beller FK, Lauritzen C: Geburtshilfe und Gynäkologie. Physiologie und Pathologie der Reproduktion. Springer, Berlin 1989

23 Kurjak A: Transvaginal color Doppler: a comprehensive guide to transvaginal color Doppler sonography in obstetrics and gynecology. Parthenon, Carnforth 1991

24 Kurjak A, Kupesic-Urek S, Schulman H, Zalud I: Transvaginal color flow Doppler in the assessment of ovarian and uterine blood flow in infertile women. Fertil. Steril. 56 (1991) 870–873

25 Long MG, Boultbee JE, Hanson ME, Begent RHJ: Doppler time velocity waveform studies of the uterine artery and uterus. Brit. J. Obstet. Gynaecol. 96 (1989) 588–593

26 McParland P, Pearce JM: Doppler studies in pregnancy. Rec. Adv. Obstet. Gynaecol. 16 (1990) 43–73

27 Mo LYL, Bascom PAJ, Ritchie K, McCowan LME: A transmission line modelling approach to the interpretation of uterine Doppler waveforms. Ultrasound Med. Biol. 14 (1988) 365–376

28 Noyes RW, Hertig AT, Rock J: Dating the endometrial biopsy. Fertil. Steril. 1 (1950) 3–25

29 Ottesen B, Gerstenberg T, Ulrichsen H, Manthorpe T, Fahrenkrug J, Wagner G: Vasoactive intestinal polypeptide (VIP) increases vaginal blood flow and inhibits uterine smooth muscle activity in women. Eur. J. Clin. Invest. 13 (1983) 321–324

30 Ottesen B, Gram GR, Fahrenkrug J: Neuropeptides in the female genital tract: effect on vascular and non-vascular smooth muscle. Peptides 4 (1983) 387–392

31 Owman Ch, Alm P, Sjöberg N-O: Pelvic autonomic ganglia: structure, transmitters, function and steroid influence. In: Elfvin L-G (eds.): Autonomic Ganglia. J. Wiley & Sons Ltd., New York 1983, 125–143

32 Owman Ch, Stjernquist M, Helm G, Kannisto P, Sjöberg N-O, Sundler F: Comparative histochemical distribution of nerve fibres storing noradrenaline and neuropeptide Y (NPY) in human ovary, fallopian tube, and uterus. Med. Biol. 64 (1986) 57–65

33 Oyesanya OA, Parsons JH, Collins WP, Campbell S: Prediction of oocyte recovery rate by transvaginal ultrasonography and color Doppler imaging before human chorionic gonadotropin administration in in vitro fertilization cycles. Fertil. Steril. 65 (1996) 806–809

34 Paulson RJ, Sauer MV, Lobo RA: Embryo implantation after human in vitro fertilization: importance of endometrial receptivity. Fertil. Steril. 53 (1990) 870–875

35 Pourcelot L: Applications cliniques de l'examen Doppler transcutane. Inserm. Colloq. 34 (1974) 213–240

36 Salle B, Bied-Damron V, Benchaib M, Desperes S, Gaucherand P, Rudigoz RC: Preliminary report of an ultrasonography and colour Doppler uterine score to predict uterine receptivity in an in-vitro fertilization programme. Hum. Reprod. 13 (1998) 1669–73

37 Scholtes MCW, Wladimiroff JW, van Rjien HJM, Hop WCJ: Uterine and ovarian flow velocity waveforms in the normal menstrual cycle: a transvaginal Doppler study. Fertil. Steril. 52 (1989) 981–985

38 Smith B, Porter R, Ahuja K, Craft I: Ultrasonic assessment of endometrial changes in stimulated cycles in an in vitro fertilization and embryo transfer program. J. Vitro Fert. Embryo Transfer 1 (1984) 233–238

39 Spernol R, Hecher K, Schwarzgruber J, Szalay St: Doppler-Flow-Messungen in der A. uterina: Ein Prognosefaktor für den Erfolg bei der Behandlung durch IVF? Ultraschall in Med. 14 (1993) 175–177
40 Steer CV, Campbell S, Pampiglione JS, Kingsland CR, Mason BA, Collins WP: Transvaginal colour flow imaging of the uterine arteries during the ovarian and menstrual cycles. Hum. Reprod. 5 (1990) 391–395
41 Steer CV, Campbell S, Tan SL et al.: The use of transvaginal color flow imaging after in vitro fertilization to identify optimum uterine conditions before embryo transfer. Fertil. Steril. 57 (1992) 372–376
42 Sterzik K, Dallenbach Ch, Schneider V, Sasse V, Dallenbach-Hellweg G: In vitro fertilization: The degree of endometrial insufficiency varies with the type of ovarian stimulation. Fertil. Steril. 50 (1988) 457–462
43 Sterzik K, Grab D, Sasse V, Hütter W, Rosenbusch B, Terinde R: Doppler sonographic findings and their correlation with implantation in an in vitro fertilization program. Fertil. Steril. 52 (1989) 825–828
44 Sterzik K, Grab D, Schneider V, Strehler EJ, Gagsteiger F, Rosenbusch BE: Lack of correlation between ultrasonography and histologic staging of the endometrium in in vitro fertilization (IVF) patients. Ultrasound Med. Biol. 23 (1997) 165–170
45 Taylor KJW, Burns PN, Wells PNT, Conway DI, Hull MGR: Ultrasound doppler flow studies of the ovarian and uterine arteries. Brit. J. Obstet. Gynaecol. 92 (1985) 240–246
46 Tekay A, Martikainen H, Jouppila P: Doppler parameters of the ovarian and uterine blood circulation in ovarian hyperstimulation syndrome. Ultrasound Obstet. Gynecol. 6 (1995) 50–53
47 Tekay A, Martikainen H, Jouppila P: Blood flow changes in uterine and ovarian vasculature, and predictive value of transvaginal pulsed colour Doppler ultrasonography in an in-vitro fertilization programme. Hum. Reprod. 10 (1995) 688–693
48 Tekay A, Martikainen H, Jouppila P: Comparison of uterine blood flow characteristics between spontaneous and stimulated cycles before embryo transfer. Human Reprod. 11 (1996) 364–368
49 Tenmoku S, Ottesen B, O'Hare MMT et al.: Interaction of NPY and VIP in regulation of myometrial blood flow and mechanical activity. Peptides 9 (1988) 269–275
50 Traurig H, Papka RE: Autonomic efferent and visceral sensory innervation of the female reproductive system: special reference to the functional roles of nerves in reproductive organs. In: Maggi CA (eds.): Autonomic Nervous System, Bd. 6: Nervous Control of the urogenital system. Harwood Academic Publishers, London 1992
51 Wentz A: Endometrial biopsy in the evaluation of infertility. Fertil. Steril. 33 (1980) 121–124
52 Zaidi J, Campbell S, Pittrof R, Tan SL: Endometrial thickness, morphology, vascular penetration and velocimetry in predicting implantation in an in vitro fertilization program. Ultrasound Obstet. Gynecol. 6 (1995) 191–198
53 Zaidi J, Jurkovic D, Campbell S, Okokon E, Tan SL: Circadian variation in uterine artery blood flow indices during the follicular phase of the menstrual cycle. Ultrasound Obstet. Gynecol. 5 (1995) 406–410
54 Zaidi J, Pittrof R, Shaker A, Kyei-Mensah A, Campbell S, Tan SL: Assessment of uterine blood flow of human chorionic gonadotropin administration by transvaginal color Doppler ultrasound in an in vitro fertilization programme. Fertil. Steril. 65 (1996) 377–381

7 Pulsed Doppler and Color Duplex Sonography in the Assessment of Tubal Patency

B. Hüneke, A. Kleinkauf-Houcken, C. Lindner, and W. Braendle

Applications of Doppler Sonography in Reproductive Medicine

Since Fitzgerald and Drumm introduced the Doppler principle into obstetric and gynecological diagnosis in 1977[6], the applications of this ultrasound technique have spread from obstetric medicine to the fields of gynecology and reproductive medicine[3, 7, 15]. The hemodynamics of the vessels of the lesser pelvis and especially of the internal genital organs are investigated by continuous-wave (CW) Doppler, pulsed-wave (PW) Doppler, and color-flow mapping (CFM). These techniques can be used, for example, to study the dependence of blood flow changes on the ovarian cycle or the hemodynamic effects of pregnancy or neoplasia.

Tubal patency. To date, only a few reports have been published on the assessment of tubal patency with spectral Doppler ultrasound[3a]. Fallopian tube disease is one of the most frequent causes of infertility, accounting for approximately 25% of cases. The detection of tubal patency, then, is an important element in the evaluation of infertility. Conventional methods of testing such as pertubation, radiographic hysterosalpingography, and laparoscopic chromopertubation have drawbacks that include lack of precision, radiation exposure, and invasiveness. One alternative is the sonographic assessment of tubal patency[9]. The first such experience at our center was gained with the transvaginal scanning of the passage of saline solution through the oviducts. When a suitable ultrasound contrast medium became available (Echovist, Schering), we were able to improve the quality of B-mode visualization[10]. Subsequent studies were done to determine whether the use of pulsed Doppler and transvaginal color duplex scanning could furnish additional information in the noninvasive evaluation of tubal patency[9a].

Assessment of Tubal Patency

Patients and Method

Clinical population. Since the advent of ultrasound tubal imaging, 210 women have been examined with B-mode and Doppler ultrasound at our infertility clinic. The average patient age was 33 years. Two-thirds of the patients had primary infertility, and one-third had secondary infertility (Table 7.**1**). Only 20% of the patients had been previously evaluated for a tubal factor, but nearly all of the women had received prior infertility treatment.

Table 7.**1** Patients enrolled in the study

Characteristics of the patients examined	
Number (*n*)	210
Age (years)	32.5 (25–41)
History of infertility (years)	3.4 (2–10)
Primary infertility (%)	65
Secondary infertility (%)	35
Duration of study (months)	60

Timing of the examination. The examination was performed in an initial series before a planned laparoscopy in patients with stimulated cycles. The patients were eventually assigned to a gamete intrafallopian transfer (GIFT) program or an in-vitro fertilization (IVF) program based on tubal imaging results and laparoscopic findings. Written informed consent was obtained. The ultrasound assessment of tubal patency was performed in the late follicular phase, as this offers the best conditions for passing an intrauterine catheter atraumatically through the canal of the estrogen-primed cervix.

Preparations for the examination. Prior to the outpatient procedure, acute genital infection was excluded by the determination of erythrocyte sedimentation rate (ESR), hemoglobin, and white blood cell count and by gynecological examination with a speculum and bimanual palpation. The ultrasound study was performed in a gynecological chair with no premedication or general anesthesia. After careful aseptic preparation of the vagina, a pediatric Foley catheter 2.7 mm in diameter (see Chapter 8) was passed through the cervical canal into the uterine cavity using sterile technique, and the catheter cuff was inflated with 2 ml of saline solution. In a few cases (e.g., patients with a previously operated cervix), it was necessary to grasp the cervix with a tenaculum or dilate the cervical canal. Next the endovaginal probe was introduced.

Ultrasound equipment. The following ultrasound systems were used in the examinations:

- Mechanical rotary scanner with a forward scan direction and 240° wide-angle view, a B-mode frequency of 5 and 7.5 MHz, a pulsed Doppler frequency of 4.5 MHz, and a pulse repetition frequency (PRF) of 3.9–15.6 kHz (Kretztechnik Combison 320–5 and 410, Zipf, Austria). The active Doppler beam could be freely steered over the entire B-mode area (240°), and the Doppler sample volume (1–15 mm) could be positioned at any depth over the 17 cm range of the scanner (Fig. 7.**1**).
- Electronic curved linear array with a forward-directed 150° beam, a B-mode frequency of 5–9 MHz, a pulsed Doppler frequency of 5 MHz, color duplex Doppler, and a penetration depth of 8–10 cm (Ultramark 9 HDI ESP, ATL, Bothell, WA, USA).

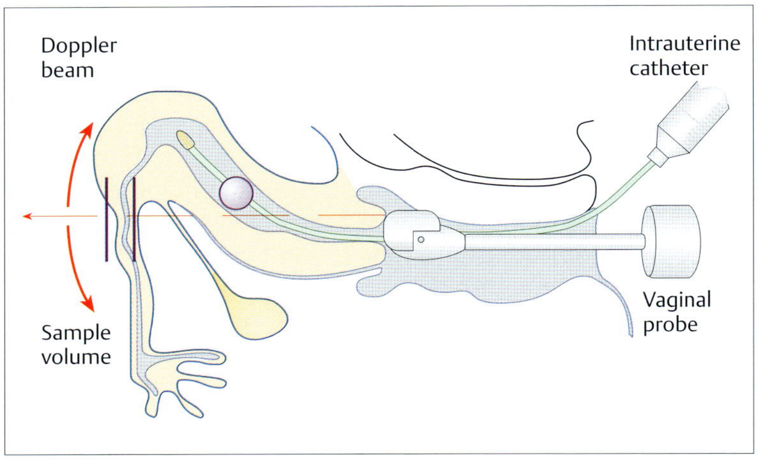

Fig. 7.**1** Instrumentation used for tubal imaging with Doppler ultrasound. The catheter for contrast instillation is inside the uterus, and the endovaginal probe is in the posterior fornix. The Doppler sample volume has been positioned over the intramural and near-proximal tubal segment.

Contrast medium. The contrast medium consisted of a freshly prepared suspension of galactose microparticles (Echovist R, Schering, Berlin). Microbubbles adsorbed to the surface of the particles provide enhanced reflection of the ultrasound waves and increased sonodensity.

Examination Technique

The uterine fundus is imaged in a B-mode reference plane in which the echogenic catheter tip and inflated cuff can be identified, and then a small contrast bolus is injected (Fig. 7.**2**).

B-mode analysis. Contrast enhancement can be seen in the uterine cavity, across the uterotubal junction, in the intramural part of the tube, and even in distal tubal segments in some cases. The flow of contrast medium can be tracked in the real-time image. The contrast agent is administered in small, pulsatile injections for optimum B-mode visualization of the flow dynamics.

a
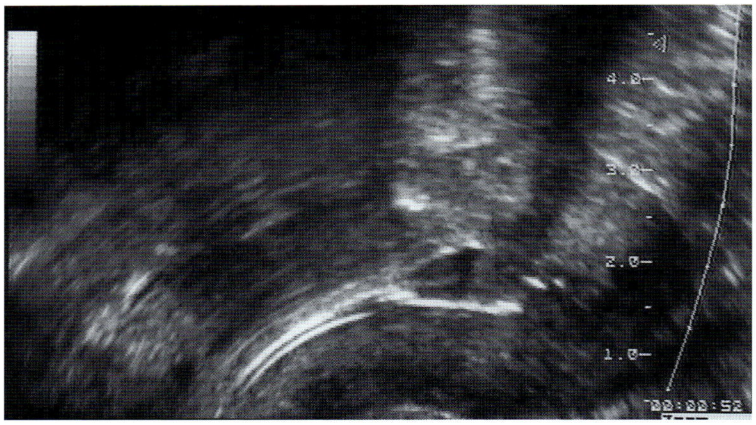

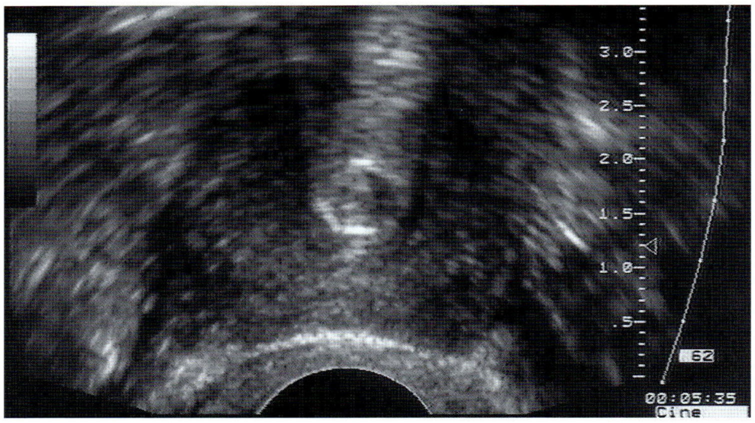
b

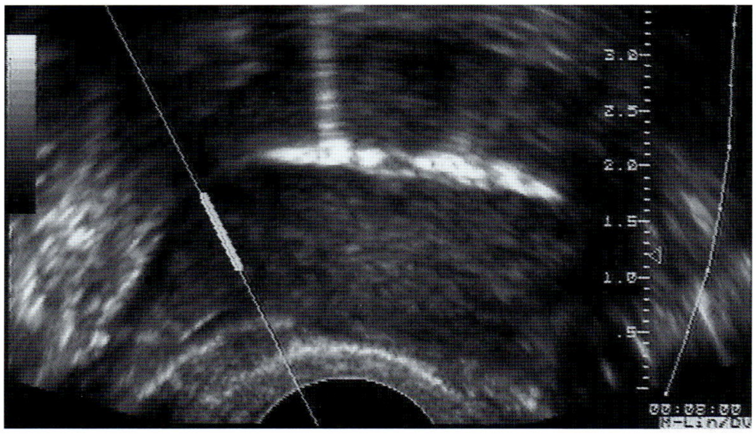
c

Fig. 7.**2** Appearance of the inflated catheter cuff within the uterus.
a Longitudinal scan.
b Transverse scan.
c Initial contrast appearance in the uterine fundus. The sample volume is positioned over the proximal part of the right fallopian tube.

Doppler mode. Following the initial B-mode analysis, the Doppler sample volume is positioned over the intramural and proximal tubal segment, using the maximum available sample length (15 mm). The machine is switched to Doppler mode, and the contrast medium is again administered in short pulsatile injections.

Based on the visual and acoustic analysis of the Doppler frequency shift of the fluid bolus passing through the fallopian tubes, the tubes are classified as patent, partially obstructed, or completely obstructed (Fig. 7.**3**).

Patent tube. If the tube is patent, the pulsatile contrast injection flows into the intramural and adjacent proximal tubal segment, where the Doppler sample volume has been placed. The inflow phase is brief and characterized by a sharp rise in the Doppler frequency shift. Unobstructed distal outflow is characterized by a slow and steady decline in the Doppler shift over time (Fig. 7.**3a**).

Partial obstruction. With a partially obstructed tube, the Doppler spectrum shows a sharp initial peak caused by intratubal turbulence. This is followed by a brief stoppage of contrast flow, with absence of a Doppler shift, caused by increased peripheral resistance (Fig. 7.**3b**). If the contrast volume is able to surmount this resistance, the Doppler trace shows an additional phase of lower-frequency shifts as a sign of decreased outflow against a resistance. This phase is always of lower amplitude and shorter duration than in a patent tube.

Complete obstruction. If the tube is completely obstructed, the contrast medium enters the sample volume in the intramural part of the tube, where it produces short, intense Doppler shifts of low amplitude. There are no additional signals following these initial peaks, signifying an absence of outflow distal to the sample volume (Fig. 7.**3c**).

Color flow. When the flow of the contrast medium is additionally analyzed by color-flow imaging, the already intense signals from the Echovist solution may produce conspicuous noise and superimposed artifacts in the patent tube unless the sensitivity of the system for flow velocities is set extremely low. Color-flow analysis (flow direction, turbulence) is indicated only if the B-mode and spectral findings are equivocal or if an obstruction is suspected. Color flow is often the only technique that can detect the spillage of contrast medium from the fimbriated end of the tube.

Documentation. When Doppler scanning is added to the qualitative B-mode analysis, it becomes possible to make a rough quantitative evaluation of flow through the fallopian tubes. The entire examination is recorded on videotape, and selected images are documented with hard copies.

Figures 7.**4**–7.**8** show typical frequency spectra that illustrate the categories of findings along with original color-flow images. After the Doppler study is completed, a final B-mode assessment is made to check for fluid collections in the tubes or cul-de-sac.

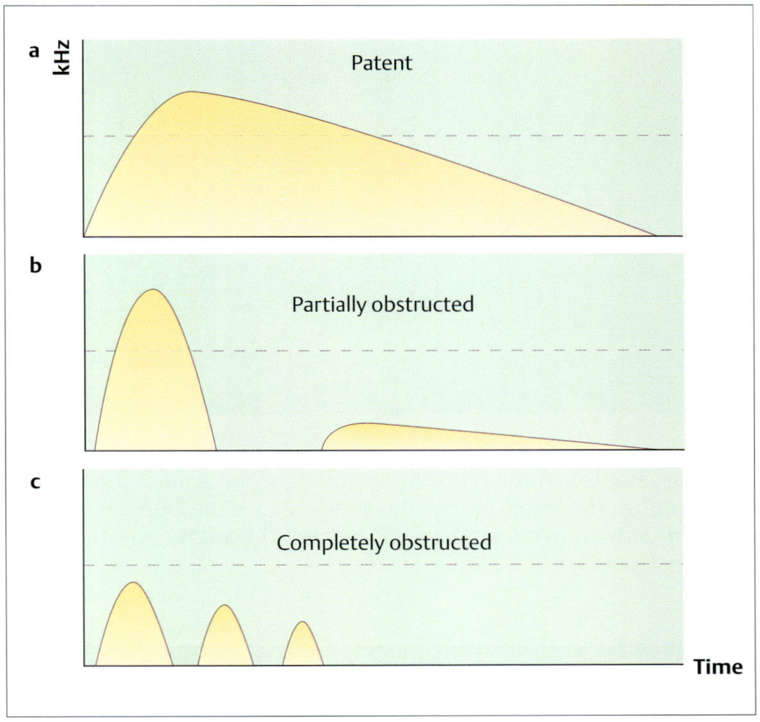

Fig. 7.**3** Doppler shift of contrast medium injected into the uterine fundus (schematic diagram).
a Patent tube.
b Partial obstruction.
c Complete obstruction.

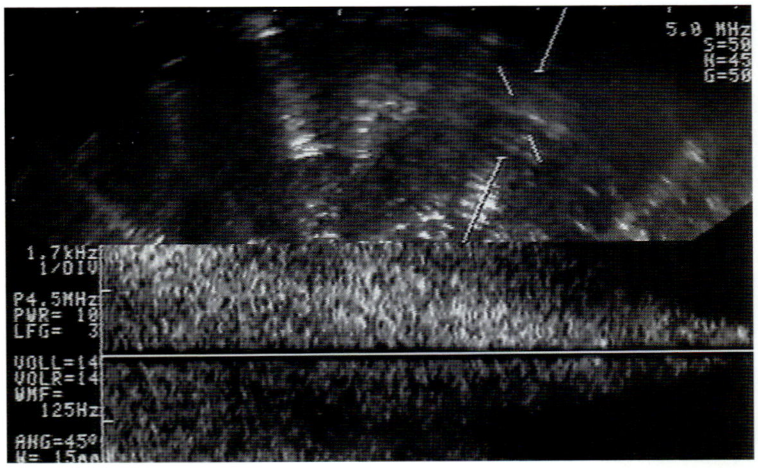

Fig. 7.**4** Original image with the Doppler sample volume placed over the proximal part of the left tube. The lower part of the image shows a uniform decline in Doppler shift associated with an unobstructed tubal passage.

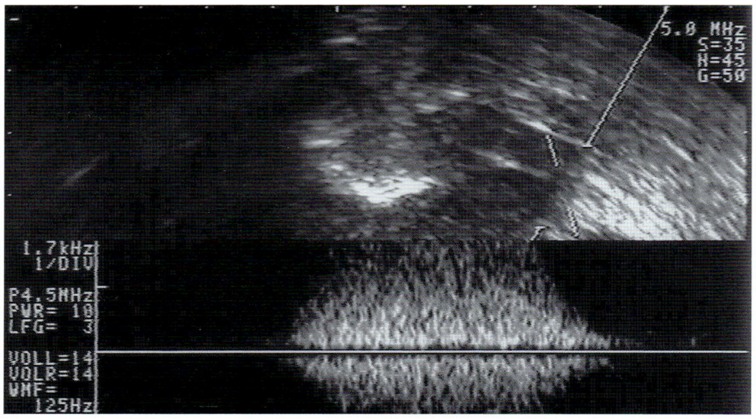

Fig. 7.**5** Original image showing contrast medium in the uterine fundus. The sample volume is positioned over the intramural segment of the left tube. The lower part of the image shows the Doppler shift associated with contrast arrival and decreased outflow due to partial tubal obstruction.

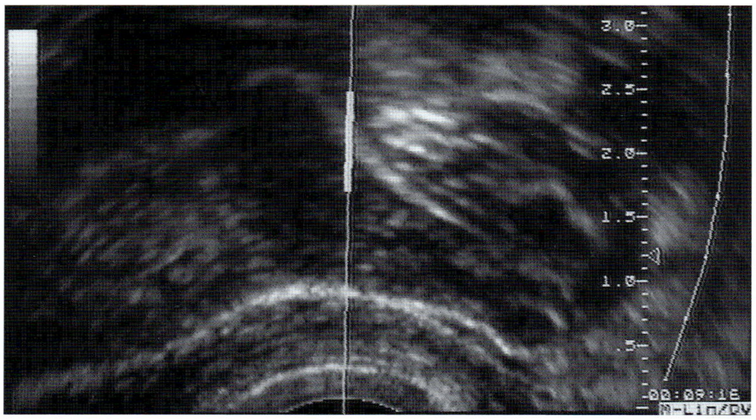

Fig. 7.**6** Original image showing contrast medium in the intramural part of the left tube. The proximal and middle thirds of the tube can be identified in the high-resolution B-mode image.

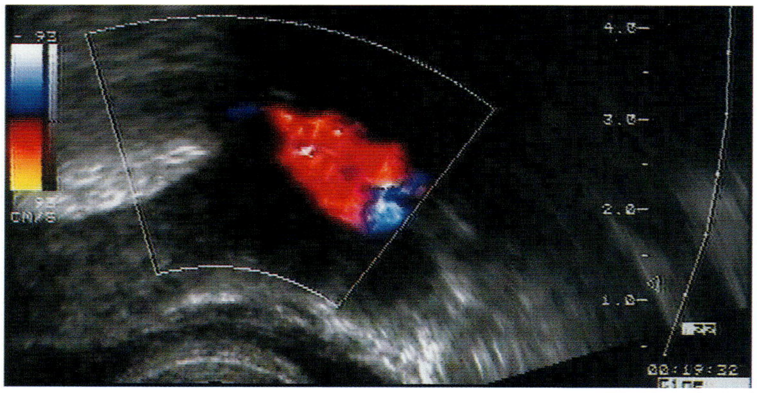

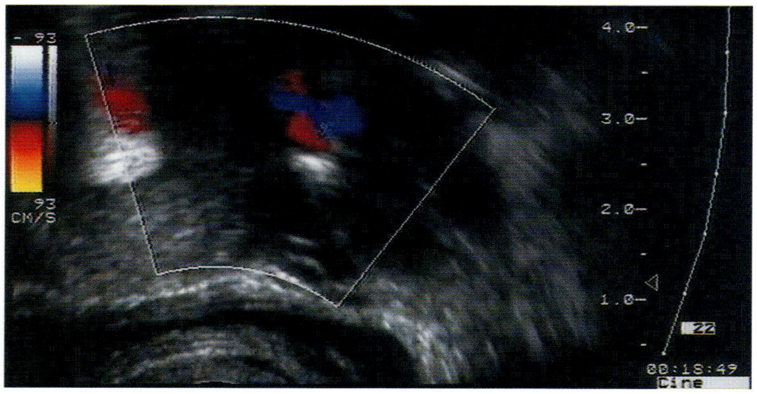

Fig. 7.**7** Original images.
a Color Doppler appearance of contrast medium entering the proximal and middle thirds of the left tube.

b Outflow of contrast medium from the fimbriated end of the left tube (turbulent zones are encoded in blue and red next to a polyfollicular left ovary).

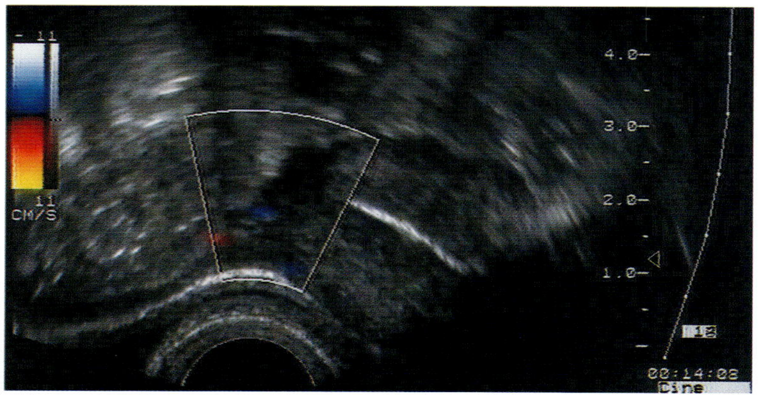

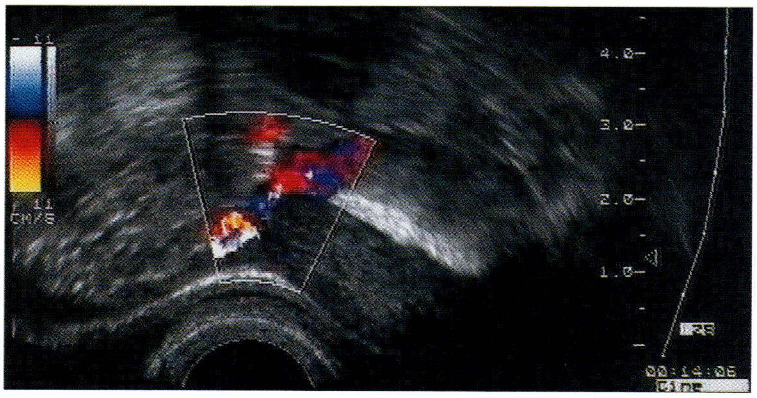

Fig. 7.**8** Original images.
a The sample volume is positioned over the proximal third of the right tube.

b Initial inflow of the contrast medium is marked by strong turbulence in the tubal lumen.

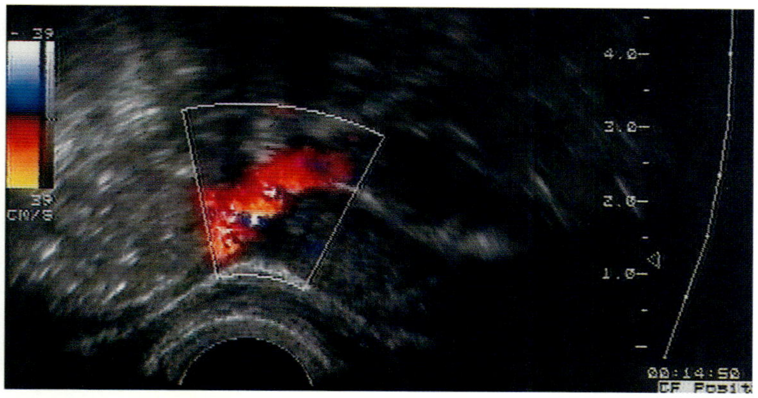

Fig. 7.**8 c** Maximum turbulent flow in the tubal lumen (red = flow toward the transducer).

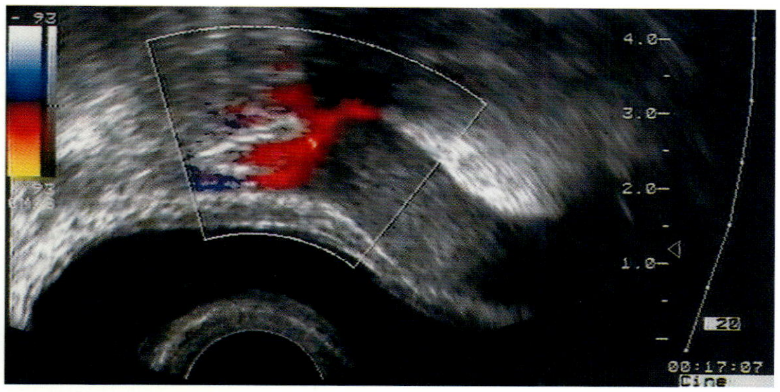

Fig. 7.**8 d** Outflow of contrast medium from the proximal into the distal segment.

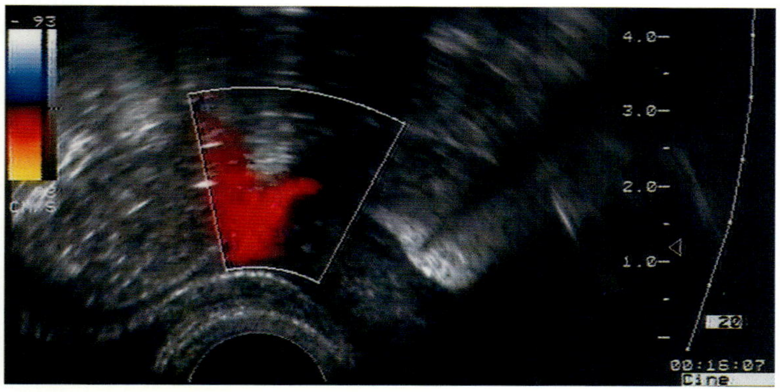

Fig. 7.**8 e** Contrast spillage from the fimbriated end of the right tube.

Results

Comparison of B-mode and Doppler. A total of 404 fallopian tubes in 210 patients were sonographically assessed for patency using B-mode and Doppler flow analysis. The results were determined separately for the proximal and distal tubal segments (Table 7.**2**). Using B-mode in the proximal tubal segment, patency was diagnosed in 92% (373 tubes) and partial or complete obstruction in 8% (31 tubes). When Doppler analysis was added, the results were more differentiated: 87% of the tubes (350) were patent, 6% (24 tubes) were partially obstructed, and 7% (30 tubes) were completely obstructed. Using B-mode in the distal tubal segment (Table 7.**2b**), we classified 72% (292 tubes) as patent and 28 % (112 tubes) as obstructed. Subsequent Doppler analysis indicated patency in 73% (297 tubes), partial obstruction in 7% (30 tubes), and complete obstruction in 20% (77 tubes).

Table 7.**2** Results of Doppler ultrasound hysterosalpingography of 404 tubes in 210 patients

a Proximal segments

Diagnostic method	Patent	Obstruction	
		Partial	Complete
B-mode	373 (92%)	31 (8%)	
B-mode and pulsed Doppler	350 (87%)	24 (6%)	30 (7%)

b Distal segments

Diagnostic method	Patent	Obstruction	
		Partial	Complete
B-mode	292 (72%)	112 (28%)	
B-mode and pulsed Doppler	297 (73%)	30 (7%)	77 (20%)

Comparison with laparoscopy. These sonographic findings in 404 tubes were checked by laparoscopy in 62% of all cases (252 tubes) (Table 7.**3**). To determine the rate of agreement between the two methods, we subdivided the whole population (404 tubes) into three sonographic groups:

- Patent (297 tubes)
- Partial obstruction (35 tubes)
- Complete obstruction (72 tubes)

The rates of agreement between sonographic and laparoscopic findings for the three groups were 95%, 95%, and 84%. Between 60% and 65% of the tubes had been examined endoscopically in all three of the groups.

The overall rate of agreement for the three subgroups was 92% (233 of 252 tubes examined).

Treatment. The patients diagnosed by ultrasound tubal imaging were either referred for ovarian stimulation therapy, followed in some cases by intratubal gamete transfer, or admitted to the IVF program.

Table 7.**3** Results of laparoscopic control examinations compared with Doppler ultrasound hysterosalpingography of 404 tubes in 210 patients, rates of agreement between ultrasound and endoscopy

Result	Total	Checked by laparoscopy	Agreement between ultrasound and laparoscopy
Patent	(297)	184 (62%)	174 (95%)
Partially obstructed	(35)	21 (60%)	20 (95%)
Completely obstructed	(72)	47 (65%)	39 (84%)
All categories	(404)	252 (62%)	233 (92%)

Table 7.4 Side-effects associated with ultrasound contrast imaging of the fallopian tubes

Side-effects	Patients (210 total)
Vasovagal reaction, pain	14 (7%)
Adnexitis	4 (2%)
Allergic reactions	0

Side-effects. Table 7.4 lists the side-effects that were registered in the study. Fourteen of the 210 patients (7%) experienced a vasovagal reaction after several milliliters of the absorbable contrast medium had been injected, necessitating discontinuation of the examination. Four patients (2%) developed mild pelvic inflammatory disease, which resolved quickly in response to antibiotic treatment. Other complications, especially allergic reactions, were not observed.

Discussion of the Value of the Test Procedures

B-mode imaging. Advances in ultrasound technology, including the development of high-resolution endovaginal transducers, have enabled anatomically precise visualization of the female internal genital organs[2]. With their proximity to the organs of interest, these transducers can be operated at higher frequencies to provide better resolution of very small details. As a result, we can evaluate the uterus and ovaries and appreciate their cycle-dependent changes on plain ultrasound images[1,8]. B-mode is a sensitive, noninvasive, ambulatory method for the detection of internal genital malformations in infertile patients and in women with palpable abnormalities[12]. At this stage of the infertility workup, ultrasound imaging and especially transvaginal sonography can already provide an effective, noninvasive screening method (sensitivity 43%, specificity 98%)[12]. Menstrual function studies such as folliculometry and endometrial testing have also become established ultrasound procedures in infertility patients[1,8].

Radiographic hysterosalpingography and laparoscopic chromopertubation. Tubal infertility has a prevalence of 20–30%, which is currently increasing owing to the rising incidence of salpingitis, ectopic pregnancy, and their conservative management [5]. In the past, the only reliable and established methods available to the fertility specialist in the pretherapeutic investigation of an unexplained tubal factor were radiographic hysterosalpingography (HSG) and laparoscopic chromopertubation[11,15,17] Both of these methods have drawbacks such as radiation exposure, possible contrast allergies, pain, invasiveness, and risks relating to general anesthesia, bleeding, and infection. There has been a pressing need for a noninvasive, risk-free screening method for tubal patency that can be performed in an outpatient setting

Abdominal sonography after uterine fluid instillation. Plain ultrasound examination of the lesser pelvis has been unable to provide information on the dynamics of tubal transit. Only abnormal fluid collections (hydrosalpinx, sactosalpinx) could be visualized. Richman et al. (1984) and Randolph et al. (1986) reported on the assessment of tubal patency with abdominal ultrasound following intrauterine fluid instillation[13,14]. This method was basically indirect, however, relying on the detection of free fluid in the cul-de-sac to confirm tubal patency. It also required the use of large amounts of saline solution (up to 200 ml), leading to an increased risk of infection. The examination was performed under general anesthesia before a planned laparoscopy or HSG.

Transvaginal hysterocontrast sonography. Deichert et al. (1988, 1989) introduced hysterocontrast sonography as a new diagnostic procedure for the differentiation of intrauterine and myometrial findings[4]. They used saline solution for examination of the uterine cavity and a galactose microparticle suspension for contrast imaging of the fallopian tubes. These studies, unlike our own, were performed under general anesthesia. Turbulent flow within the tube, contrast spillage from the fimbriated end, and increased fluid in the cul-de-sac were the criteria used for the B-mode assessment of tubal patency[5]. The authors reported 65% agreement with laparoscopy and radiographic HSG, with a 33% rate of partial agreement.

Doppler sonography. After Fitzgerald and Drumm introduced the Doppler principle into obstetric medicine in 1977[6], Taylor et al. (1985) were the first to perform blood flow measurements of the ovarian and uterine arteries during the menstrual cycle, thus applying the principle to the field of reproductive medicine[16]. They used abdominal PW sector transducers as well as endovaginal CW transducers under general anesthesia before a scheduled laparotomy. Since the early 1990s, authors have reported on a variety of results using transvaginal B-mode and Doppler imaging techniques in almost all areas of reproductive and prenatal medicine[9c]. With the high-resolution endovaginal transducers currently available, flow signals can be selectively recorded from the lesser pelvis using spectral and color Doppler techniques down to penetration depths of 10–20 cm. But because some internal genital structures (fallopian tubes) generally cannot be evaluated in their anatomy or function by plain imaging, it is necessary to incorporate contrast administration into the ultrasound protocol[3a].

Doppler sonography with contrast medium. After our group had established B-mode contrast visualization of the fallopian tubes in good agreement with classic methods[10], we attempted to achieve an even more accurate, dynamic, noninvasive assessment of tubal patency by combining this technique with the pulsed Doppler principle[9a]. The examination can be done on an ambulatory basis, but for safety reasons we combined it with one night of hospital observation in the early phase of the study. The method is well accepted by patients as it does not involve premedication, general anesthesia, or pain. If the cervical canal is stenotic or poorly estrogen-primed, or if the patient has a uterine position anomaly, the cervix may be grasped with a tenaculum to straighten the uterus, or a special intrauterine catheter may be used that has greater stability and a smaller lumen (Zinnanti Instruments, Chatsworth, CA, USA).

Contrast medium. The absorbable contrast medium incited a vasovagal response in a few patients (7%), requiring the examination to be terminated. Four patients developed mild pelvic inflammatory disease, which responded well to antibiotics. We observed no serious allergic reactions like those associated with iodinated contrast media.

Standard protocol. The examination should follow a standard protocol using designated planes of section. Following contrast instillation, the uterine cavity is first surveyed in longitudinal and transverse sections. Gross uterine malformations and abnormalities (bicornuate uterus, subseptate uterus, leiomyomas, etc.) should be excluded. This is followed by a dynamic examination of both fallopian tubes, starting with a B-mode analysis and proceeding to Doppler spectral analysis. The tubes cannot be fully visualized with B-mode because of their tortuosity, but portions of the tubes can be identified, even peripherally, and the flow pattern of the contrast medium can be analyzed by proper placement of the sample volume. This can significantly increase the quality and information content of the B-mode examination[9b] (Table 7.**2**).

Color Doppler analysis. Color Doppler analysis can further add to the diagnostic capabilities of the ultrasound examination[3,7]. Multiple color boxes are placed all over the B-mode image, allowing flow to be measured from one pixel to the next over the entire sectional image. Flow toward the transducer is encoded in red, flow away from the transducer in blue. Turbulent flow produces a mixed pattern of green and yellow pixels. In contrast to unidimensional analysis with spectral Doppler, flow phenomena can be imaged simultaneously over the entire scanned region. This requires a considerably more time-consuming adjustment of the color sensitivity of the ultrasound system, and the increase in examination time can be justified only in cases where the spectral Doppler findings are equivocal. Moreover, the contrast visualization of flow in the tubal segments with a 9 MHz transducer can show details in the B-mode image that are comparable to those revealed by color Doppler. Generally, however, only color-flow imaging can demonstrate fluid spillage from the fimbriated end of the tube.

With the Doppler technique described above, it is not possible to evaluate the mobility of the infundibulum or peritubal region, even when ultrasound confirms tubal patency. This explains why postsonographic laparoscopic findings were used in admitting patients to the IVF program. The 92% concordance between the studies (233 of 252 tubes examined in 210 patients) (Table 7.**3**) agrees with the rates of 80–91% reported in a review by Campbell et al.[3a] and identifies the method as a serious alternative to the invasive procedures of radiographic HSG and diagnostic laparoscopy.

Comments. The assessment of tubal patency by transvaginal pulsed Doppler scanning and in selected cases by color duplex sonography provides an easy-to-use, noninvasive screening method for excluding tubal obstruction. It is superior to radiographic HSG in the concomitant evaluation of uterine malformations and anatomical variants. It shows good agreement with the classic diagnostic tests and is potentially safer owing to high patient tolerance of the contrast medium. Done as an outpatient procedure with no premedication or general anesthesia, it is less costly and more comfortable for the patients.

If the findings are equivocal, the classic procedures of radiographic HSG and laparoscopic chromopertubation should be employed. The use of color-flow imaging will improve diagnostic accuracy only in selected cases with an inconclusive spectral Doppler analysis.

Summary

Tubal pathology is a major cause of infertility, and so the assessment of tubal patency is an important element in the diagnostic workup of infertile patients. Conventional methods such as pertubation, radiographic hysterosalpingography, and laparoscopic chromopertubation have disadvantages that include lack of precision, radiation exposure, and invasiveness. An alternative is the sonographic assessment of tubal patency. After initial positive experience with transvaginal ultrasound hysterosalpingography following the injection of contrast medium, the results could be improved by adding transvaginal pulsed and color duplex sonography to the examination. In our series of 210 infertility patients, tubal patency was evaluated by transvaginal sonography following the intrauterine injection of an absorbable galactose-based contrast medium (Echovist, Schering, Berlin). Depending on the result of the examination (tube patent, partially obstructed, or completely obstructed), the finding was checked by laparoscopy in 62%, 60%, and 65% of the cases. The agreement rates between ultrasound and laparoscopy were 95%, 95%, and 84%, respectively. The contrast medium was very well tolerated. The method provides an effective, noninvasive screening test for the exclusion of tubal obstruction. The addition of pulsed Doppler ultrasound provides a level of objectivity that cannot be achieved with the use of B-mode imaging alone.

References

1 Bald R, Hackelöer BJ: Ultraschalldarstellung verschiedener Endometriumformen. In: Otto R, Jann FX (eds.): Ultraschalldiagnostik. Thieme, Stuttgart 1983

2 Bernaschek G: Vorteile der endosonographischen Diagnostik in Gynäkologie und Geburtshilfe. Geburtshilfe Frauenheilkd. 47 (1987) 471–476

3 Becker R, Fobbe F, Schlief R, Wolf KJ, Hammerstein J: Prüfung der Durchgängigkeit der Tubae uterinae durch farbcodierte Duplexsonographie mittels eines Ultraschallkontrastmittels (Echovist). Ultraschall Klin. Prax. Suppl.1 (1988)

3a Campbell S, Bourne TH, Tan SL, Collins WP: Hysterosalpingo contrast sonography (HyCoSy) and its future role within the investigation of infertility in Europe. Ultrasound Obstet. Gynecol. 4 (1994) 245–253

4 Deichert U, van de Sandt M, Lauth G, Daume E: Die transvaginale Hysterokontrastsonographie (HKSG) – Ein neues diagnostisches Verfahren zur Differenzierung intrauteriner und myometraler Befunde. Geburtshilfe Frauenheilkd. 48 (1988) 835–844

5 Deichert U, Schlief R, van de Sandt M, Juhnke I: Transvaginal hystero-salpingo-contrast-sonography (Hy-Co-Sy) compared with conventional tubal diagnostics. Human Reproduction 4 (1989) 418–424

6 Fitzgerald DE, Drumm JE: Non-invasive measurement of human fetal circulation using ultrasound: a new method. Brit. Med. J. 2 (1977) 1450–1451

7 Hata T, Hata K, Senoh D et al.: Transvaginal Doppler Color Flow Mapping. Gynecol. Obstet. Invest. 27 (1989) 217–218

8 Hackelöer BJ, Nitschke S, Daume E, Sturm G, Buchholz R: Ultraschalldarstellung von Ovarveränderungen bei Gonadotropinstimulierung. Geburtshilfe Frauenheilkd. 37 (1977) 185–190

9 Henkel B, Schlief R: Die ambulante Hysterokontrastsonographie (HKSG) – eine frühzeitige Selektionsmethode in der Diagnostik der mechanisch bedingten Sterilität. Ultraschall Klin. Prax. 1 (1986) 1–11

9a Hüneke B, Lindner Ch, Braendle W: Untersuchung der Tubenpassage mit der vaginalen gepulsten Kontrastmittel-Doppler-Sonographie. Ultraschall Klin. Prax. 4 (1989) 192–198

9b Kleinkauf-Houcken A, Hüneke B, Lindner Ch, Braendle W: Combining B-mode ultrasound with pulsed wave Doppler for the assessment of tubal patency. Human Reproduction 12 (1997) 2457–2460

9c Kurjak A: An Atlas of Transvaginal Color Doppler. Parthenon Publishing Group, London 1994

10 Lindner Ch, Braendle W, Schlief R, Bispink L, Luckhardt M, Bettendorf G: Die sonographische Hysterosalpingographie. Alete Wissenschaftliche Reihe, 100. Tagung der Nordwestdeutschen Gesellschaft für Gynäkologie und Geburtshilfe (1988) 101–104

11 Maathuis JB, Horbach JGM, Van Hall EV: A comparison of the results of hysterosalpingography and laparoscopy in the diagnosis of fallopian tube dysfunction. Fertil. Steril. 23 (1972) 428–431

12 Nicolini U, Belotti M, Bonazzi B, Zamberletti D, Candiani GB: Can ultrasound be used to screen uterine malformations? Fertil. Steril. 47 (1987) 89–93

13 Randolph JR, Ying YK, Maier DB, Schmidt CL, Riddick DH: Comparison of real-time ultrasonography, hysterosalpingography, and laparascopy/hysteroscopy in the evaluation of uterine abnormalities and tubal patency. Fertil. Steril. 46 (1986) 828–832

14 Richman TS, Viscomi GN, de Cherney A, Polan ML, Alcebo LO: Fallopian tubal patency assessed by ultrasound following fluid injection. Radiology 152 (1984) 507–510

15 Sanfilippo JS, Yussman MA, Smith O: Hysterosalpingography in the evaluation of infertility: a six-year review. Fertil. Steril. 30 (1978) 636–643

16 Taylor KJW, Burns PN, Wells PNT, Conway DI, Hull MGR: Ultrasound Doppler flow studies of the ovarian and uterine arteries. Brit. J. Obstet. Gynecol. 92 (1985) 240–246

17 Tristant H, Benmussa M: Atlas der Hysterosalpingographie. Enke, Stuttgart 1984

8 Abnormalities of Corpus luteum Function

S. Kupesic, A. Kurjak, and T. Zodan

Many authors have described the clinical importance of normal corpus luteum function in the initiation of a normal pregnancy[12]. An inadequate luteal phase is defined as a delay of more than two days in the histological development of the endometrium relative to the calculated date. It is often a direct result of hormonal dysfunction of the corpus luteum. This dysfunction can have a variety of causes, most notably decreased levels of follicle-stimulating hormone (FSH) in the follicular phase, inadequate secretion of luteinizing hormone (LH), decreased LH and FSH levels at the time of ovulation, and poor endometrial responsiveness to progesterone.

Morphology and Biochemistry of the Corpus luteum

Cell types and hormone production. The differentiation of the corpus luteum is an important event in the ovarian cycle and the critical factor in sustaining an early pregnancy. After ovulation has occurred, vessels sprout from the theca to form a network of blood vessels, marking the start of corpus luteum formation (Fig. 8.**1**). Ultimately the corpus luteum contains various cells: K cells in addition to large and small luteal cells. The large luteal cells develop from the granulosa cells, and the small luteal cells from the thecal cells. The large luteal cells produce more progesterone than the small luteal cells, but the latter appear to be more sensitive to stimulation by LH and chorionic gonadotropin. Also, the small luteal cells appear to produce various angiogenesis factors, and this may occur independently of the production of steroid hormones. Production of the prostaglandins I_2, E_2, and $F_{2\alpha}$ has been demonstrated in cell cultures. The prostaglandins, whose formation is influenced by lipoxygenase products of arachidonic acid such as 5-HETE (hydroxyeicosatetraenoic acid) and is independent of chorionic gonadotropin, act directly on progesterone metabolism. Prostaglandin I_2 and prostaglandin E_2 promote progesterone formation, while prostaglandin $F_{2\alpha}$ has a luteolytic action. Besides being regulated by the hypothalamic–pituitary axis (FSH, LH), the corpus luteum has its own paracrine regulatory mechanism whose details require further investigation.

Luteal phase. The luteal phase begins with the release of the oocyte and the formation of the corpus luteum, accompanied by a significant rise of LH and FSH. The small luteal cells increasingly produce LH receptors, which stimulate progesterone production. The midluteal phase is characterized by peak levels of circulating LH and progesterone and by the lowest resistance index (RI) in the corpus luteum blood vessels, as Kupesic et al.[11] demonstrated by transvaginal pulsed color Doppler sonography (Fig. 8.**2**). The rising progesterone concentration finally suppresses the secretion of gonadotropins, the LH and progesterone levels fall, and the RI in the luteal blood vessels increases. A condition known as luteal phase defect (LPD) can result from faulty "internal" regulation as well as adverse external factors (e.g., strenuous exercise or ovulation-stimulating medications).

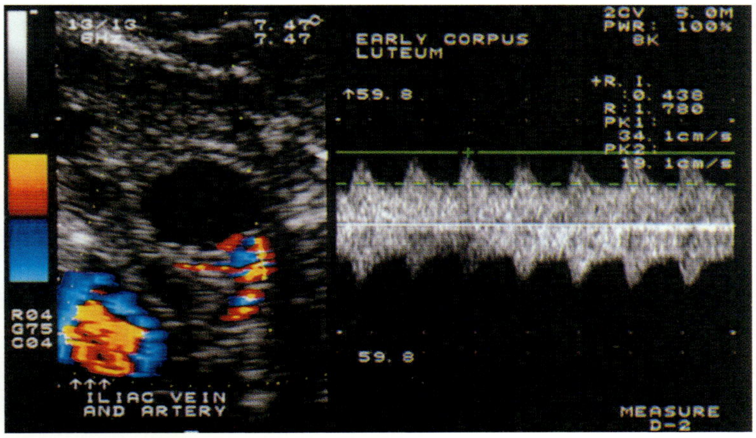

Fig. 8.**1** Transvaginal ultrasound scan of the ruptured follicle (left). An increased blood flow velocity and decreased RI (0.44) are typical signs that indicate ovulation and the start of corpus luteum formation.

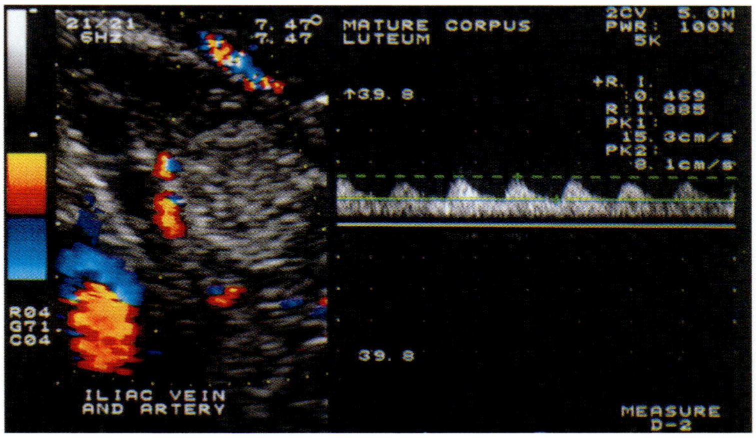

Fig. 8.**2** Increased blood flow in the mature corpus luteum (left). The Doppler trace indicates a high blood flow velocity and low RI (0.47).

Conventional Methods in the Diagnosis and Treatment of Luteal Phase Defect

Definition. Various terms have been applied to this disorder: luteal phase defect, luteal phase deficiency, short luteal phase, luteal insufficiency, and inadequate luteal phase. All these terms describe the same condition, which consists of a progesterone deficiency, a luteal phase shorter than 11 days, and a delay of 2 or more days in the secretory transformation of the endometrium.

Possible Causes of Luteal Phase Defect

Effect of LH. Zeleznik and Little-Ihrig studied the effect of LH on corpus luteum function in rhesus monkeys[20]. Gonadotropin-releasing hormone (GnRH) was administered to induce gonadotropin secretion. Various plasma LH concentrations were measured, depending on the amount of GnRH administered. It was found that an LH concentration of 50% normal was still able to sustain progesterone secretion during the late luteal phase.

These results confirm the hypothesis that regression of the corpus luteum in the nonfertile cycle is due primarily to a reduced luteal cell responsiveness to LH rather than a reduction in gonadotropin secretion.

Jones showed that an imbalance between the FSH and LH levels is responsible for inadequate folliculogenesis and for inadequate transformation of the granulosa and theca cells into the granulosa luteal cells and theca luteal cells of the corpus luteum [10]. This leads to luteal phase defect. Corpus luteum dysfunction with a normal length of the luteal phase may result from impaired granulosa cell function or from an inadequate LH surge with a fairly normal total LH secretion and theca cell response. A short luteal phase is associated with a poor LH surge and low LH secretion. There appears to be a critical LH level that must be maintained after ovulation to ensure the morphological and functional transformation of the granulosa and theca cells and the induction of enzymes for the steroidogenesis of regulatory peptides and various peptide receptors.

Strenuous exercise. Beitinis et al.[2] studied 28 female students with regular cycles who performed regular, strenuous physical exercise during a two-month training program. All subjects collected daily overnight urine samples for three months: at the start of the study, during a control cycle without exercise, and during two exercise cycles. LH, FSH, estriol, and free progesterone were determined and related to creatinine excretion. Twenty cycles with a luteal phase defect were observed in 18 participants. Four women had an inadequate luteal phase in the first month of the training program, combined with decreased free progesterone secretion and a luteal phase shorter than nine days. During the second exercise month, two inadequate and four short luteal phases were observed. Because disturbances of LH and estriol secretion were also observed in women with short luteal phases, it may be assumed that regular, strenuous exercise, which was associated with significant weight loss in 12 cases, can indeed lead to anovulation. It is interesting to note, however, that only two of the women had a luteal phase defect in both cycles.

These results show that physical exercise, change of living conditions, stress, or other extraneous factors can cause menstrual disturbances that may affect the entire cycle or only the luteal phase. It appears, however, that these effects are transient in the majority of cases. Clinical practice has shown that many women will show menstrual abnormalities when followed for several months, but that very few of these problems are permanent.

Ovarian stimulation. There is still disagreement whether ovarian stimulation causes luteal insufficiency. Hecht et al.[8] showed that luteal insufficiency is rare in clomiphene-induced cycles. On the other hand, Reshef et al.[15] found that in 30 women who had been treated with gonadotropin and hCG, 27% of the patients showed inadequate endometrial development.

Diagnosis of Luteal Phase Defect

Endometrial biopsy. One of the greatest problems for scientists and clinicians is the detection of luteal insufficiency. One option is the histological analysis of endometrial biopsies. This method is fairly precise, since the amount of progesterone produced by the corpus luteum can be estimated by the transformation that the endometrium has undergone for implantation. In a regular cycle, the biopsy is taken shortly before the start of menstrual bleeding on the basis of an assumed cycle length of 28 days and a luteal phase of 14 days duration. Because the luteal phase may last from 12 to 14 days, however, the optimum timing of the biopsy is disputed. Biopsies that are taken in the early and midluteal phase show greater histological variation than biopsies from the late luteal phase. Timing the biopsy close to menstruation will best reflect the cumulative progesterone activity.

A major problem is different biopsy interpretations rendered by different evaluators or by the same evaluator at different times.

Gibson et al.[5] obtained duplicate endometrial biopsy samples from 25 women in one sitting. Five colleagues evaluated the two biopsy series on two separate occasions at least two weeks apart. In 43.1% of the cases, the same evaluator gave exactly the same reading for duplicate slides from the same patient, and in 5% of cases the same evaluator saw a difference in endometrial transformation of three or more days in duplicate readings. Even greater inconsistencies occurred among different evaluators. The readings agreed in only 25% of the cases, and in 22% of cases the discrepancy was greater than two days.

Serum progesterone level. The serum progesterone level is also used in the diagnosis of luteal insufficiency. Minimum levels of 2.5–5.0 ng/ml in the midluteal phase indicate ovulation, while levels of 10–15 ng/ml in this phase correspond to normal corpus luteum function[19]. But because diurnal variations in progesterone secretion can cause deviations of more than 30% from the mean value, a single progesterone determination does not reflect either corpus luteum function or its effect on the endometrium.

8

Placental protein. A "placental protein 14" has been described, which is expressed by the endometrial glands and has occasionally been detected in the blood of women with anovulatory cycles[4]. A test procedure has not yet been established.

Dawood[3] states that the combined use of all test procedures is necessary in order to draw relevant conclusions.

Treatment of Luteal Phase Defect

Stimulation with clomiphene and hCG. The current treatment for luteal phase defect is based on optimum follicular development by stimulation with clomiphene and hCG (human chorionic gonadotropin) or hMG (human menopausal gonadotropin). The response to hCG depends on the age of the corpus luteum. A good increase in progesterone secretion was achieved by administering 5000 IU of hCG 8–12 days after the LH surge. A moderate increase was achieved by administering hCG four days after the LH surge, and no increase was achieved when it was administered at the time of the LH surge[19]. Based on these observations, the support of corpus luteum function should begin in the midluteal phase, approximately 7–8 days after the LH surge or 6–7 days after the rise in basal body temperature (i.e., on day 21 of a 28-day cycle)[3]. In a randomized controlled study, no significant benefit was derived from treatment with progesterone suppositories or oral dehydrogesterone compared with no treatment. Three comparative studies showed no difference with or without treatment.

FSH. Insler[9] states that luteal insufficiency is only one of a number of disorders that include an imbalance of intraovarian and extraovarian hormones, peptide interactions, and follicle formation at the wrong time of the cycle. Consequently, there are only a small number of women in whom treatment with FSH alone during the early follicular phase can prevent an endometrial delay observed in previous cycles from occurring in the next luteal phase.

LUF Syndrome

Another functional disturbance that should be addressed is the syndrome of the luteinized but unruptured follicle (LUF syndrome). In LUF syndrome, rupture of the preovulatory follicle does not occur despite luteinization. Progesterone production may be deficient in LUF syndrome or may be within normal limits. The postovulatory parameters may also be essentially normal. This can mimic a "presumed ovulatory" cycle, making it very difficult to diagnose LUF syndrome. The syndrome can have various causes: a deficient LH surge, an absence of preovulatory progesterone secretion, primary oocyte abnormalities, and changes in progesterone synthesis or other mediators leading to rupture of the follicle. The condition was initially diagnosed laparoscopically, and later its diagnosis was aided by detecting low concentrations of ovarian steroids in the cul-de-sac fluid. As ultrasound techniques are refined, it is reasonable to expect that new discoveries on this rare syndrome will be forthcoming.

Ultrasound and Doppler Sonography in the Detection of Luteal Phase Defect

No diagnostic methods used to date have been sufficiently accurate or reliable, making it necessary to introduce new methods into research. Ultrasonography is a highly promising approach, using transvaginal sonography for better identification of the corpus luteum and combining B-mode and real-time imaging (Fig. 8.**3**) with pulsed color Doppler. These techniques are opening up new capabilities for investigations of the corpus luteum, LPD, early pregnancies, and abnormal early pregnancies.

Corpus luteum volume and hormone production. Glock et al.[6] investigated whether the size of the corpus luteum as determined by ultrasound or a change in corpus luteum size in early pregnancy correlates with the serum levels of progesterone, E_2, and 17-hydroxyprogesterone or might even serve as a criterion for predicting the course of the pregnancy. They hypothesized that the corpus luteum volume in early pregnancy correlates with steroid production by the corpus luteum, that the luteal echo pattern based on the relative amount of cystic components correlates with the serum hormone concentration or pregnancy outcome, and that regression of the corpus luteum leads to pregnancy loss. In fact, their study data showed no correlation between changes in corpus luteum volume and changes in steroid production during early pregnancy. They did find, however, that a decreasing corpus luteum volume before the 8th week of pregnancy was associated with an increased abortion risk. Color Doppler sonography was used to distinguish the dominant ovary with the corpus luteum from the contralateral, nondominant ovary (Fig. 8.**4**). Early pregnancy was characterized by a low RI of 0.39–0.49 in the dominant ovary and a high RI of 0.69–1.0 on the contralateral side. In one patient the RI was 0.74 in the dominant ovary and 0.79 on the other side. This high RI on both sides was associated with failure of the pregnancy.

Intraovarian RI. Kupesic et al.[11] compared the intraovarian RI in 47 healthy subjects with ovulatory cycles with that in 28 patients with luteal phase defect (LPD) and 4 patients with LUF syndrome. They measured the follicular diameter by daily examinations and observed the rupture of the follicle at ovulation as well as the demarcation of the corpus luteum, presenting as a hypoechoic structure with an irregular border. The thickened endometrium and the presence of free fluid in the cul-de-sac were simultaneously documented. All of these findings were interpreted as evidence of ovulation. Equivocal cases (in which a corpus luteum was not visualized or follow-ups were not possible) were dropped from the ongoing study. LPD was diagnosed on the basis of progesterone levels and endometrial biopsies in the midluteal phase. The sonographic and Doppler sonographic findings were correlated with measurements of hormone levels and the histopathological data. LUF syndrome was documented by daily ultrasound examinations and hormone assays. In all four cases of LUF syndrome, normal follicular growth and normal preovulatory follicular diameters were found. The follicles maintained their size and shape during the phase in which ovulation was expected to occur. Luteinization of the unruptured follicle was manifested by increasingly high-level echoes at the follicular margin.

In the group of subjects with normal ovulatory cycles ($n = 47$), different ovarian RI values were measured at different times. Follic-

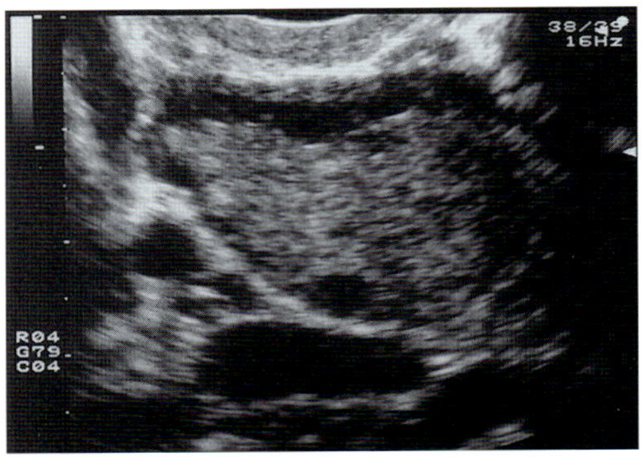

Fig. 8.3 Transvaginal ultrasound scan of an ovary that contains a corpus luteum. It was not possible with B-mode ultrasound to determine anything about the functional status of the ovary.

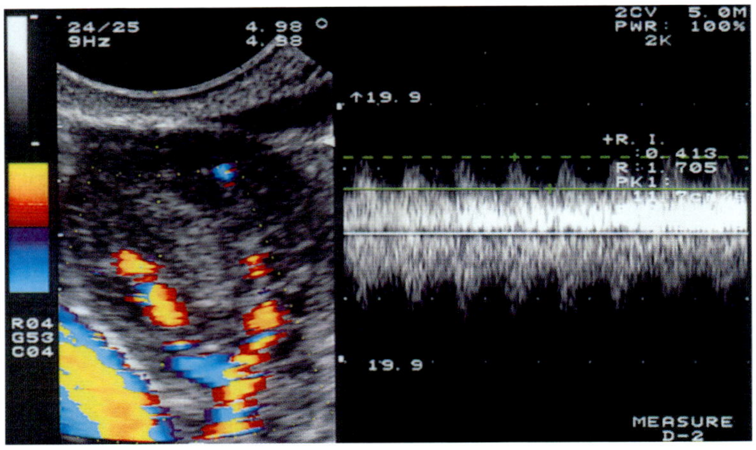

Fig. 8.4 Color Doppler clearly demonstrates the copious blood supply to the corpus luteum. The low RI (0.41) is typical of the corpus luteum.

ular growth and development were associated with moderate to high RI values (0.56 ± 0.06). A significant decrease in RI ($p < 0.001$) was observed on the day of the LH surge (RI 0.44 ± 0.04). The lowest RI values were measured in the midluteal phase (RI 0.42 ± 0.06), and these values increased again during the late luteal phase (RI 0.50 ± 0.04). Normal endometrial dating was confirmed by biopsy in 15 patients. In the group of patients with LPD ($n = 28$), no change in ovarian RI was found during the follicular phase ($p > 0.05$). The mean RI throughout the luteal phase was significantly higher (RI 0.56 ± 0.04, $p < 0.001$) than that in the women with normal cycles. No differences were seen between the early, mid- and late luteal phases ($p > 0.05$).

In the control group, the RI in the follicular and luteal phases was significantly higher on the dominant side ($p < 0.001$). In the LPD group, there were no differences ($p > 0.05$) between the dominant and nondominant sides. The mean progesterone level was significantly lower in the LPD group ($p < 0.001$) (6.9 ± 2.3 ng/ml) than in the control group (24.1 ± 11.4 ng/ml). The histopathological findings showed a delayed endometrial pattern in all of the LPD patients. Progesterone and the RI were correlated in the midluteal phase ($r = 0.09$, $p < 0.83$). The patients with LUF syndrome ($n = 4$) showed no difference in intraovarian RI values following the LH surge. Similar RI values were measured during the follicular and luteal phases (0.55 ± 0.04 and 0.54 ± 0.06). The mean progesterone level in this group was 14.1 ± 6.2 ng/ml.

Luteal conversion. Merce et al.[14] explored all aspects of transvaginal sonography—its advantages, disadvantages, current capabilities, and future outlook. In their study on ovarian blood flow during the luteal phase, they introduced the term "luteal conversion" to describe the Doppler phenomena that occur in the luteal phase: conspicuous Doppler signals, an increase in the frequency spectrum, increased blood flow turbulence with a large scatter of maximum frequencies and superimposed waveforms presenting different peak systolic velocities, and finally an increase in the intensity and area of the Doppler signals recorded over the ovary. The same authors in their LPD study observed a fall of the RI in the dominant ovary during the luteal phase compared with the follicular phase, like that also occurring in a normal cycle. A significant correlation between RI and progesterone level was not described.

LUF syndrome. Merce et al.[14] saw no decrease in the intraovarian RI after the LH surge in patients with LUF syndrome. The RI values were in the upper normal range within four days after the LH peak and during the growth and luteinization of the follicle. Later they were again similar to those during the follicular phase, indicating that the progression of RI values in LUF syndrome does not show true cyclic variations, similar to the pattern in an anovulatory cycle. "Luteal conversion" does not take place, which shows that the changes in microvascularity described in normal cycles either do not occur in LUF syndrome or occur in an altered form, possibly due to the failure of follicular rupture.

Implantation. Merce et al.[14] also noted the importance of endometrial and ovarian blood flow for implantation. They specifically recommended the use of Doppler ultrasound techniques in studies that deal with blood flow phenomena during implantation and their relationship to the course of pregnancy.

Intraovarian blood flow and progesterone levels. Glock and Brumsted[7] studied the correlation between intraovarian blood flow values and the progesterone (P) level during the course of the ovarian cycle. The mean P levels in the luteal phase of patients with LPD were significantly lower than in women with normal cycles ($p < 0.001$). The mean RI in the LPD patients was significantly increased in both the follicular and luteal phase ($p = 0.02$). The systolic and diastolic blood flow velocities were decreased in the LPD patients, although these differences were not statistically significant ($p = 0.54$, $p = 0.11$). Correlations between P level and RI were observed in all portions of the luteal phase, the highest correlation occurring in the midluteal phase (early luteal phase $r = 0.73$, $p = 0.03$; midluteal phase $r = 0.80$, $p < 0.01$; late luteal phase $r = 0.63$, $p = 0.07$). The mean RI in the dominant ovary of women with normal cycles was significantly lower throughout the cycle than in the contralateral ovary (0.50 vs. 0.65, $p = 0.001$). This was not the case in patients with LPD (0.60 vs. 0.66, $p = 0.37$). The RI in both ovaries remained elevated in one patient with an anovulatory cycle (mean value 0.76 [0.70–0.82]).

The study demonstrated a correlation between corpus luteum blood flow and progesterone level during the normal ovarian cycle. The strongest correlation exists in the midluteal phase—the period in which there is maximum neovascularization of the corpus luteum. Consistent with this finding, the authors observed a rise of vascular resistance in the late luteal phase—the period in which the corpus luteum begins to regress. These

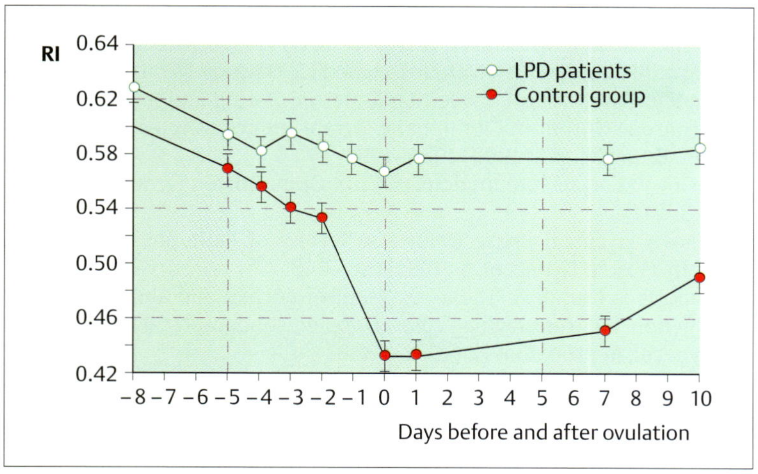

Fig. 8.**5** Changes in intraovarian blood flow before and after ovulation in luteal phase defect (LPD) patients and a control group.

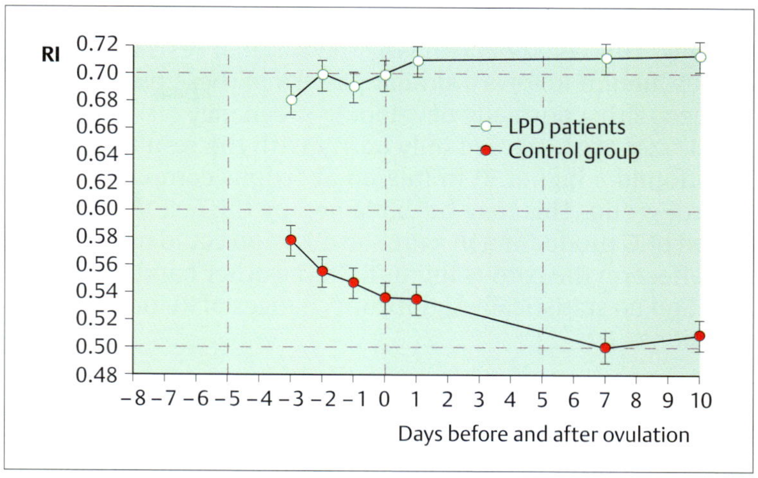

Fig. 8.**6** Blood flow changes in the spiral arteries before and after ovulation in luteal phase defect (LPD) patients and a control group.

findings suggest that measuring the RI in the corpus luteum may be a useful adjunct to the plasma progesterone assay in the evaluation of luteal function.

Tinkanen et al.[18], on the other hand, found no differences in luteal blood flow measurements between healthy subjects and infertile patients. The authors found that a shortened luteal phase was not associated with a premature regression of corpus luteum vascularization.

Strigini et al.[17] investigated the changes of vascular resistance in patients with FSH-treated cycles. The uterine pulsatility index (PI) in stimulated cycles was significantly lower than in normal cycles both before and after ovulation. This was attributed to an increased E_2 concentration in the plasma.

Doppler findings, endometrial biopsy, and hormone levels. Kupesic et al.[12] correlated Doppler flow measurements with histological findings and hormone studies. They presumed that the diagnosis of a luteal phase defect could be improved by combining ultrasound results with endometrial biopsy and hormonal markers. The mean progesterone levels in women with LPD were significantly lower than in the control group (10.2 ± 4.3 ng/ml vs. 21.0 ± 4.2 ng/ml, $p < 0.01$). The FSH/LH ratio was significantly lower in the group with a delayed endometrial pattern ($p < 0.001$). A close correlation was found between estradiol levels and follicular size on days −5 to −1. An increase in follicular diameter and endometrial thickness was observed in both groups.

No differences in intraovarian flow resistance were found in the proliferative phase ($p > 0.05$). The control group showed a significant regression of the RI on the day of the LH surge (RI = 0.45 ± 0.04, $p < 0.05$). Thereafter the RI again rose to the values that were measured during the follicular phase (RI = 0.49 ± 0.02). The mean RI was significantly higher in the LPD group (RI = 0.58 ± 0.04, $p < 0.001$) (Fig. 8.**5**) than in the control group. The RI in the dominant ovary of the control group was significantly lower than in the contralateral ovary, whereas both sides were equal in most of the LPD patients. Blood flow in the spiral arteries was also measured. The RI in the control group was 0.53 ± 0.04 in the periovulatory phase, 0.50 ± 0.02 in the midluteal phase, and 0.51 ± 0.04 in the late luteal phase (Fig. 8.**6**). In the LPD group, increased resistance values were measured in the spiral arteries during all phases (periovulatory phase RI = 0.70 ± 0.06, $p < 0.001$; midluteal phase RI = 0.72 ± 0.06, $p < 0.001$; late luteal phase RI = 0.72 ± 0.04, $p < 0.001$). A close correlation was found between plasma estradiol levels and follicular diameter.

The study shows that, in women with normal endometrial development, a decline of vascular resistance occurs in the uterine, radial, and spiral arteries from the follicular phase to the luteal phase. On the other hand, retarded endometrial development is marked by a rise of vascular resistance in the uterine vessels during the luteal phase. Since the most significant differences were found in measurements of the spiral arteries, the changes in endometrial blood flow could provide an important parameter for predicting endometrial development and the likelihood of implantation.

Blood Flow in the Corpus luteum during Early Pregnancy

Normal and abnormal early pregnancy. Salim et al.[16] compared corpus luteum blood flow in normal and abnormal early pregnancies. Their study tested the hypothesis that the absence of measurable luteal blood flow did not correlate with a normal early pregnancy. The resistance in the intraovarian vessels in women with an abnormal early pregnancy (missed abortion, incomplete or threatened abortion) was significantly higher ($p < 0.01$) than in women with a normal early pregnancy. This did not apply to patients with a molar or ectopic pregnancy, however.

The differences between the subgroups of abnormal early pregnancy may be due to the different nature of the causes. Missed abortion and incomplete abortion present as an early loss of pregnancy with no potential for further development. Threatened abortion is a similar situation. It is unclear whether decreased blood flow in the corpus luteum is a potential cause of this condition or a result. Molar and ectopic pregnancies present a different situation. In these cases the pathological state is progressive, i.e., it continues to develop and is not self-limiting. This could explain why the vascular resistance in the corpus luteum follows a similar pattern in these patients as in women with a normally developing pregnancy.

Alcazar et al.[1] agreed only partly with the results of Salim. They found a higher RI in missed abortions compared with a control group. This may be explained by the fact that the impaired hCG production in a missed abortion could have a negative effect on the corpus luteum. On the other hand, Alcazar et al. found no statistically significant changes of RI in threatened abortions.

Outlook. The true capabilities of transvaginal pulsed and color Doppler sonography in the assessment of corpus luteum and ovarian blood flow have not yet been adequately explored. With their help, however, it should be possible in the foreseeable future to better understand the physiology of reproductive processes and find ways to treat pathological conditions that are not treatable at present.

References

1 Alcazar JL, Laparte C, Lopez-Garcia G: Corpus luteum blood flow in abnormal early pregnancy. J. Ultrasound Med. 15 (1996) 645–649
2 Beitinis IZ, McArthur JW, Turnbull BA, Skrinar GS, Bullen BA: Exercise induces two types of human luteal dysfunction: Confirmation by urinary free progesterone. J. Clin. Endocrinol. Metab. 72 (1991) 1350–1358
3 Dawood MY: Corpus luteal insufficiency. Curr. Opin. Obstet. Gynecol. 6 (1994) 121–127
4 Fay TN, Jacobs IJ, Teisner B, Westergaard JG, Grudzinskas JG: A biochemical test for direct assessment of endometrial function: measurement of the major secretory endometrial protein PP14 in serum during menstruation in relation to ovulation and luteal function. Hum. Reprod. 5 (1990) 382–386
5 Gibson M, Badger GJ, Bym F, Lee KR, Korson R, Trainer TD: Error in histologic dating of secretory endometrium: variance component analysis. Fertil. Steril. 56 (1991) 242–247
6 Glock JL, Blackman JA, Badger GJ, Brumsted JR: Prognostic Significance of Morphologic Changes of the Corpus Luteum by Transvaginal Ultrasound in Early Pregnancy Monitoring. Obstet. Gynecol. 85 (1995) 37–41
7 Glock JL, Brumsted JR: Color flow pulsed Doppler ultrasound in diagnosing luteal phase defect. Fertil. Steril. 64 (1995) 500–504
8 Hecht BR, Bardawil WA, Khan-Dawood FS, Dawood MY: Luteal Insufficiency: Correlation Between Endometrial Dating and Integrated Progesterone Output in Clomiphene Citrate-Induced Cycles. Amer. J. Obstet. Gynecol. 163 (1990) 1986–1991
9 Insler V: Corpus luteum defects. Curr. Opin. Obstet. Gynecol. 4 (1992) 203–211
10 Jones GS: Luteal Phase Defect: A Review of Pathophysiology. Curr. Opin. Obstet. Gynaecol. 3 (1991) 641–648
11 Kupesic S, Kurjak A: The assessment of normal and abnormal luteal function by transvaginal color Doppler sonography. Eur. J. Obstet. Gynecol. Reprod. Biol.72 (1997) 83–87
12 Kupesic S, Kurjak A, Vujisic S, Petrovic Z: Luteal phase defect: comparison between Doppler velocimetry, histological and hormonal markers. Ultrasound Obstet. Gynaecol. 9 (1997) 1–8
13 McNeely MJ, Soules MR: The diagnosis of luteal phase deficiency: A critical review. Fertil. Steril. 50 (1988) 1–15
14 Merce LT, Garces D, De la Fuente F: Conversion lutea de la onda de velocidad de fluio ovarica: nuevo parametro ecografico de ovulacion y funcion lutea. Acta Obstet. Gynecol. Scand. (ed. Esp.) 2 (1989) 113–114
15 Reshef E, Segars JH, Hill GA, Pridham DD, Jussman MA, Colston-Wentz A: Endometrial inadequacy after treatment with human menopausal gonadotropin/human chorionic gonadotropin. Fertil. Steril. 54 (1990) 1012–1016
16 Salim A, Žalud I, Farmakides G, Schulmal H, Kurjak A, Latin V: Corpus luteum blood flow in normal and abnormal early pregnancy: Evaluation with transvaginal color and pulsed Doppler sonography. J. Ultrasound Med. 13 (1994) 971 –975
17 Strigini FAL, Scida PAM, Parri C, Visconti A, Susini S, Genazzani AR: Modifications in uterine and intraovarian artery impedance in cycles of treatment with exogenous gonadotropins: effects of luteal phase support. Fertil. Steril. 64 (1995) 76–80
18 Tinkanen H: The role of vascularization of the corpus luteum in the short luteal phase studied by Doppler ultrasound. Acta. Obstet. Gynecol. Scand. 73 (1994) 321–323
19 Yeko TR, Khan-Dawood FS, Dawood MY: Human corpus luteum: Luteinizing hormone and chorionic gonadotropin receptors during the menstrual cycle. J. Clin. Endocrinol. Metab. 68 (1989) 529–534

9 Interventional Ultrasound in Reproductive Medicine

S. Kupesic, A. Kurjak, and A. K. Ertan

Laparoscopy was the method of choice for oocyte retrieval in the earliest case reports on successful in-vitro fertilization[21, 30, 31, 45, 70]. Its greatest disadvantage, however, was that follicles maturing deep in the ovary or in an ovary concealed by matted adhesions could not be directly visualized[55]. Moreover, laparoscopy usually required general anesthesia and was associated with increased perioperative morbidity and mortality. Another problem was that the pneumoperitoneum created with carbon dioxide led to very slight, transient pH changes that could have a harmful effect on the oocytes. In the early 1980 s, the first generation of real-time scanners offered sufficiently high resolution for the percutaneous ultrasound-guided aspiration of oocytes.

Follicular Aspiration in Assisted Reproduction

Transabdominal Follicular Aspiration

Lenz et al.[40] were the first authors to describe ultrasound-guided follicular aspiration. Analgesic and sedative premedication allowed for a painless procedure with high patient acceptance[39]. General anesthesia was not used because it might induce hyperprolactinemia, which could adversely affect the microenvironment of the oocyte[60].

Equipment. Early studies described the use of linear transducers[75], whereas most operators today prefer sector scanners. High-resolution transducers have a separate biopsy attachment with corresponding software. Most authors use a 16-gauge needle with an inside diameter of 1.1 mm[43, 75]. The Teflon coating on the needle counteracts the adhesive tendency of the oocytes, while the sharp needle tip can easily pierce the ovarian tissue with minimal pressure or pain. To flush the follicle, the proximal end of the needle is connected by Teflon tubing to a syringe filled with warmed, heparinized culture medium. Flushing the follicle after it has been aspirated helps to maximize the oocyte recovery rate per punctured follicle [60]. The irrigation lumen may be parallel or concentric to the aspiration lumen. The first attempts employed manual aspiration with a 5 ml syringe. But since the uncontrolled pressure of manual aspiration could cause rupture of the zone pellucida, operators developed a foot-controlled mechanical aspiration pump with a maximum pressure of 80–100 mmHg[14]. Other equipment includes a comfortable operating table, sterile drapes and protective sleeves, sterile ultrasound gel, and a water bath or heating unit.

Patient preparation. Every patient requires thorough counseling and meticulous preparation. A large number of different stimulation protocols are used to induce the growth of multiple follicles. A combination of endocrinological and sonographic data are used to determine the timing of hCG (human chorionic gonadotrophin) administration. An ultrasound examination is advised before the start of the procedure. The goal of this examination is to determine the number and size of the follicles and compare them with the data on the day of hCG administration. The patient should drink 2 liters of liquid one hour before the procedure to distend the bladder. In some cases it may still be necessary to insert a transurethral catheter and instill Hartmann solution into the bladder, and the operator should understand the associated risks of iatrogenic infection and bladder irritation. A distended bladder permits clear visualization of the ovarian follicles and moves the bowel out of the area between the abdominal wall and ovaries. Combined analgesic and sedative premedication is given to most patients, and the needle insertion site is infiltrated with local anesthetic. Other options are peridural or spinal anesthesia. The lower abdomen is prepared with an antiseptic solution, and the operative field is packed off with sterile drapes. The endovaginal transducer is smeared with sterile ultrasound gel and covered with a sterile protective sleeve. The control panel of the ultrasound unit is covered with sterile, transparent plastic film.

Technique. The puncture needle is inserted directly into the bladder using freehand technique or a needle guide. After the needle has pierced the posterior bladder wall and ovarian capsule, the needle tip is directed into the center of the closest follicle. The follicular fluid is aspirated, and the collapsed follicle is flushed with an equal volume of culture medium. The needle tip is kept within the ovary, and all the follicles are systematically aspirated using the same technique. The operator should understand that inadequate bladder filling or pelvic adhesions can alter the position of the ovaries and that overweight and heavy scar tissue result in poor visualization of the pelvic organs. After the procedure is completed, the patient may empty her bladder and can usually be discharged home two hours later.

Transurethral Follicular Aspiration

Parsons[50] and Dellenbach[17] developed the transurethral technique of oocyte retrieval. The procedure was done on an outpatient basis to reduce costs. The patient was placed in the lithotomy position with the operator on her right, the ultrasound cart on her left, and an assistant seated in the middle between the patient's legs[9]. The vulva was aseptically prepared with chlorhexidine solution. The needle tip was introduced through the side opening of the Foley catheter and passed through the urethra into the bladder. After the bladder was filled with Hartmann solution, the catheter cuff was inflated and the needle tip was advanced out of the catheter. The operator then passed the needle tip through the posterior bladder wall under sonographic guidance and directed it into the nearest follicle. All the follicles were successively aspirated and flushed. Finally the needle was withdrawn, the bladder was emptied, and the catheter was removed. The patient drank several glasses of liquid and emptied her bladder before leaving the hospital. The overall complication rate was very low. There were no complaints of cystitis or adnexitis following the procedure.

Booker et al.[8] compared the transurethral route for ultrasound-guided follicular aspiration with the transvaginal route in a randomized prospective study. Considerably more follicles could be visualized by the transvaginal route, but there were no differences in the number of oocytes harvested, in the duration of the procedure, or in the fertility, embryo-transfer or pregnancy rates. This led the authors to conclude that the transurethral and transvaginal routes were of equal value for ultrasound-guided follicular aspiration.

Transvaginal Follicular Aspiration

Gleicher et al.[25] were the first authors to report on transvaginal follicular aspiration guided by abdominal ultrasound. A speculum is placed in the vagina, and the needle is passed through the fornix to the ovary. Subsequent experience has shown that the transvaginal approach guided with an endovaginal transducer is superior to all other ultrasound-guided techniques[22].

Advantages. The proximity of the transducer to the pelvic organs enables the use of high-frequency transducers that provide significantly better resolution and greater clinical efficiency. Owing to the natural compliance of the fornix, the tip of the transducer can be moved closer to the ovaries by applying gentle pressure. Since a full bladder is not required, the position of the pelvic organs is unchanged and the ovaries are located within the focal zone of the transducer. Overweight or adhesions do not prevent visualization of the follicles and thus are not a contraindication to this method.

Standard stimulation protocols are monitored with the aid of transvaginal sonography[32]. Additional information can be gained by hormone assays and by the color Doppler assessment of blood flow in the uterus and ovary[34, 37, 38].

Preparations. The entire treatment is done as an outpatient procedure. The patient is placed in the lithotomy position. Sedatives (such as flunitrazepam, droperidol, or pentazocine) can be given, but approximately 50% of IVF groups do not use anesthesia or sedation[23]. Since follicular aspiration takes about 10 minutes on average, most patients tolerate the procedure with no difficulties. Nevertheless, the operator should be aware of possible hypotensive reactions or queasiness. After ultrasound gel has been applied to the transducer probe, the protective sleeve (a sterile condom, surgical glove, or specially designed sheath) is placed over the gel-smeared probe, avoiding any air bubbles that might cause artifacts. The gel should not be used to lubricate the probe insertion because of its spermicidal and reputed embryotoxic properties[58]. Physiological saline solution or culture medium should be used instead.

Technique. The vagina is irrigated with isotonic saline solution or culture medium, and the transducer probe is inserted into the vagina. A sterile needle guide is used for transvaginal follicular aspiration. An automatic puncture device has been developed to avoid the potential risks of needle insertion. This device consists of a movable metal tube and a guide mechanism into which the aspiration needle is inserted and secured[22]. The device should be "locked and loaded" before it is inserted into the vagina with the transducer. After the insertion, a detailed sonographic examination is done to locate the uterus and ovaries. The probe is positioned so that the puncture line indicating the needle path is aimed precisely at the center of the nearest follicle. The operator calculates the exact distance along the puncture line to the targeted follicle and "fires" the needle into the follicle. The follicular fluid is then aspirated into a test tube connected to the aspiration pump. The collapse of the follicle can be observed during aspiration of the follicular fluid[22]. Flushing of the aspirated follicle may be done to increase the number of retrieved oocytes. This is done by infusing culture medium with heparin added through the puncture needle or through a separate line. Without removing the needle, the operator aspirates all of the follicles that are situated along the puncture line.

Complications. Feichtinger et al.[24] described a low complication rate with ultrasound-guided follicular aspiration. In 2.4% of cases the iliac vein was mistaken for a follicle and erroneously punctured. Ultrasound revealed intraperitoneal bleeding into the cul-de-sac, but this resolved spontaneously in all cases. It was shown that filling the bladder can stop the bleeding by exerting pressure on the puncture site. The use of color Doppler may provide a simple way to avoid this complication, as the iliac vessels are easily visualized with this technique.

Bleeding from the posterior vaginal wall is easy to diagnose and can be stopped by compression. Pelvic inflammatory disease (PID) is considered a rare complication of transvaginal follicular aspiration (0.14% of all patients)[24]. In most cases the infection was caused by bacterially contaminated semen and occurred in women with a prior history of PID.

GIFT (Gamete Intrafallopian Transfer) and ZIFT (Zygote Intrafallopian Transfer)

GIFT. The GIFT procedure was developed to overcome various obstacles to spermatozoon transport and the failure of the fallopian tube to capture the oocyte during ovulation[3]. Patients in this program undergo ovarian stimulation under ultrasound surveillance. The oocytes are harvested by laparoscopy, identified in the laboratory, and transferred to a catheter containing 100 000 spermatozoa collected by the swim-up technique[3]. The transfer catheter is introduced into the fimbriated end of the fallopian tube, where its contents are gently emptied. By bringing together the sperm and oocyte, this technique circumvents many of the factors that can disrupt sperm transport and fusion[46]. Because fertilization occurs in the natural milieu, the success rate is 26.5% deliveries per oocyte retrieval[61].

ZIFT. ZIFT is an advanced form of GIFT in which the oocytes are harvested by transvaginal aspiration, fertilized in vitro, and transferred the next day in the pronuclear stage into one of the fallopian tubes using the GIFT technique[62]. The transvaginal aspiration technique used in the ZIFT method is the same as that described for IVF. The overall success rate appears to be identical to that of GIFT and IVF.

Advantages and disadvantages. Disadvantages of the GIFT and ZIFT procedures are that they cannot be used in patients with tubal pathology and they require anesthesia and operating room technique. Also, neither procedure can be used in patients with male factor infertility since fertilization is uncertain. One refinement is to transfer the spermatozoa and oocytes into the fallopian tube by transcervical catheterization of the tube, but this method is associated with the same pregnancy rates with no significant reduction in costs or risks.

Embryo Transfer

Intrauterine embryo transfer represents the last critical step in in-vitro fertilization. Inadequate transfer of the embryo into the uterine cavity can be an important factor in the failure of implantation. For this reason, ultrasound-guided embryo transfer might offer significant advantages over the traditional "blind" transfer[27].

> Hurley et al.[29] performed intrauterine embryo transfer with a transcervical catheter in 94 patients using transvaginal ultrasound guidance. The culture medium that contained the embryos was injected along with air bubbles so that the embryos could be optimally placed in the uterine cavity under visual guidance. This method was particularly helpful in six cases where the catheter became "stuck" in the endocervical canal, contrary to the operator's impression that the catheter was accurately placed in the uterine cavity.

This method has also proved effective in patients with submucous leiomyomas or uterine anomalies. Moreover, the ultrasound-guided procedure allays anxiety by allowing patients to watch the transfer on the monitor screen.

Lenz et al.[41] and Parsons et al.[49] used transvaginal ultrasound for guidance during transabdominal embryo transfer through the wall of the uterus into the uterine cavity. This method could provide an alternative to transcervical transfer in patients with known cervical stenosis making it difficult to pass a catheter through the cervical canal.

Fallopian Tube Catheterization

Indications and technique. Fifteen percent of female infertility cases are caused by proximal occlusion of the fallopian tubes. The occlusions at that level may result from intraluminal cellular debris, mild adhesions, or muscular spasms[64]. Transcervical tubal catheterization can be used for both the diagnosis and treatment of infertility patients[15, 16, 33, 56, 59, 68]. The criteria used in selecting patients for this procedure are the presence of bilateral tubal occlusions or the stenosis of the remaining tube in patients who have had a unilateral salpingectomy[27]. The catheter is advanced into the fallopian tube over a guidewire to stretch open the tube. When aided by fluoroscopic guidance, tubal catheterization has a success rate of 98% for proximal occlusions in the area of the uterotubal junction but only 33% for more distal stenoses[56].

Ultrasound guidance. Transvaginal ultrasound can be used to direct the tubal catheterization, although there has been little experience with this technique to date. Lisse and Sydow[44] described tubal catheterization under transvaginal ultrasound guidance. Concomitant laparoscopy confirmed the accurate placement of the catheter in the proximal part of the tube (3–6 cm from the intramural junction). Tubal patency was restored in 85% of the patients with a proximal occlusion. Breckenridge and Schinfeld[11] and Thurmond et al.[67] recommend transabdominal ultrasound as a simpler method of guiding the catheter insertion than transvaginal ultrasound. They report that the full bladder necessary for abdominal ultrasound straightens the uterus and makes it more easily accessible.

Complications. Possible complications of transcervical tubal catheterization are perforation of the fallopian tube, vasovagal reactions during the procedure, and infection. Patients should also be informed of the increased risk of ectopic pregnancy following the procedure.

Aspiration of Ovarian Cysts

Indications and technique. Transvaginal ultrasound allows the direct visualization and aspiration of persistent follicular cysts[27]. These cysts can impair folliculogenesis either by hormone secretion or by compressing the tissue and restricting blood flow. For the aspiration of an ovarian or paraovarian cyst, the needle tip is introduced into the center of the cyst. The merits of this procedure are debated in the literature. The fear of seeding cells into the abdominal cavity from a potentially malignant ovarian process has discouraged many operators from using this method. While cytological analysis is always performed on the aspirated cyst fluid, a negative cytological report is occasionally false-negative. The high sensitivity and specificity of transvaginal color Doppler ultrasound in the difference of benign and malignant adnexal masses can apparently aid the decision of which cysts should be aspirated (Fig. 9.**1**).

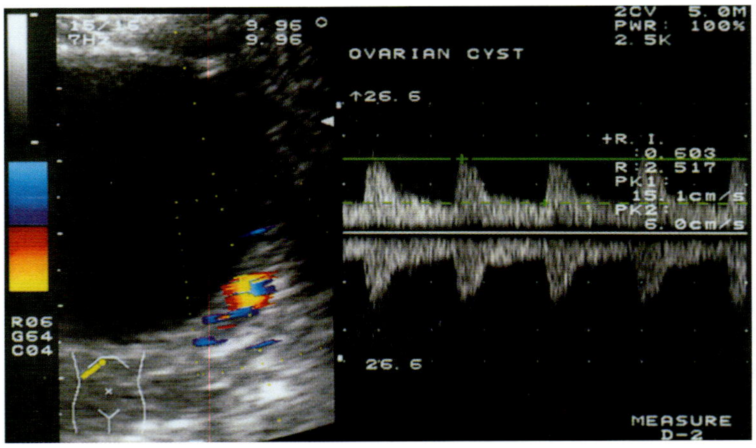

Fig. 9.**1** Transvaginal scan of an anechoic ovarian cyst. Color Doppler indicates a high RI (0.75) for blood flow in the pericystic tissue.

> **Recurrence rates.** Bret et al.[12, 13] published two reports on their experience with transvaginal ultrasound in the aspiration of ovarian cysts. They described a recurrence rate after cyst aspiration of 48% in premenopausal patients and 80% in postmenopausal women. Alcohol was injected into the aspirated cyst to prevent recurrence, but this was successful in only four of seven patients[12].

Vaegemaekers et al.[74] performed the transvaginal aspiration of 32 unilocular, hypoechoic ovarian cysts (average diameter 45 mm) in infertile patients. The authors concluded from their study that ovarian cyst aspiration in the early follicular phase might reduce the dropout rate from IVF cycles.

Endometriosis. Endometriosis is considered a relative contraindication to cyst aspiration. Aboulghar et al.[1] examined 21 patients after the transvaginal ultrasound-guided aspiration of endometriotic cysts. They found six recurrences over a 12-month follow-up period.

While the needle aspiration of endometriotic cysts is technically simple, the overall benefit and safety of this method have not yet been adequately evaluated owing to the relatively small number of patients in whom the procedure has been performed to date[42].

Drainage of Cul-de-Sac Abscesses

In 30–40% of infertile patients, occlusion or dysfunction of the fallopian tubes is the cause of the infertility. It is known that tubal occlusion is a common sequel to PID. Because PID has a high recurrence rate, it is reasonable to conclude that this cause has a high incidence in the population of infertile women.

The drainage of tubo-ovarian abscesses under sonographic guidance can accelerate the healing process and enhance the effectiveness of antibiotic therapy. The needle tip is advanced into the abscess cavity, and the fluid inside the abscess is aspirated as completely as possible. Then the needle is withdrawn or a drainage catheter is inserted[27]. Teisala et al.[66] reported on the drainage of 10 tubo-ovarian abscesses performed with antibiotic therapy and transvaginal ultrasound guidance. Only mild sedation was needed for pain control, and the procedure was well tolerated by the patients. The technique is a widely recognized alternative to laparoscopy in the treatment of tubo-ovarian abscesses.

Selective Reduction of Multiple Pregnancies

During the past 15 years, the incidence of multiple gestation has increased owing to the frequent use of ovulation-stimulating products and the more widespread use of assisted reproductive techniques. Multiple pregnancies are associated with a high risk of morbidity and mortality. The likelihood of carrying the pregnancy to term and bearing a healthy child is inversely proportional to the number of fetuses. This has prompted efforts to selectively reduce the number of embryos to improve the outcome for the remaining fetuses[6]. Selective reduction can be offered in a pregnancy with more than four embryos, usually reducing it to a twin pregnancy[27]. Normally the procedure is done after eight weeks' gestation, when the spontaneous abortion rate is very low. A group of French authors[20] were the first to perform the procedure with a needle introduced under abdominal ultrasound guidance, and other groups have adopted the same method[7, 10]. With advances in transvaginal sonography, this technique has also been used successfully in multifetal pregnancy reduction. The advantages of the transvaginal technique are the shorter needle path and the ability to direct the needle more precisely, thereby reducing the risk of injury to adjacent gestational sacs or pelvic organs.

Technique. The technique of transvaginal multifetal pregnancy reduction consists of visualizing the gestational sacs with ultrasound, carefully analyzing the fetal heart tones, introducing the needle into the heart of the selected fetus, and injecting 0.5–1 ml of a 2 mEq /ml potassium chloride solution. The fetus is then observed for 5–10 minutes to confirm cardiac asystole. The patient should be reexamined with ultrasound three hours after the procedure and again at one week. The disadvantage of transvaginal fetal reduction is that it is still uncertain at this early stage of pregnancy how many fetuses would actually continue to grow without the procedure[42].

Techniques of Ultrasound Tubal Imaging

One aspect in the diagnostic investigation of infertility has remained largely unchanged over the past 20 years: the examination of the fallopian tubes. Rubin[57] described the first attempt to evaluate tubal patency in 1954. Since then, the most widely used methods of tubal examination have been radiographic hysterosalpingography and laparoscopic chromopertubation[48].

Radiographic hysterosalpingography. For decades, hysterosalpingography (HSG) using radiographic contrast medium has been the standard procedure for demonstrating the structure of the uterus and fallopian tubes. One disadvantage of HSG is the risk posed by ionizing radiation to the oocyte, which could result in a congenital anomaly if fertilization were to occur. Also, allergy to iodinated contrast medium is considered a contraindication to radiographic HSG (Table 9.**1**). There is still disagreement as to whether an oil-based or a water-based contrast medium is better for HSG.

Hysteroscopy. Hysteroscopy should be viewed as a complementary technique to hysterosalpingography. It can be used for both the accurate investigation and treatment of endometrial polyps and submucous leiomyomas. It is also very helpful in the diagnosis and treatment of intrauterine synechiae and some congenital uterine anomalies (Table 9.**1**).

Laparoscopy. Laparoscopy has been the gold standard for the diagnostic evaluation of tubal function during the past two

Table 9.**1** Comparison of the advantages of laparoscopy, hysteroscopy, radiographic hysterosalpingography (HSG), and color Doppler hysterosalpingography in the assessment of tubal function

Criteria	Radiographic HSG	Hysteroscopy	Laparoscopy	Color Doppler HSG
Anesthesia	➤ Not required	➤ ± General anesthesia	➤ General anesthesia	➤ Not required
Risks	➤ Pelvic infection ➤ Contrast allergy	➤ Anesthesia risk ➤ Uterine perforation ➤ Bleeding ➤ Infection	➤ Anesthesia risk ➤ Intra-abdominal injuries ➤ Bleeding ➤ Infection	➤ Pelvic infection
Necessary facilities and equipment	➤ X-ray machine ➤ Fluoroscope	➤ ± Operating room	➤ Operating room	➤ Color Doppler unit
Advantages	➤ Contrast visualization (evaluation of the ampulla, demonstration of intramural and intraluminal tubal abnormalities)	➤ Visualization of uterine cavity (polyps, synechiae, leiomyomas, septa, etc.) and of tubal ostia ➤ Treatment possible in the same sitting	➤ Visualization of abdominal cavity and peritoneum (adhesions, endometriosis, etc.) ➤ Treatment possible in the same sitting	➤ No adverse reactions to contrast media or dyes ➤ Can be repeated at any time ➤ Requires active patient cooperation (patient is better informed) ➤ Permits analysis of tubal motility ➤ Findings can be recorded on videotape and discussed with the couple ➤ Best views can be printed out for documentation purposes and filed ➤ Does not require help of radiology department (lower costs)
Costs	+	++	+++	+

decades. It does require general endotracheal anesthesia, however, and there is a risk of surgical complications such as vascular and bowel injuries[2], preperitoneal emphysema, and postoperative pain. Laparoscopy permits the direct examination of the entire lesser pelvis and upper abdomen. Ovarian diseases, genital anomalies, and tubal and ovarian function can be evaluated. Laparoscopy also permits the identification and staging of endometrial lesions in the lesser pelvis. Irrigating fluid can be sampled and cultured in patients with a prior history of PID (Table 9.**1**).

Transabdominal ultrasound. Richman et al.[54] were the first to report on the assessment of tubal patency with transabdominal ultrasound. They used a special intrauterine catheter for this purpose (Harris Uterine Injector, Unimar, Canoga Park, CA, USA). At least 20 ml of the ultrasound contrast agent Hyskon (dextran dissolved in dextrose, Pharmacia, Piscataway, NJ, USA) is injected through the catheter, and the collection of this fluid in the cul-de-sac is assessed as an indicator of tubal patency.

Randolph et al.[53] used a comparable method in which 200 ml of isotonic saline solution was instilled into the uterus through a Rubin cannula. The retrouterine fluid volume determined with transabdominal ultrasound was used as an index for confirming the patency of one or both fallopian tubes, although sidedness could not be determined.

Transvaginal ultrasound. The ultrasound visualization of the pelvic organs has been significantly improved by the use of high-frequency endovaginal transducers. A full bladder is not necessary for this examination. Usually the normal fallopian tube cannot be demonstrated by transvaginal scanning unless it is surrounded by fluid for contrast. This fluid can have various sources:

- Normal peritoneal secretion, which occurs in many healthy women
- Follicular fluid during or after ovulation
- Blood
- Ascites
- Purulent discharge from an infectious focus

The lumen of a normal fallopian tube that is not filled with fluid cannot be visualized[69].

Hysterosonosalpingography

In 1989 Deichert et al.[19] developed a new transvaginal sonographic technique for evaluating the fallopian tubes, which they called transvaginal hysterosalpingo-contrast sonography or Hy-Co-Sy. The tubes were visualized and their patency was tested by the transcervical injection of the hyperechoic ultrasound contrast agent SHU 454 (Echovist; Schering, Berlin, Germany) through a Rubin cannula or No. 8 bladder catheter. Tüfecki et al.[71] simplified this procedure so that it could be done on an ambulatory basis, calling it transvaginal sonosalpingography. They evaluated tubal patency directly by the intrauterine injection of isotonic saline solution. This transvaginal examination is easily performed without anesthesia. It is safe, noninvasive, and cost-effective, and causes less patient discomfort than other conventional techniques. It also eliminates the problem of contrast allergy.

Contrast media. All materials that have a different sonodensity from human tissue can be used as ultrasound contrast media. They are divided into two categories: hypoechoic and hyperechoic. Isotonic saline, Ringer solution, and dextran solution are examples of hypoechoic media. Their instillation helps to delineate hyperechoic boundary structures. The main disadvantage of these media is that they cannot demonstrate motion- or flow-related phenomena. Hyperechoic contrast media amplify the sonographic signals, making it possible to detect flow using both B-mode and Doppler techniques. Gramiak and Shah[26] and Meltzer et al.[47] discovered that tiny gas bubbles are excellent reflectors of ultrasound waves. For this reason, all commercial ultrasound contrast agents contain microbubbles. The products Echovist and Levovist (Schering, Berlin, Germany) consist of a suspension of microbubbles that are produced from special galactose microparticles. These microparticles are dissolved either in a galactose solution (Echovist) or in sterile water (Levovist)[65].

Contraindications. Color Doppler hysterosalpingography is contraindicated by bleeding, pregnancy, or adnexal tumors detected clinically or sonographically. An ultrasound examination should be done before the procedure to assess the position of the uterus and adnexa and exclude anomalies. The procedure should be postponed if acute PID is present. Prophylactic antibiotics should be used in patients with a prior history of PID[36].

Timing and preparation. Color Doppler hysterosalpingography should be performed in the early follicular phase of the cycle after the end of the menstrual period. This will prevent the introduction of menstrual blood residues into the abdominal cavity. An examination in the immediate premenstrual phase is recommended, as uterine constriction is greatest in this phase of the cycle. The timing of the examination is of key importance in obtaining optimum results. Premedication or sedation is routinely employed; good results have been achieved with 5–10 mg of diazepam. Significant pain indicates an occlusion with possible intravascular contrast or tubal rupture, so it should not be masked by anesthesia.

Technique. After voiding, the patient is placed in the lithotomy position on a gynecological chair. The vagina and cervix are prepared with beta-isadone solution, and the cervix is exposed with a speculum so that the os is easily accessible. The anterior os is grasped with a tenaculum, and light traction is applied as the catheter is gently inserted into the endocervical canal. The uterine cavity is visualized with ultrasound, and the catheter placement is checked for accuracy (Fig. 9.**2**). The tenaculum is withdrawn, and the transducer is advanced into the posterior fornix of the vagina. Then a maximum of 5–10 ml of the negative contrast medium (sterile saline solution) is slowly injected into the uterine cavity under sonographic guidance (Fig. 9.**3**). The morphology of the uterus and the echo pattern and surface contour of the endometrium can be evaluated at this time (Figs. 9.**4**– 9.**6**). With the uterine cavity filled with anechoic contrast medium, it is not difficult to detect uterine anomalies,

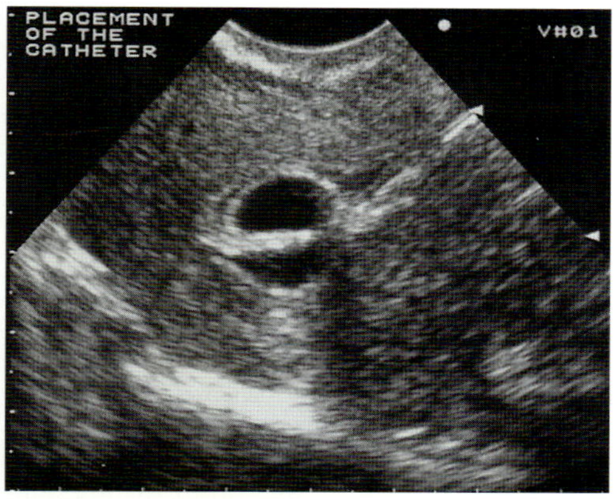

Fig. 9.**2** Transvaginal scan of the uterus after insertion of the intrauterine catheter. The catheter cuff is visible at the level of the internal os.

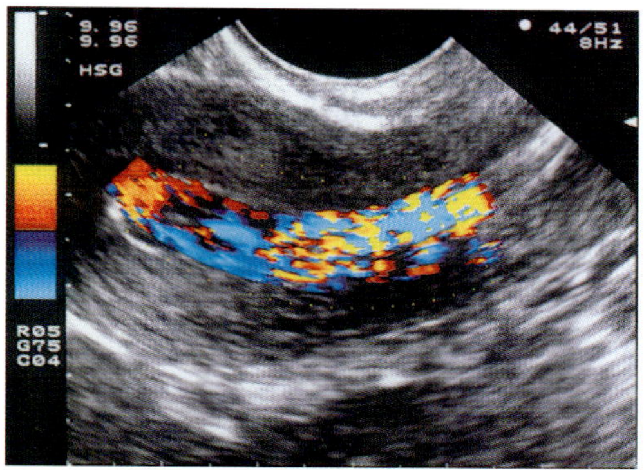

Fig. 9.**3** Transvaginal color Doppler hysterosalpingography. The contrast medium and color Doppler signals are directed by the intrauterine cannula into the uterine cavity.

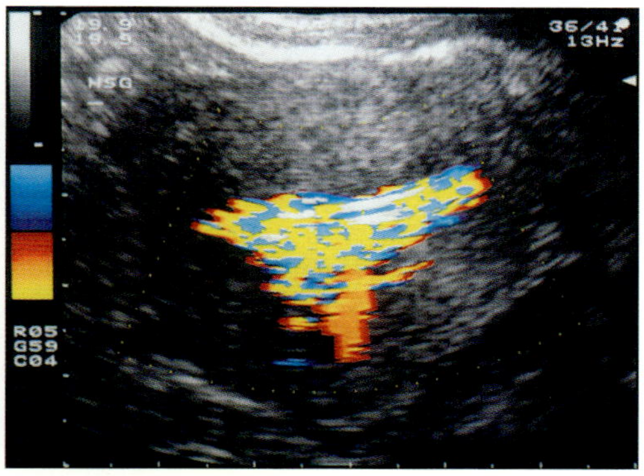

Fig. 9.**4** Transvaginal scan of the uterus after the injection of isotonic saline solution. The color Doppler signals clearly define the triangular uterine cavity.

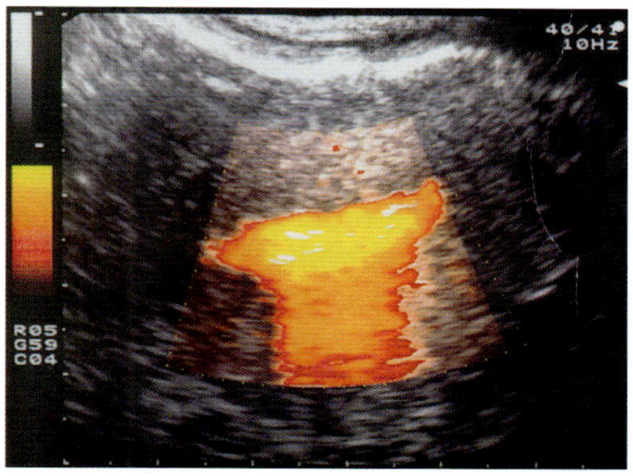

Fig. 9.**5** Power Doppler imaging makes it easier to evaluate the uterine cavity and assess tubal patency.

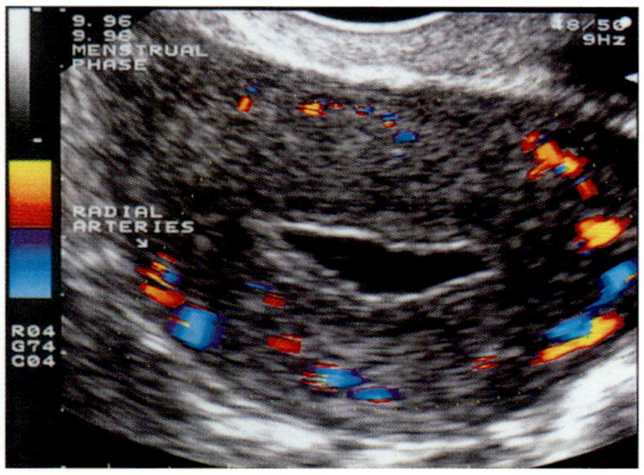

Fig. 9.**6** The regular surface of the uterine cavity is clearly demonstrated by the instillation of a negative contrast medium.

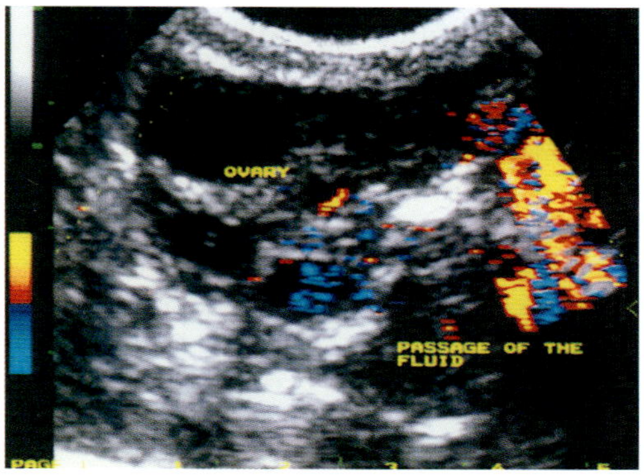

Fig. 9.**7** Color Doppler hysterosalpingography. The color-flow signals are produced by saline solution flowing through the right fallopian tube, confirming its patency. A collection of the anechoic fluid is visible in the cul-de-sac.

endometrial polyps, or submucous leiomyomas that project into the uterine cavity. Next the color Doppler beam is directed toward the tubal ostia. The use of a positive (echogenic) contrast medium will significantly increase the accuracy of tubal patency assessment: color Doppler signals within the tube indicate patency (Fig. 9.**7**), while an absence of signals is interpreted as tubal occlusion[51, 52]. The collection of contrast medium in the cul-de-sac on the side of the selective injection, demonstrated by transvaginal pulsed color Doppler scanning, is a reliable indicator of tubal patency on that side. There is no question that selective tubal injection increases the accuracy of the procedure and of the interpretation.

If the patient experiences cramping pains, the contrast injection should be halted for several minutes. If neither tube is

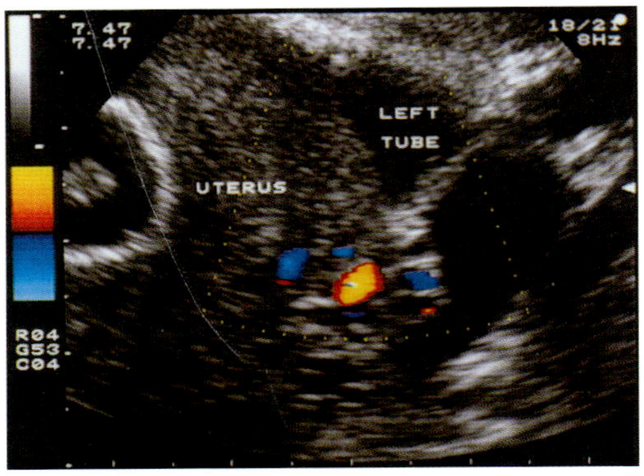

Fig. 9.**8** Transvaginal color Doppler image of the uterus and left adnexa. The intrauterine cannula and the catheter are visible on the left side. The color signals are produced by contrast medium flowing into the left fallopian tube. The tube itself appears as a septate, "sausage-like" structure owing to the collection of fluid proximal to its fimbriated end.

defined by the contrast medium, this signifies a proximal tubal occlusion at the ostial level, and the differential diagnosis should include a tubal spasm. This spasm can be prevented by administering of atropine (0.5 mg) before the examination.

Serious side-effects have not been reported during or after transvaginal Doppler hysterosalpingography. The duration of the examination is from 5 to 14 minutes. At the end of the procedure, the speculum examination is repeated and the vaginal instruments are removed. The cervix is checked for any bleeding caused by the tenaculum, and compression is applied as needed.

Problems. Problems in the diagnosis of tubal occlusion can occur in patients with hydrosalpinx, as the fluid in the distended tube can mimic tubal patency by Doppler ultrasound (Fig. 9.**8**). Also, Doppler hysterosalpingography with saline solution does not permit the assessment of tubal morphology.

Validity of the method. Using our modified technique, we compared the results of color Doppler hysterosalpingography with the results of laparoscopic chromopertubation in 47 patients[35]. The Aloka SSD 680 and 2000 color Doppler systems were used. The results of color Doppler HSG agreed with those of chromolaparoscopy in 43 of the 47 patients (91.48%). There was one patient in whom Doppler could not demonstrate patent tubes, but the presence of free fluid in the cul-de-sac gave indirect evidence of tubal patency.

Thus, our results prove the safety and efficiency of transvaginal Doppler hysterosalpingography in the assessment of tubal patency without radiation exposure and without the use of contrast medium. The costs of this method are significantly lower than those of radiographic hysterosalpingography, and the result is available immediately[54]. It is good practice to document all findings with a VCR or Polaroid camera[18] (Table 9.**1**).

To assess the accuracy of the diagnosis of tubal occlusion by color Doppler hysterosalpingography, Peters et al.[51] studied 129 infertility patients (Tables 9.**2** and 9.**3**). When the results of ultrasound HSG were compared with those of radiographic HSG and/or chromopertubation, 69 of 85 (81%) studies showed agreement, and 50 of 58 (86%) ultrasound HSG findings agreed with the results of chromopertubation. The frequency of comparable findings between radiographic HSG and chromopertubation was 75%.

Richman et al.[54] also compared their ultrasound results with those of conventional HSG. Ultrasound detected bilateral tubal occlusion with a sensitivity of 100%, and tubal patency was assessed with a specificity of 96% (Table 9. **2**).

In the 38 infertility patients studied by Tüfekci et al.[71] (Table 9.**3**), the results of transvaginal sonosalpingography and laparoscopy showed complete agreement in 29 cases (76.32%) and partial agreement in 8 cases (21.5%). Only one case (2.63%) showed a discrepant result.

Stern et al.[63] injected saline solution transcervically during transvaginal color Doppler sonography in 238 women (Table 9.**2**). The correlation between ultrasound HSG and chromopertubation in this study was 81% versus 60% between radiographic findings and chromopertubation. In 49 women who underwent all three procedures, the color Doppler results correlated with chromopertubation more often than the findings of radiographic HSG (82% versus 57%). The authors recommended that ultrasound HSG be repeated before the definitive diagnosis of a unilateral tubal occlusion.

Deichert et al.[18] sought to determine whether the additional use of pulsed Doppler sonography could improve the tubal diagnosis reached with gray-scale imaging in doubtful cases (Table 9.**3**). They concluded from their study that pulsed Doppler can be recommended as a supplement to gray-scale imaging in cases of suspected tubal occlusion where intratubal flow is detectable only over a short distance.

Allahbadia[2] reported a 92.59% rate of agreement between color Doppler examination and HSG or laparoscopy (Table 9.**3**). HSG and laparoscopy agreed in 100% of cases.

Possible therapeutic implications. The diagnostic value of this method is beyond question, and we may speculate about its possible therapeutic applications[36]. The increased pregnancy rate during the first three months after the procedure (two patients in our study) could be the result of mechanical irrigation of the uterine cavity, which can flush mucus plugs from the uterus, disrupt peritoneal adhesions, or have a stimulatory effect on the tubal cilia.

Ayida et al.[4] did a study to determine how patients would tolerate Hy-Co-Sy and HSG as outpatient tests. The procedures used to evaluate tubal patency and uterine anomalies were

Table 9.**2** Correct results of ultrasound hysterosalpingography compared with radiographic hysterosalpingography

Authors	Total number	Correct results (%)	Sensitivity (%)	Specificity (%)
Richman et al. (1984)[54]	36		100	96
Peters et al. (1991)[51]	27	19	70.37	
Stern et al. (1992)[63]	89	72	80.90	
Volpi et al. (1991)[72]	21	19	90.47	

Table 9.**3** Correct results of ultrasound hysterosalpingography compared with chromopertubation

Authors	Total number	Correct results (%)	Sensitivity (%)
Allahbadia et al. (19934)[2]	27	25	92.59
Tüfekci et al. (1992)[71]	38	37	97.37
Peters et al. (1991)[51]	58	50	86.20
Kupesic et al. (1994)[35]	47	43	91.48
Stern et al. (1992)[63]	121	99	81.82
Deichert et al. (1992)[18]	16	16	100.00
Volpi et al. (1996)[73]	29	24	82.70
Holte et al. (1995)[28]	14	12	85.70
Volpi et al. (1991)[72]	21	17	80.90

tolerated equally well. There were no significant differences in the frequency or severity of pains at different stages during and after the procedure. The analgesia requirements were also comparable.

Albunex. In a recent study, Holte et al.[28] investigated the safety and efficacy of a new ultrasound contrast medium, Albunex, for evaluating tubal patency by transvaginal sonography (Table 9.**3**). All of the women had sought medical attention because of uterine leiomyomas and excessive menstrual bleeding and were scheduled for a hysterosalpingectomy three weeks after the ultrasound examination. Albunex and saline solution were alternately injected transcervically during transvaginal ultrasound scanning. There were no serious side-effects. Agreement between hystero-contrast sonography (Hy-Co-Sy) and the postoperative testing of tubal patency was observed in 12 of 14 cases. The discrepancy between Hy-Co-Sy and postoperative findings in the other two cases was attributed to the location of the uterine myomas. The results suggest that Albunex can be safely used as a contrast medium in ultrasound investigations of tubal patency.

Recommendations and objections. Volpi et al.[73] recommended transvaginal sonography (using air bubbles and saline as contrast media) as the first step in assessing tubal patency. Equivocal or definite tubal occlusions should be investigated further by laparoscopy.

It should also be considered that normal findings in color Doppler HSG examinations cannot replace diagnostic laparoscopy in all cases. Several authors have criticized the costs and diagnostic efficacy of the method. The costs of the ultrasound system, including the extra equipment for color Doppler scanning, combined with indirect costs relating to equipment wear, depreciation, and person-hours, must be taken into account.

Unlike the authors cited above, Balen et al.[51] found that ultrasound contrast hysterosonography using a negative (sterile saline) or positive (Echovist) contrast medium was not sufficiently accurate in the evaluation of tubal patency.

Color Doppler HSG versus radiographic HSG. While radiographic HSG is the most accurate method for the diagnosis of intramural or intraluminal tubal disease, color Doppler HSG is the only noninvasive method available for the analysis of tubal motility.

The most accurate interpretation can be made during the procedure itself, as the flow of contrast medium can be tracked through the entire genital tract. To obtain as much information as possible, the procedure should be performed by an experienced examiner who is familiar with the color Doppler method, the relevant anatomy and pathology, the necessary instrumentation, and the technique of contrast injection. There is no question that the development of Hy-Co-Sy is proving to be a safe and well-tolerated option in the evaluation of tubal status.

Comments. Although ultrasound-guided procedures are most commonly used in assisted reproductive medicine, similar techniques can also be used in other clinical situations. The accuracy of the method and its high patient acceptance have contributed greatly to the more widespread use of transvaginal needle procedures. The growing use of assisted reproductive technologies has led to a marked increase in ectopic pregnancies. The use of pulsed and color Doppler sonography in the diagnosis and treatment of ectopic pregnancy is covered in Chapter 12.

References

1 Aboulghar MA, Mansour RT, Serour GI, Rizk B: Ultrasonic transvaginal aspiration of endometriotic cysts: an optional line of treatment in selected cases of endometriosis. Hum. Reprod. 6 (1991) 1408–1410
2 Allahbadia GN: Fallopian tube patency using color Doppler. Int. J. Gynecol. Obstet. 40 (1993) 241
3 Asch RH, Balmareda JP, Ellsworth LR, Wong PC: Preliminary experience with gamete intrafallopian transfer (GIFT). Fertil. Steril. 45 (1986) 366–369
4 Ayida G, Kennedy S, Barlow D, Chamberlain P: A comparison of patient tolerance of hysterosalpingo-contrast sonography (HyCoSy) with Echovist-200 and X-ray hysterosalpingography for outpatient investigation of infertile women. Ultrasound Obstet. Gynecol. 7 (1996) 201–204
5 Balen FG, Allen CM, Siddle NC, Lees WR: Ultrasound contrast hysterosalpingography-evaluation as an outpatient procedure. Br. J. Radiol. 66 (1993) 592–599
6 Berkowitz RI, Lynch L: Selective reduction: an unfortunate misnomer. Obstet. Gynecol. 75 (1990) 873–874
7 Birnholz JC, Dmowski WP, Binor Z, Radwanska E: Selective continuation in gonadotropin-induced multiple pregnancy. Fertil. Steril. 48 (1987) 873
8 Booker MW, Pampiglione JS, Parsons J: A prospective randomized comparison of the perurethral transvesical route with the transvaginal route for ultrasound directed follicle aspiration. Fertil. Steril. 50 (suppl.) (1988) 120–125
9 Booker M, Parsons J: Ultrasound directed follicle aspiration for oocyte collection by the periurethral technique. In Chervenak FA, Isaacson GC, Campbell S (eds.): Ultrasound in obstetrics and gynecology. Little, Brown and Company, London 1993, 1391–1396
10 Brandes JM, Itskovitz J, Timor-Tritsch IE: Reduction of the number of embryos in multiple pregnancy. Fertil. Steril. 48 (1987) 325–327
11 Breckenridge JW, Schinfeld JS: Technique for US-guided fallopian tube catheterization. Radiology 180 (1991) 569–570
12 Bret PM, Atri M, Guibaud L: Ovarian cysts in postmenopausal women: preliminary results with transvaginal alcohol sclerosis. Radiology 184 (1992) 661
13 Bret PM, Guibaud L, Atri M: Transvaginal US-guided aspiration of ovarian cysts and solid pelvic masses. Radiology 185 (1992) 377

14 Cohen J, Avery S, Campbell S, Mason BA, Riddle A, Sharma V: Follicular aspiration using a syringe suction system may damage the zona pellucida. J. In Vitro Fert. Embryo Transfer 3 (1986) 224–228
15 Confino E, Friberg J, Gleicher N: Transcervical balloon tuboplasty. Fertil. Steril. 46 (1986) 963–966
16 Deaton JL, Gibson M, Riddick DH, Brumsted JR: Diagnosis and treatment of cornual obstruction using a flexible guide wire. Fertil. Steril. 53 (1990) 232–236
17 Dellenbach P, Nisand I, Moreau L, Ferger B, Plumere C, Gerlinger P: Transvaginal sonographically controlled follicle puncture for oocyte retrieval. Fertil. Steril. 44 (1985) 656–661
18 Deichert U, Schlief R, van de Sandt M, Daume E: Transvaginal hysterosalpingo-contrast sonography for the assessment of tubal patency with gray scale imaging and additional use of pulsed wave Doppler. Fertil. Steril. 57 (1992) 62 –67
19 Deichert U, Schlief R, van de Sandt M, Junke I: Transvaginal Hysterosalpingo-contrast Sonography (Hy-Co-Sy) compared with conventional tubal diagnostics. Hum. Reprod. 4 (1989) 418–422
20 Dumez Y, Oury JF: Method for first trimester selective abortion in multiple pregnancy. Contrib. Gynecol. Obstet. 15 (1986) 50–53
21 Edwards RG, Steptoe PC, Purdy JM: Estalishing full-term human pregnancies using cleaving embryos grown in vitro. Br. J. Obstet. Gynaecol. 87 (1980) 737–741
22 Feichtinger W: Transvaginal oocyte retrieval. In Chervenak FA, Isaacson GC, Campbell S (eds.): Ultrasound in obstetrics and gynecology. Little, Brown and Company, London 1993, 1397–1406
23 Feichtinger W, Putz M, Kemeter P: New aspects of vaginal ultrasound in an in vitro fertilization program. Ann. N. Y. Acad. Sci. 541 (1988) 125–130
24 Feichtinger W, Putz M, Kemeter P: Four years experience with ultrasound-guided follicle aspiration. Ann. N. Y. Acad. Sci. 541 (1988) 138–145
25 Gleicher N, Friberg J, Fullan N et al.: Egg retrieval for in vitro fertilization by sonographically controlled vaginal culdocentesis. Lancet 2 (1983) 508–600
26 Gramiak R, Shah PM: Echocardiography of the aortic root. Invest Radiol. 3 (1968) 356–366
27 Hill ML, Nyberg DA: Transvaginal sonography guided procedures. In Nyberg DA, Hill LM, Bohm-Velez M, Mendelson EG (eds.): Transvaginal ultrasound. Mosby Year Book, St. Louis 1992, 319–329
28 Holte J, Rasmussen C, Morris H: First clinical experience with sonicated human serum albumin (Albunex) as an intrafallopian ultrasound contrast medium. Ultrasound Obstet. Gynecol. 6 (1995) 62
29 Hurley VA, Osborn JC, Leoni MA, Leeton J: Ultrasound-guided embryo transfer: a controlled trial. Fertil. Steril. 55 (1991) 559–562
30 Jones HW Jr, Acosta AA, Garcia JE, Sandow BA, Veeck L: On the transfer of conceptuses from oocytes fertilized in vitro. Fertil. Steril. 39 (1983) 241–245
31 Jones HW Jr, Jones GS, Andrews MD et al.: The program for in vitro fertilization at Norfolk. Fertil. Steril. 38 (1982) 14–19
32 Kemeter P, Feichtinger W: Experience with a new fixed-stimulation protocol without hormone determinations for programmed oocyte retrieval for in-vitro fertilization. Hum. Reprod. 4 (suppl.) (1989) 53–54
33 Kumpe DA, Zwerdlinger SC, Rothbarth LJ: Proximal fallopian tube occlusion: diagnosis and treatment with transcervical fallopian tube catheterization. Radiology 17 (1990) 183–187
34 Kupesic S, Kurjak A: Uterine and ovarian perfusion during the periovalutory phase assessed by transvaginal color Doppler. Fertil. Steril. 60 (1993) 439–443
35 Kupesic S, Kurjak A: Gynecological vaginal sonographic interventional procedures-what does color add? Gynaecol. Perinatol. 3 (1994) 57–60
36 Kupesic S, Kurjak A: The role of color Doppler in vaginal sonographic puncture procedures. In Kurjak A (ed): An atlas of transvaginal color Doppler. The Parthenon Publishing group Ltd, London 1994, 335–347
37 Kurjak A, Kupesic S: Ovarian senescence and its significance on uterine and ovarian perfusion. Fertil. Steril. 64 (1995) 532–537
38 Kurjak A, Kupesic S, Schulman H, Zalud I: Transvaginal color Doppler in the assessment of ovarian and uterine blood flow in infertile women. Fertil. Steril. 56 (1991) 870–873
39 Lenz S, Lauritsen JG: Ultrasonically guided percutaneous aspiration of human follicles under local anaesthesia: a new method of collecting oocytes for in vitro fertilization. Fertil. Steril. 38 (1982) 673–677
40 Lenz S, Lauritsen JG, Kjellow M: Collection of human oocytes for in vitro fertilization by ultrasonically guided follicular puncture. Lancet 1 (1981) 1163–1164
41 Lenz S, Leeton J, Rogers P, Trounson A: Transfundal transfer of embryos using ultrasound. J. In Vitro Fert. Embryo Transf. 4 (1987) 13–17
42 Lerner JP, Monteagudo A: Vaginal sonographic puncture procedures. In Goldstein SR, Timor-Tritsch IE (eds.): Ultrasound in Gynecology. Churchill Livingstone, New York 1995, 223–238
43 Lewinthal D, Mahadevan M, Pattinson HA, Taylor PJ, Persaud D: Follicular factors, serum estradiol, and outcome of pregnancy following in vitro fertilization and embryo transfer. Fertil. Steril. 48 (1987) 840–845
44 Lisse K, Sydow P: Fallopian tube catheterization and recanalization under ultrasonic observation: a simplified technique to evaluate tubal patency and open proximal obstructed tubes. Fertil. Steril. 56 (1991) 198–201
45 Lopata A, Johnston IW, Hoult IJ, Speirs AI: Pregnancy following intrauterine implantation of an embryo obtained by in vitro fertilization of a preovulatory egg. Fertil. Steril. 33 (1980) 117–121
46 Marmar JL: Idiopathic male infertility. In Garcia CR, Mastroianni L, Arnelar RD, Dubin L (eds.). Current therapy of infertility 3. BC Decker, Toronto 1988, 196–201
47 Meltzer RS, Tickner G, Sahines TP, Popp RL: The source of ultrasound contrast effect. J. Clin. Ultrasound 8 (1980) 121
48 Page H: Estimation of the prevalence and incidence of infertility in a population: a pilot study. Fertil. Steril. 71 (1989) 571–574
49 Parsons JH, Bolton VN, Wilson L, Campbell S: Pregnancies following in vitro fertilization and ultrasound-directed surgical embryo transfer by periurethral and transvaginal techniques. Fertil. Steril. 48 (1987) 691–693
50 Parsons J, Riddle A, Booker M et al.: Oocyte retrieval for in vitro fertilization by ultrasonically guided needle aspiration via the urethra. Lancet 1 (1985) 1076–1078
51 Peters JA, Coulam CB: Hysterosalpingography with color Doppler ultrasonography. Am. J. Obstet. 164 (1991) 1530–1532
52 Peters JA, Stern JJ, Coulam CB: Color Doppler Hysterosalpingography. In Jaffe R, Warsof SL (eds): Color Doppler in Obstetrics and Gynecology. Mc Graw Hill, New York 1992, 283
53 Randolph JR, Ying YK, Maier DB, Schmidt CL, Ridelick DH: Comparison of real time ultrasonography, laparoscopy and hysteroscopy in the evaluation of uterine abnormalities and tubal patency. Fertil. Steril. 46 (1986) 828–830
54 Richman TS, Bisconi GN, de Cherney A, Polan ML, Alcebo LO: Falllopian tubal patency assessed by ultrasound following fluid injection. Radiology 152 (1984) 502–504
55 Riddle AF, Sharma V, Mason BA et al.: Two year's experience of ultrasound directed oocyte retrieval. Fertil. Steril. 48 (1987) 454–457
56 Rosch J, Thurmond A, Uchida B, Sovak M: Selective transcervical fallopian tube catheterization: technique update. Radiology 168 (1988) 1–5
57 Rubin I: Differences between the uterus and tubes as a cause of oscillations recorded during uterotubal insufflation. Fertil. Steril. 5 (1954) 147–153
58 Schwimer SR, Rothman CM, Lebovic J, Oye DM: The effect of ultrasound coupling gels on sperm motility in vitro. Fertil. Steril. 42 (1984) 946–950
59 Segars JH, Herbert CM, Moore DE: Selective fallopian tube cannulation: initial experience in an infertile population. Fertil. Steril. 53 (1990) 357–359
60 Sharma V: Transabdominal oocyte recovery. In Chervenak FA, Isaacson GC, Campbell S (eds): Ultrasound in obstetric and gynecology. Little, Brown and Company, London 1993, 1379–1389
61 Society for Assisted Reproductive Technology, The American Fertility Society: Assisted reproductive technology in the United States and Canada: 1991 results from the Society for Assisted Reproductive Technology generated from the American Fertility Society Registry. Fertil. Steril. 59 (1993) 956–959

62 Speroff L, Glass RH, Kase N: Assisted reproduction. In Speroff L, Glass RH, Kase NG (eds.): Clinical gynecologic endocrinology and infertility (5th ed.). Williams and Wilkins, Baltimore 1994, 931–946
63 Stern J, Peters AJ, Coulam CB: Color Doppler ultrasonography assessment of tubal patency: a comparison study with traditional technique. Fertil. Steril. 58 (1992) 897–900
64 Sulak PI, Letterie GS, Coddington CC: Histology of proximal tubal occlusion. Fertil. Steril. 48 (1987) 437–440
65 Suren A, Puchta J, Osmers R: Fluid instillation into the uterine cavity. In Osmers R, Kurjak A (eds): Ultrasound and uterus. The Parthenon Publishing Group Ltd, London (1995)
66 Teisala K, Heinonen PK, Punnonen R: Transvaginal ultrasound in the diagnosis and treatment of tuboovarian abscess. Br. J. Obstet. Gynaecol. 97 (1990) 178–180
67 Thurmond AS, Patton PE, Hector DM, Jones MK: US-guided fallopian tube catherization. Radiology 180 (1991) 571–572
68 Thurmond A, Rosch J: Nonsurgical fallopian tube recanalization for treatment of infertility. Radiology 174 (1990) 371–374
69 Timor-Tritsch IE, Rottem S: Transvaginal ultrasonographic study of the fallopian tube. Obstet. Gynecol. 70 (1987) 424–428
70 Trounson OA, Leeton JF, Wood C, Webb J, Wood J: Pregnancies in humans by in vitro fertilization and embryo transfer in the controlled cycle. Science 212 (1981) 681–684
71 Tüfekci EC, Girit S, Bayirli MD, Durmusoglu F, Yalti S: Evaluation of tubal patency by transvaginal sonosalipingography. Fertil. Steril. 57 (1992) 336–340
72 Volpi E, De Grandis T, Sismondi P et al.: Transvaginal-sonography (TSSG) in the evaluation of tubal patency. Acta Eur. Fertil. 22 (1991) 325–328
73 Volpi E, Zuccaro A, Patriarca S, Rustichelli, Sismondi P: Transvaginal sonographic tubal patency testing air and saline solution as contrast media in a routine infertility clinic setting. Ultrasound Obstet. Gynecol. 7 (1996) 43–48
74 Waegemaekers CT, Berg-Helder A, Blankhart A, Naaktgeboren N: Transvaginal ovarian cyst puncture in the early follicular phase of an IVF cycle, indications and results. Hum. Reprod. 3 (suppl. 1) (1988) 80

Obstetric Ultrasound

10 Doppler Ultrasound Examinations from Folliculogenesis to Early Pregnancy

S. Kupesic, A. Kurjak, and W. Schmidt

Overview

The introduction of transvaginal color Doppler sonography has dramatically enhanced our understanding of early human development. This study permits the noninvasive assessment of blood flow in the corpus luteum and the changes in uterine blood flow that occur in association with pregnancy. For example, when corpus luteum blood flow was analyzed in normal and abnormal pregnancies, no difference was found between ectopic pregnancy and normal early pregnancy, but the resistance and pulsatility indices in cases of threatened abortion, incomplete abortion, and missed abortion were found to be significantly higher than in normal pregnancies.

In several studies, color Doppler imaging has shown a progressive decline in the resistance and pulsatility indices from the main branch of the uterine artery to the spiral arteries with advancing gestational age. All the vessels examined showed a decrease in flow resistance as the pregnancy progressed.

Transvaginal color Doppler ultrasound has provided a unique opportunity to study both the morphology and physiology of human life from conception to implantation. The follicular cells become luteinized after ovulation. Many biochemical, morphological, and circulatory changes take place during this transformation process, and a large proportion of these changes can be investigated by transvaginal color Doppler scanning. After fertilization of the oocyte, the embryo is transported into the uterine cavity, where it implants under favorable local and hormonal conditions and develops into a new individual. Transvaginal ultrasound with color Doppler imaging and blood flow analysis during these phases permits a detailed examination of the small vessels that supply the preovulatory follicle, corpus luteum, and endometrium.

Monitoring Folliculogenesis

The ovary and ovarian blood flow. Transvaginal ultrasound scanning also permits an increasingly precise visualization of ovarian morphology[8]. The addition of color Doppler ultrasound makes it easier to measure the sequential changes in the arteries that supply the ovaries during the course of the cycle[2,3,4,24,38]. The highest flow resistance is measured on the first day of the cycle, the lowest on the day of the LH (luteinizing hormone) peak[3,4].

Follicular blood flow. The surveillance of folliculogenesis with ultrasound has become a well-established modality in routine clinical practice. The blood flow at the periphery of the follicle can be clearly demonstrated with transvaginal color Doppler ultrasound[3,4]. Normally, a follicular flow velocity waveform can be recorded when the dominant follicle has reached a diameter of 10 mm. The resistance index (RI) is approximately 0.54 in the periovulatory period[24] (Fig. 10.**1**). The RI begins to decline two days before ovulation and reaches its lowest value when ovulation occurs (0.44 ± 0.04). Measurements taken shortly before ovulation typically show a sharp rise in the peak systolic blood flow velocity with a relatively constant resistance index. This observation may result from dilatation of the newly formed vessels between the vascularized theca cell layer and the hypoxic granulosa cell layer of the follicle[12]. Normal ovulation depends upon changes in the oxygen partial pressure and the effect of these changes on cellular function (Fig. 10.**2**). Transvaginal sonography with color Doppler imaging is becoming the method of choice for investigating the subtle blood flow changes that take place in the normally and abnormally functioning ovary (Fig. 10.**3**).

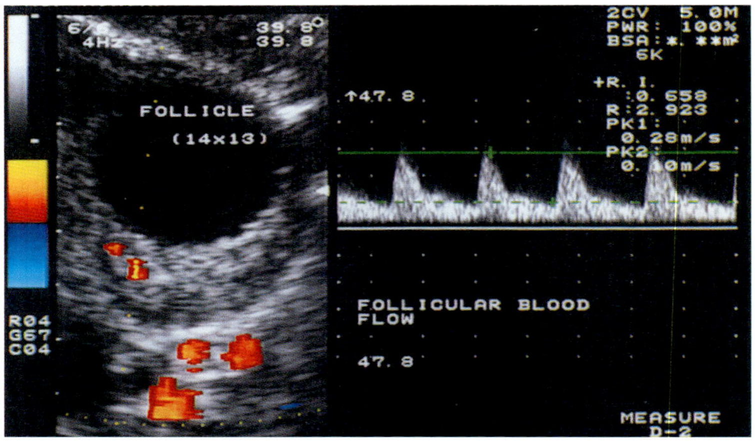

Fig. 10.**1** Transvaginal scan of the dominant follicle three days before ovulation indicates a moderate vascular resistance (RI = 0.65). Color Doppler ultrasound can define the small vessels that supply the follicle.

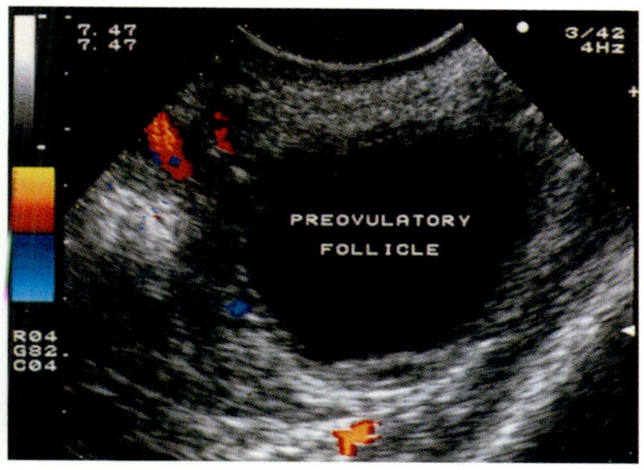

Fig. 10.**2** Triangular outline of the mature follicle.

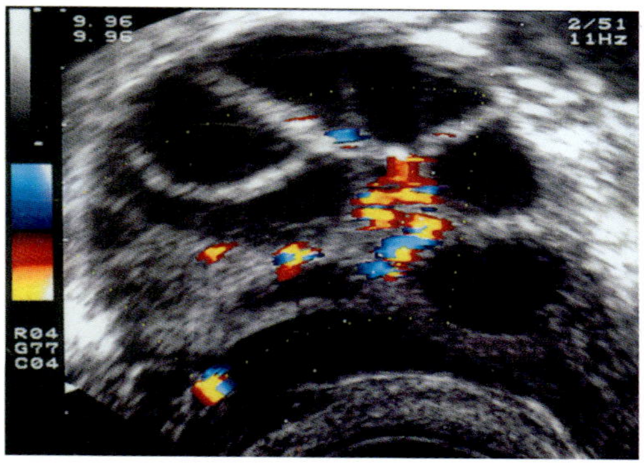

Fig. 10.**3** Image of an ovary with numerous follicles after prior ovulation induction with hCG. Each follicle is ringed by vascular structures that can be visualized with Doppler ultrasound.

Development of the Corpus luteum

Developmental stages. The cells of the follicular wall undergo structural and functional changes as they develop into the components of the corpus luteum[17]. Microscopically, the corpus luteum undergoes four developmental stages: proliferation (Fig. 10.**4**), vascularization, maturation, and regression. In the proliferative phase, the theca interna becomes infolded while its vascular channels become greatly dilated. Vessels sprout from the endothelium, penetrate the granulosa cell layer, and enter the hemorrhagic cavity of the ruptured follicle. The vascularization phase is characterized by a rapid organization of the blood-filled cavity of the ruptured follicle (Figs. 10.**5**, 10.**6**). With increasing maturation, the theca cells and the lutein cells derived from the granulosa acquire vacuoles and are physiologically active. The mature corpus luteum measures 1–3 cm in section and displays low resistance signals (average RI = 0.43). Regressive changes in the corpus luteum can be seen as early as day 23 of the cycle (Fig. 10.**7**). Typically these changes are marked by a decreased blood flow velocity and a

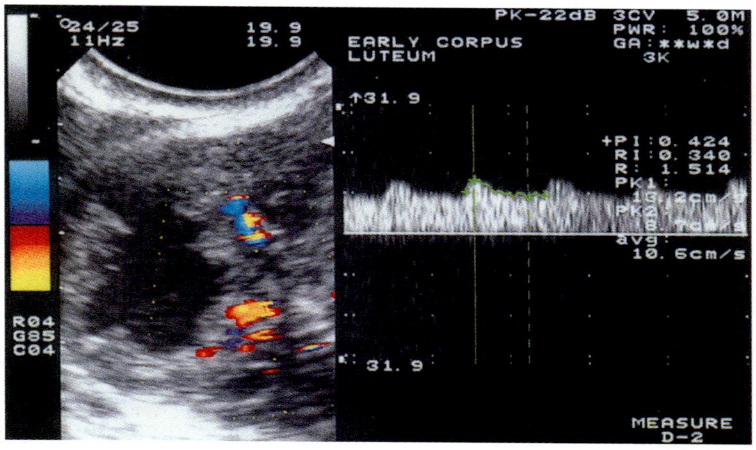

Fig. 10.**4** Early development of the corpus luteum in a transvaginal color Doppler scan. Note the dilated vessels at the periphery of the collapsed follicle (left). A decreased vascular resistance (RI = 0.40) is demonstrated (right).

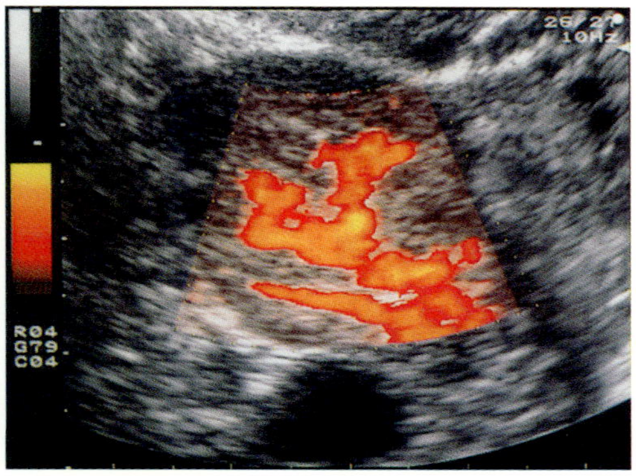

Fig. 10.**5** Transvaginal power Doppler image showing increased vascularization of the corpus luteum.

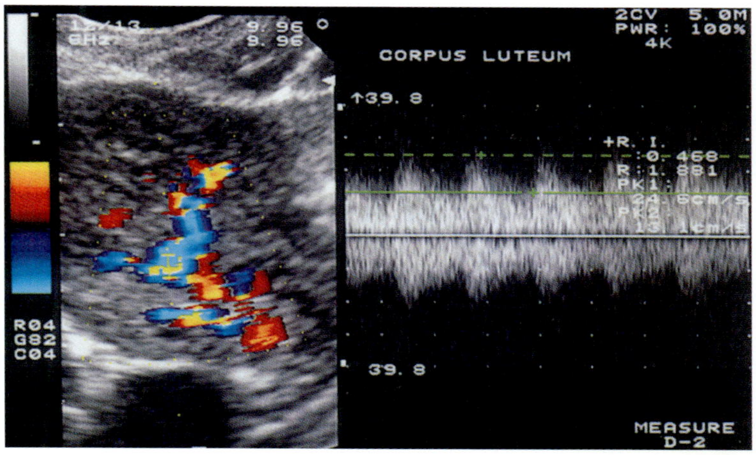

Fig. 10.**6** Typical blood flow pattern of the mature corpus luteum: high blood flow velocity and low vascular resistance (RI = 0.47).

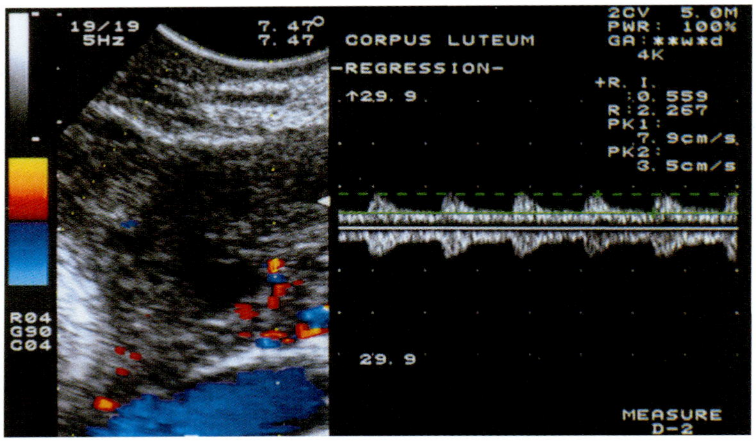

Fig. 10.**7** Transvaginal scan of the ovary in the late luteal phase (left). During the luteal regression phase, diastolic blood flow is decreased and the RI is increased (right).

rising resistance index (average RI = 0.49). If pregnancy occurs, the corpus luteum persists in response to hCG (human chorionic gonadotropin) synthesized by the trophoblasts and produces progesterone to support the developing embryo. The corpus luteum begins to regress after 10 weeks' gestation, at which time the placenta assumes the task of progesterone production. Regression of the corpus luteum is complete by the 16th week of gestation[29].

The use of Doppler ultrasound during these stages can supply interesting information[15,36]. The noninvasive Doppler assessment of blood flow velocity in the corpus luteum provides direct information on the arterial perfusion of the ovary[26,32]. Transvaginal color Doppler can demonstrate even small vessels in the ovary. The results are very precise and reproducible.

Changes in Endometrial Blood Flow

Uterine blood supply. Implantation and subsequent pregnancy in humans are dependent upon complex changes in the uterine artery and its branches in the myometrium and endometrium. The blood flow changes in the uterine vessels during a reproductive cycle are important for implantation and for the development of a uteroplacental blood supply. Transvaginal color Doppler gives a detailed view of the genital organs and of most blood flow changes that occur in these organs[19]. The color-flow signal from the uterine artery can be identified lateral to the cervix (Fig. 10.**8**). After passing through the outer third of the myometrium, the uterine arteries divide into the vascular plexus of the arcuate arteries, which are distributed within the uterus[18]. The radial arteries arise from this plexus and run toward the uterine cavity (Fig. 10.**9**). On reaching the junction of the myometrium and endometrium, they are called the spiral arteries. Transvaginal color Doppler is an efficient tool for analyzing the waveforms of the radial and spiral arteries.

It is well known that the endometrium undergoes profound structural and functional changes during the course of the cycle[9,11,18,33]. Demonstrable histological changes include the vigorous development of essential blood vessels, especially the spiral arteries, during the menstrual cycle (Fig. 10.**10**). The increased blood flow to the endometrium depends on blood flow changes that occur in the uterine arteries, in the arcuate plexus, and in the radial arteries.

Blood flow changes in the uterine arteries. Complex relationships appear to exist between the levels of circulating ovarian hormones in the peripheral venous blood and uterine blood flow parameters[13,27]. In most women, low end-diastolic blood flow is measured in the uterine arteries during the proliferative

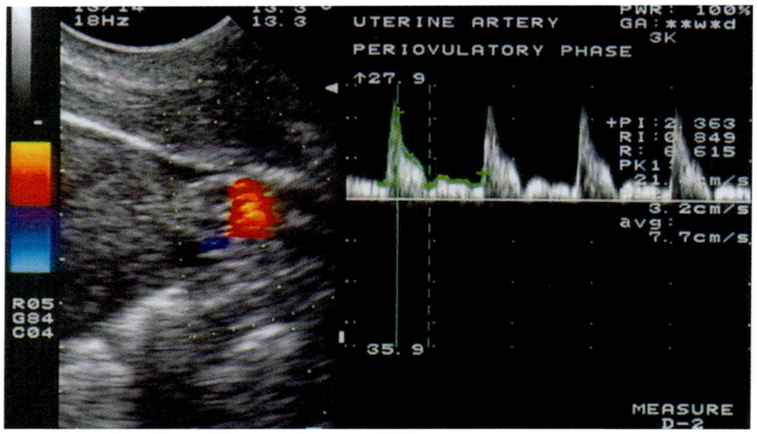

Fig. 10.**8** Scan of the uterine artery lateral to the cervix at the level of the corpus–cervix junction (left). The end-diastolic flow velocity in the uterine artery is decreased before ovulation, with RI = 0.87 (right).

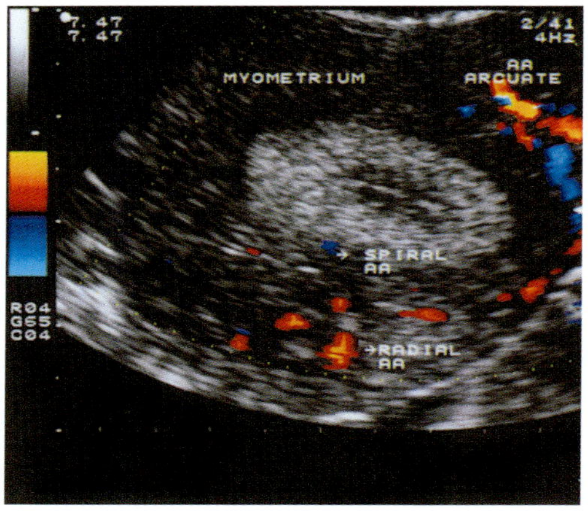

Fig. 10.**9** The radial arteries can be clearly visualized with Doppler ultrasound.

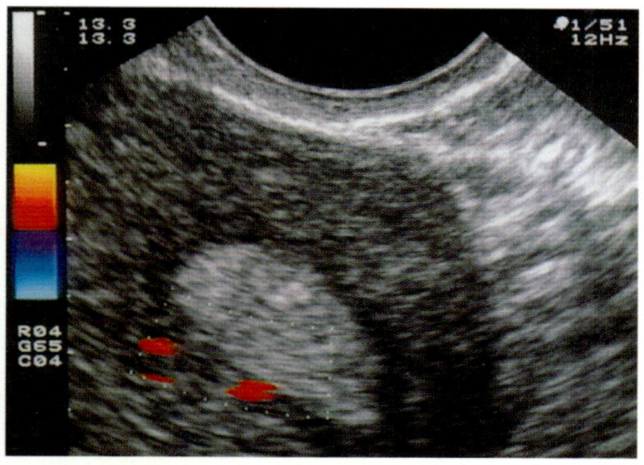

Fig. 10.**10** The detection of color-flow signals from the spiral arteries at the periphery of the hyperechoic endometrium during the luteal phase is suggestive of normal uterine receptivity.

phase[29]. The resistance index of the uterine arteries during this phase is 0.88 ± 0.04 and begins to fall on the day before ovulation. In a normal ovulatory menstrual cycle, the end-diastolic blood flow velocity shows a sharp rise between the proliferative and secretory phases. The lowest resistance index (RI = 0.84 ± 0.04) is measured on day 18 of the cycle, and this remains constant until the end of the cycle. Because the changes in blood flow velocity begin before ovulation, it is reasonable to assume that they are related both to angiogenesis and to hormonal factors. Some authors have noted the absence of end-diastolic flow in the uterine arteries of infertility patients and have related it to low fertility or infertility[13, 14, 29, 35]. However, this finding could also be a variant that disappears with the onset of pregnancy[29]. The changes described above cannot be demonstrated in anovulatory cycles.

Blood flow changes in the radial and spiral arteries. With transvaginal color Doppler sonography, it is possible to investigate blood flow changes that occur in the radial and spiral arteries under normal and abnormal conditions. A fall in the resistance and pulsatility indices across the uterine artery toward the spiral arteries can be detected in the uterus of nonpregnant women. The RI in the radial arteries during the proliferative phase is 0.78 ± 0.10. This index begins to fall significantly at midcycle and reaches its lowest value in the midluteal phase (0.68 ± 0.04). The RI in the spiral arteries is 0.54 ± 0.03 on the day before ovulation (Fig. 10.**11**) and reaches its lowest value (0.49 ± 0.05) approximately between days 16 and 18 of the cycle (Fig. 10.**12**). The changes in the blood flow parameters of the radial and spiral arteries correspond to the blood flow changes that are measured in the uterine arteries.

Therapeutic implications. Transvaginal color Doppler sonography appears to provide a means of predicting implantation rates, investigating previously unexplained infertility problems, and referring patients with impaired myometrial and endometrial blood flow for appropriate therapy. Another logical application of this technique is in testing the effects of various medications on uterine blood flow[21]. A knowledge of the changes in endometrial blood flow could play an important role in determining the optimum timing for implantation and embryo transfer. In this regard, transvaginal color Doppler can provide a tool for the noninvasive assessment of uterine receptivity. This will enable clinicians to cryopreserve embryos when uterine conditions are poor and transfer a smaller number of embryos when the conditions are optimal (Figs. 10.**13**, 10.**14**).

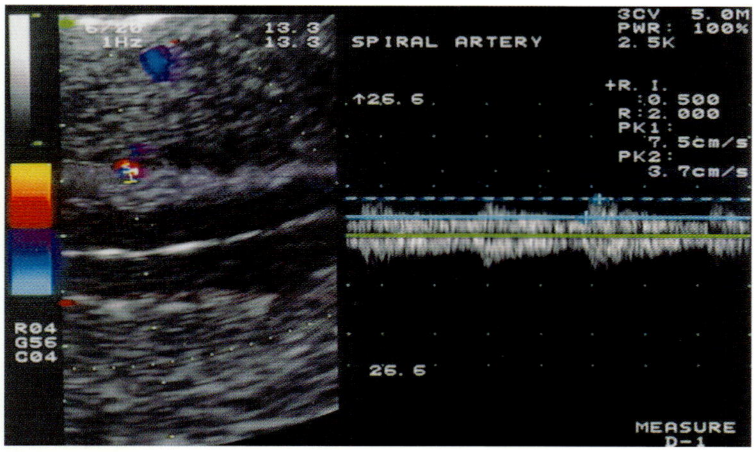

Fig. 10.**11** Spiral artery waveform in the periovulatory phase. The low RI (0.50) and increased blood flow velocity are characteristic of ovulation.

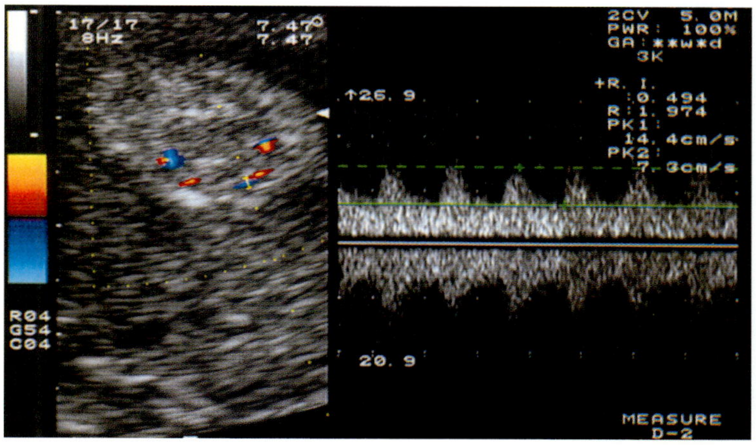

Fig. 10.**12** Transvaginal scan demonstrates a thickened endometrium. The color areas represent the spiral arteries (left). Analysis of the pulsed Doppler spectrum (right) shows a low to moderate vascular resistance, which is characteristic of normal endometrial receptivity.

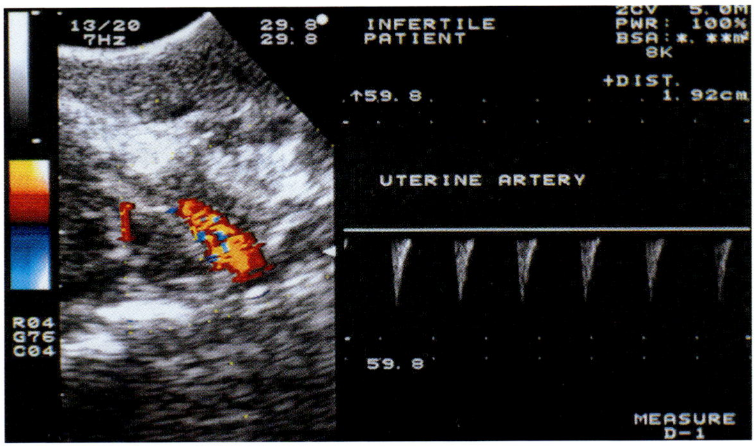

Fig. 10.**13** Uterine artery waveform in an infertile patient (left). Note the absence of diastolic blood flow (RI = 1.0) (right).

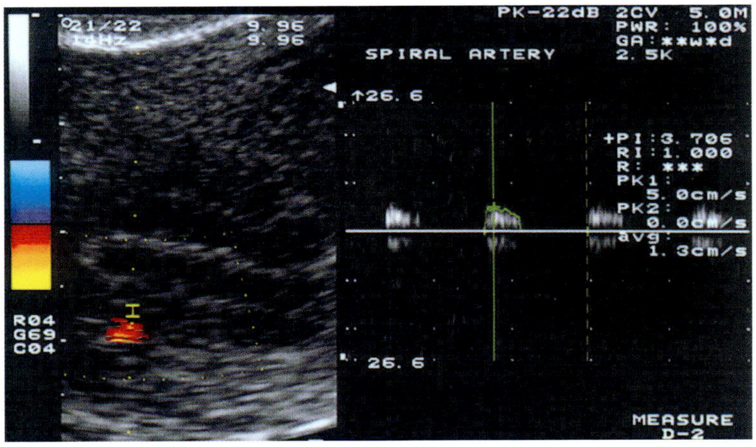

Fig. 10.**14** Absent diastolic flow in the uterine artery on the predicted day of ovulation (right). This finding may occur in association with low fertility or infertility.

Luteal Blood Flow in Normal and Abnormal Pregnancies

A total of 181 pregnancies were studied: 127 with a normal course of pregnancy and 54 with an abnormal course. The course of pregnancy in both groups was routinely evaluated with ultrasound. All of the patients had regular cycles before their pregnancy.

Material and methods. The uterus, gestational sac, embryo, and ovaries were examined with transvaginal B-mode ultrasound. The pulsed Doppler signals were acquired in a 2 mm wide sample volume using the Aloka color Doppler SSD-680 and 2000 systems (Aloka Corporation, Tokyo, Japan) with a 5 MHz transducer. Blood flows were recorded in 90° sectors at a scanning depth of 2–20 cm. The pulse repetition frequency was between 2 and 32 kHz, and 100 Hz wall filters were used to eliminate unwanted low-frequency signals. The maximum energy density was 80 mW/cm^2, which is below the FDA-prescribed limit for fetal ultrasound. The waveforms of all the sampled vessels were determined twice by each examiner. Three different cardiac cycles were recorded in each run, and the RI was computed. The average length of the examination was less than 20 minutes.

Results. None of the 127 women in the uncomplicated pregnancy group had an RI higher than 0.5 in the 5th– 12th week of gestation. The resistance and pulsatility indices showed no statistically significant changes between weeks 6 and 12. The disappearance of luteal blood flow after 12 weeks is attributed to regression of the luteal vessels. Studies of the intraovarian blood supply raise the question whether there are factors other than hCG that sustain corpus luteum function[1, 5, 10, 34, 37]. Although the hCG value in nonpregnant women is negative and is of no significance, luteal blood flow is present throughout the second half of the cycle. Kratzer et al.[20] found that corpus luteum function in early pregnancy was dependent on the rate of the hCG increase. The authors also found that this rate of increase was lower in patients with ectopic pregnancy and spontaneous abortion than in women with a normal course of pregnancy. The resistance and pulsatility indices during the luteal phase follow the same pattern in patients with ectopic pregnancy as in nonpregnant women[32]. Luteal blood flow in ectopic pregnancy corresponds to that in normal early pregnancy ($p > 0.05$). Luteal blood flow during the first trimester of pregnancy is comparable to that in a nonpregnant woman during the luteal phase. In cases of inevitable abortion, imminent abortion or missed abortion, the resistance and pulsatility indices are significantly increased relative to the normal population ($p < 0.001$). Zalud and Kurjak[38] investigated luteal blood flow in the ovary of nonpregnant women and in ectopic and intrauterine pregnancies. The RI of luteal blood flow was lowest in the nonpregnant women (0.42 ± 0.12) and highest in the women with an intrauterine pregnancy (0.53 ± 0.09). The RI in ectopic pregnancies was 0.48 ± 0.07. Luteal blood flow was present on the affected side in 86.4 % of the ectopic pregnancies.

Implications. It appears that luteal blood flow has prognostic significance in cases of threatened abortion. An elevated RI suggests a decreased chance of embryo survival, especially after eight weeks' gestation, when the critical period for survival is concluded. It seems reasonable to use luteal blood flow as a parameter for corpus luteum regression that can be correlated with other regulatory factors such as progesterone, relaxin, prostaglandin, and angiogenetic peptide growth hormone. Prostaglandin $F_{2\alpha}$ ($PGF_{2\alpha}$), also known as luteolysin[28], is produced locally by luteal cells during the regression phase. Because $PGF_{2\alpha}$ is a potent vasoconstrictor, it can increase the resistance in the corpus luteum vessels (reflected in the RI or PI). It remains unclear whether endothelial-derived relaxing factor (EDRF) also plays a role. It is also unclear whether vascular factors are important in luteal phase disturbances. Transvaginal scanning and pulsed Doppler sonography will certainly be of help in the further investigation of these questions.

Trophoblastic Invasion and Development of the Placenta

Trophoblastic invasion. After the blastocyst has established contact with the endometrium, the trophoblast penetrates deep into the endometrium and first encounters dilated venous vessels, followed by superficial arterioles and finally, in the 4th week of gestation, the spiral arteries[18, 19]. Cytotrophoblastic cells reach the boundary between the decidua and myometrium between the 8th and 12th weeks of gestation[31]. The trophoblastic invasion of the myometrium continues until the 18th week[7]. The spiral arteries undergo a morphological transformation in early pregnancy as the trophoblast partially disrupts the vascular structure, causing hypertrophy of the smooth-muscle cells of the media and thickening of the endothelium[7, 30].

Blood flow changes. The erosion of the spiral arteries and altered steroid hormone levels in the plasma cause a significant fall of uterine vascular resistance. It is common to detect low vascular resistance and turbulence, especially near the placenta[22]. Jaffe and Warsof[16] observed Doppler waveform changes in connection with placentation as early as the 5th week of gestation.

Kurjak et al.[23] described the occurrence of Doppler changes in early pregnancy even before a gestational sac could be identified. These changes consist of low peak systolic flow velocities in the periphery of the endometrium. Larger maternal vessels are eroded as the placenta continues to develop, with Doppler ultrasound showing higher blood flow velocities with an increased diastolic component.

Outlook. The uteroplacental vascular supply and blood flow during pregnancy have been the subject of extensive research[6, 25]. Color Doppler sonography, as an adjunct to transvaginal B-mode imaging, is broadening our knowledge of the functional aspects of fetal morphology and physiology. Normal development of the embryo depends upon chromosomal structure, implantation mechanism, and uterine perfusion. Poor implantation conditions and a deficient blood supply can be detected noninvasively by Doppler scanning. In the future, transvaginal color Doppler sonography could become the procedure of choice in threatened abortion cases for differentiating among the various causes of abnormal hemodynamic changes.

References

1 Adams EC, Hertig AT: Studies on the human corpus luteum I and II. Observations on the ultrastructure of luteal cells during pregnancy. J. Cell. Biol. 41 (1969) 496–498

2 Baber RJ, McSweeney MB, Gill RW et al.: Transvaginal pulsed Doppler ultrasound assessment of blood flow to the corpus luteum in IVF patients following ET. Brit. J. Obstet. Gynaecol. 95 (1988) 1226–1230

3 Bourne T, Jurkovic D, Waterstone J, Campbell S, Collins WP: Intrafollicular blood flow during human ovulation. Ultrasound. Obstet. Gynecol. 1 (1991) 53–59

4 Collins WP, Jurkovic D, Bourne TH, Kurjak A, Campbell S: Ovarian morphology, endocrine function and intrafollicular blood flow during peri-ovulatory period. Hum. Reprod. 6 (1991) 319–324

5 Csapo A, Pulkkinen M: Indispensability of the human corpus luteum in the maintenance of early pregnancy. Obstet. Gynecol. Survey 33 (1978) 69–81

6 Deutinger J, Rudelstorfer R, Bernarschek G: Vaginosonographic velocimetry of both main uterine arteries by visual vessel recognition and pulsed Doppler method during pregnancy. Amer. J. Obstet. Gynecol. 159 (1988) 1072–1074

7 De Wolf FC, De Wolf-Peeters C, Brosens I, Robertson WB: The human placental bed: electron microscopic study of trophoblastic invasion of spiral arteries. Amer. J. Obstet. Gynecol. 137 (1979) 58–70

8 Feichtinger W, Kemeter P: Transvaginal sector sonography for needle guided transvaginal follicle aspiration and other applications in gynecologic routine and research. Fertil. Steril. 45 (1989) 722–725

9 Fleischer A, Herbert CM, Hill GA, Keppler DM, Worrell JA: Transvaginal sonography of the endometrium during induced cycles. J. Ultrasound. Med. 10 (1991) 93–95

10 Flint APF, Scheldric EL: Ovarian peptides and luteolysis. In Edwards RG, Purdy JM, Streptoe PC (eds): Implantation of human embryo. Academic, London 1985

11 Glissant A, De Mouzon J, Frydman R: Ultrasound study of the endometrium during in vitro fertilization cycles. Fertil. Steril. 44 (1985) 786–790

12 Gosden RG, Byat-Smith JG: Oxygen concentration across the follicular epithelium: model, prediction and implications. Hum. Reprod. 1 (1986) 65–68

13 Goswamy RK, Steptoe PC: Doppler ultrasound studies of the uterine in spontaneous ovarian cycles. Hum. Reprod. 3 (1988) 721–724

14 Goswamy RK, Williams SG, Steptoe PC: Decreased uterine perfusion—a cause of infertility. Hum. Reprod. 3 (1988) 955–959

15 Hata K, Hata T, Senot D et al.: Change in ovarian arterial compliance during the human menstrual cycle assessed by Doppler ultrasound. Brit. J. Obstet. Gynecol. 97 (1990) 163–166

16 Jaffe R, Warsof SL: Transvaginal color Doppler imaging in the assessment of uteroplacental blood flow in the normal first trimester of pregnancy. Amer. J. Obstet. Gynecol. 164 (1991) 781–785

17 Jaffe R, Warsof LS (ed.): Color Doppler imaging in Obstetrics and Gynecology. McGraw Hill, New York 1992

18 Jauniaux E, Jurkovic D, Campbell S: In vivo investigation of the anatomy and the physiology of early human placental circulations. Ultrasound. Obstet. Gynecol. 1 (1991) 435–445

19 Jurkovic D, Jauniaux E, Kurjak A, Hustin J, Campbell S, Nicolaides KH: Transvaginal color Doppler assessment of utero-placental circulation in early pregnancy. Obstet. Gynecol. 77 (1991) 365–369

20 Kratzer PG, Taylor RN: Corpus luteum function in early pregnancies is primarily determined by the rate of change of human chorionic gonadotropin levels. Amer. J. Obstet. Gynecol. 163 (1990) 1497–1502

21 Kupesic S, Kurjak A: Uterine and ovarian perfusion during the preovulatory period assissted by transvaginal color Doppler. Fertil. Steril. 60 (1993) 439–443

22 Kurjak A (ed.): Transvaginal color Doppler. Parthenon Publishing, Carnworth 1991
23 Kurjak A, Kupesic-Urek S, Predanic M et al.: Transvaginal color Doppler in the study of early pregnancies associated with fibroids. J. Matern. Fetal. Invest. 2 (1992) 81–83
24 Kurjak A, Kupesic-Urek S, Schulman H, Zalud I: Transvaginal color flow Doppler in the assessment of ovarian and uterine blood flow in infertile women. Fertil. Steril. 56 (1991) 870–873
25 Kurjak A, Miljan M, Zalud I: Transabdominal and transvaginal color Doppler in the assessment of fetomaternal circulation during all three trimesters of pregnancy. Eur. J. Obstet. Gynecol. Reprod. Biol. 36 (1990) 240–246
26 Kurjak A, Zalud I, Jurkovic D, Alfirevic Z, Miljan M: Transvaginal color Doppler in the assessment of pelvic circulation. Acta Obstet. Gynecol. Scand. 68 (1989) 131–135
27 Long MC, Boultbee JE, Hanson ME, Begent RHJ: Doppler time velocity waveform studies of the uterine artery and uterus. Brit. J. Obstet. Gynecol. 96 (1989) 588–593
28 Nett TM, Niswander GD: Luteal blood flow and receptors for LH during PGF2-alpha induced luteolysis: productionof PGE2 and PGF2-alpha during early pregnancy. Acta. Vet. Scand. 77 (1981) 117–130
29 Nyberg DA, Hill LM, Bohm-Velez M, Mendelson EB (eds.): Transvaginal ultrasound. Mosby YearBook, St. Louis 1992
30 Pijnenborg R, Bland JM, Robertson WB, Brosens I: Uteroplacental arterial changes related to interstitial trophoblast migration in early human pregnancy. Placenta 4 (1983) 397–414
31 Pijnenborg R, Bland JM, Robertson WB, Dixon G, Brosens I: The pattern of interstitial invasion of the myometrium in early human pregnancy. Placenta 3 (1990) 19–21
32 Salim A, Kurjak A, Zalud I: Ovarian luteal flow in normal and abnormal early pregnancies. J. Matern. Fetal. Invest. 2 (1992) 119–124
33 Smith B, Porter R, Ahuja K, Craft I: Ultrasonic assessment of changes in stimulated cycles in in vitro fertilization and ET program. J. In. Vitro. Fert. Embryo. Transf 1 (1984) 233–238
34 Soules MR, Bremner WJ, Dahl KD, Rivier JE, Vale W, Clifton DK: The induction of premature luteolysis in normal women-follicular phase luteinizing hormone secretion and corpus luteum function in the subsequent cycle. Amer. J. Obstet. Gynecol. 164 (1991) 989–996
35 Steer V, Mulls CL, Campbell S: Vaginal colour Doppler assessment on the day of embryo transfer accurately predicts patients in an in vitro fertilization programe with suboptimal uterine perfusion who fail to become pregnant. Ultrasound. Obstet. Gynecol. 1 (1991) 79–80
36 Taylor KWJ, Burns PN, Wells PNE, Conway DI, Hull MGR: Ultrasound Doppler flow studies of the ovarian and uterine arteries. Brit. J. Obstet. Gynecol. 92 (1985) 240–246
37 Tulsky AS, Koff AK: Some observations on the role of corpus luteum in early human pregnancy. Fertil. Steril. 8 (1957) 118–121
38 Zalud I, Kurjak A: The assessment of luteal blood flow in pregnant and nonpregnant women by transvaginal color Doppler. J. Perinat. Med. 18 (1990) 215–221

11 Evaluation of Early Placentation and the Embryonic Circulation with Doppler Ultrasound

A. Kurjak and S. Kupesic

Implantation

There is a brief period after ovulation in which the endometrium is most receptive to implantation. During this time, a blastocyst entering the uterine cavity can establish contact with the endothelial surface and become implanted[41]. The optimum time for implantation is between the fifth and seventh day after ovulation. Thus, implantation of the blastocyst occurs on about day 21 of the cycle and is complete by day 26, when the endometrial membrane completely envelops the blastocyst. When implantation commences, the inner cells of the blastocyst face the endometrium[41]. The action of proteolytic enzymes produced by the syncytiotrophoblast causes erosion and invasion of the uterine mucosa. As implantation continues, the trophoblast erodes adjacent maternal capillaries, allowing maternal blood to come into direct contact with the embryo. Finally this lacunar system gives rise to the intervillous space of the placenta.

Development of the Intervillous Circulation

Classic Theory

Transformation of the spiral arteries. During the 4th week of gestation, the trophoblast invades the uterine wall and gradually penetrates into larger venous sinusoids and superficial arterioles. The extravillous cytotrophoblastic cells penetrate to the lumina of the spiral arteries. They transform the thick-walled arteries surrounded by smooth-muscle cells into flaccid, saclike uteroplacental vessels, which can accommodate the surge of maternal blood flow through passive dilation. The increased blood flow is necessary to supply the fetus with oxygen and support its growth[3] (Fig. 11.**1**).

Cells of the trophoblast can be found in the spiral arteries starting at about five weeks after fertilization (Fig. 11.**2**). The destructive action of the trophoblast on the muscle cells and elastic fibers of the spiral arteries has two effects:

1. The increasing blood flow causes a progressive expansion of the vessels. They are transformed into uteroplacental arteries that can accommodate the burgeoning blood supply.
2. The uteroplacental arteries are not subject to control by the autonomic nervous system[17].

Connection of the spiral arteries to the intervillous space. During the second month of gestation, the intervillous space becomes markedly larger owing to arborization of the villi. At this time, nests of cytotrophoblastic cells are found in the tips of many spiral arteries near the intervillous space. Meanwhile, numerous anastomoses form among the veins of the decidua. After 40 days (crown–rump length [CRL] = 15 mm), the spiral arteries communicate through direct openings with the intervillous space, and cytotrophoblastic cells begin to line the arterial lumina. Maternal blood reaches the intervillous space through gaps between the cells of the endovascular tropho-

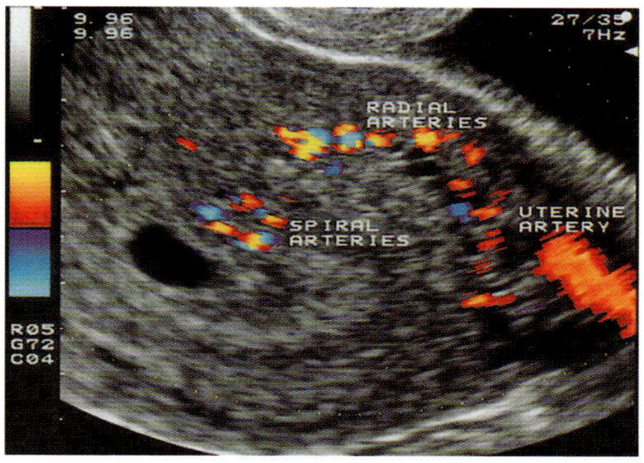

Fig. 11.**1** Transvaginal ultrasound scan of an early gestational sac. Note the eccentric position of the sac, its location in the fundus, its elliptical shape, and its double contours. Color Doppler provides an excellent view of the uterine vessels.

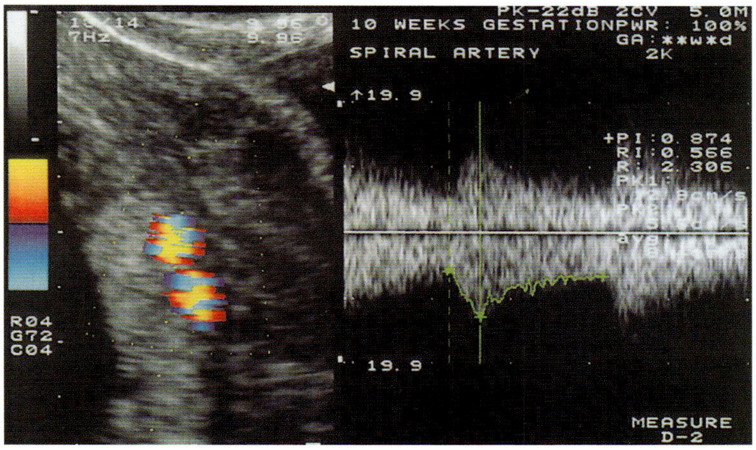

Fig. 11.**2** Transvaginal scan of an early gestational sac. The color-encoded area represents the spiral arteries (left). The Doppler spectrum (right) indicates a high flow velocity and low impedance.

blasts. The fact that the cytotrophoblastic cell columns are not flushed from the lumen of the spiral arteries suggests that the blood pressure is not high.

During the third month of gestation, the cytotrophoblastic cell columns completely occlude the tips of most of the spiral arteries, so that the spiral arteries do not terminate freely in the intervillous space. Later in this stage, the cell columns are more loosely arranged and apparently pose less of an obstacle to maternal blood entering the intervillous space.

Definitive placenta. By the end of the fourth month of gestation, the definitive placenta has developed from the chorion frondosum. The peripheral villous trees of the chorion, which are connected to the decidua capsularis, degenerate and the associated intervillous space disappears. The smooth, avascular chorion laeve[17] is formed by fusion of the chorionic plate and basal plate.

Trophoblastic infiltration of the myometrium occurs between the 8th and 18th weeks of gestation. The endovascular cytotrophoblast partially replaces the endothelium of the myometrial vessels and invades the smooth-muscle cells of these vessels. This results in increasing expansion of the radial arteries located in the myometrium.

Uteroplacental circulation. The uteroplacental circulation is a low-pressure system because the diameters of the vessels steadily increase on their way to the intervillous space. To keep the normal arterial pressure from being transmitted to the intervillous space, a considerable pressure drop occurs between the proximal, nondilated portion of the uteroplacental arterioles and the distal, dilated portion.

Objections and Alternative Theories

Opening of the spiral arteries in the 12th week of gestation. The classic theory on the origin of the uteroplacental circulation described above[4,43] was challenged for a time by the results of Hustin and Shaaps[6,7]. Their in-vivo studies of the placenta using transvaginal sonography, intervillous hysteroscopy, and phase-contrast examination of chorion villus sampling material showed that there is no true continuous blood flow in the intervillous space during the first 12 weeks of pregnancy.

> The authors studied tomograms of hysterectomy specimens from an in-situ pregnancy that were taken in the 7th, 8th and 9th weeks of pregnancy and found no contrast medium in the intervillous space. When this study was done in the 13th week of pregnancy, the placenta filled rapidly with contrast medium. Histological examinations of these hysterectomy specimens showed occlusion of the uteroplacental arteries by trophoblastic cells until the 12th week of gestation. Serial sections of spiral arteries also indicated an absence of intervillous blood flow before the 12th week. On the other hand, no trophoblastic plugs were found in the uteroplacental arteries in the 13th week, and contrast medium was found surrounding the villi in the intervillous space.

These results suggested that mainly fluid from the maternal plasma and uterine glandular secretions enter the early placenta on the maternal side. The authors assumed that intervillous blood circulation was absent or rudimentary during the first 12 weeks of gestation. According to this theory, the expansile changes in the spiral arteries take place throughout the first trimester. Finally, by about the 12th week of gestation, all of the trophoblastic cell columns in the spiral arteries are loosely arranged and displaced. This process gives the maternal blood free access to the intervillous space, clearing the way for a fully developed placental blood flow.

P_{O_2} values in the placenta. This theory was supported by a more recent study[45] using a polarographic oxygen electrode that was introduced under ultrasound guidance. This study showed that the Po_2 values in the placenta were significantly lower than in the endometrium between the 8th and 10th week of gestation, while these values were similar between the 12th and 13th week. The intraplacental P_{O_2} values rose significantly between the 8th–10th and 12th–13th weeks. These findings suggest that the rise of the placental P_{O_2} is related to the development of continuous maternal blood flow in the intervillous space at the end of the first trimester.

Color Doppler Studies

The introduction and evolution of transvaginal color Doppler sonography has made all portions of the embryonic/fetal and uteroplacental circulation accessible to hemodynamic evaluation in vivo[22,23,27,28,30–33]. Researchers have been able to focus their attention directly on the developing embryo, greatly expanding our knowledge and understanding of the embryonic circulation. Current knowledge on the anatomy and physiology of the uteroplacental blood supply is based largely on the classic early studies on this subject. We shall therefore review chronologically the theories on the origin of the intervillous circulation that have been derived from Doppler ultrasound studies.

Intraplacental flow. In 1991 and 1992, Jauniaux et al.[10,12] and Jaffe and Warsof[9] were unable to detect "intraplacental" flow before the 12th week of gestation with transvaginal color Doppler ultrasound. They did detect intraplacental flow by the 14th week of gestation, accompanied by the appearance of pandiastolic flow in the umbilical artery and a sudden rise of peak systolic flow velocity in the uterine artery. Building on the theories of Hustin and Shapps[6,7], they postulated that the simultaneous occurrence of these three phenomena could be explained by the disappearance of the trophoblastic plugs from the spiral arteries.

However, Kurjak et al.[26] detected no abrupt change in the uteroplacental circulation between the 12th and 14th week of gestation.

When a new generation of much more sensitive color Doppler scanners became available, several authors reported on the existence of intervillous blood flow during the first trimester of pregnancy (Fig. 11.**3**).

Pulsatile and continuous flow. In 1995, Kurjak et al.[25] published the first study on the combined assessment of intervillous blood flow using Doppler ultrasound and pathomorphological analysis. Two types of intervillous flow were detected in all

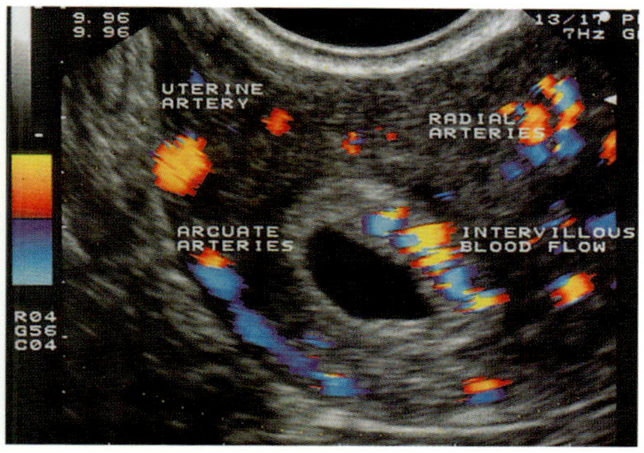

Fig. 11.**3** Color Doppler permits the simultaneous visualization of uterine, embryonic, and intervillous blood flow.

patients scanned with an endovaginal color Doppler probe: pulsatile, artery-like flow (Fig. 11.**4**) and continuous, veinlike flow (Fig. 11.**5**). In a parallel histological study, there was no time at which the spiral artery lumina were completely occluded by trophoblastic cells. These results suggest that the development of the intervillous circulation is more a continuous process than an abrupt event at the end of the first trimester.

Passage of red blood cells. Subsequently, several other groups of authors published similar results. Valentin et al.[46] did a study on uteroplacental and luteal flow combined with a pathomorphological analysis. Color Doppler measurements indicated the presence of intervillous blood flow starting as early as week 6 of a normal pregnancy. They also detected and measured the two types of Doppler signals (pulsatile and continuous flow) in more than 90% of 66 pregnancies from the 5th to the 11th gestational weeks. The authors claimed that the high blood flow velocities observed in subchorionic arteries were at odds with the concept that these arteries are completely occluded by trophoblastic plugs. On pathomorphological analysis, it was found that trophoblastic cells did not completely occlude the spiral arteries but allowed red blood cells to pass through. This led the authors to conclude that intervillous flow was present as early as the first trimester.

Merce et al.[36] published similar results based on a study of 108 normal singleton pregnancies from the 4th to the15th gestational weeks. They were able to detect intervillous blood flow as early as the 5th week (+ 6 days). This flow produced a slightly undulating, veinlike signal that tended toward higher velocities over the course of the first trimester. They also documented arterial signals in the retrochorionic segments of the uteroplacental vessels. They concluded that their results were consistent with the classic embryological concept that intervillous flow develops between the 4th and 7th weeks of gestation. According to Merce et al.[36], the uteroplacental circulation undergoes pronounced changes starting in the 4th week of pregnancy. Intervillous blood circulation and initial umbilical blood flow were detected from the 5th week onward.

Comparison with nonhuman primates. Experiments in primates, especially monkeys, have been a crucial part of research on the development of the placenta and the uteroplacental circulation. The classic study by Elisabeth Ramsey on blood circulation in the intervillous space of the primate placenta is the basis for all present-day research in this area[43,44]. Nimrod et al.[38,39] more recently reported on the assessment of early uteroplacental blood flow in the cynomolgus monkey (*Macaca fascicularis*) using color and Doppler sonography. They detected intervillous blood circulation starting on the 18th day after conception.

Despite the known differences between humans and monkeys in the depth of trophoblastic invasion of the spiral arteries, this result can be considered further evidence for the early development of intervillous blood circulation in all primate placentas, although the "analogy" argument should be applied with caution.

Normal and abnormal early pregnancies. In recently published studies, Kurjak et al. 20, 21) reported on their analysis of 60 normal pregnancies between the 6th and 12th week and 34 abnormal early pregnancies (22 missed abortions and 12 anembryonic pregnancies) between the 7th and 12th weeks. The same Doppler signals were found in the intervillous space in all the pregnancies, i.e., pulsatile artery-like signals with characteristic waveform peaks and continuous veinlike signals. The Doppler findings were similar between the women with missed abortion and the women with nor-

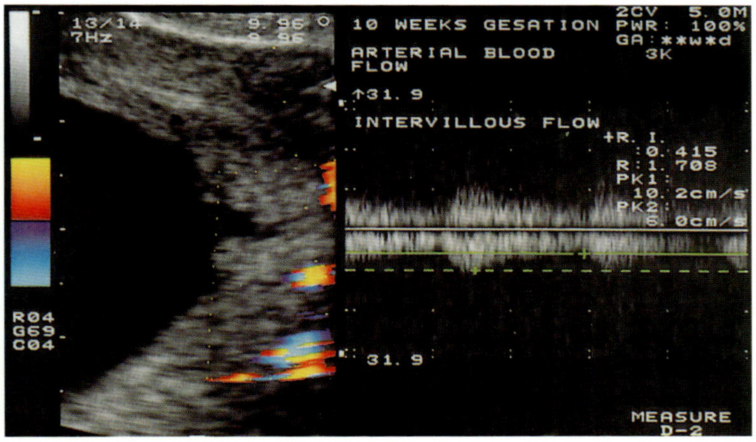

Fig. 11.**4** Pulsatile, artery-like blood flow pattern in the intervillous space (left) is characterized by a low vascular resistance (RI = 0.42).

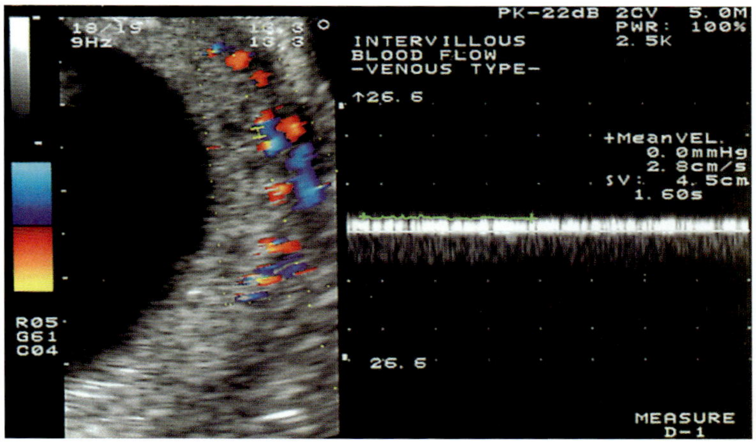

Fig. 11.**5** Continuous venous blood flow (right) is another flow pattern that can be clearly identified in the intervillous space (left).

mal pregnancies, but lower vascular resistance (as measured by the RI and PI) was found in the women with anembryonic pregnancies. These results contrast markedly with those of Jauniaux et al.[16], who found increased intervillous blood flow in 70% of all abnormal pregnancies before the 12th week. Histopathological analysis in these cases showed thinning and fragmentation of the trophoblastic shell and massive infiltration of the intervillous space by maternal blood.

The authors hypothesized that the trophoblastic plugs in the spiral arteries keep maternal blood from entering the intervillous space, thereby protecting the vulnerable villi from the high arterial pressure. According to this hypothesis, the premature entry of maternal blood into the intervillous space can disrupt the interface between maternal and embryonic tissue, leading to separation of the early placenta and possible abortion. However, this hypothesis does not exclude the presence of continuous intervillous blood flow in the first trimester. It appears that there are some areas in the spiral arteries in which the trophoblastic plugs are loosely arranged and allow intervillous blood circulation to occur. Initially there are only a few areas in which this blood flow delivers sufficient oxygen and nutrients to sustain the pregnancy. They are still accompanied by areas of restricted blood flow in which nutrients and oxygen diffuse through the intracellular fluid. The number of areas with a functioning intervillous circulation increases along with the size of the embryo and placenta in order to maintain a metabolic balance. This process concludes when the intervillous space of the mature placenta is completely formed. This hypothesis does not conflict with the concept that the entry of maternal blood into the intervillous space under an arterial pressure too high for this stage of gestation disrupts the materno-embryonic interface and may be the mechanism that precipitates abortion[16].

The use of transvaginal color and pulsed Doppler sonography provides exciting new insights into the functional development of the intervillous space, especially with regard to the blood supply and mechanisms of blood circulation.

Vascularization of the Yolk Sac and Vitelline Duct

Various ultrasound parameters such as the size of the gestational sac, embryonic growth, and the size of an intrauterine hematoma have been suggested as prognostic parameters for predicting the outcome of a pregnancy.

The secondary yolk sac is the earliest indication of the embryo. It can be seen in the gestational sac during the 5th week of pregnancy. According to Levi et al.[34], the yolk sac can be identified when the gestational sac has reached a size of 8 mm. When the outlines of the yolk sac are closely scrutinized, it should be possible to detect early embryonic heart activity in the 6th week of gestation. Between the 6th and 12th weeks of gestation, the diameter of the yolk sac gradually increases from 3.4 to 5.4 mm. If ultrasound demonstrates an abnormal size and morphology of the yolk sac, the outcome may be early pregnancy loss.

The vascular system begins to develop in the wall of the yolk sac approximately two weeks after ovulation. Because the yolk sac is the first vascular and hematopoietic organ of the embryo, our group investigated the vascularization of the secondary yolk sac and the vitelline duct by transvaginal sonography[22] within the framework of our research on the maternofetal circulation[19,31,32].

The patient group comprised 105 women between the 6th and 10th week of gestation. The first color and pulsed signals were recorded from the yolk sac between the 5th and 6th weeks. The highest detection rate, at 85.71%, was achieved in the 7th and 8th weeks (Fig. 11.**6**). All of the yolk sacs displayed a characteristic waveform marked by a low flow velocity (5.8 ± 1.7 cm/s) and the absence of diastolic flow. The PI had a mean value of 4.24 ± 0.94. While the functional activity of the yolk sac gradually declined, there was a parallel decrease in the detection rate from 78.26% in the 9th week to 61.11% in the 10th week. Color and pulsed Doppler signals could be recorded from the vitelline duct in 85.71% of the patients during the 7th week of gestation. The vessels of the vitelline duct showed similar values for peak systolic velocity and PI as the yolk sac. The detection rate of these vessels was highest (89.3%) during the 8th week of gestation. The process of vitelline duct elongation was accompanied by a declining detection rate during weeks 9 (73.9%) and 10 (55.6%).

Abnormal vascular development was found in the yolk sac and vitelline duct of pregnancies that ended in abortion. It may be that a veinlike signal pattern, irregular waveforms, or an increased diastolic component result from poor development of the embryo or even the absorption of embryonic remnants.

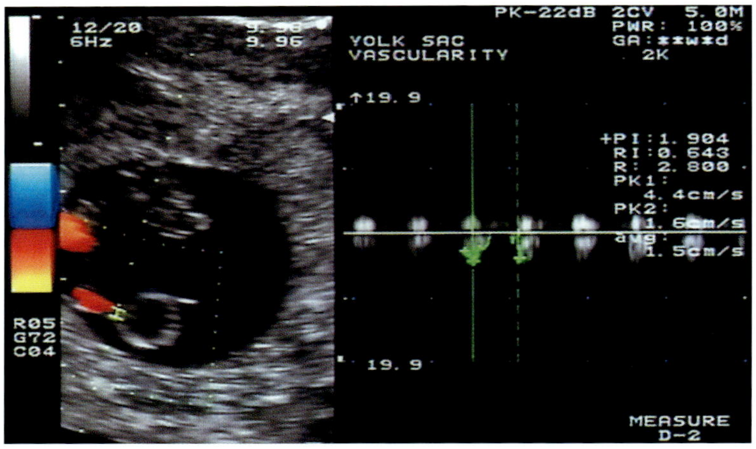

Fig. 11.**6** Vascularization of the yolk sac and yolk stalk. A low flow velocity and absent diastolic flow are typical of these structures.

Changes in Uterine Perfusion after Placentation

Anatomy. The maternal part of the uteroplacental circulation consists of the uterine arteries and their branches, which spread out in the uterus before they reach the deciduous portions of the placenta[18]. The uterine arteries arise from the iliac arteries. They run along the lateral pelvic wall before crossing the external iliac arteries and reaching the uterus at the level of the cervix. After giving off a cervical branch, they ascend in a tortuous course along the lateral wall of the uterine corpus and give off a branch that anastomoses with the ovarian artery. The uterine arteries then divide into a vascular plexus that surrounds the uterus. This plexus gives rise to smaller arteries, called the radial arteries, that run toward the lumen of the uterine cavity, where they become the basal arteries. The spiral arteries, which are a continuation of the basal arteries, supply the endometrium and can be visualized at the myoendometrial junction. The uterine blood supply in humans is rich in anastomoses [8]. Branches from the uterine arteries anastomose with branches of the ovarian and vaginal arteries to form a vascular arcade that supplies the internal genital organs.

Uterine Arteries and Spiral Arteries

Placental development requires adaptive processes in the vascular structures of the uterus. It is known from anatomical studies that the uterine vascular network elongates and dilates during pregnancy[8]. The uterine vascular changes can be demonstrated by transvaginal color and pulsed Doppler sonography.

Uterine artery. Numerous Doppler studies[9, 10, 23, 26, 28, 31, 32, 33] have shown a gradual decline of the uterine artery resistance index during the first trimester of pregnancy. Apparently this decrease persists during the second and third trimesters and can be observed in all portions of the uteroplacental circulation.

Spiral arteries. During early pregnancy, the spiral arteries are gradually transformed into amuscular, dilated, tortuous vascular channels[15, 42]. Turbulent flow with low impedance, typical of the altered spiral arteries, is often detected near the placental bed[15] (Fig. 11.**2**). Moreover, Doppler indicates higher flow velocities and an increased diastolic component caused by the invasion of larger maternal blood vessels under higher blood pressure. Jaffe and Warsof[9] investigated these vascular changes with Doppler ultrasound in the 5th week of gestation. Kurjak et al.[24] described vascular changes in early pregnancy that were detectable even before the gestational sac could be visualized.

Characteristic waveforms. The uteroplacental circulation has been studied intensively through all stages of pregnancy. Vascular resistance declines from the uterine arteries to the spiral arteries with advancing gestation, accompanied by an increase in uterine blood flow. The peak systolic flow velocity tends to decline from the uterine arteries across the arcuate arteries to the radial arteries.

The pulsed Doppler waveforms recorded from the uterine arteries are characteristic: they show a high systolic component with a typical notch in the systolic downstroke and low end-diastolic flow. The higher systolic flow velocity and lower vascular resistance in the spiral arteries compared with the rest of the uteroplacental circulation may result from the trophoblast-induced dilatation of the spiral arteries, hormonal factors, and the decreasing viscosity of the maternal blood. As the spiral arteries change their wall structure during the course of pregnancy, they exhibit hemodynamic properties that are entirely different from those of other arteries in the uteroplacental circulation.

Practical implications. It is known that early human development depends on uterine blood flow, implantation mechanisms, and chromosomal structure. Inadequate implantation and deficient uterine blood flow can be detected noninvasively by Doppler sonography. As a result, Doppler sonography could become the method of choice for determining the causes of abnormal embryonic development that relate to hemodynamic factors. Also, low trophoblastic penetration of the decidua and of the spiral arteries appears to be associated with the same chromosome abnormalities. Hence the examination of blood flow in the intervillous space and placental bed may be of value in predicting the outcome of pregnancy.

Embryonic and Fetal Circulation

Documentation of heart rate. Approximately 21 days after ovulation, corresponding to 5 weeks' menstrual age, the primitive embryonic heart begins to beat (Fig. 11.**7**). Embryonic heart activity has been documented in the uterus as early as postmenstrual day 36[35]. The heart rate rises from 80–90 bpm to 150–170 at the end of 9 weeks. Thereafter the heart rate declines to a mean value of 158 bpm by the 14th week of gestation. Some studies claim that the assessment of embryonic heart activity could be helpful in the assessment of pregnancy outcome[2, 35, 37]. A heart rate lower than 85 bpm in the initial examination between the 5th and 7th weeks or a falling heart rate in subsequent examinations are signs of impending abortion.

Fetal echocardiography. It has been shown that normal fetal cardiac anatomy can be demonstrated very early by transvaginal fetal echocardiography. This suggests that this method can also be used to diagnose major fetal cardiac anomalies in the late first trimester and early second trimester of pregnancy.

Intracardiac waveforms. Doppler flow velocities have been measured in the atrioventricular plane and at the level of the

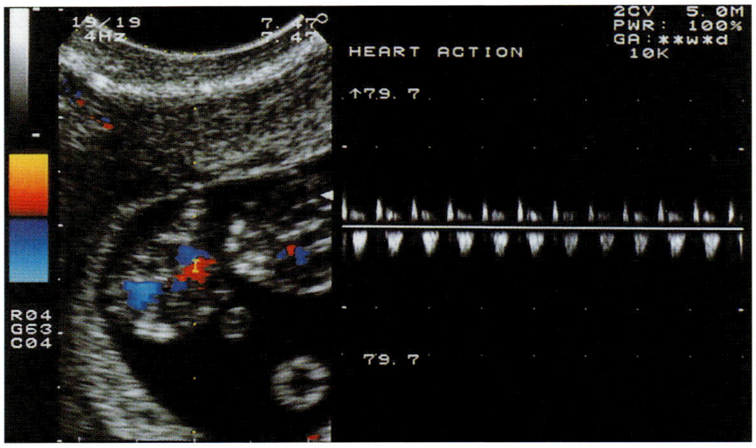

Fig. 11.**7** Transvaginal scan of an embryo in weeks 7–8 of gestation. The regular heart activity is clearly defined.

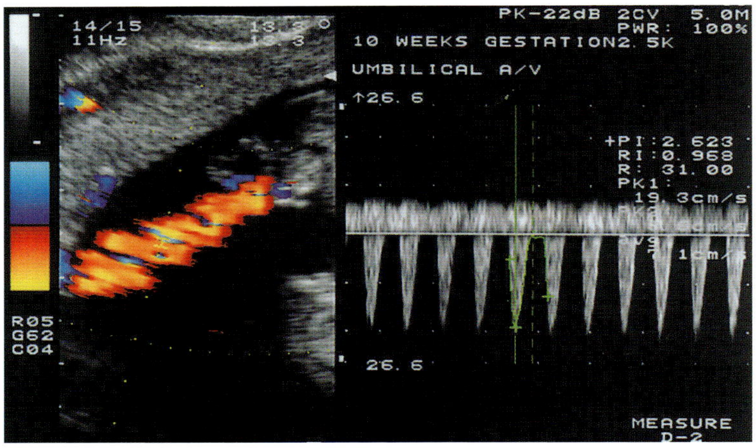

Fig. 11.**8** A free loop of umbilical cord and its insertion demonstrated by conventional Doppler in the 10th week of gestation. The pulsed Doppler signals recorded from the umbilical cord show an absence of diastolic flow, which is typical of the umbilical artery, and flow with a venous component, which is typical of the umbilical vein.

outflow tract, and characteristic intracardiac waveforms have been identified[48]. E-waves (early diastolic filling) and A-waves (atrial contraction) can be measured over the mitral valve and over the tricuspid valve. The E/A ratio expresses the relationship of the passive and active phases of ventricular filling. This ratio increases from 0.5 in the first trimester of pregnancy to 0.9 at term, which may reflect the increasing ventricular compliance during the course of pregnancy.

Fetal Vessels

Fetal vessels that are commonly examined to evaluate fetal condition are the fetal aorta, the umbilical artery, the carotid arteries, and the middle cerebral artery. Pulsations of the fetal aorta and umbilical artery can be detected as early as the 6th week of gestation.

Umbilical artery. End-diastolic flow is not detected in the umbilical artery before the end of 10 weeks' gestation (Fig. 11.**8**). End-diastolic flow is first detectable from the 11th to the 14th week, although this initial flow is inconstant or incomplete. Constant end-diastolic flow velocities are consistently detected after 14 weeks' gestation[1,18,32].

Arteries of the chorion and intraplacental branches of the umbilical artery can also be demonstrated in a large number of pregnancies. Vascular resistance declines from the umbilical artery in the direction of its branches[5,13].

Intracranial vessels. Intracranial blood flow can be visualized as early as the 7th week of gestation. At this time subtle pulsations of the internal carotid artery can be detected at the skull base. Color signals that encode blood flow can be recorded in the anterolateral quadrant of the skull base between the 9th and 10th weeks. Starting in the 9th week, arterial pulsations can be identified in transverse scans lateral to the mesencephalon. However, often it is not possible to distinguish between the internal carotid artery and middle cerebral artery. A characteristic spectral pattern featuring a prominent systolic component and absent end-diastolic components can be seen from the 7th to the 10th weeks, signifying a high vascular resistance at the fetal and umbilical level in comparison with late pregnancy. An end-diastolic flow component is not consistently present from the 11th to 12th weeks, but end-diastolic flow is consistently observed in the middle cerebral artery starting in the 12th week.

Cerebral autoregulatory mechanism. A significant fall in the PI of the intracranial vessels has been documented with advancing gestational age. This decline was noted two weeks earlier than in other parts of the fetal circulation[27,47,49]. End-diastolic flow velocities are also observed earlier in the cerebral vessels than in the fetal aorta and umbilical artery. This indicates a low vascular resistance in the fetal brain that is independent of the resistance changes in the fetal trunk or in the uteroplacental circulation. This autoregulatory mechanism serves to ensure an adequate blood supply for the growing fetal brain. After 12 weeks' gestation, end-diastolic flow components also gradually appear in the umbilical artery and descending aorta, signifying a decrease in fetal vascular resistance. Recently, aided by an improved instrument, we were able to record continuous diastolic flow in the middle cerebral artery as early as weeks 9 and 10 (Fig. 11.**9**).

Choroid plexus. Transvaginal color Doppler can also be used to investigate blood flow in the fetal choroid plexus[29]. Vessels of the choroid plexus can be clearly identified in the 9th week as faint color signals situated along the inner border of the choroid plexus of the lateral ventricle. Besides venous signals, arterial blood flow without a diastolic component can also be clearly visualized. Low RI values are measured in these vessels after 11 weeks. The vascular network of the choroid plexus is best demonstrated in the 13th week. Thereafter the visualization rate declines in association with the morphological development of the plexus. Like other cerebral vessels, the arteries of the choroid plexus show a steady decrease in resistance and an increase in blood flow with advancing gestational age.

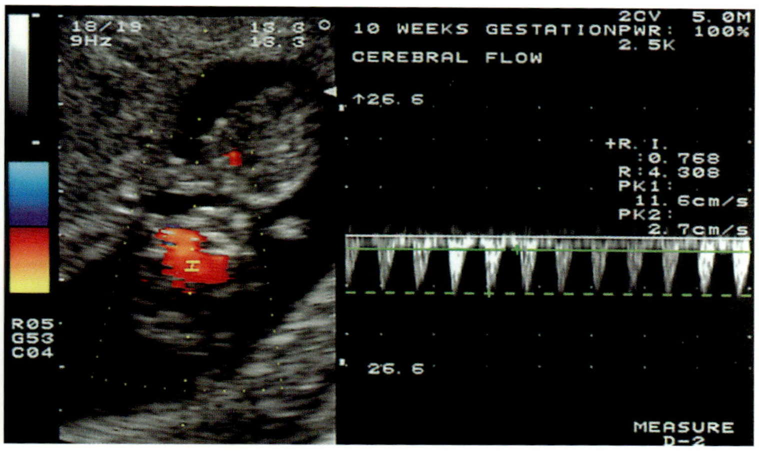

Fig. 11.**9** Transvaginal color Doppler scan of the middle cerebral artery in the 10th week of gestation (left). Pulsed Doppler signals indicate continuous diastolic flow and a lower resistance to blood flow (RI = 0.77) than in other vessels.

Summary

Not too long ago, the only reason to perform an ultrasound examination in early pregnancy was to verify the continuation of the pregnancy and detect heart activity. With the advent of transvaginal color Doppler ultrasound, it became possible to examine the maternal uterine circulation and the fetal circulation during the first trimester of pregnancy. The transvaginal technique allows better access to the fetus and ultimately provides higher image resolution. Color Doppler sonography is very helpful in the localization of arterial blood flow as well as intracardiac and venous flow during early pregnancy. It is not surprising, therefore, that this new technique has assumed great popularity within a short time.

References

1 Arduini D, Rizzo G: Umbilical artery velocity waveforms in early pregnancy: a transvaginal color Doppler study. J. Clin. Ultrasound 19 (1991) 335–339

2 Birnholz JC: First trimester fetal arrhythmias. Fetal. Diagn. Ther. 8 (Suppl. 2) (1990) 6

3 Fox H: Current topic: trophoblastic pathology. Placenta 12 (1991) 479–486

4 Hamilton WJ, Boyd JD, Mossman MW: Human embryology. Heffer, Cambridge 1972

5 Hsieh FJ, Kuo PL, Ko TM, Chang FM, Chen HY: Doppler velocimetry of intraplacental fetal arteries. Obstet. Gynecol. 77 (1991) 478–482

6 Hustin J, Shaaps JP: Echographic and anatomic studies of the maternotrophoblastic border during the first trimester of pregnancy. Amer. J. Obstet. Gynecol. 157 (1987) 162–168

7 Hustin J, Shaaps JP, Lambotte R: Anatomical studies of the uteroplacental vascularization in the first trimester of pregnancy. Troph. Res. 3 (1988) 49–60

8 Itskovitz J, Lindenbaum ES, Brandes JM: Arterial anastomosis in the pregnant human uterus. Obstet. Gynecol. 1 (1980) 3–19

9 Jaffe R, Warsof S: Transvaginal color Doppler imaging in the assessment of uteroplacental blood flow in the normal first-trimester pregnancy. Amer. J. Obstet. Gynecol. 164 (1991) 781–785

10 Jauniaux E, Jurkovic D, Campbell S: In vivo investigations of anatomy and physiology of early human placental circulations. Ultrasound Obstet. Gynecol. 1 (1991) 435–445

11 Jauniaux E, Jurkovic D, Campbell S: Pathophysiology and diagnosis of early pregnancy complications. In Barnea ER, Check JH, Grudzinskas JG, Maruo T (eds.): Implication and Early Pregnancy in Humans. The Parthenon Publishing, London1 992, 465–485

12 Jauniaux E, Jurkovic D, Campbell S, Hustin J: Doppler ultrasonographic features of the developing placental circulation: correlation with anatomic findings. Amer. J. Obstet. Gynecol. 166 (1992) 585–587

13 Jauniaux E, Jurkovic D, Campbell S, Kurjak A. Hustin J: Investigation of placental circulations by color Doppler ultrasound. Amer. J. Obstet. Gynecol. 164 (1991) 486–488

14 Jauniaux E, Jurkovic D, Henriet Y: Development of the secondary yolk sac: correlation of sonographic and anatomic features. Hum. Reprod. 6 (1991) 1160–1165

15 Jauniaux E, Jurkovic D, Kurjak A, Hustin J: Assessment of placental development and function. In Kurjak A (ed.): Transvaginal color Doppler. Parthenon Publishing, New Jersey 1991

16 Jauniaux E, Zaidi J, Jurkovic D, Campbell S, Hustin J: Comparison of color Doppler features and pathohistological finding in complicated early pregnancy. Hum. Reprod. 9 (1994) 2432–2437

17 Kong TY, Pearce JM: Development and investigation of the placenta and its blood supply. In Lavery JP (ed.): The Human Placenta. Clinical Pespectives. Aspen Publishers, Rockville 1987, 25–45

18 Kurjak A, Chervenak F, Zudenigo D, Kupesic S: Early fetal hemodynamics assessed by transvaginal color Doppler. In Kurjak A, Chervenak F (eds.): The Fetus as a Patient. Parthenon Publishing, London 1994, 435–457

19 Kurjak A, Crvenkovic G, Salihagic A, Zalud I, Miljan M: The assessment of normal early pregnancy by transvaginal color Doppler ultrasonography. J. Clin. Ultrasound 21 (1993) 3–8

20 Kurjak A, Kupesic S: Doppler assessment of the intervillous blood flow in normal and abnormal early pregnancy. Obstet. Gynecol. 2 (1997) 252–256

21 Kurjak A, Kupesic S, Kos M, Latin V, Zudenigo D: Early hemodynamics studied by transvaginal color Doppler. Prenat. Neonat. Med. 1 (1996) 38–49

22 Kurjak A, Kupesic S, Kostovic Lj: Vascularization of yolk sac and vitelline duct in normal pregnancies studied by transvaginal color Doppler. J. Perinat. Med. 22 (1994) 433–440

23 Kurjak A, Kupesic S, Predanic M, Salihagic A: Transvaginal color Doppler assessment of the uteroplacental circulation in normal and abnormal early pregnancy. Early Hum. Dev. 29 (1992) 385–389

24 Kurjak A, Kupesic-Urek S, Predanic M, Zudenigo D, Matijevic R, Salihagic A: Transvaginal color Doppler in the study of early pregnancies associated with fibroids. J. Matern. Fetal. Invest. 2 (1992) 81–87

25 Kurjak A. Laurini R, Kupesic S, Kos M, Latin V, Bulic K: A combined Doppler and morphopathological study of intervillous circulation. In The Fifth World Congress of Ultrasound in Obstetrics and Gynecology. Vol. 6 (Suppl. 2) (1995) 116

26 Kurjak A, Predanic M, Kupesic-Urek S: Transvaginal color Doppler in the assessment of placental blood flow. Eur. J. Obstet. Gynecol. Reprod. Biol. 49 (1993) 29–32

27 Kurjak A, Predanic M, Kupesic S, Funduk-Kurjak B, Demarin V, Salihagic A: Transvaginal color Doppler study of middle cerebral artery blood flow in early normal and abnormal pregnancy. Ultrasound Obstet. Gynecol. 2 (1992) 424–428

28 Kurjak A, Predanic M, Kupesic S, Zudenigo D, Matijevic R, Salihagic A: Transvaginal color Doppler in the study of early normal pregnancies and pregnancies associated with uterine fibroids. J. Matern. Fetal Invest. 3 (1992) 81–85

29 Kurjak A, Schulman H, Predanic M, Kupesic S, Zalud I: Fetal choroid plexus vascularization assessed by color and pulsed Doppler. J. Ultrasound Med. 13 (1994) 841–844

30 Kurjak A, Zalud I, Salihagic A, Crvenkovic G, Matijevic R: Transvaginal color Doppler in the assessment of abnormal early pregnancy. J. Perinat. Med. 19 (1991) 155–165

31 Kurjak A, Zudenigo D, Funduk-Kurjak B, Shalan H, Predanic M, Sosic A: Transvaginal color Doppler in the assessment of the uteroplacental circulation in normal early pregnancy. J. Perinat. Med. 21 (1993) 25–34
32 Kurjak A, Zudenigo D, Predanic M, Kupesic S: Recent advances in the Doppler study of early fetomaternal circulation. J. Perinat. Med. 22 (1994) 419–439
33 Kurjak A, Zudenigo D, Predanic M, Kupesic S, Funduk-Kurjak B: Transvaginal color Doppler study of fetomaternal circulation in threatened abortion. Fetal. Diagn. Ther. 9 (1994) 341–347
34 Levi CS, Lyons EA, Lindsay DJ: Early diagnosis of nonviable pregnancy with endovaginal ultrasound. Radiology 167 (1988) 383–387
35 May DA, Sturtevant NV: Embryonic heart rate as a predictor of pregnancy outcome: A prospective analysis. J. Ultrasound Med. 10 (1991) 591–593
36 Merce LT, Barco MJ, Bau S: Color Doppler sonographic assessment of placental circulation in the first trimester of normal pregnancy. J. Ultrasound Med. 15 (1996) 135–142
37 Merchiers EH, Dhont M, De Sutter PA, Beghin CJ, Vandekerckhove DA: Predictive value of early embryonic cardiac activity for pregnancy outcome. Amer. J. Obstet. Gynecol.165 (1991) 11–14
38 Nimrod C, Simpson N, De Vermette R, Fournier J: Placental and early fetal haemodynamics: the suitability of the monkey model. The Fetus as a Patient. XII International Congress, Grado, Italy, May 1996 68
39 Nimrod C, Simpson N, Hafner T et al.: Assessment of early placental development in the cynomolgus monkey (Macaca fascicularis) using color and pulsed wave Doppler sonography. J. Med. Primat. 25 (1996) 106–111
40 Pijnenborg R, Bland JM, Robertson WB, Brosens I: Uteroplacental arterial changes related to interstitial trophoblast migration in early human pregnancy. Placenta 4 (1983) 397–414
41 Pijnenborg R, Bland JM, Robertson WB, Dixon G, Brosens I: The pattern of interstitial invasion of the myometrium in early human pregnancy. Placenta 2 (1981) 303–316
42 Pijnenborg R, Dixon G, Robertson WB, Brosens I: Trophoblastic invasion of human decidua from 8 to 18 weeks of pregnancy. Placenta 1 (1980) 3–19
43 Ramsey EM: Circulation in the intervillous space of the primate placenta. Amer. J. Obstet. Gynecol. 84 (1962) 1649–1663
44 Ramsey EM, Chez RA, Doppman JL: Radioangiographic measurement of the internal diameters of the uteroplacental arteries in Rhesus monkeys and man. Carnegie Inst. Contrib. Embriol. 38 (1979) 59–70
45 Rodesh F, Simon P, Donner C, Jauniaux E: Oxygen measurements in endometrial and trophoblastic tissues during early pregnancy. Obstet. Gynecol. 80 (1992) 283–285
46 Valentin L, Sladkevicius P, Laurini R, Sodeberg H, Marsal K: Uteroplacental and luteal circulation in normal first trimester pregnancies. Doppler ultrasonographic and morphologic study. Amer. J. Obstet. Gynecol. 174 (1996) 768–775
47 Van Zalen-Sprock MM, Van Vugt JMG, Colenbrander G, Geijn HP: First trimester uteroplacental and fetal blood flow velocity waveforms in normally developing fetuses: a longitudinal study. Ultrasound Obstet. Gynecol. 4 (1994) 284–288
48 Wladimiroff JW, Huisman TWA, Stewart PA: Cardiac Doppler flow velocities in the late first trimester fetus: a transvaginal Doppler study. J. Amer. Coll. Cardiol. 17 (1991) 1357–1359
49 Wladimiroff JW, Huisman TWA, Stewart PA: Intracerebral, aortic and umbilical artery flow velocity waveforms in the late-first-trimester fetus. Amer. J. Obstet. Gynecol. 166 (1992) 46–49

12 Color Doppler Sonography in Ectopic Pregnancy

H. J. Voigt

The following principle has proved useful in the prospective, risk-oriented care of patients in early pregnancy. Until an intact intrauterine pregnancy has been definitely confirmed, it is always in the best interests of the patient to consider the possibility of an ectopic pregnancy ("think ectopic!"). The exclusion of an abnormal early pregnancy, then, basically dictates the diagnostic efforts prior to 10 weeks' gestation, before the start of the established three-point screening program.

Three main diagnostic advances have improved the early detection of ectopic pregnancy during the past 10 years:

- The development of rapid, highly sensitive methods for detecting human chorionic gonadotropin and its beta-subgroup (β-hCG)
- The higher resolution achieved with transvaginal sonography
- Laparoscopy (minimally invasive laparoscopic surgery)

Importance of Transvaginal Sonography and Serum hCG

β-hCG. Various studies have consistently shown that ectopic pregnancy can be diagnosed by transvaginal ultrasound in 80–95% of cases[2,9]. Nyberg et al.[7] found associated β-hCG levels of 1000 IU/l or more. Our own studies showed that the chorion could be visualized at β-hCG levels of 500 IU/l in favorable cases and consistently visualized at levels in the range of 1000–2000 IU/l[14]. This means that ectopic pregnancy should always be considered in asymptomatic patients with a serum β-hCG level of 1000 IU/l or more who do not have a demonstrable intrauterine chorion, and that further tests should be done to confirm or exclude the diagnosis.

Routine vaginal scans. Rein et al.[8] state that a vaginal ultrasound examination should be performed routinely in all early pregnancies, since the early diagnosis of ectopic pregnancy can significantly reduce the maternal risk and improve the chances for an organ-conserving laparoscopy. This particularly applies to high-risk patients with a history of infertility or chronic recurring inflammations, IUD (intrauterine device) wearers, and of course women with a prior history of ectopic pregnancy.

Typical sonographic findings. Today the following vaginal ultrasound findings are considered to be definite or compelling evidence of an early ectopic pregnancy (assuming the patient has a positive pregnancy test):

- Exclusion of an intrauterine chorion-type structure
- Presence of an extrauterine and extraovarian chorion-type structure
- Demonstrable embryonic heart activity and movements within the structure (<5% of cases)
- Enlarged uterus with a thickened endometrium of high echogenicity
- Free fluid in the cul-de-sac and paracolic gutters with clot formation and fibrin strands (hemoperitoneum)

The greater the number of these suggestive findings, the higher the index of suspicion for an ectopic pregnancy.

Differential diagnosis. It is difficult to distinguish a nonintact ectopic pregnancy from an adnexal tumor, because the chorionic cavity cannot always be differentiated from a cystic-solid mass. In this case the clinical presentation and β-hCG progression should be used to confirm the diagnosis.

With its higher resolution, transvaginal sonography does not demonstrate the classic "pseudogestational sac" of ectopic pregnancy that is seen with transabdominal ultrasound. If a ringlike structure is detected in the uterine cavity with transvaginal ultrasound, the differential diagnosis will include a blighted ovum or incomplete abortion, especially if there is associated hemorrhage.

Despite the improved capabilities of current routine studies, ectopic pregnancy continues to pose a diagnostic and therapeutic challenge in both asymptomatic and symptomatic cases.

Application of new techniques. There is no question that improved ultrasound techniques have helped to increase preoperative diagnostic accuracy and reduce the number of unnecessary surgical procedures. Transvaginal color Doppler sonography is an innovative technique that is being tested for its ability to identify extrauterine chorion-like structures based on their increased blood flow and typical peritrophoblastic blood flow patterns.

Transvaginal Color Doppler Sonography

Typical Findings in Normal and Ectopic Pregnancy

Typical Doppler features of the peritrophoblastic region. With the increasing clinical use of transvaginal color Doppler sonography, studies began to be published in the early 1990 s dealing with the circulatory patterns that are typically found in association with early placentation[5]. Taylor et al.[12] found that the peritrophoblastic region displayed typical Doppler features characterized by a high systolic flow velocity, low impedance, and a high diastolic flow velocity. This suggested the presence of a high pressure gradient between a maternal artery and a perfused area under low pressure, the intervillous space. The normal histomorphological correlate is found in the maternal vessels at the site of placentation. With an ectopic implantation, corresponding signs of early placentation are found until the limitations of the abnormal implantation site put an end to the process and (barring onset of an acute disease process) regressive changes become predominant.

Decreased pulsatile flow in the fallopian tube. When the blastocyst implants into the tubal mucosa, the trophoblast infiltrates the lamina propria of the mucosa and the muscular layer and grows chiefly between the tubal lumen and the serosa, enlarging in the circumferential and longitudinal directions. The vasotropic growth and the invasion of the surrounding vessels, the branches of the uterine and ovarian arteries, leads to greatly increased blood flow and/or intratubal and extratubal hemorrhage. As a result of this, the ectopic pregnancy reduces the normally pulsatile blood flow with a high perfusion resistance in the fallopian tube. These hemodynamic changes can be demonstrated by selective color Doppler imaging of the tube. Of course, these resistance changes can also be detected by pulsed Doppler sonography alone, but without direct vascular imaging the Doppler velocimetry scan is time-consuming, fraught with uncertainties, and therefore impractical in urgent cases. Color Doppler can provide a rapid and precise "probe." If a suspicious anatomical structure has been detected in the lower pelvis with transvaginal ultrasound, the color-flow image will yield further information on the hemodynamics at the site and permit the targeted sampling of flow-velocity waveforms and their analysis using Doppler indices.

Diagnostic Efficiency

An analysis of the diagnostic efficiency of current methods for detecting ectopic pregnancy has shown that the added use of color Doppler improves the diagnostic efficiency of transvaginal ultrasound from 84–95%[4, 11] to 87–96 %[3, 13]. The gain appears small, and a comparison of the study conditions used by different authors makes it difficult to draw a clear positive conclusion.

True-positive and false-positive diagnoses. An important clinical issue is the confidence with which a suspected ectopic pregnancy can provide a primary indication for laparoscopy. The accuracy of the preoperative presumptive diagnoses compared with the actual findings is expressed by the rate of true-positive and false-positive cases. True validation is not possible with this type of analysis, because true-negative cases do not lead to laparoscopy, while false-negative cases take time to produce overt clinical signs, providing at best a secondary indication for intervention. These include the rather frequent cases in which curettage is done for a presumed nonviable early pregnancy (incomplete abortion or blighted ovum) and a trophoblast is not found in the curettage material. They also include cases of early pregnancy loss, ectopic pregnancies that are not initially recognized, and the occasional cases with a poorly vascularized, involuting ectopic pregnancy[1]. Meyers et al. studied the degree of vascularity as a function of trophoblastic activity as measured by the serum hCG level. They found a statistical association between avascularity and hCG levels that were low or fell below 1000 IU/l[6]. This observation should prompt further prospective comparative studies. Highly acute disease states that warrant immediate action without comprehensive prior tests are also inaccessible to validation.

Author's Studies

In a kind of historical comparison, we reviewed the quality of indications for laparoscopy in patients with ectopic pregnancy. Our retrospective analysis covered 263 patients who underwent laparoscopy during a four-year period before the advent of color Doppler-assisted vaginal sonography. These cases were compared with 60 laparoscopies for ectopic pregnancy during a one-year period in which the primary workup included color Doppler sonography. The results of the comparison are shown in Fig. 12.**1** and Table 12.**1**. An improvement in positive predictive value from 91.6% to 95% indicates a slight increase in the confidence level for laparoscopy referral. The rate of false-positive diagnoses fell from 8.4% to 5 %, representing only a slight improvement.

The corpus luteum as a major source of error. Critical analysis reveals that the causes of misdiagnoses have remained the same. The ultrasound visualization and interpretation of the corpus luteum led to errors in both the old and new study populations. Profuse luteal hemorrhage, torsion, ruptures with intraperitoneal blood collections, and painful sonopalpation of the adnexal mass and blood-filled cul-de-sac are findings that strongly bias the clinical decision-making process toward operative intervention.

Kurjak and his group[10] investigated luteal blood flow in patients with intrauterine and ectopic pregnancies and in nonpregnant subjects during the second phase of the menstrual cycle. They could detect no qualitative difference in perfusion. In 86.4% of the patients with ectopic pregnancy, they found that the corpus luteum and pregnancy were on the same side of the adnexa. While this observation can be helpful in locating the ectopic pregnancy, it should also serve as a warning against false security because of the potential for internal and external migration of the blastocyst.

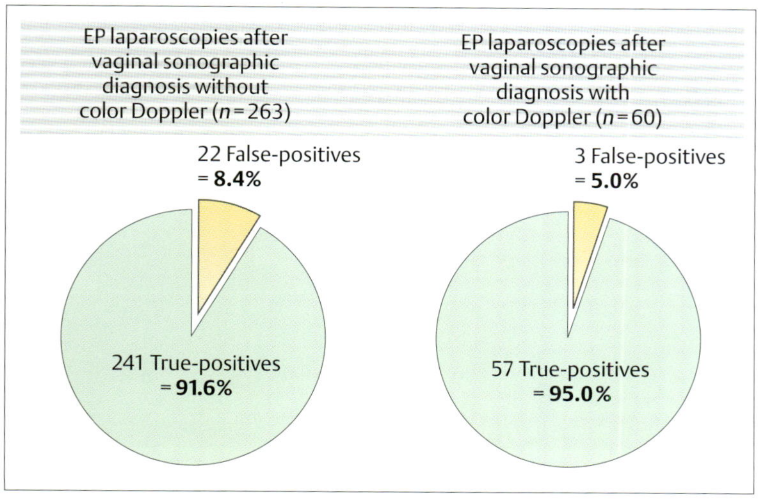

Fig. 12.1 Diagnostic quality of transvaginal sonographic examinations without color Doppler and with color Doppler in cases of suspected ectopic pregnancy (EP).

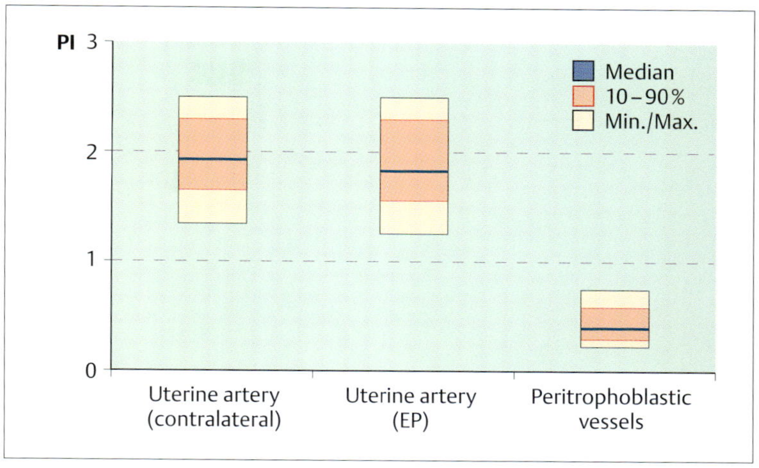

Fig. 12.2 Perfusion resistance of uterine and peritrophoblastic tubal vessels in ectopic pregnancy.

Table 12.1 Diagnostic quality of transvaginal sonographic examinations without color Doppler and with color Doppler

Vaginal ultrasound without color Doppler		Vaginal ultrasound with color Doppler	
True positive, 241 cases	91.6%	True positive, 57 cases	95%
False-positive, 22 cases	8.4%	False-positive, 3 cases	5%
Partial/complete torsion with large corpus luteum of pregnancy	4	Early pregnancy loss, hemorrhagic corpus luteum	1
Large hemorrhagic corpus luteum	7	Early pregnancy loss, corpus luteum	1
Early pregnancy loss, corpus luteum	7	Early pregnancy, not diagnosed	1
Early pregnancy, not diagnosed	4		
Positive predictive value	91.6%	Positive predictive value	95%

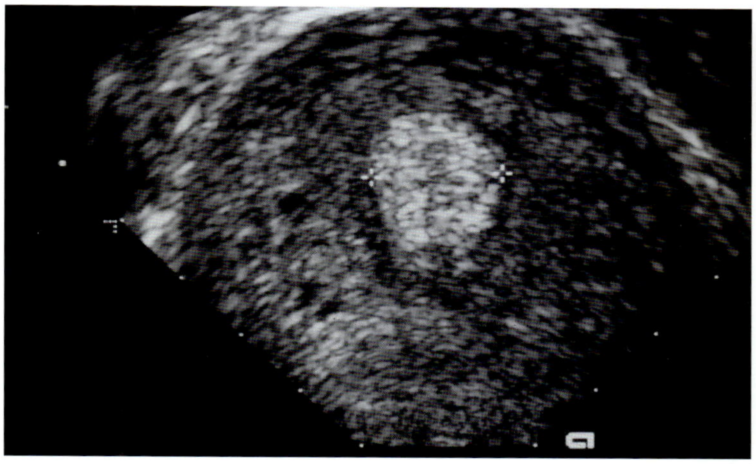

Fig. 12.3 Transvaginal ultrasound scan shows a hyperechoic endometrium due to a decidual reaction, with no chorionic ring in the uterus. Serum hCG = 914 IU/l.

It is important to note that the suspicious adnexal structure must be extraovarian in its location for the acquisition of useful color flow signals. The very rare cases of ovarian implantation (<1%) should also be considered in this regard.

Pulsatility index. We also determined perfusion resistance based on the pulsatility index (PI) of the uterine arteries and, where possible, color-flow signals recorded from the suspicious adnexal structure. Our goal was to assess and compare the flow impedance of the uterine artery on the side of the ectopic pregnancy and on the opposite side and also in the peritrophoblastic vessels. The results are shown in Fig. 12.2. We found no significant difference between the resistance indices of the uterine arteries. The mean PI values, at 1.8 and 1.9, were not significantly different. The median PI value in the peritrophoblastic tubal tissue was 0.38 and thus significantly lower than in the uterine arteries ($p < 0.01$). The indices for the uterine arcade and spiral arteries in intact intrauterine pregnancies are in the same range, signifying comparable perfusion conditions. We did not find these values in the intrauterine color-flow signals from ectopic pregnancies. This observation may be helpful as a differentiating sign.

Clinical examples. Figures 12.3– 12.6 illustrate a typical case of an asymptomatic tubal ectopic pregnancy that was detected early and managed by a tube-conserving procedure.

Figures 12.7– 12.9 are from a case with a false-positive diagnosis caused by a richly vascularized corpus luteum. Multiple, thorough searches were necessary to find a small intrauterine chorionic structure signifying an early pregnancy. A history of irregular cycles with prolonged amenorrhea had caused a three-week error in dating the pregnancy.

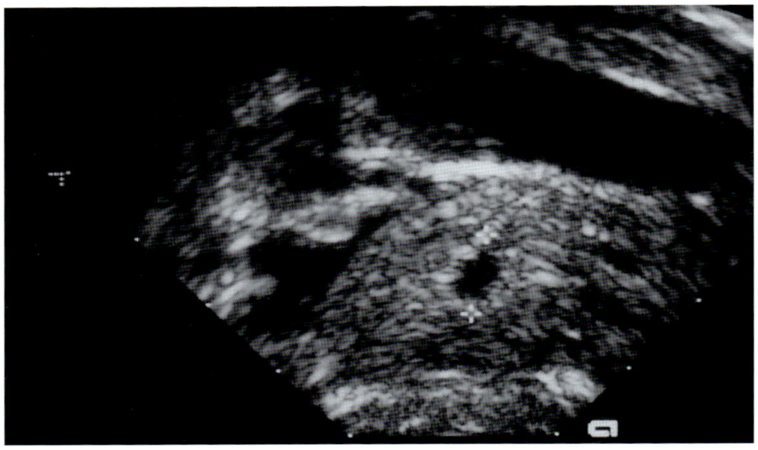

Fig. 12.**4** A chorionic structure is distinguishable from the left ovary in the left adnexal region.

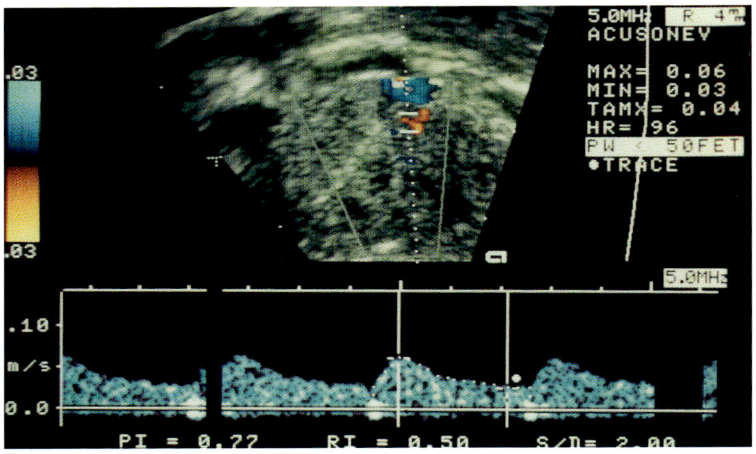

Fig. 12.**5** Intense color Doppler signals are seen at the periphery of the extrauterine chorionic structure. The color scale was adjusted to display low flow velocities (slow-flow imaging). The sampled Doppler spectrum (triplex mode) shows low indices (PI = 0.7, RI = 0.5, S/D ratio = 2) characteristic of a low perfusion resistance at the sampled site. There is a marked similarity to the blood flow patterns of utero-placental vessels.

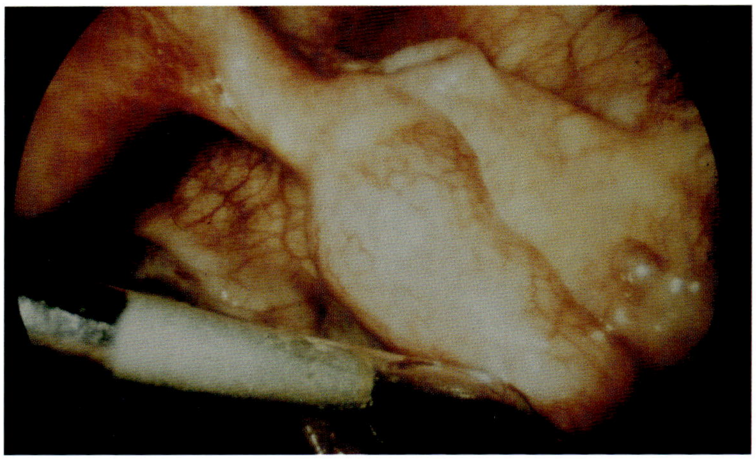

Fig. 12.**6** Laparoscopic view of the asymptomatic ectopic pregnancy (same patient as in Figs. 12.**3**– 12.**5**) shows a fusiform swelling of the right tube with no hemoperitoneum or signs of rupture.

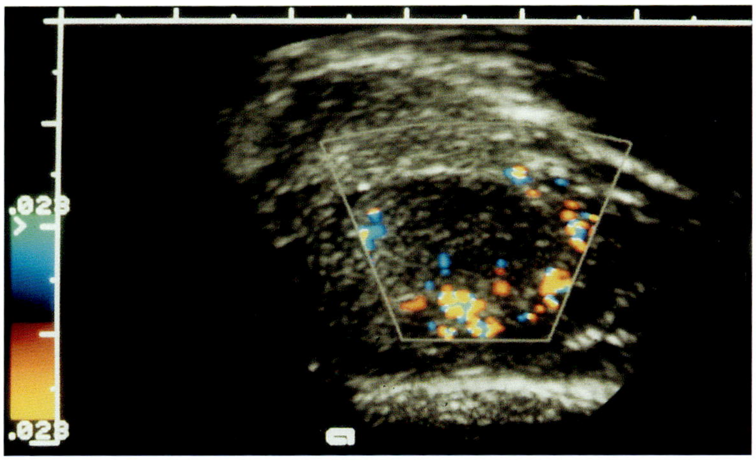

Fig. 12.**7** Right adnexal mass in a patient with a positive pregnancy test and an "empty" uterus by transvaginal sonography. The assumed gestational age is 7 weeks. Color Doppler shows heavy vascularization with numerous color-flow signals (slow-flow imaging, low signal detection threshold < 0.028 m/s).

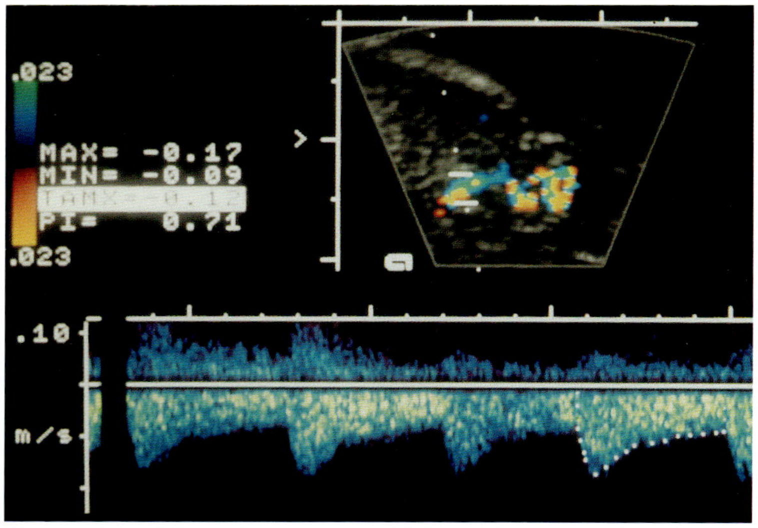

Fig. 12.**8** Selectively acquired Doppler spectrum indicates low perfusion resistance at the sampled site.

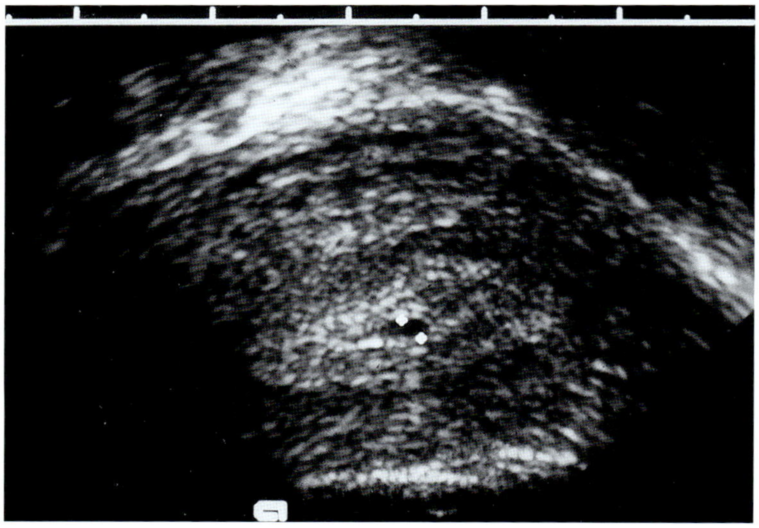

Fig. 12.**9** Initially missed intrauterine chorion consistent with the 5th week of gestation (same patient as in Figs. 12.**7** and 12.**8**).

Assessment of the Method

Little extra effort. It would be an overstatement to claim that color Doppler vaginal sonography is indispensable for the early detection of ectopic pregnancy, since an experienced sonographer using quality equipment can diagnose an ectopic pregnancy with reasonable confidence even without color Doppler. It is known that a radical improvement in diagnostic accuracy can be achieved only through tremendous effort. Today, however, the widespread availability of color-flow techniques gives many examiners the opportunity to increase their diagnostic confidence level with little extra effort, even if only to a statistically moderate degree. With its protean clinical manifestations, it is rare for an ectopic pregnancy to exhibit a typical course. Color Doppler sonography thus offers a valuable aid to differential diagnosis.

Triplex mode. There are no compelling, fundamental arguments against the use of color Doppler sonography, but the examiner should be aware of the high energy outputs that are associated with imaging in the triplex mode. Since the uterus may contain a very early pregnancy that is still below the resolution limit of the ultrasound system, the uterine cavity should be shielded from triplex scanning to eliminate exposure as a potential issue. The case in Figs. 12.**7**– 12.**9** illustrates the importance of this aspect.

Summary

The introduction of transvaginal sonography has significantly improved diagnostic accuracy in patients with a suspected ectopic pregnancy. The added use of color Doppler ultrasound and its capabilities have been addressed and studied by numerous groups of authors. All have documented a modest increase in diagnostic efficiency, and our own experience has confirmed this. The positive predictive value has improved from approximately 90% to 95% while the false-positive rate has decreased from 8.4 % to 5%. A potential source of error is the corpus luteum, which can have a great diversity of clinical presentations. For safety reasons, the uterus should not be exposed to prolonged ultrasound scanning in the triplex mode.

The typical candidate for transvaginal color Doppler evaluation has an "empty" uterus and a positive pregnancy test or a serum hCG level of 1000 IU/l or more with no clinical symptoms. The early detection of ectopic pregnancy helps to reduce morbidity and increase the prospects for an organ-conserving laparoscopic procedure.

References

1 Bonilla-Musoles FM, Ballester MJ, Tarin JJ, Raga F, Osborne NG, Pellicer A: Does transvaginal color Doppler sonography differentiate between developing and involuting ectopic pregnancys? J. Ultrasound Med. 3 (1995) 175
2 De Crespigny LC: Demonstration of ectopic pregnancy by transvaginal ultrasound. Brit. J. Obstet. Gynaecol. 95 (1988) 1253
3 Emerson DS, Cartier MS, Altieri LA et al.: Diagnostic efficacy of endovaginal color Doppler flow imaging in an ectopic pregnancy screening program. Radiology 183 (1992) 413
4 Kivikosky AI, Martin CM, Smeltzer JS: Transabdominal and transvaginal ultrasonography in the diagnosis of ectopic pregnancy: a comparative study. Amer. J. Obstet. Gynecol. 163 (1990) 123
5 Kurjak A, Zalud J, Jurkovic D, Alfirevic Z, Miljan M: Transvaginal color Doppler for the assessment of pelvic circulation. Acta Obstet. Gynecol. Scand. 68 (1989) 131
6 Meyers M, Feyock A, Holland S, Taylor KJW: Correlation of duplex Doppler and HCG levels in ectopic pregnancy. Radiology 173 (1989) 247
7 Nyberg DA, Hill LM: Normal early intrauterine pregnancy: sonographic development and hCG correlation. In: Transvaginal ultrasound. Mosby, St. Louis 1992, 65–84
8 Rein MS, Di Salvo DN, Friedman AJ: Heterotopic pregnancy associated with in vitro fertilization and embryo transfer: possible role for routine vaginal ultrasound. Fertil. Steril. 51 (1989) 1057
9 Rempen A.: Vaginal sonography in ectopic pregnancy: A prospective evaluation. J. Ultrasound Med. 7 (1988) 381
10 Salim A, Zalud J, Farmakides G, Schulman H, Kurjak A, Latin V.: Corpus luteum blood flow in normal and abnormal early pregnancy: evaluation with transvaginal colour and pulsed Doppler sonography. J. Ultrasound Med. 13 (1994) 971
11 Schurz B, Wenzel R, Eppel W, Schon HJ, Reinold E: Early detection of ectopic pregnancy by transvaginal ultrasound. Arch. Gynecol. Obstet. 248 (1990) 25
12 Taylor KJW, Ramos IM, Feyock AL et al.: Ectopic pregnancy: duplex Doppler evaluation. Radiology 173 (1989) 93
13 Taylor KJW, Meyer WR: New techniques in the diagnosis of ectopic pregnancy. Obstet. Gynecol. Clin. N. Am. 18 (1991) 39
14 Voigt HJ: Pathologie der Frühschwangerschaft. Gynäkologe 29 (1996) 165

13 Physiology of Doppler Flow in Maternal Vessels during Pregnancy

T. Golaszewski, J. Deutinger, and G. Bernaschek

Conditions of Intrauterine Life

The introduction of Doppler flow measurements has enhanced our ability to investigate the conditions of intrauterine fetal life. For more than 20 years, fetal Doppler ultrasound has provided a simple, noninvasive method for measuring blood flow velocities in the fetal vessels[23]. Doppler measurements of uterine blood flow, or the maternal part of the fetomaternal circulation, were first reported in 1983[3].

Since the advent of transvaginal sonography and the development of color-encoded pulsed Doppler technology, Doppler flow measurements of the uterine artery and arcuate arteries have been substantially improved. The color mapping of blood flow is particularly advantageous for examining the maternal circulation owing to the many small-caliber vessels in the maternal system.

Physical Principles

Doppler effect. Christian Doppler (born in Salzburg in 1803) described the effect by which sound waves are perceived as having a higher or lower frequency, depending on whether the source of the sound is moving toward or away from the listener. The amount of this frequency shift is directly related to the velocity of the sound source. By applying this effect, it was possible to measure the frequency shift between the ultrasound emitted by a transducer and the echoes returning from red blood cells and thus determine the velocity of blood flow in fetal and maternal vessels. We know from hemodynamics that the blood flow velocity (i.e., the velocity of the moving red cells) in a vessel is directly proportional to the blood pressure and indirectly proportional to the vascular resistance.

Continuous-wave transducers. Various types of Doppler transducer are available. Continuous-wave (CW) transducers consist of a transmitter that continuously emits sound and a receiver that continuously receives the frequency-shifted echoes. One disadvantage of CW Doppler is that it does not discriminate echoes according to their depth. In theory, multiple vessels along the path of the beam can lead to superimposed signals and summation effects.

Pulsed-wave transducers. Pulsed-wave (PW) transducers emit ultrasound for a fraction of a second and then switch to the receive mode. These instruments can provide depth discrimination by measuring the echo delay time. Only the echoes that return from a specified tissue depth are analyzed for their frequency shift. This selected depth, called the Doppler gate or sample volume, can be positioned precisely within the vessel lumen, allowing signals to be acquired from a uniquely defined anatomical location.

Color Doppler. A color Doppler system displays not only the magnitude of the frequency shift but also the direction of blood flow (either toward or away from the transducer) by means of a color-encoding system (red = toward the transducer, blue = away from the transducer). This makes it easier to locate small vessels such as the arcuate arteries.

Indices. Several angle-independent indices can be used to analyze the frequency-shift data: the S/D ratio of Stuart (S = maximum frequency shift in systole, D = minimum frequency shift in diastole), the resistance index of Pourcelot ($RI = (S-D)/S$), and the pulsatility index ($PI = (S-D)/\text{mean value}$).

Anatomical and Physiological Principles

Uteroplacental blood supply. The uterus derives most of its blood supply from the two uterine arteries, which arise from the iliac artery. A small amount of blood is also supplied by anastomoses with the ovarian vessels, which branch directly from the abdominal aorta. The two uterine artery trunks divide within the myometrium into 10–15 arcade vessels on both sides. These vessels encircle the uterus and form anastomoses near the anterior and posterior midline. The radial arteries arise from these arcade vessels and pass more deeply into the uterus. The radial arteries in turn give rise to the basal arteries, which supply the basal endometrium and the spiral arteries with blood. The tortuous, helixlike spiral arteries supply blood to the placenta; this blood enters the intervillous space from the decidua basalis through broad openings in the spiral arteries.

Adaptive Processes during Pregnancy

Spiral arteries. At one time it was believed that the ends of the spiral arteries bore constrictions that produced a "jet effect." If this were true, a high resistance would be present in the spiral arteries. In fact, invading trophoblasts destroy the muscular portion of the spiral artery wall within the decidua by the 20th week of gestation, causing the lumen to expand in the direction of the placenta and creating a pressure difference of only 8 mmHg between the spiral artery and the amniotic cavity[25].

Uterine enlargement. Normally the uterus increases greatly in size during the course of pregnancy. The weight of a gravid uterus at term is 20 times the uterine weight outside of pregnancy (1000 g versus 50 g). The weight increase is not due mainly to an increased number of muscle fibers but to muscular hyperplasia, i.e., a marked enlargement of the muscle fibers of the myometrium. The blood supply during pregnancy adapts accordingly.

Arterial dilatation. The uterine arteries dilate to approximately 3 times their original caliber, the arcuate arteries to 10 times, and the spiral arteries to 30 times[15]. By the end of pregnancy, blood is flowing through the intervillous space at a rate of approximately 60 ml/min[28]. The systemic blood pressure prevails to the level of the arcuate arteries. Beyond that, the pressure dwindles to the spiral artery openings, resulting in a low pressure gradient relative to the intervillous space (perfusion pressure 15–20 mmHg).

It is apparent, then, that Doppler ultrasound scans of the uterine vessels are representative only to the level of the arcuate arterial bed.

Placental blood supply. Because of the numerous anastomoses between the two uterine arteries, both vessels supply the placenta. The uterine arteries each supply a different number of cotyledons, depending on the placental location, and this accounts for the normal difference between the two sides that is recorded with Doppler flowmetry.

Perfusion in normal and abnormal pregnancies. In a normal pregnancy, the maternal uterine arteries reflect the perfusion of the entire uterofetoplacental circulation. Owing to the tremendous increase in terminal vascular branches, the peripheral flow resistance declines, causing a rise in diastolic flow velocity. This results in part from progesterone-induced vasodilation and from the development of the intervillous space.

In an abnormal pregnancy, vasoconstriction may occur in the uterine circulation (hypertension, preeclampsia) producing a high S/D ratio, or there may be an inadequate development of collateral channels causing an abnormal discrepancy between the sides. Normally a postsystolic notch in the Doppler waveform does not persist beyond the 24th week of gestation. If it persists after 24 weeks, the possibility of an impending pregnancy complication should be considered.

The following values are normal for the third trimester:

- S/D ratio of the uterine arteries < 3, difference between the sides < 1
- S/D ratio of the arcuate arteries < 2

Technique of Transvaginal Pulsed Doppler Flowmetry

Patient position. In all vaginal ultrasound examinations, it is advantageous to place the patient in a lithotomy position as this allows greater mobility of the endovaginal probe. This position does not compromise uterine blood flow, eliminating the possibility of false-positive findings[34].

Transducer placement. Contact gel is applied to the ultrasound probe, which is then covered with a condom-like sterile rubber sheath. A lubricating gel is applied, and the probe is carefully inserted into the vagina. It is oriented to provide a sagittal scan of the lesser pelvis. A standard routine should be followed for Doppler flowmetry. In all of our examinations the right side of the monitor is caudal in the sagittal scan, and the left side is cranial. The probe is rotated 90° to obtain a coronal or transverse scan, in which case the right side of the monitor corresponds to the patient's left side, and vice versa[7].

Adjusting the Doppler beam and sample volume. After making an initial survey of the lesser pelvis and identifying key landmarks (cervix, gestational sac, ovary, pelvic wall), the examiner positions the ultrasound probe in the lateral fornix of the vagina. Rotating the probe 90° makes it easier to locate the uterine artery in the parametrium, which can be examined in coronal section. The echogenic parametrium is surveyed by slowly moving the scanner anteriorly from the sacrum. With some practice, the uterine artery can be consistently identified as an elongated, pulsating vascular echo 2–4 mm in diameter. Besides its typical course, which can be defined over a variable length, the pulsations of the uterine artery are its most important distinguishing feature.

The examination is done with a minimum amount of probe pressure. When the vessel has been located in the real-time scan, the Doppler beam and sample volume are adjusted so that the Doppler frequency shift can be received and recorded. It takes from 5 to 10 minutes to examine both uterine arteries and interpret their spectra. The total examination time depends partly on gestational age.

In late pregnancy it can be difficult or impossible to position the probe in the lateral fornix owing to the low position of the presenting fetal part. This is not a problem, however, because Doppler velocimetry of the uterine arteries is used mainly for the early detection of pregnancy complications and is rarely performed after 36 weeks.

Authors' Studies

Instruments. We use an end-fire vaginal probe (KretzTechnik, Zipf, Austria) with a 240° field of view. This instrument can cover all of the lesser pelvis, making orientation easier. We use a transducer frequency of 7.5 MHz and a Doppler beam frequency of 4.5 MHz. The optimum pulse repetion frequency (PRF) setting for most examinations is 5.2 kHz with a penetration depth of approximately 7–8 cm and a 125 Hz filter settings. According to the manufacturer, the power output of the scanner is less than 100 mW/cm^2.

Parameters. In principle, all of the angle-independent parameters mentioned above can be used for waveform analysis. However, because the maternal vessels exhibit diastolic flow even when the peripheral resistance is extremely high (early pregnancy, hypertension), it is sufficient to use simple indices such as the S/D ratio or RI. The qualitative analysis of Doppler waveforms would mainly be useful for detecting a postsystolic notch and has not found wide application[12, 36].

Uterine Perfusion in a Normal Pregnancy

Establishing normal values. We performed transvaginal Doppler velocimetric studies of the uterine artery in singleton pregnancies from 7 to 40 weeks' gestation in order to establish normal values[6]. We analyzed the data from 88 women. The criteria for enrollment in the study were an ultrasound examination in early pregnancy for accurate dating, an uncomplicated course of pregnancy, and a birthweight within normal limits.

Early pregnancy. Figure 13.**1** shows a coronal scan through the parametrium demonstrating the uterine artery and a Doppler waveform pattern that is typical of early pregnancy, i.e., high pulsatility with high systolic flow velocities and low end-diastolic velocities. The further progression of pregnancy is marked by an increase in the flow velocities, especially during diastole, causing a significant decrease in the S/D ratio (Fig. 13.**2**).

Second and third trimesters. The greatest changes in our study were observed at the start of the second trimester. We saw no significant increase in diastolic flow velocity after the end of the second trimester. The mean S/D ratio was 5.44 in the first trimester and 2.20 in the third trimester, while the pulsatility index fell from 2.59 to 1.32 (Figs. 13.**3** and 13.**4**).

Differences between the sides. When we compared the right and left uterine arteries by their respective contributions to the uterine blood supply, we noted some marked differences between the two sides in early pregnancy, which were also responsible for the large scatter of individual values measured in the first and second trimesters. When the mean values for both vessels were compared, however, no differences were found. The difference in the S/D ratio between the right and left uterine arteries decreased during the third trimester and averaged only 0.4–0.3. In the third trimester of a normal pregnancy, the S/D ratio in both uterine arteries was less than 3, while the difference between the sides was less than 1.

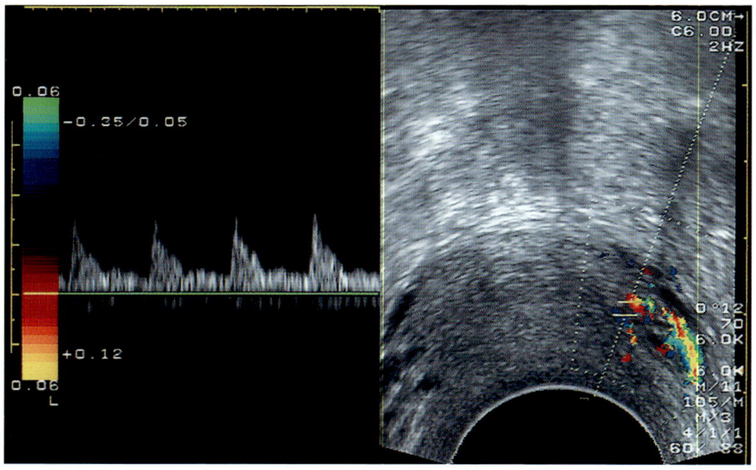

Fig. 13.1 Example of an original transvaginal pulsed Doppler scan of the left uterine artery in the first trimester of pregnancy. Transverse scan through the uterine wall at the level of the isthmus shows numerous color-encoded vessels of small caliber. The Doppler sample volume is positioned in the descending branch of the left uterine artery. The peripheral resistance is still high at this stage, and consequently the diastolic flow velocities are low.

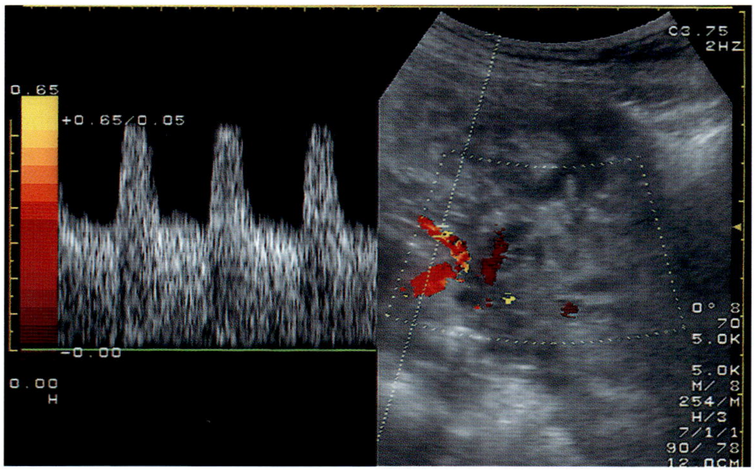

Fig. 13.2 Transabdominal Doppler velocimetry of the right uterine artery in the second trimester of pregnancy. The high flow velocity in diastole is caused by the low peripheral resistance, which is considered a sign of adequate placentation and uterine blood flow.

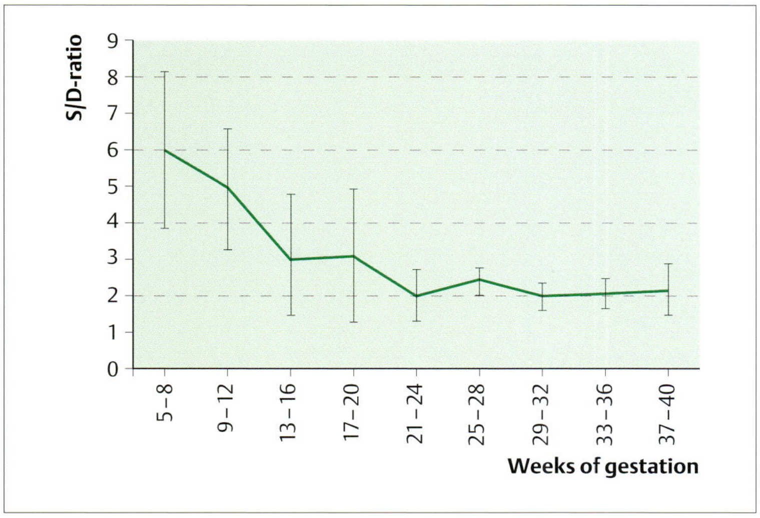

Fig. 13.**3** Fall of the S/D ratio during the course of pregnancy.

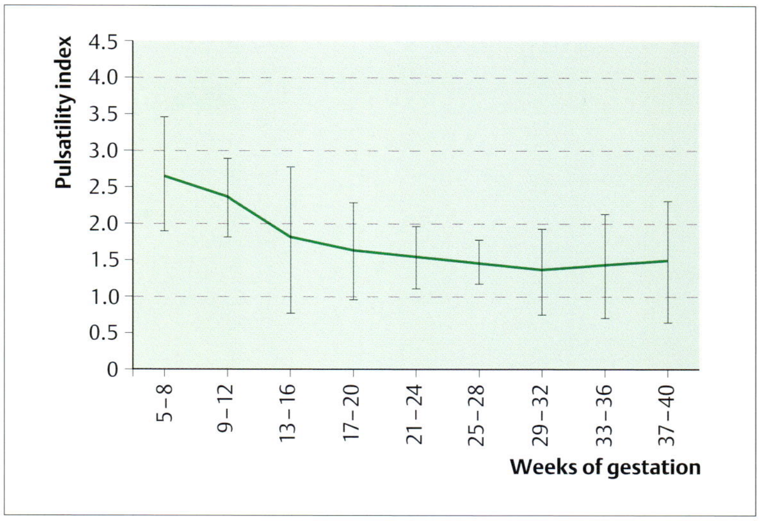

Fig. 13.**4** Fall of the pulsatility index during the course of pregnancy.

Uterine Perfusion in an Abnormal Pregnancy

In another study, uterine artery Doppler velocimetry was performed in 176 women with singleton pregnancies (114 primiparae and 62 multiparae) between the 27th and 40th weeks of gestation[8].

The women were enrolled in the study for the following reasons:

- Suspected fetal growth retardation (94)
- Risk history (30)
- Pregnancy-induced hypertension (23)
- Diabetes mellitus (16)
- Chronic hypertension (13)

S/D ratio and PI. The S/D ratio was used for waveform analysis. Three parameters were used for the classification of uterine perfusion, drawing on our tables of normal values: the S/D ratio in the right uterine artery, the S/D ratio in the left uterine artery, and the left–right difference. Analysis of the fetal circulation was based on the S/D ratio and the pulsatility index (PI) of the umbilical artery, aorta, and internal carotid artery. For further analysis we classified the calculated values as normal or abnormal by referring to published normal values (S/D ratio in the umbilical artery < 3[31]; PI of the fetal aorta < 2.2[37]; and PI of the internal carotid artery > 1.3[40]).

Six groups of uterine perfusion. We were able to distinguish six different groups of uterine perfusion based on the calculated S/D ratios and left–right differences in the uterine arteries (Table 13.**1**). They were distributed as follows: in 113 of the 176 patients (64%), the S/D ratios and left–right differences were within our normal values. These cases served as the control group (group I). In 63 patients (36%) we found values that were outside our normal range. These patients were subdivided into five groups (II–VI) with abnormal uterine perfusion (Fig. 13.**5**).

Table 13.**1** Definition of groups I–VI in relation to uterine perfusion

Left-right difference	< 1	> 1
Both S/D ratios < 3	I	IV
One S/D ratio > 3	II	V
Both S/D ratios > 3	III	VI

Fetal circulation. The results of Doppler flow measurements in fetal arteries were grouped in relation to uterine perfusion and compared with the control group. We found that most cases with abnormal flow velocities in fetal vessels belonged to groups III, V and VI. In groups III and VI, all of the fetuses had an abnormal waveform in one of the three vessels examined (umbilical artery, aorta, internal carotid artery). Most of the fetuses in group V (35 of 38) had an abnormal flow velocity in at least one of the vessels examined (Table 13.**2**).

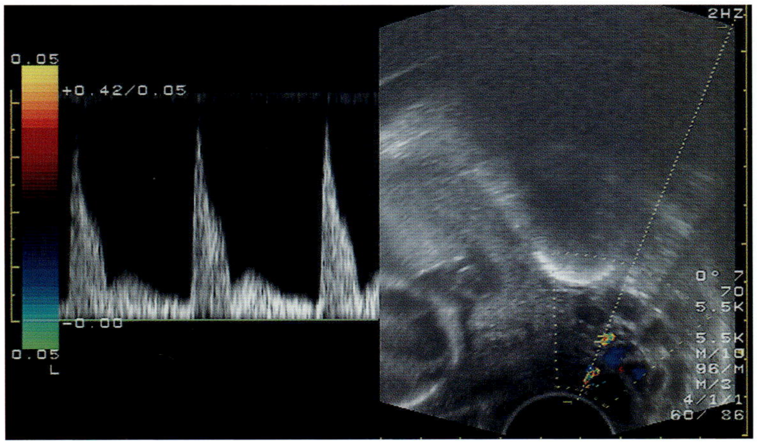

Fig. 13.**5** Transvaginal pulsed Doppler scan of the left uterine artery in the 25th week of gestation, showing an elevated S/D ratio. The waveform also contains a postsystolic notch. In this case the pregnancy was complicated by preeclampsia.

Table 13.2 Normal and abnormal Doppler waveforms (n = number of patients) of fetal vessels in relation to uterine perfusion (groups I–VI)

Group	I	II	III	IV	V	VI
n	113	10	4	5	38	6
Normal ratios in all fetal vessels	39	4	0	2	3	0
%	34	40	0	40	8	0
Abnormal	74	6	4	3	35	6
%	66	60	100	60	92	100
UA + aorta + ICA	–	–	2	–	–	–
UA + aorta	15	2	1	–	13	2
UA + ICA	–	1	–	–	3	–
Aorta + ICA	3	–	–	1	–	1
UA	14	3	–	–	5	–
Aorta	34	–	1	1	12	3
ICA	8	–	–	1	2	–

UA = umbilical artery, ICA = internal carotid artery

Uterine Perfusion on Medication or following Uterine Manipulation

We also did a study to determine how amniocentesis, cervical cerclage, or the administration of beta-mimetics would affect uterine blood flow[9].

Six groups. We examined a homogeneous population of 117 women (Table 13.3) who were randomly divided into four groups. Groups 1 and 2 consisted of women who underwent genetic amniocentesis for advanced maternal age. These patients either received no beta-mimetics (group 1) or prophylactic oral beta-mimetics (group 2). Groups 3 and 4 consisted of women who underwent prophylactic cervical cerclage. Group 3 did not receive i.v. beta-mimetic prophylaxis, while group 4 did. Further clinical data on the patients are presented in Table 13.3. The oral beta-mimetics were administered in six individual doses of 10 mg each. The i.v. beta-mimetics were administered at a rate of 0.2 mg/min.

Table 13.3 Demographic data on the patients examined

	Group 1	Group 2	Group 3	Group 4
Medication	None	Oral	None	i.v.
Surgical procedure	Amniocentesis	Amniocentesis	Cerclage	Cerclage
n	30	30	28	29
Age (years) ± SD	37.8 ± 2.3	38.2 ± 2.4	26.4 ± 2.5	25.1 ± 2.9
Primiparae	6	4	9	10
Multiparae	24	26	19	19

Doppler velocimetry. Transvaginal Doppler velocimetry was performed before and after surgery or medication between the 14th and 16th weeks of pregnancy. In groups 1 and 2 we performed Doppler flowmetry before and 24 hours after amniocentesis. In groups 3 and 4 the measurements were taken before and six hours after cervical cerclage. We analyzed the mean values of both measurements and then evaluated the total perfusion of the uterus. We based our analysis on the S/D ratio, pulsatility index, and maternal heart rate.

Results. The diastolic blood flow velocity rose significantly after cervical cerclage and the postoperative administration of i.v. beta-mimetics. This was caused mainly by a fall in peripheral resistance. This phenomenon was observed only in group 4. Amniocentesis (group 1) and cerclage alone (group 3) had no effect on uterine perfusion. Oral treatment with beta-mimetics (group 2) also caused no change in uterine perfusion. In the patients who received i.v. beta-mimetics (group 4), we found a significant decrease in the S/D ratio and pulsatility index of the uterine arteries and a rise in the maternal heart rate. Thus, only the i.v. administration of beta-mimetics led to an increased perfusion rate of the pregnant uterus in the second trimester (Table 13.4).

Table 13.4 Hemodynamic parameters of uterine perfusion before and after surgery and/or prophylactic beta-mimetics

	Group 1	Group 2	Group 3	Group 4
Medication	None	Oral	None	i.v.
Surgical procedure	Amniocentesis	Amniocentesis	Cerclage	Cerclage
S/D preoperative	2.63 ± 0.70	2.67 ± 0.90	2.62 ± 1.14	2.69 ± 1.17
S/D postoperative	2.66 ± 0.76	2.54 ± 0.86	2.61 ± 0.76	2.11 ± 0.44*
PI preoperative	1.53 ± 0.44	1.67 ± 0.46	1.54 ± 0.46	1.62 ± 0.47
PI postoperative	1.54 ± 0.40	1.58 ± 0.75	1.50 ± 0.48	1.21 ± 0.30*
HR preoperative	85.3 ± 11.6	81.4 ± 11.4	84.0 ± 12.6	82.6 ± 11.1
HR postoperative	90.2 ± 13.2	89.7 ± 18.5	83.9 ± 15.9	99.4 ± 15.7*

S/D = S/D ratio; PI = pulsatility index; HR = maternal heart rate
* $p < 0.05$

Discussion

Uterine Perfusion in a Normal Pregnancy

Early Pregnancy

Arcuate arteries. Methods of flowmetry on the maternal side of the fetomaternal circulation continue to be controversial. Two decades ago, Campbell et al.[3] were able to show that changes in the flow pattern of the arcuate arteries could be recorded much earlier than with any other known parameters[3]. There is a normal, early decline in the resistance and pulsatility indices of the radial and spiral arteries that can be detected in the first trimester with color Doppler sonography [24]. Disadvantages of this method are its poor reproducibility, uncertainty over the precise anatomical source of the measurement, and the fact that each uterine artery divides into 12–15 arcuate arteries. The shape of the waveforms depends on whether the measurements are taken on the inside or outside of the uterine wall. Also, the pulsatility index depends on the maternal heart rate and is therefore highly variable. The best parameters for Doppler flow analysis at the level of this vascular bed are the S/D ratio and the resiatnce index[5]. Spiral arteries are consistently detectable between 17 and 20 weeks' gestation at the center of the placenta, where the RI and PI are significantly lower than at the periphery[22].

The increase in uterine blood flow has been correlated with the size of the conceptus, underscoring the importance of the physiology of uterine perfusion in the first trimester. Uterine blood flow shows a linear increase up until the 9th week of gestation, and thereafter it rises exponentially[10]. Color Doppler sonography during the first trimester can already show a difference of uterine blood flow in patients with a threatened abortion or molar pregnancy compared with women with a normal pregnancy[18,20].

Uterine arteries. The physiological changes in early pregnancy chiefly affect the uterine vessels and are a result of trophoblastic invasion and enlarged vessel lumina, whereas rheological factors such as decreased maternal blood viscosity are of no importance[26]. It has long been known that preeclampsia can develop owing to deficient trophoblastic invasion and transformation of the spiral arteries into a low-pressure system[4]. Thus, screening for later pregnancy complications can be done during the first trimester by checking for an abnormal PI in the uterine arteries[19,27].

Reports on transabdominal uterine artery sampling illustrate the difficulty of this type of flowmetry. In many cases it is not possible to define the uterine artery in the B-mode image[16]. The vaginal approach eliminates a source of error in these measurements owing to the proximity of the sampled vessel and appears to be superior to transabdominal scanning. The transvaginal scanning of both uterine arteries can be done successfully in early pregnancy and provides a simple method of examination that is well accepted by patients who have received the necessary information.

Compared with velocimetry of the arcuate arteries, Doppler sonography of the uterine arteries permits a more general evaluation of uterine perfusion since the uterine arteries represent the sum of the flow conditions in all the arcuate arteries. This is particularly advantageous in patients with extensive placental infarctions and in smokers. Doppler flowmetry that is confined to a single, circumscribed area of the placental bed can lead to erroneous results in some circumstances.

Late Pregnancy

So far, the Doppler velocimetry of maternal vessels has not yielded convincing results in terms of fetal surveillance[13]. This is unlikely from a physiological standpoint, because uterine artery velocimetry tends to be unrewarding unless there is absent or deficient trophoblastic invasion of the spiral arteries. The assessment of fetal condition at term or the detection of fetal compromise is accomplished more effectively with fetal heart rate monitoring[35].

Uterine Perfusion in an Abnormal Pregnancy

Fetal growth retardation and/or pregnancy-induced hypertension. Abnormal waveforms in the uterine vessels show a strong correlation with the presence or subsequent development of fetal growth retardation and preeclampsia[33]. Abnormal PI values in the uterine artery show a significantly higher association with premature delivery and low birthweight compared with normal flow patterns[30].

In our study of high-risk pregnancies, we examined one group of patients with normal uterine perfusion and five groups with varying degrees of abnormal uterine perfusion. In cases with pregnancy-induced hypertension or fetal growth retardation, many of the spiral arteries are left out of the transformation into a low-impedance vascular bed or are resistant to such a change. This is manifested by an increase in the S/D ratio by Doppler flowmetry. It is our opinion that group VI had the most severe degree of abnormal uterine perfusion. All of the fetuses in that group responded to the deficient blood supply with abnormal flow velocities in at least one vessel.

One abnormal perfusion parameter. The mildest form of abnormal uterine perfusion appears to exist when only one parameter (the S/D ratio in one uterine artery or a difference between the right and left sides) is abnormal. Hemodynamically, we can interpret this finding as follows: only one uterine artery has undergone normal expansion and transformation into a low-impedance vessel with high diastolic flow, while the contralateral artery has not undergone this change or has failed to form normal collateral channels. This type of pathology appears to have only a moderate adverse effect on the fetal circulation, however. Apparently there has been adequate fetal cardiovascular compensation for the decrease in uterine perfusion.

Two or three abnormal perfusion parameters. When two or three parameters of uterine perfusion were abnormal (increased S/D ratio in one or both uterine arteries, with or without an increased difference between the two sides), we interpreted this as a severe form of abnormal uterine perfusion. These groups (III, V, and VI) had severe fetal circulatory changes that affected all the fetal vessels, suggesting that the presence of two or three abnormal perfusion parameters leads to serious fetal compromise [7]. For this reason, Doppler velocimetry of both uterine arteries is recommended for the clinical assessment of uterine perfusion.

Postsystolic notch. A postsystolic notch in uteroplacental waveforms during the second half of pregnancy indicates an abnormal pulse-wave reflection in the spiral arteries—presumably a result of deficient trophoblastic invasion during placentation. The presence of a postsystolic notch is the Doppler sign of preeclampsia[3] and has a very strong association with intrauterine growth retardation. The notch should no longer be detectable after 24 weeks' gestation; otherwise there is a 70 % likelihood of a pregnancy complication with hypertension[14].

Uterine Perfusion on Medication or after Uterine Manipulation

Mechanisms of action of intravenous beta-mimetics. The increased perfusion rate of the gravid uterus in response to intravenously administered beta-mimetics is based on two mechanisms: an increase in cardiac output and a lowering of peripheral vascular resistance. Since we also observed a fall in the pulsatility index, it is reasonable to assume that the more important mechanism is peripheral vasodilation. The calculated pulsatility index is less affected than the S/D ratio by an increased diastolic blood flow velocity caused by the increase in cardiac output.

No effect of amniocentesis, cervical cerclage, or oral beta-mimetics. In our study population, which was homogeneous for gestational age, we found that amniocentesis, cervical cerclage, and orally administered beta-mimetics produced no change in uterine perfusion that was measurable by vaginal sonography. The intravenous administration of beta-mimetics led to a fall in the S/D ratio and pulsatility index.

Our findings confirm the clinical experience that none of the foregoing procedures hampers blood flow to the pregnant uterus and that the intravenous administration of beta-mimetics even appears to improve it.

It is interesting to note that nifedipine administration produces similar results[29], but this effect has not been seen in response to indomethacin therapy[21].

Doppler Flowmetry of Maternal Vessels as a Screening Test?

The Doppler flowmetry of maternal blood vessels is an important step in evaluating the blood flow of the fetomaternal unit[11, 32]. Besides its use for risk assessment in the first trimester, the transvaginal pulsed Doppler velocimetry of both uterine arteries can also be recommended as a screening test at the end of the second trimester. Placentation is already complete by this time, so that the physiological processes of trophoblastic invasion will not produce false-positive findings (difference between the sides).

Sensitivity for lateral hypertension. When both uterine arteries are examined and the resistance indices are calculated, Doppler velocimetry of the uterine arteries can be done as early as the 22nd week of gestation to screen for pregnancy complications. In the study by Valensise et al.[39], this screening test had a sensitivity of 50% for predicting later pregnancy-induced hypertension, 88% for preeclampsia, and 100% for intrauterine growth retardation.

Negative predictive value. Abnormal uterine artery waveforms can have clinical implications when detected as early as the 24th week of gestation. The negative predictive value of this method is impressive: with a normal finding, there is a 96 % likelihood that a pregnancy complication will not occur[33]. It is sufficient in these cases to schedule follow-up examinations at about four-week intervals. It should be emphasized, however, that not all pregnancy complications can be predicted and diagnosed by Doppler flowmetry. It has been shown that this method is most rewarding in cases of pregnancy-induced hypertension or intrauterine growth retardation, and therefore the test should be performed in patients who are considered from their history to be at risk.

False-positive results. On the other hand, there is a possibility of false-positive test results, and the measurements cannot always be interpreted with complete certainty. As a result, uterine artery velocimetry cannot presently be recommended as a general screening test for all pregnancies[2, 17].

Summary

Doppler velocimetry of maternal vessels is one tool that is used for evaluating the perfusion of the fetomaternal unit. The most rewarding vessels are those that are directly or indirectly responsible for sustaining the pregnancy. Abnormalities in the waveforms of major uterine vessels show a strong correlation with the presence or subsequent development of fetal growth retardation and preeclampsia. Flowmetry of the arcuate arteries can sometimes lead to false results in patients with extensive placental infarcts and in smokers because the examination is confined to a circumscribed, terminal portion of the vascular bed.

Doppler velocimetry in a normal early pregnancy demonstrates high systolic and low diastolic flow velocities combined with a high S/D ratio and a high pulsatility index. The diastolic flow velocity increases with advancing gestation, leading to a decline in the S/D ratio and pulsatility index.

References

1 Arabin B: Die Bedeutung von Perzentilenwerten utero-plazentarer und fetaler Blutflußparameter für die klinische Praxis. Ultraschall Klin. Prax. Suppl. 1 (1987) 51

2 Bewley S, Cooper D, Campbell S: Doppler investigation of uteroplacental blood flow resistance in the second trimester: a screening study for pre-eclampsia and intrauterine growth retardation. Brit. J. Obstet. Gynecol. 98 (1991) 871–879

3 Campbell S, Griffin DR, Pearce JM, Wilson K, Teague MJ: New Doppler technique for assessing uteroplacental blood flow. Lancet 1 (1983) 675–677

4 Campbell S, Pearce JM, Hackett G, Cohen-Overbeek T, Hernandez C: Qualitative assessment of uteroplacental blood flow: early screening test for high-risk pregnancies. Obstet. Gynecol. 68 (1986) 649–653

5 Chaoui R, Hoffmann H, Bollmann R, Halle H, Zienert A, Metzner A: Erfassung der uteroplazentaren Durchblutung ungestörter Schwangerschaften mittels gepulstem Doppler. Ultraschall. Zentralbl. Gynäkol. 112 (1990) 11–18

6 Deutinger J, Rudelstorfer R, Bernaschek G: Vaginosonographic velocimetry of both main uterine arteries by visual vessel recognition and pulsed Doppler method during pregnancy. Amer. J. Obstet. Gynecol. 159 (1988) 1072–1076

7 Deutinger J: Vaginosonographische gepulste Doppler-Strömungsmessungen in Gefäßen des kleinen Beckens. Enke, Stuttgart 1990

8 Deutinger J, Rudelstorfer R, Bernaschek G: Vaginosonographic Doppler velocimetry in both uterine arteries: elevated left-right differences and relationship to fetal haemodynamics and outcome. Early Hum. Develop. 25 (1991) 187–196

9 Deutinger J, Rudelstorfer R, Pattermann A, Bernaschek G: Vaginosonographic velocimetry in uterine arteries before and after administration of beta-mimetics. Brit. J. Obstet. Gynaecol. 99 (1992) 417–421

10 Dickey RP, Hower JF: Ultrasonographic features of uterine blood flow during the first 16 weeks of pregnancy. Hum. Reprod. 10 (1995) 2448–2452

11 Fendel H, Fendel M, Warnking R: Fehlermöglichkeiten der gepulsten Dopplermethode zur Blutflußmessung am Feten. Z. Geburtsh. Perinat. 187 (1983) 83–87

12 Fendel H, Fendel M, Jörn H, Funk A, Docker B, Meyer W: Doppler-Score zur Beurteilung des perinatalen Risikos. Ultraschall Klin. Prax. 5 (1990) 68–73

13 Fischer RL, Kuhlman KA, Depp R, Wapner RJ: Doppler evaluation of umbilical and uterine-arcuate arteries in the postdates pregnancy. Obstet. Gynecol. 78 (1991) 363–368

14 Fleischer A, Schulman H, Farmakides G: Uterine artery Doppler velocimetry in pregnant women with hypertension. Amer. J. Obstet. Gynecol. 154 (1996) 806–813

15 Fuller EO, Galletti PM, Takeuchi T: Major and collateral components of blood flow to pregnant sheep uterus. Amer. J. Physiol. 229 (1973) 279–285

16 Hanretty KP, Whittle M: Doppler uteroplacental waveforms in pregnancy induced hypertension: A re-appraisal. Lancet 1 (1988) 850–852

17 Jacobson S, Imhof R, Manning N et al.: The value of Doppler assessment of the uteroplacental circulation in predicting preeclampsia or intrauterine growth retardation. Amer. J. Obstet. Gynecol. 162 (1990) 110–114

18 Jauniaux E, Zaidi J, Jurkovic D, Campbell S, Hustin J: Comparison of color Doppler features and pathological findings in complicated early pregnancy. Hum. Reprod. 9 (1994) 2432–2437

19 Kurjak A, Predanic M, Kupesic-Urek S: Transvaginal color Doppler in the assessment of placental blood flow. Eur. J. Obstet. Gynecol. Reprod. Biol. 49 (1993) 29–32

20 Kurjak A, Zalud I, Predanic M, Kupesic S: Transvaginal color and pulsed Doppler study of uterine blood flow in the first and early second trimesters of pregnancy: normal versus abnormal. J. Ultrasound Med. 13 (1994) 43–47

21 Mari G, Kirshon B, Wassersturm N, Moise KJ, Deter RL: Uterine blood flow velocity waveforms in pregnant women during indomethacin therapy. Obstet. Gynecol. 76 (1990) 33–36

22 Matijevic R, Meekins JW, Walkinshaw SA, Neilson JP, McFadyen IR: Spiral artery blood flow in the central and peripheral areas of the placental bed in the second trimester. Obstet. Gynecol. 86 (1995) 289–292

23 McCallum WD, Williams CS, Nagel S, Daigle RE: Fetal blood flow velocity waveforms. Amer. J. Obstet. Gynecol. 132 (1978) 425–429

24 Merce LT, Barco MJ, Bau S: Color Doppler sonographic assessment of placental circulation in the first trimester of normal pregnancy. J. Ultrasound Med. 15 (1996) 135–142

25 Moll W, Nienartowicz A, Hes H, Lenz A: Blood flow regulation in the uteroplacental arteries. Trophoblast Res. 3 (1988) 83–96

26 Oosterhof H, Wichers G, Fidler V, Aarnoudse JG: Blood viscosity and uterine artery flow velocity waveforms in pregnancy: a longitudinal study. Placenta 14 (1993) 555–561

27 Olofson P, Laurini RN, Marsál K: A high uterine artery pulsatility index reflects a defective development of placental bed spiral arteries in pregnancies complicated by hypertension and fetal growth retardation. Eur. J. Obstet. Gynecol. Reprod. Biol. 49 (1993) 161–168

28 Palmer SK, Zamudio S, Coffin C: Quantitative estimation of human uterine artery blood flow and pelvic blood flow redistribution in pregnancy. Obstet. Gynecol. 80 (1992) 1000–1006

29 Pirhonen JK, Erkkola RU, Ekblad UU: Uterine and fetal flow velocity waveforms in hypertensive pregnancy: the effect of a single dose of nifedipine. Obstet. Gynecol. 76 (1990) 37–41

30 Ruckhäberle KE, Faber R, Robel R, Viehweg B: Diagnostik und Therapie gestörter Hämodynamik. Ein Beitrag zum Management bei Schwangerschaften mit drohender Frühgeburt. Z. Geburtsh. u. Perinat. 196 (1992) 152–158

31 Schulman H, Fleischer A, Stern W, Farmakides G, Jagani N, Blattner P: Umbilical velocity wave ratios in human pregnancy. Amer. J. Obstet. Gynecol. 148 (1984) 985–990

32 Schulman H, Fleischer A, Farmakides G, Bracero L, Rochelson B, Grunfeld L: Development of uterine artery compliance in pregnancy as detected by Doppler ultrasound. Amer. J. Obstet. Gynecol. 155 (1986) 1031–1036

33 Schulman H: The clinical implications of Doppler ultrasound analysis of the uterine and umbilical arteries. Am. J. Obstet. Gynecol. 156 (1987) 889–893

34 Sohn C, Kesternich P, Fendel H: Der Einfluß der Körperposition auf die uterine Durchblutung im 3. Schwangerschaftstrimenon. Ultraschall 10 (1989) 10–14

35 Sohn C, Stolz W: Untersuchung zur Entwicklung der Dopplerparameter in fetalen und mütterlichen Gefäßen 10 Tage vor bis 10 Tage nach errechnetem Geburtstermin. Geburtsh. u. Frauenheilk. 54 (1994) 102–107

36 Thaler I, Manor D, Itskovitz J et al.: Changes in uterine blood flow during human pregnancy. Amer. J. Obstet. Gynecol. 162 (1990) 121–125

37 Tonge HM, Wladimiroff JW, Noordam MJ, van Kooten C: A study on fetal blood flow velocity waveforms in cases of intrauterine growth retardation. Obstet. Gynecol. 67 (1986) 851–855

38 Trudinger BJ, Giles WB, Cook CM: Uteroplacental blood flow velocity-time waveforms in normal and complicated pregnancy. Brit. J. Obstet. Gynaecol. 92 (1985) 39–45

39 Valensise H, Bezzeccheri V, Rizzo G, Tranquilli AL, Garzetti G, Romanini C: Doppler velocimetry of the uterine artery as a screening test for gestational hypertension. Ultrasound Obstet. Gynecol. 3 (1993) 18–22

40 Wladimiroff JW, Tonge HM, Stewart PA: Doppler ultrasound of cerebral blood flow in the human fetus. Brit. J. Obstet. Gynaecol. 93 (1986) 471–475

14 Color Doppler Sonography of the Uterine Arteries in Early Pregnancy to Screen for Preeclampsia and Uteroplacental Insufficiency

A. Funk

Applications of Color Doppler Sonography during Pregnancy

Placental insufficiency and preeclampsia in the third trimester. Doppler sonography of the uterofetoplacental unit in the third trimester is part of the standard diagnostic workup of patients with suspected placental insufficiency and preeclampsia. At this late stage of gestation, however, it is possible only to confirm the presence of a chronic perfusion deficit. The Doppler findings, along with B-mode findings and biophysical and biochemical tests, are helpful in assessing the degree of fetal compromise and directing clinical management.

Early detection of decreased uterine perfusion. The efforts of modern medicine are also aimed at early detection, clinical intervention, and prevention, and these goals have prompted the use of transvaginal ultrasound scanning to assess blood flow early in the pregnancy.

Changes in the spiral arteries. The central question is whether Doppler sonography is able to confirm the normal development of blood flow and also detect decreased blood flow during the first half of pregnancy. Another key issue is whether Doppler evidence of decreased uterine perfusion actually correlates with the development of pregnancy-induced hypertension (PIH) and intrauterine growth retardation.

Normally a massive increase of blood flow occurs during the first half of pregnancy, and this requires a morphological transformation of the spiral arteries as described by Brosens et al.[3]. Trophoblastic cells invade and erode the musculoelastic media of the spiral arteries, transforming the vessels into large, saclike tubes with a funnel-like expansion at their junction with the intervillous space. These changes take place between the 14th and 20th weeks of menstrual age and are completed no later than the 24th week, even when maturation is delayed. The calibers of the spiral arteries are increased by a factor of 30, the arcuate arteries by a factor of 10, and the uterine arteries by a factor of 1.5–3[2].

Blood circulation in the intervillous space. The blood flow changes that occur during the first trimester are more difficult to classify. Hustin and Shaaps[7] performed hysteroscopic, radiological, and morphological studies showing that intervillous blood circulation does not occur before the 12th week of gestation. This led them to conclude that there is no contact between the chorionic villi and maternal blood during the first trimester and that the embryo and placenta are independent of the maternal environment. They distinguished between a period in which implantation, organogenesis, and placentation occur and a second period that begins with active maternal blood flow through the intervillous space toward the end of the first trimester.

It is reasonable to conclude that uterine blood flow in the first trimester is controlled in part by the rising estrogen levels in the maternal serum. This can be inferred from the fact that cyclic changes in uterine blood flow correlate with the estradiol level[4, 12], that estrogen receptors have been identified in muscle cells of the uterine arteries[9], and that animal studies have shown dilatation of the terminal segments of the uterine vessels prior to trophoblast invasion[10].

Technique of Transvaginal Doppler Sonography

Transducer placement. With the patient in the lithotomy position, the ultrasound probe, sheathed in a condom and smeared with contact gel, is inserted into the vagina until it is in contact with the anterior fornix (anteflexed uterus) or posterior fornix (retroflexed uterus). An initial survey is carried out, and the probe is positioned in the lateral fornix so that it is in direct contact with the uterine attachment of the parametrium through the vaginal wall. The proximity to the target organs makes it possible to use transducer frequencies of 5 MHz or higher, which are excellent for blood flow studies.

Vascular anatomy. After arising from the internal iliac artery, each uterine artery, surrounded by the uterine veins, curves over the ureter about 2 cm from the cervix and reaches the uterus at the level of the isthmus. There it divides into a descending cervical branch and an ascending branch. The latter branch ascends tortuously along the side of the uterus and anastomoses with the ovarian artery via the tubal branch. The best sites for recording Doppler signals are the main trunk of the uterine artery and the origin of the ascending branch. Both vessels can be clearly identified in those areas and can be scanned at an optimum beam–vessel angle.

Advantages of transvaginal scanning. According to our studies, the transvaginal approach has several major advantages over transabdominal scanning, especially in early pregnancy examinations[5]:

- Accurate overall assessment of uterine perfusion
- Greater pulsatility of blood flow, allowing better recognition of impedance criteria in the Doppler scan
- Smaller range of error owing to a more favorable beam–vessel angle

Normal Development of Uterine Artery Doppler Spectra

The Doppler spectra of the uterine arteries display characteristic changes from early pregnancy until the conclusion of trophoblastic invasion in midpregnancy (Fig. 14.**1**).

Doppler spectra in early pregnancy. At the start of pregnancy, the Doppler spectra of the uterine arteries do not differ from those recorded in the periovulatory period. Highly variable patterns can occur. The spectra typically have a sharp systolic upstroke and downstroke with a sharp, narrow systolic peak and a low maximum frequency shift. A variable notch may appear in the systolic downstroke. The diastolic phase of the cardiac cycle consistently shows a very low frequency shift whose pattern changes more characteristically with advancing gestation than the systolic peak. During the first trimester the sharp systolic downslope typically ends in an early diastolic notch, which may convert to reverse flow during the initial weeks. The postsystolic notch is followed by a diastolic peak that shows the highest velocities in the diastolic part of the cardiac cycle during the first trimester. The end-diastolic frequency shift is low in the first trimester, and absent or reverse flow is seen in rare cases. These normal features during early pregnancy are useful for the visual interpretation of Doppler waveforms in cases where abnormalities develop later in the pregnancy.

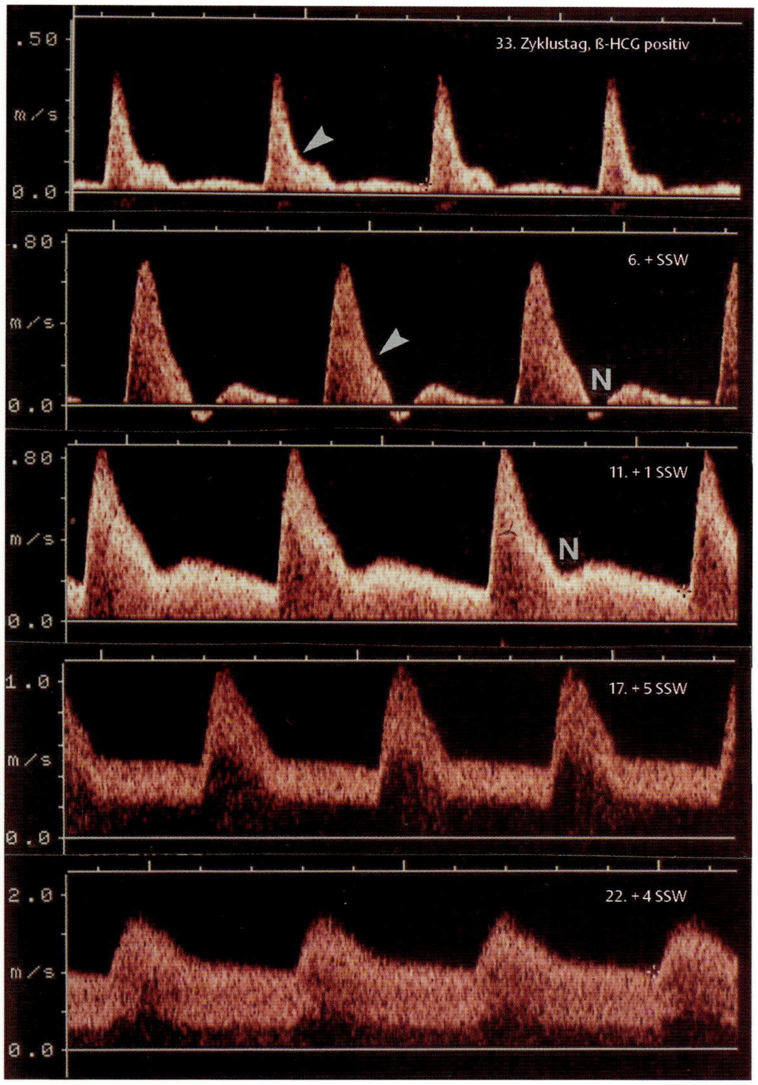

Fig. 14.**1** Comparison of uterine Doppler spectra recorded during the first half of pregnancy. The maximum frequency shift in each spectrum can be read on the adjacent scale. Arrows = systolic notch; N = early diastolic notch.

Changes after the first trimester. The characteristic changes past the first trimester are a continued increase in the systolic maxima with broadening of the systolic peak and an increasingly high diastolic frequency shift, with disappearance of the early diastolic notch. The increase in flow velocities is relatively greater at end diastole than in systole. The angle between the systolic downstroke and the maximum diastolic frequencies becomes greater, and there is an overall flattening of the spectral waveform. Generally this development is completed in the 20th week of gestation or by the 24th week at the latest.

Normal Values in Early Pregnancy

Progression of values from weeks 4 to 24. In a prospective longitudinal study, the progression of normal values for Doppler ultrasound parameters was documented from the 4th to 24th weeks of menstrual age. Of 257 pregnancies examined, 79 met the criteria for inclusion in the study (confirmed dates at examination, no significant uterine anomalies, uncomplicated singleton pregnancy, term delivery of a healthy eutrophic infant) based on a retrospective postpartum review. A maximum of 117 individual examinations of the uterine vessels could be evaluated. In addition to the standard indices, the early diastolic notch frequency and maximum diastolic frequency were also determined. The latter quantities were used to calculate an index, the notch/peak (N/P) ratio, which characterizes the presence or absence of an early diastolic notch in the spectral waveform. The data from both uterine arteries were used in calculating all the indices. The following instruments were available for the study: Combison 320–5 combined with the D 300 Doppler unit (Kretztechnik, 5 MHz vaginal probe) and the 128 XP/10 color Doppler unit (Acuson Corp., 5 MHz vaginal probe). SAS software from the Department of Medical Information and Biometry of Aachen Medical College was used for statistical analysis. The progressions of normal values for the S/D ratio, resistance index (RI), pulsatility index (PI), and N/P ratio are summarized in Table 14.**1**.

Table 14.**1** Normal values of uterine artery Doppler indices from the 4th to 24th weeks menstrual age

Weeks	*n*	Mean	SD	P5	P10	P50	P90	P95
S/D ratio (*n* = 100)								
4– 6	14	12.18	8.14	4.87	6.44	10.26	15.45	38.50
7– 9	13	9.54	3.68	4.92	5.53	7.92	13.90	16.77
10–12	8	5.97	4.62	2.86	2.86	4.04	16.75	16.75
13–15	16	3.89	1.59	2.15	2.50	3.47	7.08	7.20
16–18	24	*2.87	1.32	1.97	2.06	2.43	3.70	6.14
19–21	13	2.38	0.37	1.79	1.82	2.37	2.80	2.83
22–24	12	2.14	0.21	1.64	1.99	2.12	2.39	2.45
RI (*n* = 104)								
4– 6	15	0.89	0.04	0.79	0.84	0.89	0.95	1.00
7– 9	13	*0.87	0.04	0.79	0.80	0.86	0.93	0.93
10–12	8	0.76	0.09	0.65	0.65	0.74	0.93	0.93
13–15	16	0.69	0.07	0.56	0.59	0.70	0.81	0.81
16–18	26	*0.61	0.08	0.49	0.51	0.60	0.72	0.75
19–21	14	0.56	0.06	0.44	0.45	0.56	0.63	0.64
22–24	12	0.52	0.05	0.39	0.49	0.53	0.57	0.58
PI (*n* = 102)								
4– 6	15	3.32	0.78	1.85	2.36	3.39	4.37	4.80
7– 9	13	2.85	1.07	0.87	1.94	2.53	4.52	4.76
10–12	8	2.12	0.92	1.17	1.17	1.93	3.93	3.93
13–15	16	1.86	0.50	0.85	1.02	2.04	2.29	2.59
16–18	25	*1.25	0.61	0.73	0.79	1.09	1.94	2.93
19–21	13	1.06	0.24	0.66	0.77	1.02	1.44	1.52
22–24	12	0.87	0.23	0.52	0.72	0.82	0.99	1.49
N/P-Ratio (*n* = 99)								
4– 6	15	0.595	0.18	0.10	0.44	0.67	0.74	0.80
7– 9	13	0.628	0.17	0.30	0.41	0.67	0.80	0.85
10–12	8	0.871	0.08	0.71	0.71	0.89	0.99	0.99
13–15	14	0.924	0.10	0.76	0.78	0.94	1.07	1.07
16–18	26	1.03	0.15	0.69	0.86	1.06	1.18	1.22
19–21	12	1.05	0.08	0.91	0.96	1.05	1.15	1.16
22–24	11	1.07	0.07	0.95	0.98	1.11	1.14	1.15

* Significant difference relative to previous value

Early Doppler Examination of Uteroplacental Blood Flow in Abnormal Pregnancy

Patients

Inclusion and exclusion criteria. This series consists of 25 pregnancies that were examined by transvaginal Doppler sonography before the 25th week of gestation and had an abnormal course or outcome. Inclusion criteria were confirmed dates at the time of the examination, a singleton pregnancy, the development of PIH (diastolic blood pressure >90 mmHg in multiple readings) with or without proteinuria (>300 mg/day), perinatal asphyxia based on histologically confirmed placental insufficiency, and the delivery of a growth-retarded infant (<10th weight percentile after Hohenauer). Pregnancies with premature delivery, fetal and/or chromosome anomalies, and early pregnancy loss ≤ 16 weeks were excluded from the study.

Analysis. The methodology and statistical tools were the same as described above. Statistical significance measured by the 10th and 90th quantiles was determined with the independent *t*-test with a significance level of $\alpha = 0.05$. The positive and negative predictive values were of limited validity because of the high prevalence in the selected at-risk population.

Results

Maternal parameters such as age and number of pregnancies and deliveries did not differ from those in the normal group. Significant differences were seen mainly in the length of pregnancy ($x = 31.8$ weeks, SD = 8.0 weeks), birthweight ($x = 2009$ g, SD = 1285 g), and in the one- and five-minute Apgar scores.

Doppler parameters. None of the Doppler parameters showed significant differences from a normal pregnancy during the first trimester. Table 14.**2** shows the data on specificity, sensitivity, and predictive values from weeks 13–15 to weeks 22–24. It is not until week 16, and thus after the end of the first trimester, that individual values for the S/D ratio, RI, and PI exceeded the 90th quantile or individual values for the N/P ratio fell below the 10th quantile for a normal pregnancy.

PI. With advancing gestation, the Doppler findings increasingly reflected the delayed maturation of uterine perfusion. The PI provided the best discrimination: 4 of 9 cases in weeks 16–18, 5 of 7 cases in weeks 19–21, and all 8 cases in weeks 22–24 were above the 90th percentile. Starting in week 19, the PI showed a sensitivity of 87 % and a specificity of 92 %. The results for the S/D ratio and RI were slightly lower.

N/P ratio. The N/P ratio, which expresses whether an early diastolic notch is present, had a mean value >1 starting in weeks 16–18 of uncomplicated pregnancies. This parameter also showed increasing sensitivity after 16 weeks. Reaching 83 % by week 19 ($n = 15$), the sensitivity of the N/P ratio was slightly lower than that of the standard Doppler indices.

Table 14.**2** Accuracy of uterine artery Doppler examination from the 13th to 24th weeks of menstrual age in predicting the development of pregnancy-induced hypertension or uteroplacental insufficiency

Weeks	*n*	Sensitivity (%)	Specificity (%)	Positive predictive value (%)	Negative predictive value (%)
S/D ratio					
13–15	2	0	93	0	88
16–18	8	37.5	92	60	92
19–21	6	67	92	80	86
22–24	7	100	92	87.5	100
≥ 19	13	85	92	85	92
RI					
13–15	2	50	93	50	93
16–18	8	25	92	50	79
19–21	6	33	92	33	67
22–24	7	100	92	87	100
≥ 19	13	69	92	82	85
PI					
13–15	2	0	94	0	88
16–18	9	44.5	92	67	83
19–21	7	71.5	93	83	87
22–24	8	100	92	89	100
≥ 19	15	87	92	87	92
N/P ratio					
13–15	2	50	93	50	93
16–18	9	44.5	92	67	83
19–21	7	57	92	80	78
22–24	8	87.5	91	87.5	91
≥ 19	15	83	91	84	85

Discussion

Opening of the spiral arteries. The visible waveform changes in the uterine vessels, like the progression of the calculated indices, reflect the morphological processes in the terminal vascular bed that are causing progressive dilatation of the spiral arteries. The variability of the frequency spectra seen during the initial weeks may be attributable to differences in endovascular trophoblastic invasion and to individual factors such as hormone levels and vascular calibers[6]. Our results show that it is not possible to distinguish between normal and delayed or impaired perfusion during the first trimester. The significant difference from weeks 13–15 to weeks 16–18 indicates that a disproportionately large decrease in vascular resistance occurs at the start of the second trimester. This may reflect increasing uterine perfusion after the spiral arteries have opened toward the intervillous space[7].

Secondary trophoblastic invasion. Secondary trophoblastic invasion is crucial for the development of uteroplacental blood flow. The characteristic Doppler changes that occur during this period are a rise in maximum flow velocities throughout the cardiac cycle with a relative predominance in diastole and a flattening or disappearance of the early diastolic notch (in weeks 16–18 on average). A key observation is that there was no case in which uterine blood flow developed normally at first and then became abnormal secondarily. At the same time, it was common to find a late onset of normal blood flow with disappearance of the early diastolic notch by the 24th week of gestation. Aristidou et al.[1] attribute this to a delayed but otherwise normal process of trophoblastic invasion and placental maturation.

Correlation of Doppler findings with pregnancy disorders. The study results indicate that Doppler sonography is a proven method for the early detection of decreased uterine perfusion. The results also show that pregnancies that are at risk for developing a hypertensive disorder and/or uteroplacental insufficiency can be predicted with increasing confidence in early pregnancy and that a definite correlation exists with the development of uterine perfusion during the period of secondary trophoblastic invasion. The results of Steel et al.[11] in particular indicate that the sensitivity of early Doppler ultrasound screening correlates with the severity of the developing pregnancy disorder.

An abnormal uterine artery Doppler finding may also reflect the presence of chromosomal and/or fetal anomalies, however (reference 1 and authors' studies). Kuhlmann et al.[8] claim that the cause is placental undervascularization resulting from arrested or delayed angiopoiesis.

How to proceed when decreased uterine perfusion is found. The diagnosis of decreased uterine perfusion warrants further testing and also a therapeutic response. When anomalies have been excluded, treatment may consist of a recommendation for rest, early hemodilution, or aspirin therapy. There are still no definitive data on the most effective treatment options.

We recommend that the following steps be taken when decreased uterine perfusion is diagnosed in the 19th or 20th week of gestation:

- Exclude a chromosomal and/or fetal anomaly.
- Recommend physical rest.
- Treatment with aspirin, hemodilution, or NO donors may be tried in controlled studies.
- Reexamine the patient at 24 weeks. If uterine perfusion is still decreased, continue treatment with close-interval pregnancy surveillance.

Screening. It remains unclear whether Doppler sonography should be used as a routine screening test. It seems prudent to include examination of the uterine vessels in routine screening at 19–22 weeks, even in patients who have no obvious risk factors.

The early diagnosis of pronounced uteroplacental insufficiency brings us closer to the goal of reducing the high morbidity and mortality of early prematurity by prolonging the pregnancy. Effective treatment options still need to be developed and evaluated.

Summary

Characteristic changes in the Doppler sonogram. Doppler sonography of the uterine vessels is proving to be a suitable method for assessing the development of the terminal uterine vascular branches to a low-impedance system and for indirectly following the progression of morphological changes, especially in the spiral arteries. The characteristic changes in the Doppler sonogram consist of a relatively greater increase in frequency shifts during the diastolic phase of the cardiac cycle compared with systole. This causes a decrease in Doppler indices until the middle of the pregnancy and a disappearance of the early diastolic notch (on average, during postmenstrual weeks 16–18).

Given the high variability of Doppler waveforms during the first trimester, it is not possible to discriminate between normal and abnormal perfusion until the start of the second trimester.

Diagnostic accuracy. The diagnostic accuracy of Doppler sonography increases with advancing gestation and is maximal between weeks 19 and 24. At this time a definite correlation exists between an abnormal Doppler sonogram and the subsequent development of preeclampsia and/or uteroplacental insufficiency.

The diagnostic value of early Doppler scanning of the uterine arteries is that pregnancy complications can be detected at an early stage and therapeutic measures can be instituted to prolong the pregnancy. The ultimate goal is to reduce the high morbidity and mortality of very early prematurity. Effective treatment options still need to be developed and evaluated, however.

References

1 Aristidou A, van den Hof MC, Campbell S, Nicolaides K: Uterine artery Doppler in the investigation of pregnancies with raised maternal serum alphafetoprotein. Brit. J. Obstet. Gynaecol. 97 (1990) 431

2 Bieniarz J, Yoshida T, Romero-Salinas G, Curuchet E, Caldeyro-Barcia R, Crottogini JJ: Aortacaval compression by the uterus in late human pregnancy. IV. Circulatory homeostatis by preferential perfusion of the placenta. Amer. J. Obstet. Gynecol. 103 (1969) 19

3 Brosens I, Robertson VB, Dixon HG: The physiological response of the vessels of the placental bed to normal pregnancy. J. Pathol. Bacterial 93 (1967) 569

4 De Ziegler D, Bessis R, Frydman R: Vascular resistance of uterine arteries: physiological effects of estradiol and progesterone. Fertil. Steril. 55 (1991) 775

5 Funk A, Jörn H, Fendel H: Abdominale versus transvaginale Doppler-Sonographie der uterinen Gefäße. Ultrasch. Klin. Prax. 7 (1992) 264

6 Goswamy RK, Steptoe PC: Doppler ultrasound studies of the uterine arteries in spontaneous ovarian cycles. Hum. Reprod. 3 (1988) 721

7 Hustin J, Schaaps J-P: Echocardiographic and anatomic studies of the maternotrophoblastic border during the first trimester of pregnancy. Amer. J. Obstet. Gynecol. 157 (1987) 162

8 Kuhlmann RS, Werner AL, Abramowicz J, Warsoff SL, Arrington J, Levy DL: Placental histology in fetuses between 18 and 23 weeks› gestation with abnormal karyotyp. Amer. J. Obstet. Gynecol. 163 (1990) 1264

9 Perrot-Applanat M, Groyer-Picart MT, Garcia E, Lorenzo F, Milgram E: Immunocytochemical demonstration of estrogen and progesterone receptors in muscel cells of uterine arteries in rabbits and humans. Endocrinology 123 (1988) 1511

10 Rudolph AM, Heyman MA: Circulatory changes during growth in fetal lamb. Circ. Res. 26 (1970) 289

11 Steel SA, Pearce JM, Mc Parland J, Chamberlain GVP: Early doppler ultrasound screening in prediction of hypertensive disorders of pregnancy. The Lancet 335 (1990) 1548

12 Weiner Z, Thaler I, Levron J, Lewit N, Itskovity-Eldor J: Assessment of ovarian and uterine blood flow by transvaginal color Doppler in ovarian-stimulated women correlation with the number of follicles and steroid hormone levels. Fertil. Steril. 59 (1993) 743

15 Normal Fetomaternal Doppler Indices in the Second and Third Trimesters of Pregnancy

A. K. Ertan, H. J. Hendrik, I. Tossounidis, and W. Schmidt

Establishing Normal Curves

The goal of defining normal values for fetomaternal perfusion is to be able to discriminate between normal and abnormal pregnancies on the basis of Doppler ultrasound findings. This requires the establishment of reference values and normal values for physiological conditions of flow[8].

"Ideal" course of pregnancy. Not infrequently, normal values are based on data from pregnancies that must meet a set of normal criteria and thus represent more of an "ideal" case. Because of this selection effect, the defined cutoff limits may be too narrow for at best they describe the variability of "ideal values" rather than the original biological scatter[27].

Dependence on gestational age. Perfusion changes have been described with advancing gestation due to maturation of the trophoblastic tissue, and these changes must be considered in the interpretation of reference curves. It is best, therefore, to use gestational-age-based charts to express the normal values for particular indices. Measurements should commence at a stage of the pregnancy when the results of the measurements can still influence obstetric management. The current consensus is that this is the mid-second trimester of pregnancy, when fetal viability is reached. There are also cases in which the analysis of maternal arterial waveforms can be rewarding in earlier weeks of pregnancy[16].

Cross-sectional studies. In a cross-sectional study design, the measurements should be spaced as evenly as possible over the study period, obtaining an adequate number of measurements for the various weeks of pregnancy, such as more than 15 per week[13]. It is best to avoid multiple measurements in any one pregnancy as this might narrow the tolerance ranges by reducing the variability relative to the natural biological scatter[9].

The goal of establishing normal reference curves is to offer a more favorable prognosis for pregnancies that are at risk by their history and/or findings but exhibit normal blood flow patterns. This also requires the analysis of populations with abnormal pregnancies[20]. The curves presented below are designed to show the distribution of normal values for the Doppler indices that are most commonly used to characterize fetomaternal blood flow.

Methodology

Defining the Normal Population

Our normal population consisted of patients whose dates had been confirmed by an ultrasound examination before the 20th week of pregnancy. We excluded cases that had a primary cesarean section for placental insufficiency and cases with an abnormal fetal heart rate (FHR) trace or impending intrauterine asphyxia. The birthweights in the normal population were between the 10th and 90th percentiles for gestational age as defined as Roemer et al.[18]. The 5- and 10-minute Apgar scores were 7 or higher, and the umbilical artery pH was 7.20 or higher. We also excluded multiple pregnancies and cases with fetal anomalies.

Plotting Quantile Curves

Indices. The qualitative analysis of recorded Doppler waveforms has proved to be a more useful clinical tool than the more complex calculation of volume flow. This is reasonable when we consider that a quantitative analysis requires the measurement of small vascular cross sections, and this adds a potential source of error.

Numerous indices have been devised for the qualitative analysis of Doppler waveforms (review in reference 8). The following are the most commonly used (Fig. 15.**1**):

- The S/D ratio of Stuart (1980)[22]
- The resistance index (RI) of Pourcelot (1974)[17]
- The pulsatility index (PI) of Gosling and King (1977)[11]

The following quantities are used in analyzing the Doppler sonogram and calculating the indices:

S = maximum peak systolic frequency
D = maximum end-diastolic frequency
Tav = temporal average of maximum frequency
D = instantaneous spatial average frequency
E = temporal average of spatial average frequencies

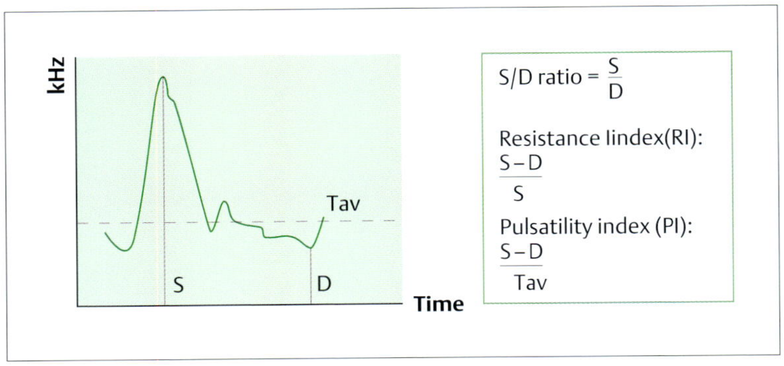

Fig. 15.1 Parameters used for qualitative waveform analysis. S = maximum systolic frequency, D = maximum end-diastolic frequency, Tav = temporal average of maximum frequency.

The indices are calculated from the maximum systolic (S) and end-diastolic (D) frequency shifts and, in the case of the PI, from the maximum frequency shift averaged over the entire cardiac cycle (Tav) using these formulas (Fig. 15.**1**):

$$\text{S/D ratio} = \frac{S}{D}$$

$$\text{RI} = \frac{S-D}{S}$$

$$\text{PI} = \frac{S-D}{\text{Tav}}$$

In some ultrasound systems, the temporal average of the maximum frequencies (E) is used in calculating the PI. This results in correspondingly higher values.

In the studies published to date, no definite advantage has been found for any single parameter[6,8,9–11]. Because all the indices express a ratio, they are largely independent of the insonation angle. The S/D ratio is very easy to determine, and the RI is easy to interpret: values close to 0 signify a very low flow resistance, while a value of 1 indicates cessation of flow. The PI, however, is the only index that can provide differentiated information on an absence of end-diastolic flow, because the average maximum frequency shift contains information on the cardiac cycle as a whole, while the S/D ratio tends toward infinity and the RI equals 1 in cases with absent end-diastolic flow.

Quantiles. Our calculation of normal values was based on 602 pulsed Doppler flow measurements in 370 patients who met the normal population criteria defined above. An average of 1.62 measurements were acquired per patient. Quan tiles were determined as a function of gestational age, since the Doppler values did not show a normal distribution. The study period was divided into intervals of two consecutive gestational weeks each, and the 5th, 10th, 50th, 90th, and 95th quantiles were calculated and plotted graphically for each interval. The curves were smoothed by cubic regression. In the first example given below, both the smoothed and original curves are shown. Since the smoothing did not significantly alter the curve shape, only the smoothed curves are presented in the figures below. Th e 5th, 10th, 50th, 90th, and 95th quantiles are shown for each of the measured indices.

Results

Figures 15.**2**– 15.**14** show the reference curves obtained for the S/D ratio, PI, and RI of the umbilical artery, fetal aorta, middle cerebral artery, and uterine artery. Figure 15.**15** shows the reference curves for the cerebroplacental ratio.

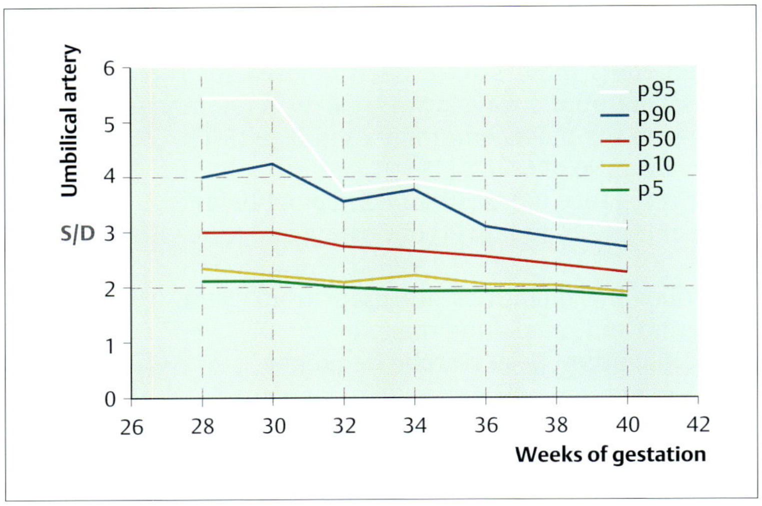

Fig. 15.2 Reference curves for the **S/D ratio of the umbilical artery**, based on 600 measurements in normal pregnancies and grouped by even gestational weeks.

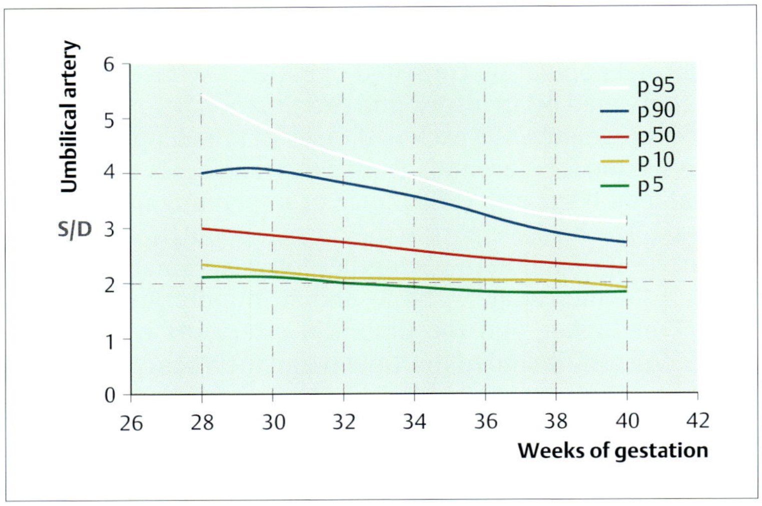

Fig. 15.3 Reference curves for the **S/D ratio of the umbilical artery**, based on 600 measurements in normal pregnancies and grouped by even gestational weeks. The curves were smoothed by cubic regression.

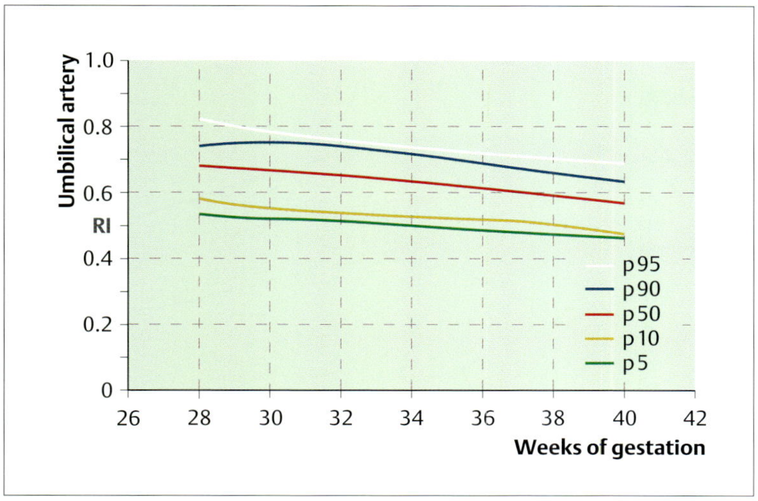

Fig. 15.**4** Reference curves for the **RI of the umbilical artery**, based on 600 measurements in normal pregnancies and grouped by even gestational weeks. The curves were smoothed by cubic regression.

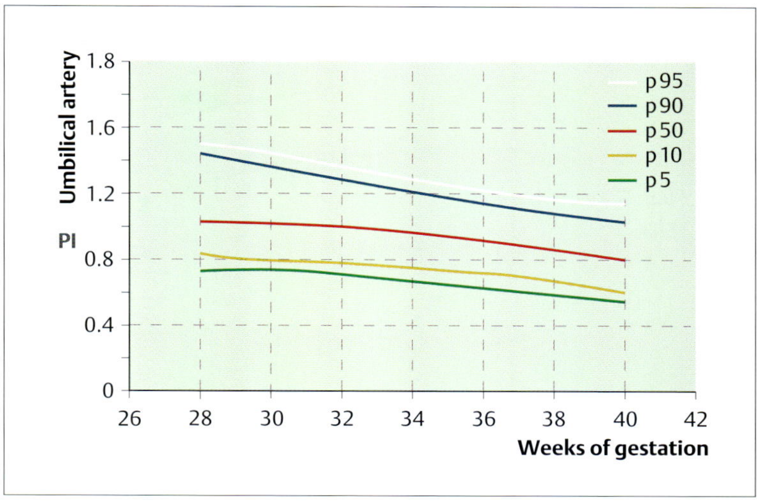

Fig. 15.**5** Reference curves for the **PI of the umbilical artery**, based on 510 measurements in normal pregnancies and grouped by even gestational weeks. The curves were smoothed by cubic regression.

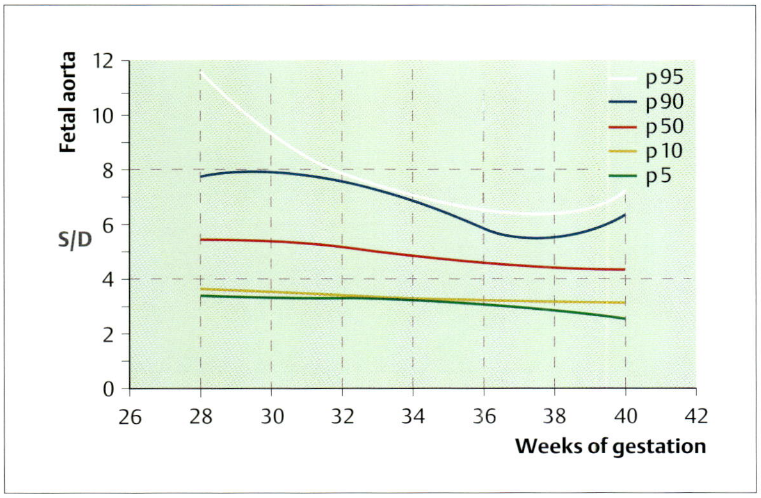

Fig. 15.**6** Reference curves for the **S/D ratio of the fetal aorta**, smoothed by cubic regression.

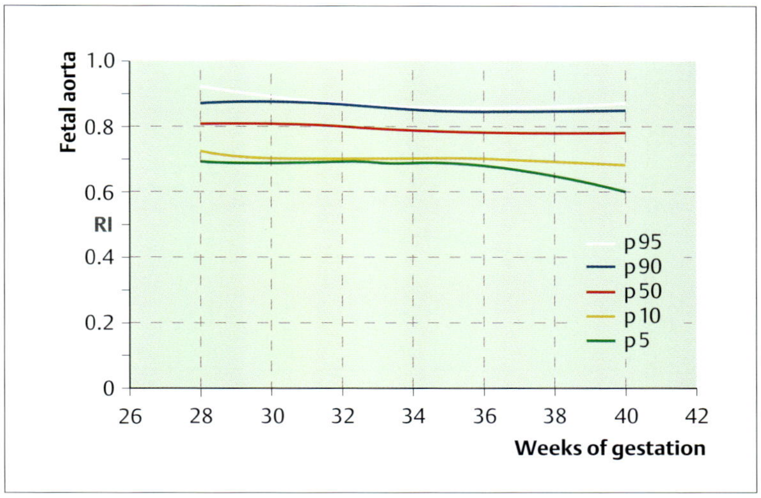

Fig. 15.**7** Reference curves for the **RI of the fetal aorta**, smoothed by cubic regression.

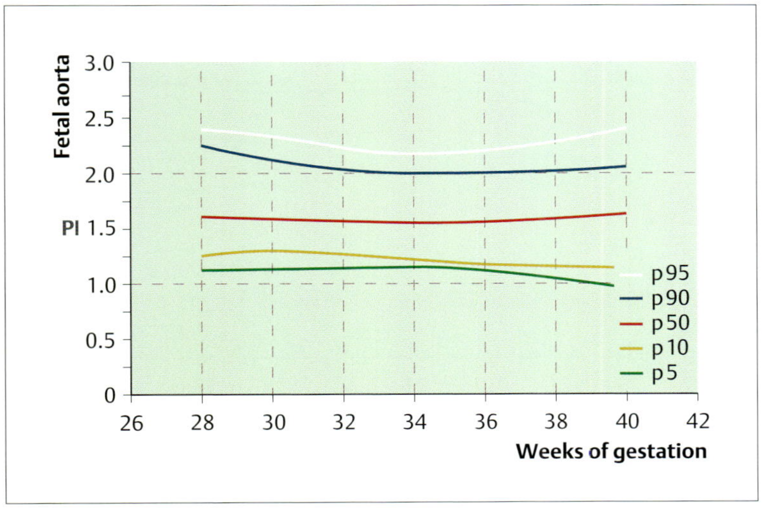

Fig. 15.**8** Reference curves for the **PI of the fetal aorta**, smoothed by cubic regression.

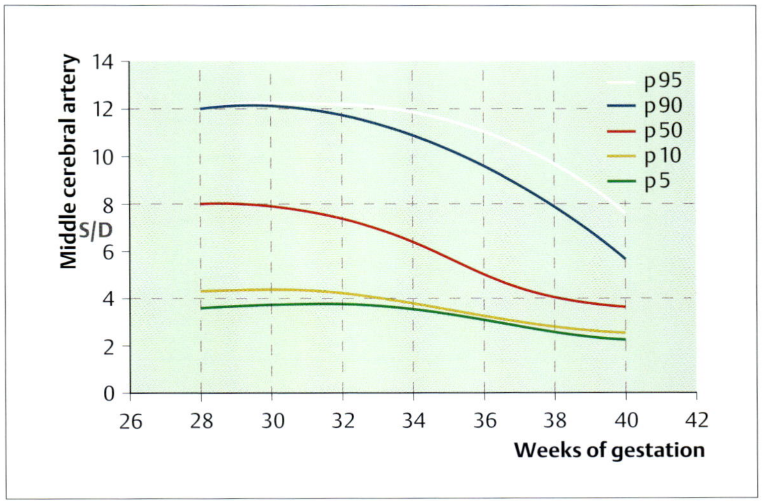

Fig. 15.**9** Reference curves for the **S/D ratio of the middle cerebral artery**, smoothed by cubic regression.

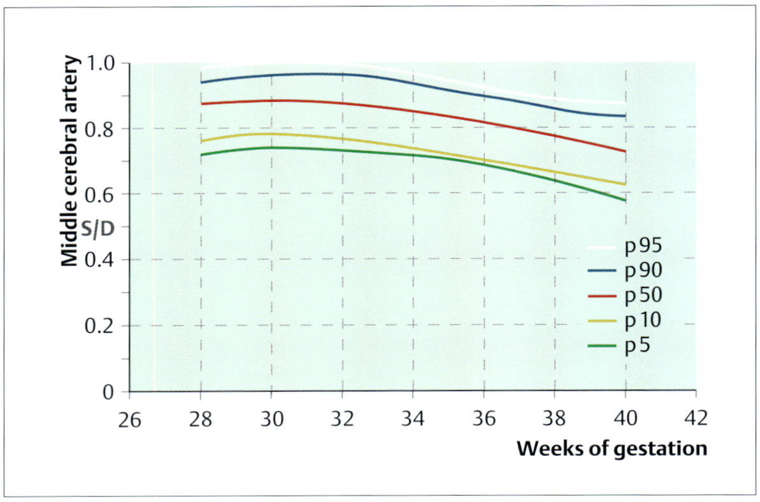

Fig. 15.10 Reference curves for the **RI of the middle cerebral artery**, based on 581 measurements in normal pregnancies and grouped by even gestational weeks. The curves were smoothed by cubic regression.

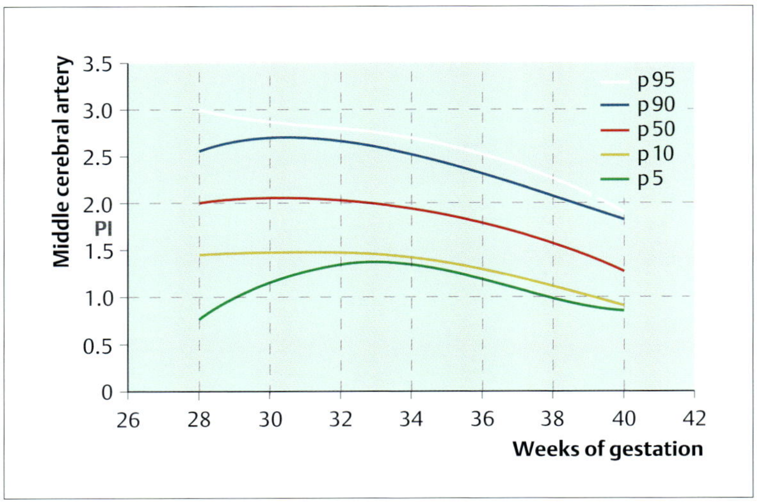

Fig. 15.11 Reference curves for the **PI of the middle cerebral artery**, smoothed by cubic regression.

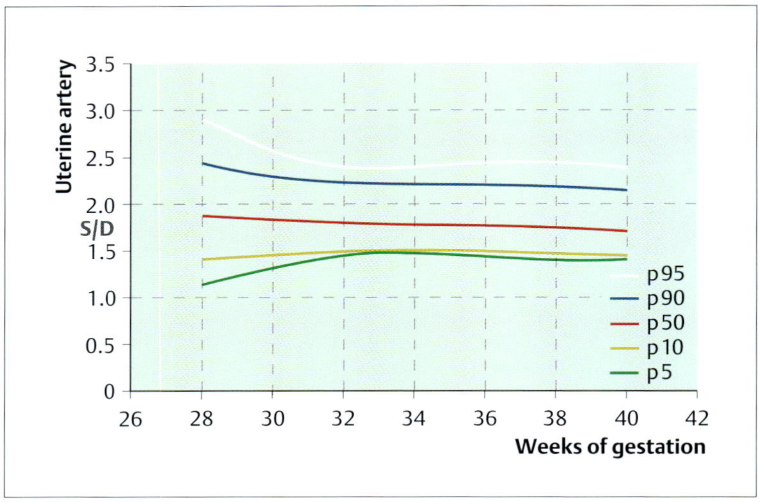

Fig. 15.12 Reference curves for the **S/D ratio of the uterine artery**, smoothed by cubic regression.

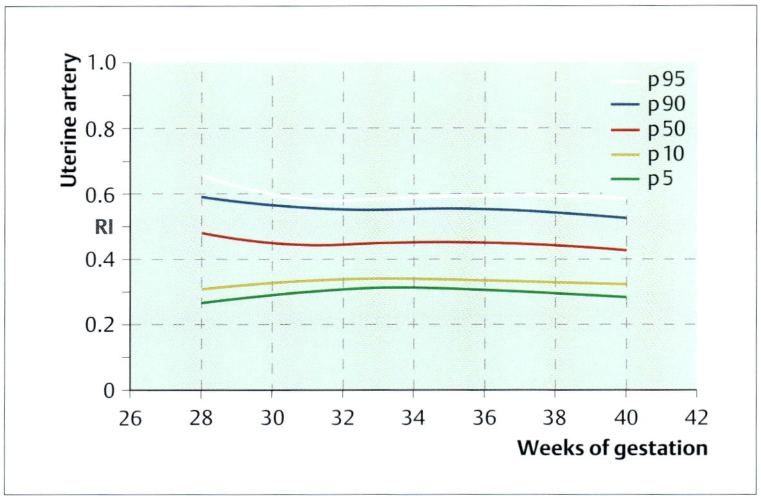

Fig. 15.13 Reference curves for the **RI of the uterine artery**, smoothed by cubic regression.

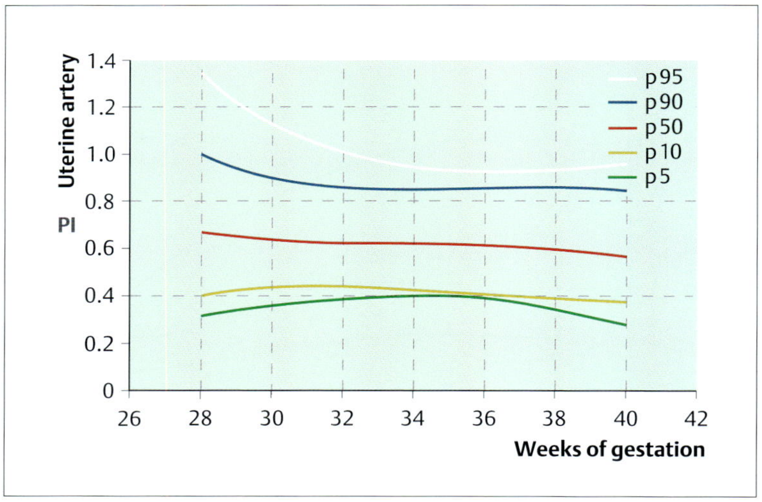

Fig. 15.14 Reference curves for the **PI of the uterine artery**, smoothed by cubic regression.

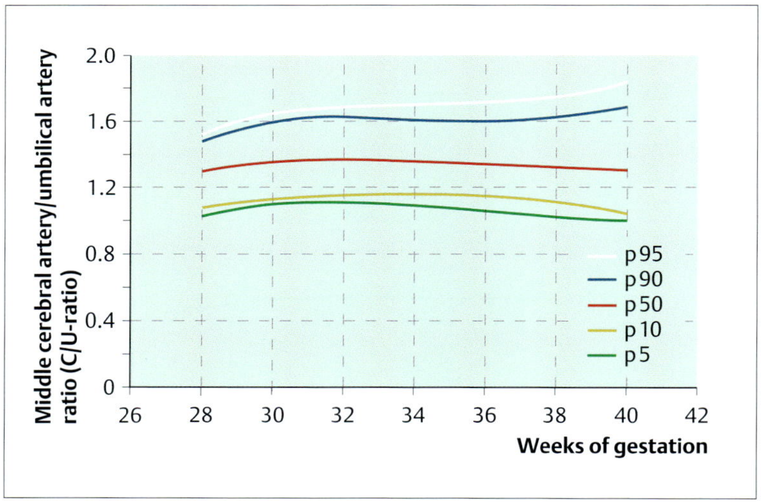

Fig. 15.15 Reference curves for the **cerebroplacental ratio**, based on 580 measurements of the **RI** of the middle cerebral artery and umbilical artery in normal pregnancies and grouped by even gestational weeks. The curves were smoothed by cubic regression.

Discussion

The advent of Doppler sonography has opened up a new functional dimension in diagnostic ultrasound. For the first time, we are able to evaluate both the physiology and the pathophysiology of uterofetoplacental hemodynamics during the course of pregnancy using a noninvasive, largely standardized method. The evaluation of fetal well-being is an important prerequisite for obstetric risk assessment.

ROC curves and cutoff values. The goal of establishing normal reference curves is to be able to classify the course of pregnancy as normal or abnormal on the basis of measured values. By establishing normal values, we are able to create receiver operating characteristic (ROC) curves to define cutoff values that discriminate between normal and abnormal pregnancies with an optimum degree of specificity and sensitivity. This can be accomplished with the reference curves shown above. From a formal statistical standpoint, however, it must be considered that the curves shown are regression-optimized graphic interpolations, and so the curves and their deviation values cannot be used as mathematical equations.

Problems with repeat measurements. Because the database is that of a cross-sectional study and not a longitudinal study, our methodology does not allow us to evaluate a change that occurs in repeat measurements. This can lead to difficulties in Doppler flowmetry, the purpose of which is to measure a change in the functional properties of the uteroplacental unit. The problem is that it cannot be determined whether the measured change reflects an individual physiological progression or a steady decline in the functional capacity of the fetoplacental unit with pathophysiological significance.

Progression of Doppler Indices during the Course of Pregnancy

The Doppler indices recorded from maternal and fetal vessels show a variable pattern of progression during the course of pregnancy as a result of anatomical and histological changes.

Umbilical artery. The S/D ratio of the umbilical artery falls continuously with advancing gestation owing to the increase in arterial blood flow during the course of pregnancy. This leads to persistent flow throughout the cardiac cycle with a steady increase in end-diastolic flow. Comparative data from two longitudinal studies by Fogarty et al.[10] and Hünecke et al.[15] and numerous cross-sectional studies[1,2,21,23] indicate a similar progression. Trudinger et al.[25] attributed this to the following mechanisms:

- Continuous maturation of the fetal placental villous system
- Increase in the cross-sectional area of the fetal placental vessels, causing a progressive decline of fetoplacental vascular resistance
- Increase in fetal cardiac output
- The changing compliance and resistance of the vessel wall
- Rise of fetal blood pressure

Interestingly, these highly individual developmental processes lead to a very large scatter of normal values by the start of the third trimester. This variability is strongly reflected in the S/D ratio (Figs. 15.**2**, 15.**3**) and less so in the PI (Fig. 15.**5**). It has little effect on the RI, which cannot exceed 1, and so the range of variance of this index remains almost constant from the 28th week of gestation until term (Fig. 15.**4**).

Fetal descending aorta. It has become routine practice to scan the fetal descending aorta in addition to the umbilical artery. The S/D ratio of the fetal aorta shows a slight, insignificant decline with advancing gestational age. Our results in this area are comparable to those of Hecher et al.[12]. The waveform of the fetal aorta exhibits forward flow throughout the cardiac cycle, but the diastolic flow velocity is lower in relation to the systolic flow velocity than in the umbilical artery. As a result, higher normal values are generally measured for the S/D ratio of the fetal aorta than for the umbilical artery. The total cross-sectional area of all the vessels increases tremendously during the course of pregnancy. This is accompanied by a falling peripheral resistance and a rising diastolic flow velocity, although there is only a slight resulting decrease in the S/D ratio of the fetal aorta (Fig. 15.**6**). This progression is weakly reflected in the PI and RI and, as in the umbilical artery, the scatters of these two indices are considerably smaller and more stable over the course of the pregnancy (Figs. 15.**7**, 15.**8**).

Middle cerebral artery. The middle cerebral artery is the most commonly selected cerebral supply artery for antenatal Doppler investigation. The cerebral supply arteries show a large range of biological variability, as these vessels reflect the degree of activity of the individual fetus. The resistance indices decline with advancing gestation[26] (Figs. 15.**9**– 15.**11**). The diastolic flow in the cerebral supply arteries is normally low but increases toward the end of pregnancy. Vigorous fetal movements can cause a false elevation of diastolic flow, which can also result from increased intrauterine pressure (e.g., hydramnios) or from extrinsic pressure on the fetal skull[28].

Uterine artery. In contrast to the arcuate arterial system, the measurement of uteroplacental perfusion in the uterine artery permits an overall assessment of uterine perfusion[6]. Higher values are measured than in the arcuate system, however, and so the site of the measurement should be carefully noted when the reference curves are used. This is easily accomplished with the aid of color-flow imaging.

The blood flow conditions depend on the location of the placenta and on gestational age[20]. With a lateralized placenta, the ipsilateral uterine artery reflects the flow conditions in the distal vascular bed and is therefore preferred for clinical evaluation. The differences between the sides are quite pronounced in the early weeks of pregnancy. But in the final trimester, the difference between the S/D ratios of the right and left uterine arteries decreases and averages only 0.4–0.3[6]. The detection of abnormal waveforms in both uterine arteries signifies a high risk for the development of preeclampsia[4] or of intrauterine growth retardation when accompanied by abnormal perfusion of the fetal vessels.

The uteroplacental vessels at the start of pregnancy show a high pulsatility with high systolic flow velocities and very low end-diastolic velocities[6]. With increasing trophoblastic invasion and transformation of the uteroplacental vascular bed, a high-impedance vascular system is transformed into a low-impedance system after the second trimester[5]. After 20 weeks the parameters show largely constant values with a stable biological scatter from the 30th week on (Figs. 15.**12** – 15.**14**).

The early diastolic notch that is normally present in the uterine artery waveform before the 24th week is attributed to a pulse-wave reflection in the vascular periphery, relating to the fact that adaptation of the uteroplacental vascular bed is still incomplete. If the notch persists after 24 weeks' gestation, this should be considered pathognomonic for pregnancy-induced hypertension[3, 14, 25].

Summary. In normal pregnancies, constant values of uteroplacental blood flow are found from the middle of the second trimester until term. Meanwhile, the fetal vessels continue to exhibit changes as the pregnancy progresses. Thus, the S/D ratio of the umbilical artery declines with advancing gestation, as does the S/D ratio of the middle cerebral artery. The values in the fetal descending aorta tend to remain constant, but like the other fetal vessels they show a decreasing scatter with advancing gestational age. It is important, therefore, to develop gestational-age-based reference curves for all of these vessels and to evaluate them for the definition of cutoff limits that can discriminate between normal and abnormal pregnancies.

References

1 Arabin B, Bergmann PL, Saling E: Simultaneous assessment of blood flow velocity waveforms in uteroplacental vessels, the umbilical artery, the fetal aorta and the fetal common carotid artery. Fetal Therapy 2 (1987) 17–26
2 Arduini D, Rizzo G: Normal values of pulsatility index from fetal vessels: A cross-sectional study on 1556 healthy fetuses. J. Perinat. Med. 18 (1990) 165–172
3 Campbell S, Pearce JMF, Hackett G, Cohen-Overbeek T, Hernandez C: Qualitative assessment of uteroplacental blood flow: Early screening test for high-risk pregnancies. Obstet. Gynecol. 68 (1986) 649–653
4 Bower S, Schuchter K, Campbell S: Doppler ultrasound screening as part of routine antenatal scanning: Prediction of pre-eclampsia and intrauterine growth retardation. Brit. J. Obstet. Gynaecol. 100 (1993) 989–994
5 Brosens I, Dixon HG, Robertson WB: Fetal growth retardation and the arteries of the placental bed. Brit. J. Obstet. Gynaecol. 84 (1977) 656–664
6 Deutinger J: Physiologie des Doppler-Flusses in maternalen Gefäßen während der Schwangerschaft. Gynäkologe 25 (1992) 284–291
7 Ertan AK, Rühle W, Gnirs J, Schmidt W: Doppler-Sonographie; Vergleich der Aussagekraft von "A/B-Ratio", "RI" und "PI" bei Aorta fetalis und Nabelarterien. Berichte Gynäkologie Geburtshilfe. 128(8) (1991) 611
8 Erz W, Franz HBG, Gonser M: Dopplersonographie des utero- und fetoplazentaren Kreislaufs: Plazentalateralität, Normalwerte und Referenzkurven. Ultrasch. Med. 19 (1998) 108–113
9 Fendel H, Giani G, Fendel M, Jung H: Die Bestimmung des Gestationsalters mit der Scheitelsteißlänge und dem biparietalen Kopfdurchmesser in der ersten Schwangerschaftshälfte – Gegenüberstellung zweier Methoden. Z. Geburtsh. u. Perinat. 188 (1984) 161–166
10 Fogarty P, Beattie B, Harper A, Dornan J: Continuous wave Doppler flow velocity waveforms from the umbilical artery in normal pregnancy. J. Perinat. Med. 18 (1990) 51–57
11 Gosling RG, King DH: Arterial assessment by Doppler shift ultrasound. Proc. Roy. Soc. Med. 67 (1977) 447–449
12 Hecher K, Spernol R, Szalay S, Stettner H, Ertl U: Referenzwerte für den Pulsatilitätsindex und den Resistanceindex von Blutflusskurven der Arteria umbilicalis und der fetalen Aorta im dritten Trimenon. Ultrasch. Med. 10 (1989) 226–229
13 Hendrik HJ: Sonographische Untersuchungen – erweiterte fetale Biometrie und semiquantitative Bestimmung der Fruchtwassermenge. Inaug. Diss. Med. Fak. Univ. Heidelberg 1988
14 Hoffmann H, Chaoui R, Bollmann R, Bayer H: Klinische Anwendungsmöglichkeiten des Doppler-Ultraschalls in der Geburtshilfe. Zentralbl. Gynäkol. 111 (1989) 1277–1284
15 Hünecke B, Holst A, Schröder HJ, Carstensen MH: Normalbereiche für die relativen Doppler-Indizes A/B-Ratio, Resistance-Index und Pulsatilitäts-Index der Arteria uterina und Arteria umbilicalis bei ungestörter Schwangerschaft. Geburtsh. u. Frauenheilk. 55 (1995) 616–622
16 Mires GJ, Christie AD, Leslie J: Are notched uterine arterial waveforms of prognostic value for hypertensive and growth disorders of pregnancy? Fetal Diagn. Ther. 10 (1995) 111–118
17 Pourcelot L: Application clinique de l‹examen Doppler transcutane. In Peronneau P (ed.): Velocimetrie ultrasonore Doppler. Inserm. 34 (1974) 213–240
18 Roemer VM, Bühler K, Kieback DG: Gestationszeit und Geburtsgewicht. Z. Geburtsh. u. Perinat. 194 (1990) 241
19 Rühle W, Graf von Ballestrem CL, Ertan AK, Schmidt W: Doppler-Sonographie der fetalen Gefäße – Optimierung der Aussagekraft durch ein Kombinationsdiagramm. Z. Geburtsh. u. Perinat. 197 (1993) 95–98
20 Schneider KTM: Standards in der Perinatalmedizin – Dopplersonographie in der Schwangerschaft. Der Frauenarzt 38 (1997) 452–458
21 Schulman H, Fleischer A, Stern W, Farmakides G, Jagani N, Blattner P: Umbilical velocity wave ratios in human pregnancy. Amer. J. Obstet. Gynecol. 148 (1984) 985–990
22 Stuart B, Drumm J, Fitzgerald DE, Duignan NM: Fetal blood velocity waveforms in normal pragnancies. Brit. J. Obstet. Gynecol. 87 (1980) 780–785
23 Thompson RS, Trudinger BJ, Cook CM: Doppler ultrasound waveform indices: A/B-Ratio, pulsatility index and Pourcelot Ratio. Brit. J. Obstet. Gynaecol. 95 (1988) 589
24 Trudinger BW, Giles WB, Cook CM: Uteroplacental blood flow velocity-time waveforms in normal and complicated pregnancy. Brit. J. Obstet. Gynaecol. 92 (1985) 39–45
25 Trudinger BW: Umbilical Artery Blood Flow. In Chervenak FA, Isaacson GC, Campbell S (eds.): Ultrasound in Obstetrics and Gynecology. Vol. 1. Little, Brown and Company, Boston 1993, 597–604
26 Vetter K, Gonser M, Gasiorek-Wiens A: Dopplersonographie in der Schwangerschaft. In Sohn C, Holzgreve W (eds.): Ultraschall in Gynäkologie und Geburtshilfe. Thieme, Stuttgart 1995, 501–540
27 Voigt M, Schneider KTM, Jährig K: Analyse des Geburtengutes des Jahrgangs 1992 der Bundesrepublik Deutschland. Teil 1: Neue Perzentilwerte für die Körpermaße von Neugeborenen. Geburtsh. u. Frauenheilk. 56 (1996) 550–558
28 Vyas S, Campbell S, Bower S, Nicolaides KH: Maternal abdominal pressure alters fetal cerebral blood flow. Brit. J. Obstet. Gynaecol. 97 (1990) 740–747

16 Venous Doppler Sonography

F. Bahlmann

Historical Development

Since its initial description by Fitzgerald in 1977[13], the Doppler ultrasound assessment of the fetomaternal vascular system has become widely utilized in the surveillance of high-risk pregnancies. In the past, scientific and clinical attention was focused primarily on examination of the uteroplacental vascular system and on the arterial system of the fetus. In recent years, there have been increasing numbers of Doppler ultrasound studies of the physiological and pathophysiological changes in the fetal heart and circulatory system, with special emphasis on the cardiac and venous vascular systems. Although Doppler studies of volume flow in the intrahepatic umbilical vein were already being done in the early 1980 s, this method has not become widely established because of its poor reproducibility[11, 12, 18, 19, 20, 35, 36, 41]. With technical advances in ultrasound instrumentation and especially the advent of color Doppler sonography, it became possible to make a detailed evaluation of the fetal venous system. Besides the umbilical vein, inferior vena cava, and hepatic veins, Doppler examination of the ductus venosus has become a particular object of current scientific interest.

Physiology

Streaming effect. The ductus venosus represents the first of three specific shunts in the fetal circulation and appears to be an important regulator in the distribution of oxygen-enriched blood. Studies in experimental animals have demonstrated the existence of two different directions of blood flow (streaming effect) in the thoracic portion of the inferior vena cava[9]. Studies in fetal lambs showed that approximately 50% of oxygenated umbilical venous blood flows through the ductus venosus. The narrow ductal lumen causes marked acceleration of the blood flow, creating a stream in the left dorsal portion of the inferior vena cava with the preferential direction of flow through the foramen ovale into the left atrium and left ventricle. This streaming mechanism ensures an optimum oxygen supply to the brain and myocardium. The considerably slower blood flow in the distal inferior vena cava forms a stream in the right ventral portion of the inferior vena cava that is directed into the right atrium and then into the right ventricle (Fig. 16.**1**).

This streaming effect appears to be supported by the crest-like caudal border of the foramen ovale[9]. Similar findings were made in color Doppler studies of human fetuses[42]. An approximately 70% increase of blood flow through the ductus venosus was documented in response to hypoxemia in fetal lambs[10]. The exact pathophysiological mechanism of this effect is still uncertain.

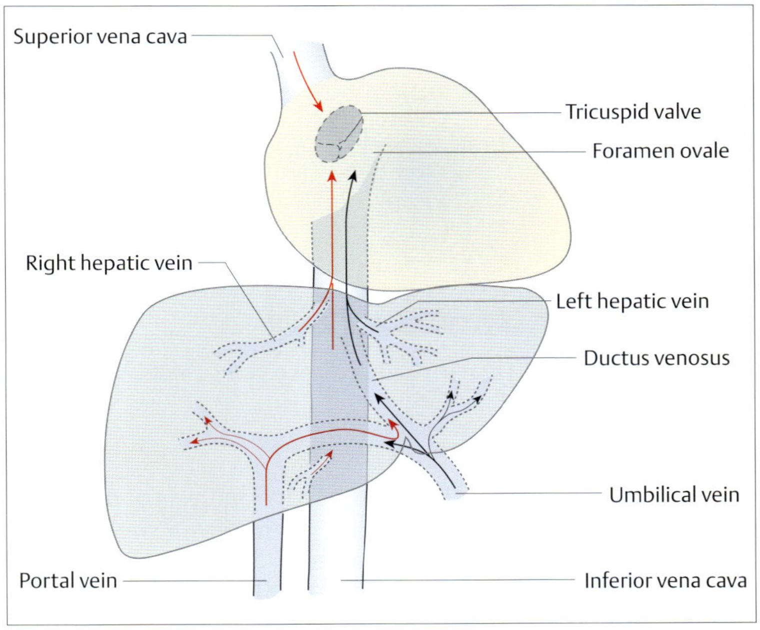

Fig. 16.**1** "Streaming effect" demonstrated by experimental studies in fetal sheep[9].

Umbilical Vein

Quantitative determination of blood flow. Eik-Nes and Gill[11, 18] used Doppler ultrasound to evaluate fetal blood volume in the early 1980 s. They found a relatively constant mean volume flow of 110–125 ml/(kg min) in the third trimester, which decreased to 90 ml/(kg min) toward the end of pregnancy[11, 12, 18] (Fig. 16.**2**). The following formula (equation 1) is used for the quantitative determination of blood flow in the umbilical vein[47]:

$$Q_{VU} = V \times (D/2)^2 \times \pi \times 0.6 \text{ ml/min} \qquad (1)$$

Venous umbilical blood flow in this equation (Q_{VU}) is dependent on the blood flow velocity (V) and the vascular cross-

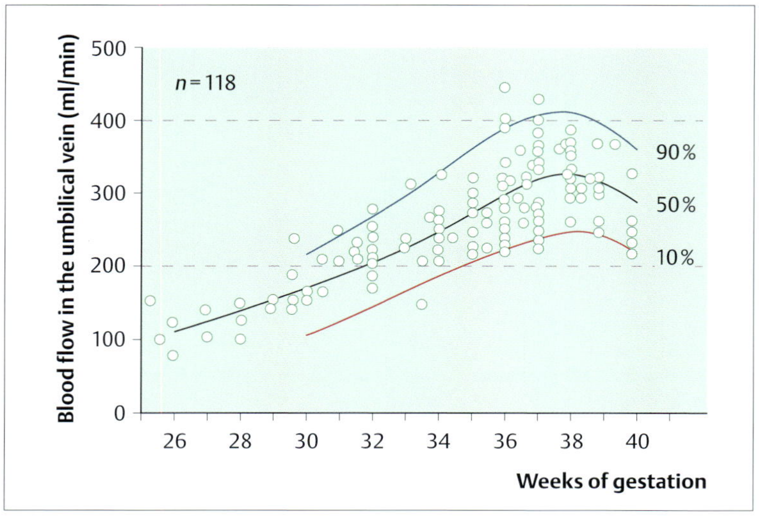

Fig. 16.**2** Mean volume flow in the umbilical vein during pregnancy[19].

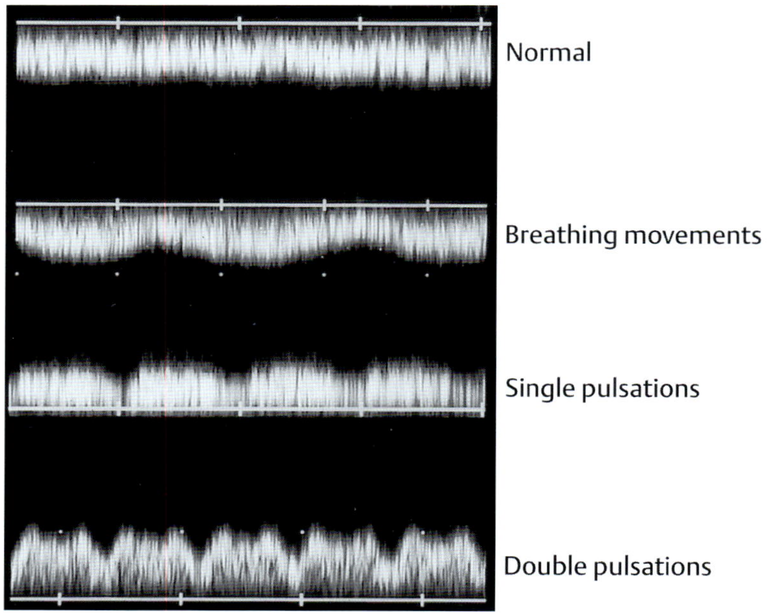

Fig. 16.**3** Normal and abnormal umbilical venous Doppler spectra.

sectional area $(D/2)^2 \times \pi$. The quantitative analysis of umbilical blood flow is problematic, however, in that even slight deviations in the vessel diameter lead to a large scatter of flow values[47]. Because of its poor reproducibility, this quantitative method of blood volume determination is not widely utilized[12].

Locating the umbilical vein. Doppler spectra are recorded from either the intra-amniotic or intrahepatic segment of the umbilical vein. The intrahepatic segment is preferred for its better reproducibility[11,12,18]. It is sampled by imaging the fetal abdomen in a transverse plane and placing the sample volume at an acute angle ($<30°$) in the central part of the intrahepatic umbilical vein[11].

Findings. The umbilical venous waveform generally shows a monophasic pattern with a mean flow velocity of 10–15 cm/s. Under physiological conditions, umbilical vein pulsations occur until the end of the first trimester or in response to fetal breathing movements[33,58] and normally are no longer seen after 13 weeks' gestation. Umbilical vein pulsations in the second or third trimester may signify a cardiac anomaly or congestive heart disease, or they are commonly associated with absent end-diastolic flow in the umbilical artery as a result of chronic placental insufficiency[22,49] (Fig. 16.**3**). Umbilical vein pulsations in these cases show a temporal correlation with atrial systole and are an expression of myocardial insufficiency. Pulsations in the umbilical vein may take the form of single pulsations, double pulsations, or a triphasic Doppler spectrum[1,3,22,49]. A markedly increased mortality rate of 50–60% is reported when these flow patterns are detected[22,33,49].

Ductus venosus

Locating the ductus venosus. The ductus venosus and its flow pattern can be demonstrated with two-dimensional real-time ultrasound and also with color Doppler (Fig. 16.**4**). Of all the precardial veins, the ductus venosus yields the best and most reliable information on the myocardial hemodynamics and function of the fetal heart, with good reproducibility of the Doppler spectra[61]. Doppler signals are recorded most easily and quickly when the fetus is in a dorsoposterior lie. For rapid venous orientation, the intrahepatic portion of the umbilical vein should be imaged first. The best view is obtained in either a midsagittal plane or oblique transverse plane of the fetal abdomen[24,30,42,43]. The entry of the intrahepatic umbilical vein into the ductus venosus is located in continuity with the vein. The diameter of the ductus venosus rarely exceeds 2 mm and shows a slight funnel-shaped expansion up to 20 mm long[44].

A marked difference in blood flow velocities between the umbilical vein and ductus venosus can be recognized with color Doppler. The 3–4 times higher flow velocity in the ductus venosus causes aliasing, which appears as an area of color reversal (Figs. 16.**5**, 16.**6**). Flow signals are acquired by positioning the Doppler sample volume directly at the entrance to the ductus venosus, at the site of the color reversal[24,42] (Fig. 16.**7**). The width of the sample volume should just cover the vessel lumen to avoid detecting signals from the adjacent hepatic veins and umbilical vein. The use of color Doppler makes it much easier to locate the ductus venosus and accurately position the sample volume. Owing to the funnel shape of the ductus venosus, the flow velocities directly at the entrance are higher than at the outlet[53]. An insonation angle less than 30° (or 50°) is recommended to obtain an optimum waveform[44]. The wall filter should be set as low as possible—between 125 Hz and 100 Hz, depending on the type of equipment used.

Waveforms and indices. A sonographer experienced in Doppler ultrasound can obtain clear Doppler signals from the ductus venosus in 94% of cases[24]. The normal velocity waveform of the ductus venosus exhibits continuous, triphasic forward flow throughout the cardiac cycle. Absent or low pulsatility is noted in 3% of cases and is considered a normal variant[44]; a brief waiting period is recommended in these cases. The maximum flow velocities in the ductus venosus are the highest in the venous system and appear to be responsible for the "streaming

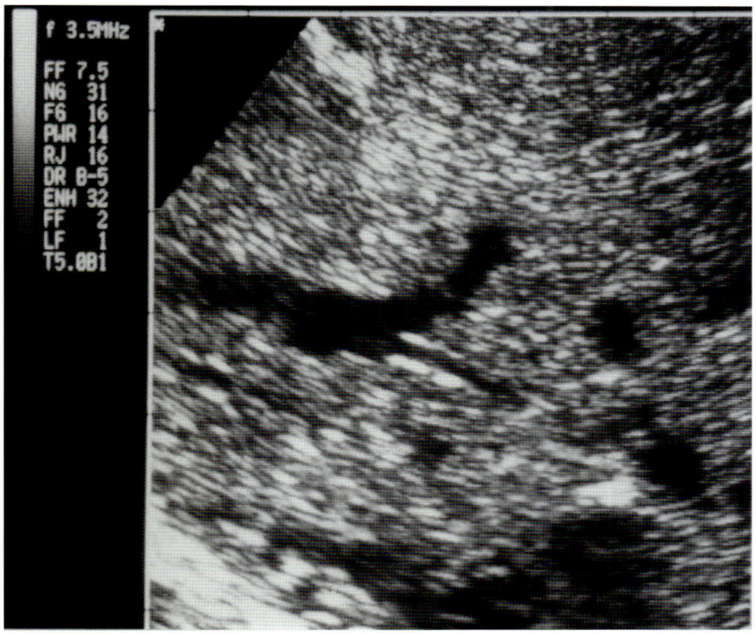

◁ Fig. 16.**4** B-mode image of the portal sinus with the origin of the ductus venosus. Dorsoposterior lie (27th week).

effect." As in the arterial system, the flow velocities in the ductus venosus are dependent upon gestational age, fetal breathing and body movements, and the fetal heart rate [44]. The maximum blood flow velocity rises on average from 65 cm/s to 75 cm/s between 18 and 40 weeks' gestation[44]. The blood flow velocity in the ductus venosus can rise to double or triple the normal value during inspiration, depending on the intensity of the breathing movements.

The quantities that are determined for ductus venosus waveform analysis are the maximum flow velocities during ventricular systole (S), ventricular early diastole (D), and ventricular late diastole (atrial contraction [a]) (Fig. 16.**8**). Hemodynamically, these phases reflect the rapidly time-varying pressure gradient between the umbilical vein and right atrium. The highest pressure gradient between the ductus ve-

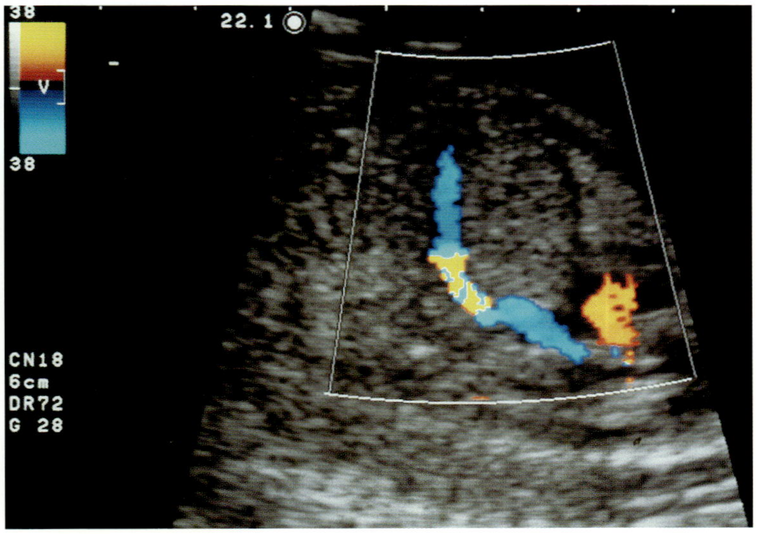

Fig. 16.**5** Midsagittal scan through the fetal abdomen demonstrates the course of the umbilical vein (blue) and its junction with the ductus venosus (yellow). The color reversal (aliasing) marks the site where the flow velocities are highest. Dorsoposterior lie (30th week).

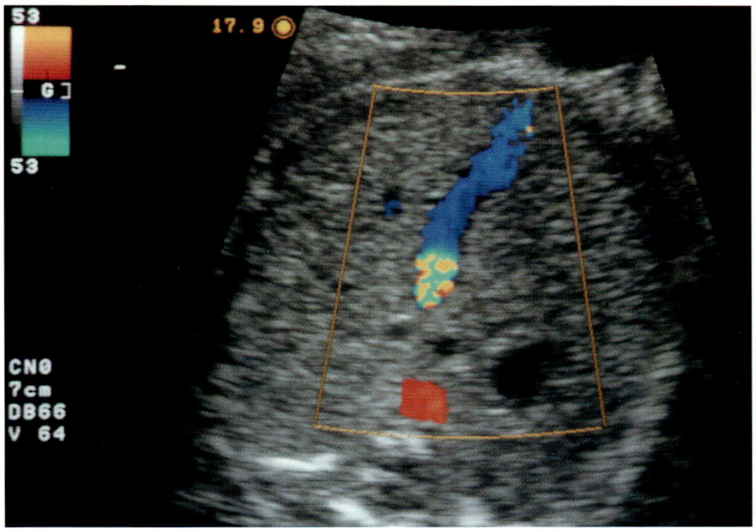

Fig. 16.**6** Oblique transverse scan through the fetal abdomen (34th week) demonstrates the course of the umbilical vein (blue) and its junction with the ductus venosus (yellow).

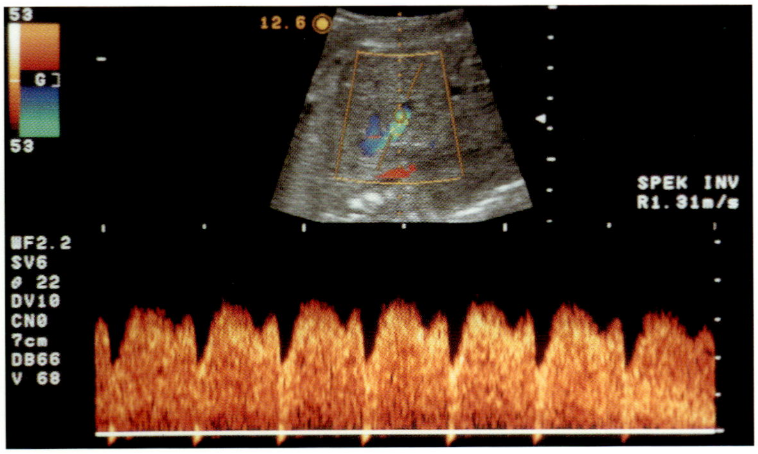

Fig. 16.**7** Color recording of the Doppler spectrum at the entry of the umbilical vein into the ductus venosus shows the typical triphasic, antegrade flow pattern.

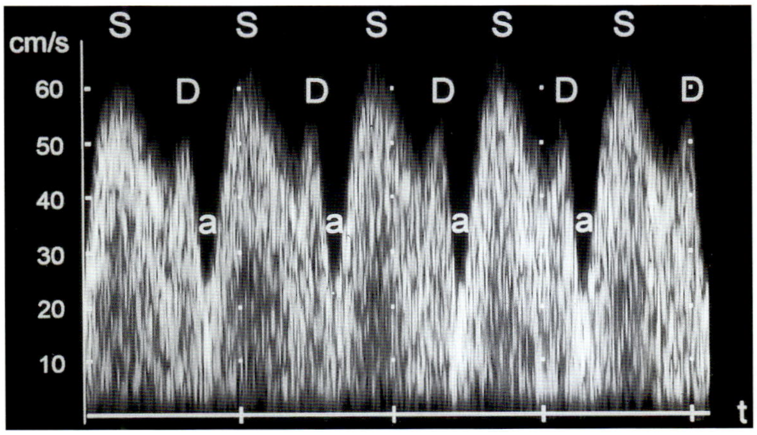

Fig. 16.**8** Normal triphasic waveform of the ductus venosus: maximum antegrade blood flow velocities during ventricular systole (S), early ventricular diastole (D), and atrial contraction (a).

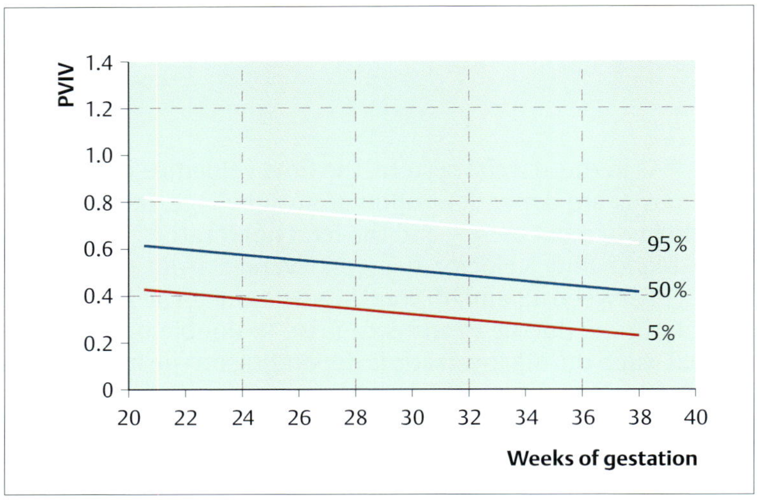

Fig. 16.**9** Ductus venosus. Preload index (S–a)/D of Hecher (1994). PVIV = peak velocity index for veins.

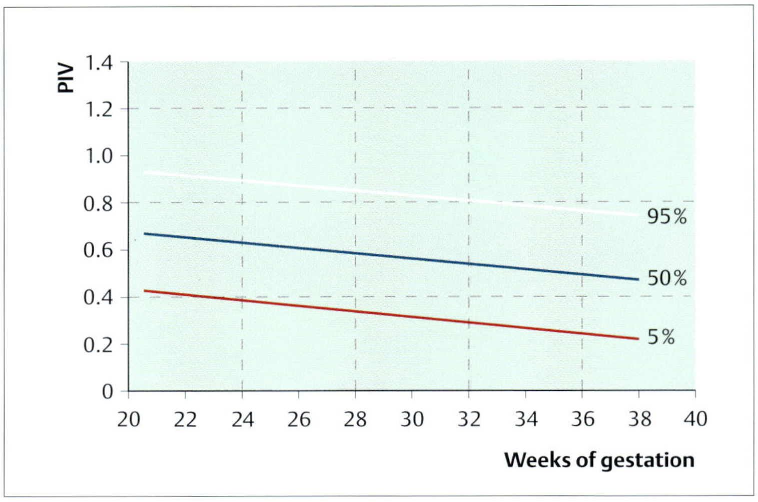

Fig. 16.**10** Ductus venosus. Preload index (S–a)/Tamx of Hecher (1994). PIV = pulsatility index for veins.

nosus and right atrium occurs during ventricular systole. This gradient is produced by the descent of the atrioventricular (AV) valve plane, resulting in antegrade flow that fills the atria. Subsequent early diastole is marked by opening of the AV valves and passive filling of the ventricles. This phase corresponds to the E component of the biphasic atrioventricular waveform. During atrial contraction, which coincides with the A component of the atrioventricular waveform, the foramen ovale closes and the rest of the atrial blood volume is actively pumped into the right ventricle. By analyzing these patterns, the sonographer can obtain information on the end-diastolic right ventricular pressure and the central venous pressure. The angle-independent parameters listed in Table 16.**1** are used for the qualitative assessment of pulsatility. These parameters can be used to evaluate the cardiac preload[8].

Normal and abnormal changes during the course of pregnancy. The end-diastolic ventricular pressure declines with advancing gestational age as a result of placental maturation processes that lower the placental resistance. This decrease is manifested by a decline of venous pulsatility and falling preload indices and is due mainly to the rising flow velocity during atrial contraction (Figs. 16.**9**, 16.**10**). A rise in the preload indices reflects an increased end-diastolic ventricular pressure in the heart. In a healthy fetus, the umbilical venous pressure during atrial contraction is higher than the central venous pressure. When there is severe centralization of the fetal circulation due, for example, to chronic placental insufficiency or hypovolemia, a hypoxia-induced myocardial insufficiency develops, leading to a rise of central venous pressure in the fetal right heart. The result is a fall of maximum flow velocities in the ductus venosus or even reverse flow during atrial contraction (Fig. 16.**11**)[23, 25, 27, 46, 62]. Complete cardiac exhaustion and decompensation are manifested by sinus bradycardia, which is marked by a decrease in systolic and early diastolic antegrade flow velocities and an increase in retrograde late diastolic flow velocities.

Table 16.**1** Doppler indices in the venous vascular system for qualitative evaluation of the cardiac preload and central venous pressures

Index	Author
a/S	Kanzaki (1990)[37]
S/D	Huisman (1991)[29]
(S–a)/S	De Vore (1993)[8]
(S–a)/D	Hecher (1994)[24]
(S–a)/Tamx	Hecher (1994)[24]

S = systole, D = diastole, a = atrial contraction, Tamx = time-averaged maximum velocity.

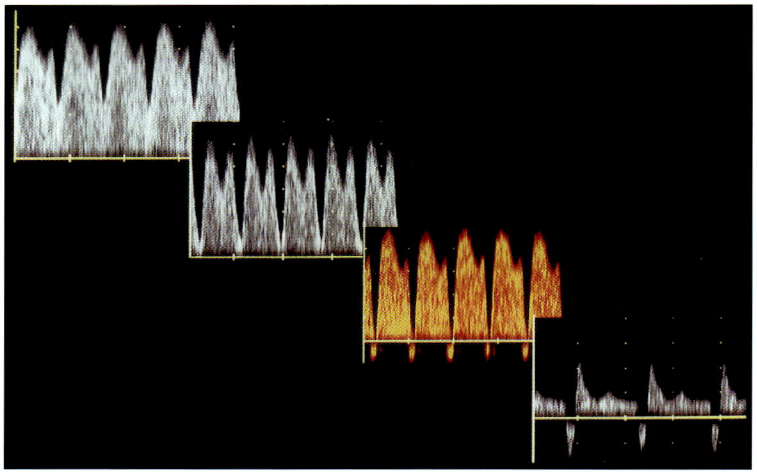

Fig. 16.**11** Normal and abnormal ductus venosus Doppler spectra in fetuses with severe asymmetrical intrauterine growth retardation due to chronic placental insufficiency. The increase in pulsatility caused by a slowing or reversal of flow during atrial contraction reflects the increasing degree of fetal compromise.

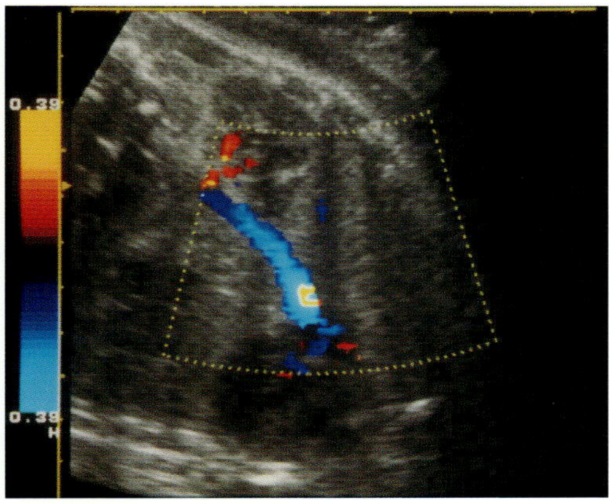

Fig. 16.**12** Color-flow image of the inferior vena cava in a parasagittal plane.

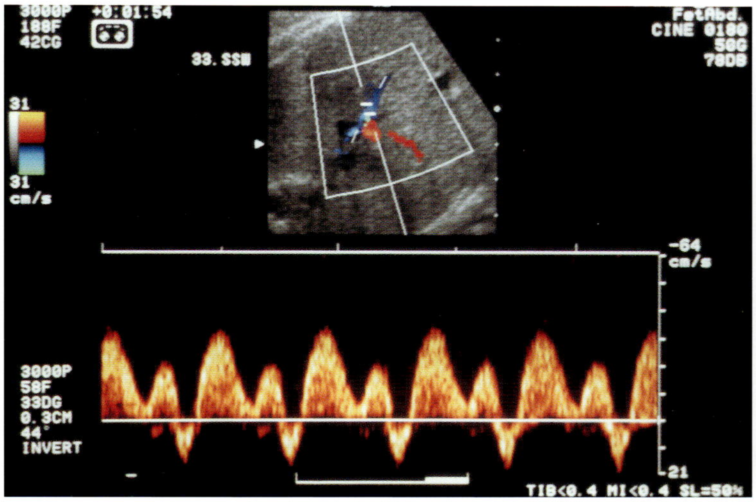

Fig. 16.**13** Color Doppler scan of the inferior vena cava (blue) below the ductus venosus (red). Note the bidirectional flow velocity waveform with typical antegrade S- and D-phases and a retrograde a-phase.

Inferior Vena Cava

Locating the inferior vena cava. There is disagreement in the literature concerning the best site for sampling flow velocity waveforms from the inferior vena cava[24, 29, 37, 54, 59]. Flow sampling directly below the right atrium appears to adversely affect the inferior vena cava waveform owing to venous flows merging in different directions from subdiaphragmatic tributaries[31]. Rizzo et al. compared various sampling sites for recording flow velocity waveforms from the inferior vena cava[59]. In this study, sites between the renal vessels and subdiaphragmatic venous tributaries and also below the ductus venosus provided the best reproducibility, the most favorable insonation angle, and the least variation [59]. With this technique the inferior vena cava is imaged in a parasagittal longitudinal plane, and the sample volume is positioned at the lowest possible angle ($<30°$) (Fig. 16.**12**). Another option is to place the sample volume just below the entrance to the right atrium[54], but this leads to greater variability of the velocity waveforms[29]. Fetal body and breathing movements in particular can lead to marked changes in the Doppler frequency spectrum[32, 34].

Waveforms and indices. The flow velocity waveform of the inferior vena cava, like that of the ductus venosus, reflects the systolic and diastolic phases of the cardiac cycle and thus the intracardiac pressure changes[57]. Unlike the ductus venosus, the inferior vena cava exhibits a bidirectional, triphasic flow pattern with a retrograde component during atrial contraction and 2–3 times lower flow velocities (Fig. 16.**13**). As in the ductus venosus, the indices shown in Table 16.**1** are used for qualitative analysis of the inferior vena cava waveform.

Normal and abnormal changes during the course of pregnancy. The percentage of retrograde flow during atrial contraction declines with advancing gestational age[29, 60, 65]. This is associated with a decrease in pulsatility and related indices and is attributable to both a decrease in fetoplacental resistance and an increasing differentiation of diastolic ventricular function. Increased pulsatility and retrograde flow are commonly found in hypoxemic fetuses with a severe centralization of blood flow, congestive heart disease, and cardiac arrhythmias (Fig. 16.**14**). Simultaneous Doppler sampling of the inferior vena cava and descending aorta can be performed to evaluate fetal arrhythmia[4, 38, 54]. For this purpose, the sample volume is enlarged to detect the flow velocities in both vessels.

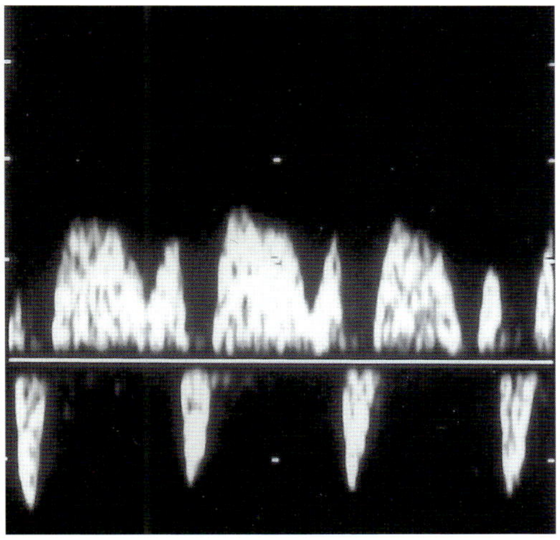

Fig. 16.**14** Increased retrograde flow during atrial contraction in a fetus with significant pulmonic stenosis (29th week, no hydrops).

Hepatic Veins

To date, only a few reports have been published on Doppler ultrasound studies of the hepatic veins and their clinical significance[24, 48, 64]. As in the inferior vena cava, the flow velocity waveform of the hepatic veins shows a triphasic, bidirectional pattern (Fig. 16.**15**). The maximum flow velocities are somewhat lower, however. The left hepatic vein is most easily located in a parasagittal plane just below the diaphragm. As in all other fetal vessels, the Doppler spectrum of the hepatic veins is modulated by fetal body and breathing movements (Fig. 16.**16**).

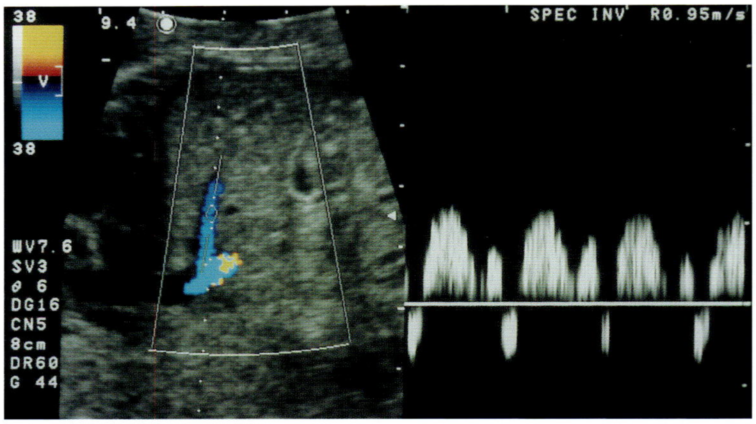

Fig. 16.**15** Color-flow image and corresponding Doppler spectrum of the left hepatic vein. The color reversal (yellow) reflects the area of the ductus venosus.

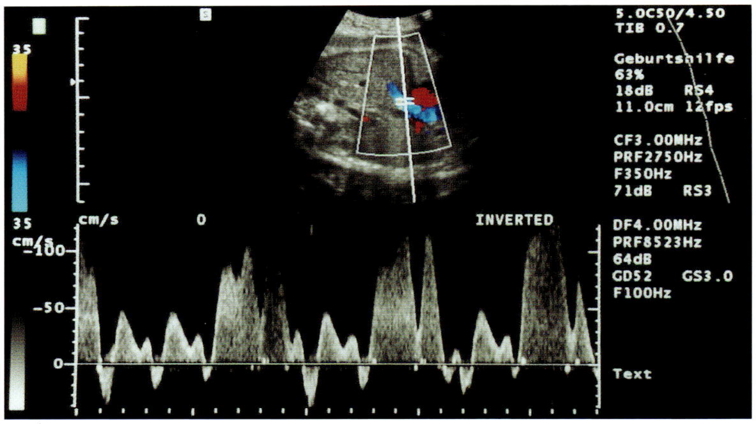

Fig. 16.**16** Effect of fetal breathing movements on the Doppler spectrum of the left hepatic vein. The maximum velocities are two to three times higher during inspiration.

Clinical Applications

Intrauterine Growth Retardation Due to Chronic Placental Insufficiency

Fetal compensatory mechanisms. Hypoxemia and acidemia trigger a compensatory centralization of the fetal circulation in which peripheral vascular resistance rises and blood flow is routed preferentially to the brain, myocardium, and adrenal glands [5,10,47,63]. In a setting of chronic placental insufficiency, it appears that a growth-retarded fetus can compensate for its chronic hypoxemic state for some time through specific adaptive processes that include an increase in maximum myocardial blood flow, the formation of new blood vessels in the myocardium, and changes in myocardial energy metabolism[2]. The hemodynamic changes, along with increased erythropoiesis and increasing anaerobic glycolysis, are logical adaptive responses by the fetus aimed at the prevention of serious injury. When this compensatory capacity has reached its limit, a further increase in acidemia will cause irreversible damage that is reflected in high rates of perinatal morbidity and mortality[27,39,62].

Centralization of the fetal circulation. Rising resistance to blood flow in the placental vascular bed evokes an increasing centralization of the fetal circulation. The Doppler manifestations of this oxygen-sparing response include elevated resistance indices in the umbilical artery and fetal aorta and a fall of the cerebral resistance indices. As resistance in the fetal aorta continues to rise (increased afterload), there is a corresponding rise in the end-diastolic right ventricular pressure,

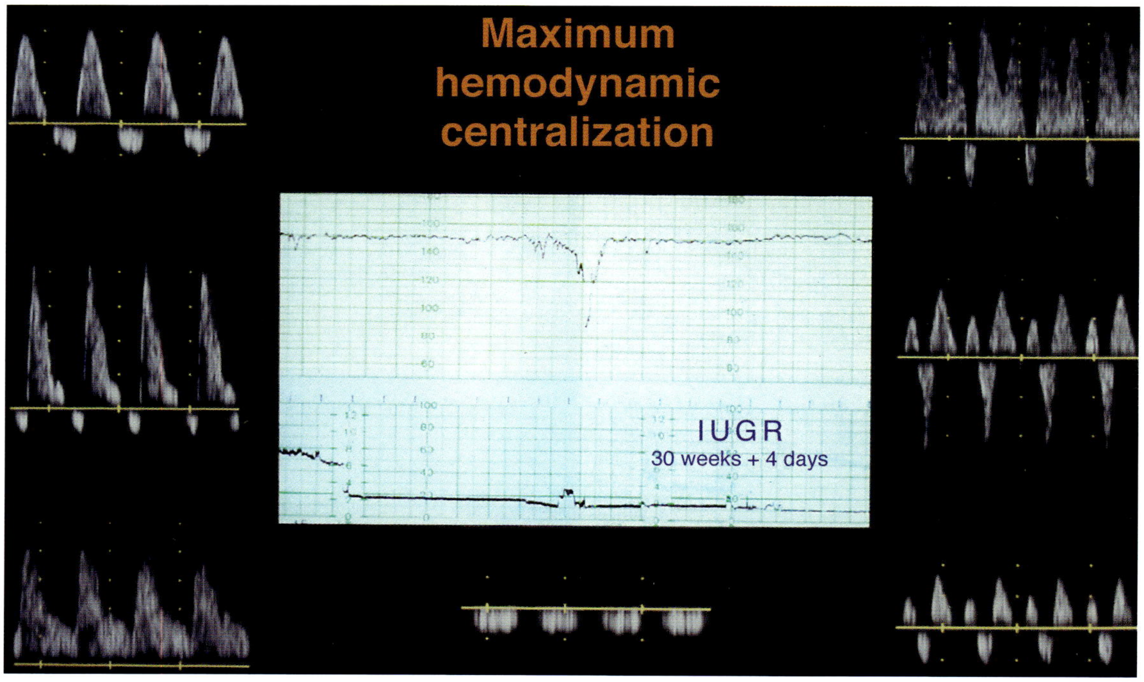

Fig. 16.**17** Severely abnormal Doppler waveforms in the arterial (left side) and venous (right side) vascular systems in a severely growth-retarded fetus at 30 weeks 4 days. The FHR trace shows a silent pattern with a late deceleration.

whose persistence leads to greater pulsatility in the venous vessels and the occurrence of umbilical vein pulsations[55,56]. These Doppler changes reflect a hypoxia-induced myocardial insufficiency and are especially pronounced in cases with progressive acidemia. It appears that abnormally increased preload indices in the ductus venosus and inferior vena cava as well as umbilical vein pulsations precede definite changes in fetal heart rate (FHR) tracings[6,25] (Fig. 16.**17**).

Use of venous Doppler sonography. Venous Doppler sonography is particularly indicated in cases with absent or reverse end-diastolic flow in the umbilical artery[22,25,46,62]. The goal of this examination is to gain additional, noninvasive information on fetal cardiac performance so that the optimum timing of the delivery can be determined. This appears to be particularly crucial in severely growth-retarded fetuses below 30 weeks in a setting of chronic placental insufficiency [21]. Clinical studies are currently underway to determine the extent to which these findings can be confirmed and clinical management principles can be derived from them.

Growth Discordance in Multiple Pregnancy

In cases where malformations and chromosome abnormalities can be excluded, a growth discordance in a multiple pregnancy is most commonly related to intrauterine growth retardation based on chronic placental insufficiency or a fetofetal transfusion syndrome.

Fetofetal transfusion syndrome. Although the pathogenesis of fetofetal transfusion syndrome is not yet fully understood, the chronic shunting of blood from one twin (donor) to the other twin (recipient) through placental vascular anastomoses appears to be the cause of the syndrome. This leads to increasing hypervolemia in the recipient and varying degrees of hypovolemia in the donor. The result is increased diuresis in the recipient and decreased diuresis in the donor, promoting the development of polyhydramnios or oligohydramnios ("stuck twin"). When this process outstrips the compensatory capacity of the recipient, the hypervolemia leads to congestive heart failure. The resulting elevation of central venous pressure in the heart leads to cardiomegaly, which in turn is reflected in elevated preload indices in the venous vessels or even pulsations in the umbilical vein (Figs. 16.**18**, 16.**19**). When congestive failure has become established in the recipient, tricuspid insufficiency can be demonstrated in 56% of cases and increased preload indices in 40%[66]. By contrast, increased venous preload indices are found in the donor twin in only 8% of cases. They may be an expression of hypovolemia and increased placental resistance. Tricuspid insufficiency is not invariably associated with increased pulsatility in the venous vessels. This is due to the time difference between the valvular and venous flows: whereas the regurgitation in tricuspid insufficiency occurs mainly during ventricular systole, reverse flow (a) in venous vessels occurs during late ventricular diastole or atrial contraction. Depending on the adaptive capacity of the heart, hydrops fetalis can develop in 8% of cases, usually in the recipient twin. Interestingly, these cases often display normal resistance indices in the arterial vessels[28].

Hydrops fetalis

Some fetuses with nonimmune hydrops are found to have a low right ventricular stroke volume, which is manifested by elevated preload indices in the venous vessels and the presence of umbilical vein pulsations[22]. The examination of venous vessels can be helpful in these cases for the etiological and prognostic evaluation of the various forms of fetal hydrops.

Fetal Arrhythmias

Supraventricular extrasystoles. Supraventricular extrasystoles (premature beats) may occur singly or at regular intervals and account for approximately 95% of all fetal arrhythmias. This type of arrhythmia is considered to have a very good overall intrauterine and neonatal prognosis. Only about 10% of extrasys-

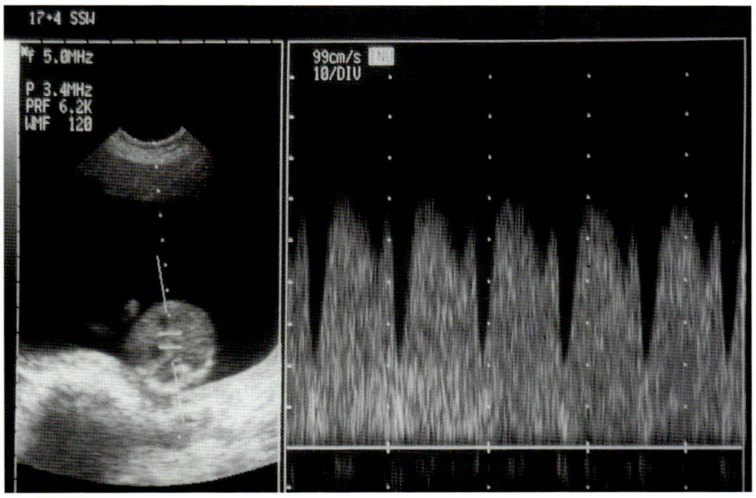

Fig. 16.**18** Fetofetal transfusion syndrome at 17 weeks 4 days. A normal Doppler spectrum is recorded from the ductus venosus of the donor twin.

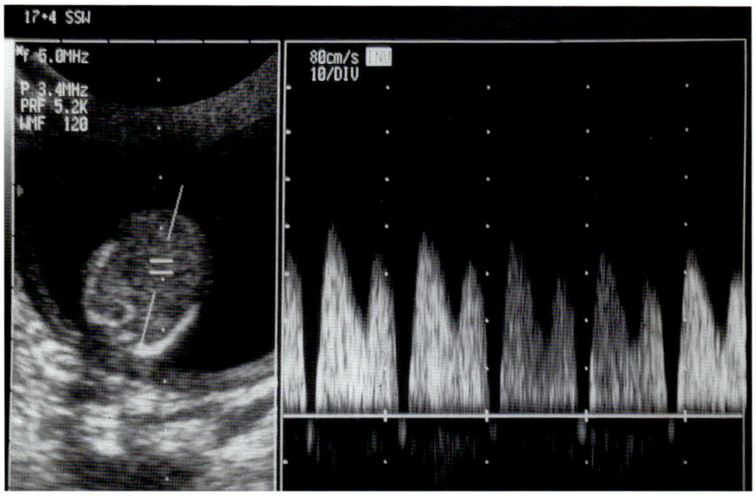

Fig. 16.**19** Fetofetal transfusion syndrome at 17 weeks 4 days. An abnormal Doppler spectrum is recorded from the ductus venosus of the recipient twin.

Obstetric Ultrasound

toles can still be detected during the first weeks of life. A detail fetal echocardiographic examination is recommended to exclude morphological cardiac abnormalities, which are present in 1–8% of cases. The cause of supraventricular extrasystoles is an immaturity of the impulse formation and conduction system of the heart. When an extrasystole occurs, the action potential arising from the sinus node (Keith–Flack node) encounters a still-refractory AV node (Aschoff–Tawara node), leading to ventricular bradycardia. Increased atrial filling occurs during this bradycardiac phase, characterized by a decrease in flow velocities and a rise of central venous pressure, especially during the subsequent atrial contraction (a) (Fig. 16.**20**). Because of the Frank–Starling mechanism, the increased blood volume is handled in the next cardiac cycle by an increase in the ventricular stroke volume. As a result of this compensatory mechanism, the central venous pressure returns to normal, and even frequent runs of extrasystoles are unable to produce cardiac failure or hydrops fetalis. It should be noted, however, that extrasystoles may progress to paroxysmal supraventricular tachycardia in approximately 1–2% of cases.

Supraventricular tachycardia. Fetal tachyarrhythmias, especially supraventricular tachyarrhythmias, atrial fibrillation, and atrial flutter, may be associated with fetal heart failure and the development of nonimmune fetal hydrops, polyhydramnios, and placental hydrops when they are of long duration.

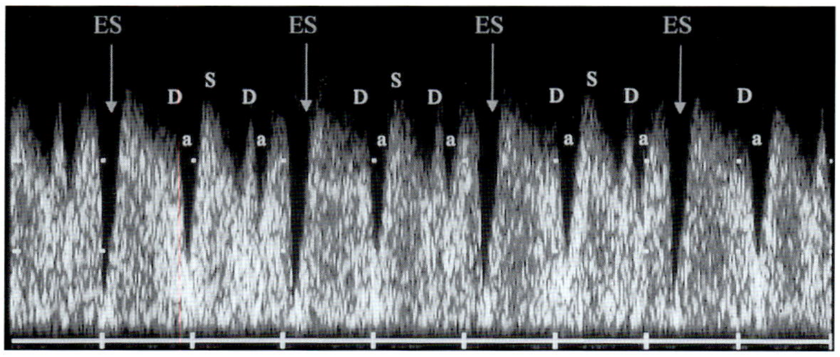

Fig. 16.**20** Supraventricular bigeminal extrasystoles in a fetus. The extrasystoles (ES) occur during ventricular systole.

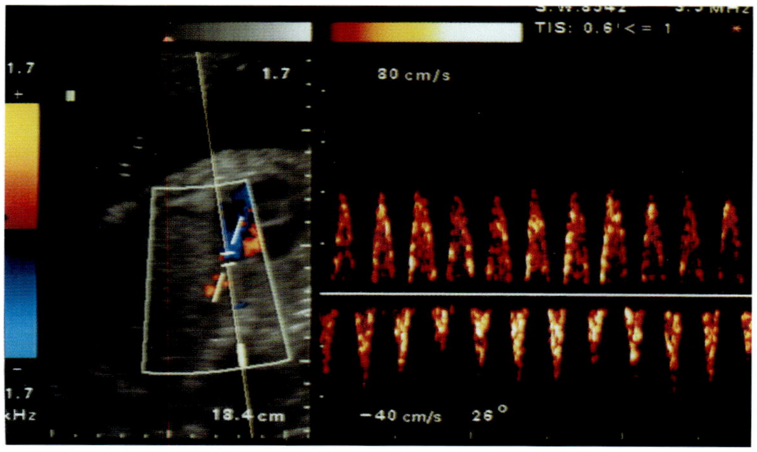

Fig. 16.**21** Monophasic, bidirectional Doppler spectrum of the inferior vena cava in a fetus with supraventricular tachycardia (220 bpm).

Supraventricular tachycardia is marked by uncoordinated electrical excitation processes in the heart (reentry circuits) that disrupt the normal hemodynamics of ventricular filling and emptying. It has been shown in animal studies that an atrial rate of 300–320 bpm immediately causes monophasic, bidirectional blood flow in the venous vessels, resulting in a 75% rise of venous pressures in the inferior vena cava[17]. Hydrops developed in these cases within 4–48 hours. The critical heart rate in fetal sheep appears to be approximately 310 bpm. Unlike the normal Doppler spectra in venous vessels, forward diastolic flow is absent in supraventricular tachycardia owing to the shortened diastolic phase, resulting in pulsatile, monophasic, holodiastolic reverse flow (Fig. 16.**21**). This retrograde flow correlates with the ventricular diastolic phase of the cardiac cycle, and its primary cause is not atrioventricular valvular reflux during ventricular systole[15,17]. Instead, the retrograde flow appears to be caused by tachycardia-induced changes in pressure and volume loads and a shortened relaxation phase during diastole. Another factor appears to be a lack of coordination between atrial contraction and the opening of the atrioventricular valves, so that the blood volume propelled by the atrial contraction encounters atrioventricular valves that are essentially closed.

Development of hydrops fetalis. If supraventricular tachycardia persists, there is a deterioration of myocardial perfusion, which occurs mainly during ventricular diastole, and cardiomyopathy will eventually develop owing to a trophic and oxidative breakdown of the myocardial energy metabolism. The cardiomyopathy leads to cardiomegaly with the development of atrioventricular valve incompetence. If cardioversion cannot be achieved with medication, the sustained tachycardia will culminate in a generalized hydrops fetalis and possible intrauterine fetal death[14]. The critical heart rate for fetuses appears to be 210 bpm[15]. If cardioversion is successful, the venous waveforms quickly return to normal and the hydrops clears.

Fetal Anemia

Severe acute anemia in experimental animals induces vasoconstriction in the splanchnic and renal vascular beds, resulting in an increased oxygen supply to the brain, heart, and adrenal glands. By contrast, fetuses with severe chronic anemia, due for example to blood group isoimmunization or a fresh parvovirus B19 infection, are able to maintain an adequate oxygen supply for a certain period by means of specific humoral, hemorheological, and cardiovascular adaptive processes.

Hyperdynamic circulation. Although there is no evidence of overall impedance changes in the fetoplacental vessels, the anemic fetus responds with a progressive rise in cardiac output and blood flow velocities[7,40,41]. These hemodynamic changes result from increased contractility of the heart and the low viscosity of the blood. This cardiovascular response to anemia is also known as a "hyperdynamic circulation"[51]. While increased flow velocities can be measured in essentially all fetal vessels, Doppler examinations of the middle cerebral artery, descending aorta, and ductus venosus appear to be the most rewarding (Fig. 16.**22**). As in the arterial system, the pulsatility in

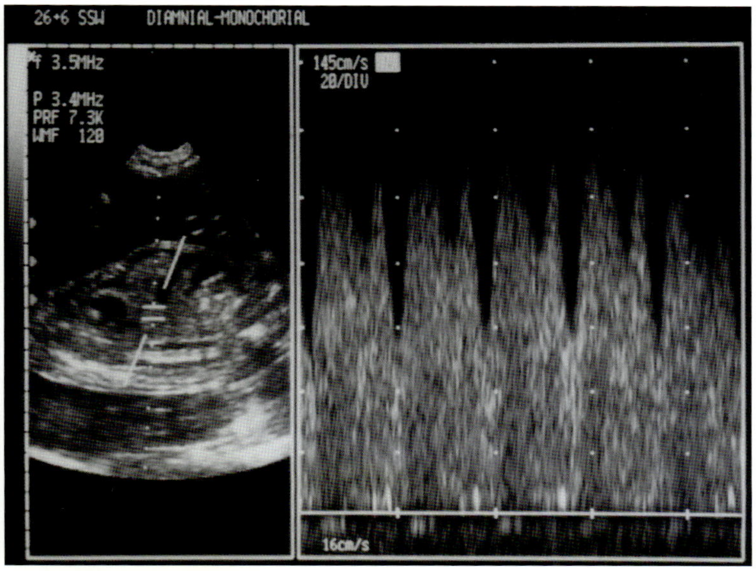

Fig. 16.**22** Doppler spectrum of the ductus venosus in a twin pregnancy (26 weeks 6 days) with a fresh parvovirus B19 infection. Note the high maximum velocities and normal waveform pulsatility. The Hb value determined by cordocentesis was 1.9 g/dl.

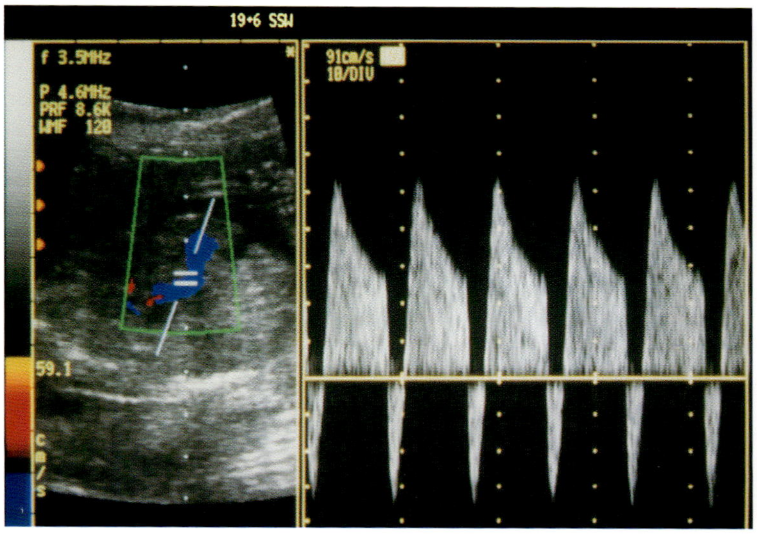

Fig. 16.**23** Doppler spectrum of the ductus venosus in a hydropic fetus with endocardial fibroelastosis. Note the extensive retrograde flow during atrial contraction.

the venous vessels is not increased[26,51]. Consequently, the resistance indices of the arterial vessels and the preload indices of the venous vessels are not useful in predicting the degree of anemia.

Development of hydrops fetalis. These findings are supported by experimental studies in fetal lambs with induced chronic anemia[7,40]. It was shown, for example, that the right ventricular pressure does not rise in the presence of increasing anemia and concomitant hydrops fetalis. Instead, there is an increase in the right ventricular stroke volume and in myocardial blood flow[7]. Thus, the hydrops does not appear to result primarily from congestive heart failure but more from changes in the colloid osmotic pressure, permeability changes, and hypoxemia-induced endothelial lesions of the fetal vessels. Cardiovascular decompensation does not occur until the hematocrit falls below 10%, at which point the dwindling cardiac stroke volume and resulting high cardiac pressures lead to congestive heart failure[7]. In the monitoring of intrauterine intravascular blood transfusion, Doppler examination of the ductus venosus can furnish information on hemodynamic changes during or immediately after the transfusion[50].

Other Diseases

Arteriovenous anastomoses. Abnormal venous Doppler spectra recorded in fetuses with a sacrococcygeal teratoma or vein of Galen aneurysm, for example, may be caused by the presence of arteriovenous anastomoses. Congestive heart failure can develop in these cases as a result of the frequent high shunt volumes.

Endocardial fibroelastosis. A progressive increase of pulsatility in the venous vessels can also develop in the setting of endocardial fibroelastosis (Fig. 16.**23**). The progressive stiffening of the myocardium and the resultant loss of contractility lead to a dwindling cardiac output and elevated central venous pressures. As cardiac compliance continues to decline, hydrops fetalis develops with a poor prognosis.

Congenital heart disease. Abnormal venous flow patterns can also be found in association with certain congenital heart defects, particularly those associated with anomalies of the ventricular inflow or outflow tracts[45]. In rare cases, hydrops fetalis may also correlate with agenesis of the ductus venosus[60].

Conclusion

In the hands of an experienced examiner, Doppler evaluation of the venous vascular system is an important adjunct in the surveillance and prognostic assessment of high-risk fetuses. The use of venous preload indices permits a detailed evaluation of fetal cardiac function and its physiological and pathophysiological changes.

References

1 Abithol MM, Rochelson BL, Monheit AG, Ryland SJ, Baumann AL, Stern W: Use of an indwelling Doppler probe to study acute changes in umbilical vein waveforms in the fetal sheep. Gynecol. Obstet. Invest. 34 (1992) 6–11
2 Baschat AA, Gembruch U, Reiss I, Gortner L, Dietrich K: Demonstration of fetal coronary blood flow by Doppler ultrasound in relation to arterial and venous flow velocity waveforms and perinatal outcome—the "heart-sparing effect". Ultrasound Obstet. Gynecol. 9 (1997) 162–172
3 Baschat AA, Gembruch U: Triphasic umbilical venous blood flow with prolonged survival in severe intrauterine growth retardation: a case report. Ultrasound Obstet. Gynecol. 8 (1996) 201–205
4 Chan FY, Woo SK, Ghosh A, Tang M, Lam C: Prenatal Diagnosis of congenital fetal arrhythmias by simultaneous pulsed Doppler Velocimetry of the fetal abdominal aorta and inferior vena cava. Obstet. Gynecol. 76 (1990) 200–204
5 Cohn HE, Sacks EJ, Heymann MA, Rudolph AM: Cardiovascular responses to hypoxemia and acidemia in fetal lambs. Amer. J. Obstet. Gynecol. 120 (1974) 817–824
6 Damron DP, Chaffin DG, Anderson CF, Reed KL: Changes in umbilical arterial and venous blood flow velocity waveforms during late decelerations of the fetal heart rate. Obstet. Gynecol. 84 (1994) 1038–1040
7 Davis LE, Hohimer AR: Hemodynamics and organ blood flow in fetal sheep subjected to chronic anemia. Amer. J. Physiol. 261 (1991) R1542–R1548
8 De Vore GR, Horenstein J: Ductus venosus index: a method for evaluating right ventricular preload in the second-trimester fetus. Ultrasound Obstet. Gynecol. 3 (1993) 338–342
9 Edelstone DI, Rudolph AM: Preferential streaming of ductus venosus blood to the brain and heart in fetal lambs. Amer. J. Physiol. 237 (1979) H724–H729
10 Edelstone DI, Rudolph AM, Heymanm MA: Effects of hypoxemia and decreasing umbilical flow on liver and ductus venosus blood flows in fetal lambs. Amer. J. Physiol. 238 (1980) H656–H663
11 Eik-Nes SH, Brubakk AO, Ulstein MK: Measurement of human fetal blood flow. BMJ (1980) 283–284
12 Erskine RLA, Ritchie JWK: Quantitative measurement of fetal blood flow using Doppler ultrasound. Brit. J. Obstet. Gynecol. 92 (1985) 600–604
13 Fitzgerald DE, Drumm DE: Non invasive measurement of human fetal circulation using ultrasound: a new method. Brit. J. Obstet. Gynecol. 2 (1977) 1450–1451
14 Gembruch U, Redel DA, Bald R, Hansmann: Longitudinal study in 18 cases of fetal supraventricular tachycardia: Doppler echocardiographic findings and pathophysiologic implications. Am. Heart J. 125 (1993) 1290–1301
15 Gembruch U, Krapp M, Baumann P: Changes of venous blood flow velocity waveforms in fetuses with supraventricular tachycardia. Ultrasound Obstet. Gynecol. 5 (1995) 394–399
16 Gembruch U, Baschat AA, Caliebe A, Gortner L: Prenatal diagnosis of ductus venosus agenesis: a report of two cases and review of the literature. Ultrasound Obstet Gynecol 11 (1998) 185–189
17 Gest AL, Martin CG, Moise AA, Hansen TN: Reversal of venous blood flow with atrial tachycardia and hydrops in fetal sheep. Pediatr. Res. 28 (1990) 223–226
18 Gill RW, Trudinger BJ, Gerrit WJ, Kossow G, Warren PS: Fetal umbilical venous flow measured in utero by pulsed Doppler and B-mode ultrasound. Amer. J. Obstet. Gynecol. 139 (1981) 720–725
19 Gill RW, Kossoff G, Warren PS, Garrett WJ: Umbilical venous flow in normal and complicated pregnancy. Ultrasound Med. Biol. 10 (1984) 349–363
20 Giles WB, Lingman G, Marsal K, Trudinger BJ: Fetal volume blood flow and umbilical artery flow velocity waveform analysis: a comparison. Brit. J. Obstet. Gynecol. 93 (1986) 461–465
21 Gonser M, Goelz R, Erz W, Mielke G: Dopplersonographie des Ductus venosus bei Hochrisikoschwangerschaften zwischen 23 und 26 Wochen. Geburtshilfe Frauenheilkd 55 (1995) 661–663
22 Gudmundson S, Hutha JC, DC Wood, Tulzer G, Cohen AW, Weiner S: Doppler ultrasonography in the fetus with nonimmune hydrops. Amer. J. Obstet. Gynecol. 164 (1991) 33–37
23 Gudmundson S, Tulzer G, Hutha JC, Marsal K: Venous Doppler in the fetus with absent end-diastolic flow in the umbilical artery. Ultrasound Obstet. Gynecol. 7 (1996) 262–267
24 Hecher K, Campbell S, Snijders R, Nicolaides K: Reference ranges for fetal venous and atrioventricular blood flow parameters. Ultrasound Obstet. Gynecol. 4 (1994) 381–390
25 Hecher K, Campbell S, Doyle P, Harrington K, Nicolaides K: Assessment of fetal compromise by Doppler ultrasound investigation of the fetal circulation. Arterial, intracardiac, and venous blood flow velocity studies. Circulation 91 (1995) 129 –138
26 Hecher K, Snijders R, Campbell S, Nicolaides K: Fetal venous, arterial, and intracardiac blood flows in red blood cell isoimmunization. Obstet. Gynecol. 85 (1995) 122–128
27 Hecher K, Snijders R, Campbell S, Nicolaides K: Fetal venous, intracardiac, and arterial blood flow measurements in intrauterine growth retardation: Relationship with fetal blood gases. Amer. J. Obstet. Gynecol. 173 (1995) 10–15
28 Hecher K, Ville Y, Snijders R, Nicolaides K: Doppler studies of the fetal circulation in twin-twin transfusion syndrome. Ultrasound Obstet. Gynecol. 5 (1995) 318–324
29 Huisman TWA, Stewart PA, Wladimiroff JW: Flow velocity waveforms in the fetal inferior vena cava during the second half of normal pregnancy. Ultrasound Med. Biol. 17 (1991) 379–382
30 Huisman TWA, Stewart PA, Wladimiroff JW: Ductus venosus blood flow velocity waveforms in the human fetus – A Doppler study. Ultrasound Med. Biol. 18 (1992) 33–37
31 Huisman TWA, Gittenberger-De Groot AC, Wladimiroff JW: Recognition of a fetal subdiaphragmatic venous vestibulum essential for fetal venous Doppler assessment. Pediatr. Res. 32 (1992) 338–341
32 Huisman TWA, van den Eijnde SM, Stewart PA, Wladimiroff JW: Changes in inferior vena cava blood flow velocity and diameter during breathing movements in the human fetus. Ultrasound Obstet. Gynecol. 3 (1993) 26–30
33 Indik J, Chen V, Reed KL: Association of umbilical venous with inferior vena cava blood flow velocities. Obstet. Gynecol. 77 (1991) 551–557
34 Indik JH, Reed KL: Correlations in umbilical blood flow Doppler velocities in the human fetus during breathing. Ultrasound Obstet. Gynecol. 4 (1994) 361–366
35 Jouppila P, Kirkinen P: Umbilical vein blood flow in the human fetus in cases of maternal and fetal anemia and uterine bleeding. Ultrasound Med. Biol. 10 (1984) 365–370
36 Jouppila P, Kirkinen P: Umbilical vein blood flow as an indicator of fetal hypoxia. Brit. J. Obstet. Gynecol. 91 (1984) 107–110
37 Kanzaki T, Chiba Y: Evaluation of the preload condition of the fetus by inferior vena cava blood flow pattern. Fetal Diagn. Ther. 5 (1990) 168–174
38 Kanzaki T, Murakami M, Kobayashi H, Chiba Y: Characteristic abnormal blood flow patterns of the inferior vena cava in fetal arrhythmias. J. Matern. Fetal. Invest. 1 (1991) 35–39
39 Karsdorp VHM, van Vugt JMG, Kostense PJ, Arduini D, Montenegro N, Todros T: Clinical significance of absent or reversed end diastolic velocity waveforms in umbilical artery. Lancet 344 (1994) 1664–1668
40 Kilby MD, Swarc R, Benson LN, Morrow RJ: Left ventricular hemodynamics in anemic fetal lambs. J. Perinat. Med. 26 (1998) 5–12
41 Kirkinen P, Jouppila P: Umbilical vein blood flow in rhesus-isoimmunization. Brit. J. Obstet. Gynecol. 90 (1983) 640–643
42 Kiserud T, Eik-Nes SH, Blaas HG, Hellevik LR: Ultrasonographic velocimetry of the fetal ductus venosus. Lancet 338 (1991) 1412–1414
43 Kiserud T, Eik-Nes SH, Blaas HG, Hellevik LR: Foramen ovale: An ultrasonographic study of its relation to the inferior vena cava, ductus venosus and hepatic veins. Ultrasound Obstet. Gynecol. 2 (1992) 389–396
44 Kiserud T, Eik-Nes SH, Hellevik LR, Blaas HG: Ductus venosus – a longitudinal Doppler velocimetric study of the human fetus. J. Matern. Fetal. Invest. 2 (1992) 5–11

45 Kiserud T, Eik-Nes SH, Hellevik LR, Blaas HG: Ductus venosus blood velocity changes in fetal cardiac diseases. J. Matern. Fetal. Invest. 3 (1993) 15–20
46 Kiserud T, Eik-Nes SH, Blaas HG, Hellevik LR, Simensen B: Ductus venosus blood velocity and the umbilical circulation in the seriously growth-retarded fetus. Ultrasound Obstet. Gynecol. 4 (1994) 109–114
47 Künzel W: Das fetale Schocksyndrom. Z. Geburtsh. u. Perinat. 190 (1986) 177–184
48 Mari G, Uerpairojkit B, Copel JA: Abdominal venous system in the normal fetus. Obstet. Gynecol. 86 (1995) 729–733
49 Nakai Y, Miyazaki Y, Matsuoka Y: Pulsatile umbilical venous flow and its clinical significance. Brit. J. Obstet. Gynecol. 99 (1992) 977–980
50 Oepkes D, Vanderbussche FP, van Bel F, Kanhai HHH: Fetal ductus venosus blood flow velocities before and after transfusion in red-cell alloimmunized pregnancies. Obstet. Gynecol. 82 (1993) 237–241
51 Oepkes D, Brand R, Vandenbussche FP, Meerman RH, Kanhai HHH: The use of ultrasonography and Doppler in the prediction of fetal haemolytic anaemia: a multivariate analysis. Brit. J. Obstet. Gynecol. 101 (1994) 680–684
52 Ozcan T, Sbracia M, Dáncona RL, Copel JA, Mari G: Arterial and venous Doppler velocimetry in the severely growth-restricted fetus and associations with adverse perinatal outcome. Ultrasound Obstet. Gynecol. 12 (1998) 39–44
53 Pennati G, Bellotti M, Ferrazzi E, Rigano S, Garberi A: Hemodynamic changes across the human ductus venous: a comparison between clinical findings and mathematical calculations. Ultrasound Obstet. Gynecol. 9 (1997) 183–391
54 Reed KL, Appleton CP, Anderson CF, Shenker L, Sahn DL: Doppler studies of vena cava flows in human fetuses. Circulation 81 (1990) 498–505
55 Reed KL, Chaffin DG, Anderson CF: Umbilical venous Doppler velocity pulsations and inferior vena cava pressure elevations in fetal lamb. Obstet. Gynecol. 87 (1996) 617–620
56 Reed KL, Chaffin DG, Anderson CF, Newman AT: Umbilical venous velocity pulsations are related to atrial contraction pressure waveforms in fetal lambs. Obstet. Gynecol. 89 (1997) 953–956
57 Reuss ML, Rudolph AM, Dae MW: Phasic blood flow patterns in the superior and inferior venae cavae and umbilical vein of fetal sheep. Amer. J. Obstet. Gynecol. 145 (1983) 70–78
58 Rizzo G, Arduini D, Romanini C: Umbilical vein pulsations: A physiologic finding in early gestation. Amer. J. Obstet. Gynecol. 167 (1992) 675–677
59 Rizzo G, Arduini D, Caforio L, Romanini C: Effects of sampling sites on inferior vena cava flow velocity waveforms. J. Matern. Fetal. Invest. 2 (1992) 153–156
60 Rizzo G, Arduini D, Romanini C: Inferior vena cava flow velocity waveforms in appropriate- and small-for-gestational-age fetuses. Amer. J. Obstet. Gynecol. 166 (1992) 1271–1280
61 Rizzo G, Capponi A, Arduini D, Romanini C: Ductus venosus velocity waveforms in appropriate and small for gestational age fetuses. Early Hum. Develop. 39 (1994) 15–26
62 Rizzo G, Capponi A, Talone PE, Arduini D, Romanini C: Doppler indices from inferior vena cava and ductus venosus in predicting pH and oxygen tension in umbilical blood at cordocentesis in growth-retarded fetuses. Ultrasound Obstet. Gynecol. 7 (1996) 401–410
63 Saling E: Die O_2-Sparschaltung des fetalen Kreislaufes. Geburtsh. u. Frauenheilk. 26 (1966) 413–419
64 van Splunder IP, Huisman TWA, Stijnen T, Wladimiroff JW: Presence of pulsations and reproducibility of waveform recording in the umbilical and left vein in normal pregnancies. Ultrasound Obstet. Gynecol. 4 (1994) 49–53
65 Wladimiroff JW, Huisman TW, Stewart PA: Normal fetal Doppler inferior vena cava, transcuspid, and umbilical artery flow velocity waveforms between 11th and 16th week's gestation. Amer. J. Obstet. Gynecol. 166 (1992) 921–924
66 Zikulnig L, Hecher K, Bregenzer T, Bäz E, Hackelöer BJ: Doppler sonography and prognostic factors in severe twin-twin transfusion syndrome with endoscopic laser therapy. Ultrasound Obstet. Gynecol. 12(Suppl.1) (1998) 9

Specific Obstetric Problems

17 Color Doppler Sonography in the Diagnosis of Nuchal Cord

A. K. Ertan, H. J. Hendrik, and W. Schmidt

Importance of Nuchal Cord

Possible sequelae of nuchal cord. Umbilical cord complications are by far the most common cause of severe intrapartum fetal hypoxic states[18]. Looping of the umbilical cord around the fetal neck (nuchal cord, NC) is observed in 20–33 % of all deliveries[1,42] and in up to 48% of breech-presenting fetuses[7]. Rare but serious complications of NC are an increased incidence of acidosis[2,10,24], neonatal hypovolemic shock[42], and abnormalities of neurological development[1]. Ultimately, NC can also be a cause of intrauterine fetal death[1,17]. It can also lead to dystocic problems in the form of prolonged cervical dilatation and hypotonic labor[10] (Table 17.**1**).

Table 17.**1** Possible complications of nuchal cord

- More frequent occurrence of variable decelerations, FHR changes
- Prolonged dilation period, hypotonic labor
- Higher rate of neonatal asphyxia
- Hypovolemic shock
- Neurologic abnormalities
- Intrauterine fetal death

FHR = fetal heart rate.

Number of loops. Birnholz[1] notes that NC is associated with an increased risk of asphyxia-induced prenatal brain injury and that the degree of risk increases with the number of entangling loops. There are varying reports on the incidence of multiple cord loops[19,20]. Kan-Pun-Shui and Eastman[16] found a 20.6% incidence of double loops, a 2.5% incidence of triple loops, and a 0.1% incidence of quadruple loops[16]. There have been isolated reports of up to nine cord loops[22].

Color Doppler Study on the Diagnosis of Nuchal Cord

Until recently, the only methods available for the diagnosis of NC were B-mode ultrasound imaging[4,6,7,11,33,36] and fetal heart rate (FHR) monitoring[23]. In recent years there have been increasing reports on the use of pulsed Doppler ultrasound in the diagnosis of NC[1,29,31].

A prospective study was conducted at the Department of Obstetrics and Gynecology, Homburg/Saar University, to assess the role of high-resolution color Doppler sonography in the antenatal diagnosis of NC. A series of 254 pregnant women underwent ultrasound examinations on admission and for a suspicious fetal heart rate to check for NC. The prenatal ultrasound results were then compared with the intrapartum and postpartum findings. All the examinations were performed with high-resolution real-time color Doppler units (Acuson Computed Sonography 128 XP/10 with a 5 MHz curved array, Picker CS 192 Integral Color Doppler, and Siemens Elegra with a 3.5 MHz curved array). Videotape and video hard-copy documentation were used. When fetal biometry was completed, the fetal neck region was imaged and systematically scanned with a color Doppler probe to check for signals from umbilical vessels. All results were entered on a documentation sheet. Positional drawings were made in special situations, such as when ultrasound showed umbilical cord signals near the fetal neck but did not show definite cord encirclement.

Examination Technique

Longitudinal scan. First the fetal neck region was examined with conventional B-mode (gray-scale) ultrasound for the presence of umbilical vessels. This was followed by a color Doppler examination. Longitudinal scanning of the fetal neck provides a cross-sectional view of the umbilical vessels (Fig. 17.**1**). Depending on the fetal lie, the neck region can be explored in sagittal sections (dorsoanterior or dorsoposterior lie) or coronal sections (fetal spine to the right or left). If a suspected umbilical cord signal is found near the fetal neck, the diagnosis can be confirmed in equivocal cases by identifying typical frequency shifts in the umbilical arteries and veins with pulsed Doppler.

Transverse scan. A definitive diagnosis of NC can be made only by demonstrating direct encirclement of the fetal neck by the umbilical cord with color Doppler. This view of the umbilical vessels is obtained by imaging the neck region in transverse section (Fig. 17.2**a**). Often it is not possible to define a complete loop of cord around the neck, especially near term (owing to oligohydramnios, a low presenting part, etc.). For this reason, at least the anterior and lateral portions of the neck should be closely scrutinized (Fig. 17.**2b**). This is the only way to avoid a false presumptive diagnosis of NC in cases where a "tangle" of tortuous umbilical cord is seen abutting the side of the neck.

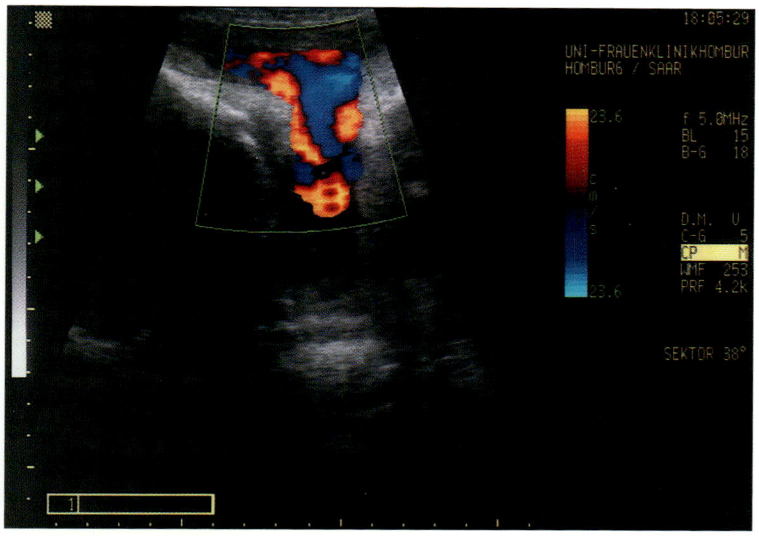

Fig. 17.**1** Umbilical cord signals in a longitudinal color Doppler scan of the fetal neck.

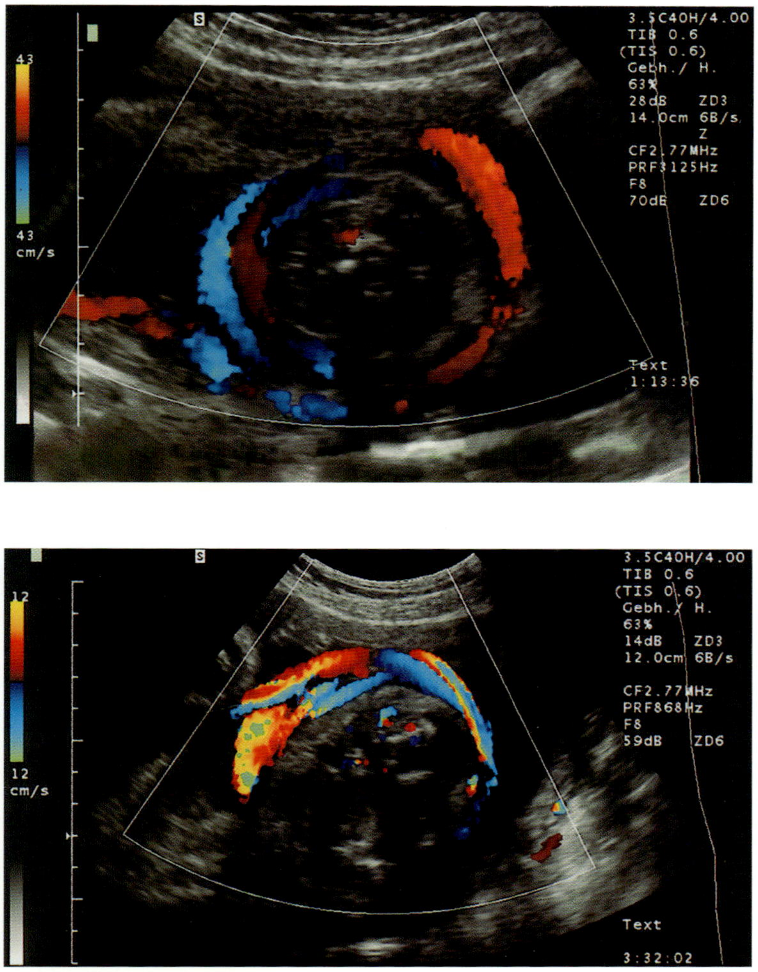

Fig. 17.**2** Umbilical cord signals in a transverse color Doppler scan of the fetal neck.
a Umbilical cord signals encircling the neck.
b Cord segments abutting the front and sides of the neck.

Results

We achieved a 97 % sensitivity rate in the diagnosis of NC by antepartum color Doppler examination, with a 6 % false-positive rate and a 1.5 % false-negative rate. The specificity was 88 %. The positive predictive value was 89 %, the negative predictive value 96 %. The efficiency was 93 %.

Sources of error. Using color flow alone, it is easy to obtain a false-positive impression of NC owing to the close proximity of the maternal pelvic vessels. The fetal and maternal vessels can be differentiated, however, by their distinctive frequency-shift patterns. NC is also difficult to diagnose in the presence of oligohydramnios, where it is almost impossible to make a correct diagnosis by B-mode imaging alone.

FHR monitoring. In 21 % of cases, an abnormal antepartum FHR trace provided the indication for targeted color Doppler imaging. In the group with an antenatal diagnosis of NC, variable heart rate decelerations were noted in 72 % of cases, early decelerations in 7 %, and late decelerations in 21 %. In the remaining cases without NC, half of the cases showed late decelerations, two cases had variable decelerations, and three cases had early decelerations.

An abnormal intrapartum FHR trace was recorded in approximately one-fourth of all the cases examined. The percentage of spontaneous deliveries was 61 %. A primary cesarean delivery was performed in 17 % of the cases, a secondary cesarean delivery in 10 %. Twelve percent of the patients had an operative vaginal delivery. In the cases that had an indication for operative delivery, the percentage of abnormal FHR traces (impending intrauterine asphyxia) was significantly higher in the group with NC (43 %) than in the group without NC (8 %, $p < 0.001$).

Doppler examination of fetal vessels. Doppler velocimetry of the fetal aorta was abnormal (S/D ratio > 7) in 30 % of the cases diagnosed with NC. There was an 11.3 % rate of abnormal findings in the umbilical arteries (S/D ratio > 4) in the NC cases. In 12 % of the cases with NC, the Doppler velocimetry findings deteriorated over the course of the pregnancy.

Umbilical cord length. Postpartum umbilical cord length was measured in all patients (minimum 28 cm, maximum 101 cm). The average umbilical cord length in the group with NC was 65 ± 10 cm, versus 54 ± 9 cm in the group without NC ($p < 0.01$). Excessive umbilical cord length, defined as >70 cm (>90th percentile), was 7 times more prevalent in the group with NC than in newborns without NC (Table 17.**2**).

Table 17.**2** Distribution of umbilical cord length in groups with and without nuchal cord

Umbilical cord length	With nucal cord	Without nucal cord
35–70 cm (normal)	79 %	97 %
> 70 cm (too long) ($p < 0.001$)	21 %	3 %

Importance of Nuchal Cord Diagnosis in the Biophysical (ABCD) Profile

Doppler sonography, especially color Doppler sonography, has gained an important and established role in pregnancy surveillance using biophysical methods. The antepartum detection of fetal compromise can be significantly improved by combining Doppler ultrasound with other biophysical methods of examination (see Chronic Placental Insufficiency, Chapter 18)[13].

Comparison of Doppler velocimetry and other biophysical parameters. We did another prospective study to determine the importance of the diagnosis of fetal NC with prenatal Doppler within the context of the biophysical (ABCD) profile. We particularly wanted to determine how the other biophysical parameters behaved in relation to Doppler velocimetry findings and assess the impact of NC on perinatal outcome in the complex biophysical evaluation. In 380 evaluations of the ABCD profile, NC was detected by color Doppler in 128 cases (34%)[12]. In 73 cases (57%) the pregnancy had no additional clinical risk factors. This group was compared with 144 pregnancies without NC and without additional clinical risk factors (Table 17.3).

ABCD profile. The ABCD profile consists of extended fetal biometry, the amniotic fluid index (AFI or EFI), kinetocardiotocography (KCTG), and Doppler examination of the fetal aorta and umbilical artery. Only singleton pregnancies were evaluated.

Comparing the biophysical profile variables in groups with and without NC in the absence of other clinical risk factors, we find only marginal differences with regard to biometry, amniotic fluid volume, and Doppler velocimetry (Table 17.4).

Table 17.**3** Relationship of ABCD profile to fetal outcomes with and without nuchal cord

	Without nuchal cord	With nuchal cord
Cases	*n* = 144	*n* = 73
Weeks' gestation at delivery	39	40
Birthweight in g (mean value, SD)	3204 (366)	3296 (240)
Mode of delivery		
Spontaneous	82%	75%
Primary cesarean section	11%	9%
Secondary cesarean section	7%	8%
Operative vaginal	–	8%
Apgar score 1 min**		
< 7	2%	16%
≥ 7	98%	84%
Apgar score, 5 min		
< 7	0%	0%
≥ 7	100%	100%
Umbilical cord arterial pH (mean, SD)	7.28 (0, 8)	7.28 (0, 9)
Intrapartum complications		
No	76%	64%
Yes	24%	36%

** $p < 0.01$

Similarly, we find no significant differences in individual KCTG criteria between the groups. At most, the number of accelerations was lower in the cases with NC than in the cases without NC (Table 17.5). Fetal heart rate parameters based on the Fischer score were the same in both groups.

A differentiated analysis of Doppler velocimetry findings showed a slight decrease of perfusion in the aorta and umbilical artery and no differences in the middle cerebral artery (Table 17.6).

Table 17.**4** ABCD profile findings in groups with and without a Doppler diagnosis of nuchal cord

	Without nuchal cord	With nuchal cord	*p*
Biometry			
Normal	72%	79%	NS
Abnormal	28%	21%	
Amniotic fluid volume			
Normal/increased	88%	92%	NS
Reduced	12%	8%	
Doppler velocimetry			
Normal	75%	70%	NS
Abnormal	25%	30%	
Movements (mean block length)			
Normal	83%	78%	NS
Shortened	17%	22%	
FHR (Fischer score)			
> 7	96%	96%	NS
≤ 7	4%	4%	

Table 17.**5** Kinetocardiotocography findings associated with nuchal cord

	Without nuchal cord	With nuchal cord	*p*
% Movements (mean value, SD)	18 (9,9)	18 (8,3)	NS
Mean block length (mean value, SD)	5.77 (2.6)	5.06 (1.2)	NS
Number of accelerations (mean value, SD)	8.6 (5.5)	6.0 (4.8)	< 0.05
Fischer score			
8–10	96%	96%	NS
5–7	4%	4%	

Table 17.**6** Individual color Doppler findings in the ABCD profile in cases with and without nuchal cord

Vessel	Evaluation	Without nuchal cord	With nuchal cord	*p*
Aorta	Normal	94%	78%	< 0.1
	Abnormal/ borderline	6%	22%	
Umbilical artery	Normal	75%	70%	< 0.1
	Abnormal	25%	30%	
Middle cerebral artery	Normal	84%	82%	NS
	Abnormal	16%	15%	

Perinatal outcome. In pregnancies with no additional clinical risk factors, there were no significant differences in perinatal outcomes between the groups with and without NC. The cases with NC had a slightly increased percentage of vaginal operative deliveries. The 1-minute Apgar score was markedly poorer than in the cases with NC, but the 5-minute scores were the same in both groups. There was no difference in arterial umbilical cord pH. Overall intrapartum complications were 12% more frequent in cases with an antepartum diagnosis of NC.

Interpretation. In summary, we may conclude that NC is of minor importance in the complex biophysical evaluation of uncomplicated pregnancies. The perfusion values in the fetal vessels are mildly impaired without causing a significant long-term perfusion deficit. The fetal outcome does show the typical effects of acute changes caused by NC, but these are easily recognized and treated within the framework of ordinary obstetric management.

Role of Doppler Sonography in NC

Nuchal cord can have a variety of antepartum and intrapartum complications ranging from sporadic, variable FHR decelerations to intrauterine fetal death[1, 10, 14, 15, 17, 25, 35, 37, 40].

High diagnostic accuracy. In a prospective study, we found that NC could be diagnosed prenatally with a 97% sensitivity by using high-resolution real-time color Doppler sonography. The 89% positive predictive value, 97% negative predictive value, and 93% efficiency confirm the high diagnostic accuracy of this method. Especially with findings that are equivocal by B-mode and color-flow imaging, the use of pulsed Doppler velocimetry can further increase the detection rate based on the typical frequency shift patterns that are observed in the umbilical arteries and vein.

Umbilical cord length. Little attention has been paid to the issue of umbilical cord length. The mean reported umbilical cord length in term deliveries is 50–60 cm. It is generally acknowledged that NC may be more common in pregnancies where the umbilical cord is too long. This contrasts with the effects of a short umbilical cord, which is associated with constrained fetal movements and abnormalities of central nervous system development. As early as 1750, Smellie reported cases of intrauterine fetal death caused by too short an umbilical cord[38]. Besides a “relatively” short umbilical cord (e.g., caused by looping around the fetal neck), there may be an “absolute” short cord whose length is less than 35 cm[8, 30, 32]. Reportedly, a short cord is present in 0.43–0.78% of pregnancies and appears to correlate with chromosome abnormalities including Down syndrome[5].

The frequent occurrence of excessive umbilical cord length in newborns with NC was confirmed in our study. A cord length >70 cm was 7 times more common in cases with NC than in cases without NC (21% versus 3%, $p < 0.001$).

Correlation with FHR trace. According to Kubli and Schmidt[18], peracute fetal hypoxic states occur without prior warning, probably as a result of occult umbilical cord compression, in approximately 0.1–0.2% of all deliveries [18]. This means that an abnormal prenatal FHR trace is an important indication for antenatal NC screening. Abnormal antepartum and intrapartum heart rate changes are found more commonly in fetuses with NC[10, 26, 28]. For example, Goldkrand et al.[9] found a 74% incidence of variable FHR decelerations in cases with certain umbilical cord abnormalities such as knotted cord and nuchal cord, while Tejani et al.[41] found an 89% incidence. In our study using the biophysical profile, NC in the absence of other clinical risk factors was manifested only by a decreased number of accelerations in the FHR. The presence of NC did not affect the Fischer scores or quantitative motion analysis by kinetocardiotocography.

It is generally accepted that variable decelerations in the FHR can be caused by a fetal vagal reflex. Umbilical cord compression appears to be a factor in this process both before and during delivery[43].

Perinatal findings. Twenty-one percent of our study patients were selected for antenatal NC screening based on abnormal antenatal FHR findings. By contrast, the FHR abnormalities detected in the group without NC were due chiefly to other factors (e.g., suspected intrauterine growth retardation). The mean Apgar score and umbilical cord pH values were also significantly lower in infants with NC, and the Pco_2 levels were higher. The incidence of neonatal asphyxia was significantly higher in cases with repeated variable decelerations than in cases without (Apgar score < 7, 35.3% versus 3.8%; and pH < 7.20, 44.1% versus 6.5%). It is reported in the literature that greenish amniotic fluid is 3–4 times more common in NC[39], and it was three times more common in the patients that we examined.

Prevention of neuromotor deficits. The current main focus of obstetric efforts, besides lowering prenatal mortality, is the prevention of perinatal morbidity with the potential risk of permanent neuromotor deficits[18]. As a general rule, the FHR trace, by monitoring a single functional quantity (the fetal heart rate), cannot cover all aspects of a potentially multifactorial disturbance, and the antenatal FHR trace offers no typical warning signs that would indicate a subsequent impairment of cerebral development[34]. FHR monitoring is more a screening test whose ultimate goal is the early prediction of subsequent (intrapartum) compromise of the fetal circulation, which is always stressed by the delivery. Greater attention is now being given to prenatal abnormalities as distinguished from intrapartum insults[27, 34]. Mallard et al.[21] found that isolated umbilical cord occlusion for 10 minutes (in fetal sheep) caused severe transient asphyxia in addition to hypotension, bradycardia, and increased cortical impedance (EEG), but the principal effect was a loss of neurons in the hippocampus[21]. Hippocampal damage in humans is manifested chiefly by memory deficits.

Clapp et al.[3] described similar results. They reported on histologically confirmed cerebral damage in 89% of cases in which central hypoxia had been induced in fetal lambs by intermittent partial occlusion of the umbilical circulation[3].

Practical aspects of Doppler screening. In the study presented here, the practical aspects of antenatal screening for NC were described in detail. Of particular importance is the color Doppler visualization of umbilical cord signals in the transverse neck scan, as this can provide a vivid color-flow image of umbilical vessels encircling the neck (Fig. 17.**2**). The 6% false-positive rate is due partly to the fact that while color and pulsed Doppler could detect typical umbilical cord signals by the fetal neck, a view in the circumferential plane was not possible on account of various factors (e.g., a deeply engaged fetal head at delivery) (Fig. 17.**3**).

The number of cord loops also appears to be an important factor. A single loop of cord can be readily distinguished from, say, three or four loops in the color Doppler image (Fig. 17.**4**).

Follow-up. It is also possible to detect loops of cord around the fetal neck, limbs, or body as well as knots of the cord during the early weeks of gestation and then monitor their progression during the rest of the pregnancy (Figs. 17.**5** , 17.**6**). This enables the physician to increase antenatal care and surveillance and intervene quickly if there is any functional deterioration of fetal condition. The combined use of biophysical evaluation methods such as the ABCD profile can help to detect functional changes at an early stage and avert their sequelae through appropriate obstetric action. On the other hand, the diagnosis of NC in an otherwise normal pregnancy (with no clinical risk factors) is a benign finding, and the patient with a normal ABCD profile can be given reassurance.

Umbilical cord prolapse. Color Doppler visualization of the umbilical cord in front of the fetal presenting part in the lower uterine segment (if necessary, using an endovaginal probe) can facilitate the prenatal diagnosis of umbilical cord prolapse.

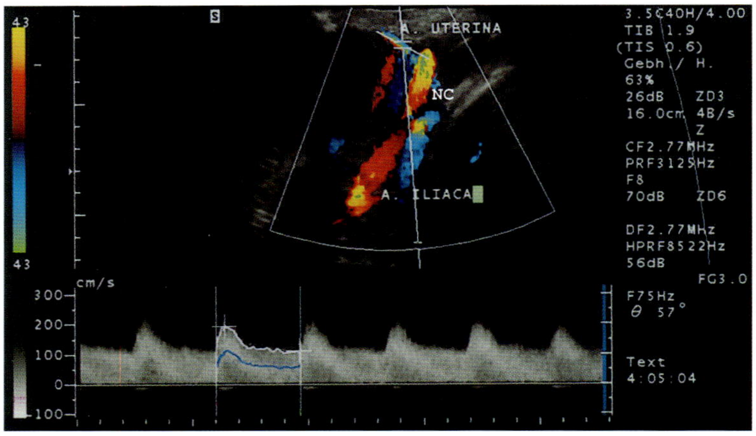

Fig. 17.**3** Pitfalls in umbilical cord identification and the targeted use of pulsed Doppler sonography. NC = nuchal cord.

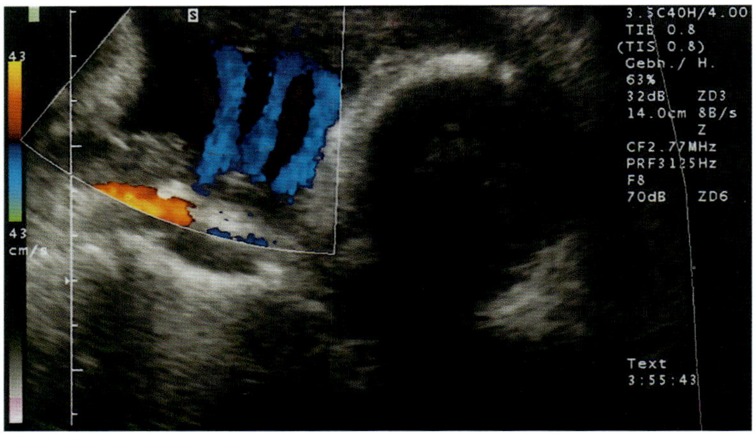

Fig. 17.**4** Double loops of umbilical cord around the fetal neck.

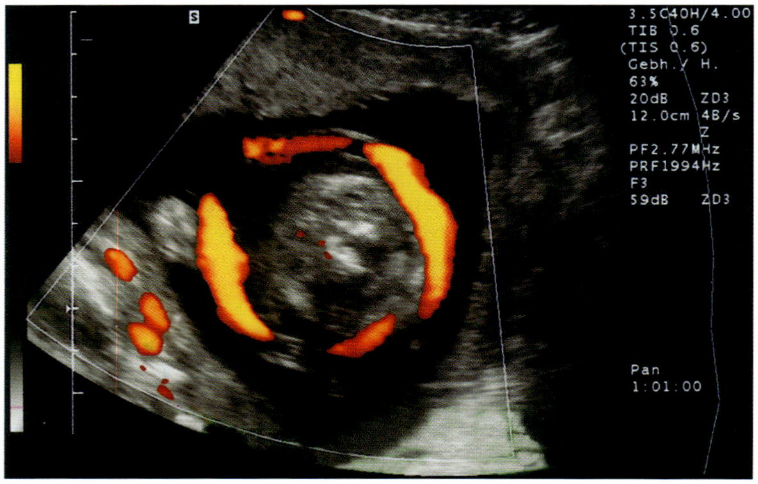

Fig. 17.**5** Nuchal cord in the 22nd week of gestation.

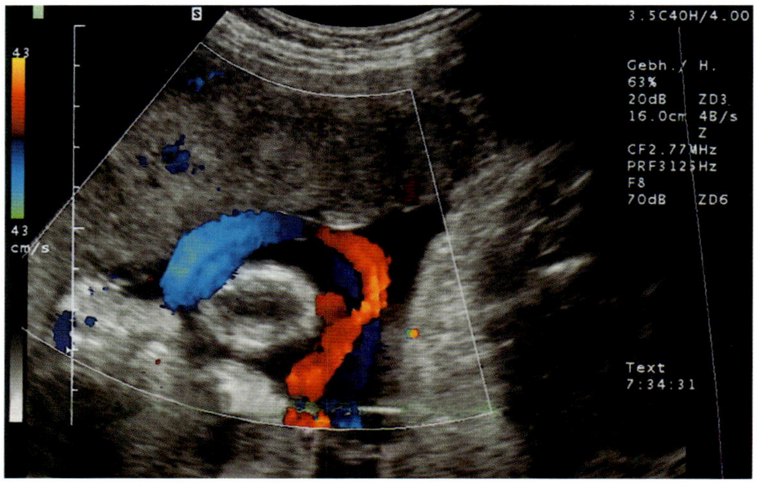

Fig. 17.**6** Loop of umbilical cord encircling the fetal wrist.

17

Summary

Previous experience with this method can be summarized as follows.

The antenatal detection of NC by color Doppler sonography can be accomplished in nearly all cases. Once this method has been learned, the presence or absence of NC can be quickly and reliably established. It may be possible to avoid intrapartum complications through the prompt antenatal detection of NC. Problems can arise when this diagnosis is communicated to the patient. Insufficient or improper counseling can provoke confusion and anxiety in many patients. For this reason, the pregnant woman should understand that NC is a frequent finding, even as an isolated condition. When antepartum FHR abnormalities are noted (e.g., unexplained variable decelerations), color Doppler can be used to determine the possible cause. The high accuracy of color Doppler in the diagnosis of NC can aid in the prevention of serious perinatal and intrapartum complications in many cases. When other abnormalities have been excluded, these patients can be managed further by intensive surveillance, which may include use of the biophysical (ABCD) profile.

References

1 Birnholz JC: Ecologic Physiology of the Fetus. Radiol. Clin. North. Am. 28 (1990) 179–188
2 Bretscher J, Saling E: pH values in the human fetus during labor. Am. J. Obstet. Gynecol. 97 (1967) 906–911
3 Clapp JF, Peress NS, Wesley M, Mann LI: Brain damage after intermittent partial cord occlusion in the chronically instrumented fetal lamb. Am. J. Obstet. Gynecol. 159 (1988) 504–509
4 Collins JH: Nuchal cord type A and type B. Am. J. Obstet. Gynecol. 177 (1997) 94
5 Ente G, Penzer PH: The umbilical cord: normal parameters. J. R. Soc. Health. 111 (1991) 138–140
6 Feinstein SJ, Lodeiro JG, Vintzileos AM, Weinbaum PJ, Campbell WA, Nochimson DJ: Intrapartum ultrasound diagnosis of nuchal cord as a decisive factor in management. Am. J. Obstet. Gynecol. 153 (1985) 308–309
7 Giacomello F: Ultrasound determination of nuchal cord in breech presentation. Am. J. Obstet. Gynecol. 159 (1988) 531–532
8 Gardiner JP: The umbilical cord. Normal length; length in cord complications; etiology end frequency of coiling. Surg. Gynecol. Obstet. 34 (1922) 254–256
9 Goldkrand JW, Speichinger JP: "Mixed cord compression", fetal heart rate pattern, and its relation to abnormal cord position. Am. J. Obstet. Gynecol. 122 (1975) 144–150
10 Hankins GDV, Snyder RR, Hauth JC, Gilstrap III LC, Hamm-Mond T: Nuchal Cords and Neonatal Outcome. Obstet. Gynecol. 70 (1987) 687–691
11 Hansen HS, Hillersborg B: Case report; Antepartum looping of the umbilical cord. Acta. Obstet. Gynecol. Scand. 67 (1988) 475–476
12 Hendrik H-J, Laub K, Ertan AK, Holländer M, Schmidt W: Bedeutung des Biophysikalischen Profils (ABCD) bei der Nabelschnurumschlingung in unkomplizierten Schwangerschaften. Arch. Gynecol. Obstet. 261(Supp.1) (1998) 50
13 Hendrik H-J, Tossounidis I, Boos R, Schmidt W: Neuentwicklung eines fetalen biophysikalischen Profils unter Verwendung verschiedener sonographischer Parameter, Doppler-Flow und der Kinetocardiotokographie. Bildgebung/Imaging 61(Supp.2) (1994) 92
14 Horwitz ST, Finn WF, Mastrota VF: A study of umbilical cord encirclement. Am. J. Obstet. Gynecol. 89 (1964) 970–974
15 Jauniaux E, Ramsay B, Peellaerts C, Scholler Y: Perinatal features of pregnancies complicated by nuchal cord. Am. J. Perinatol. 12(4) (1995) 255–258
16 Kan-Pun-Shui, Eastman NJ: Coiling of the umbilical cord around the foetal neck. J. Obstet. Gynaecol. Brit. Emp. 64 (1957) 227–228
17 Kubli F, Rüttgers H: Die antepartale fetale Herzfrequenz. Verhalten von Grundfrequenz, Fluktuation und Dezelerationen bei antepartalem Fruchttod. Z. Geburtsh. u. Perinatol. 176 (1972) 309
18 Kubli F, Schmidt W: Zustandsdiagnostik des Feten. In von Bachmann KD, Ewerbeck H, Kleihauer E, Rossi E, Stalder G (eds.): Pädiatrie in Praxis und Klinik. Band I. Thieme, Stuttgart 1989, 79–94
19 Larson JD, Rayburn WF, Crosby S, Thurnau GR: Multiple nuchal cord entanglements and intrapartum complications. Am. J. Obstet. Gynecol. 173(4) (1995) 1228–1231
20 Larson JD, Rayburn WF, Harlan VL: Nuchal cord entanglements and gestational age. Am. J. Perinatol. 14(9) (1997) 555–557
21 Mallard EC, Gunn AJ, Williams CE, Johnston BM, Gluckman PD: Transient umbilical cord occlusion causes hippocampal damage in the fetal sheep. Am. J. Obstet. Gynecol. 167 (1992) 1423–1430
22 McCaffrey LE: Umbilical cord encircling neck and its relation to intrapartum complications. Am. J. Obstet. Gynecol. 13 (1927) 104–108
23 Mendez-Bauer C, Troxell RM, Roberts JE et al.: A clinical test for diagnosing nuchal cords. J. Reprod. Med. 32 (1987) 924–927
24 Nelson KB, Grether JK: Potentially asphyxiating conditions and spastic cerebral palsy in infants of normal birth weight. Am. J. Obstet. Gynecol. 179(2) (1998) 507–513
25 Osak R, Webster KM, Bocking AD, Campbell MK, Richardson BS: Nuchal cord evident at birth impacts on fetal size relative to that of the placenta. Early Hum. Dev. 49(3) (1997) 193–202
26 Paul RH, Krosnick G: Fetal Heart Rate in Excessive Cord Encirclement. Obstet. Gynecol. 28 (1966) 646–649
27 Paul RH, Yonekura ML, Cantrell CJ, Turkel S, Pavlova Z, Sipos L: Fetal injury prior to labor: Does it happen? Am. J. Obstet. Gynecol. 154 (1986) 1187–1193
28 Pelosi MA: Antepartum ultrasonic diagnosis of cord presentation. Am. J. Obstet. Gynecol. 162 (1990) 599–601
29 Pilu G, Falco P, Guazzarini M, Sandri F, Bovicelli L: Sonographic demonstration of nuchal cord and abnormal umbilical artery waveform heralding fetal distress. Ultrasound Obstet. Gynecol. 12(2) (1998) 125–127
30 Rayburn WF, Beynen A, Brinkman DL: Umbilical Cord Length and Intrapartum Complications. Obstet. Gynecol. 57 (1981) 450–452
31 Romero R, Pilu G, Jeanty P, Ghidini A, Hobbins JC: The Umbilical Cord. In Prenatal diagnosis of congenital anomalies. Appleton & Lange, Norwalk 1988 pp. 385–402
32 Rosen RH: The short umbilical cord. Am. J. Obstet. Gynecol. 66 (1953) 1253–1259
33 Schaefer M, Laurichesse-Delmas H, Ville Y: The effect of nuchal cord on nuchal translucency measurement at 10–14 weeks. Ultrasound Obstet. Gynecol. 11(4) (1998) 271–273
34 Schmidt W: Zur perinatalen Hirnschädigung – Aussagekraft des Kardiotokogramms. Arch. Gynecol. Obstet. 250 (1991) 1089–1095
35 Shepherd AJ, Richardson CJ, Brown JP: Nuchal Cord as a Cause of Neonatal Anemia. AJDC 139 (1985) 71–73
36 Sherer DM, Onyeije CI, Binder D, Bernstein PS, Divon MY: Uncomplicated baseline fetal tachycardia or bradycardia in postterm pregnancies and perinatal outcome. Am. J. Perinatol. 15(5) (1998) 335–338
37 Sherer DM, Menashe M, Sadovsky E: Severe fetal bradycardia caused by external vibratory acoustic stimulation. Am. J. Obstet. Gynecol. 159 (1988) 334–335
38 Smellie W: Treatise on the theory and practice of midwifery: cases 174 and 175. New Sydeham Society, London (1750) 237
39 Spellacy WN, Gravem H, Fisch RO: The umbilical cord complications of true knots, nuchal coils, and cords around the body. Am. J. Obstet. Gynecol. 94 (1966) 1136–1142

40 Stembera ZK, Horska S: The Influence of Coiling of the Umbilical Cord around the Neck of the Fetus on its Gas Metabolism and Acid-Base Balance. Biol. Neonate. 20 (1972) 214–225
41 Tejani NA, Mann LI, Sanghavi M, Bhakthavathsalan A, Weiss RR: The Association of Umbilical Cord Complications and Variable Decelerations With Acid-Base Findings. Obstet. Gynecol. 49 (1977) 159–162
42 Vanhaesebrouck P, Vanneste K, De Praeter C, Van Trappen Y, Thiery M: Tight nuchal cord and neonatal hypovolaemic shock. Arch. Disease. Childh. 62 (1987) 1276–1277
43 Weiss E, Hitschold T, Berle P: Umbilical artery blood flow velocity waveforms during variable decelerations of the fetal heart rate. Am. J. Obstet. Gynecol. 164 (1991) 534–540

18 Chronic Placental Insufficiency

H. J. Hendrik, A. K. Ertan, and W. Schmidt

Definitions

Normal pregnancy. The uncomplicated human pregnancy lasts from initial cell division of the fertilized ovum to the birth of a mature neonate with an average birthweight of 3450 g at 266 days[43]. Of course, not all pregnancies culminate in the delivery of a normal-weight infant at term.

Intrauterine growth retardation. The concept of intrauterine growth retardation was formulated in the early 1960s and was clearly distinguished from prematurity. Until that time, birthweight was considered the only relevant quantity, and infants weighing less than 2500 g were classified as premature.

Chronic placental insufficiency. Chronic placental insufficiency is one aspect of fetal disorders that are classifiable, as shown in Table 18.**1**.

Perinatal mortality. The efficacy of obstetric care is documented by performance markers such as perinatal mortality (all infant deaths before, during, and in the first week after birth). During the past 30 years, there has been a marked improvement in the obstetric standard in the Federal Republic of Germany (as measured by perinatal mortality). The perinatal mortality rate in Germany has fallen to 0.6%. Various reasons can be cited for this decline (Table 18.**2**).

Table 18.**1** Spectrum of fetal abnormalities and diseases[49]

- Malformations
- Specific fetal diseases (immunological, infectious, endocrinological, etc.)
- Placental insufficiency
- Abnormal length of gestation (prematurity, postmaturity)
- Trauma (antepartum, intrapartum)

Table 18.**2** Reasons for the decline of perinatal mortality in Germany during the past 30 years

- Systematic program of prenatal care, standard formulation of risks
- Technological progress: biophysical examination methods
- Systematic intrapartum monitoring
- Systematic organizational structures: perinatal centers, three-level ultrasound screen
- Advances in neonatal intensive care
- Better understanding of pathophysiological relationships at the fetal and neonatal level

Antepartum and intrapartum mortality. Interestingly, the most significant decline in mortality has occurred in liveborn infants. The percentage of antenatal deaths (approximately 60% of perinatal mortality) has remained constant despite all measures and advances[51]. The intrapartum mortality rate is low, accounting for 9% of perinatal deaths.

Even in countries with very low perinatal mortality, progressive chronic placental insufficiency is the major cause of perinatal losses next to lethal congenital anomalies[21]. Strategies are therefore needed for the management of pregnancies with chronic placental insufficiency in order to avoid indeterminate risks and reduce the incidence of deaths and permanent disabilities.

Definition and Incidence of Chronic Placental Insufficiency

The concept of placental insufficiency can be viewed from various aspects.

Acute, subacute, and chronic forms. According to Gruenwald[34], it is important to distinguish among acute, subacute, and chronic forms of placental insufficiency. These forms may also occur in combination, as they exist on a continuum. Chronic placental insufficiency lasts for weeks or months. Subacute chronic placental insufficiency lasts for a period of days or weeks, and acute placental insufficiency poses an immediate threat to the fetus. Since acute placental insufficiency affects immediate oxygen exchange, it is a life-threatening emergency that requires an entirely different mode of obstetric management.

Chronic placental insufficiency. Chronic placental dysfunction is based on a permanent reduction of uteroplacental blood flow that has functional sequelae. It is helpful to recall Kubli's definition[46]: "Chronic placental insufficiency is a condition in which the functional performance of the placenta is no longer adequate to maintain fetal homeostasis, normal and timely fetal development, or a duration of pregnancy that is sufficient to reach fetal maturity."

In this definition, placental function is characterized by the adequacy of its performance in relation to embryofetal demands: even a damaged placenta is not insufficient as long as it can sustain normal fetal development and homeostasis[48].

Table 18.**3** Criteria for the diagnosis of chronic placental insufficiency

Antepartum	• Clinical (symphysis-fundus height) • Ultrasound biometry • Amniotic fluid volume • Fetomaternal Doppler flowmetry • Fetal movements • Fetal behavioral states • Fetal heart rate (nonstress test) • Amnioscopy • Oxytocin stress test • Cordocentesis (Astrup, SO_2, lactate, etc.)
Intrapartum	• Fetal heart rate • Amniotic fluid color ➤ SO_2 ➤ Fetal blood gases
Postpartum	• Birthweight centile • Clifford rating • Ponderal index • Placental histology • Apgar, umbilical artery pH • Normoblasts, reticulocytes, lymphocytes, platelets • Erythropoietin • Imaging studies to detect an asphyxiating brain lesion (ultrasound, MRI)

MRI = magnetic resonance imaging

Pathoanatomical changes. Qualitative signs of placental insufficiency can be formulated at a pathoanatomical level, but often these signs do not correlate with the clinical presentation. Insufficiency of the placental membrane as an exchange organ is caused by an obstruction or reduction of the surface area available for exchange and/or by a broadening or obstruction of the diffusion pathway. These changes are manifested at the electron-microscopic level as a decrease in the proliferation of cytotrophoblasts, stromal fibrosis, and slight villous angiogenesis in the form of straight, unbranched capillaries of the placental terminal villi[56].

Effects and causes. The effects of chronic placental insufficiency in the fetus depend on its duration and extent and on the affected (partial) function of the organism. They may include disturbances in the nutritive, respiratory, and endocrine systems of the fetoplacental unit[47].

The precise pathophysiological mechanisms are still poorly understood. The difficulties are compounded by the fact that there are different types of chronic placental insufficiency with a number of different causes, and that assigning an individual case to a specific type of chronic placental insufficiency (see below) is more like a jigsaw puzzle than a logical chain of pathophysiological events.

Efforts have been made, therefore, to define chronic placental insufficiency on the basis of its observable features, without reaching a general consensus on the underlying criteria[52].

Clinical features. Conceivable criteria are clinical features such as weight, the ratio of weight to body length, skin turgor, and subcutaneous fat.

Chemical values. Chemical values such as the human placental lactogen (HPL) production rate, estriol production rate, respiratory acidosis in the cord blood, hematocrit, normoblast count[6], and erythropoietin[39] in the cord blood can also be used as criteria.

Histological evaluation. Histological criteria such as placental weight, infarctions, villous maturation defects, vessel wall hyperplasia, villous stromal fibrosis, etc. can also be used in the definition of chronic placental insufficiency (Table 18.**3**).

Intrauterine Growth Retardation

One manifestation of chronic placental insufficiency is a slower than normal rate of intrauterine growth. If the birthweight is below a threshold value (e.g., the 10th, 5th, or 3rd percentile, or 2 or 3 standard deviations below the mean) in a corresponding reference population, then intrauterine growth retardation (IUGR) is said to be present. Other terms are also in common use (Table 18.**4**).

Reference Curves and Thresholds

The usual practice since 1974 is to view the birthweight in relation to gestational age and fetal gender and to classify newborns as eutrophic, hypotrophic, or macrosomic. Dunn[22] and Chiswick[18] reviewed and commented upon the current WHO definitions.

Percentile curves. The percentile curves describe the normal distribution of birthweights by sex and length of gestation in a reference population. These curves show that even children below a certain threshold level may still be included in the normal population. Growth-restricted infants may also fall below this threshold, of course, but they do not necessarily do so. There are growth disturbances within a subtle clinical range that cause an arrest of growth in a genetically large child but do not cause the birthweight to fall below the reference value or may even affect placental functions other than cell growth. The definitive birthweight represents only one aspect of intrauterine growth dynamics.

Thus the terms "chronic placental insufficiency" and "intrauterine growth retardation" are not synonymous, and the limits of their definitions can become an issue, especially when function-testing procedures like Doppler sonography are used.

Reference population. There is no universal rule for selecting a particular reference population, because different ethno-

Table 18.**4** Synonyms for intrauterine growth retardation

- Fetal growth restriction
- Small-for-date infant
- Small-for-date baby
- Fetal malnutrition
- Chronic fetal distress
- Pseudoprematurity
- Dysmaturity

graphic conditions lead to differences in absolute values. Since ethnic, genetic, medical, socioeconomic, geographic and climatic factors cause differences in prenatal growth patterns, population-based (national) standards are required for the classification of newborns.

Different investigators. For a variety of reasons, the data from individual investigators have not been widely utilized in Germany[87]. It is important to cite the particular reference curve that is being used (e.g., Thomson et al. [82] or Voigt et al. [87]). It is preferable to use norm tables that include an adequate number of cases in the early viable weeks of gestation (i.e., from 24 weeks on), although births at this gestational age bias the overall weight distribution and can therefore distort the curve (perhaps toward lower weight classes).

Importance of Pregnancy Dating

Accurate pregnancy dating is necessary in order to assign a measured parameter to a percentile class. It is known that estimating gestational age from the reported last menstrual period (LMP) is accurate in only about 30–40% of cases[65]. In all other cases the examiner must rely on embryonic or fetal dating measurements obtained with ultrasound. This provides an accuracy of ±4–7 days, depending on the parameter used. It is best to determine the delivery date, and thus the current gestational age, before 20 completed weeks of gestation; otherwise unacceptable dating errors are likely to result[36]. It is important to consider the rare case of very early intrauterine growth retardation prior to 20 weeks, usually due to a genetic cause or the presence of other fetal malformations.

Etiology and Historical Risk Factors

Most of the etiologically important factors in intrauterine growth retardation have already been identified[42]. They are listed below in descending order of importance:

- Nicotine abuse or passive smoking (especially in countries with a high percentage of smoking mothers)
- Scant maternal weight gain during pregnancy
- Low body mass index before pregnancy
- Primiparity
- Preeclampsia
- Maternal short stature
- Non-white ethnicity
- Other genetic factors
- Alcohol and drug abuse during pregnancy

Effects and severity. Recent studies confirm that the different factors correlate in varying degrees with the severity of the growth disturbance and that they produce different effects at different stages in the pregnancy. For example, parity and maternal anthropometric factors have greater effects on mature fetuses with IUGR, suggesting that these effects are mediated by placental blood flow and/or nutritional effects late in the third trimester of pregnancy. On the other hand, pregnancy-induced hypertension appears to affect growth at a much earlier stage, making it a far more important factor in determining the postnatal prognosis. The results of the study by Kramer et al.[45] suggest that the severity and prognosis of intrauterine growth retardation are dependent on the cause. This should be considered during the application and interpretation of diagnostic procedures.

Types of Intrauterine Growth Retardation

Two types of intrauterine growth retardation are distinguished on the basis of fetal body proportions: symmetrical and asymmetrical. Gruenwald[34] states that the type of growth retardation documents the timing of the insult to fetal growth.

Symmetrical IUGR. In symmetrical IUGR, the growth restriction affects the dimensions of the head and trunk equally. The injurious factors act early in the pregnancy during general organ growth. The causes include genetic anomalies, malformations, intrauterine infections (e.g., TORCH), and other toxic insults (alcohol, nicotine, drugs, medications).

Asymmetrical IUGR. Asymmetrical IUGR is characterized by a reduction of fat and glycogen stores, causing the fetal abdominal circumference to be disproportionately small in relation to the head circumference. This type of growth restriction is usually associated with late-onset idiopathic placental insufficiency and also with preeclampsia[64].

Prognosis

Antenatal measurements. The antenatal sonographic measurement of various fetal body dimensions makes it possible to determine the type of growth retardation that is present so that the necessary conclusions can be drawn[77]. However, the degree of asymmetry does not tell us the severity of fetal malnutrition, because prolonged malnutrition can also affect the growth of the head[44]. Some authors claim on the basis of statistical analyses that the antenatal diagnosis of intrauterine growth retardation has little impact on the fetal outcome and pediatric prognosis owing to frequent postnatal rebound growth, with most children showing normal long-term development. Recent evidence suggests the value of a more active policy, however, in the expectation that the morbidity and mortality in specific cases can be reduced through appropriate management. This includes the targeted use of fetomaternal Doppler ultrasound, which enables us to assess the individual dynamics of the supply deficit.

Factors. The prognosis of fetal growth retardation depends on several factors:

- The severity of the growth restriction
- The onset of malnutrition
- The duration of chronic hypoxia
- Perinatal management[73]

The prognosis is also influenced by the ability of the fetus to compensate for antenatal insults.

Morbidity and mortality. Even when infants are delivered at a specialized center, severe intrauterine growth retardation accounts for approximately one-third of perinatal mortality, with an incidence of 0.12–0.18%[10,73]. Increasing attention is being focused on morbidity problems in growth-retarded infants.

The majority of growth-retarded children develop normal intelligence during their preschool and school years and do not manifest significant disability. But there are also reports of a slightly increased incidence of cerebral palsy, mental retardation, and minimal cerebral dysfunction[2]. This is particularly true in children who are both premature and growth-retarded[20].

Fetoplacental perfusion deficit. Recent studies show that, besides gestational age, the degree of the fetoplacental perfusion deficit detected with Doppler ultrasound has a critical impact on prognosis[25]. The criteria of interest in this regard are intermediate- and long-term neuromotor development and isolated performance deficits. Follow-up studies in school-age children have shown gross motor disturbances in very small (often premature) children, while abnormalities of fine motor skills or delays in speech development tended to occur in hypotrophic children[11,73].

These effects of chronic placental insufficiency are significantly more common in association with abnormal and severely abnormal perfusion patterns, especially in cases where the fetus was no longer able to compensate for cerebral hypoxia through a brain-sparing redistribution of blood flow[24]. It is uncertain how Doppler velocimetric findings in the borderline or low-abnormal range correlate with long-term sequelae. Based on the results reported to date, however, Doppler sonography appears to be useful for analyzing the dynamics of chronic placental insufficiency with respect to long-term morbidity. This means that it is also one of several useful criteria (along with gestational age, etc.) that can be an aid to obstetric decision-making.

Diagnosis of Chronic Placental Insufficiency

Diagnostic Systems

Diagnostic systems for chronic placental insufficiency may be intrauterine or postnatal and may establish the presence of growth retardation as a single-point assessment or determine abnormal growth as a baseline for further evaluations. The diagnostic system may be evaluated in terms of birthweight percentiles, clinical complications before, during or after delivery, laboratory values, or postpartum criteria such as the Clifford rating, Ponderal index, etc. Intrauterine diagnostic systems generally use postpartum values, often in the form of birthweight percentiles. It should be noted in these cases that there is, by definition, a 5% incidence of idiopathically small (i.e., healthy) children that are below the 5th birthweight percentile. The success of the diagnosis, moreover, depends on the circumstances of the examination, i.e., whether the method is used for screening or specifically to confirm or exclude a diagnosis of placental insufficiency.

Clinical evaluation. Clinical evaluation by abdominal palpation or measurement of the symphysis-fundus height bears only a tenuous relationship to fetal weight[55,72,67]. As a result, clinical examination has potential value only as a screening method.

Ultrasound biometry. Symmetrical or asymmetrical growth retardation can be detected in almost 90% of cases by means of (extended) ultrasound biometry, although there is relatively high false-positive rate that exceeds 20%[71]. The accuracy rate of ultrasound biometry for IUGR in nonselected populations is considerably lower. In a meta-analysis published by Schneider, a positive predictive value of only 62% was achieved[78]. It is not possible to differentiate between size and growth with a single-point measurement obtained in the third trimester. Measurements on two separate occasions (e.g., an initial measurement before 20 weeks' gestation followed by a measurement at 34–36 weeks) can increase the detection rate of a small-for-gestational-age (SGA) baby but does not improve the fetal outcome[63]. Also, it is difficult to determine the exact timing of the repeat measurement, which may be too early or too late to prevent adverse sequelae in any given fetus [28]. Diagnosis may be improved by multiple follow-up measurements from the 25th week on and by an individual analysis of the growth rate[27] so that fetuses with increasing compromise can be referred for appropriate surveillance.

Hormone levels. Serial biometry can provide an indirect means of evaluating placental function. A more direct method is to measure the hormone levels of estriol and HPL. However, this method has been abandoned because it is costly and time-consuming, and has a low predictive value[12,72].

Amniotic fluid volume. A decreased amniotic fluid volume is found more frequently in pregnancies with chronic placental insufficiency[53]. After premature rupture of the membranes and a fetal malformation (e.g., of the urogenital tract) have been excluded, the decreased amniotic fluid may reflect chronic hypoxia secondary to disordered placental function. If we regard the postdate pregnancy as a model case of chronic placental insufficiency[69], we can find parallels in the fetomaternal consequences: perinatal morbidity and mortality rise with decreasing amniotic fluid volume and/or increasing postmaturity[61].

The usefulness of amniotic fluid volume assessment in the diagnosis of intrauterine growth retardation is highly variable. In a meta-analysis on the diagnosis of IUGR, the sensitivity of this test ranged from 24% to 80% and its specificity from 72% to 98%. The positive predictive value ranged from 21% to 55%, and the negative predictive value from 92% to 99%[66].

Amniotic fluid volume assessment. Amniotic fluid volume assessment by the semiquantitative four-quadrant technique[68] is more accurate and reproducible than the purely qualitative method[3]. An AFI (amniotic fluid index) ≤ 5.0 cm in term fetuses was associated with a higher rate of cesarean delivery for fetal distress and lower 5-minute Apgar scores in a meta-analysis[17], but there was no direct evidence of fetal compromise (e.g., umbilical arterial pH) in cases with a decreased amniotic fluid volume. There are, however, ultrastructural signs of local fetoplacental physiological adaptive processes to vascular compromise in oligohydramnios placentas[4,80]. It appears that traditional methods of fluid measurement (single deepest pocket diameter, AFI) are unable to measure an abnormal amniotic fluid volume with sufficient accuracy[58]. It may be better to use the "expanded amniotic fluid index" (EFI), a modified technique that has shown a significantly higher correlation with the actual amniotic fluid volume (quantified by 3D volumetry) than the AFI[38].

Doppler sonography. Chronic placental insufficiency was a major research application for Doppler ultrasound during the early clinical testing phase of this modality. Various fetal vessels were investigated for their significance in chronic placental insufficiency, initially using qualitative flow analysis and later using quantitative methods. It was common to find decreased flow rates or at least alterations in the Doppler waveforms. Trudinger[84] showed in an initial review that 79% of growth-retarded fetuses, who often had a complicated postpartum course, were found to have abnormal flow velocity waveforms in the umbilical artery. At about the same time, definite correlations were discovered between abnormalities of placental development and abnormal umbilical artery waveforms[29]. Fetal growth retardation is significantly more common in association with abnormal Doppler indices[15]. Abnormal Doppler flow patterns can predict which growth-retarded fetuses are at greater risk for a poor fetal outcome[5,8]. In cases with poor to very poor fetomaternal perfusion, Doppler sonography is very accurate in predicting the risk of long-term disturbances of neuromotor development[23]. Prognostic accuracy can be further improved by observation of the "brain-sparing effect" in critical situations of chronic placental insufficiency.

It should be noted that Doppler sonography is not suitable for the primary diagnosis of fetal growth retardation in most circumstances, but it provides an excellent adjunctive study for risk differentiation in suspicious cases.

Indications for Doppler Sonography

Use in high-risk pregnancies. Since 1995, fetomaternal Doppler ultrasound has been included in the official Prenatal Care Guidelines in Germany as a special examination. So far, prospective randomized studies of fetal vessels and their collective evaluation have shown no benefit of Doppler ultrasound screening in a nonselected population[79]. In high-risk pregnancies, however, there is definite evidence that the use of Doppler sonography can significantly reduce the number of antenatal examinations and the number of necessary inductions and cesarean deliveries for suspected fetal distress. When Doppler ultrasound was included in clinical management, there was a reduction in perinatal mortality, the number of elective deliveries, the incidence of intrapartum distress, and the occurrence of hypoxic encephalopathy[1]. In this regard, a definite benefit has been established for the use of Doppler scanning in selected cases (Table 18.**5**).

Regarding the indications for obstetric Doppler sonography, a basic distinction is drawn between examinations of the maternal and fetal vascular systems. It is also important to consider whether abnormalities in a particular compartment may have effects on the opposing side of the fetomaternal vascular system of the placenta.

Table 18.**5** Indications for Doppler examination based on Prenatal Care Guidelines[14]

- Suspicion of intrauterine growth retardation
- Pregnancy-induced hypertension (PIH), preeclampsia, eclampsia
- High index of suspicion for a fetal malformation or disease
- Multiple pregnancy with discordant growth
- Investigation of suspected cardiac anomaly or heart disease
- Abnormalities of fetal heart rate
- Preexisting maternal diseases with vascular relevance
- Prior history of IUGR or intrauterine fetal death
- Prior history of preeclampsia

IUGR = intrauterine growth retardation

Maternal Diseases

Pregnancy-induced hypertension (PIH), preeclampsia, and eclampsia. This complex of conditions poses a significant pregnancy risk. Doppler examination of the uteroplacental and fetal vascular systems makes it possible to evaluate current perfusion in both compartments and improves risk assessment. It appears that hypertensive complications during the further course of pregnancy can be predicted by scanning the uterine arteries in the second trimester[81]. Beyond antihypertensive therapy, the prophylactic and therapeutic implications of this examination are still considered experimental at the present time.

Chronic maternal diseases. Significant risk factors in the maternal history, such as chronic hypertension, diabetes mellitus, and vascular-related autoimmune disorders, can cause significant perfusion deficits on the fetal side. The disease itself may be manifested in the maternal vascular pattern, constituting a special risk situation that requires appropriate surveillance[88].

Fetal Malformations and Diseases

(Color) Doppler sonography has gained an established place in the diagnosis and follow-up of fetal malformations and diseases[16,83].

Presence, location, and function of organs. Chronic placental insufficiency is more common in fetuses with congenital malformations, especially when associated with chromosomal defects. When oligohydramnios is present, perfusion studies can distinguish between a fetal organ malformation and a placental abnormality as the cause. Color Doppler sonography can identify the vascular structures of specific organs, thereby confirming or excluding the presence, location, and function of the organ. This particularly applies to fetal echocardiography, which may be indicated on the basis of the history and clinical manifestations (e.g., fetal arrhythmia). It should be noted, however, that the above diagnostic methods are complementary rather than competitive, since congenital heart disease and chronic placental insufficiency may coexist.

Fetal anemia. Fetal diseases that may be associated with an overload of the right heart provide a clear indication for Doppler ultrasound surveillance. The hydrops that develops in fetal anemia (e.g., in the setting of blood group isoimmunization) is a form of right-sided heart failure. The indirect assessment of the degree of fetal anemia has an important bearing on further management. Fetal anemia cannot be recognized from sonographic suggestive signs until it has reached an advanced stage[19], and treatment at this time can involve far greater hemodynamic risks than in earlier phases where the anemia is still compensated. Once sensitization has occurred, the use of invasive diagnostic procedures also runs the risk of inducing additional irregular maternal antibodies of different types[57]. This underscores the advantage of reliable, indirect methods for diagnosing fetal anemia. The hypercirculatory state that is induced by fetal anemia can be detected by Doppler sonography and has been evaluated in terms of predicting the degree of anemia. Different studies have yielded contradictory results[35,62], but it has been consistently found that the peak systolic velocity in both arterial and venous vessels is elevated in cases of progressive fetal anemia with no evidence of hydrops. Based on a large volume of data, Mari et al.[62] were able to predict moderate to severe fetal anemia with a sensitivity of 100% and a false-positive rate of 12% by analyzing the peak velocity in the middle cerebral artery (the positive and negative predictive values were 65% and 100%) and comparing it with normal values for gestational age.

Discordant growth in twins. Discordant twin growth is an indication for Doppler scanning in the second half of the pregnancy and especially from the 25th week on. Discordant growth is observed in approximately one-third of twin pregnancies, while growth retardation of both twins is noted in approximately one-fifth of cases when current birthweight percentiles (e.g., Thomson et al.) are used[41]. Using the indices for singleton pregnancies, an individual analysis of fetal flow patterns can clarify important differential diagnostic aspects of discordant and hypotrophic growth. Some of the hemodynamic problems that occur in monochorionic twin pregnancies (fetofetal transfusion syndrome, acardia, twin reversed arterial perfusion) can be evaluated with Doppler ultrasound[40].

Doppler examination of the venous system in these cases can show evidence of a volume overload in the recipient twin. Venous Doppler also has applications in other disorders[33] (Table 18.**6**).

Table 18.**6** Indications for fetal venous Doppler examination

- Fetal arrhythmias
- Suspected fetofetal transfusion syndrome
- Nonimmune hydrops fetalis
- Suspected stenoses in the cardiac outflow tract
- Congenital heart disease
- Severe centralization of the fetal circulation (brain sparing)
- Suspicious FHR trace

Clinical Management of Chronic Placental Insufficiency Suspected from Doppler Findings

Antenatal Fetal Heart Rate Monitoring

According to conventional recommendations, antenatal electronic fetal heart rate (FHR) monitoring is the most reliable instrument available for detecting acute alterations in high-risk pregnancies. Künzel and Hohmann[50] were able to show that fetuses with chronic placental insufficiency were more likely to have an abnormal nonstress test, especially before the 37th week, than fetuses without chronic placental insufficiency. The differences were less pronounced in infants delivered at term (i.e., cases that did not require early delivery).

Criteria. What are the most conspicuous FHR criteria that are seen in cases with chronic placental insufficiency and impending decompensation? In a study by Gnirs and Schmidt[30], normal pregnancies were compared with pregnancies with severe chronic placental insufficiency (> 5th percentile). A significantly higher average FHR was found in cases with impending decompensation—apparently an adaptive response to the chronic perfusion deficit and hypoxia. Silent fluctuations were much more common in these cases. Significantly fewer accelerations were observed per minute, and the mean duration of the accelerations was half that in cases with a normal course of pregnancy.

Disadvantages. The main disadvantage of FHR monitoring is the high rate of false-positive findings. A normal FHR trace generally means that acute fetal distress is not present at the time of the examination. An abnormal trace may signify hypoxia but does not necessarily do so. One solution is to obtain frequent FHR traces, but this is a very costly process. Göschen[32] suggests examinations at intervals from once weekly to several times daily, depending on the initial situation. Following this regimen would entail an unacceptably high frequency of FHR examinations and would often necessitate hospitalization. Given these practical difficulties and the limited financial resources of our health care industry, it is essential that we optimize the use of available surveillance methods. This includes the use of concomitant studies that have a sound pathophysiological rationale.

Pathological Changes in Organ Systems

Times of occurrence. Visser[86] presented data mainly from animal studies concerning the effects of chronic hypoxia on various organ functions. He observed successive, dynamic phases of subtle to latent supply deficits followed by varying degrees of chronic hypoxemia, with subsequent progression to acidemia. The results of different studies indicate different times of occurrence of pathological changes due to chronic hypoxia in the various measurable biophysical fetal variables. Thus, even with latent chronic placental insufficiency, we find umbilical arterial waveform changes that may also be accompanied by qualitative changes in fetal movements. After a time the fetal heart rate responds with various changes, some subtle, and this culminates in a terminal FHR pattern that is no longer accompanied by observable fetal movements. The effects of chronic placental insufficiency can sometimes last for weeks before they actually produce overt, severely abnormal changes in the FHR.

Compensatory reserves. The deleterious effects of chronic asphyxia on various organs depend on the duration and extent of the insult. At the same time, we must concede that the organ functions have a recovery capacity that can range from the recruitment of replacement functions to a complete restoration of normal function. It is also reasonable to assume that the fetus has compensatory reserves that help to withstand acute or chronic alterations.

Disturbances of neurological development. It is generally agreed that abnormalities of neurological development due to perinatal asphyxia are caused by intrapartum events in only 10% of cases. Some 30–40% of these disturbances arise well before birth and are unrelated to the delivery, and an equal percentage are based on postpartum complications. Despite this awareness, the definitive timing of the insult is generally uncertain.

Fetal movements. A prolonged decrease in placental perfusion leads to fetal growth retardation with increasingly abnormal Doppler waveforms. As the stress situation persists, renal perfusion is diminished and less amniotic fluid is produced. Nevertheless, these fetuses may continue to exhibit normal movements. With a further reduction in placental perfusion, fetal movements are also reduced until acute asphyxia supervenes, marked by an abnormal FHR combined with abnormal fetal movement patterns.

Biophysical Profile

Combining various biophysical methods of pregnancy surveillance in a "biophysical profile" is a relatively old idea that has yielded good results in selected patients[60].

New diagnostic methods. Besides technical advances (high-resolution ultrasound, pulsed Doppler ultrasound, kineto-cardiotocography), new discoveries in fetal physiology and pathophysiology (e.g., principles of neuromotor coordination development in fetuses[86]) have suggested the concept of combining the new methods into a system that can improve the strategy for diagnosing and monitoring compromised fetuses.

Doppler spectra and FHR monitoring. In a study by our group[74], we presented abnormal Doppler spectra and FHR findings in patients with markedly decreased placental perfusion. Our criteria were drastic perfusion changes in the fetal vessels and abnormalities in the maternal vascular supply. This resulted in a study population in which hypotrophic fetuses below the 10th weight percentile were present in 94% of cases and fetuses with very severe growth retardation were present in 68%. As expected, we found prepathological or pathological FHR changes in 94% of these cases. The abnormal FHR findings were frequently inconstant (40%), however, and in 17% of the cases with severely abnormal Doppler spectra, a change from normal to abnormal FHR patterns did not occur until one week before delivery. The average interval from the detection of a severely abnormal Doppler spectrum to the appearance of acute asphyxia signs in the FHR trace was 13 days (median). In cases with absent end-diastolic flow, the interval was shortened to 8 days.

Behrens et al.[7] reported comparable figures in a recently published study. More than 80% of the FHR findings recorded on initial diagnosis of absent end-diastolic flow or reverse flow were suspicious or abnormal.

Fetal movements. Based on the pathophysiological assumptions mentioned earlier and the screening study by our group[70] showing markedly decreased fetal movements in association with chronic placental insufficiency, we expected that the fetal movement pattern could provide an indicator of fetal compromise. This is analogous to the findings of postnatal follow-up studies[23] showing that increased pathology in antenatal flow measurements of the fetomaternal system is associated with an increase in neurological abnormalities after 2 years of age (Table 18.**7**). Thus we attempted to assess the value of fetal movements as a connecting link between intermediate- and long-term compromise.

In their biophysical profile, Manning et al.[59] placed considerable emphasis on the analysis of fetal movements. Vintzileos et al.[85] observed increasing acidosis in the umbilical artery by antenatal cordocentesis in fetuses with a nonreactive nonstress test or an absence of movements (Table 18.**8**).

Table 18.**7** Occurrence of neuromotor developmental abnormalities after 2 years of age (gestational age-adjusted) as a function of antenatal Doppler flow findings[23]

Doppler flow findings	Percentage of neuromotor development abnormalities
Normal	15.3%
Abnormal S/D ratio	22.2%
Absent end-diastolic flow, reverse flow	31.7%

Table 18.**8** Relationship between abnormal biophysical parameters in the Manning biophysical profile and acidosis in the umbilical arterial blood at cordocentesis[85]

Biophysical parameter	Arterial pH	P_{O_2} (mmHg)	P_{CO_2} (mmHg)	Base excess (mmol/l)
Nonreactive nonstress test	7.27 ± 0.09	19.9 ± 8.0	49.5 ± 9.7	–3.0 ± 4.1
Absence of breathing movements	7.27 ± 0.10	20.0 ± 8.4	50.3 ± 10.5	–2.8 ± 4.7
Absence of fetal movements	7.17 ± 0.07	16.6 ± 5.9	55.1 ± 14.3	–7.0 ± 3.6
Absence of fetal tone	7.14 ± 0.06	13.6 ± 7.2	57.2 ± 18.6	–7.8 ± 3.3

Table 18.**9** Prediction of intrapartum complications (fetal distress) by KCTG recording up to 7 days before the delivery of hypotrophic and eutrophic neonates. European Multicenter Study on the Evaluation of KCTG[37]

KCTG parameters	Birthweight < 10th centile with intrapartum complications (*n* = 61)	Birthweight > 10th centile without intrapartum complications (*n* = 121)
Number of accelerations/h	10.05 ± 6.81	10.13 ± 6.27
Mean block length (s)	5.21 ± 1.46	5.92 ± 1.71**
Number of movement blocks/h	118.73 ± 41.64	117.65 ± 43.39
Absolute duration of movement blocks/h (s)	637.19 ± 335.88	716.16 ± 388.76
Nonreactive nonstress test	33%	14%

** $p < 0.01$
KCTG = kinetocardiotocography

Kinetocardiotocography (KCTG). The disadvantage of the biophysical profile is that the evaluation of the variables is strongly examiner-dependent. This led us to test the value of the kinetocardiotocogram developed by our group[75]. In this test more than 90% of fetal movement activities are documented and quantitatively evaluated with regard to the duration and number of movement units[37]. The duration of a fetal movement sequence appears to play an important role [31]. We were able to confirm this observation by taking the mean length of a "movement block" as our main parameter. We theorized that the motor competence of a growth-retarded fetus is expressed in a shorter mean duration of a movement block, reflecting an overall decline in physical fitness. This criterion also appears to be important for the prediction of intrapartum complications in fetuses with chronic placental insufficiency (Table 18.**9**).

ABCD profile. A prospective study is currently underway at our center to determine the value of combined qualitative biophysical test methods in diagnosing chronic placental insufficiency, assessing its dynamics, and predicting intrapartum complications. For simplicity, we refer to this project as the "ABCD profile": A for amniotic fluid index, B for extended biometry, C for kinetocardiotocography, and D for Doppler examination of the fetomaternal system.

With the combination of methods in the ABCD profile, we have been able to enhance our understanding of chronic placental insufficiency as a dynamic disease process and assess individual fetal compromise based on the activity level of the fetus. The evaluations to date have yielded the following results:

- In cases with chronic placental insufficiency, fetuses with abnormal biometry had intrapartum complications in 46% of cases. The complication rate was 67% in cases with abnormal Doppler spectra within 7 days of delivery and 69% in cases with abnormally decreased fetal movements.
- No complications were found in cases with a small fetus plus normal Doppler flow, but the complication rate was 68% in cases with a small fetus plus abnormal Doppler flow. When this was combined with abnormal fetal movements, the complication rate rose to 82% (versus 50% in cases with normal fetal movements) (Table 18.**10**).

Thus, in cases with observable dynamics of placental insufficiency, we can use the combined biophysical variables to better evaluate the danger to the fetus, define follow-up intervals more precisely, reduce hospitalizations, and optimize the management of the delivery and postnatal care.

Table 18.**10** Combination of different biophysical variables in sonographically hypotrophic fetuses and the prediction of intrapartum complications

Biophysical variables	Intrapartum complications present	Intrapartum complications absent
Abnormal biometry plus		
Normal flow	0%	100%
Abnormal flow	68%	32%
Abnormal flow plus		
Normal Movements (KCTG)	50%	50%
Abnormal movements (KCTG)	82%	18%

Summary

Identifying Cases with IUGR

Based on these experiences, gleaned mostly from individual studies, an effective combination of clinical and biophysical methods is available for identifying cases with IUGR[76,77]. Besides the history and clinical examination, major emphasis is placed on accurate dating of the pregnancy, precise fetal biometry, and sonographic evaluation of the amniotic fluid volume. If growth retardation is suspected, hypoxia assessment should be done with the aid of dynamic tests (fetal movements, Doppler velocimetry, fetal heart rate). Of course, further tests may be needed to exclude other causes of biometric growth restriction. Prime examples are genetic aberrations and fetal malformations that can be recognized by their sonographic features.

Obstetric Management

Induction of fetal lung maturation. The cardinal goals of obstetric management are to improve placental perfusion, promote fetal lung maturation (e.g., with corticosteroids), and avoid additional stressors. In compromised fetuses with chronic placental insufficiency, the obstetrician must often weigh various opposing aspects of his or her actions (intrauterine hypoxia versus prematurity) against one another. The decision to proceed with elective delivery is very difficult before fetal lung maturity is reached. The latest results of the European GRIT study (Growth Restriction Intervention Trial) show that the long-term neurological development of fetuses with a severe perfusion deficit will benefit more from an expectant approach and the induction of lung maturation than from early delivery immediately following the Doppler ultrasound diagnosis[13].

Other therapeutic options such as the intra-amniotic or direct fetal infusion of amino acids or glucose, maternal low-dose aspirin therapy or prophylaxis, maternal oxygen administration, etc. have so far yielded contradictory results or have failed to benefit fetal growth. The main goals of antenatal surveillance are to assess the current disease process in the mother and fetus, test therapeutic options, and prolong the gestation until fetal lung maturity is reached while avoiding hypoxic injury.

Doppler sonography. Initial longitudinal studies confirm that the changes in Doppler flow indices that occur in cases of severe, chronic placental insufficiency resemble a progressive cascade of pathology that culminates in the delivery of a severely compromised, hypotrophic infant[26]. When the above instruments are used, good clinical judgment combined with flexibility and individualization are the key elements in achieving the best fetal outcome[54]. The mode of delivery depends on various criteria such as prematurity, degree of fetal compromise, maternal indication, and the prediction of fetal intrapartum stress. Doppler sonography and other biophysical assessments can provide a good impression of the stress tolerance of the compromised fetus (see above), which may still allow for a vaginal delivery in many cases. An awareness of the preexisting deficits in placental function can optimize perinatal management and help prevent fetal injuries[9]. The best setting for the delivery thus depends on the expected degree of intrapartum and postpartum fetal compromise. Delivery at a perinatal center is generally recommended.

Doppler sonography, then, is an effective instrument for case selection and for optimizing the course of the pregnancy and the perinatal outcome.

References

1 Alfirevic Z, Neilson JP: Doppler ultrasonography in high-risk pregnancies: systematic review with meta-analysis. Amer. J. Obstet. Gynecol. 172 (1995) 1379–1387

2 Allen MC: Developmental outcome and follow up of the small for gestational age infant. Sem. Perinatol. 8 (1984) 123

3 American College of Obstetricians and Gynecologists: Ultrasonography in pregnancy. The College Technical Bulletin. Washington 1993, No. 187

4 Apodaca CC, Moore K, Hume RF et al.: Expression of Adrenomedullin and its Receptor in the Human Placenta: Normal and Pathologic Pregnancy. Amer. J. Obstet. Gynecol. 182 (2000) 636

5 Arduini D, Rizzo G: Prediction of fetal outcome in small for gestational age fetuses: Comparison of Doppler measurements obtained from different fetal vessels. J. Perinat. Med. 20 (1992) 29–38

6 Axt R, Ertan K, Hendrik J, Wrobel M, Mink D, Schmidt W: Nucleated red blood cells in cord blood of singleton term and post-term neonates. J. Perinat. Med. 27 (1999) 376–381

7 Behrens O, Wedeking-Schohl H, Goeschen K: Prognostischer Wert der Kardiotokographie bei Schwangerschaften mit pathologischem Dopplerbefund. Geburtsh. u. Frauenheilk. 56 (1996) 272–277

8 Berkowitz GS, Mehalek KE, Chitkara U, Rosenberg J, Cogswell C, Berkowitz RL: Doppler velocimetry in the prediction of adverse outcome in pregnancies at risk for intrauterine growth retardation. Obstet. Gynecol. 71 (1988) 742–746

9 Blair E: A research definition for "birth asphyxia"? Dev. Med. Child. Neurol. 35 (1993) 449–452

10 Bolte A, Fuhrmann U, Hamm W, Kusche M, Schlensker K-H, Stenzel B: Geburtshilfliches Management bei schwerer fetaler Wachstumsretardierung. Geburtsh. u. Frauenheilk. 47 (1987) 518–524

11 Bolte A, Eibach HW, Gladtke E et al.: Die kindliche Entwicklung nach schwerer intrauteriner Wachstumsretardierung – Ergebnisse von Follow-up Studien. Geburtsh. u. Frauenheilk. 47 (1987) 525–532

12 Bonatz G, Lorcher E, Weisner D, Schulz V, Semm K: Stellenwert der pulsatilen Dopplersonographie zur Diagnostik bei Verdacht auf Plazentainsuffizienz im Vergleich zu anderen Untersuchungsparametern. Zentralbl. Gynäkol. 118 (1996) 129–135

13 Bröcker J, Schneider KTM, Gnirs J, Schelling M, Thornton J: Optimierung des Entbindungszeitpunktes bei kompromittierten IUGR-Feten. Z. Geburtsh. Neonatol. 203 (1999) (Suppl.II) 69

14 Bundesärztekammer: Indikationen zur Doppler-Untersuchung nach den Mutterschaftsrichtlinien. Dtsch. Ärztebl. 92 (1995) 311–313

15 Campbell S, Pearce JMF, Hackett G, Cohen-Overbeck T, Hernandez C: Qualitative assessment of uteroplacental blood flow: Early screening test for high-risk pregnancies. Obstet. Gynecol. 66 (1986) 649–653

16 Chaoui R: Dopplersonographie bei fetalen Fehlbildungen. Arch. Gynecol. Obstet. 260 (1997) 159–161

17 Chauhan SP, Sanderson M, Hendrix NW, Magann EF, Devoe LD: Perinatal outcome and amniotic fluid index in the antepartum and intrapartum periods: A meta-analysis. Amer. J. Obstet. Gynecol. 181 (1999) 1473–1478

18 Chiswick ML: Commentary on current World Health Organisation definitions used in perinatal statistics. Brit. J. Obstet. Gynaecol. 93 (1986) 1236–1238

19 Chitkara U, Wilkins I, Lynch L, Mehalek K, Berkowitz RL: The role of sonography in assessing severity of fetal anemia in Rh- and Kell-isoimmunized pregnancies. Obstet. Gynecol. 71 (1988) 393–398

20 Comney JO, Fitzhardinge PM: Handikap in the pre-term small-for-gestational-age infant. J. Pediatr. 94 (1979) 779–784

21 de Galan-Roosen AE, Kuijpers JC, Mackenbach JP: Perinatal mortality in Delft and the surrounding area, 1983–1992: a possibility for further reductions through targeted attention to lethal congenital abnormalities and placental insufficiency. Ned. Tijdschr. Geneeskd. 143 (1999) 152–157

22 Dunn P: The search for perinatal definitions and standards. Acta Paediatr. Scand. Suppl. 319 (1985) 7–16

23 Ertan AK, Jost W, Hendrik J, Lauer S, Uhrmacher S, Schmidt W: Perinatal events and neuromotoric development of children with zero flow in the fetal vessels during the last trimester. In: Cosmi EV, Di Renzo GC (eds.). 2nd World Congress of Perinatal Medicine, Monduizzi Editore, Rome 1993, 1049–1052

24 Ertan AK, Hendrik HJ, Schmidt W: Perinatale Auffälligkeiten und fetal outcome bei Fällen mit hochpathologischen Doppler-Flow-Befunden (enddiastolischer Block und Reverse Flow) in A. umbilicalis und fetaler Aorta. In: Merz E (ed.): Sonographische Diagnostik in Gynäkologie und Geburtshilfe. Band 2: Geburtshilfe. 2. Auflage. Thieme, Stuttgart, im Druck

25 Ertan AK, He JP, Hendrik HJ, Holländer M, Limbach HG, Schmidt W: Reverse Flow in den fetalen Gefäßen – Klinische Bedeutung und perinatale Auffälligkeiten. Z. Geburtsh. Neonatol. 204 (2000)

26 Ferrazzi E, Bellotti M, Bozzo M et al.: The temporal sequence of changes in fetal velocimetry indices for growth restricted fetuses. Amer. J. Obstet. Gynecol. 182 (2000) (Suppl.1) 17

27 Gardosi J: Prediction of birth weight and fetal weight gain. Contemp. Rev. Obstet. Gynaecol. 6 (1994) 122–127

28 Gardosi JO, Mongelli IM, Mul T: Intrauterine growth retardation. Bailliere's Clin. Obstet. Gynaecol. 9 (1995) 445–463

29 Giles WB, Trudinger BJ, Blaird P: Fetal umbilical artery flow velocity waveforms and placental resistance: Pathologic correlation. Brit. J. Obstet. Gynaecol. 92 (1985) 31–38

30 Gnirs J, Schmidt W: Kinetokardiotokogramm – fetale Bewegungsaktivität als zusätzliches Beurteilungskriterium antepartaler CTG-Registrierungen. In: Schmidt W (ed.): Jahrbuch der Gynäkologie und Geburtshilfe 1991. Biermann, Zülpich 1991, 190–197

31 Gnirs J, Schneider KTM: Fetales Verhalten und Bewegungsaktivität. Gynäkologe 27 (1994) 136–145

32 Göschen K: Überwachung der Schwangerschaft aus forensischer Sicht. Gynäkologe 27 (1994) 197–207

33 Gonser M, Erz W: Indikationen zur venösen Dopplersonographie. Arch. Gynecol. Obstet. 260 (1997) 161–162

34 Gruenwald P: Chronic fetal distress and placental insufficiency. Biol. Neonat. 5 (1963) 215–265

35 Hecher K, Snijders R, Campbell S, Nicolaides K: Fetal venous, arterial, and intracardiac blood flows in red blood cell isoimmunization. Obstet. Gynecol. 85 (1995) 122–128

36 Hendrik HJ: Sonographische Untersuchungen – erweiterte fetale Biometrie und semiquantitative Bestimmung der Fruchtwassermenge. Inaugural Diss. Med. Fak. Univ. Heidelberg 1988

37 Hendrik HJ, Uhrmacher S, Schmidt W: Ergebnisse der europäischen Multicenterstudie zur klinischen Evaluierung des Kinetocardiotokogramms (KCTG). Arch. Gynecol. Obstet. 255 (1994) (Suppl.1) 35

38 Hendrik HJ: Der erweiterte Fruchtwasserindex (EFI), ein neues Verfahren zur Objektivierung der Beurteilung der Fruchtwassermenge. Vortrag Mittelrheinische Gesellschaft für Geburtshilfe und Gynäkologie, Mainz, 28.–30.4.1995

39 Hendrik H-J, Mink D, Laufhütte S, Schmidt W: Erythropoetin im Nabelschnurblut bei normalem Schwangerschafts- und Geburtsverlauf. 169. Tagung der Mittelrheinischen Gesellschaft für Geburtshilfe und Gynäkologie, Homburg/Saar, 6.–8.6.1997

40 Holländer M, Hendrik HJ, Ertan AK, Uzlu S, Tossounidis I, Schmidt W: Zur Rolle der Dopplersonographie in der Überwachung von Mehrlingsgraviditäten. Arch. Gynecol. Obstet. 261 (1998) (Suppl.1) 21

41 Ilkhanipur I: Geminigravidität. Schwangerschafts- und geburtshilfliches Management einer Risikogravididtät. Inaugural. Diss. Med. Fak. Univ. Heidelberg 1998

42 Keirse MJNC: Epidemiology and aetiology of the growth retarded baby. Clin. Obstet. Gynecol. 11 (1984) 415–436

43 Kloostermann GJ: On intrauterine growth. Int. J. Obstet. 8 (1970) 895–912

44 Kramer MS, Olivier M, Mclean FH: Impact of intrauterine growth retardation and body proportionality on fetal and neonatal outcome. Pediatrics 86 (1990) 707–713

45 Kramer MS, Platt R, Yang H, McNamara H, Usher RH: Are all growth-restricted newborns created equal(ly)? Pediatrics 103 (1999) 599–602

46 Kubli F: Die chronische Plazentainsuffizienz. Gynäkologe 1 (1968) 53–60

47 Kubli F: Gestose und Plazentainsuffizienz. In: Rippmann ET (ed.): Die Spätgestose (EPH-Gestose). Schwabe, Basel 1970

48 Kubli F, Wernicke K: Plazentainsuffizienz. In: Becker V, Schiebler TH, Kubli F. (eds.): Die Plazenta des Menschen. Thieme, Stuttgart 1981, 391–477

49 Kubli F, Arabin B: Frühgeburt. In: Dudenhausen JW (ed.): Praxis der Perinatalmedizin. Thieme, Stuttgart 1984, S. 148–180

50 Künzel W, Hohmann M: Grundlagen der kardiotokographischen Überwachung des Feten während der Schwangerschaft. Gynäkologe 22 (1989) 156–161

51 Künzel W, Hohmann M: Das biophysikalische Profil des Feten. Indikationen zur antepartalen Überwachung des Feten. Gynäkologe 27 (1994) 117–122

52 Kuss E: Was ist das „Plazenta-Insuffizienzsyndrom"? Geburtsh. u. Frauenheilk. 47 (1987) 664–670

53 Leucht W, Schmidt W, Hendrik HJ, Rabe D, Kubli F: Schwangerschafts- und Geburtsverlauf nach ultrasonographisch auffälliger Fruchtwassermenge. In: Dudenhausen JW, Saling E (eds.): Perinatale Medizin. Band 10. Thieme, Stuttgart 1984, 200–201

54 Lin CC, Santolaya-Forgas J: Current concepts of fetal growth restrictions: Part II. Diagnosis and management. Obstet. Gynecol. 93 (1999) 140–146

55 Loeffler FE: Clinical foetal weight estimation. J. Obstet. Gynaecol. Br. Cwlth. 74 (1967) 675–677

56 Macara L, Kingdom JCP, Kaufman P et al.: Structural analysis of placenta terminal villi from growth-restricted pregnancies with abnormal umbilical artery Doppler waveforms. Placenta 17 (1996) 37–48

57 MacGregor SN, Silver RK, Sholl JS: Enhanced sensitization after cordocentesis in a rhesus-isoimmunizied pregnancy. Amer. J. Obstet. Gynecol. 165 (1991) 382–383

58 Magann EF, Chauhan SP, Barrilleaux PS, Withworth NS, Martin JN: Neither Amniotic Fluid Index nor Single-Deepest Pocket Methods accurately reflect abnormal amniotic fluid volumes. Amer. J. Obstet. Gynecol. 182 (2000) 340

59 Manning FA, Platt LD, Sipos L: Antepartum fetal evaluation: Development of a fetal biophysical profile. Amer. J. Obstet. Gynecol. 136 (1980) 787–796

60 Manning FA, Baskett TF, Morrison I, Lange I: Fetal biophysical profile scoring: A prospective study in 1184 high-risk patients. Amer. J. Obstet. Gynecol. 140 (1981) 289–294

61 Manning FA: Ultrasound in perinatal medicine. In: Creasy R, Resnick R (eds.): Maternal Fetal Medicine. Saunders, Philadelphia 1984, p. 203

62 Mari G: Noninvasive diagnosis by doppler ultrasonography of fetal anemia due to maternal red-cell alloimmunization. (For the collaborative group for doppler assessment of the blood velocity in anemic fetuses). New Engl. J. Med. 342 (2000) 9 –14

63 Neilson JP, Munjana SP, Whitfield CR: Screening for small for dates fetuses: a controlled trial. Brit. Med. J. 289 (1984) 1179–1182

64 Niswander KR: Management of growth retardation with a view to preventing neuromotor dysfunction and mental handicap. In: Kubli F, Patel N, Schmidt W, Linderkamp O (Hrsg.): Perinatal Events and Brain Damage in Surviving Children. Springer, Berlin 1988, 106–118

65 Ott WJ: Accurate gestational dating. Obstet. Gynecol. 65 (1985) 460–464

66 Ott WJ: Sonographic diagnosis of intrauterine growth restriction. Clin. Obstet. Gynecol. 40 (1997) 787–795

67 Persson B, Stangenberg M, Lunell NO, Brodin U, Holmberg NG, Vaclavinkova V: Prediction of size of infants at birth by measurements of the symphysis-fundus height. Brit. J. Obstet. Gynaecol. 93 (1986) 206–211

68 Phelan JP, Smith CV, Broussard P, Small M: Amniotic fluid volume assessment with the four-quadrant technique at 36–42 week›s gestation. J. Reprod. Med. 32 (1987) 540–542

69 Runge H: Über einige besondere Merkmale der übertragenen Frucht. Zentralbl. Gynäkol. 31 (1942) 756–764

70 Schmidt W, Garoff L, Heberling D et al.: Überwachung der fetalen Bewegungsaktivität mit Real-time Ultraschall und deren Bedeutung für den Schwangerschaftsverlauf. Geburtshilfe Frauenheilkd. 41 (1981) 601–606

71 Schmidt W, Hendrik HJ, Gauwerky J, Junkermann H, Leucht W, Kubli F: Diagnose der intrauterinen Wachstumsretardierung durch die erweiterte Ultraschallbiometrie. Geburtsh. u. Frauenheilk. 42 (1982) 543–548

72 Schmidt W, Kubli F, Garoff L, Hendrik HJ, Leucht W, Runnebaum B: Diagnose der intrauterinen Wachstumsretardierung – Vergleich von Klinik, Gesamtöstrogenbestimmung aus dem 24-h-Urin und Ultraschallbiometrie (Distanzmessungen, biparietaler Kopfdurchmesser und thorako-abdominaler Querdurchmesser) unter Berücksichtigung des antepartalen und subpartalen CTGs. Geburtsh. u. Frauenheilk. 42 (1982) 709–716

73 Schmidt W, Hendrik J: Frühgeburt und Mangelgeburt. In: Bolte A, Wolff F (eds.): Hochrisikoschwangerschaft – Diagnose, Therapie, Prognose für Mutter und Kind. Steinkopff, Darmstadt 1989, 17–30

74 Schmidt W, Graf von Ballestrem CL, Ertan AK, Rühle W, Gnirs J, Boos R: Pathologische Doppler-Flow-Befunde und kardiotokographische Ergebnisse. Geburtsh. u. Frauenheilk. 51 (1991) 523–531

75 Schmidt W, Gnirs J: Das KCTG – erste klinische Erfahrungen beim Einsatz des Kinetokardiotokogramms. Geburtsh. u. Frauenheilk. 51 (1991) 437–442

76 Schmidt W, Hendrik HJ: Biophysikalische Überwachungsmethoden in der Geburtshilfe unter besonderer Berücksichtigung der klinischen Bedeutung des Kinetocardiotokogramms (KCTG). In: Hillemanns HG (ed.): Geburtshilfe – Geburtsmedizin. Eine umfassende Bilanz zukunftsweisender Entwicklungen am Ende des 20. Jahrhunderts. Springer, Berlin 1995, 351–363

77 Schneider KTM: IUGR – Probleme der Diagnostik. Zentralbl. Gynäkol. 113 (1991) 467–474

78 Schneider KTM: (1993), zitiert nach Gnirs J, Schneider KTM: Dopplersonographie bei der intrauterinen Wachstumsretardierung. Arch. Gynecol. Obstet. 260 (1997) 154–156

79 Schneider KTM: Standards in der Perinatalmedizin – Dopplersonographie in der Schwangerschaft (für die Standardkommission der AGDMFM). Geburtsh. u. Frauenheilk. 56 (1996) M69–M73

80 Schweikert S, Incerpi M, Pavlova Z: Placental Pathology in Gravidas With Oligohydramnios. Amer. J. Obstet. Gynecol. 182 (2000) 461

81 Steiner H: Dopplersonographie bei schwangerschaftsinduzierter Hypertonie. Arch. Gynecol. Obstet. 260 (1997) 156–157

82 Thomson AM, Billewicz JW, Hytten FE: The assessment of fetal growth. J. Obstet. Gynaecol. Br. Cwlth. 75 (1968) 903–916

83 Tossounidis I, Hendrik HJ, Ertan K, Holländer M, Schmidt W: Bedeutung des Biophysikalischen Profils (ABCD) bei chromosomalen und strukturellen Fehlbildungen. Arch. Gynecol. Obstet. 261 (1998) (Suppl.1) 50

84 Trudinger BJ: The umbilical circulation. Semin. Perinatol. 11 (1987) 311–321

85 Vintzileos AM, Fleming AD, Scorza WE et al.: Relationship between fetal biophysical activities and umbilical cord blood gas values. Amer. J. Obstet. Gynecol. 165 (1991) 707–713

86 Visser GHA: Fetale Herzfrequenz und Bewegungsmuster in Abhängigkeit von der Sauerstoffversorgung und vom neurologischen Status des Neugeborenen. In: Bolte A, Wolff F (eds.): Hochrisikoschwangerschaft – Diagnose, Therapie, Prognose für Mutter und Kind. Steinkopff, Darmstadt 1989, 89–95

87 Voigt M, Schneider KTM, Jährig K: Analyse des Geburtengutes des Jahrgangs 1992 der Bundesrepublik Deutschland. Geburtsh. u. Frauenheilk. 56 (1996) 550–558

88 Wisser J: Dopplersonographie bei anamnestischen Schwangerschaftsrisiken. Arch. Gynecol. Obstet. 260 (1997) 1539–1541

19 Severely Abnormal Doppler Findings and Perinatal Abnormalities

A. K. Ertan, H. J. Hendrik, and W. Schmidt

Surveillance of Compromised Fetuses

Diseases that cause fetal compromise. Chronic placental insufficiency is considered the prototype of a disease that causes fetal compromise. Often it can be detected early through the combined use of antenatal surveillance methods, making it possible to save the life of the fetus and reduce permanent disability[30]. Increasing experience in the use of these surveillance methods and new discoveries in fetal pathophysiology and placental pathoanatomy[26] have shown that chronic placental insufficiency has an individual dynamic, the recognition and understanding of which appear to be important for successful obstetric management. Important factors besides the severity of the placental changes are the gestational age, and thus organ maturity, and also the ability of the fetus to mobilize compensatory reserves[12]. While the classic definition of a growth-retarded fetus is one with a birthweight below the 5th or 10th percentile, there are a number of pregnancies with an intermediate to long-term imbalance between placental supply and fetal demand in which significant compromise can develop in situations that require increased fetoplacental performance.

Surveillance methods. Thus we can evaluate available surveillance methods in terms of their ability to detect chronic subtle fetal hypoxia (e.g., fetal biometry), impending hypoxia (e.g., fetal movements, amniotic fluid volume, fetal Doppler waveforms), or acute hypoxia (e.g., nonstress test)[6]. There is a tradition of using a combination of different test methods to evaluate fetal compromise[33,48]. New developments such as Doppler sonography[17] and kinetocardiotocography[47] have gained an established place in modern obstetrics and are commonly used.

Capabilities of Doppler sonography. Doppler sonography can be used to identify high-risk cases in obstetrics and provide them with appropriate surveillance. It is widely agreed that Doppler ultrasound can help to optimize perinatal management, especially in the surveillance of high-risk pregnancies. A causal relationship has been established between abnormal Doppler waveforms and adverse fetal outcome[8,44,50,52]. A chronic increase of flow impedance in the placenta causes a state of chronic fetal hypoxia, which in turn leads to growth retardation and altered fetal hemodynamics[59,61]. Abnormal flow patterns are an indicator of hypoxemic fetal compromise caused by impaired gas exchange in the placenta [42,59]. Thus, a decreased intrauterine fetal oxygen supply can be diagnosed at an early stage by detecting an abnormal flow pattern[24,59].

Absent End-Diastolic Flow (AEDF) and Reverse Flow

Technical Aspects in the Diagnosis of Absent End-Diastolic Flow and Reverse Flow

Definitions. Figures 19.**1** and 19.**2** show normal flow velocity waveforms recorded from the umbilical artery and the fetal aorta, respectively, under optimum examination conditions. For comparison, Figs. 19.**3** and 19.**4** illustrate abnormal waveforms recorded from the same vessels, characterized by decreased but still-present end-diastolic flow (causing an elevated resistance index). Increasing pathology of fetal perfusion can lead to a complete absence of blood flow during the end-diastolic phase. This finding is referred to in the literature as absent end-diastolic flow (AEDF). Figure 19.**5** shows a typical example of AEDF in the umbilical artery. In cases where AEDF is present, the further deterioration of intrauterine condition can lead to retrograde blood flow, also known as reverse flow. Figure 19.**6** illustrates end-diastolic reverse flow in the umbilical artery.

Technical factors. The detection of AEDF or reverse flow in the fetal vessels requires a meticulous examination technique, since technical factors can easily produce false-positive findings and prompt obstetric decisions with potentially far-ranging consequences. Whenever possible, the diagnosis should be confirmed by a second, independent examiner.

Several factors should be given special consideration when the images are read. Besides an optimum beam–vessel angle (must be $< 60°$; the greater the insonation angle, the smaller the frequency shift and the lower the diastolic velocity), the wall filter setting should not exceed 50–100 Hz. If the high-pass filter is accidentally set too high, it may truncate the end-diastolic flows and create a false impression of AEDF (Fig. 19.**7**).

Gestational age-dependent factors. Gestational age-dependent factors should also be considered. For example, when the gestational age is less than 28 weeks at the time of examination, decreased end-diastolic flows are within physiological limits. Fetal body movements, which may be very pronounced

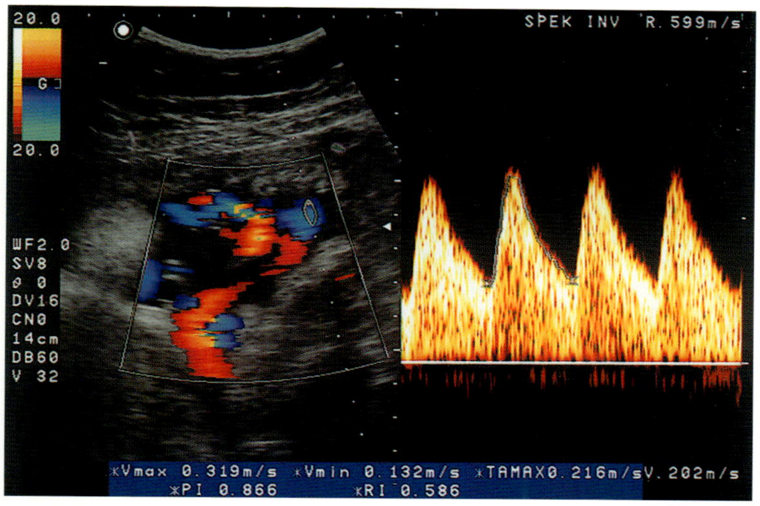

Fig. 19.**1** Normal flow velocity waveform of the umbilical artery.

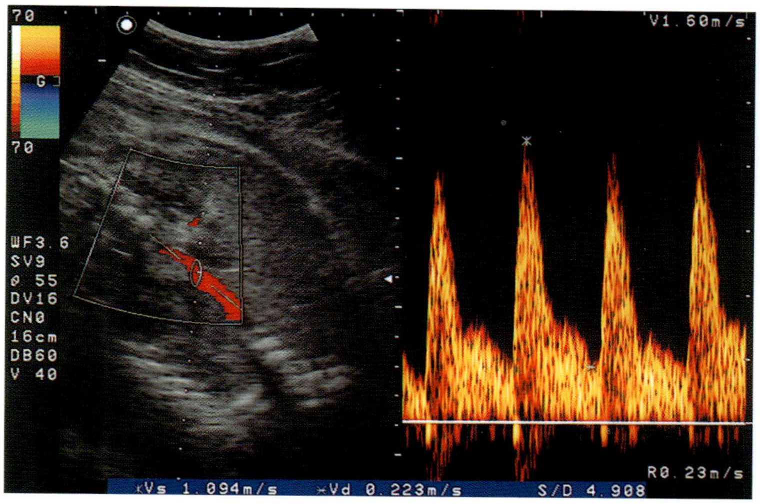

Fig. 19.**2** Normal flow velocity waveform of the fetal aorta.

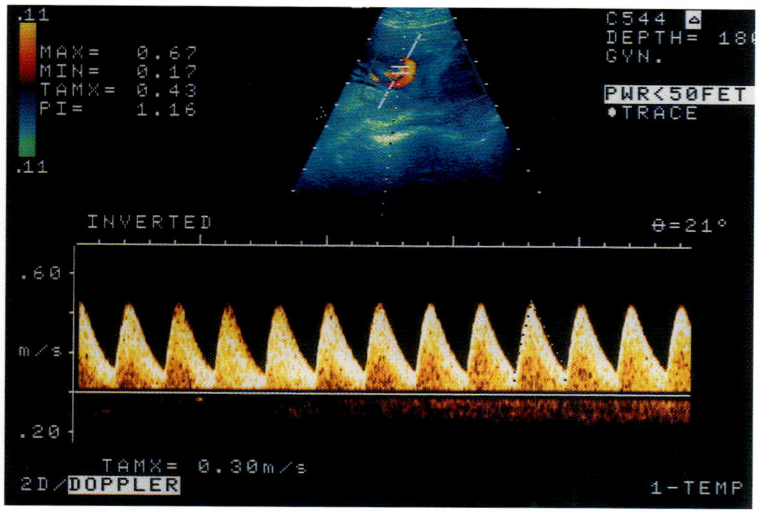

Fig. 19.**3** Abnormally high S/D ratio in the umbilical artery.

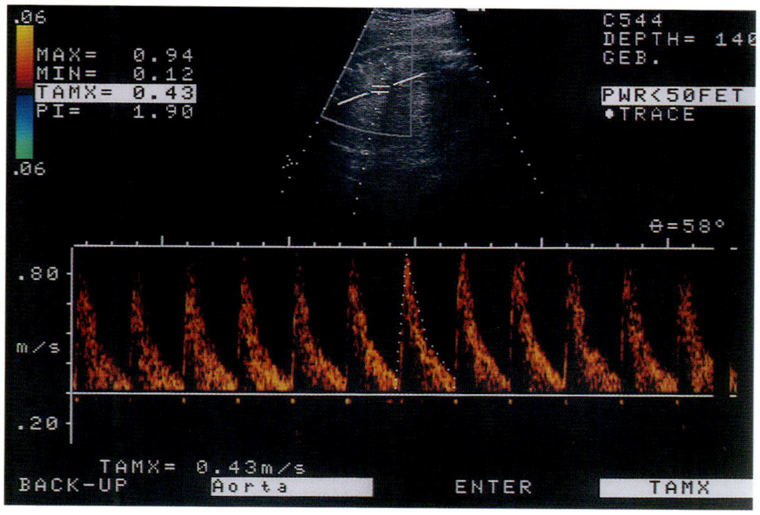

Fig. 19.**4** Abnormally high S/D ratio in the fetal aorta.

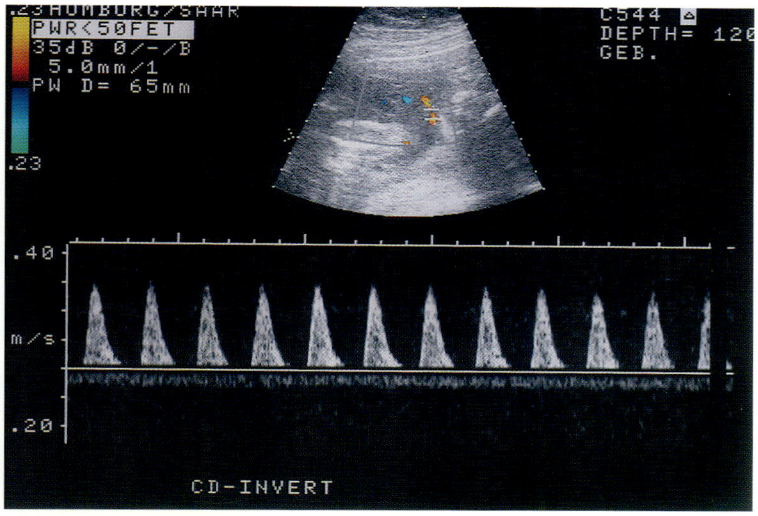

Fig. 19.**5** Absent end-diastolic flow in the umbilical artery.

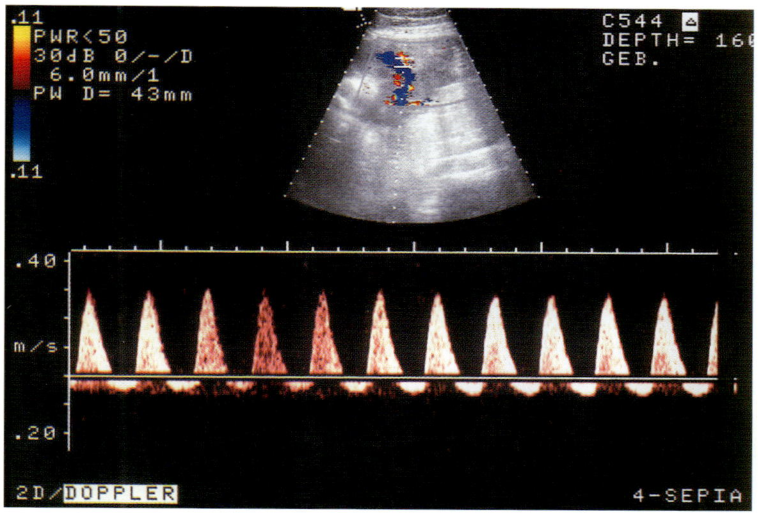

Fig. 19.**6** Typical appearance of reverse flow in the umbilical artery.

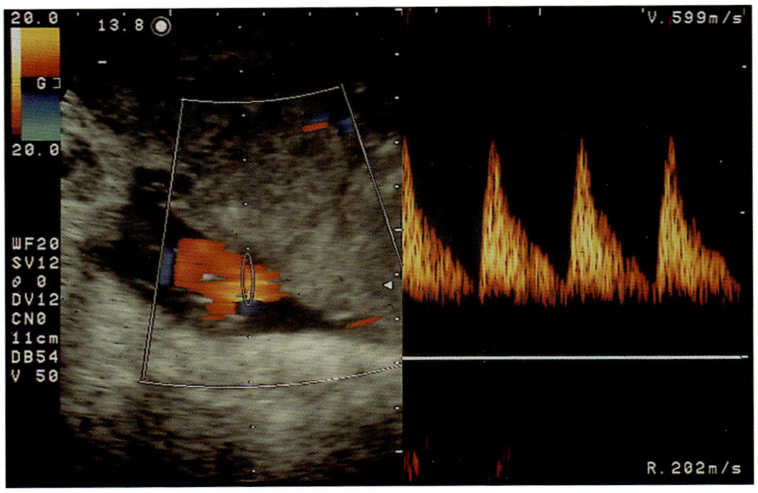

Fig. 19.**7** Pitfall: apparent AEDF caused by setting the high-pass filter too high.

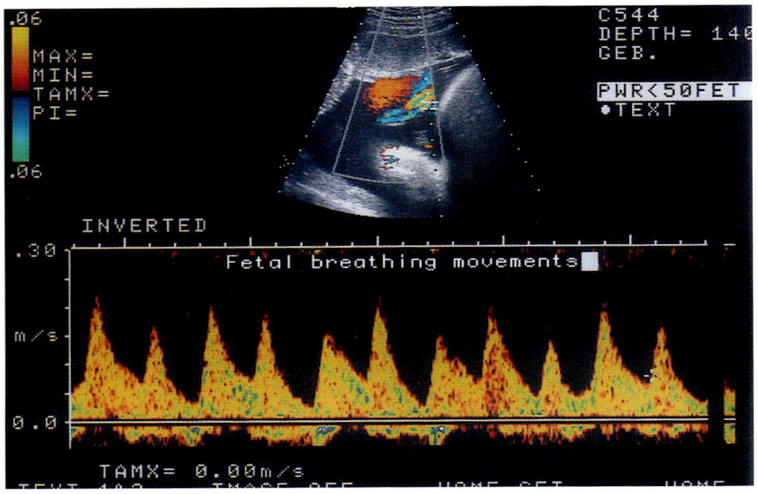

Fig. 19.**8** Pitfall: episodes of absent and reverse end-diastolic flow caused by fetal breathing movements.

at this stage of gestation, as well as fetal breathing movements can cause a transient decrease in diastolic flow ranging to AEDF or even reverse flow (Fig. 19.**8**). End-diastolic flow is also subject to other fluctuations creating a condition of "partial AEDF" or "partial reverse flow."

Absent End-Diastolic Flow in the Umbilical Artery and/or Fetal Aorta

Cases in which Doppler ultrasound shows an absence of end-diastolic flow in the umbilical artery or fetal aorta are included in the obstetric high-risk population, where their incidence is between 2% and 8%. Various authors have reported on the presence of AEDF and presume that it is related to intrauterine hypoxia[19,37,41,52].

Brain-sparing effect. A redistribution of blood flow is a common occurrence in fetuses with AEDF. This centralization of blood flow with decreased perfusion of the peripheral vessels and autoregulation of the cerebral vessels is called the brain-sparing effect[2,54,57,60]. In the literature it has been associated with increased rates of cesarean section, preterm delivery, and neonatal ICU admission as well as increased morbidity and mortality[49,52]. By contrast, relatively little is known about long-term developmental abnormalities in these infants. Some authors point to increased neonatal morbidity rates with permanent neuromotor deficits[14,15,53].

Reverse Flow in the Umbilical Artery and/or Fetal Aorta

There is ample reason to believe that the functional impairment of the fetoplacental unit is a continuous, progressive process and that this progression (after loss of the compensatory reserves) may be reflected in increasingly abnormal Doppler flow indices. In this respect the severity of Doppler abnormalities correlates with the degree of intrauterine fetal compromise. For example, it is likely that serious perinatal problems will arise when end-diastolic reverse flow is detected in the umbilical artery and/or fetal aorta[8,10,26,49].

High-risk situation. This type of finding is associated with a perinatal mortality rate between 50% and 100%[4,8,45,49]. This Doppler finding thus reflects a hazardous situation for the fetus. Most fetuses with reverse flow in the fetal vessels may die in utero within a few days[10,63]. Often a cesarean section must be performed due to suspicion of fetal distress (e.g., an abnormal FHR trace)[8,10,45]. The morbidity in these high-risk infants is particularly high.

Obstetric management. The relationship between fetal outcome and the presence of reverse flow and its causes is still uncertain due to the low prevalence of this finding (approximately 0.3–1%). According to the literature, there is also uncertainty as to the pathophysiological mechanisms and optimum obstetric management of cases with reverse flow. The question of how to proceed when reverse flow is detected in early pregnancy remains unanswered. Although the clinical population with reverse flow is very small, these fetuses warrant very close attention due to the high morbidity and mortality rates.

Clinical Results of AEDF or Reverse Flow in the Umbilical Artery and/or Fetal Aorta

Long-term study. We followed 120 fetuses with absent end-diastolic flow and 30 fetuses with reverse flow in the umbilical artery or fetal aorta over a 10-year period. Besides perinatal abnormalities, we analyzed perinatal outcome and long-term neuromotor development in children who had these severely abnormal Doppler findings in the fetal vessels during the third trimester. The purpose of this long-term study was to help us filter out and identify patterns of antenatal injury independent of perinatal problems. We used the Munich Functional Developmental Score for this purpose[21,28]. Additionally, any special circumstances were identified by questioning the parents and reviewing the children's medical files. Of the surviving children in this high-risk population, 30 cases with absent end-diastolic flow (AEDF) were examined postnatally to assess neuromotor development. The perinatal abnormalities and disturbances of neuromotor development in these children were compared with those in a matched-pair group of comparable gestational age with no Doppler abnormalities ($n = 30$ children). Each child's developmental status was assessed with regard to gross and fine motor skills, perception, independence, speech, language comprehension, and social age.

Absent End-Diastolic Flow

Perinatal results. The mean gestational age at the time of delivery in this group of 120 children was 32 weeks + 5 days, and the mean birthweight was 1385 g. The incidence of severely dystrophic infants (< 5th percentile) was 69%, with a perinatal mortality of 18%. In 97% of cases the liveborn infants were admitted to neonatal ICU (Table 19.**1**).

It is particularly noteworthy that 80% of the infants with AEDF had an abnormal S/D ratio ("oxygen sparing") in the middle cerebral artery, as opposed to only 7% in the group with normal Doppler findings.

Neuromotor development. To evaluate long-term morbidity following severely abnormal antenatal Doppler findings, the neuromotor development of these children was prospectively studied in two parallel groups matched by gestational age at delivery. Thirty children with normal Doppler findings in the fetal vessels (group 1) were compared with 30 children with AEDF in the umbilical artery and/or fetal aorta (group 2). The age of the children was between 9 and 36 months at the time of neuromotor examination. For each functional category, the developmental age was determined and the deviation from the adjusted age was calculated in months.

All of the examined children with AEDF lagged behind the equal-age children without placental dysfunction in their average neuromotor development: 32% of the children with AEDF showed impairment of neuromotor development, compared with only 17% of the children with normal Doppler findings (Fig. 19.**9**).

The deviations from the adjusted age mainly involved gross motor skills, perception processing, and speech.

Other developmental parameters. When the two groups were compared by weight, longitudinal growth, and postpartum head circumference, significant differences were found in both U1 and U7.

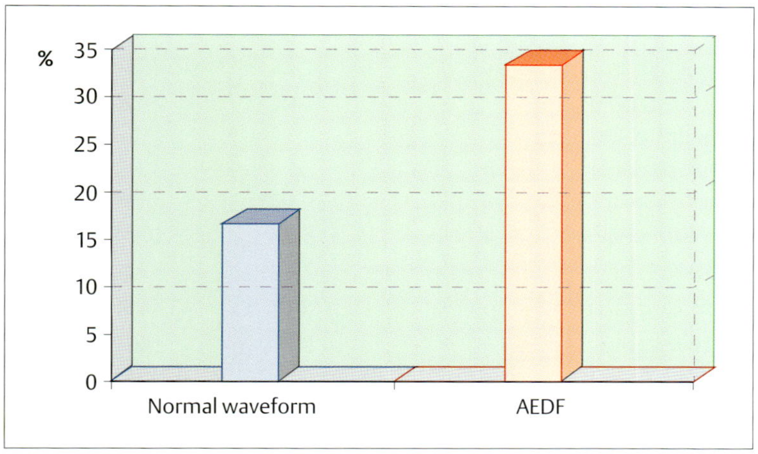

Fig. 19.**9** Frequency of neuromotor abnormalities associated with normal Doppler waveforms ($n = 30$) and with AEDF in the fetal vessels ($n = 30$).

Table 19.**1** Perinatal abnormalities in infants with AEDF in the umbilical artery and/or fetal aorta ($n = 120$ patients)

Perinatal abnormality	Incidence
Pregnancy-induced hypertension	62%
Oligohydramnios	60%
Abnormal FHR trace (Fischer score < 5)	70%
Gestational age at delivery	32 weeks + 5 days
Preterm delivery < 37 weeks	85%
Preterm delivery < 33 weeks	49%
Primary cesarean section	84%
Birthweight (average)	1385 g
5-min Apgar score	11%
pH (average)	7,24
Dystrophy (< 5th percentile)	69%
Perinatal mortality	18%
Congenital anomalies	22%

Reverse Flow

Fetuses with absent end-diastolic flow constitute a high-risk population with serious perinatal problems and a markedly increased risk for neuromotor handicap. In some cases with protracted AEDF segments (e.g., significant pregnancy-induced hypertension [PIH] with preeclampsia), we also observed reverse flow in the fetal vessels over a period of several days. Cases with reverse flow already present in the umbilical artery or fetal aorta at the time of examination were also referred to our center. To compare the perinatal abnormalities associated with AEDF and reverse flow, two gestational age-matched groups of 30 cases each were identified at delivery. Besides the prenatal surveillance methods, neonatal neurosonographic and echocardiographic studies were included in our evaluation.

Perinatal results. Reverse flow in the fetal vessels was diagnosed in 30 cases at an average gestational age of 30 weeks + 1 day. The risk factors of preeclampsia, placental insufficiency, oligohydramnios, and nicotine abuse were significantly more common in cases with reverse flow than in cases with AEDF. The mean gestational age at delivery was 30 weeks + 6 days in both groups. For comparable modes of delivery, a higher acidosis rate (pH ≤ 7.2) was found in association with reverse flow (31.3%) than with AEDF (8.8%). Severe intrauterine growth retardation (< 5th percentile) was demonstrated in 86% of children with reverse flow (odds ratio 9.7) and in 63% of cases with AEDF. Intrauterine death occurred in 43% of the fetuses with reverse flow (odds ratio 22.7), 67% of these fetuses had chronic placental insufficiency, and 25% had a congenital anomaly on pathoanatomical examination. By contrast, intrauterine death occurred in only 3.3% of the cases with AEDF. Thus, the perinatal mortality associated with reverse flow (29%) is markedly higher than in fetuses with AEDF (7%).

Neonatal morbidity. The neonatal morbidity in cases with reverse flow, at 81%, was quite high compared with the 63% incidence in AEDF. Postpartum ultrasound imaging revealed a cerebral abnormality (e.g., cysts, ventricular dilatation or hemorrhage) in 44% of the cases with reverse flow, compared with 31% of the children with AEDF. The incidence of cerebral hemorrhage in the surviving neonates with antepartum reverse flow was 25% (versus 17% in AEDF). Four of 10 infants with cerebral hemorrhage died during the neonatal period. No intracerebral hemorrhages were found in gestational age-matched infants that did not have severely abnormal Doppler findings. Based on the available data, it is our opinion that fetuses with end-diastolic reverse flow have a markedly higher incidence of perinatal problems and a poorer prognosis compared with AEDF, and that consequently these fetuses should not be treated as a common group.

Significance of Severely Abnormal Doppler Findings

Doppler velocimetry of the fetoplacental unit is a very promising technique that has significantly enhanced our ability to evaluate intrauterine fetal well-being.

Comparison with FHR findings. In an immature fetus with a suspicious nonstress test, Doppler sonography can qualify the FHR findings and help to direct clinical management for the benefit of the fetus[18, 45, 49]. In our experience, the average interval from the appearance of a severely abnormal Doppler waveform to the appearance of an abnormal FHR trace is approximately 12 days[49]. Other authors report 4 to 21 days[6, 10, 24]. In many cases an abnormal FHR trace is already present when severely abnormal flow is first detected[49]. In our series, the FHR trace was already abnormal (Fischer score ≤ 4) in 50% of cases that had reverse flow at initial Doppler velocimetry but in only 17% of the cases with AEDF. Various authors have stressed the advantage of Doppler velocimetry over FHR recordings in the early detection of fetal compromise[2, 6, 46]. It may be that the altered hemodynamics in the fetal umbilical artery leads to autoregulation of the cerebral artery, causing a change in the central control of the fetal heart rate. The earliest finding is a biphasic change of blood flow in the middle cerebral artery, followed by a loss of vasodilation of the artery and a decrease in left cardiac output. This is followed in turn by a change in the variability of the fetal heart rate[3].

Neuromotor abnormalities due to failure of the brain-sparing effect. Absent end-diastolic flow should be interpreted as a serious clinical sign[5, 11, 37, 40, 45, 58, 62]. It is associated with increased perinatal morbidity and mortality. In our studies, neuromotor impairment was found in 33% of the children with AEDF that were assessed by the Munich Functional Developmental Score[15].

Cerebral Doppler findings, especially the brain-sparing effect, also had an important bearing on neurological development in our series. This effect is characterized by the presence of end-diastolic frequencies and a decreased S/D ratio or pulsatility index in the cerebral vessels[2, 3, 43, 57]. It results from a redistribution of blood flow in growth-retarded fetuses favoring the brain. In the literature, the brain-sparing effect is interpreted as a mechanism to protect the fetal brain from hypoxia[43]. When this mechanism fails, terminal symptomatology can develop in fetuses with AEDF before the 30th week of gestation[54].

In our study on absent or reverse flow in the umbilical artery and/or fetal aorta, AEDF was observed only in the cerebral vessels of fetuses that later displayed abnormalities. It is reasonable to assume that the brain-sparing effect failed in these fetuses. The condition of the fetuses was so poor that a centralization of blood flow could no longer be achieved. An apparent normalization of abnormal cerebral flow velocity waveforms is also described in the literature[9, 16, 57, 59].

Intrauterine growth retardation. Surprisingly, the children with and without neuromotor abnormalities in our study showed only very minor differences in immediate postnatal data such as Apgar score, pH, and blood gases (Table 19.**2**). Thus, perinatal asphyxia led to an increase in perinatal morbidity but did not cause a permanent impairment of development. A longitudinal study identified fetal growth retardation as a predisposing factor for subsequent learning deficits at 9 to 11 years of age but was unable to relate the deficits to factors of perinatal morbidity[32]. Thus, the problem of neuromotor retardation in later childhood does not appear to arise during the delivery. It is reasonable to assume that the developmental disturbance has a predominantly antenatal cause, i.e., an adverse effect on brain development due to a deficient intrauterine supply. The greater influence of the antenatal period is also emphasized in the literature[39, 51]. Other adverse prognostic factors for childhood development in our series were prematurity, a birthweight below the 3rd percentile, a head circumference below the 3rd percentile, and a low placental weight in relation to birthweight (Table 19.**3**). Dystrophy and immaturity are also cited in the literature as causes of perinatal problems in fetuses with AEDF[1, 30, 41, 63, 64]. Twenty-four percent of normals and 38% of abnormals were still severely dystrophic in their size and weight when seen at follow-up. Other authors also report that infants who lag behind in weight and length do not catch up in their later development[32, 38, 55]. Vohr and Oh consider head circumference at one year of age to be the critical prognostic factor for development[55].

Intracerebral hemorrhage. Neurological abnormalities at birth can influence further development[34]. Cerebral hemorrhage occurred in 10% of infants with AEDF, while Weiss et al. reported 15%[54, 62]. The high incidence of cerebral hemorrhage can be attributed in part to increased cerebral blood flow due to the brain-sparing effect[10]. Cerebral hemorrhage occurred in 25% of children with neuromotor abnormalities in the present study,

Table 19.**2** Comparison of perinatal abnormalities in children with normal Doppler findings (group 1 = 30 children) and with AEDF in the fetal vessels (group 2 = 30 children) for comparable gestational age at birth

Perinatal abnormalities	Normal (group 1)	AEDF (group 2)
Oligohydramnios	14%	23%
Abnormal FHR trace (Fischer score < 5)	14%	29%
Gestational age at birth	34 weeks + 0 days	33 weeks + 3 days
Preterm delivery < 37 weeks	82%	100%
Preterm delivery < 33 weeks	31%	53%
Primary cesarean section	44%	84%
Birthweight (average)	2570 g	1460 g
1-min Apgar < 7	27%	47%
pH (mean)	7,26	7,29
Dystrophy (< 10th percentile)	23%	53%
Congenital anomalies	9%	24%
Transfer to neonatal ICU	57%	93%

Table 19.**3** Review of the literature on abnormalities associated with AEDF

Authors	Year	Cases	Gestational age at delivery (weeks)	Cesarean section rate (%)	IUGR (%)	Birthweight (g)	Abnormalities (%)
Reed	1987	14	33	80	79	1227 g	29
Rochelson	1987	15	34	80	60	1851 g	27
Ombelet	1988	21	31	100	95	924 g	–
Johnstone	1988	24	32	83	92	1282 g	–
Kirkinen	1988	84	33 + 5	72	–	–	9
Arabin	1988	30	33	100	100	–	–
Rochelson	1989	10	34	80	60	1581 g	–
Jouppila	1989	84	33 + 5	72	–	1820 g	9
Gutmundsson	1990	14	37	100	86	2086 g	–
Pillai and James	1990	4	32 + 1	100	100	1285 g	–
Wenstrom	1991	22	29	–	45	1077 g	45
Trudinger	1991	96	31 + 1	91	81	1198 g	9
Pattinson	1993	21	31 + 4	–	17	1014 g	–
Ashmead	1993	5	33	–	–	1710 g	–
Valcamonico	1994	26	31 + 4	–	100	1172 g	8
Rizzo	1994	192	30 + 6	61	–	1124 g	13
Poulain	1994	62	–	86	39	–	16
Ulrich	1994	68	31 +	–	56	1225 g	–
Weiner	1994	10	32 + 3	90	–	1258 g	–
Karsdorp (multicenter study)	1994	178	31 + 4	96	–	1209 g	–
Zelop	1996	32	31 + 1	94	–	1139 g	–
Average values for all studies		52	32	72,1	87,5	1343 g	18,3
Our results	1998	120	32 + 5	84	69	1385 g	22

IUGR = intrauterine growth retardation.

compared with 4.5% of normal children, and thus it can be identified as one cause of the developmental disturbances. Only Scherjon et al. described fewer cerebral hemorrhages in children with later abnormalities[43]. Ulrich et al. found a significantly higher incidence of cerebral hemorrhage and pronounced neurological impairment in AEDF infants than in a corresponding group of preterm infants with normal Doppler waveforms. Thirty-one percent of the AEDF infants exhibited abnormalities of neurological and psychomotor development[53]. This figure is comparable to the 33% rate of developmental delays found in our study.

Gross and fine motor skills and perception were the areas of development that were most strongly affected. Other authors also found a preponderance of fine and gross motor abnormalities[7,36,56] or of motor and perceptual dysfunction [35] in premature infants. Autonomy was least affected.

Reappearance of positive end-diastolic flow. Brar and Platt reported that positive end-diastolic flow was subsequently found in approximately 15% of fetuses with AEDF[8]. These cases were found to have a better fetal outcome. This finding may relate to a change in placental blood flow. Bell et al. found in their study that 11 of 40 (27.5%) fetuses with AEDF regained positive end-diastolic blood flow during the course of the pregnancy. The interval from the detection of AEDF to delivery, gestational age at delivery, and birthweight were greater in these fetuses, and neonatal mortality was lower. It was postulated that the outcome of fetuses with AEDF may improve following the reappearance of positive end-diastolic blood flow. Weiss and Berle found that the rate of fetal acidosis and the number of necessary emergency cesarean sections were higher when there was a short interval between the initial diagnosis and delivery, and that a better fetal prognosis could be achieved with conservative management[60]. Improvement of umbilical artery blood flow and the reappearance of end-diastolic frequencies have been reported in growth-retarded fetuses following maternal oxygen therapy[4,27].

Karsdorp et al. found that antihypertensive medication and hemodilution could improve the rheology of the material circulation[26]. Positive end-diastolic umbilical arterial flow reappeared in all seven pregnancies with AEDF, but the AEDF persisted in the seven pregnancies without hemodilution. The fetal outcome varied considerably. Five of seven fetuses in which positive end-diastolic flow reappeared after AEDF survived, compared with only one of seven fetuses that consistently had AEDF. It might have been possible to improve placental blood flow with treatment in these latter fetuses.

Research on causes of reverse flow. Although high fetal mortality has been observed in cases with end-diastolic reverse flow, the pathophysiology leading to reverse flow is still uncertain. Because of small case numbers, adequate epidemiological data have not yet been published on reverse flow. Nearly all authors studied fewer than 30 cases and usually analyzed the data on AEDF and reverse flow together ("ARED" = absent or reverse end-diastolic flow), without differentiating between them. An exception is the study by Karsdorp et al., who analyzed the data from nine perinatal centers[26] (Table 19.**4**).

Many factors that can affect fetal and maternal hemodynamics alter the flow pattern in the umbilical artery or fetal aorta[8]. Absent and/or reverse end-diastolic flow is significantly more common in pregnancies with intrauterine growth retardation (IUGR) (odds ratio 3.1), with PIH, or with IUGR and PIH (odds ratio 7.4)[5,13,26,45,64]. In our study, we found that cases with IUGR were at high risk for the development of reverse flow (odds ratio 22.6). In cases with severe IUGR, we found either a small placenta or massive intervillous fibrin deposits in the placenta. This could account for the occurrence of abnormal flow patterns.

Pregnancy-induced hypertension. Patients diagnosed with PIH in our study had an increased risk for end-diastolic reverse flow (odds ratio 3.8). A number of authors suggest that a close correlation exists between severely abnormal Doppler waveforms and PIH[13,64]. An increased production of prostacyclin and/or endothelium-derived relaxing factors has been detected in patients with PIH. This is believed to decrease placental blood flow due to the reduction of active renin and angiotensin II in the peripheral circulation or to increased activity of the renin–angiotensin system in the uteroplacental circulation. In turn, this leads to local hypoxemia in the placenta and also induces the production of oxygen free radicals, which are known to mediate local vasoconstriction of the placental vessels in PIH. Apparently this could explain the increased placental vascular resistance that is found in patients with PIH.

Nicotine abuse. An increased incidence of reverse flow was also documented in women who smoke more than 10 cigarettes per day (odds ratio 9.4). These results conflict with those reported in the literature. Karsdorp et al. found no relationship between the risk of reverse flow and maternal smoking habits[26]. Another study found that nicotine did not affect hemodynamics prior to the appearance of vascular damage with morphological changes[20]. Nicotine can induce the vasoconstriction of placental vessels, however, causing a reduction of uteroplacental blood flow. These changes are well documented by the available literature on placental studies[25,42].

Intrauterine fetal death. The increased incidence of intrauterine death is a problem in fetuses with reverse flow[5,8,45,58]. Seven authors in the literature reported an average incidence of 47%, meaning that almost half of fetuses with reverse flow died in utero (Table 19.**5**). The interval from initial diagnosis to fetal death ranged from one day to several days. In our own study, intrauterine death occurred in 40% of the fetuses with reverse flow. The average interval between the detection of reverse flow and the occurrence of intrauterine fetal death was only 2.5 days. Overall, 92% of the intrauterine fetal deaths occurred within one week after diagnosis.

Todros et al. reported that absent or reverse end-diastolic flow had a high predictive value for an adverse fetal outcome. They therefore consider that a definitive diagnosis of ARED is an indication for immediate delivery.

Table 19.**4** Review of the literature on mortality rates associated with reverse end-diastolic flow

Authors	Year	Cases	Gestational age at delivery (weeks)	Cesarean section rate (%)	IUFD (%)	Perinatal mortality	Postpartum mortality
Brar	1988	12	30 + 1	50	33	50	18
Illyes	1988	5	32 + 2	100	100	100	–
Schmidt	1991	4	30 + 4	100	75	50	25
Fouron	1993	5	28 + 3	60	60	–	–
Valcamonico	1994	5	30 + 1	–	20	40	20
Karsdorp (multicenter study)	1994	67	29 + 0	75	24	–	51
Zelop	1996	24	29 + 1	–	17	33	–
Average values for all studies		17	30	77	47	55	29
Our results	1998	30	30 + 6	53	40	27	22

IUFT = intrauterine fetal death

Table 19.5 Comparison of perinatal abnormalities

Perinatal abnormalities	Reverse flow (authors = 8)		AEDF (authors = 21)		Reverse flow/AEDF (authors = 16)	
Total cases	152		1062		560	
	Average values	Median	Average values	Median	Average values	Median
Cases	19	9	51	30	35	32
Gestational age at delivery (weeks)	30.1	30.1	31.9	31.6	31.2	31.2
Total mortality	73.0	67.5	32.2	34	43.9	40
Perinatal mortality	50.0	45	28.7	22	35.4	38
Neonatal mortality	27.1	22	19.4	20	20.0	10
IUFD (%)	46.0	36.5	16.7	11	25.1	16
Anomalies (%)	21.8	23	18.2	15	21	21
IUGR (%)	100	100	88.3	91	90	94
Birthweight (g)	997	983	1337	1225	1114	1037
Cesarean delivery (%)	93	96	72.7	80	74.4	75
1-min Apgar ≤ 7 (%)	84.3	78	63.8	66	68.5	72
pH ≤ 7.2 (%)	41.5	41.5	19.0	19	26.0	26

IUFD = intrauterine fetal death.
IUGR = intrauterine growth retardation.

Summary

Prognostic factor for high-risk pregnancies. Doppler velocimetry is an important method in perinatal diagnosis for the evaluation of high-risk pregnancies. Since severely abnormal findings are often noted in the initial Doppler examination, high-risk pregnancies should undergo Doppler evaluation as early as possible. A severely abnormal Doppler finding in the fetal vessels, such as absent end-diastolic flow (AEDF) or reverse flow, appears to be a significant additional prognostic factor for the high-risk pregnancy.

Absent and reverse flow. Because of small case numbers, absent and reverse end-diastolic flow have often been assigned to a common group in the previous literature ("ARED flow"). This practice should be discontinued in the future. If we define "hypovascularization of the placenta" as the underlying cause of these flow patterns, we must regard both phenomena as the end points on a continuum of placental malperfusion. Generally the clinical consequences of AEDF are more "benign" than in cases with reverse flow.

Doppler sonography and ABCD profile. It is essential that we continue to research the underlying principles of placental malperfusion. We should also direct our efforts toward learning to assess the dynamics of placental insufficiency in order to prevent antenatal intrauterine insults to the maturing organs, as these insults appear to be chiefly responsible for long-term morbidity. Proficient Doppler scanning of the fetomaternal vessels will continue to be the principal instrument for making this assessment. It may also be necessary to apply further biophysical tests within the context of a biophysical profile (ABCD profile)[22, 23] to meet this important clinical challenge.

Therapeutic implications. There are occasional cases in which end-diastolic flow can be improved with appropriate treatment in fetuses who manifest absent or reverse flow. This means that immediate delivery may be unnecessary in some fetuses that have a severely abnormal flow pattern but receive optimum, intensive perinatal care. A prompt cesarean delivery is unavoidable in many cases, however.

Postnatal development. These children have markedly increased mortality and serious morbidity in the neonatal period as well as a markedly increased risk of later neuromotor abnormalities compared with children of comparable gestational age with normal prenatal Doppler findings.

Long-term sequelae can also result from hypoxia-induced intracerebral hemorrhage, which is common in cases with a brain-sparing redistribution of blood flow. Fetuses that are delivered in a compromised condition (abnormal umbilical cord pH, abnormal Apgar score) have a markedly higher risk of hemorrhage than fetuses that have received optimum perinatal management. Optimum management includes early referral to a perinatal center, where the timing of the delivery and perinatal management can be decided in close cooperation with a neonatologist and tailored to the individual situation following a detailed consultation with the parents.

References

1 Adiotomre PN, Johnstone FD, Laing IA: Effect of absent end diastolic flow velocity in the fetal umbilical artery on subsequent outcome. Arch. Dis. Child Fetal Neonatal. 76(1) (1997) 35–38

2 Arabin B, Saling E: Die „Sparschaltung“ des fetalen Kreislaufs dargestellt anhand von eigenen quantitativen Doppler-Blutflussparametern. Z. Geburtshilfe Perinatol. 191 (1987) 213–218

3 Arduini D, Rizzo G: Prediction of fetal outcome in small for gestational age fetuses: Comparision of Doppler measurements obtained from different fetal vessels. J. Perinat. Med. 20 (1992) 29–38

4 Ariyuki Y, Hata T, Kitao M: Reverse end-diastolic umbilical artery velocity in a case of intrauterine fetal death at 14 weeks' gestation. Amer. J. Obstet. Gynecol. 169(6) (1993) 1621–1622

4a Babajan A: Die intrauterine fetale Wachstumsretardierung; Ätiologie und Diagnostik. Dissertationsschrift, Universität des Saarlandes 1998

5 Battaglia C, Artini PG, Galli PA, D'ambrogio G, Droghini F, Genazzani AR: Absent or reversed end-diastolic flow in umbilical artery and severe intrauterine growth retardation. Acta. Obstet. Gynecol. Scand. 72 (1993) 167–171

6 Bekedam DJ, Visser GHA, van der Zee AGJ, Snijders RJM, Poelmann-Weesjes G: Abnormal velocity waveforms of the umbilical artery in growth-retarded fetuses: Relationship to antepartum late heart rate decelerations and outcome. Early Hum. Dev. 24 (1990) 79–89

7 Bjerre F, Hansen E: Psychomotor development and school adjustment of 7-year-old children with low birthweight. Acta. Paediatr. Scand. 65 (1976) 88–96

8 Brar HS, Platt LD: Reverse end-diastolic flow velocity on umbilical artery velocimetry in high-risk pregnancies: An ominous finding with adverse pregnancy outcome. Amer. J. Obstet. Gynecol. 159 (1988) 559–561

9 Chandran R, Serra-Serra V, Sellers SM, Redman CWG: Fetal middle cerebral artery flow velocity waveforms – a terminal pattern. Brit. J. Obstet. Gynaecol. 98 (1991) 937–938

10 Chaoui R, Hoffmann H. Zienert A, Bollmann R, Halle H, Grauel EL: Klinische Bedeutung und fetal outcome beim enddiastolischen Flowverlust in der A. umbilicalis und/oder fetale Aorta: Analyse von 51 Fällen. Geburtsh. u. Frauenheilk. 51 (1991) 532–539

11 Comas C, Carrera M, Devesa R et al.: Early detection of reversed diastolic umbilical flow: should we offer karyotyping? Ultrasound Obstet. Gynecol. 10 (1997) 400–402

12 Edelstone DI: Fetal compensatory response to reduced oxygen delivery. Sem. Perinat. 8 (1984) 184–191

13 Ertan AK, He JP, Tossounidis I, Schmidt W: Einfluss der EPH-Gestose auf perinatale Faktoren bei Fällen mit „Reverse Flow“ bzw. „enddiastolischem Block“ in den fetalen Gefäßen. In Künzel W (ed.), Arch. Gynecol. Obstet. 258 (1996) 113

14 Ertan AK, Jost W, Mink D, Schmidt W: Neuromotoric development of children after ARED-Flow during pregnancy. In Kurjak A, Latin V, Rippmann E (eds.): Advances on the pathophysiology of pregnancy. CIC Edizioni internazionali 1995, 55–62

15 Ertan AK, Jost W, Hendrik J, Lauer S, Uhrmacher S, Schmidt W: Perinatal events and neuromotoric development of children with zero flow in the fetal vessels during the last trimester. In Cosmi EV, Di Renzo GC: 2nd World Congress of Perinatal Medicine. Monduzzi Editore 1993, 1049–1052

16 Erz W, Gonser, M: Dopplersonographie der fetalen Arteria cerebri media: Präfinale Normalisierung des zerebralen Blutflusses? Geburtshilfe Frauenheilkd. 55 (1995) 407–409

17 Fitzgerald DE, Drumm JE: Non-invasive measurement of human fetal circulation using ultrasound: a new method. Brit. Med. J. 2 (1977) 1450–1451

18 Göschen K: Überwachung der Schwangerschaft aus forensischer Sicht. Gynäkologe 27 (1994) 197–207

19 Gudmundsson S, Tulzer G, Huhta JC, Marsal K: Venous Doppler in the fetus with absent end-diastolic flow in the umbilical artery. Ultrasound Obstet. Gynecol. 7(4) (1996) 262–267

19a He JP: Reverse Flow; perinatologische Auffälligkeiten. Dissertationsschrift, Universität des Saarlandes 1998

20 He JP, Ertan AK, Reitnauer K, Mink D, Schmidt W: Pathomorphologische Veränderungen der Endzotten: Vergleich bei Fällen mit „Reverse Flow“ / enddiastolischem Block bzw. mit normalen Doppler-Flow-Befunden in den fetalen Gefäßen. Im Druck 2000

21 Hellbrügge T, Lajosi F, Menara D, Schamberger R, Rautenstrauch T: Münchner Funktionelle Entwicklungsdiagnostik. Erstes Lebensjahr. Hansisches Verlagskontor, Lübeck 1978

22 Hendrik H-J, Ertan AK, Schmidt W: Die Überwachung der Risikoschwangerschaft mit einem neuen biophysikalischen Profil (ABCD-Profil). Im Druck 2000

23 Hendrik H-J, Tossounidis I, Boos R, Schmidt W: Neuentwicklung eines fetalen biophysikalischen Profils unter Verwendung verschiedener sonographischer Parameter, Doppler-Flow und der Kinetocardiotokographie. Ultraschalldiagnostik, Dreiländertreffen, Bildgebung/Imaging 61(Suppl. 2) (1994) 92

23a Hitschold T, Ulrich S, Kalder M, Müntefering H, Berle P: Blutströmungsprofile der A. umbilicalis. Korrelation zur Plazentamorphologie und zu klinisch-geburtshilflichen Daten im Rahmen der Präeklampsie. Z. Geburtsh. u. Neonat. 199 (1995) 8–12

24 Jouppila P, Kirkinen P: Increased vascular resistance in the descending aorta of the human fetus in hypoxia. Brit. J. Obstet. Gynaecol. 91 (1984) 853–856

25 Karsdorp VH, Dirks BK, van der Linden JC, van Vugt JM, Baak, JP, van Geijn HP: Placenta morphology and absent or reversed end diastolic flow velocities in the umbilical artery: a clinical and morphometrical study. Placenta 17 (1996) 393 –399

26 Karsdorp VH, van Vugt JM, van Geijn HP et al.: Clinical significance of absent or reversed end diastolic velocity waveforms in umbilical artery. Lancet 17(344) (1994) 1664–1668

27 Kingdom JC, Kaufman P: Oxygen and placental villous development: origins of fetal hypoxia. Placenta 18 (1997) 613–621

28 Köhler G, Egelkraut H: Münchner Funktionelle Entwicklungsdiagnostik für das zweite und dritte Lebensjahr. Eigenverlag der Aktion Sonnenschein 1984

29 Krebs C, Macara LM, Leiser R, Bowman AWF, Greer IA, Kingdom JCP: Intrauterine growth restriction and absent diastolic flow velocity in umbilical artery is associated with maldevelopment of the terminal placental villous tree. Amer. J. Obstet. Gynecol. 175 (1996) 1534–1542

30 Kubli F, Schmidt W: Zustandsdiagnostik des Feten. In Bachmann KD, Ewerbeck H, Kleihauer E, Rossi E, Stalder G (eds.): Pädiatrie in Klinik und Praxis. Bd. 1. Thieme, Stuttgart (1987) 79–94

31 Kurkinen-Raty M, Kivela A, Jouppila P: The clinical significance of an absent end-diastolic velocity in the umbilical artery detected before the 34th week of pregnancy. Acta. Obstet. Gynecol. Scand. 76 (1997) 398–404

31a Lauer S: Enddiastolischer Block und neuromotorische Entwicklung. Dissertationsschrift, Universität des Saarlandes 1997

32 Low JA, Handley-Derry MH, Burke SO et al.: Association of intrauterine fetal growth retardation and learning deficits at age of 9 to 11 years. Amer. J. Obstet. Gynecol. 167 (1992) 1499–1505

33 Manning FA, Morrison I, Lange IR: Fetal biophysical profile scoring. A prospective study in 1184 high-risk patient. Amer. J. Obstet. Gynecol. 140 (1981) 289–293

34 Marlow N, Hunt LP, Chiswick ML: Clinical factors associated with adverse outcome for babies weighing 2000 g or less at birth. Arch. Dis. Child 63 (1988) 1131–1136

35 Matilainen R, Heinonen K, Siren-Tiusanen H, Jokela V, Launiala K: Neurodevelopmental screening of in utero growth-retarded prematurely born children before school age. Eur. J. Pediatr. 146 (1987) 453–457

36 Nickel RE, Bennett FG, Lawson FN: School performance of children with birthweights of 1000 g or less. Amer. J. Dis. Child 136 (1982) 105–110

37 Poulain P, Palaric JC, Milon J et al.: Absent end diastolic flow of umbilical artery Doppler: pregnancy outcome in 62 cases. Eur. J. Obstet. Gynecol. Reprod. Biol. 53(2) (1992) 115–119

38 Richter T, Lietz R, Beyreiss K: Gewichts- und Längenentwicklung ehemals hypotroph geborener Kinder in Abhängigkeit vom Schweregrad der intrauterinen Retardierung. Kinderärztl. Praxis 59 (1991) 341–345

39 Riegel K: Die Entwicklung des Kindes nach Schwangerschafts- und Geburtsrisiken. Diagnostik 14 (1981) 493–500

40 Rizzo G, Pietropoll-Li A, Capponi A, Arduini D, Romanini C: Chromosomal abnormalities in fetuses with absent end-diastolic velocity in umbilical artery: analysis of risk factors for an abnormal karyotype. Amer. J. Obstet. Gynecol. 171(3) (1994) 827–831

41 Rochelson B: The clinical significance of absent end-diastolic velocity in the umbilical artery waveforms. Clin. Obstet. Gynecol. 32 (1989) 692–702

42 Salafia CM, Pezzullo JC, Minior VK, Divon MY: Placental pathology of absent and reversed end-diastolic flow in growth-restricted fetuses. Obstet. Gynecol. 90(5) (1997) 830–836

43 Scherjon SA, Smolders-De Haas H, Kok JH, Zondervan HA: The "brain-sparing" effect: Antenatal cerebral Doppler findings in relation to neurologic outcome in very preterm infants. Amer. J. Obstet. Gynecol. 169 (1993) 169–175

44 Schmidt W, Ertan AK: Dopplersonographie in der Geburtsmedizin. Geburtshilfliches Management bei hochpathologischen Doppler-Flow-Befunden. In Hillemans HG (ed.): Geburtshilfe – Geburtsmedizin. Eine umfassende Bilanz zukunftsweisender Enwicklungen am Ende des 20. Jahrhunderts. Springer, Heidelberg 1995, 317–325

45 Schmidt W, Ertan AK, Rühle W, von Ballestrem CL, Gnirs J, Boos R: Dopplersonographie: „Enddiastolischer Block bzw. Reverse Flow" – Perinatologische Daten und geburtshilfliches Management. Jahrbuch der Gynäkologie und Geburtshilfe. Biermann Verlag, Zülpich 1991, 99–106

46 Schmidt W, Graf-von-Ballestrem CL, Ertan AK, Rühle W, Gnirs J, Boos R: Pathologische Doppler-Flow-Befunde und kardiotokographische Ergebnisse. Geburtsh. u. Frauenheilk. 51 (1991) 523–531

47 Schmidt W, Gnirs J: Fetale Bewegungsaktivität und akustische Stimulation. Gynäkologe 23 (1990) 289–297

48 Schmidt W, Kubli F, Garoff L et al.: Diagnose der intrauterinen Wachstumsretardierung – Vergleich von Klinik, Gesamtöstrogenbestimmung aus dem 24 h Urin und Ultraschallbiometrie (Distanzmessungen, biparietaler Kopfdurchmesser, thorako-abdominaler Querdurchmesser) unter Berücksichtigung des antepartalen und subpartalen CTGs. Geburtshilfe Frauenheilkd. 42 (1982)

49 Schmidt W, Rühle W, Ertan AK, Boos R, Gnirs J: Doppler-Sonographie – Perinatologische Daten bei Fällen mit enddiastolischem Block bzw. Reverse Flow. Geburtshilfe Frauenheilkd. 51 (1991) 288–292

50 Schulman H, Fleischer A, Stern W, Farmakides G, Jagani N, Blattner P: Umbilical velocity wave ratios in human pregnancy. Amer. J. Obstet. Gynecol. 148 (1984) 985–990

51 Taylor DJ, Howie PW: Fetal growth achievement and neurodevelopmental disability. Brit. J. Obstet. Gynaecol. 96 (1989) 789–794

52 Trudinger BJ, Cook CM, Giles WB, Connelly A, Thompson RS: Umbilical artery flow velocity waveforms in high-risk pregnancy. Randomised controlled trial. Lancet 1 (1987) 188–190

53 Ulrich S, Ernst JP, Kalder M, Weiss E, Berle P: Neurologische Spätmorbidität von Frühgeburten mit intrauterin diagnostiziertem Null- oder Negativflow der Nabelarterien. Z. Geburtshilfe Perinatol. 198 (1994) 100–103

54 Ulrich S, Weiss E, Kalder M, Hitschold T, Berle P: Doppler sonographic flow measurements of the middle cerebral artery in end-diastolic zero flow in the umbilical arteries in relation to fetal outcome. Z. Geburtsh. Neonatol. 200(1) (1996) 21 –24

55 Vohr B, Oh W: Growth and development in preterm infants small for gestational age. J. Pediatr. 103 (1983) 941–944

56 Vohr B, Garcia-Coll C: Neurodevelopmental and school performance of very low-birthweight infants: a seven year longitudinal study. Pediatrics 76 (1985) 345–350

57 Vyas S, Nicolaides KH, Bower S, Campbell S: Middle cerebral artery flow velocity waveforms in fetal hypoxaemia. Brit. J. Obstet. Gynecol. 97 (1990) 797–803

58 Wang KG, Chen CP, Yang JM, Su TH: Impact of reverse end-diastolic flow velocity in umbilical artery on pregnancy outcome after the 28 th gestational week. Acta. Obstet. Gynecol. Scand. 77(5) (1998) 527–531

59 Weiner Z, Farmakides G, Schulman H, Penny B: Central and peripheral hemodynamic changes in fetuses with absent enddiastolic velocity in umbilical artery: Correlation with computerized fetal heart rate pattern. Amer. Obstet. Gynecol. 170 (1994) 509–515

60 Weiss E, Berle P: Clinical management of fetuses with diastolic zero or reverse flow of the umbilical arteries: Duration of clinical surveillance and fetal outcome. Z. Geburtsh. Perinat. 195 (1991) 37–42

61 Weiss E, Hitschold T, Müntefering H, Berle P: Dopplersonographie der Art. umbilicalis: Differenzierte Diagnostik bei der intrauterinen Mangelentwicklung. Geburtsh. u. Frauenheilk. 49 (1989) 466–471

62 Weiss E, Ulrich S, Berle P: Condition at birth of infants with previously absent or reverse umbilical artery end-diastolic flow velocities. Arch. Gynecol. Obstet. 252 (1982) 37–43

63 Woo JSK, Liang ST, Lo RLS: Significance of an absent or reversed end diastolic flow in doppler umbilical artery waveforms. J. Ultrasound Med. 6 (1987) 291–297

64 Zelop CM, Richardson DK, Heffner LJ: Outcomes of severely abnormal umbilical artery doppler velocimetry in structurally normal singleton fetuses. Obstet. Gynecol. 87(3) (1996) 434–438

20 Fetal Doppler Findings in Late Pregnancy

K. Vetter

Physiological Findings in Late Pregnancy

Late pregnancy and delivery are characterized by a constant rise in fetal demand with very little concomitant growth of the placental supply organ.

Uterine artery blood flow. The blood flow rate in the uterine arteries rises steadily during the course of pregnancy, increasing from about 190 ml/min before pregnancy to approximately 680 ml/min in late pregnancy[1,35]. The mean diameter of the uterine artery expands from 1.6 mm before pregnancy to 3.7 mm near term. The ratio of the peak systolic velocity to the maximum end-diastolic velocity—a measure of flow impedance—decreases from a mean value of 5.3 (RI = 0.81) before pregnancy to 2.3 (RI = 0.57) near term[35].

Placental weight and blood flow. The mass of the placenta grows steadily from approximately 6 g at 6 weeks menstrual age to more than 500 g at term[3,40]. The fetus grows considerably faster than the placenta. As a result, the placental-to-fetal weight ratio of 1.16 at 16 weeks menstrual age declines to 0.13 at term[40]. Very early Doppler data show that the mean placental blood flow increases during the course of pregnancy from approximately 100 ml/min at 22 weeks to a maximum of more than 320 ml/min at 37–38 weeks. It then falls to about 300 ml/min during the final two weeks. The rate of placental blood flow relative to fetal weight is a constant 120 (ml/min)/kg at 36–37 weeks. This may decline to 90 (ml/min)/kg during the final weeks[13]. Meanwhile the resistance index falls from 0.6 at 28 weeks to 0.5 at term[26].

The amniotic fluid volume dwindles as term approaches, and finally the vernix caseosa disappears from the fetal skin as a sign of decreased placental nutrition.

Maternal adaptation. The signs of maternal adaptation to pregnancy gradually regress: the hematocrit rises again, the maternal blood pressure returns to its original level, AT-II sensitivity returns to normal, and pregnancy-related edema may clear.

These are the boundary conditions for the responses of the fetal circulation: a relative reduction in the amount of placental tissue available for nutrient supply and gas exchange, and an absolute decrease in fetoplacental blood flow.

Aorta: Quantitative Analysis

In the early years of obstetric Doppler ultrasound, quantitative blood flow measurements were performed on the uteroplacentofetal unit. The largest easily accessible vessel is the fetal aorta, and one of the first reported quantitative Doppler findings was a flow rate of 185 ± 7.6 (ml/min)/kg estimated fetal weight in the descending aorta[7,8].

Values during the course of pregnancy. In subsequent years, various groups of authors determined and published values during the course of pregnancy[9–11,15–17,20,21,24,36,37]. The mean blood flow velocity (TASAV = temporal average of spatial average velocity) in the aorta ranged from 26.5 to 34.6 cm/s, with a median value of 29.0 cm/s. The relative mean blood flow was between 169 and 246 (ml/min)/kg, with a median value of 220 (ml/min)/kg. The peak systolic velocity (Mv_{max}) was between 70 and 118 cm/s, with a median value of 100 cm/s. The reported mean aortic stroke volume at term was 5.4 ± 1.6 ml, with a relative stroke volume of 1.8 ± 0.5 ml/kg. The mean systolic diameter of the aorta was 7.3 ± 1.1 mm; the mean diastolic diameter was 6.3 ± 1.1 mm. The average effective aortic diameter (determined by echo tracking) was 7.0 ± 1.1 mm[19].

In a study of blood flow in the descending aorta by the author, the values showed the greatest development between 24 weeks and delivery, with a considerable range of variation[38].

It is notable that the quantitative assessment of blood flow shows substantial variability in both normal and abnormal pregnancies. One reason for this is the difficulty of accurately measuring fetal vascular diameters with ultrasound. Clinicians were disappointed when Doppler ultrasound did not provide the accuracy of blood flow measurements that they had hoped for.

Aorta: Qualitative Analysis

Waveform analysis. Analysis of the envelope curve of the Doppler spectrum, either alone or as an adjunct to quantitative measurements, has provided an effective impetus for prenatal Doppler scanning. Flow impedance and pulse-wave reflections have become more important in the interpretation of Doppler sonograms than comparisons of flow volumes. Despite technical advances, qualitative results are easier to obtain than quantitative data—one reason being that they are more or less independent of the insonation angle. Qualitative waveform analysis has made it possible to evaluate blood flow even in small vessels. Over time, therefore, the focus of Doppler scanning has shifted away from the fetal aorta and more toward the cerebral arteries in addition to other peripheral vessels such as the renal arteries.

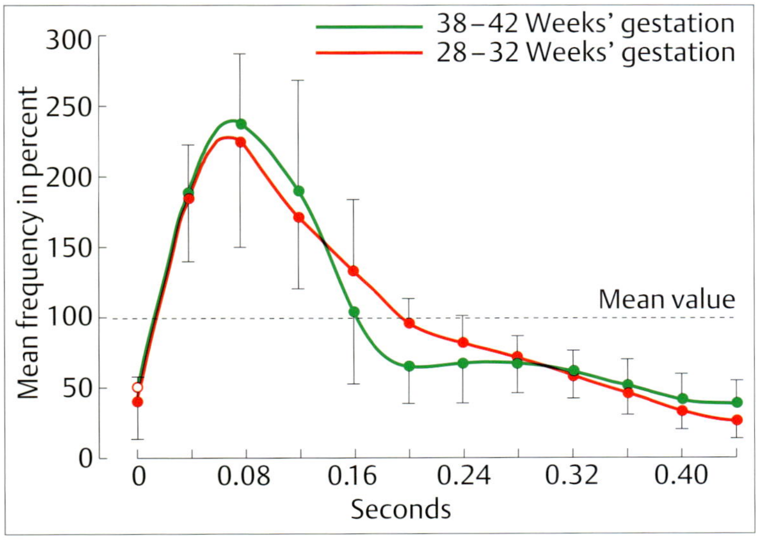

Fig. 20.1 Frequency index profile (FIP). Nomogram ± 2 SD for normalized descending aorta waveforms in the early and late third trimester. A deepening notch appears in the late third trimester as compared with the early third trimester. (From Griffin et al. 1983[15].)

Indices. Numerous methods have been devised for analyzing the Doppler spectrum of the aorta. Most analyses in the literature are based on two-value indices: the S/D ratio (Stuart[33]) and the resistance index RI (Pourcelot[28]). The three-value pulsatility index PI (Gosling[14]) was occasionally used by virtue of its greater sensitivity to waveform changes between the maximum and minimum. The frequency index profile (FIP[6]) is a very interesting but complicated analytical method (Fig. 20.**1**).

Values during the course of pregnancy. A longitudinal study of blood flow in the thoracic aorta during the third trimester of pregnancy showed no significant changes in relation to gestational age: the peak velocity was 115.6 ± 19.0 cm/s, the pulsatility index 1.96 ± 0.31, the acceleration time percentage 19.2 ± 2.2%, the rising slope 25.7 ± 5.6, and the descending slope 4.5 ± 0.9. The values in the abdominal aorta showed a similar lack of change: peak velocity 99.7 ± 18.8 cm/s, pulsatility index 1.68 ± 0.28, acceleration time percentage 19.1 ± 2.2%, rising slope 29.9 ± 4.9, and descending slope 5.3 ± 0.9[21].

The mean blood flow velocity in the aorta shows a definite increase from week 17 to week 32 of gestation, remains constant until the due date, and then falls again until week 42[4,38]. The PI in the aorta remains constant throughout the pregnancy[4]. The most important quantitative and qualitative indices of aortic blood flow after 24 weeks' gestation are reviewed in Table 20.**1**.

Cerebral Arteries

The brain is a very important region in terms of the blood flow conditions that prevail there. The common carotid artery, internal carotid artery, circle of Willis arteries, and the anterior, middle and posterior cerebral arteries are accessible to sonographic evaluation. With the offset Doppler scanners that were originally used, it was much easier to analyze the carotid arteries[25] than the circle of Willis arteries. Today, however, sector scanners with integrated Doppler make these vessels much easier to interrogate—especially the middle cerebral artery, which can usually be scanned at an optimum beam–vessel angle that approaches 0°.

Common carotid artery. The flow velocity in the common carotid artery increases throughout the pregnancy. The PI shows a sharp decline after 32 weeks' gestation[4].

Middle cerebral artery. The Doppler velocity waveforms of the middle cerebral artery (MCA) typically show a biphasic pattern with continuous forward flow during diastole. The RI values show a significant decrease at the end of pregnancy (Fig. 20.**2**)[18,31]. In another study, the S/D ratio showed a significant decline from 6.89 ± 1.48 at 25 weeks to 4.23 ± 0.67 at term[42].

Table 20.**1** Quantitative and qualitative blood flow indices in the fetal descending aorta after 24 weeks in five gestational-age groups

Parameter	Unit	Weeks of gestation				
		26	**30**	**34**	**38**	**> 40**
Heart rate	beats/min	146	143	142	145	145
TASAV	cm/s	26	30	31	30	28
Diameter	mm	3.9	5.2	5.8	6.6	7.4
Blood flow	ml/min	204	400	480	638	694
MV_{max}	cm/s	78	91	97	100	92
RI		0.77	0.80	0.79	0.77	0.79
PI		1.73	1.77	1.62	1.59	1.66

TASAV = temporal average of spatial average velocities = ?????
Mv_{max} = peak systolic velocity

Renal Arteries

The PI of the renal arteries shows a linear decline between 18 and 42 weeks' gestation, decreasing from an average of 3 to 2. This reflects a substantial decrease in impedance and may therefore signal an increase in renal perfusion[41].

Femoral Arteries

Unlike most other arterial sonograms during pregnancy, the PI of the femoral arteries shows a linear rise from 1.8 at 15 weeks to 5.0 at 42 weeks[23]. Reverse flow in the femoral arteries is considered a normal finding in the third trimester.

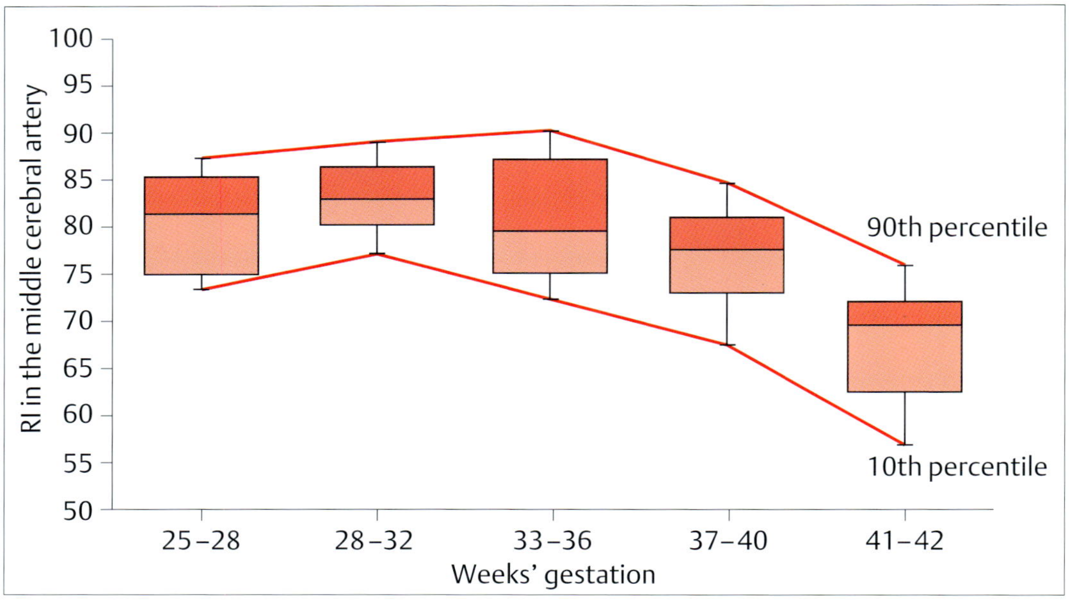

Fig. 20.2 Resistance indices (RI values) of fetal intracranial arterial waveforms in normal pregnancies. A box plot (25th, 50th and 75th percentiles) and whiskers plot (10th and 90th percentiles) is shown for each gestational age. (From Kirkinen et al. 1987[18].)

Changes in Findings at Term and in Postterm Pregnancies

A variety of changes occur in the fetal circulation when the pregnancy reaches term. As noted in a recent publication, "we conclude that postterm pregnancy may mimic a mild fetal growth restriction"[2].

Fetoplacental blood flow. The first report challenging the traditional view of a continuous progression of fetal values during pregnancy was published in 1981. It was observed that fetoplacental blood flow in the umbilical artery increased until 36 weeks, was maximal between 37 and 38 weeks, and then decreased during the last two weeks of pregnancy. The blood flow relative to fetal weight was constant until 36–37 weeks of pregnancy, when a reduction occurred (Fig. 20.**3**)[13].

Term Effect

Fetal aorta. When blood flow in the fetal descending aorta was evaluated with pulsed Doppler ultrasound, it was found that the flow showed a steady decline after term, with very little concomitant change in the RI[29]. Besides a decreased mean flow velocity, we observed a very sharp velocity drop in late systole, which in some cases was so pronounced that a notch appeared in the Doppler velocity waveform. This phenomenon occurred about two weeks before the start of spontaneous delivery (Fig. 20.**4**). Significant dilatation of the fetal aorta was noted at the same time[39]. A later study of pregnancies carried beyond term showed that notching of the aortic waveform was a normal

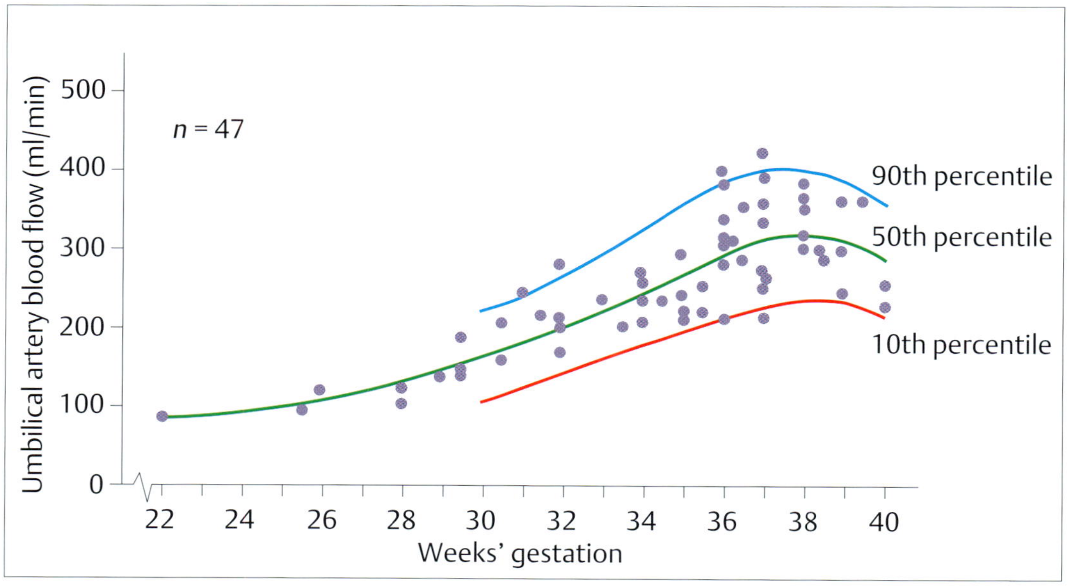

Fig. 20.3 Fetal umbilical cord blood flow in normal pregnancies as a function of gestational age. The 10th and 90th percentile values are shown. (From Gill et al. 1981[13].)

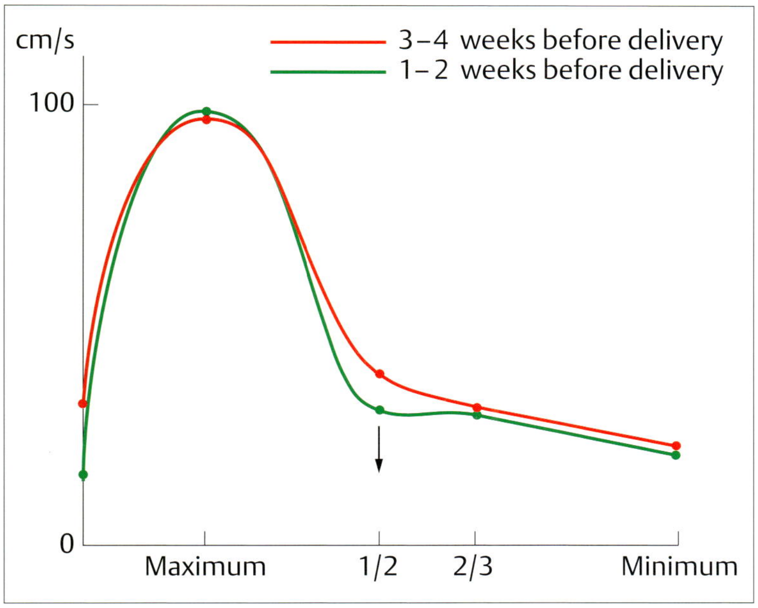

Fig. 20.**4** The term effect. Median values of the fetal descending aorta waveform recorded 3–4 weeks before delivery compared with the median values recorded during the last two weeks before delivery. The arrow indicates the postsystolic notch. (From Vetter et al. 1989[39].)

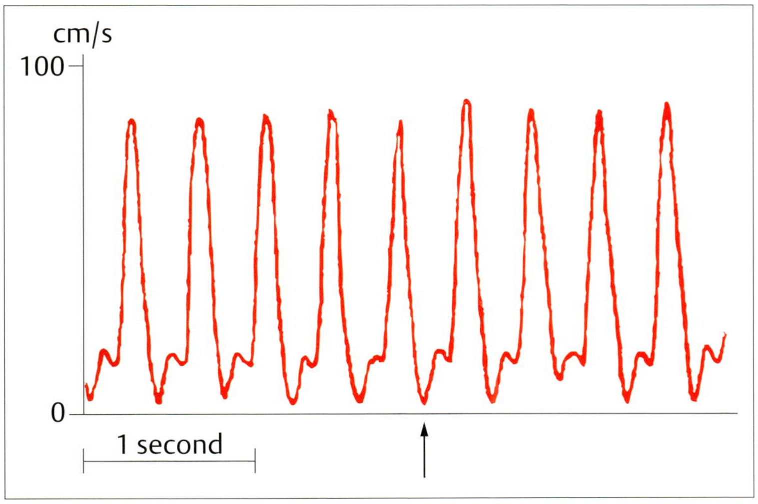

Fig. 20.**5** Postterm pregnancy. The waveform of the fetal descending aorta exhibits a notch in early diastole. (From Malcus et al. 1991[22].)

finding in prolonged pregnancies (Fig. 20.**5**)[22]. The compliance of the aorta appears to decrease around term on the basis of increased pulse-wave velocities determined by measurements using an echo-tracking technique[32]. We refer to these changes in the aorta as the "term effect"[39].

Aortic isthmus. It a Doppler study of the aortic isthmus, forward flow was recorded throughout the cardiac cycle before 20 weeks' gestation, and the diastolic deceleration phase was gradual and smooth. After 20 weeks, a notch appeared at end-systole that progressively increased, and by 30 weeks' gestation a brief phase of reverse diastolic flow was consistently recorded. Experimental and clinical observations have shown that an increase of resistance in the fetoplacental circulation initially causes changes in the waveforms of the aortic isthmus before any significant changes appear in the umbilical artery waveforms. The aortic isthmus waveform was used to calculate an index called the balance index (BI), equal to S–D divided by the difference between the forward-and reverse-flow velocity integrals. The BI increased slowly throughout gestation. It could be shown by color Doppler that the reverse flow observed late in gestation in the aortic isthmus was coming from the ductus arteriosus[12].

Table 20.**2** Doppler ultrasound findings at term and in postterm pregnancies

- Decreased placental blood flow
- Unchanged or decreased impedance in the umbilical arteries
- Increased impedance in the femoral arteries
- Decreased impedance in the renal arteries
- Decreased mean blood flow velocity in the dilated aorta
- Increased pulsatility with a notch in the descending aorta (normal finding)
- Decreased impedance in the aorta
- Reverse diastolic flow in the aortic arch and elevated balance index (BI)
- Increased diastolic flow velocities in the cerebral arteries

Table 20.**2** reviews the hemodynamic situation at term and beyond term as it may be demonstrated by Doppler scanning of the large fetal vessels.

Circulatory Balance

In a simplified scheme, pulsatile blood flow reflects the driving forces of the heart on the one hand and the moderating forces of peripheral vascular resistance on the other. Several intervening factors are also present:

- Vessel wall compliance
- Effects of branching vessels
- Differences of impedance and flow resistance in the vascular regions supplied by the examined vessel

Thus, the Doppler sonogram of a vessel represents the balance of all these factors.

Redistribution of blood flow. In a normal pregnancy, the region with the lowest flow resistance is the placenta, followed by the cerebral arteries. Assuming a stable combined cardiac output, the differential distribution of the blood follows the path of least resistance. For this reason, most of the blood is directed into the placenta. At the end of the pregnancy, the requirements of the brain increase both absolutely and in relation to other organs. One way to redistribute the blood flow (in this case, to meet the increased cerebral demand) is to open up circulatory channels to the brain. This is accomplished by lowering the flow resistance in the cerebral arteries. The increased supply to the brain must not occur at the expense of the placenta, however. Because blood flow to the placenta cannot be directly regulated, another way to redistribute the flow (to the placenta in this case) is to close some peripheral channels by raising their flow resistance. This is illustrated by the significant rise of impedance that occurs in the lower extremity arteries. Both mechanisms—the opening of some channels and the closing of others—can be observed near term. The result is a redistribution of blood flow that favors the brain without diminishing the placental supply.

Changes in the aortic arch. These redistribution mechanisms lead to significant changes in the vascular connections between the parallel circulatory systems that are perfused by the left and right cardiac ventricles[5]. The "link" between the two systems is located in the aortic arch[34] and functions as a watershed (Fig. 20.**6**)[27]. Under normal circumstances, blood flows forward through the aortic arch, but near term it is not uncommon to see notches in the Doppler waveform or periods of reverse flow. These partial or complete shifts follow either an increased peripheral resistance in the vascular bed of the aorta or decreased resistances in the cerebral circulation—or both.

Interpretation of Doppler findings near term. Many changes take place within a few days as the pregnancy reaches term. Some findings can mimic serious pregnancy complications[2], making it more difficult to interpret the Doppler sonogram. Not all changes occur synchronously, despite isolated reports, and therefore the overall result is extremely difficult to predict in any given case. Our present knowledge appears to be too limited to draw simple conclusions from individual unexpected findings, especially with regard to changes in the aorta. Thus, before definitive conclusions are drawn from Doppler ultrasound findings, all available information should be gathered together and interpreted on the basis of broad physiological and pathophysiological knowledge. The major difference between the physiological changes that occur near term and abnormal placental changes is the continuation of normal fetoplacental blood flow at term in the normal pregnancy.

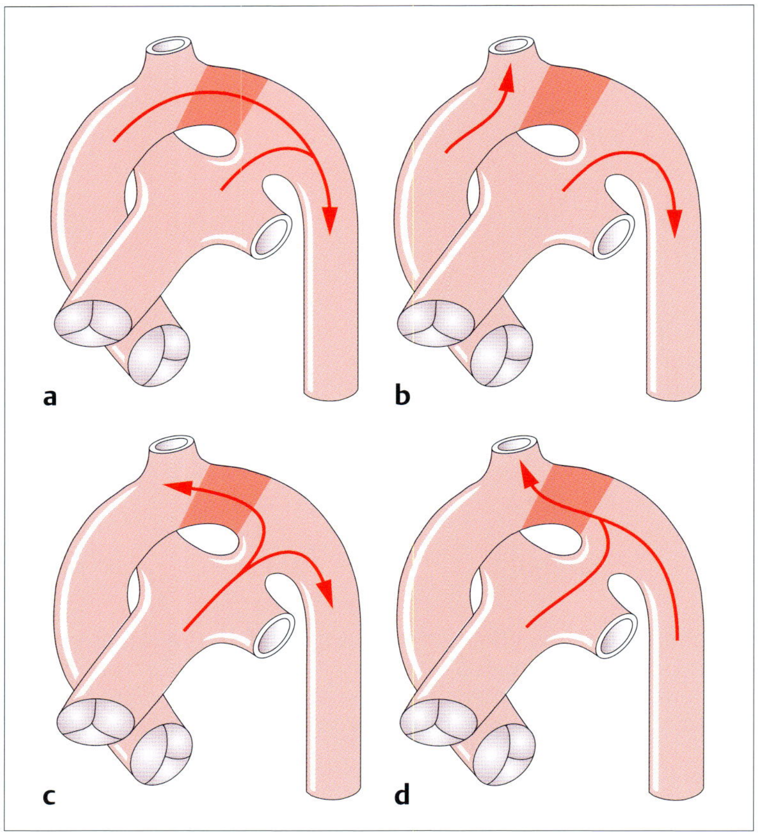

Fig. 20.**6** Diastolic flow patterns in the aortic isthmus associated with various degrees of resistance to placental blood flow.
a Normal resistance.
b Slight increase.
c Moderate increase.
d Severe increase.
(From Teyssier et al. 1993[34].)

Summary

A redistribution of blood flow can occur at the end of pregnancy to correct for the discrepancy that may arise between fetal demand and placental supply. Flow to the placenta is increased by an elevation of peripheral resistance in the lower half of the body, while flow to the brain is augmented by a fall of impedance in the cerebral vessels. The effects of these mechanisms can be seen either in the affected vascular territories or in the central vessel, the aorta, where they can be observed in all segments. Thus, the term effect in the fetal aorta is the net result of all the redistribution mechanisms that occur in the arteries of the extremities, kidneys, and brain.

References

1 Adamsons K, Myers RE: Circulation in the intervillous space; obstetrical considerations in fetal deprivation. In Grünwald P (ed.): The placenta and its maternal supply line. MT Press, Lancaster 1975, 158–177

2 Battaglia C, Artini PG, Ballestri M et al.: Hemodynamic, hematological and hemorheological evaluation of post-term pregnancy. Acta. Obstet. Gynecol. Scand. 74 (1995) 336–340

3 Benirschke K, Kaufmann P: Pathology of the human placenta. Springer, New York 1995

4 Bilardo CM, Campbell S, Nicolaides KH: Mean blood velocities and flow impedance in the fetal descending thoracic aorta and common carotid artery in normal pregnancy. Early Hum. Dev. 18 (1988) 2–3

5 Bonnin P, Fouron JC, Teyssier G, Sonesson SE, Skoll A: Quantitative assessment of circulatory changes in the fetal aortic isthmus during progressive increase of resistance to umbilical blood flow. Circulation 88 (1993) 216–222

6 Campbell S, Diaz-Recasens J, Griffin DR et al.: New Doppler technique for assessing uteroplacental blood flow. Lancet 32 (1983) 675–677

7 Eik-Nes SH, Marsál K, Brubakk AO, Kristofferson K, Ulstein M: Ultrasonic measurement of human fetal blood flow. J. biomed. Engng. 4 (1982) 28–36

8 Eik-Nes SH, Maršál K, Kristofferson K, Vernersson E: Noninvasive Messung des fetalen Blutstromes mittels Ultraschall. Ultraschall 2 (1981) 226–231

9 Eldridge MW, Berman W Jr, Greene ER: Serial echo-Doppler measurements for human fetal abdominal aortic blood flow. J. Ultrasound Med. 4 (1985) 453–458

10 Erskine RL, Ritchie JW: Umbilical artery blood flow characteristics in normal and growth-retarded fetuses. Brit. J. Obstet. Gynaecol. 92 (1985) 605–610

11 Fendel H, Fendel M, Warnking R: Fehlermöglichkeiten der gepulsten Dopplermethode zur Blutflußmessung am Feten. Z. Geburtsh. u. Perinat. 187 (1983) 83–87

12 Fouron JC, Zarelli M, Drblik SP, Lessard M: Flow velocity profile of the fetal aortic isthmus through normal gestation. Amer. J. Cardiol. 74 (1994) 483–486

13 Gill RW, Trudinger BJ, Garrett WJ, Kossoff G, Warren PS: Fetal umbilical venous flow measured in utero by pulsed Doppler and B-mode ultrasound. I. Normal pregnancies. Amer. J. Obstet. Gynecol. 139 (1981) 720–725

14 Gosling R, Dunbar G, King DH et al.: The quantitative analysis of occlusive peripheral arterial disease by a non-intrusive ultrasonic technique. Angiology 22 (1971) 52–55

15 Griffin D, Cohen Overbeek T, Campbell S: Fetal and utero-placental blood flow. Clin. Obstet. Gynaecol. 10 (1983) 565–602

16 Jouppila P, Kirkinen P: Blood velocity waveforms of the fetal aorta in normal and hypertensive pregnancies. Obstet. Gynecol. 67 (1986) 856–860

17 Kaesemann H, Trenkel K: Kombinierte B-Bild-Doppler-Sonographie – Messungen am In-vitro-Modell und in fetalen Gefäßen. Z. Geburtsh. u. Perinat. 190 (1986) 43–48

18 Kirkinen P, Müller R, Huch R, Huch A: Blood flow velocity waveforms in human fetal intracranial arteries. Obstet. Gynecol. 70 (1987) 617–621

19 Lingman G, Gennser G, Maršál K: Ultrasonic measurements of the blood velocity and pulsatile diameter changes in the fetal descending aorta. In Rolfe P (ed.): Fetal physiological measurements. Vol. 56. Butterworths, London 1986, 206 –210

20 Lingman G, Maršál K: Fetal central blood circulation in the third trimester of normal pregnancy – A longitudinal study. I. Aortic and umbilical blood flow. Early Hum. Dev. 13 (1986) 137–150

21 Lingman G, Maršál K: Fetal central blood circulation in the third trimester of normal pregnancy – A longitudinal study. II. Aortic blood velocity waveform. Early Hum. Dev. 13 (1986) 151–159

22 Malcus P, Maršál K, Persson P-H: Fetal and uteroplacental blood flow in prolonged pregnancies. Ultrasound Obstet. Gynecol. 1 (1991) 40–45

23 Mari G: Arterial blood flow velocity waveforms of the pelvis and lower extremities in normal and growth-retarded fetuses. Amer. J. Obstet. Gynecol. 165 (1991) 143–151

24 Maršál K, Lindblad A, Lingman G, Eik-Nes SH: Blood flow in the fetal descending aorta; intrinsic factors affecting fetal blood flow, i.e. fetal breathing movements and cardiac arrhythmia. Ultrasound Med. Biol. 10 (1984) 339–348

25 Maršál K, Lingman G, Giles WB: Evaluation of the carotid, aortic and umbilical blood velocity waveforms in the human fetus. XIth Annual Conference of the Society for the Study of Fetal Physiology. Oxford 1984; C33

26 Pal A, Kurjak A: Examination of the blood flow in the umbilical artery by ultrasound. Zentralbl. Gynäkol. 108 (1986) 724–729

37 Patton DJ, Fouron JC: Cerebral arteriovenous malformation: Prenatal and postnatal central blood flow dynamics. Pediatr. Cardiol. 16 (1995) 141–144

28 Pourcelot L: Applications cliniques de l'éxamen Doppler transcutane. INSERM Colloq. 34 (1974) 213–240

29 Rightmire DA, Campbell S: Fetal and maternal Doppler blood flow parameters in postterm pregnancies. Obstet. Gynecol. 69 (1987) 891–894

30 Rudolph AM: Distribution and regulation of blood flow in the fetal and neonatal lamb. Circ. Res. 57 (1985) 811

31 Satoh S, Koyanagi T, Hara K, Shimokawa H, Nakano H: Developmental characteristics of blood flow in the middle cerebral artery in the human fetus in utero, assessed using the linear-array pulsed Doppler method. Early Hum. Dev. 17 (1988) 2–3

32 Stale H, Gennser G, Maršál K: Blood flow velocity and pulsatile diameter changes in the fetal descending aorta: A longitudinal study. Amer. J. Obstet. Gynecol. 163 (1990) 26–29

33 Stuart B, Drumm J, FitzGerald DE, Duignan NM: Fetal blood velocity waveforms in normal pregnancy. Brit. J. Obstet. Gynaecol. 87 (1980) 780–785

34 Teyssier G, Fouron JC, Maroto E, Sonesson S-E, Bonnin P: Blood flow velocity profile in the fetal aortic isthmus: A sensitive indicator of changes in systemic peripheral resistances. I. Experimental studies. J. Mat. Fetal Invest. 3 (1993) 213 –218

35 Thaler I, Manor D, Itskovitz J et al.: Changes in uterine blood flow during human pregnancy. Amer. J. Obstet. Gynecol. 162 (1990) 121–125

36 Tonge HM, Stewart PA, Wladimiroff JW: Fetal blood flow measurements during fetal cardiac arrhythmia. Early Hum. Dev. 10 (1984) 1–2

37 Van Lierde M, Oberweis D, Thomas K: Ultrasonic measurement of aortic and umbilical blood flow in the human fetus. Obstet. Gynecol. 63 (1984) 801–805

38 Vetter K: Dopplersonographie in der Schwangerschaft. VCH Verlagsgesellschaft, Weinheim 1991

39 Vetter K, Favre Y, Suter T, Huch R, Huch A: Doppler-sonographisch ermittelte spezifisch hämodynamische Veränderungen im Kreislauf von Feten in den letzten 4 Wochen vor Geburt. Z. Geburtsh. u. Perinat. 193 (1989) 215–218

40 Vogel M: Atlas der morphologischen Plazentadiagnostik. Springer, Berlin 1992

41 Vyas S, Nicolaides KH, Campbell S: Renal artery flow-velocity waveforms in normal and hypoxemic fetuses. Amer. J. Obstet. Gynecol. 161 (1989) 168–172

42 Woo JSK, Liang ST, Lo RLS, Chan FY: Middle cerebral artery Doppler flow velocity waveforms. Obstet. Gynecol. 70 (1987) 613–616

21 Intrapartum Fetal Heart Rate Changes and Doppler Sonography

E. Weiss

The surveillance of high-risk pregnancies with Doppler ultrasound is concerned with chronic or at most subacute changes of blood flow in the uterofetoplacental system. Fetal and uterine waveforms are recorded when the fetus is at rest and the uterus is not contracting.

A very different situation exists when Doppler waveforms are recorded during labor, especially when scanning is done in cases with intrapartum fetal heart rate changes. Here it is important to consider the entirely different conditions of uterine blood flow that exist during and between uterine contractions. In the umbilical cord and intrafetal vessels, direct mechanical effects due to vascular compression must be distinguished from other causes of acute impedance changes.

This chapter explores the diagnostic capabilities of intrapartum Doppler sonography based on personal results with uterine waveforms recorded during labor and with umbilical waveforms recorded during decelerations in the fetal heart rate (FHR) at delivery.

Pathophysiology and Technical Problems

Causes of FHR decelerations. Uterine contractions lead to an acute reduction of uterine blood flow, causing a decreased maternal oxygen supply in the intervillous space[3]. Similarly, an acute interruption of fetoplacental blood flow can occur as a result of umbilical cord compression during labor, especially if the membranes have ruptured[26]. Compression of the fetal head causes a rise of intracranial pressure, which in turn alters the blood flow patterns of intracranial arteries [48]. The FHR rarely shows accelerations in response to these acute intrapartum events. It is more common to observe decelerations, which cannot always be referred to a specific pathophysiological cause[28]. Even when a causal diagnosis is attempted, its prognostic significance is disputed[15,27]. A slowing of the FHR, with a corresponding prolongation of the fetal cardiac cycle, leads basically to a lengthening of diastole with a consequent decrease in the maximum end-diastolic frequency shift in the Doppler waveforms[53].

Intrapartum Doppler velocimetry is noninvasive and is well tolerated by patients, making it a useful study for investigating the pathophysiological mechanisms that underlie FHR alterations. It may also allow for a more accurate prognostic assessment in some cases and can help to determine whether acute fetal asphyxia, with its dramatic effects on fetal hemodynamics[21,23,29,30], leads to altered blood flow patterns in the fetal and/or maternal vessels.

Technical problems. To date, only a few authors have performed intrapartum Doppler velocimetry examinations even though such studies are of extraordinary help in understanding pathophysiological mechanisms during delivery. This is most likely due to the formidable technical problems that arise when Doppler velocimetry is attempted during labor. The most significant problems and obstacles are listed below.

- Uterine contractions during the dilation stage lead to changes in maternal circulatory parameters.
- Maternal respiratory excursions, especially at the height of contractions, hamper the continuous recording of Doppler signals.
- The amniotic fluid volume is usually reduced at term or may be absent due to amniotomy, greatly increasing the difficulty of examining the fetal vessels with Doppler ultrasound.
- Uterine contractions alter the position of the transducer, the maternal abdominal wall, and the fetus, often resulting in loss of signal.
- The deeply engaged fetal head makes it difficult to record Doppler waveforms from intracranial vessels with an abdominal transducer.
- At present, on-line analysis of Doppler waveforms correlated with the FHR trace is practically impossible or requires very costly, nonroutine protocols.

Changes in Uterine Artery Waveforms during Labor

Method of measurement. When the ascending branch of the uterine artery has been identified with color Doppler (Fig. 21.**1**), the frequency spectra are continuously recorded with pulsed Doppler during spontaneous or induced uterine contractions (Fig. 21.**2**). The reduction of uterine perfusion achieved during a contraction can be quantified by measuring the area under the uterine artery waveform. When the angle between the vessel and Doppler beam is kept constant, the area under the waveform is proportional to uterine perfusion, assuming that the low flow velocities contributing to the spectrum are uniformly distributed and can thus be disregarded when determining a relative change in perfusion. If the area integral between contractions is taken as the initial value, then the relative reduction in blood flow can be stated for each individual contraction. An additional angle-independent measure of uterine impedance is the resiatnce index (RI) of the scanned uterine artery, which is easily determined.

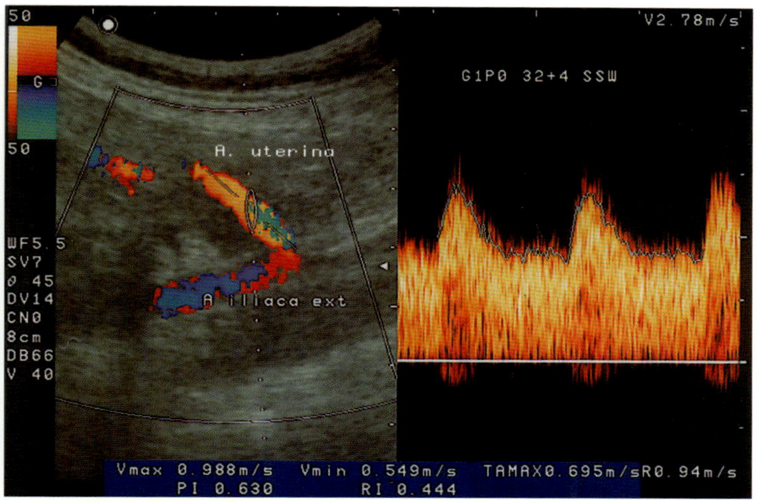

Fig. 21.**1** Color-flow image of the ascending branch of the right uterine artery at its apparent "crossover" with the iliac artery.

Our Results

Reduction in uterine blood flow during contractions. A total of 68 uterine contractions in 16 pregnancies were studied during oxytocin-induced contractions. Within limits, the results of these "Doppler-controlled" contraction stress tests are also applicable to the intrapartum situation. The area under the induced uterine waveforms showed a considerable variability of flow reduction during contractions, with values ranging from 14% to 85% of the initial value (median 54 ± 20 %), despite comparable tocodynamometry (TKD) traces. The example in Fig. 21.**3** shows inadequate quantification by external TKD. This is particularly evident with a positive contraction stress test, where the continuous uterine arterial waveform indicates the degree of uterine flow reduction during the contraction and also explains the decrease in hypoxic FHR decelerations occurring at contractions 2, 3, and 4. Contraction 5, which is especially prominent in the TKD trace, appears to have the least impact on uterine perfusion and does not induce a late deceleration in this example. We cannot offer a satisfactory explanation for the absence of FHR decelerations at contraction 1 despite synchronous uterine velocimetry.

Reduction in uterine blood flow during premature labor. An unphysiological reduction in uterine blood flow may occur during labor as well as during prenatal contractions. The following example shows a case in which premature contractions occurred at exactly 31 weeks' gestation (Fig. 21.**4**). The uterine contractions were recorded by external TKD. The patient did not perceive the contractions as painful (hemodynamically stable, semilateral position), but the uterine artery waveforms showed a massive, bilateral decrease in blood flow velocities with reverse diastolic flow that practically canceled out the forward flow. As a result of this, an abnormally high reduction of blood flow in the uterine bed occurred during uterine contractions, evoking a hypoxic response in the FHR. After tocolytic therapy, the eutrophic fetus was delivered at 32 weeks 4

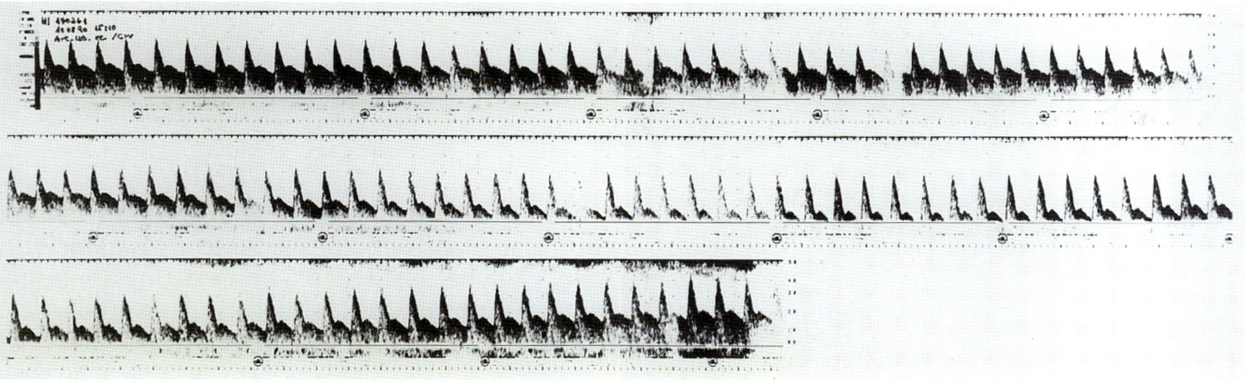

Fig. 21.**2** Blood flow pattern of the ascending branch of the uterine artery with oxytocin-induced contractions at 37 weeks.

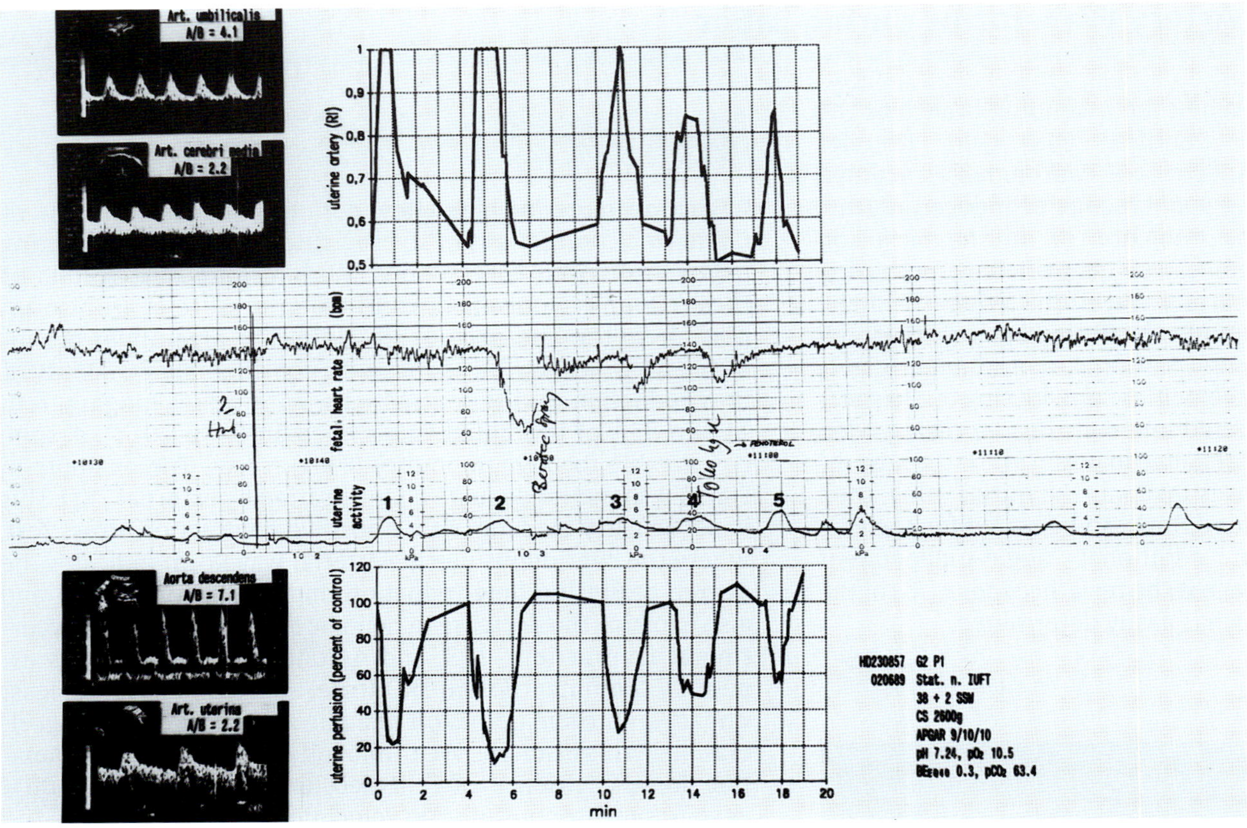

Fig. 21.3 Synchronous displays of the uterine artery RI, the FHR in beats/min, uterine contractions recorded by external tocodynamometry, and relative uterine blood flow in the examined vessel, given as a percentage of the area integral under the waveform in relation to the value measured between contractions during the oxytocin stress test. Gestational age 38 weeks + 2 days, possible placental insufficiency.

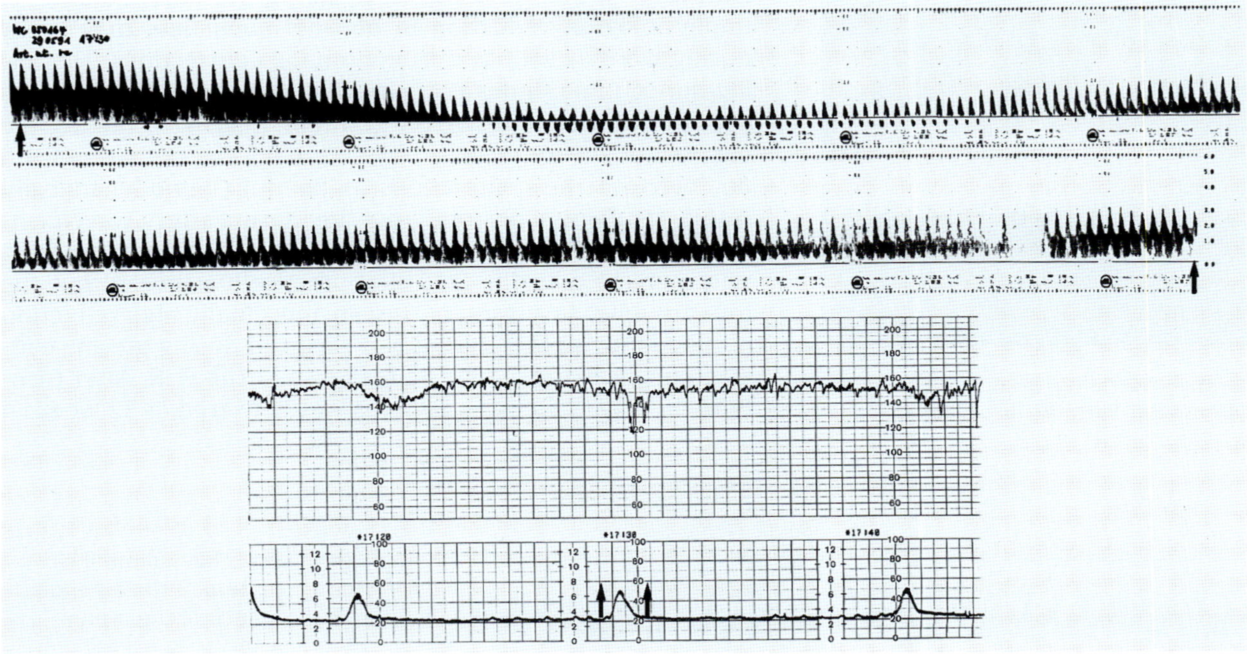

Fig. 21.4 Doppler waveform recorded continuously from the ascending branch of the uterine artery during a uterine contraction in premature labor (31 weeks).

days by cesarean section due to a breech presentation with increasing cervical dilation. The placenta showed no histomorphological abnormalities (H. Müntefering, personal communication), and the acid–base balance in the umbilical cord blood was normal.

Physiological and unphysiological reduction in uterine blood flow. The typical uterine waveform recorded continuously during a contraction under physiological conditions is compared in Fig. 21.**5** with an unphysiological response recorded during premature labor and with the response in a patient with pregnancy-induced hypertension and a preexisting abnormal uterine blood flow pattern. We see that in the physiological case, the maximum systolic and diastolic flow velocities are initially reduced to an equal degree until the height of the contraction, at which point we observe a massive shift in the RI (Fig. 21.**5a**). With an unphysiological flow reduction during premature labor, the strong decrease in systolic flow velocities

Fig. 21.**5** Uterine artery waveforms.
a Physiological response of continuously recorded uterine waveforms to an oxytocin-induced contraction.
b Unphysiological response of uterine waveforms in premature labor.
c Response of uterine waveforms in a patient with pregnancy-induced hypertension and a preexisting early diastolic notch.

is accompanied by holosystolic reverse flow (Fig. 21.**5b**). An entirely different pattern is seen when there is preexisting pregnancy-induced hypertension and an early diastolic notch in the uterine artery waveform (Fig. 21.**5c**). While the uterine contraction has almost no effect on maximum systolic velocities, end-diastolic reverse flow occurs at the height of the contraction, and the waveform pattern is similar to that of the external iliac artery. The relative reduction in blood flow, calculated as the area under the waveform, is approximately 50–60% of the initial value in the physiological case and in pregnancy-induced hypertension. The initial level is assumed to be different in both cases, however.

Discussion of Uterine Doppler Changes during Labor

Marked reduction of diastolic flow. The first Doppler measurement of blood flow in the arcuate arteries was described in 1983[6]. The first uterine artery waveforms recorded during labor[11, 13] showed a reduction in the mean blood flow velocity to approximately 40% of the initial value for a maximum intrauterine pressure of approximately 60 mmHg. The maximum systolic flow velocity decreased by only 25%, while the detectable diastolic flow velocities were absent or very low[13]. This can be explained by the markedly reduced perfusion pressure of uterine blood flow at the height of the contraction. The perfusion pressure during systole is approximately 60–70 mmHg when the intrauterine pressure is 50–60 mmHg (systolic perfusion pressure = maternal systolic blood pressure minus the intrauterine pressure minus the maternal central venous pressure). In diastole, a perfusion pressure is no longer present when the intrauterine pressure exceeds the maternal arterial diastolic blood pressure (disregarding the maternal central venous pressure of 3–8 mmHg). Fendel et al.[12] showed a definite correlation between the diastolic reduction of Doppler flow and the strength of the uterine contraction, with diastolic flow dropping to zero when the intrauterine pressure is 80 mmHg or higher. Systolic flow can still be demonstrated up to an intrauterine pressure of 130 mmHg[12].

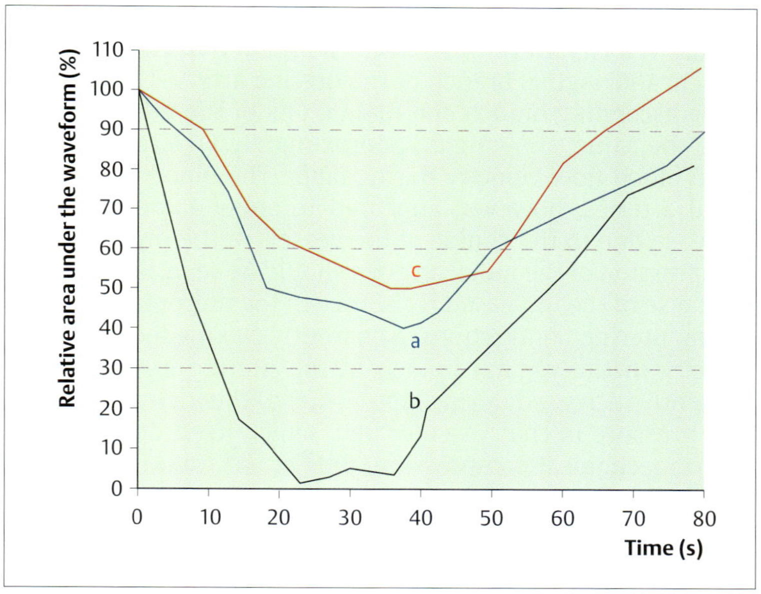

Fig. 21.**5 d** Relative reduction in blood flow determined for cases **a–c** (calculated as the relative area under the uterine waveforms).

Table 21.1 Doppler measurements of uteroplacental perfusion during labor and in the dilation stage

Authors	*n*	UP	RM	COD	Flow parameters	Vessel	Velocimetry
Brar et al. (1988)[4]	27	Int	+	?	S/D ratio	Uterine artery	Up to 60 torr, inverse correlation between S/D ratio and intrauterine pressure
Janbu et al. (1985)[20]	19	Ext	±	?	TASAV	Radial arteries	40–100% reduction
						Uterine artery	37% reduction
						Ascending uterine artery	60% reduction
						Vaginal artery	100% increase
Fendel et al. (1984)[11]	10	Int	+	?	TASAV	Uterine artery	58% reduction
Fendel et al. (1987)[12]	7	Int	+	?	TASAV	Uterine artery	43% reduction
Fendel et al. (1989)[13]	?	Int	+	?	TASAV	Uterine artery	Absent diastolic flow at 100 mmHg or higher
Fleischer et al. (1987)[16]	12	Int	+	?	S/D ratio	Uterine artery	Absent diastolic flow at 36 mmHg or higher
Our results (CST)	16	Ext	–	1–2	TAMV	Uterine artery	Reduction to 54% ± 20% of initial value (68 contractions, range 14–85%)

UP = uterine pressure; RM = rupture of membranes; COD = cervical os diameter,Int = internal pressure measurement; Ext = external pressure measurement; TASAV = time-averaged spatial average velocity over one cardiac cycle; TAMV = time-averaged maximum velocity over one cardiac cycle; CST = contraction stress test.

Compression of the radial arteries. The Doppler studies of uterine blood flow during labor that have been published to date are summarized in Table 21.**1**. The data confirm the findings reported by Fendel et al.[11]. Janbu et al.[20] recorded transvaginal velocity waveforms from the main branch, ascending branch, and vaginal branch of the uterine artery. The patterns in the ascending branch and main trunk of the uterine artery were the same, showing an approximately 60% reduction of mean blood flow velocity during labor. The blood flow in the radial arteries, however, appeared to show a much greater decrease due to the course of the vessels in the myometrium. Compression of the radial arteries should be seen as the essential cause of the increased resistance in the uteroplacental bed during uterine contractions. Apparently the mean blood flow velocity in the vaginal branch of the uterine artery is subsequently increased during labor[20] due to the increased perfusion pressure in that vascular bed, which itself does not undergo mechanical compression. Brar et al.[4] found an inverse correlation between intrauterine pressure during labor and the ratio of the maximum systolic and end-diastolic velocities (S/D ratio) in the examined arteries, with absent diastolic flow occurring when the intrauterine pressure reaches 60 mmHg.

Omitting the contraction stress test. Our own studies were performed during induced uterine contractions. We found that the reduction in mean uterine blood flow velocity corresponded to the reductions measured during spontaneous labor. Our studies confirmed the principle of a transient reduction in uterine blood flow, and thus of the oxygen supply in the intervillous space, that underlies the contraction stress test. It is also clear, however, that the inability to quantify the findings by external TKD is responsible for the poor sensitivity and the high percentage of false-positive findings. Moreover, the improvement of FHR monitoring by the assessment of fetal movements as well as intrafetal arterial and venous Doppler sampling can provide an accurate picture of fetal and placental condition, suggesting that the contraction stress test is unnecessary in cases with a suspicious FHR and could actually increase the risk of fetal deterioration in cases with abnormal hemodynamic findings by Doppler ultrasound. In postterm pregnancies as well, we have abandoned the oxytocin stress test at our center due to its poor sensitivity and specificity in favor of a differentiated induction strategy with prostaglandins.

Indication for tocolysis. Uterine artery velocimetry during normal or premature labor can detect an unphysiological impedance increase in cases with suspected uterine hyperactivity, providing an indication for tocolysis. Systolic blood flow is present up to an intrauterine pressure of approximately 130 mmHg under physiological conditions[12] and guarantees a certain minimum blood flow to the intervillous space.

Findings in pregnancy-induced hypertension. The results in patients with pregnancy-induced hypertension (PIH) and a preexisting early diastolic notch in the uterine artery waveform represent isolated cases that are certainly not representative. They demonstrate, however, that diastolic flow disappears in these cases when there is an increase in uterine tone, and that a reflected resonance wave appears in early diastole due to the high peripheral resistance. The blood supply to the intervillous space during contractions is partially maintained almost entirely by the unchanged systolic flow velocities, apparently as a result of the raised perfusion pressure due to maternal hypertension. These findings and considerations should caution against the indiscriminate use of antihypertensive medical therapy in patients with PIH. In cases with a preexisting abnormal uterine waveform, Doppler scanning shows a greater rise of impedance during labor compared with the physiological situation, indicating a greater danger to these fetuses. Olofsson et al.[40] found a significantly greater rise of uterine artery flow resistance in cases with a positive oxytocin challenge test than in OCT-negative cases. The authors suggest that continued susceptibility to vasopressor stimuli in the uterine resistance vessels based on impaired trophoblastic invasion may be the cause.

Intrapartum Waveform Changes in Umbilical and Intrafetal Vessels

Even more interesting than the effect of labor on uterine artery waveforms is its effect on umbilical and intrafetal waveforms, owing to the potential interactions between mechanical and hypoxic factors. Some of the relationships and interactions are so complex that they allow for different interpretations of the observed phenomena. The material is organized below by subheads in order to make the diverse flow changes easier to understand.

Umbilical Cord Doppler during Labor

No change in blood flow velocities during the dilation stage of labor. When the fetal membranes are intact and an adequate amniotic fluid volume is present, the uterus can be regarded as a hollow sphere[30]. In this model, proposed by Künzel, the pressure in both the umbilical vein and the umbilical arteries rises in response to uterine contractions. As a result, there is no change of perfusion pressure in the placental vascular tree. To a degree, this is also true after rupture of the membranes has occurred, since the fetal head seals off most of the cervical orifice during labor. At the same time, direct mechanical compression of the umbilical cord is much more frequent after the membranes have ruptured[30]. Stuart et al. reported on Doppler velocimetry of the umbilical arteries during spontaneous labor in 1981[44]. When the FHR was normal, it was found that the velocity waveforms of the umbilical arteries were not altered by uterine contractions during the dilation stage, nor were they affected by artificial rupture of the membranes or the infusion of oxytocin. Several other authors confirmed these results in subsequent years (Table 21.**2**). Figure 21.**6** illustrates this with a case from our files. The slight changes in end-diastolic flow velocities are caused entirely by fluctuations of the FHR and the associated change in the length of diastole.

Heart-rate effect. Only a few authors to date have described Doppler velocimetry of the umbilical arteries during intrapartum decelerations of the FHR in human patients. Fairlie[9] found that unspecified decelerations in the FHR during uterine contractions were associated with a marked decrease in end-diastolic blood flow velocities. Kirkinen[24] described a case with variable decelerations in the FHR and absent end-diastolic flow during the decelerations. The flow velocity waveforms were almost unchanged, however, suggesting that a heart-rate effect may have been involved. This effect is caused by the lengthening of diastole that occurs when the FHR is slowed. It results in a protracted fall of diastolic flow velocities to a lower end-diastolic value, with a corresponding change in Doppler indices. Various authors have described the relationship between end-diastolic flow velocity and the fetal heart rate[35,36,49].

Table 21.**2** Doppler measurements of fetoplacental perfusion during labor

Authors	*n*	Umbilical artery waveforms
Stuart et al. (1981)[44]	10	No change in waveforms during contraction No effect from amniotomy or oxytocin
Fendel et al. (1987)[12]	15/7	No change in waveforms during contraction
Fleischer et al. (1987)[16]	12	No change in waveforms during contraction
Brar et al. (1988)[4]	27	No change in waveforms during contraction
Kirkinen et al. (1988)[24]	?	Absent diastolic flow during decelerations
Fairlie et al. (1989)[9]	43	Heart-rate effect during decelerations with no significant waveform change
Weiss et al. (1989, 1991)[50,51]	8	Variable decelerations in FHR Reverse diastolic flow in 6 fetuses, heart-rate effect in 2 fetuses
Cruz et al. (1988)[7]	71	Rupture of fetal membranes had no effect on S/D ratio (measured between contractions)
Sarno et al. (1989)[43]	109	No relationship between intrapartum asphyxia and S/D ratio (measured between contractions)
Feinkind et al. (1989)[10]	273	No relationship between P_{O_2}, P_{CO_2}, O_2 saturation, and velocity waveforms (measured between contractions)

FHR = fetal heart rate.

Our Results

Simultaneous recording of Doppler waveforms and the FHR. A total of 130 decelerations of the FHR were studied in eight term fetuses with ruptured membranes and variable decelerations by simultaneous Doppler measurement of umbilical artery flow and internal FHR monitoring. The pregnant women were between 38 weeks + 1 day and 40 weeks + 6 days' gestation, and the fetuses were eutrophic. Typical, repetitive variable FHR decelerations were recorded in all cases. Fetal-scalp blood gas analysis, done before Doppler velocimetry, was normal in seven cases and preacidotic in one case. The umbilical artery RI between contractions was within the normal range of our physiological reference curves in all the fetuses.

Technical problems. The technical problems in Doppler velocimetry were related to an absence of amniotic fluid and increased maternal respiratory excursions during labor, which often caused the selected fetal umbilical artery to move out of

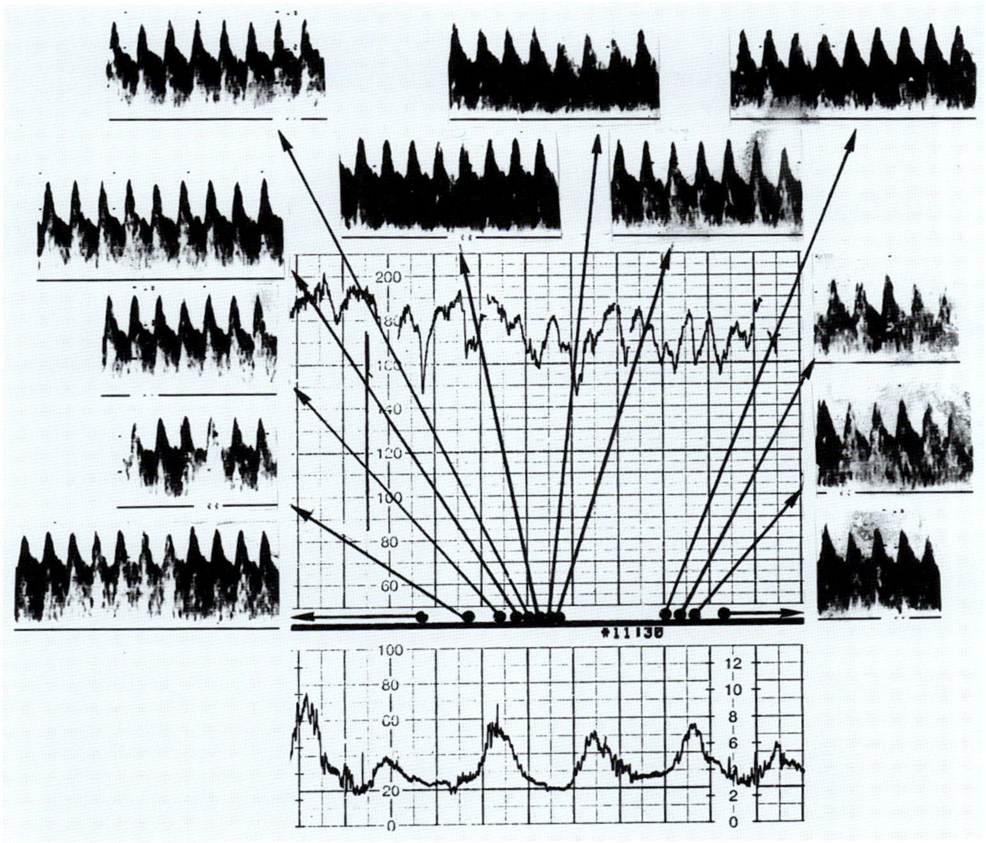

Fig. 21.**6** Flow velocity waveforms recorded from an umbilical artery during labor (cervical dilation 5 cm), with mild fetal tachycardia.

the Doppler sample volume. In 45 contractions the signals were lost for more than 10 seconds per contraction, and in 50 contractions they were lost for 5–10 seconds per contraction. Thirty-five contractions were recorded with a signal loss of less than 5 seconds, and 22 with no signal loss at all. At least two recordings with losses of less than 5 seconds could be obtained for each case. The Doppler waveforms were recorded on videotape and analyzed later.

Analysis. The video recordings were digitized, and the area integral under the Doppler envelope curve was determined with special PC analytical software. For exact synchronization of the flow patterns and FHR traces, the duration of each cardiac cycle was measured in the flow spectra to calculate the beat-to-beat heart rate. The FHR curve was then plotted with a graphic program and expanded according to the paper feed rate of the Doppler traces (4 cm/s). In this way the FHR tracing could be synchronized with the waveforms.

Reverse diastolic flow. Although typical umbilical cord decelerations were present in the FHR traces based on the criteria of Fischer[15], two distinctly different waveform changes were observed in the umbilical arteries. The waveforms in two fetuses showed only a heart-rate effect consisting of a prolonged decline of diastolic flow velocities in bradycardia (Fig. 21.**7a**). The slope of the diastolic velocity decline was not significantly altered, however, so no change was found in the impedance parameter (RI) when the measurement was adjusted for heart rate. By contrast, the umbilical artery waveforms in six fetuses changed dramatically during the decelerations, showing a reversal of diastolic flow (Fig. 21.**7b**). This finding cannot be explained by the decrease in FHR or by changes in the beam–vessel angle. The reverse diastolic flow pattern in these cases correlates with an acute rise of impedance in the fetoplacental circulation caused by the occlusion of umbilical blood flow. When Abitbol et al.[1] and Fouron et al.[17] experimentally occluded the umbilical veins in fetal sheep, they observed Doppler patterns of reverse diastolic flow in the umbilical arteries like those we observed during human labor[50,51].

Delayed appearance of deceleration. We related the timing of the drop in the FHR and the appearance of absent end-diastolic flow in the umbilical arteries using the complicated method outlined above (detailed description in reference 50). The synchronous display in Fig. 21.**8** shows that when end-diastolic flow disappears, the flow changes can be detected approximately 6–8 seconds prior to the FHR deceleration. This finding is consistent with the effects of umbilical vein occlusion in experimental animals[26]. This time delay was evident in all cases with absent end-diastolic umbilical flow, although the range of variation was from 4 to 10 seconds. Figure 21.**8** also shows that the FHR deceleration may still be present at a point where almost normal diastolic flow velocities reappear in the umbilical arteries.

Relative perfusion of the fetal placenta with reverse diastolic flow. To obtain a measure of the actual flow reduction in cases with absent end-diastolic umbilical flow and umbilical cord decelerations, we determined the area under the umbilical artery waveforms for each individual cardiac cycle and compared it with the average area in the interval between contractions (mean value of 10 waveforms before the next contrac-

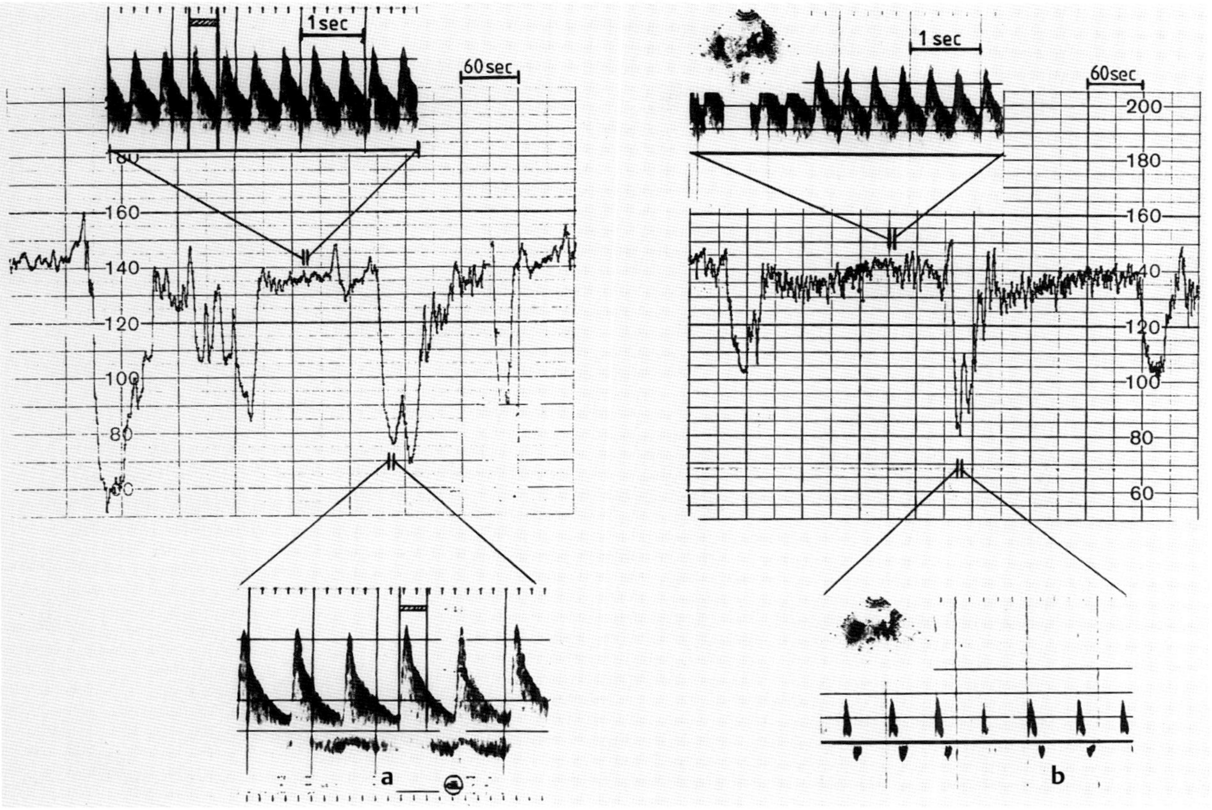

Fig. 21.**7** Umbilical artery velocity waveforms recorded before and during variable decelerations in the FHR. **a** Heart-rate effect associated with fetal bradycardia. Measurement of the end-diastolic flow velocity after an interval corresponding to the cardiac cycle

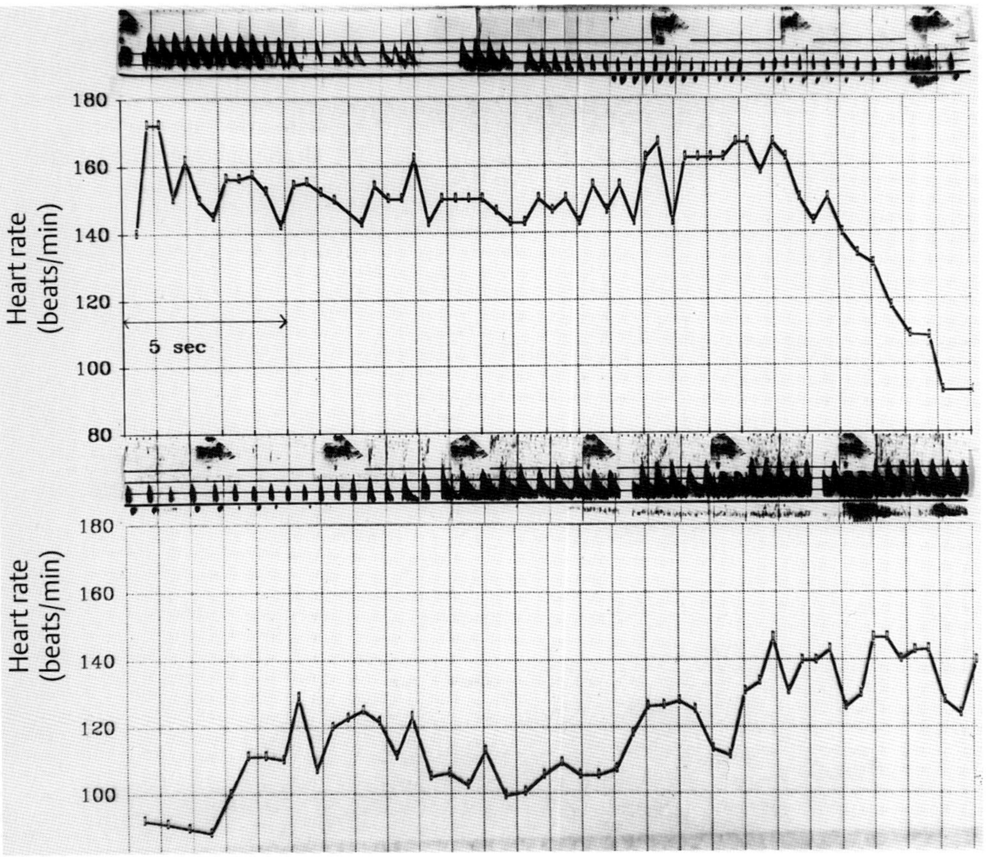

Fig. 21.**8** Synchronous display of continuously recorded Doppler waveforms and FHR traces during the appearance of absent end-diastolic flow.

tion). This procedure assumes a constant beam–vessel angle as well as a uniform distribution of all flow velocities over the vessel cross section (ideal laminar flow with a parabolic profile) in determining the area integral under the waveform. This type of flow is essentially present in fetal vessels when a normal hematocrit is assumed [47]. Other authors have described a close correlation with actual volume-flow changes in area-under-the-waveform determinations[42]. The possibility of a change in beam–vessel angle cannot be completely ruled out, but it cannot account for the occurrence of reverse diastolic flow. Also, repeat measurements taken at different umbilical cord sites in the same fetus showed almost identical flow changes, and so this error is considered slight. When the relative perfusion of the fetal placenta is calculated by determining the relative flow velocity (Fig. 21.**9**), we find an almost complete cessation of fetoplacental perfusion during the deceleration with absent end-diastolic flow in the umbilical artery.

Relative perfusion of the fetal placenta with a heart-rate effect. When the same calculation is performed during a deceleration and shows only a heart-rate effect on the umbilical artery waveforms (Fig. 21.**10**), we observe a decrease in the relative flow per minute, which paralleled the deceleration in the FHR (Fig. 21.**11**). But the fall in FHR from approximately 135 to 70 bpm causes only a 30% decrease in relative blood flow, because an increase in the area under the waveform due to the prolonged cardiac cycle in bradycardia represents an increased stroke volume. It may be assumed, then, that in cases that show a heart-rate effect only, the fetal blood volume that is

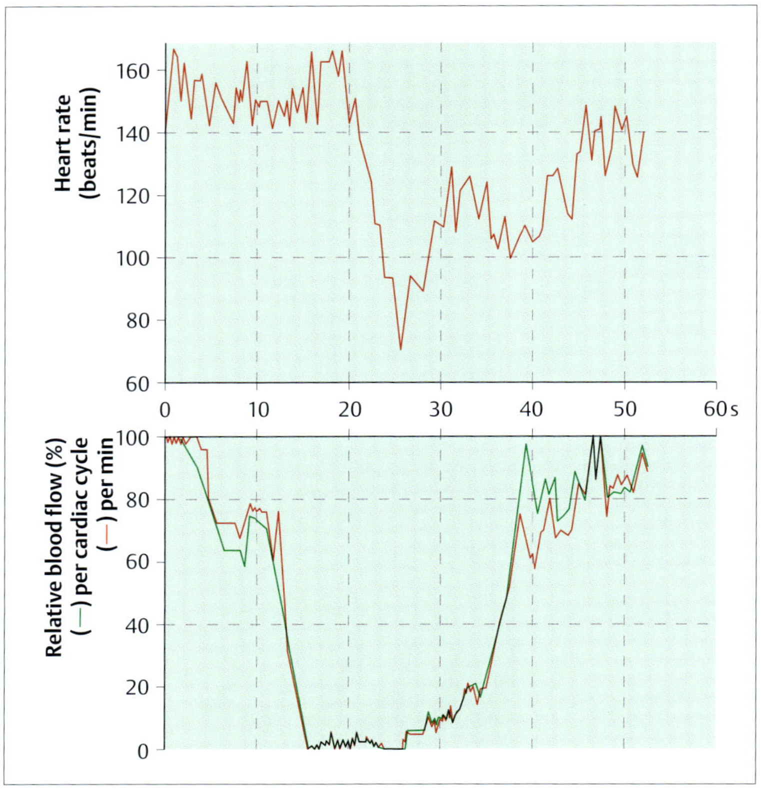

Fig. 21.**9** FHR trace (top) and relative blood flow (bottom) during FHR deceleration with reverse diastolic flow in the umbilical artery. The area integral under the waveform in relation to the mean area integral between contractions was used as a measure of relative blood flow.

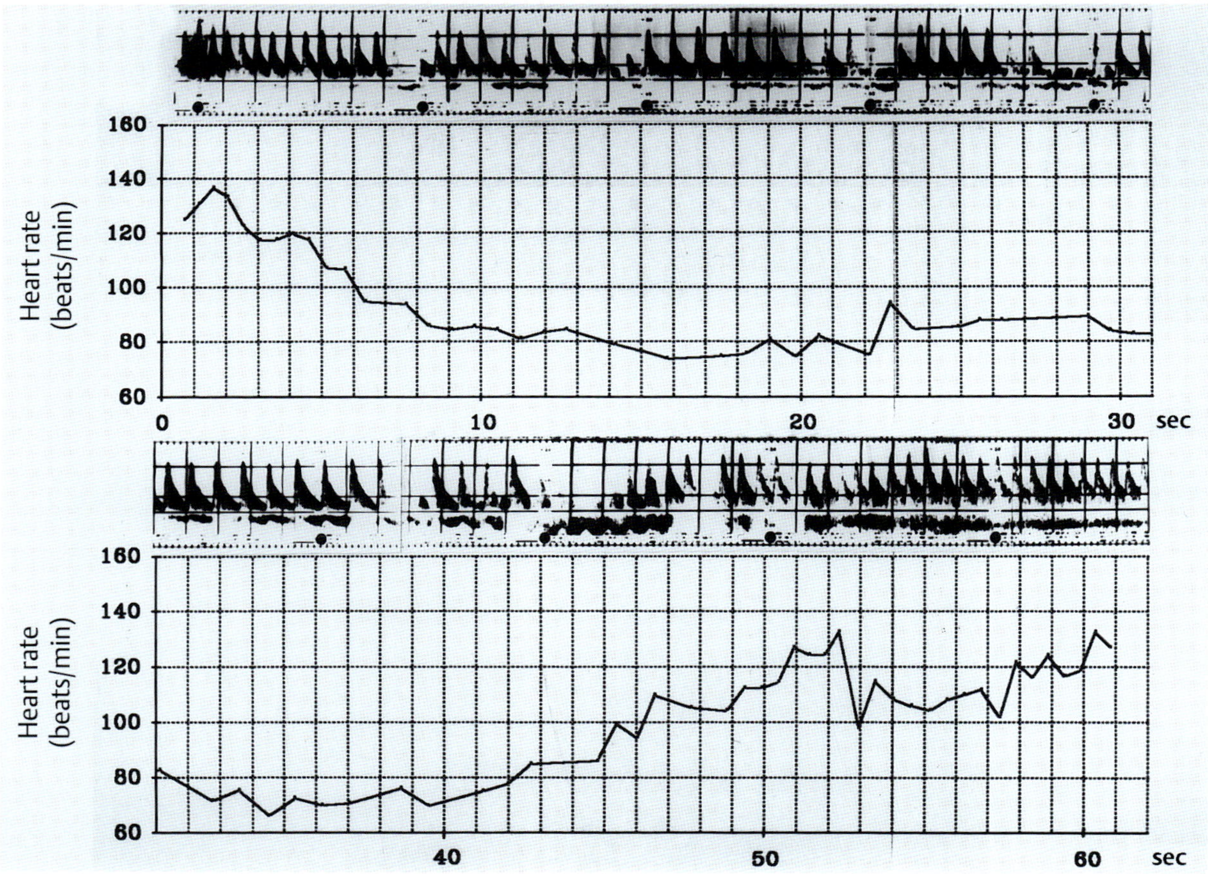

Fig. 21.**10** Synchronous display of continuously recorded Doppler waveforms and FHR traces, with a heart-rate effect on the fetal umbilical artery waveform.

21

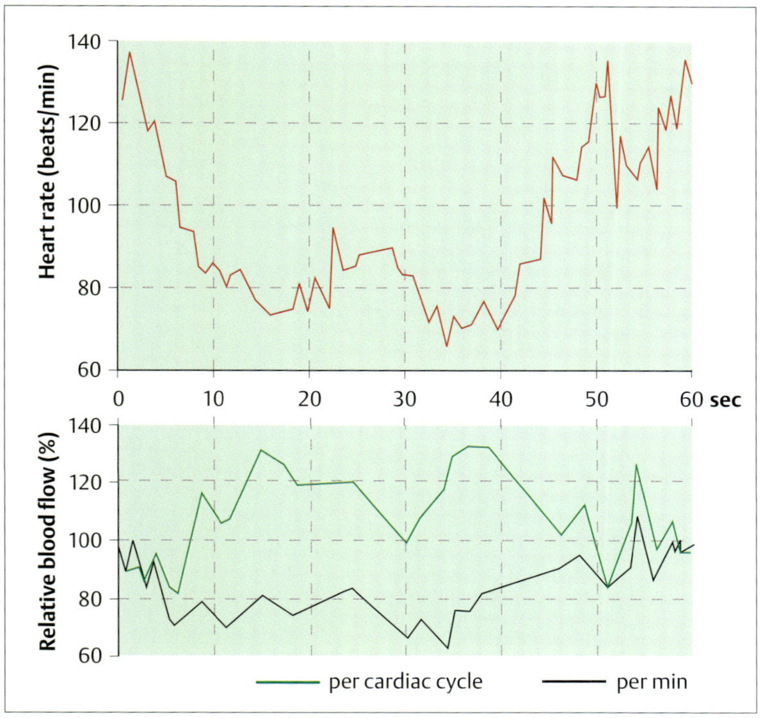

Fig. 21.**11** FHR trace (top) and relative perfusion of the scanned umbilical artery (bottom) during the deceleration in Fig. 21.**10**, with a heart-rate effect on the velocity waveforms. Calculated as in Fig. 21.**9**.

delivered into the placenta per heartbeat is increased and partially compensates for the loss of perfusion due to bradycardia.

Different reverse flow patterns. In cases with acute reverse diastolic flow during the deceleration, we were able to identify different types of reverse flow pattern (Fig. 21.**12**). The early diastolic form with brief reverse flow, occurring in some cases during the previous flow phase, can be interpreted in terms of the double-expansion-chamber ("windkessel") model of Moll[37] as resonant flow in the presence of an extremely high resistance located just past the sampling site and may well reflect an occlusion of the umbilical artery. According to this model, holodiastolic reverse flow that tends to increase at the end of diastole would be interpreted as compensatory flow from the peripheral expansion chamber (i.e., the placenta) in the presence of an umbilical vein occlusion. The question whether an occlusion of the umbilical vein alone or of both the umbilical vein and artery underlies the cases with acute diastolic reverse flow is of minor importance in terms of placental perfusion. In both cases the result is a cessation of fetoplacental perfusion, as there is only an ineffectual to-and-fro movement of the fetal blood column in the umbilical arteries.

Clinical evaluation of variable decelerations. Since animal experiments have shown that the oxygen supply to the fetus remains almost constant when umbilical blood flow is reduced to approximately 50% of normal but that the fetal oxygen supply declines exponentially when umbilical flow is reduced below 50%, intrapartum Doppler velocimetry is a valuable tool in the clinical evaluation of variable decelerations in the FHR,

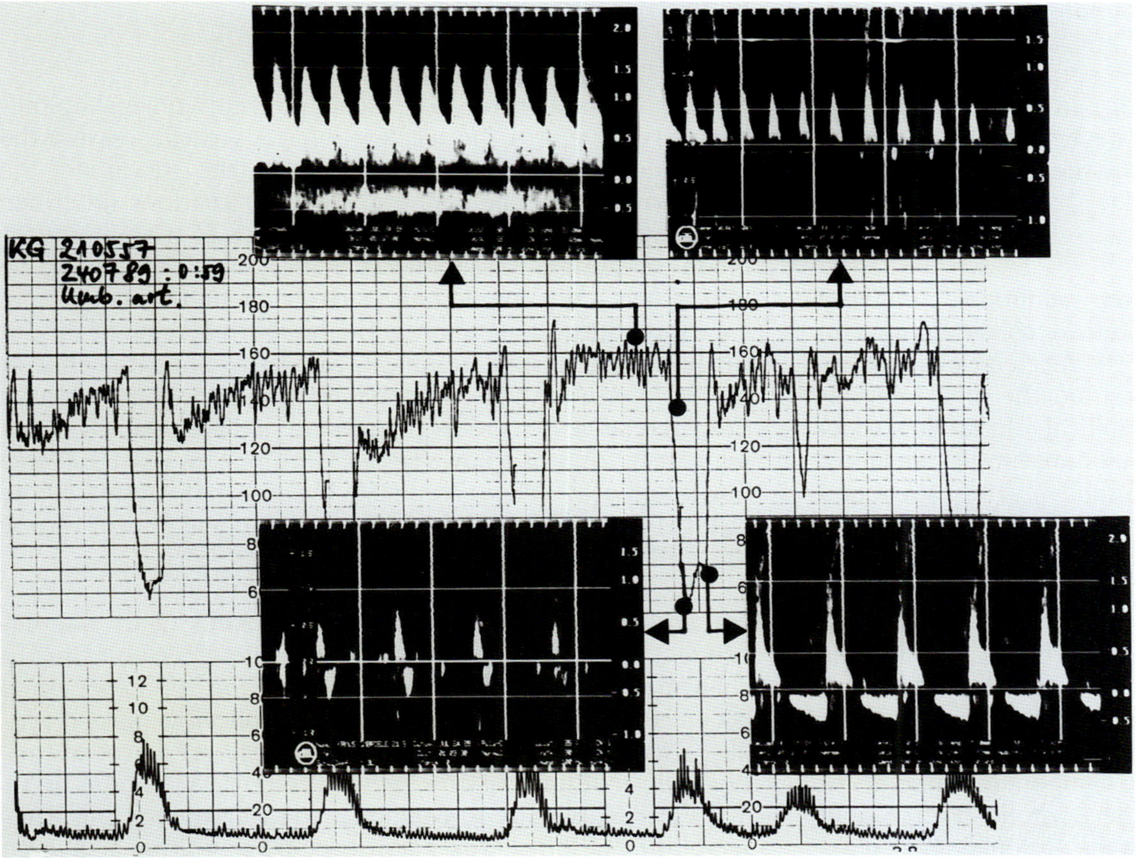

Fig. 21.**12** Early diastolic resonant flow and holodiastolic reverse flow during a variable deceleration in the FHR.

which are associated with the appearance of reverse diastolic flow with a significant transient reduction in fetal oxygen supply. Time and again in recent years we have benefited from the technical ability to make a gross assessment of fetal oxygen supply in the delivery room.

Effect of Intrapartum FHR Decelerations on Quantitative Parameters of Umbilical Blood Flow

Fetal bradycardia. Within the physiological range of FHR (120–160 bpm), the effects of heart rate on umbilical artery end-diastolic velocity waveforms with a normal impedance are not clinically significant[5, 25, 36, 39, 52]. But because the FHR correlates with the length of diastole, fetal bradycardia of less than 100 bpm is associated with a significant prolongation of the compensatory diastolic flow described by Moll[37] and thus of the end-diastolic frequency shift. Whereas direct resistance changes during uterine contractions often radically alter the umbilical artery waveform even before the FHR decelerates (see above) and the heart-rate effect is "added on," the presence of fetal bradycardia in cases where impedance is unchanged by the uterine contraction has only an indirect effect on the umbilical artery waveform. When the technique for area-under-the-curve determination described above is used for the quantitative assessment of the relative perfusion of the umbilical arteries in these cases, we find that these areas have major pathophysiological significance. For example, fetal bradycardia in a normal umbilical arterial waveform (Fig. 21.**13a**) leads to an increase in the area under the waveform per cardiac cycle. As a result, the decrease in perfusion caused by the low heart rate is partially offset by blood still flowing at a relatively high velocity during the prolonged diastolic phase. This effect is barely evident in an abnormal umbilical waveform (Fig. 21.**13b**) and is imperceptible in a waveform with absent diastolic flow (Fig. 21.**13c**). In cases with reverse diastolic blood flow (Fig. 21.**13d**), the reverse flow volume would be theoretically increased due to the prolonged diastole, and in extreme cases this would quickly cause the net quantitative blood flow to approach zero.

Fetal bradycardia in normal and abnormal umbilical waveforms. We tested these theoretical considerations in practice. In cases with intrapartum FHR decelerations with no direct impedance changes, we measured the area integrals under the waveform for a normal umbilical arterial flow spectrum, for reduced end-diastolic flow, for absent diastolic flow, and for reverse diastolic flow in the umbilical artery. We found that FHR deceleration occurring in a normal Doppler spectrum was associated with a marked increase in the area under the waveform relative to the length of the cardiac cycle (Figs. 21.**10**, 21.**11**). This effect was much less pronounced when there was a primary decrease in end-diastolic flow. With absent diastolic flow, we observed no heart rate-dependent change in the area integral under the waveform. It is particularly interesting when FHR deceleration occurs in cases with negative diastolic flow. While the systolic forward flow component remains unchanged during the deceleration, the diastolic reverse flow component increases in proportion to the prolongation of diastole (Fig. 21.**14**). Contrary to the conditions in a normal umbilical artery waveform, the deceleration leads to a decrease in the net blood flow per beat. If we calculate the area under the waveform (in cm^2 per minute) for the situations described above, which provides a relative measure of the amount of blood perfusing the scanned vascular segment per minute, we find a less steep reduction in the area under the normal waveform during fetal bradycardia, a slightly steeper decline in the initially abnormal umbilical waveform, a linear decline in the waveform with absent diastolic flow, and an exponential drop in relative blood flow per minute in the waveform showing reverse diastolic flow (Fig. 21.**15**). The area in the reverse flow channel at 60 bpm is equal to the area in the forward flow channel during systole, resulting in a net blood flow of zero at this FHR.

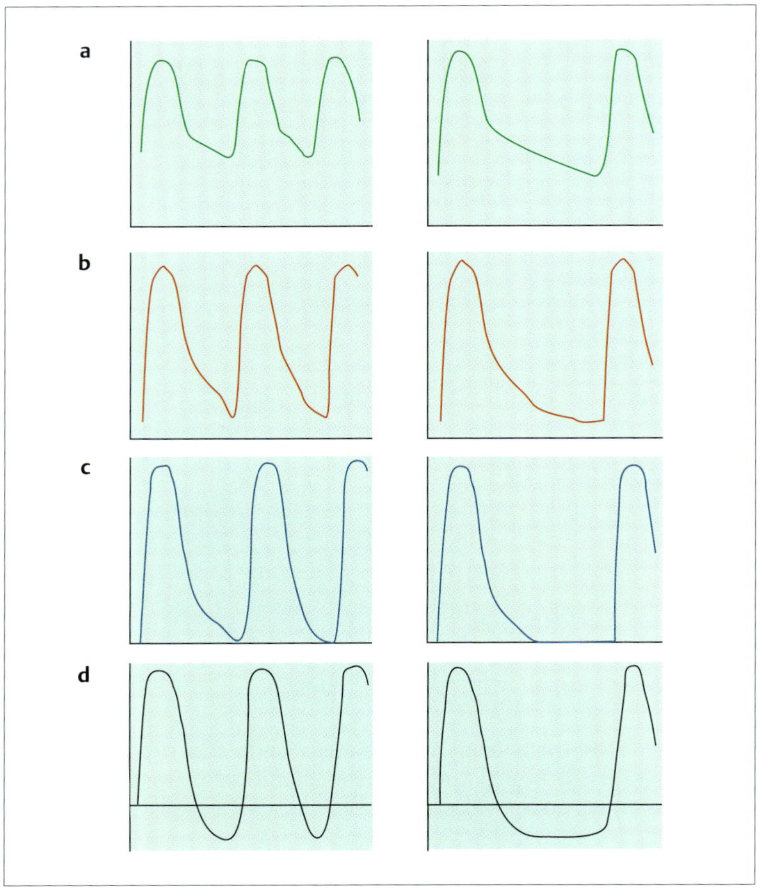

Fig. 21.**13** Waveform patterns in the fetal umbilical artery. Left column: normal FHR. Right column: fetal bradycardia.
a With normal impedance.
b With primary increase in impedance.
c With absent end-diastolic flow.
d With reverse diastolic flow.

Practical implications. The markedly greater heart rate-dependent decrease in quantitative blood flow per unit time in abnormal umbilical arterial waveforms and the exponential decline in waveforms with negative diastolic flow make it very likely that these fetuses, already in jeopardy, are placed at considerable additional risk for hypoxia when decelerations occur in the FHR. This helps us to understand the frequent clinical observation that some fetuses can develop a significant acid–base

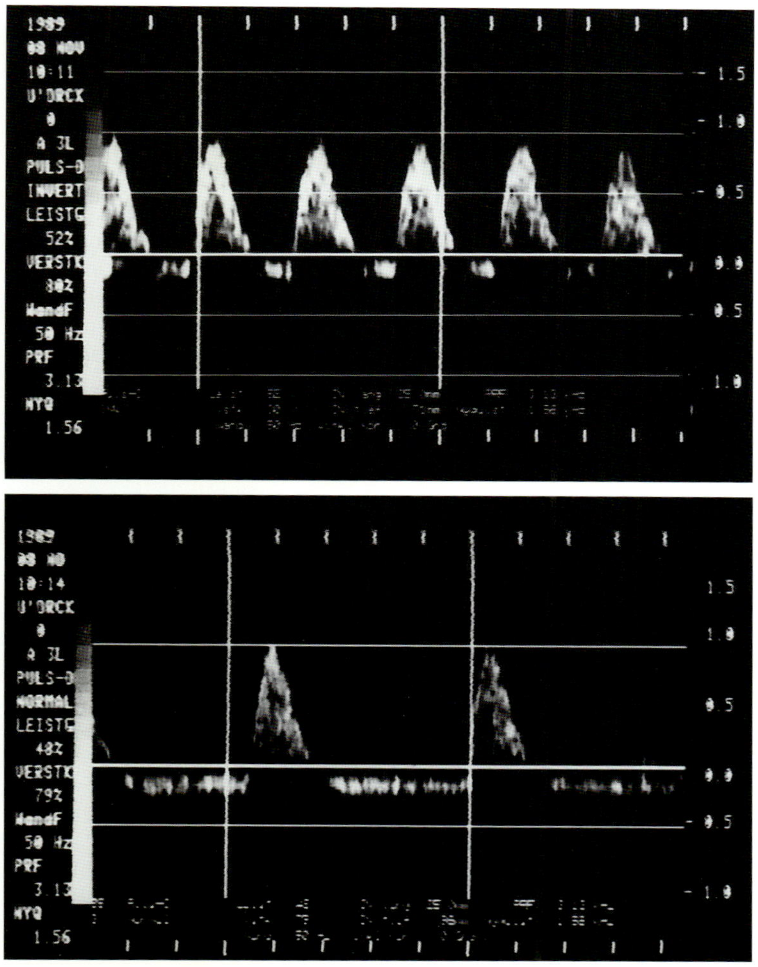

Fig. 21.**14** Severely abnormal umbilical artery waveform with reverse diastolic flow (top). Waveform from the same fetus during a deceleration in the FHR (bottom).

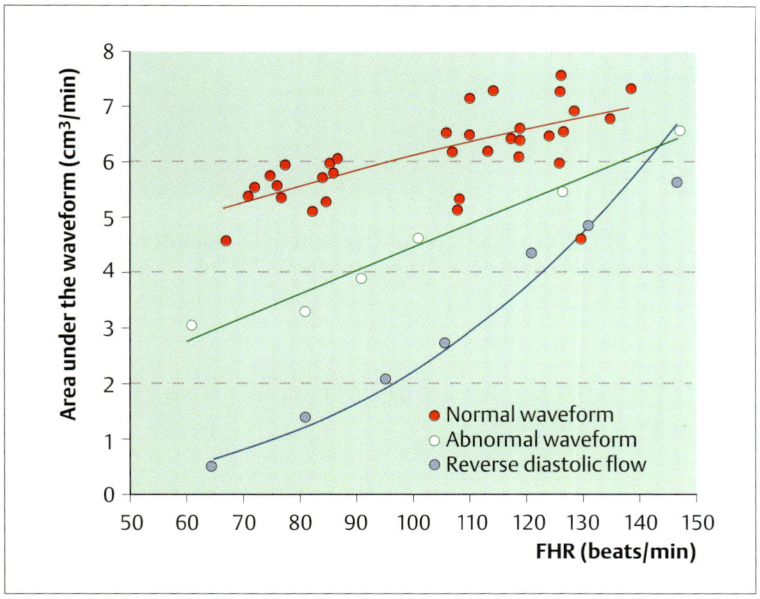

Fig. 21.**15** Area under the umbilical artery waveform (flow volume/min) as a function of the FHR. Top: normal waveform. Center: reduced diastolic flow. Bottom: reverse diastolic flow.

problem in the form of acute respiratory acidosis after just a few FHR decelerations, while other fetuses continue to exhibit normal acid–base values even after repeated decelerations. Animal studies have also documented the heightened susceptibility of growth-retarded fetuses to an additional hypoxic stress[2].

Because the fetal oxygen supply declines significantly when umbilical blood flow falls below 50% of the initial value[19], every deceleration in the FHR should be considered a serious threat to fetuses with reduced diastolic flow in the umbilical arteries and especially to fetuses with absent or reversed umbilical diastolic flow. One clinical implication for cases with intrapartum decelerations is that "basic Doppler" of the umbilical arteries between contractions or on admission to the delivery room can provide valuable information for estimating the hypoxic risk to the fetus from FHR decelerations. Based on the discoveries to date, spontaneous or induced labor (oxytocin test) should be avoided in fetuses that exhibit absent or reverse flow. These contractions could trigger a potentially damaging cascade of oxygen deprivation in the intervillous space and FHR deceleration causing an exponential decline of blood flow in the umbilical arteries, culminating in severe hypoxia.

Direct Effect of Intrapartum Fetal Hypoxia or Hypoxemia on Blood Flow Patterns in the Umbilical Arteries and Vein

Normal umbilical waveforms in acute hypoxemia. While the effect of reduced umbilical perfusion on the fetal oxygen supply is obvious, it is important to consider the direct effects of fetal asphyxia on blood flow in the umbilical arteries. According to Jensen[22], one minute of asphyxia leads to an approximately 35% reduction in umbilical blood flow, due mainly to the decrease in FHR (Fig. 21.**11**). Placental blood flow at the level of the initial value has been found after brief, recurring periods of asphyxia despite centralization of the fetal circulation, but with a normal FHR[22]. Acute or chronic hypoxic states can be induced in fetal sheep by restricting the oxygen supply to the mother or by occluding the blood supply to the intervillous space. Umbilical blood flow remains unchanged in these studies even when there are marked changes in the fetal arterial oxygen partial pressure[41]. Doppler velocimetry was unable to demonstrate acute fetal hypoxia or asphyxia in experimental animals[17,38,45]. These results are not surprising when we consider that the perfusion resistance in the fetal placenta probably cannot change acutely due to the lack of a nerve supply to the arterial resistance vessels of the placenta, although it might be affected by the fetal epinephrine level. The elevated fetal blood pressure in asphyxia[29] leads to an increased perfusion pressure, however, and could mask a hypothetical epinephrine effect. The clinical example in Fig. 21.**16** shows a normal umbilical artery waveform during acute fetal asphyxia in the expulsion stage despite fetal respiratory acidosis and hypoxemia. A slight increase in the RI is not seen until terminal bradycardia owing to the heart-rate effect described above.

Abnormal Doppler indices associated with a chronic supply deficit. The significant correlation between abnormal Doppler

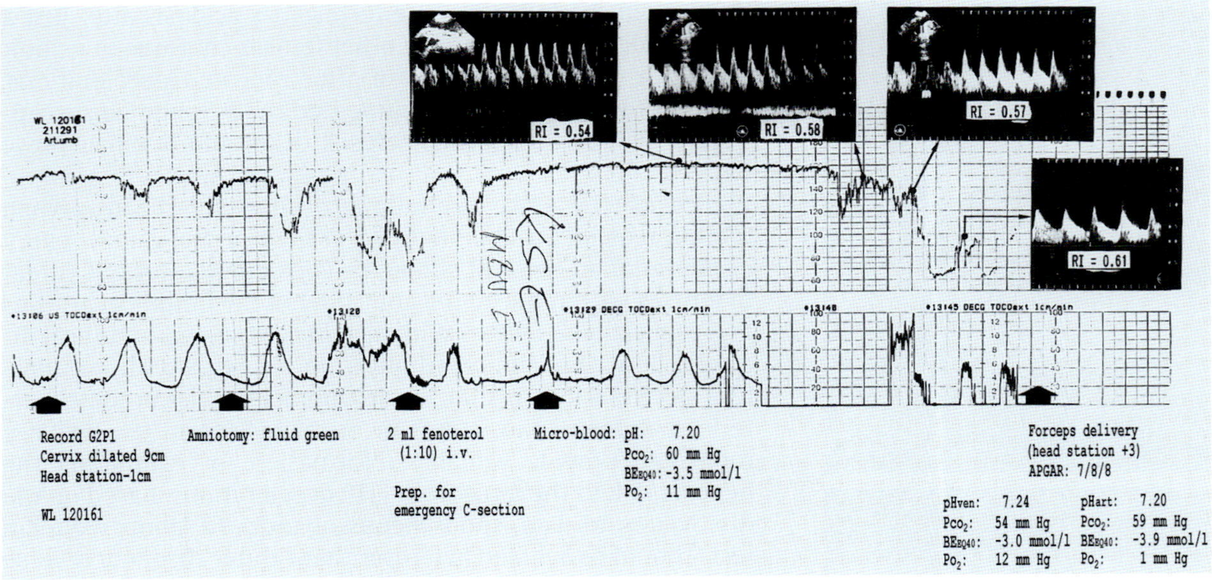

Fig. 21.**16** Abnormal FHR tracing in a patient admitted at term with a fully dilated cervix. A prolonged, W-shaped deceleration was recorded after amniotomy, which yielded a greenish amniotic fluid, and intrauterine resuscitation was carried out. Preparations were made for an emergency cesarean section. Microblood test showed respiratory acidosis. Fetal asphyxia cannot be detected by Doppler velocimetry of the umbilical arteries.

indices in the umbilical arteries and a poor fetal outcome in a high-risk pregnancy does not contradict these findings, as the abnormal waveform in these cases reflects the chronic resistance increase in the fetoplacental circulation due to deficient formation or secondary occlusion of the vascular tree of the fetal placenta[18]. This helps us to understand the studies that found a poor correlation between the umbilical artery Doppler findings obtained between contractions and the fetal outcome[10,43]. Besides the risks to the fetus that are incurred by the delivery process itself, this is due mainly to the fact that the prevalence of abnormal Doppler findings in term fetuses at delivery is relatively small, since fetuses with an abnormal umbilical artery flow pattern already manifest clinical abnormalities at an earlier time.

Late decelerations and an elevated S/D ratio. In cases with late decelerations in the FHR, Brar et al.[5] found that Doppler velocimetry of the umbilical arteries could detect an elevated S/D ratio in fetuses that developed signs of hypoxemia during the delivery. These measurements were obtained between contractions, however, and our experience indicates that the S/D ratio is not affected during late decelerations, aside from heart-rate effects. This was confirmed in a study by Damron et al.[8], who recorded umbilical artery waveforms between and also during contractions. Fetuses with late decelerations were found to have a higher S/D ratio than fetuses with a normal heart rate in measurements performed between contractions. Measurements at the height of contractions yielded indices that corresponded to the values between contractions.

Umbilical vein pulsations. Ninety percent of the fetuses with late decelerations showed umbilical vein pulsations during contractions, signifying a possible short-term overload of the right heart. In cases with a normal intrapartum FHR, the authors found constant, undisturbed blood flow in the umbilical vein, which is consistent with the earlier results of other authors[12,16,44]. The umbilical vein recording in Fig. 21.**17** also shows that the acute hypoxic response of the fetus can be clearly recognized in the flow pattern of the venous system.

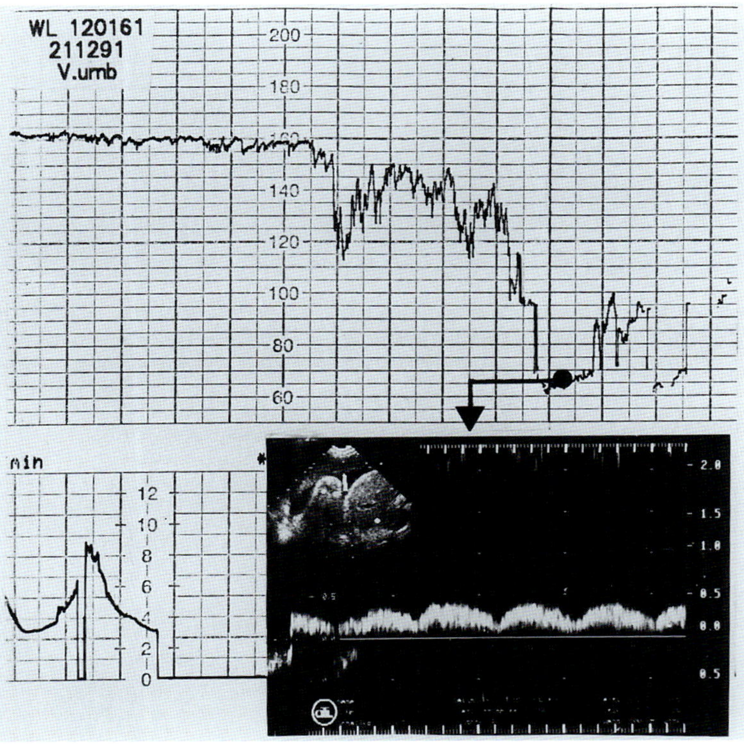

Fig. 21.**17** Blood flow pattern of the umbilical vein during terminal bradycardia before delivery of the fetus in Fig. 21.**16**.

Intrapartum Blood Flow Patterns in the Fetal Aorta

With a normal intrapartum FHR, uterine contractions do not have an appreciable effect on the Doppler waveforms of the fetal aorta[12]. Decelerations lead to a decrease in end-diastolic flow velocities due to a heart-rate effect[13]. Occlusion of the umbilical cord can induce reverse diastolic flow in the fetal aorta of experimental animals[33]. These results have been confirmed by other authors in animal studies using isolated compression of the umbilical vein[17] and correspond to the findings in the umbilical arteries on occlusion of the umbilical vein. It is reasonable to conclude that acute fetal hypoxia does not have a direct effect on the waveform of the fetal aorta. On the other hand, a redistribution of the fetal cardiac output in favor of cerebral perfusion could have an indirect effect on the fetal aortic waveform, as acute animal experiments have demonstrated[22,23].

Quantitative blood flow measurements. Quantitative measurements of blood flow in the fetal aorta showed no significant change in volume flow during labor[12]. Other authors measured a slightly significant increase in quantitative flow from 200 to 245 (ml/min)/kg between contractions[32]. The reliability of these quantitative measurements is uncertain, however, since the determination of aortic diameters involves considerable errors[31] that become exponentially large in calculations.

Intrapartum Blood Flow Patterns in Fetal Cerebral Vessels

Compression of the fetal head. Animal studies have demonstrated a reduction of cerebral blood flow in response to fetal head compression as a cause of decelerations in the FHR[30]. The fetal anterior, middle, and posterior cerebral arteries and the internal and common carotid arteries are accessible to Doppler velocimetry[54]. Transducer pressure on the fetal head is sufficient to lower end-diastolic flow velocities in cerebral vessels, and strong transducer pressure can even cause reverse diastolic flow in the middle cerebral artery (MCA)[48]. Similar changes in the internal carotid artery flow pattern have been demonstrated in oligohydramnios[46].

Table 21.**3** Doppler measurements of cerebral perfusion during labor

Authors	*n*	Flow pattern during dilation stage (DS)
Fendel et al. (1990)[14]	7	Early DS: S/D ratio slightly increased Late DS: S/D ratio markedly increased Breech presentation: S/D ratio not affected
Maesel et al. (1990)[34]	15	No waveform change during contractions (COD 4 – 9 cm, membranes ruptured)
Our results	8	No waveform change during contractions with a normal FHR (COD 2 – 9 cm), absent diastolic flow during early deceleration

DS = dilation stage; COD = cervical os diameter; FHR = fetal heart rate

Early and late dilation stage. Only a few studies on intrapartum measurements have been published to date. With a normal FHR, the blood flow pattern of the MCA is apparently unchanged during uterine contractions[34]. Other authors[14] found a slightly increased S/D ratio in the internal carotid artery during the early dilation stage of labor. But as the fetal head descends during the late dilation stage, the S/D ratio rises from 3.5 to 5.5 during contractions, reflecting the increased resistance to cerebral perfusion with increasing head compression. Our own studies (Table 21.**3**) have shown no effect on the MCA waveforms with a normal FHR. In the presence of decelerations, which may be synchronous or delayed in relation to uterine contractions, the absence of diastolic flow in the MCA may well be an expression of fetal head compression (Fig. 21.**18**). Thus, Doppler has confirmed the concept of a reduction in cerebral blood flow by fetal skull compression as the cause of these relatively early-onset decelerations during labor.

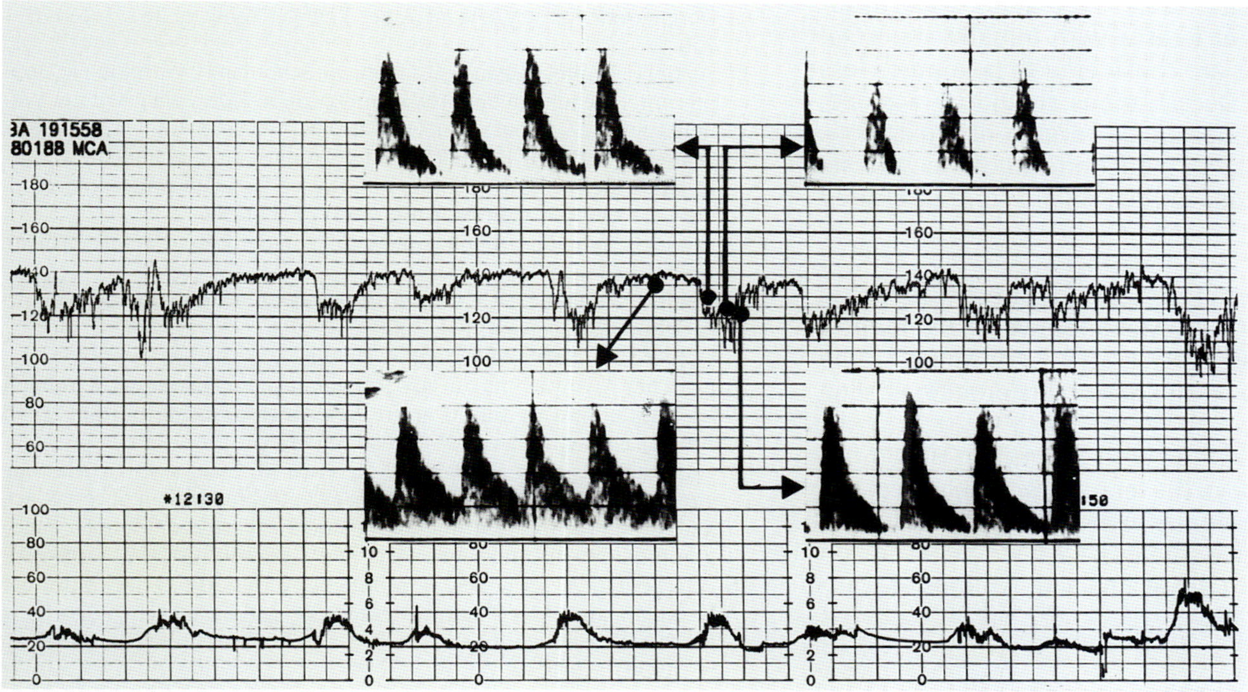

Fig. 21.**18** Absent diastolic flow in the middle cerebral artery (MCA) during labor as an expression of fetal head compression with typical early decelerations in the FHR.

Summary

Doppler velocimetry during labor, as opposed to measurements in the resting fetus and uterus, can detect waveform changes in uterine and fetal vascular regions that occur in a matter of seconds.

Uterine arteries. Contraction of the uterine muscles leads to a dramatic rise of impedance in the uterine arteries, with a corresponding reduction in quantitative blood flow. The extent of these changes depends on the strength of the uterine contractions and does not show a reliable correlation with external tocometry in the contraction stress test. With a preexisting abnormal uterine artery waveform in a patient with pregnancy-induced hypertension, Doppler velocimetry during a uterine contraction shows forward-directed systolic flow but reverse diastolic flow.

Umbilical arteries. In examinations of the umbilical and intrafetal vessels, direct mechanical effects due to compression are distinguished from heart-rate- and/or hypoxia-related factors that can alter the flow patterns. In cases with a normal FHR, uterine contractions do not alter the blood flow pattern in the umbilical arteries. In cases with umbilical cord compression during labor, Doppler can demonstrate an arrest of umbilical perfusion caused by bidirectional movement of the blood column due to occlusion of the umbilical vessels. If decelerations due to a different cause occur during labor, the heart rate-induced waveform changes in deceleration with a normal initial flow pattern lead only to a slight reduction of mean blood flow velocity in the umbilical arteries owing to the relatively high diastolic flow velocities. This ensures that a volume flow of approximately 70% of the initial value is maintained even if the FHR falls below 60 bpm. But in cases with primarily abnormal waveforms and especially with preexisting reverse diastolic flow in the umbilical arteries, a deceleration in the FHR will cause a significantly sharper or even exponential fall of relative volume flow in the umbilical circulation. Thus, intrapartum FHR decelerations in fetuses with initially abnormal umbilical artery waveforms are considered to be a much less favorable prognostic sign. Decelerations should be avoided, therefore, in fetuses with absent or reverse diastolic flow. Based on available results, labor induction and challenge tests that may induce decelerations in the FHR are contraindicated in these cases.

Fetal hypoxemia. Fetal hypoxemia does not directly affect the umbilical artery waveforms during labor. However, if umbilical vein pulsations occur during a deceleration or during a uterine contraction, this appears to be an unfavorable prognostic sign.

Fetal aorta. Uterine contractions with a normal FHR do not alter the blood flow patterns in the fetal aorta. When decelerations occur, the changes are similar to those seen in the umbilical arteries. No representative studies have been published on this subject, however, due to the technical difficulties of scanning the fetal aorta during labor.

Cerebral vessels. The rise of intrauterine pressure during labor does not affect the velocity waveforms of the cerebral vessels. However, the mounting external pressure on the descending fetal head during labor leads to a decrease in diastolic flow velocities as an expression of the increased intracranial pressure.

References

1 Abitbol MM, Monheit AG, Rochelson BL, Stern W, Blyakher L, Saraf V: The use of an indwelling Doppler probe to study acute changes in umbilical artery waveforms in the fetal sheep. Amer. J. Obstet. Gynecol. 161 (1989) 1324–1331
2 Block BSB, Llanos AJ, Creasy RK: Responses of the growth-retarded fetus to acute hypoxemia. Amer. J. Obstet. Gynecol. 148 (1984) 878–885
3 Borell V, Fernström I, Ohlson L, Wiquist N: Influence of uterine contractions on the uteroplacental blood flow at term. Amer. J. Gynecol. Obstet. 93 (1965) 44–57
4 Brar HS, Platt LD, DeVore GR, Horenstein J, Medearis AL: Qualitative assessment of maternal uterine and fetal umbilical artery blood flow and resistance in laboring patients by Doppler velocimetry. Amer. J. Obstet. Gynecol. 158 (1988) 952 –956
5 Brar HS, Platt LD, Paul RH: Fetal umbilical blood flow velocity waveforms using Doppler ultrasonography in patients with late decelerations. Obstet. Gynecol. 73 (1989) 363–366
6 Campbell S, Diaz-Recasens J, Griffin DR, Cohen-Overbeek TE, Pearce JM, Willson K: New Doppler technique for assessing uteroplacental blood flow. Lancet i (1983) 675–677
7 Cruz AC, Frentzen BH, Gomez KJ, Allen G, Tyson-Thomas M: Continuous-wave Doppler ultrasound and decreased amniotic fluid volume in pregnant women with intact or ruptured membranes. Amer. J. Obstet. Gynecol. 159 (1988) 708–714
8 Damron DP, Chaffin DG, Anderson CF, Reed KL: Changes in umbilical arterial and venous blood flow velocity waveforms during late decelerations of the fetal heart rate. Obstet. Gynecol. 84 (1994) 1038–1040
9 Fairlie FM, Lang GD, Sheldon CD: Umbilical artery flow velocity waveforms in labour. Brit. J. Obstet. Gynaecol. 96 (1989) 151–157
10 Feinkind L, Abulafia O, Delke I, Feldmann J, Minkoff H: Screening with Doppler velocimetry in labor. Amer. J. Obstet. Gynecol. 161 (1989) 765–770
11 Fendel H, Fendel M, Pauen H, Liedtke B, Schonlau H, Warnking R: Doppleruntersuchungen des arteriellen uterinen Flows während der Wehentätigkeit. Z. Geburtsh. u. Perinat. 188 (1984) 64–67
12 Fendel H, Fettweis P, Billet P et al.: Doppleruntersuchungen des arteriellen utero-feto-plazentaren Blutflusses vor und während der Geburt. Z. Geburtsh. u. Perinat. 191 (1987) 121–129
13 Fendel H, Sohn Ch: Dopplersonographie in der Geburtshilfe. Springer, Berlin 1989
14 Fendel H, Funk A, Jörn H, Gans A: Zerebraler Blutfluß unter der Geburt. Z. Geburtsh. u. Perinat. 194 (1990) 272–274
15 Fischer WM: Kardiotokographie. Thieme, Suttgart 1981
16 Fleischer A, Anyaegbunam AA, Schulman H, Farmakides G, Randolph G: Uterine and umbilical artery velocimetry during normal labor. Amer. J. Obstet. Gynecol. 157 (1987) 40–43
17 Fouron JC, Teyssier G, Maroto E, Lessard M, Marquette G: Diastolic circulatory dynamics in the presence of elevated placental resistance and retrograde diastolic flow in the umbilical artery: A doppler echographic study in lambs. Amer. J. Obstet. Gynecol. 164 (1991) 195–203
18 Hitschold T, Weiss E, Beck T, Müntefering H, Berle P: Beeinflußt die Vaskularisation der Placenta fetalis die enddiastolischen Blutflußgeschwindigkeiten in den Nabelarterien? Geburtshilfe Frauenheilkd. 50 (1990) 623–627

19 Itskovitz J, LaGamma EF, Rudolph AM: The effect of reducing umbilical blood flow on fetal oxygenation. Amer. J. Obstet. Gynecol. 145 (1983) 813–818

20 Janbu T, Koss KS, Nesheim B-I, Wesche J: Blood velocities in the uterine artery in humans during labour. Acta. Physiol. Scand. 124 (1985) 153–161

21 Jensen A, Hohmann M, Künzel W: Redistribution of fetal circulation during repeated asphyxia in sheep: effects on skin blood flow, transcutaneous PO2, and plasma catecholamines. J. Dev. Physiol. 9 (1987) 41–45

22 Jensen A: Das Schocksyndrom des Feten. Med. Welt 38 (1987) 1072–1083

23 Jensen A, Hohmann M, Künzel W: Dynamic changes in organ blood flow and oxygen consumption during acute asphyxia in fetal sheep. J. Dev. Physiol. 9 (1987) 543–559

24 Kirkinen P, Jouppila P, Huch R, Huch A: Blood flow velocity waveforms at late pregnancy and during labor. Arch. Gynecol. Obstet. 244 (1988) 19–23

25 Kofinas AD, Espeland M, Swain M, Penry M, Nelson LH: Correcting umbilical artery flow velocity waveforms for fetal heart rate is unnecessary. Amer. J. Obstet. Gynecol. 160 (1989) 704–707

26 Künzel W, Mann LI, Bhakthavathsalan A, Airomlooi J, Liu M: The effect of umbilical vein occlusion on fetal oxygenation, cardiovascular parameters, and fetal electroencephalogramm. Amer. J. Obstet. Gynecol. 128 (1977) 201–208

27 Künzel W, Kurz CS, Kastendieck E: Die Variabilität der fetalen Herzfrequenzreaktion auf die Reduktion der uterinen Durchblutung. Z. Geburtsh. u. Perinat. 185 (1981) 343–350

28 Künzel W, Hohmann M: Interpretation der fetalen Herzfrequenz während der Schwangerschaft und Geburt. Gynäkologe 17 (1984) 255–261

29 Künzel W: Das fetale Schocksyndrom. Z. Geburth. u. Perinat. 190 (1986) 177–184

30 Künzel W: Überwachung des Feten während der Geburt. In Wulf KH, Schmidt-Matthiesen H (eds.): Klinik der Frauenheilkunde und Geburtshilfe. Bd 7/I. Urban & Schwarzenberg, München 1990, 91–134

31 Künzel W, Jovanovic V, Grüßner S: Der Blutfluß in der Vena und Arteria umbilicalis während der Schwangerschaft. Geburtshilfe Frauenheilkd. 51 (1991) 513–522

32 Lindblad A, Bernow J, Marsál K: Obstetric analgesia and fetal blood flow during labour. Brit. J. Obstet. Gynaecol. 94 (1987) 306–311

33 Lingman G, Marsál K, Rosén K-G, Kjellmer I: Blood flow measurements in exteriozized lamb fetuses during asphyxia. In Jung H, Fendel H (eds.): Doppler technics in obstetrics. Thieme, Stutgart 1986, 36–40

34 Maesel A, Lingman G, Marsál K: Cerebral blood flow during labor in human fetus. Acta. Obstet. Gynecol. Scand. 69 (1990) 493–495

35 Mansouri H, Gagnon R, Hunse C: Relationship between fetal heart rate and umbilical blood flow velocity in term human fetuses during labor. Amer. J. Obstet. Gynecol. 160 (1989) 1007–1012

36 Mires G, Dempster J, Patel NB, Crawford JW: The effect of fetal heart rate on umbilical artery flow velocity waveforms. Brit. J. Obstet. Gynaecol. 94 (1987) 665–669

37 Moll W: Strömungsgeschwindigkeit und Doppler-Shift in fetalen und maternen Gefäßen. Gynäkologe 25 (1992) 278–28

38 Morrow RJ, Adamson SL, Bull SB, Ritchie JWK: Hypoxic acidemia, hyperviscosity, and maternal hypertension do not affect the umbilical arterial velocity waveform in fetal sheep. Amer. J. Obstet. Gynecol. 163 (1990) 1313–1320

39 Newnham J, Patterson L, James I, Reid S: The effect of heart rate on Doppler flow velocity systolic/diastolic ratios in umbilical and uteroplacental arterial waveforms. Early Hum. Dev. 21 (1990) 21–29

40 Olofsson P, Thuring-Jönsson A, Marsál K: Uterine and umbilical circulation during the oxytocin challenge test. Ultrasound Obstet. Gynecol. 8 (1996) 247–251

41 Peeters LLH, Sheldon RE, Jones MD, Makowski EL, Meschia G: Blood flow to fetal organs as a function of arterial oxygen content. Amer. J. Obstet. Gynecol. 135 (1979) 637–646

42 Rosenkrantz TS, Oh W: Cerebral blood flow in infants with polycythemia and hyperviscosity: Effect of partial exchange transfusion with plasmanate. J. Pediatr. 101 (1982) 94–98

43 Sarno AP, Ahn MO, Brar HS, Phelan JP, Platt LD: Intrapartum Doppler velocimetry, amniotic fluid volume, and fetal heart rate as predictors of subsequent fetal distress. Amer. J. Obstet. Gynecol. 161 (1989) 1508–1514

44 Stuart B, Drumm J, FitzGerald DE, Duignan NM: Fetal blood velocity waveforms in uncomplicated labour. Brit. J. Obstet. Gynaecol. 88 (1981) 865–869

45 van Huisseling H, Hasaart THM, Ruissen CJ, Muijsers GJJ, de Haan J: Umbilical artery flow velocity waveforms during acute hypoxemia and the relationship with hemodynamic changes in the fetal lamb. Amer. J. Obstet. Gynecol. 161 (1989) 1061–1064

46 van den Wijngaard JAGW, Wladimiroff JW, Reuss A, Stewart PA: Oligohydramnios and fetal cerebral blood flow. Brit. J. Obstet. Gynaecol. 95 (1988) 1309–1311

47 Vetter K: Dopplersonographie in der Schwangerschaft. Ed. Medizin VCH, Weinheim 1991

48 Vyas S, Campbell S, Bower S: Maternal abdominal pressure alters fetal cerebral blood flow. Brit. J. Obstet. Gynaecol. 97 (1990) 740–747

49 Weiss E, Hundemer H-P, Berle P: Dopplersonographische Untersuchung der Art. umbilicalis: Abhängigkeit von der fetalen Herzfrequenz. In Dudenhausen JW, Saling E (eds.): Perinatale Medizin. Bd XII. Thieme, Stuttgart 1988, 170 –171

50 Weiss E, Hitschold T, Berle P: Untersuchungen zur Reduktion der umbilikalen Durchblutung bei Feten mit variablen Dezelerationen der Herzfrequenz mittels Dopplersonographie der Nabelarterie. Z. Geburtsh. u. Perinat. 193 (1989) 60–67

51 Weiss E, Hitschold T, Berle P: Umbilical artery blood flow velocity waveforms during variable decelerations of the fetal heart rate. Amer. J. Obstet. Gynecol.164 (1991) 534–540

52 Weiss E: Die Blutflußmessung in der Schwangerschaft. Bedeutung der Dopplersonographie in der klinischen Anwendung. Habilitationsschrift, Johannes Gutenberg Universität Mainz 1993

53 Weiss E: Fetale Herzfrequenz und Dopplersonographie. Gynäkologe 27 (1994) 146–150

54 Wladimiroff JW, Tonge HM, Stewart PA: Doppler ultrasound assessment of cerebral blood flow in the human fetus. Brit. J. Obstet. Gynaecol. 93 (1986) 471–475

22 Color Doppler Ultrasound in Fetal Echocardiography

A. Lindinger

Congenital Heart Disease—Incidence and Risk Factors

Congenital heart disease has a reported incidence of 0.4–0.8%[2,3,8]. The etiology of most congenital heart defects is unknown but is believed to be multifactorial. A genetic defect can be demonstrated in approximately 5–10% of cases. Teratogenic agents can be identified in a small percentage of cases[4,9].

It is estimated that a cardiac anomaly is present:

- In approximately two-thirds (35–99%) of cases where a fetal chromosome abnormality exists.
- In approximately 25% of cases with extracardiac anomalies.
- In fetuses with a structural cardiac malformation, a chromosome abnormality is present in approximately 30% of cases and an extracardiac anomaly in 50%[6,12,14,15].

Heart defects account for a very high percentage of fetal deaths in the early stage of pregnancy. This percentage declines with advancing gestation (Table 22.**1**)[5,7,16].

Table 22.**2** lists the most frequent chromosome abnormalities and genetic defects with the most important associated cardiac anomalies and their incidences.

Table 22.**3** lists the maternal diseases and risk factors and fetal findings that are commonly associated with congenital heart disease and thus warrant an examination of the fetal heart.

The risk of recurrence in patients with a family history of congenital heart disease is between 2% and 20%, depending on the degree of the relationship and the number of affected relatives (Table 22.**4**)[1,10,11,13,18].

There is a high risk of recurrence for heart defects that involve left ventricular outflow tract obstruction such as severe aortic stenosis and especially hypoplastic left heart syndrome, for which recurrence rates up to 20% have been reported [10,18].

Thus, a fetal ultrasound examination should be recommended in the 20th or 21st week of gestation in patients with a positive family history of congenital anomalies, patients with maternal metabolic diseases, patients on chronic medication, and patients with pregnancy complications that involve one of the known risk factors.

Table 22.**1** Percentage of fetal deaths due to heart defects at various gestational ages. (After reference 7)

Gestational age	Percentage of fetal deaths due to heart disease
< 10th week of gestation	40
Weeks 11–15	18
Weeks 16–20	8
Weeks 21–25	7

Table 22.**2** Types and incidences of cardiac anomalies in the setting of chromosome abnormalities and genetic defects

Chromosome abnormality/genetic defect	Cardiac disease	Incidence (%)
Down syndrome (trisomy 21)	Complete AV septal defect, tetralogy of Fallot	40
Pätau syndrome (trisomy 13)	Atrial or ventricular septal defect, patent ductus arteriosus	90
Edwards syndrome (trisomy 18)	Conotruncal defects	100
CATCH 22 (microdeletion 22q11)	Conotruncal defects	90
Marfan syndrome (defect in fibrillin-1 gene, chromosome 15q21.1)	Dilatation of aortic root and ascending aorta, mitral valve prolapse	90
Williams–Beuren syndrome (deletion at 7q11.23)	Supravalvular aortic stenosis, peripheral pulmonary stenosis	100
Bourneville–Pringle disease, tuberous sclerosis (9q34, 16p13.3)	Multiple cardiac rhabdomyomas	50
Turner syndrome (45 XO)	Coarctation of the aorta	30

Table 22.**3** Maternal and fetal risk factors and diseases that are an indication for ultrasound evaluation of the fetal heart

Maternal risk factor	Incidence (%)
Infections (e.g., rubella)	up to 70
Alcohol abuse	30–50
Metabolic disorders:	
Poorly controlled diabetes mellitus	3
PKU: phenylalanine level > 15 mg/dl	15
Medications: hydantoin, lithium, thalidomide, sex hormones, retinoids	2–10
Polyhydramnios/oligohydramnios	10/20
Fetal risk factors	
Extracardiac anomalies	2–50
Single umbilical artery	1–2
Fetal arrhythmias:	
Supraventricular extrasystoles	1–2
Supraventricular tachycardia	5–10
Third-degree AV block	40
Fetal effusions, hydrops	10–20
Intrauterine growth retardation	10
Monochorionic twins	2

Table 22.**4** Risk of recurrence for congenital heart disease in patients with a positive family history

Affected relatives	Recurrence risk (%)
One affected sibling	2–4
Two affected siblings	6–12
Affected father or mother	4–15

General Introductory Remarks on Color Doppler Sonography of the Fetal Heart

Indications for Color Doppler Sonography—Advantages and Limitations

In most cases a structurally normal fetal heart can be adequately examined with B-mode and Doppler ultrasound. The use of color Doppler sonography is advantageous for the rapid localization of extracardiac structures (e.g., the aortic arch) and for evaluating the function of the cardiac valves. In the presence of a heart defect, color Doppler always provides a valuable adjunct by color-encoding the direction of blood flow (e.g., in cases with valvular insufficiency or reversed flow in the ductus arteriosus) and the velocity of the flow (e.g., turbulence).

The limitations of color Doppler ultrasound are based on technical factors and are a particularly important consideration in fetal cardiac examinations. As in Doppler velocimetry, blood flow is optimally depicted only when the ultrasound beam is roughly parallel to the flow, i.e., when the beam axis forms an angle less than 20° relative to the flow direction (20° is the maximum permissible angle for the Doppler sampling of blood flow in the heart or an adjacent vessel; there is no need for angle correction within this range[17]).

The preselected velocity range of the Doppler spectrum should be matched to the velocities of the sampled blood flow, and the color intensity should be finely adjusted. This will eliminate the majority of false-positive and false-negative artifacts in color Doppler imaging (e.g., extraluminal "color bleed").

Special Features of Fetal Echocardiography

Limits are generally imposed by scanning at greater depths and by the frequent need for image magnification. These limitations result in low signal amplitudes with decreased lateral and depth resolution and low frame rates. Image formation can be optimized under these conditions by using the smallest possible sector angle.

Ultrasound Examination of the Fetal Heart

Normal Findings

The procedure for ultrasound examination of the fetal heart consists of the following components, which are illustrated in Figs. 22.**1**–22.**11**.

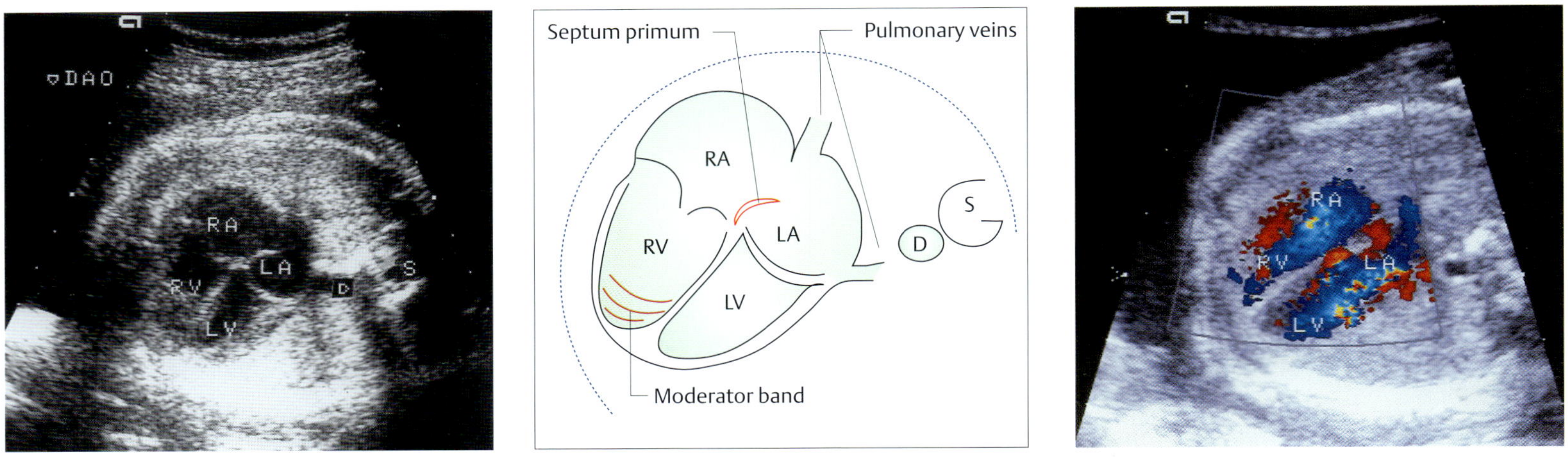

a b c

Fig. 22.**1** Fetal heart in a transverse scan through the thorax.
a, b The descending aorta (DAO) appears anterior to the spine (S). In front of the descending aorta are the left atrium (LA) and left ventricle (LV). The cardiac apex points toward the left side. The four-chamber view demonstrates both atria (RA, LA) and both ventricles (RV, LV). The right ventricle is distinguished morphologically by a transverse muscle bundle at the apex (the moderator band) and abuts the right chest wall. The site of entry of the pulmonary veins into the left atrium can be identified. The septum primum flap is deflected into the left atrium.
c Color Doppler image in the same plane as **a** shows blood, encoded in blue, flowing from the atria into the ventricles.

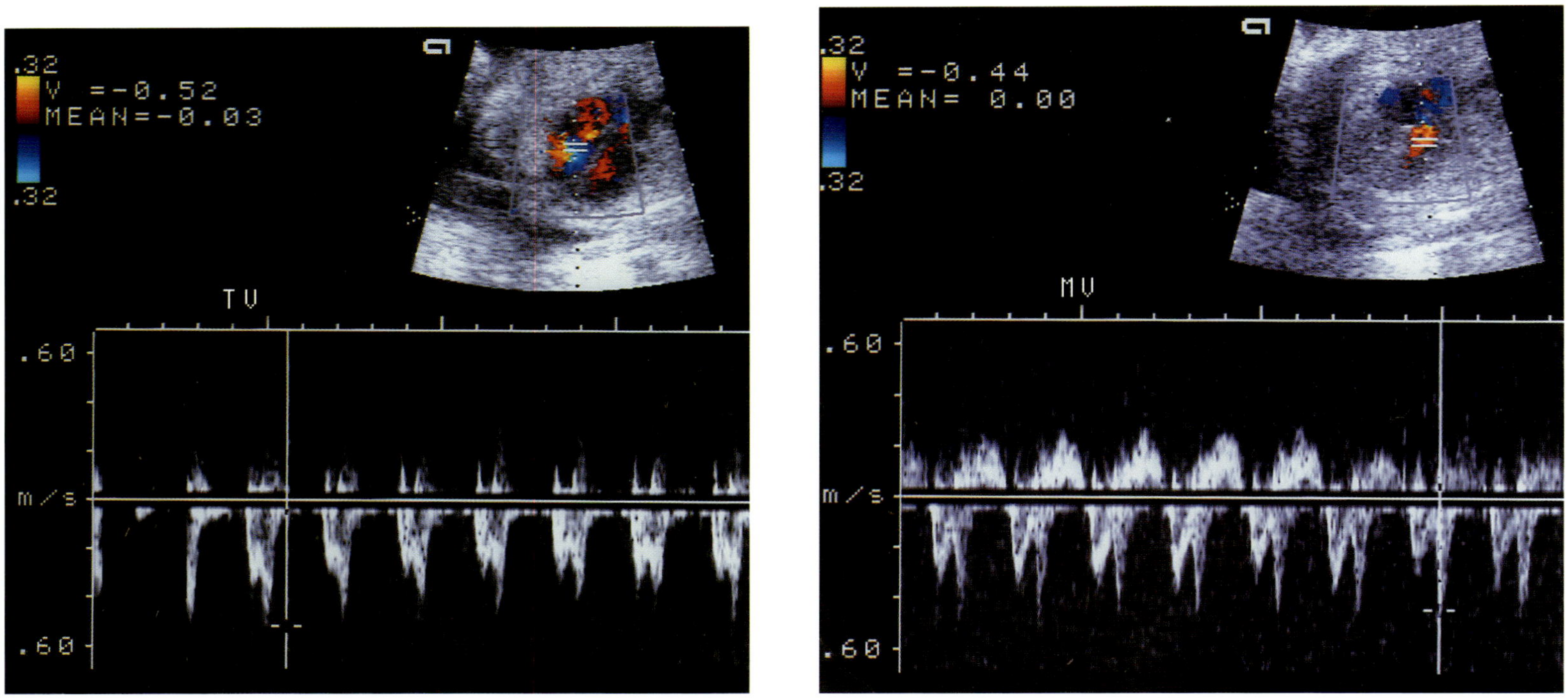

a b

Fig. 22.**2** Four-chamber view with color Doppler image of blood flowing from both atria into the ventricles. The Doppler spectra across the tricuspid valve (TV) and mitral valve (MV) show the typical diastolic flow pattern that occurs across the AV valves: the flow velocity during the e-wave (early diastolic inflow) is slower than during the a-wave (late diastolic inflow through atrial contraction).
a Tricuspid valve.
b Mitral valve.

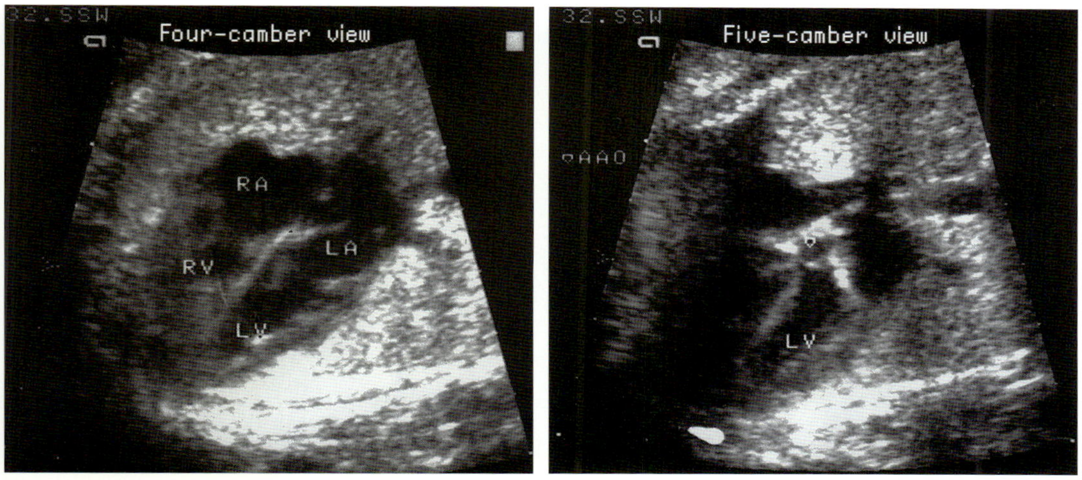

a b

Fig. 22.**3** The scan plane is tilted from the four-chamber view (**a**) to the five-chamber view (**b**) by anterior angulation of the transducer. This brings the left ventricle (LV) and the base of the aortic valve (AAo) into view. RA = right atrium, LA = left atrium, RV = right ventricle.

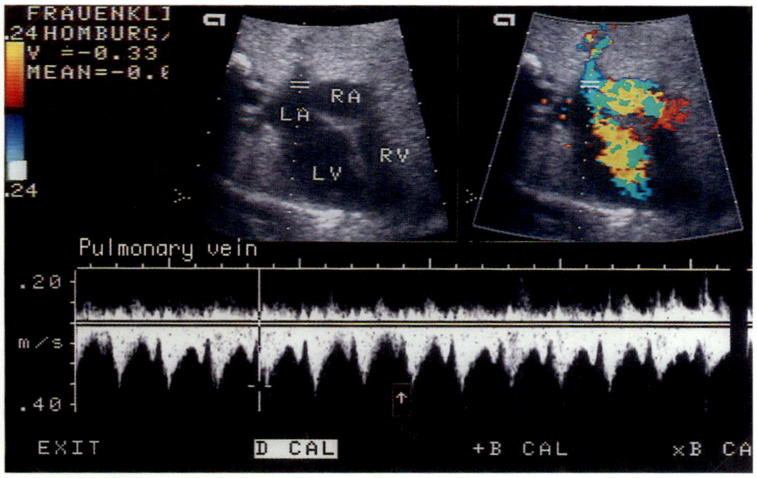

Fig. 22.**4** Upper left: four-chamber view demonstrating the sites of entry of the right and left pulmonary veins into the left atrium (LA). The sample volume has been placed in the right pulmonary vein. Upper right: color Doppler image of the entry of the right pulmonary vein and blood flow from the right atrium (RA) into the left atrium and from there into the left ventricle (LV). RV = right ventricle. Below: typical Doppler spectrum of pulmonary venous flow with a brief dip during atrial contraction (↑).

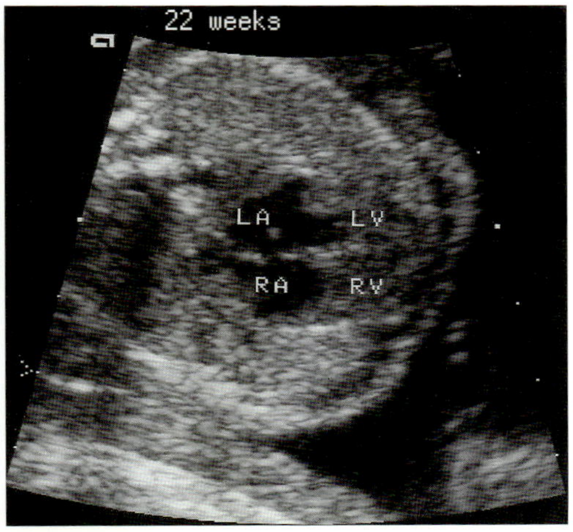

a

Fig. 22.**5** Four-chamber view demonstrating the foramen ovale.
a The septum primum flap is deflected into the left atrium (LA).
b Color Doppler image of blood flow from the right atrium (RA) into the left atrium (LA), with a typical Doppler spectrum. RV = right ventricle, LV = left ventricle.

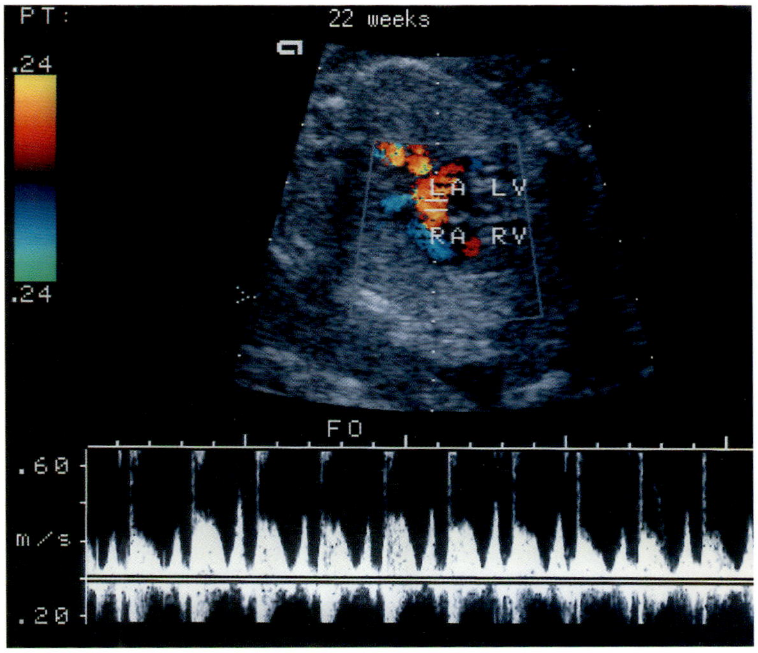

b

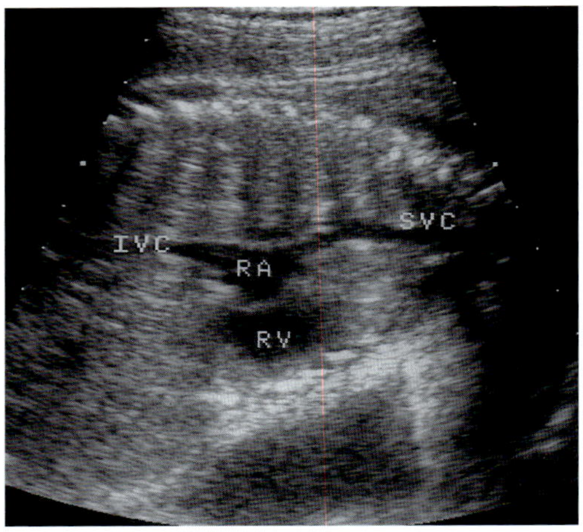

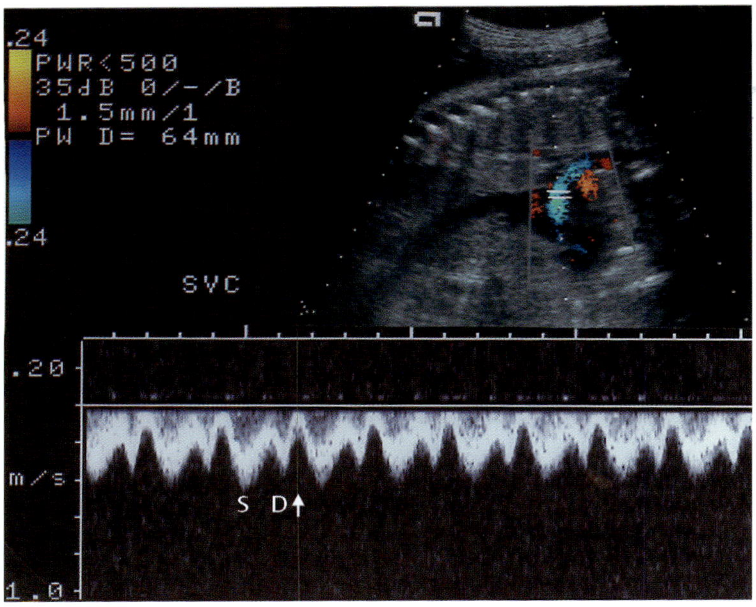

Fig. 22.**6 a** Site of entry of the superior vena cava (SVC) and inferior vena cava (IVC) into the right atrium (RA), with the spine above. RV = right ventricle.

Fig. 22.**6 b** Color Doppler image of superior vena cava flow. The typical Doppler spectrum displays two peaks representing ventricular systole (S) and diastole (D). Flow is decreased during atrial systole (↑).

Fig. 22.**7 a** Junction of the superior vena cava (SVC) with the right atrium (RA). Next to the vena cava is the ascending aorta (AO), which arises from the left ventricle (LV). The scan plane also cuts the pulmonary artery (PA) just in front of the aorta.

Fig. 22.**7 b** Color Doppler image in the same plane as **a**. The blood flow from the superior vena cava into the right atrium is encoded in blue, while flow in the ascending aorta and pulmonary artery is encoded in red.

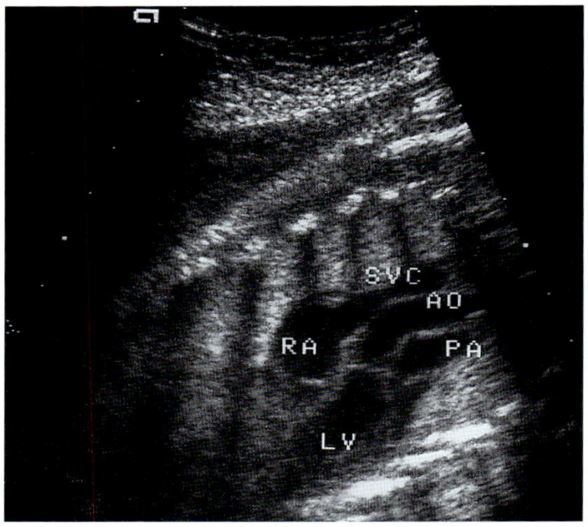

a

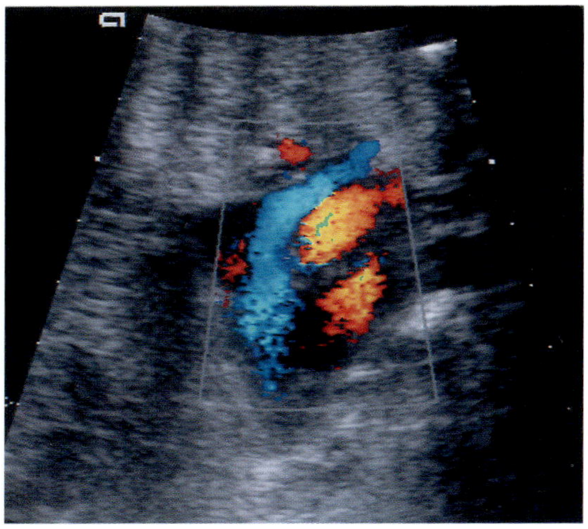

b

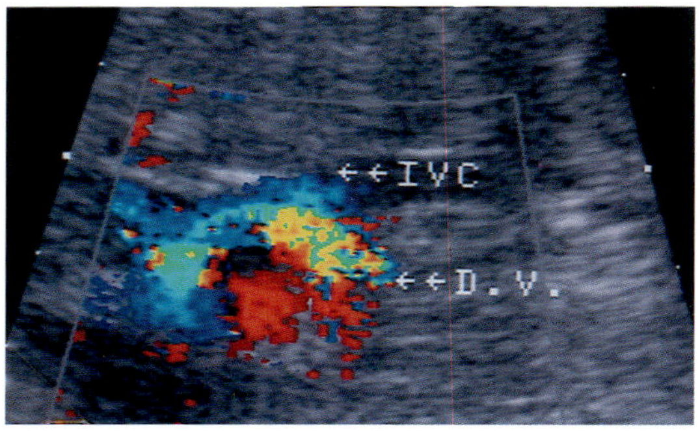

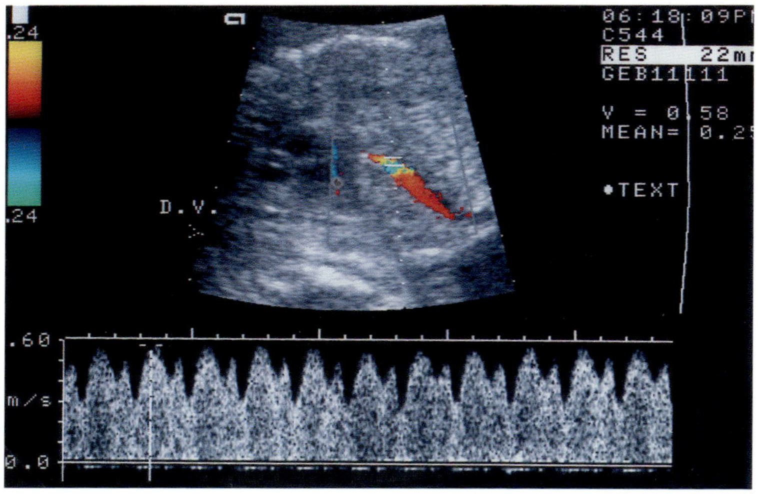

Fig. 22.**8 a** Site of entry of the ductus venosus (DV) into the inferior vena cava (IVC) just before its entry into the right atrium. Color Doppler demonstrates the parallel arrangement of the two blood flows. The blood in the ductus venosus is encoded in yellow.

Fig. 22.**8 b** Typical Doppler spectrum of the ductus venosus, showing a systolic and diastolic peak and a reduction in flow during atrial contraction.

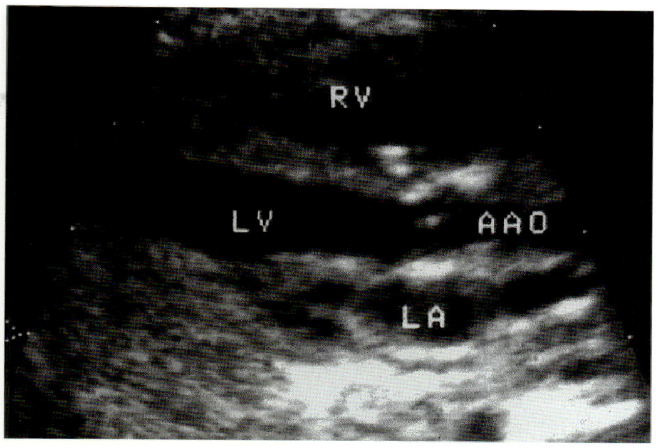

Fig. 22.**9 a** Long-axis view of the heart. The right ventricle (RV) is anterior. The posterior chambers are the left atrium (LA) and left ventricle (LV), from which the aorta (AAo) arises.

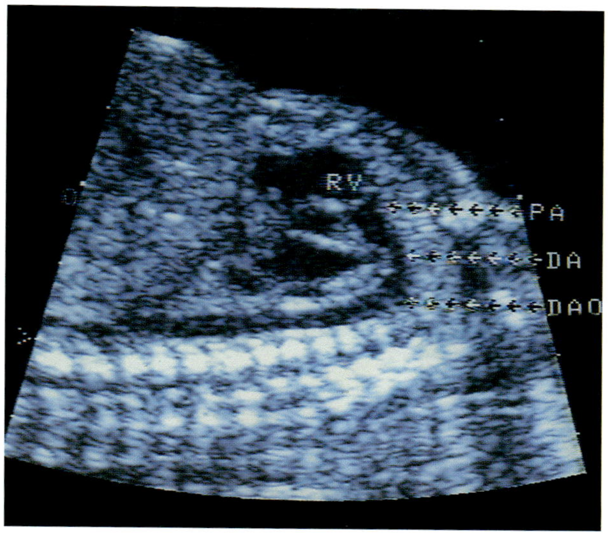

Fig. 22.**9 b** Rotating the transducer clockwise demonstrates the origin of the pulmonary artery (PA) from the right ventricle (RV). The pulmonary artery "crosses" the aorta in this projection. The ductus arteriosus (DA) connects the pulmonary artery to the descending aorta (DAO) via the ductal arch. Below the aorta is the fetal spine.

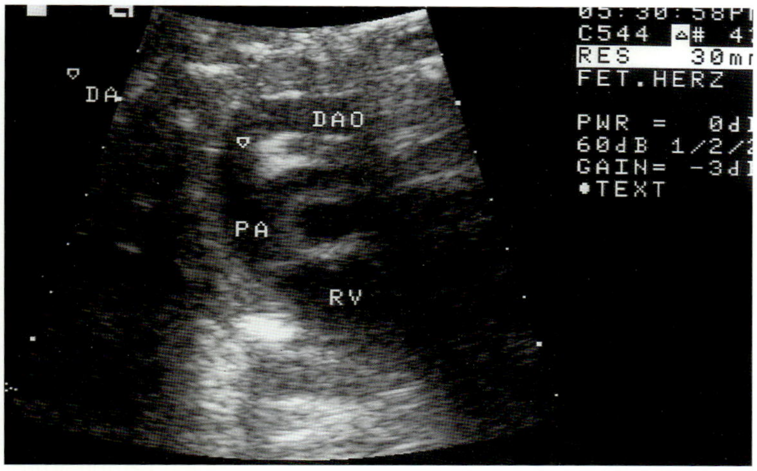

Fig. 22.**10 a** The ductal arch (with the fetal spine above). The pulmonary artery (PA), ductus arteriosus (DA), and descending aorta (DAO) are in direct continuity. In front of the descending aorta is the right ventricle (RV), from which the pulmonary artery arises. The ductus arch is broad and does not give rise to any brachiocephalic vessels.

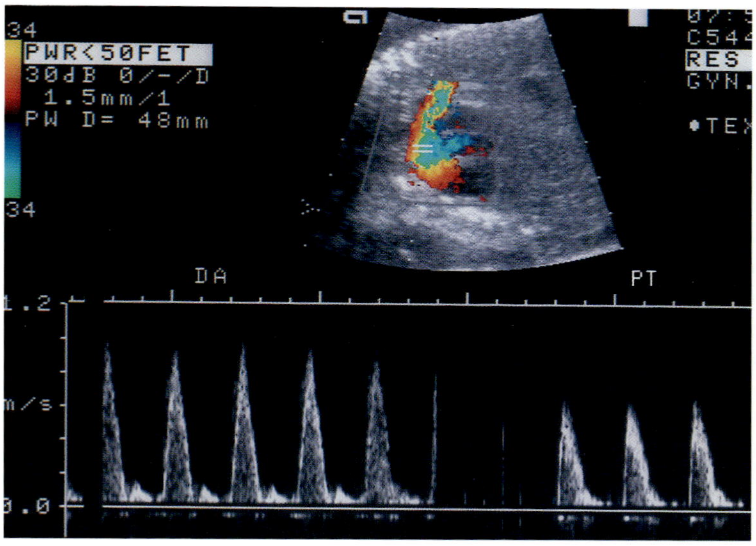

Fig. 22.**10 b** Color Doppler image in the same plane. The sample volume is placed first in the ductus arteriosus (DA, left spectrum) and then more proximally in the pulmonary trunk (PT, right spectrum). The ductus arteriosus exhibits a higher flow velocity with a brief early diastolic dip followed by low diastolic flow.

Position of the Fetal Heart in Abdominal and Thoracic Scans

Horizontal transverse scan. The position of the abdominal aorta and inferior vena cava is evaluated in the horizontal transverse scan. The descending aorta runs anterior to the spinal column on the left side, the inferior vena cava on the right side.

Thoracic transverse scan. The spinal column also provides the orientation landmark in the thoracic transverse scan. The thoracic part of the descending aorta is located anterior to the spinal column on the left side. In front of the aorta is the left atrium, into which the pulmonary veins drain. The fetal heart is located in the anterior part of the chest, with the cardiac apex and the interventricular septal axis pointing toward the left side (in the four-chamber view). The right ventricle abuts the right anterior chest wall. It is distinguished by the presence of the moderator band, a transverse muscle bundle located at the right ventricular apex. The left ventricle lies anterior to the left atrium, which is identified by its relationship to the aorta.

Morphology of the Fetal Heart

Atria and ventricles. The right atrium and right ventricle of the fetal heart are each slightly larger than the left atrium and left ventricle.

Foramen ovale. The foramen ovale is an opening between the right atrium and left atrium. It occupies approximately one-third of the interatrial septum and measures about two-thirds the diameter of the aortic root. Consistent with the right-to-left shunt through the foramen ovale, the free edge of the septum primum is deviated to the left.

Cardiac valves. An atrioventricular (AV) valve is located between each atrium and the corresponding ventricle, and the normal function of these valves should be confirmed by Doppler sonography. The same applies to the semilunar valves of the pulmonary and aortic arteries. The aortic valve is located at the center of the heart where the atrioventricular septum and AV valve plane intersect (five-chamber view). In the long-axis view, the pulmonary artery is anterior, and the ductus arteriosus (DA) connects the pulmonary trunk with the descending aorta by way of the ductal arch. The aortic arch and ductal arch can both be identified in the long-axis view and its variations. Normally the two great arteries cross each other after leaving the heart (crossing of the outflow tracts).

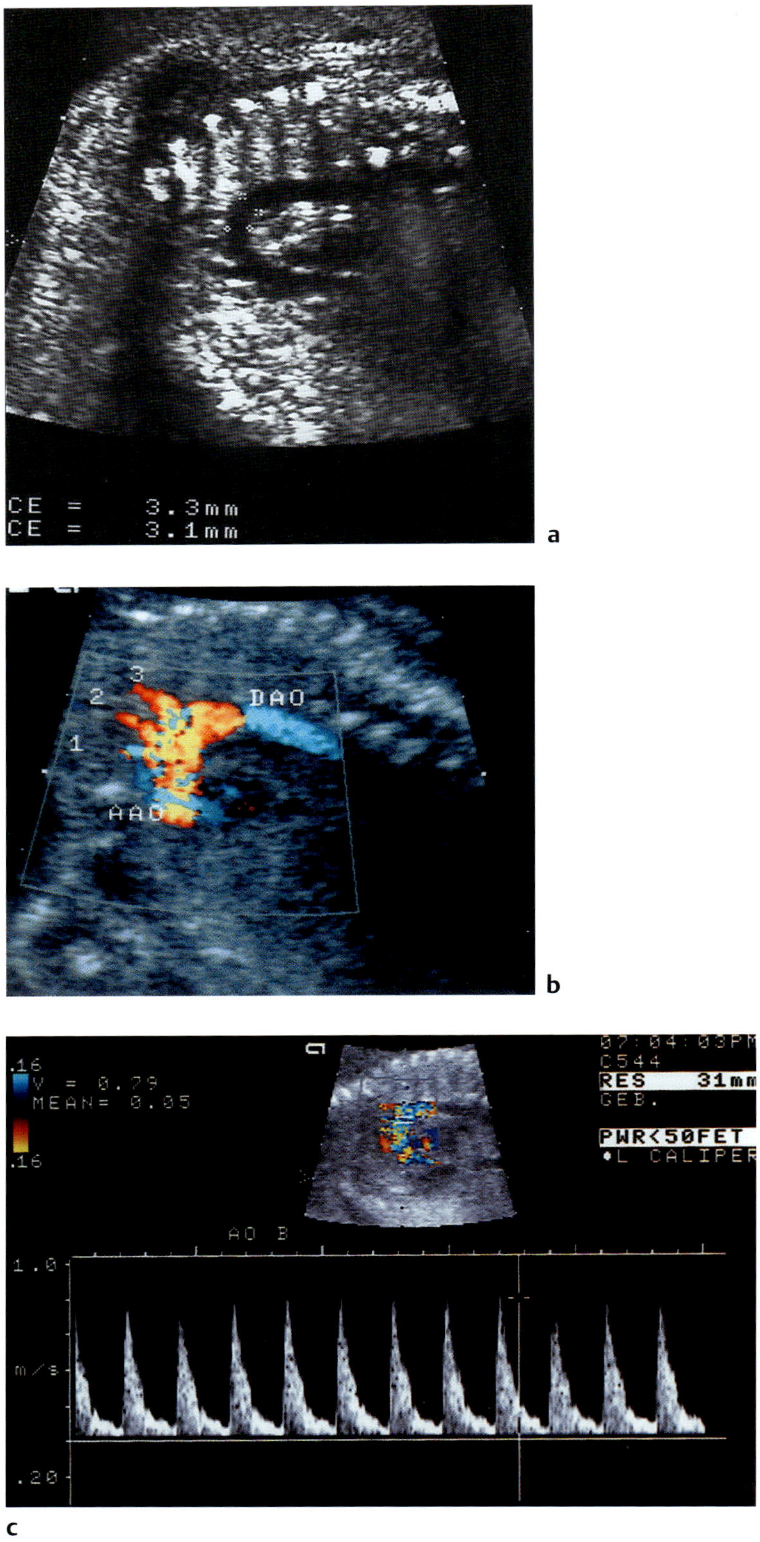

Fig. 22.**11 a** The aortic arch (with the fetal spine above). The three brachiocephalic vessels are seen arising from the aortic arch.
b Color Doppler image of the aortic arch in the same plane demonstrates the origins of the brachiocephalic trunk (1), left carotid artery (2), and left subclavian artery (3). Note the associated color reversal between the ascending aorta (AAO) and descending aorta (DAO).
c Typical Doppler spectrum of the aortic arch, indicating a low diastolic flow velocity.

Color Doppler in the Diagnosis of Congenital Heart Disease

This section reviews the echocardiographic features of specific congenital heart defects and their associated anomalies.

The distinctive features of the heart defects and the diagnostic information that is gained by the use of color Doppler are highlighted in boxes thus:

Distinctive features of the heart defect.

Color Doppler

Diagnostic information gained by adding color Doppler.

Defects Involving Abnormal Shunts

Atrial Septal Defect of the Ostium secundum Type (ASD II)

Only larger defects can be distinguished antenatally from the foramen ovale, which normally occupies one-third of the atrial septum (Fig. 22.**12**). The following differentiating features are helpful:

- Defect located at the center of the atrial septum
- Absence of the septum primum flap
- (Volume overload on the right atrium and ventricle)

Color Doppler

Demonstrates broad shunting of blood across the atrial septal defect.

Total Anomalous Pulmonary Venous Return

Three main types of total anomalous pulmonary venous return are recognized, as described below. Mixed forms may also occur.

➤ **Supracardiac anomalous pulmonary venous return.** The pulmonary veins join to form a confluence behind the left atrium that drains through a left ascending vertical vein into the anonymous vein and from there into the superior vena cava.

➤ **Cardiac type.** The pulmonary veins drain directly into the right atrium or into the lower part of the superior vena cava or coronary sinus.

➤ **Infracardiac type.** Blood drains from the pulmonary venous confluence into a vessel that descends through the diaphragm and opens into the ductus venosus or inferior vena cava.

All three types have several features in common:

- The pulmonary veins do not drain into the left atrium at the usual location.
- There is an increased volume load on the right atrium and right ventricle, which are markedly larger than the left atrium and left ventricle.
- An atrial septal defect is usually present.

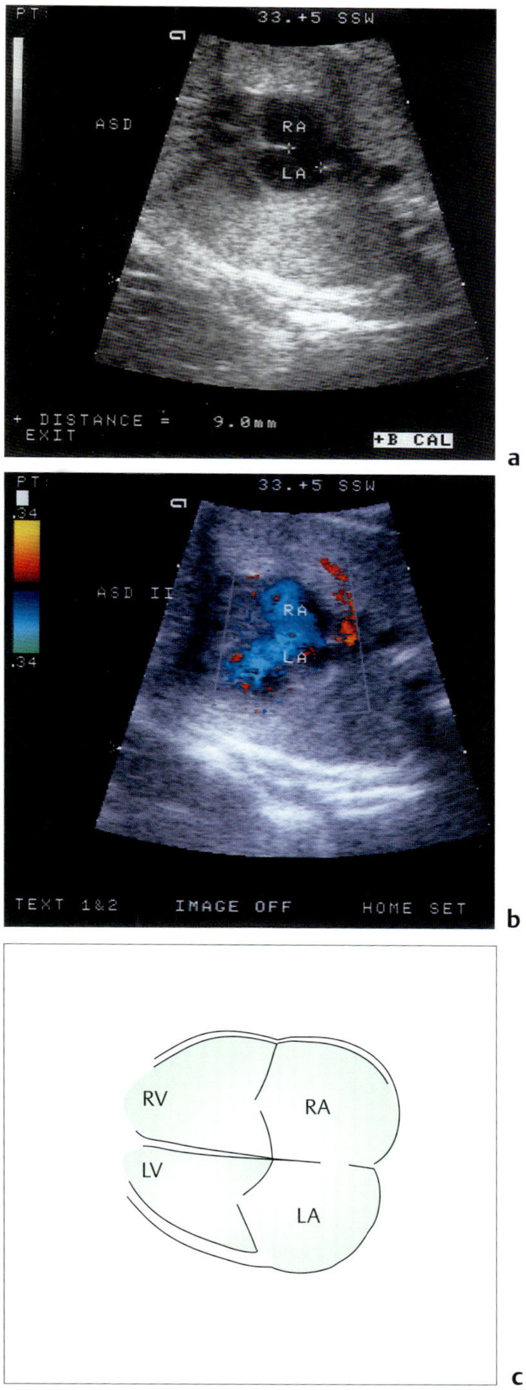

Fig. 22.**12** Large atrial septal defect in the ostium secundum position (ASD II), 27th week of gestation.
a–c Pictured are the two atria (RA, LA) with the AV valve plane and the entrances to both ventricles. The atrial septum is absent except for a small portion near the AV valves. RV = right ventricle, LV = left ventricle.

Color Doppler

Demonstrates the venous anomalies.

Atrial Septal Defect in the Ostium primum Position (ASD I, Partial Atrioventricular Septal Defect)

- This defect is located in the lower part of the interatrial septum, extending to the AV valve plane (Fig. 22.**13**).

- The defect is often combined with AV valve regurgitation, most commonly affecting the mitral valve.

Color Doppler

Demonstrates the shunt across the atrial septal defect and AV valve regurgitation.

Ventricular Septal Defect (VSD)

VSD is the most common congenital heart defect, with a relative incidence of 30–35%.

➤ Defect located in the perimembranous area.
  - Usually this defect is relatively easy to define with ultrasound (Fig. 22.**14**).

Associated anomalies: tetralogy of Fallot, AV canal, transposition of the great arteries, etc.

➤ Defect located in the muscular ventricular septum (middle and apical septal region).
  - This VSD is usually small, and so it cannot always be detected prenatally.

Color Doppler

Color flow is not always rewarding with an isolated VSD because the antenatal pressures in the two ventricles are approximately equal.

Atrioventricular Septal Defect

In this anomaly, an ASD I is combined with an inlet VSD and AV valve regurgitation (Fig. 22.**15**).
- Ultrasound demonstrates a large defect extending from the atrium to the center of the ventricular septum and a large "common" AV valve.
- Occasional hypoplasia of one of the ventricles.

Color Doppler

Demonstrates shunt across the atrioventricular septal defect.

Common in trisomy 21 and other chromosome abnormalities, less common in heterotaxic syndromes.

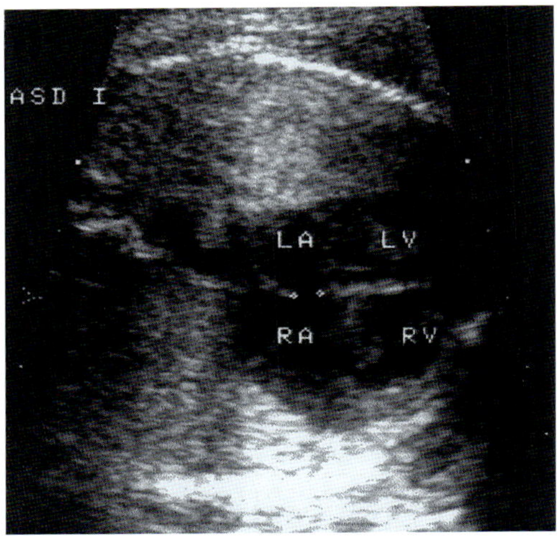

Fig. 22.**13** Atrial septal defect in the ostium primum position (ASD I, 31st week of gestation). The lesion consists of a 3.3 mm defect in the inferior part of the interatrial septum above the AV valve plane. RV = right ventricle; LV = left ventricle; LA = left atrium; RA = right atrium. The pregnant woman herself had been operated for an ASD I.

Ductus Arteriosus (DA)

- The ductus arteriosus normally connects the pulmonary artery to the descending aorta in the fetus with right to left shunting.

Defects involving the ductus may cause shunt reversal: Pulmonary atresia with an intact ventricular septum or VSD Tricuspid atresia with an absence of antegrade pulmonary artery flow

Color Doppler

Color Doppler demonstrates the flow direction across the DA, which is confirmed in the Doppler spectrum.

Constriction of the ductus arteriosus. This can result from the maternal use of prostaglandin synthesis inhibitors (e.g., ASS doses higher than 100 mg/kg per day or indomethacin). An increased flow velocity can be detected in the duct (systolic velocity > 1.40 m/s), and there is overload of the right heart.

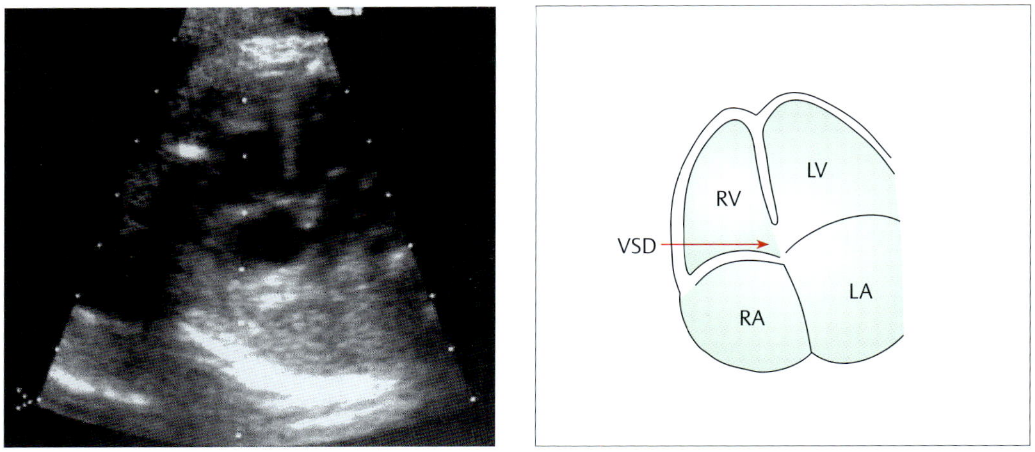

a b

Fig. 22.**14** Perimembranous ventricular septal defect (VSD), 30th week of gestation. Ultrasound was done to investigate fetal arrhythmia.
a, b Four-chamber view demonstrates a moderately large ventricular septal defect in the perimembranous area. No additional anomalies were found. RV = right ventricle; LV = left ventricle; LA = left atrium; RA = right atrium.

22

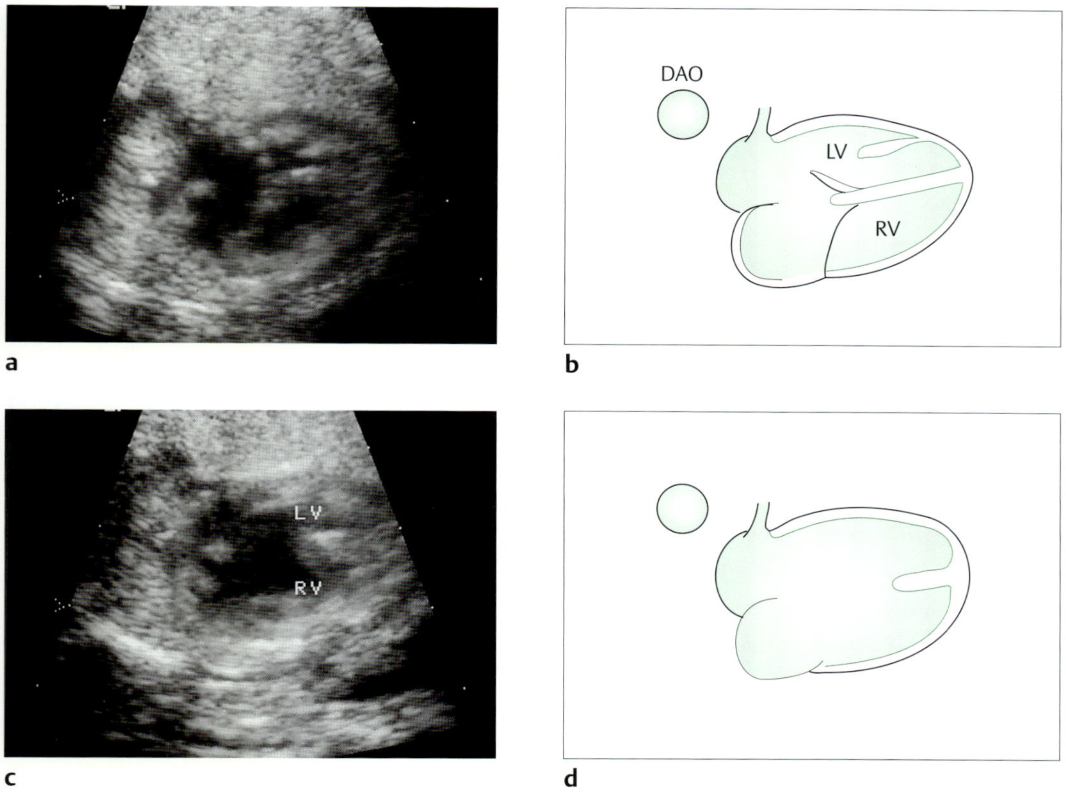

Fig. 22.**15** Complete atrioventricular septal defect, 24th week of gestation.
a, b Four-chamber view with the AV valves closed. The septum primum defect (ASD I) can be identified.
c, d In a four-chamber view with the AV valves open, the large composite defect (ASD I plus VSD) can be seen. DAO = descending aorta' RV = right ventricle; LV = left ventricle.

Defects with Obstruction of the Left Heart

Valvular Aortic Stenosis (Fig. 22.**16**)

The left ventricle in this condition may exhibit a variety of morphological criteria.

➤ **Dilated left ventricle.** Poor pump function, mitral insufficiency.

➤ **Normal-size left ventricle.** Usually there is only moderate impairment of pump function.

➤ **Small left ventricle.** Narrow ventricular cavity, myocardial hypertrophy, endocardial fibroelastosis.

- Base of aortic valve and ascending aorta are narrow or of normal width.
- Aortic valve is thickened.
- Left atrium is dilated (especially with mitral insufficiency).
- Left-to-right shunt through the foramen ovale or intact interatrial septum with severe stenoses.
- Doppler flow velocity in the ascending aorta is usually increased, but may be normal or reduced if there is severe aortic stenosis with impaired left ventricular function.

Color Doppler

Turbulent color pattern in the ascending aorta. Color flow may be absent, though, if there is minimal blood flow in the ascending aorta due to a critical stenosis.
Demonstration of mitral insufficiency.

Coarctation of the Aorta

The degree of this malformation is variable. The most severe form is interruption of the aortic arch. Coarctation of the aorta is rarely diagnosed before birth. The complete form is usually detected postnatally after the ductus arteriosus has closed.

- Diagnosis is aided by hypoplasia of the aortic arch and a narrow isthmus.
- The left ventricle and ascending aorta may be hypoplastic.

Color Doppler

Color flow does not add diagnostic information in the isthmic region.

Associated ventricular septal defect in 25% of cases. Associated chromosome abnormalities (45 XO, trisomy 13 or 18).

Hypoplastic Left Heart Syndrome (Fig. 22.**17**)

- Hypoplasia, high-grade stenosis, or atresia of the mitral valve and/or aortic valve
- Hypoplasia of the left atrium, left ventricle, aortic valve, ascending aorta
- Hyperechoic left ventricle (endocardial fibroelastosis)
- Reverse flow in the ascending aorta from the DA due to aortic valve atresia
- Increased flow velocity in the ductus and descending aorta (total cardiac output)
- Left-to-right shunt through the foramen ovale or intact interatrial septum (due to increased pressure in the left atrium)

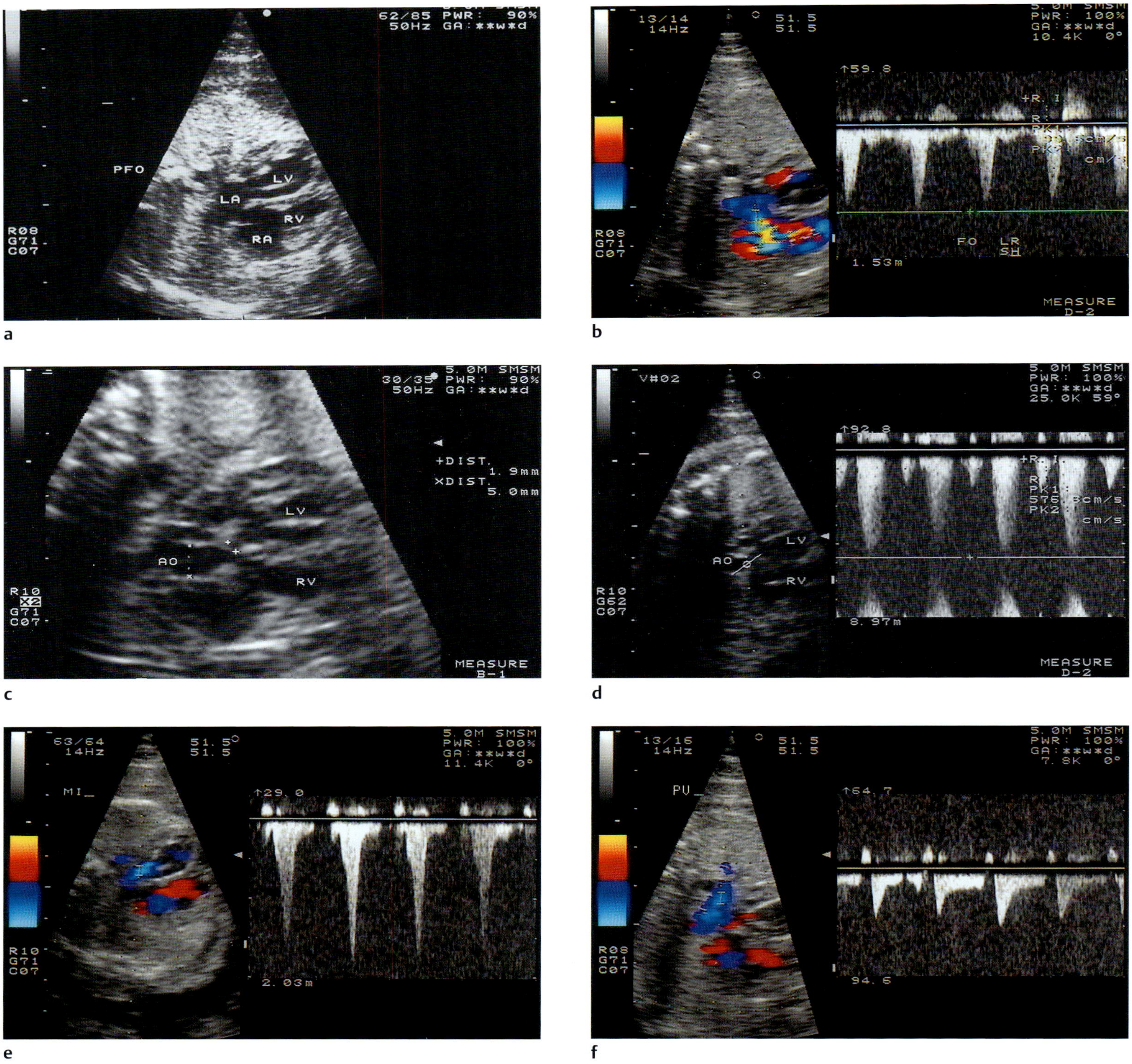

Fig. 22.**16** Valvular aortic stenosis (26th week of gestation).
a In the four-chamber view, the left ventricle (LV) is not enlarged but is hypocontractile with an echogenic endocardium (endocardial fibroelastosis). The septum secundum is deflected into the right atrium (RA) by the increased pressure in the left atrium (LA). RV = right ventricle.
b Color flow through the foramen ovale (FO) and the Doppler spectrum indicate a left-to-right shunt resulting from the increased pressure in the left atrium.
c The base of the aortic valve is narrow (1.9 mm), and the valve is thickened and shows decreased motion. The ascending aorta (AO) shows poststenotic dilatation (5 mm).
d The maximum systolic Doppler gradient across the aortic valve is 130 mmHg.
e Mitral insufficiency because of pressure load on the left ventricle. The flow is encoded in blue, indicating a flow direction from the left ventricle into the left atrium.
f Color-encoded flow from the left pulmonary vein. Doppler indicates brief reverse flow (encoded in blue) during atrial contraction.

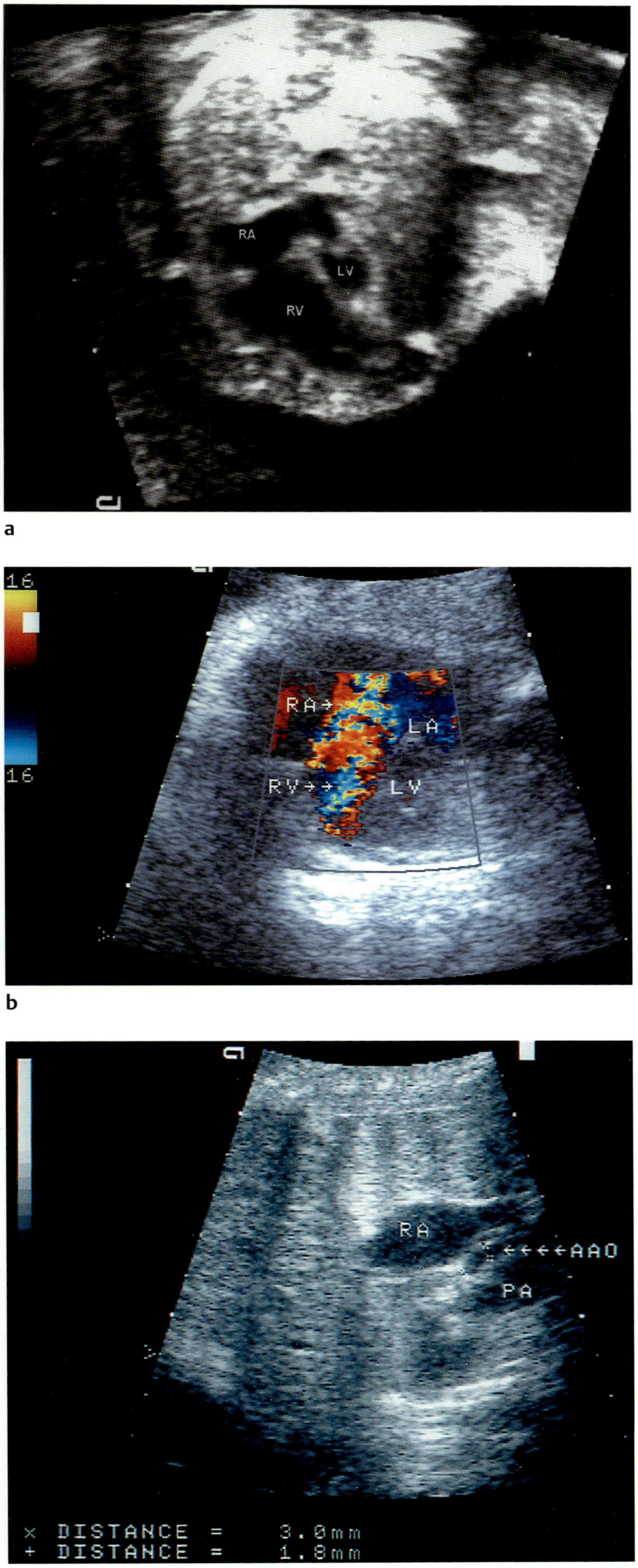

◁ Fig. 22.**17** Hypoplastic left heart syndrome (37th week of gestation).
a Four-chamber view: the right atrium (RA) is markedly dilated, and the right ventricle (RV) is slightly enlarged. The cavity of the left ventricle (LV) is hypoplastic with an atretic mitral valve.
b Color Doppler image confirms mitral valve atresia.
c Hypoplastic ascending aorta (AAO, diameter 3 mm) with aortic valve atresia (valve diameter 1.8 mm). Antegrade flow is not demonstrated by spectral or color Doppler. PA = pulmonary artery.

Color Doppler

Absence of blood flow into the left ventricle.
Absence of antegrade blood flow in the ascending aorta.
Possible reverse flow in the ascending aorta (from the DA).
Left-to-right shunt across the foramen ovale.

Rarely, the syndrome is combined with VSD or transposition of the great arteries. Relatively high familial risk of recurrence (10–20 %).

All defects involving a left ventricular outflow tract obstruction are an indication for serial ultrasound examinations, as the defect often does not become fully developed until the second half of the pregnancy. Approximately 7 % of these heart defects are associated with fetal hydrops.

Defects Involving Obstruction of the Right Heart

In all heart defects that are associated with stenosis of the pulmonary valve and/or hypoplasia of the pulmonary trunk, it is common to find that the degree of stenosis increases mainly in the third trimester of pregnancy. It can be difficult to differentiate complete valvular atresia from a high-grade stenosis.

Valvular Pulmonary Stenosis

- Thickened pulmonary valve
- Hypertrophy of the right ventricle (depending on the degree of stenosis)

Color Doppler

Turbulence in the pulmonary trunk due to accelerated flow across the pulmonary valve.

Critical Pulmonary Stenosis, Pulmonary Atresia with an Intact Ventricular Septum (Fig. 22.**18**)

- No opening of the atretic pulmonary valve
- Hypertrophic right ventricle, usually with a small cavity
- Frequent hypoplastic tricuspid valve, tricuspid insufficiency

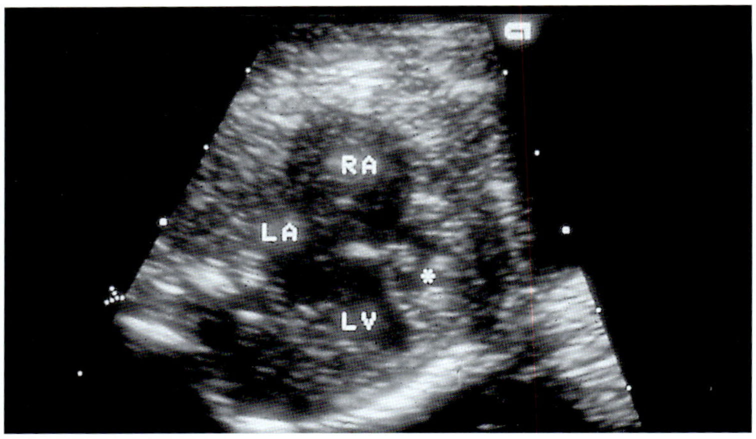

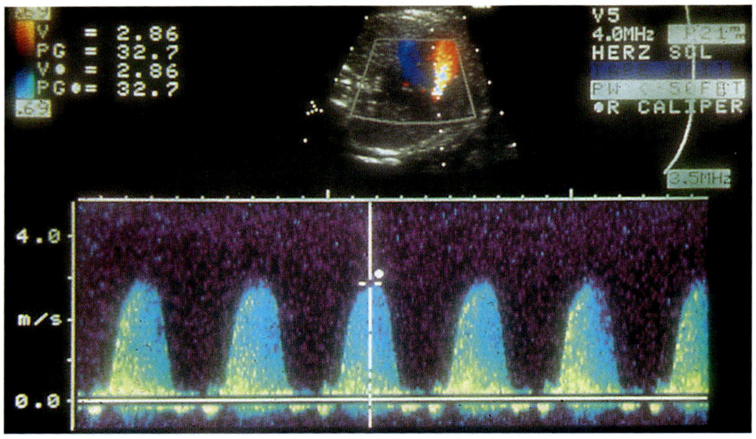

Fig. 22.**18** Pulmonary atresia with an intact ventricular septum.
a Four-chamber view: the right atrium (RA) is markedly enlarged, and the tricuspid valve and right ventricle (*) are hypoplastic. LA = left atrium; LV = left ventricle.

b Color flow shows pronounced tricuspid insufficiency with a jet directed onto the lateral wall of the right atrium (with kind permission of Professor Michael Hofbeck, Department of Pediatric Cardiology, Tübingen University Children's Hospital).

Color Doppler

Reverse flow in the pulmonary trunk via the DA with pulmonary atresia.

Tetralogy of Fallot (Fig. 22.**19**)

- Both ventricles are of equal size.

Color Doppler

Ventricular septal defect with a broad aorta overriding the ventricular septum defect; the degree of pulmonary stenosis and hypoplasia of the pulmonary trunc are often increasing during the third trimester.
Turbulence in the pulmonary trunk.

Association with chromosome abnormalities:
CATCH 22 (acronym for cardiac defects, *a*bnormal facies, *t*hymic hypoplasia, cleft palate, *h*ypocalcemia, and chromosome *22*).
Associated with DiGeorge syndrome, velocardiofacial syndrome, and conotruncal facial anomaly syndrome.
Coexists with atrioventricular septal canal defect in trisomy 21.

Pulmonary Atresia with Ventricular Septal Defect

- Like the tetralogy of Fallot, but with an atretic pulmonary valve
- Variable hypoplasia of the pulmonary trunk

Color Doppler

Reverse flow in the pulmonary trunk through the DA.

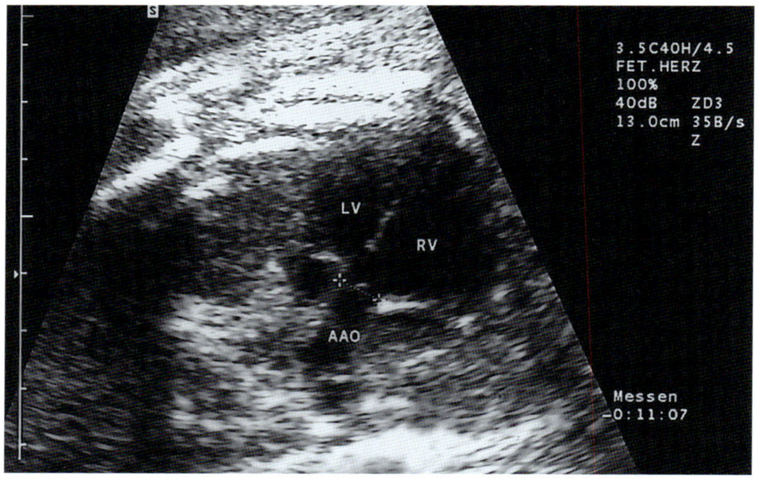

a

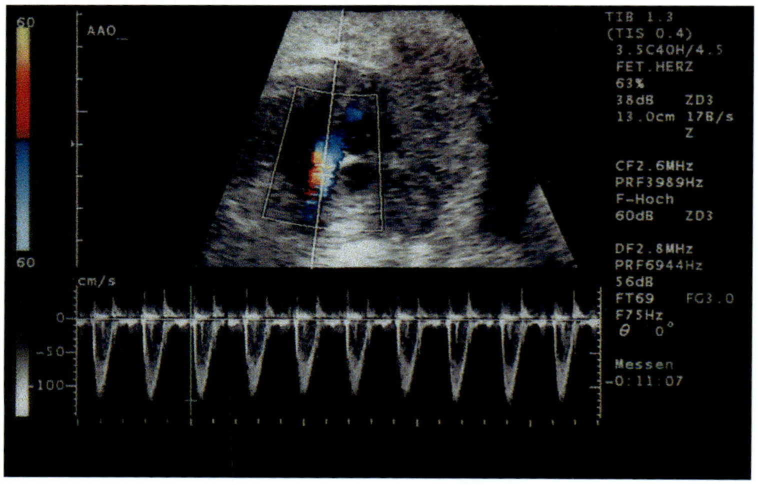

b

Fig. 22.**19** Tetralogy of Fallot (35th week of gestation).
a Four-chamber view with a subaortic ventricular septal defect. There is apparent marked dextroposition of the aorta. RV = right ventricle; LV = left ventricle; AAO = ascending aorta.

b Aortic flow demonstrated by color and spectral Doppler. The aortic flow velocity is 120 cm/s.

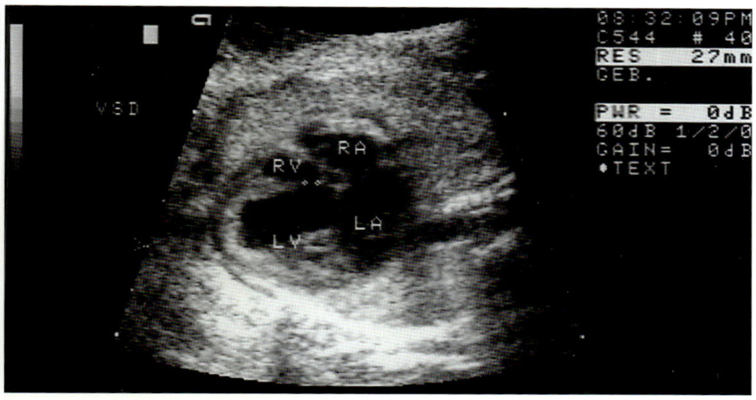

a

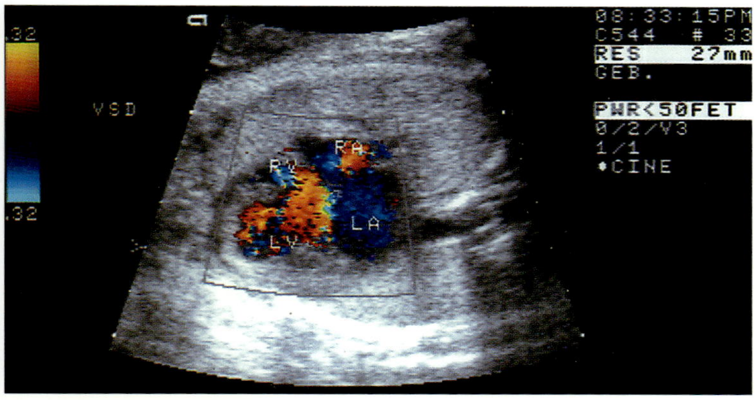

b

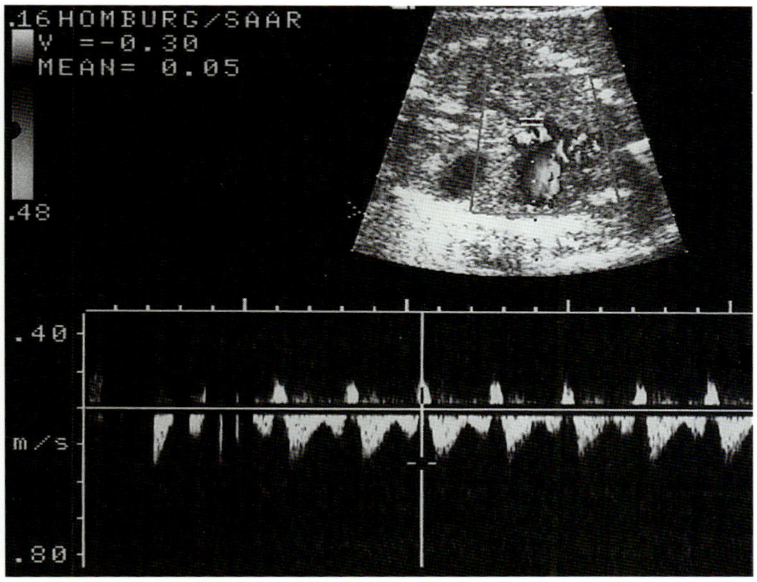

c

Fig. 22.**20** Tricuspid atresia with a ventricular septal defect (32nd week of gestation).
a The four-chamber view shows a large left ventricle (LV) that communicates with a hypoplastic, anteriorly situated right ventricle (RV) through a 2.6 mm ventricular septal defect (VSD). The tricuspid valve is atretic. LA = left atrium; RA = right atrium.
b Color Doppler image demonstrates flow from the left ventricle into the right ventricle across the ventricular septal defect (encoded red) and flow from the right atrium into the left atrium across the atrial septal defect (encoded blue).
c Doppler spectrum of flow in the inferior vena cava shows a brief period of reverse flow during atrial contraction (indicating right-sided inflow stasis).

Tricuspid Atresia (Fig. 22.**20**)

- Variable hypoplasia of the right ventricle, hypertrophy of the ventricular myocardium, small right ventricular cavity
- Enlarged right atrium
- Frequent ventricular septal defect of variable size, with a tendency to occlude

Color Doppler

With a ventricular septal defect: antegrade flow in the pulmonary trunk.
With pulmonary atresia: reverse flow across the DA.
Absence of inflow through the tricuspid valve into the right ventricle.

Rarely, combined with transposition of the great arteries. These cases have a hypoplastic ascending aorta.

Single Ventricle

Single ventricle (univentricular heart) is the term for a very heterogeneous group of anomalies whose varying pathomorphological features determine the therapeutic approach and the prognosis. In all forms, only one functionally competent ventricle is present; the second ventricle is usually present but hypoplastic. The classification of this anomaly is based on the configuration of the AV valves and ventricles and on the ventriculoatrial connection (Fig. 22.**21**).

➤ **Atrioventricular connection**
- Two AV valves open into the main chamber (double-inlet configuration)
- One common AV valve
- One absent or atretic AV valve

➤ **Ventricular morphology**
- Morphological right ventricle with a left ventricular outlet chamber
- Morphological left ventricle with a right ventricular outlet chamber
- Common ventricle with a right and left ventricular component; the ventricular septum is (almost completely) absent
- Single ventricle of indeterminate morphology

➤ **Ventriculoatrial connection**
- Normal position of the great arteries
- Transposition of the great arteries
- Frequent hypoplasia, stenosis, or atresia of one of the two arteries

The principal associated cardiac anomalies, besides atrial septal defect, are systemic venous anomalies (heterotaxy syndrome) and abnormalities of cardiac position and of thoracic and abdominal situs.

The most common variants of this complex anomaly are tricuspid atresia, mitral atresia or hypoplastic left heart syndrome, double-inlet left ventricle, and double-outlet right ventricle. Complete absence of the second ventricle is rare. Factors that critically affect peri- and postnatal hemodynamics and the therapeutic approach are the interatrial communication, persistence of the ductus arteriosus, and possible obstruction of the right or left ventricular outflow tract or the associated arterial vessel.

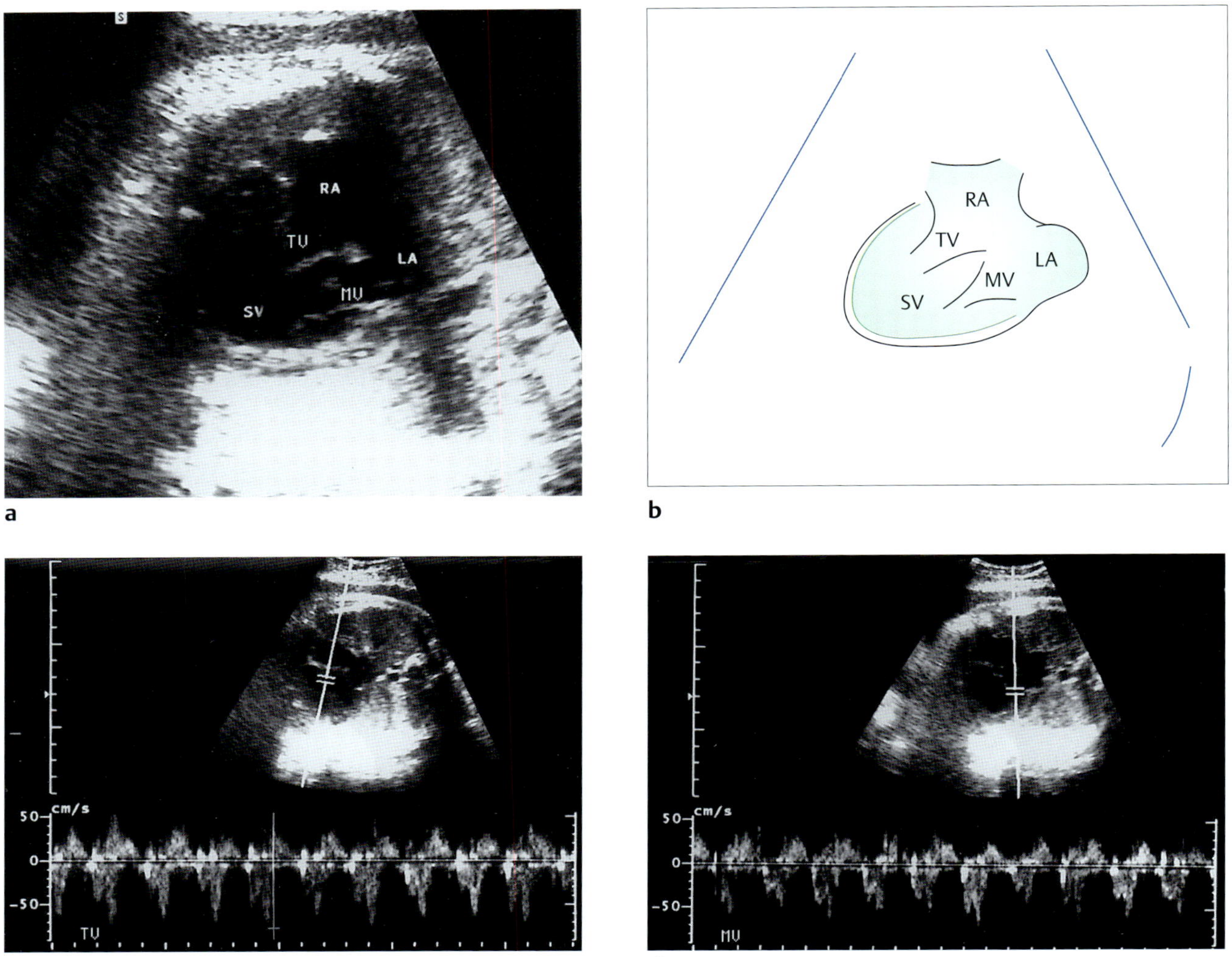

a b c d

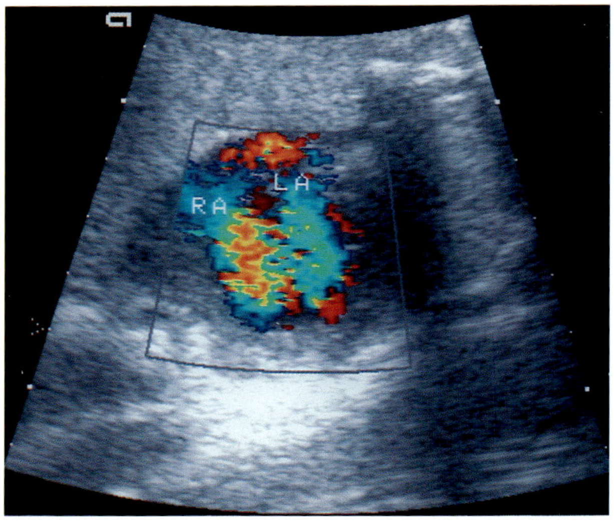

e

Fig. 22.**21** Single ventricle (30th week of gestation) with a double-inlet left ventricle, a right ventricular outlet chamber, VSD, and normal position of the great arteries.
a, b The right and left atria (RA, LA) drain respectively through a tricuspid valve (TV) and mitral valve (MV) into the single ventricle (SV).
c, d Doppler scan confirms the normal function of the tricuspid valve (TV) and mitral valve (MV).
e Color Doppler shows blood flowing from the atria into the ventricle. There is only partial mixing of the flows within the single ventricle.

Fig. 22.**21 f** and **g** ▷

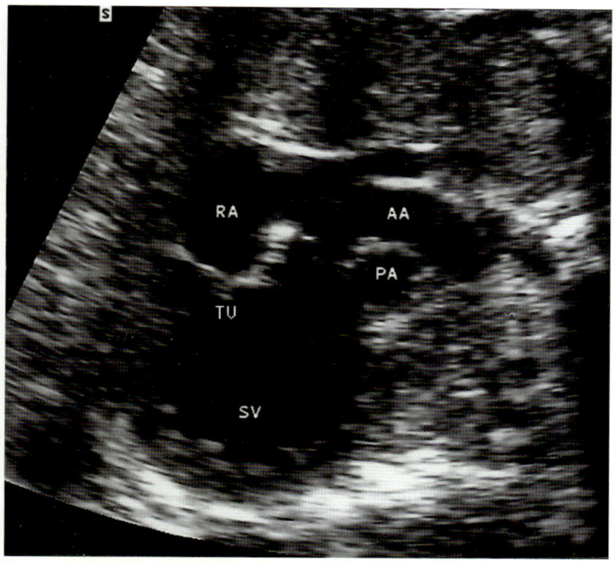

f

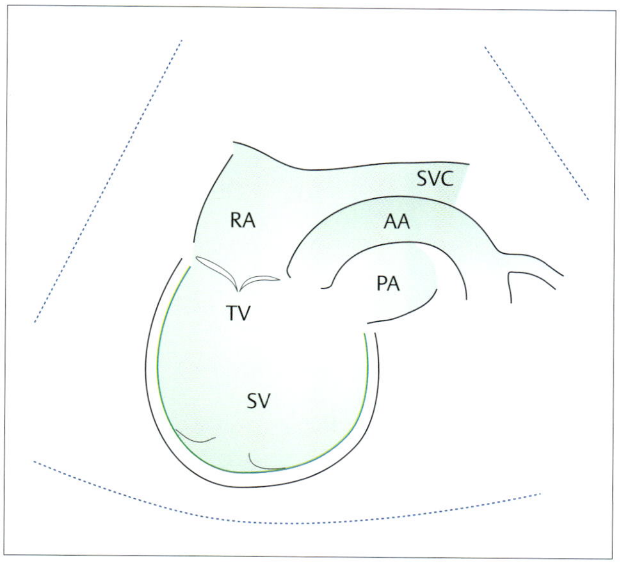

g

Fig. 22.**21 f, g** The aorta arises from the single ventricle, which has a left ventricular morphology. The aortic arch (AA) can be identified. The pulmonary artery trunk (PA) is imaged in cross section. RA = right atrium; TV = tricuspid valve; SV = single ventricle; SVC = superior vena cava.

Anomalies of the Great Arteries (Fig. 22.**22**)

D(Dextro)-Transposition of the Great Arteries (Fig. 22.**22a–d**)

Transposed position of the great arteries with:
- Atrioventricular concordance and ventriculoatrial discordance
- An anterior, right-sided aorta arising from the right ventricle
- A posteriorly situated pulmonary artery arising from the left ventricle
- A normal-appearing four-chamber view, but absence of a centrally located aortic valve in the five-chamber view!

Color Doppler

Both great arteries leave the heart in a parallel arrangement; they do not cross!
Combined with ventricular septal defect, rarely with pulmonary stenosis or coarctation of the aorta.

L(Levo)-Transposition of the Great Arteries (Fig. 22.**22e–h**)

There is ventricular inversion with atrioventricular and ventriculoatrial discordance:
- The right atrium is connected to a right-sided morphological left ventricle, which empties into the pulmonary artery.
- The left atrium is connected to a left-sided morphological right ventricle, which empties into the aorta.

Color Doppler

The ascending aorta is located anterior to the pulmonary artery (the aorta arises from the right ventricle at a site anterior and to the left of the pulmonary artery).
May be combined with ventricular septal defect and/or pulmonary stenosis.
Third-degree AV block is rarely present.

Associated with abnormalities of cardiac position.

Common Truncus arteriosus (Fig. 22.**23**)

- Two ventricles of equal size.

Color Doppler

A common, large arterial trunk overrides the ventricular septal defect.
The pulmonary artery arises posteriorly from the truncus arteriosus (with a common trunk or separate origin of both pulmonary branches).

Frequently associated with CATCH 22 syndrome.

Tricuspid Valve Anomalies

➤ **Tricuspid valve dysplasia** (Figs. 22.**24**, 22.**25**)
  - Valve leaflets are dysplastic, chordae are lengthened or shortened.

➤ **Ebstein anomaly**
  - Functional tricuspid valve plane is displaced into the right ventricle.
  - There is resultant arterialization of the right ventricular inlet.
  - Tricuspid valve is malformed and usually incompetent.

  Both cardiac anomalies are highly variable in their degree.

Color Doppler

Tricuspid regurgitation with an enlarged right atrium.
Hypoplastic pulmonary artery trunk with reduced flow.
Right-sided inflow stasis with dilatation of the venae cavae and hepatic veins.

Severe forms are usually associated with ascites, pleural effusion, hydrops, lung hypoplasia, and intrauterine death.

The two forms are difficult to distinguish from each other antenatally.

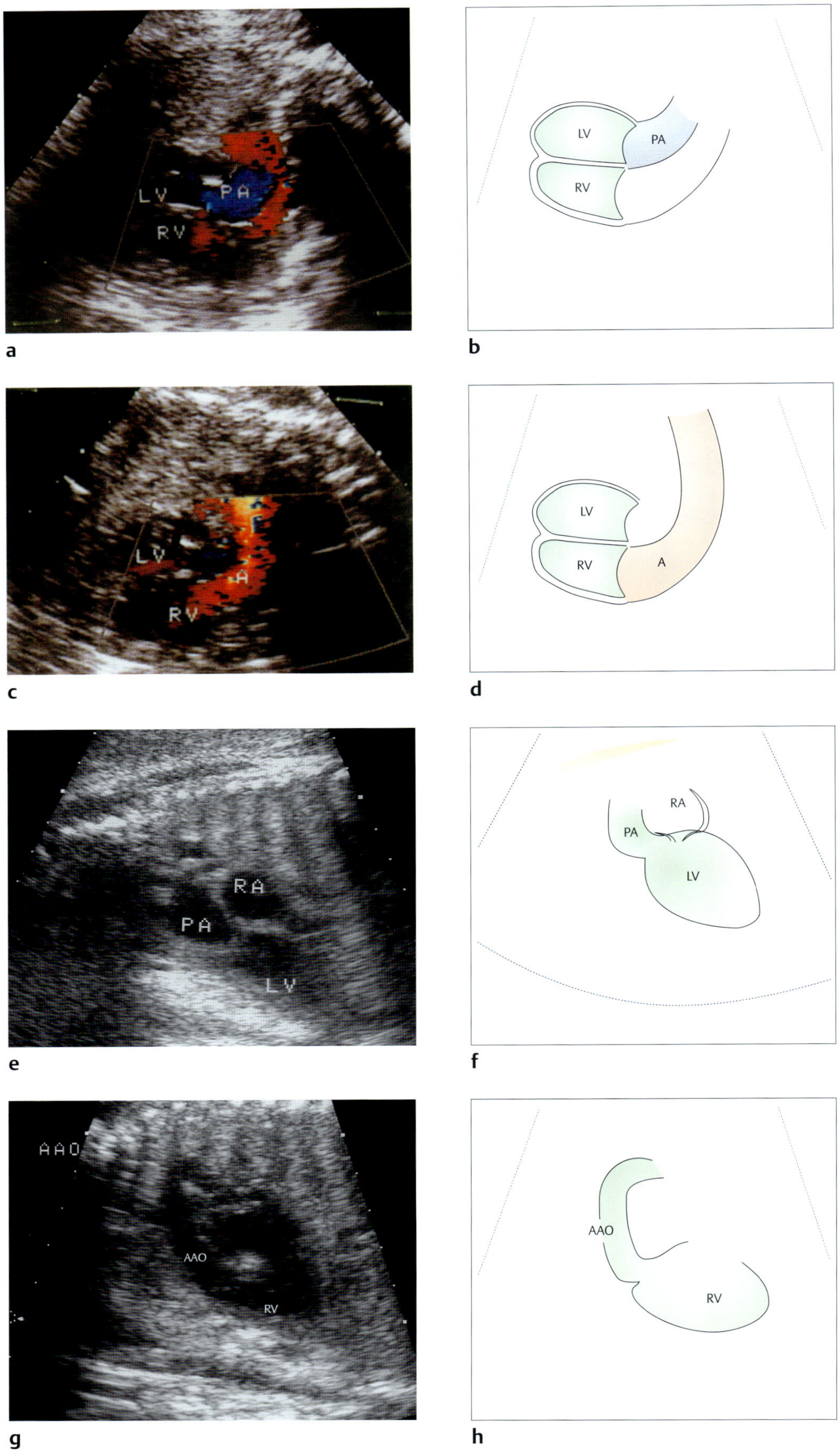

Fig. 22.**22** Transposition of the great arteries.
a, b D-transposition of the great arteries. The pulmonary artery (PA) arises from the posteriorly located left ventricle (LV).
c, d D-transposition of the great arteries. The aorta (A) arises from the anteriorly located right ventricle (RV).
e, f L-transposition of the great arteries (36th week of gestation). The right atrium (RA) drains into the left ventricle (LV), from which the pulmonary artery arises.
g, h L-transposition of the great arteries. The aorta (AAO) arises anteriorly from the right ventricle (RV).
(**a–d** with kind permission of Dr. V. Fesslova, Department of Pediatric Cardiology, Azienda Ospedaliera, Instituti Clinici di Perfezionamento, Milan, Italy.)

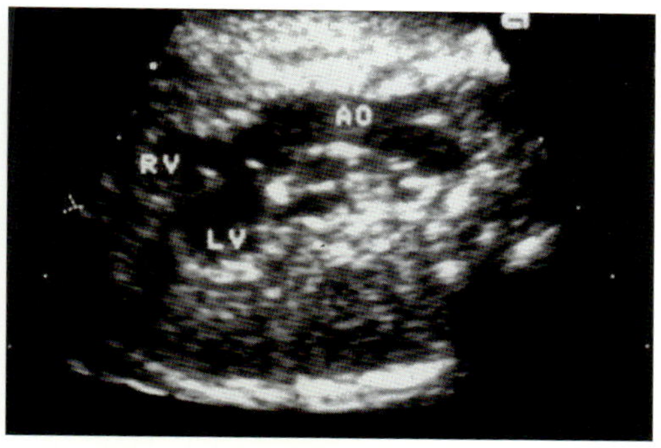

a

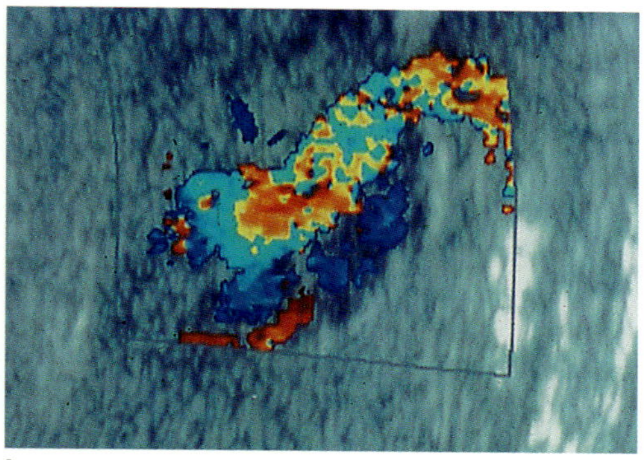

b

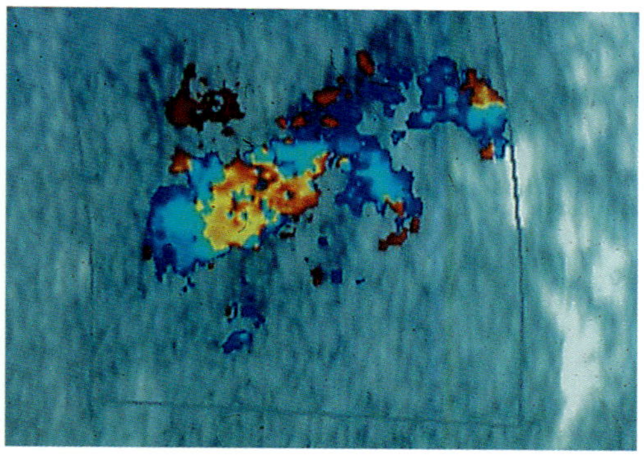

c

Fig. 22.**23** Common truncus arteriosus (32nd week of gestation).
a B-mode cardiac scan demonstrates both ventricles and a large arterial vessel. Both ventricles communicate through a large ventricular septal defect straddled by an arterial trunk (AO). A pulmonary vessel is not seen arising from the right ventricle. RV = right ventricle; LV = left ventricle.
b Same situation as in **a**: color Doppler image of the common truncus arteriosus.
c The pulmonary artery arises from the posterior part of the trunk just above the valve plane.

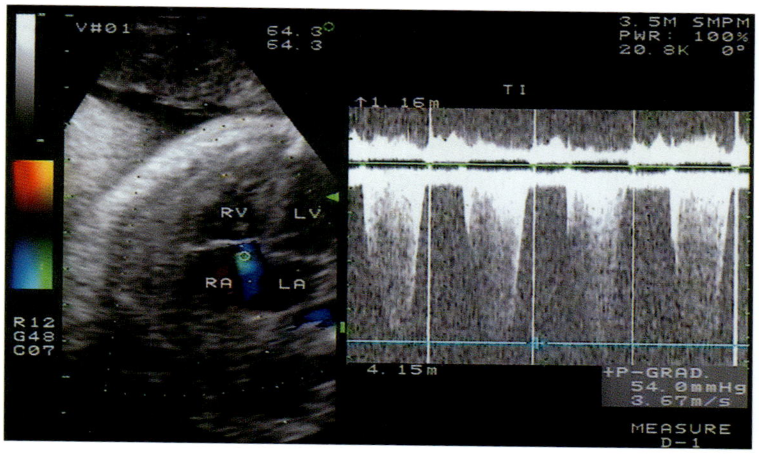

Fig. 22.**24** Tricuspid insufficiency (38th week of gestation). The four-chamber view shows a very mildly dilated right atrium (RA) with color flow indicating mild to moderate tricuspid regurgitation (blue jet across the tricuspid valve into the right atrium, directed toward the interatrial septum). LA = left atrium; RV = right ventricle; LV = left ventricle. The Doppler spectrum (right side of image) indicates a maximum systolic flow velocity of 3.67 m/s, corresponding to a pressure gradient of 54 mmHg.

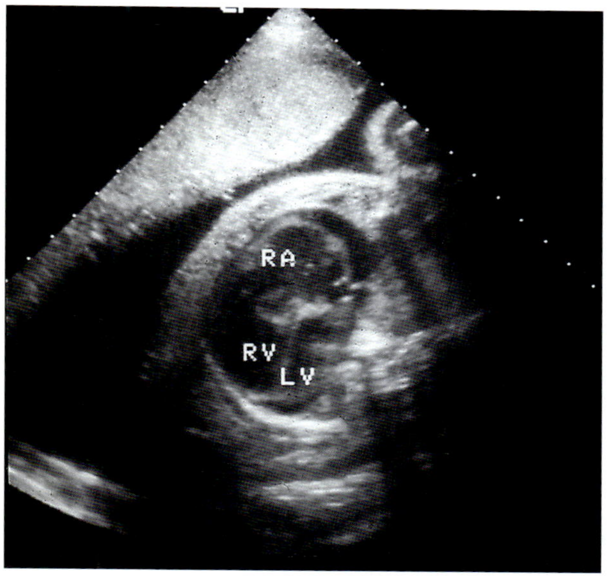

Fig. 22.**25** Tricuspid valve dysplasia with fetal hydrops (28th week of gestation). Polyhydramnios, fetal pericardial effusion, and fetal cutaneous edema are present. The four-chamber view shows a greatly enlarged right atrium (RA) with a dysplastic tricuspid valve. Otherwise the right and left ventricles (RV, LV) are of normal size. There was marked general enlargement of the heart leading to bilateral pulmonary hypoplasia, with severe postnatal breathing problems.

Heterotaxy Syndromes

These syndromes involve complex anomalies of the heart, vessels, and thoracoabdominal organs (synonym: Ivemark syndrome). Right atrial isomerism (asplenia syndrome) is distinguished from left atrial isomerism (polysplenia syndrome).

- Anomalous positions of the upper abdominal organs. The liver is often centered in the abdomen, with the stomach on the right side.
- Multiple small spleens (polysplenia) or absence of the spleen (asplenia).
- Anomalous course of the inferior vena cava (on the same side of the spine as the descending aorta) or duplication of the superior vena cava.

Color Doppler

Cardiac anomalies: single ventricle, common atrium, AV valve anomalies, pulmonary stenosis or atresia, malposition of the great arteries, anomalous pulmonary venous return, third-degree AV block.

Affected infants show normal cerebral development postnatally, but often the cardiac anomalies cannot be fully corrected.

Cardiac Tumors

Rhabdomyomas. The most common tumors are rhabdomyomas, which are usually attached to the interior walls of the heart (ventricular walls, interventricular septum, occasionally the atrial walls) and are usually multiple. The tumors frequently enlarge during the course of pregnancy. Hemodynamic complications depend on the size and location of the rhabdomyomas (e.g., valvular obstruction). After birth, the tumors usually show a gradual reduction in size over a period of several years. Bourneville–Pringle disease (tuberous sclerosis) is present as an underlying disorder in most cases (Fig. 22.**26**).

Hemangiomas and teratomas. Rarely, hemangiomas can occur in the pericardium and may cause pericardial effusion. Teratomas are also rare.

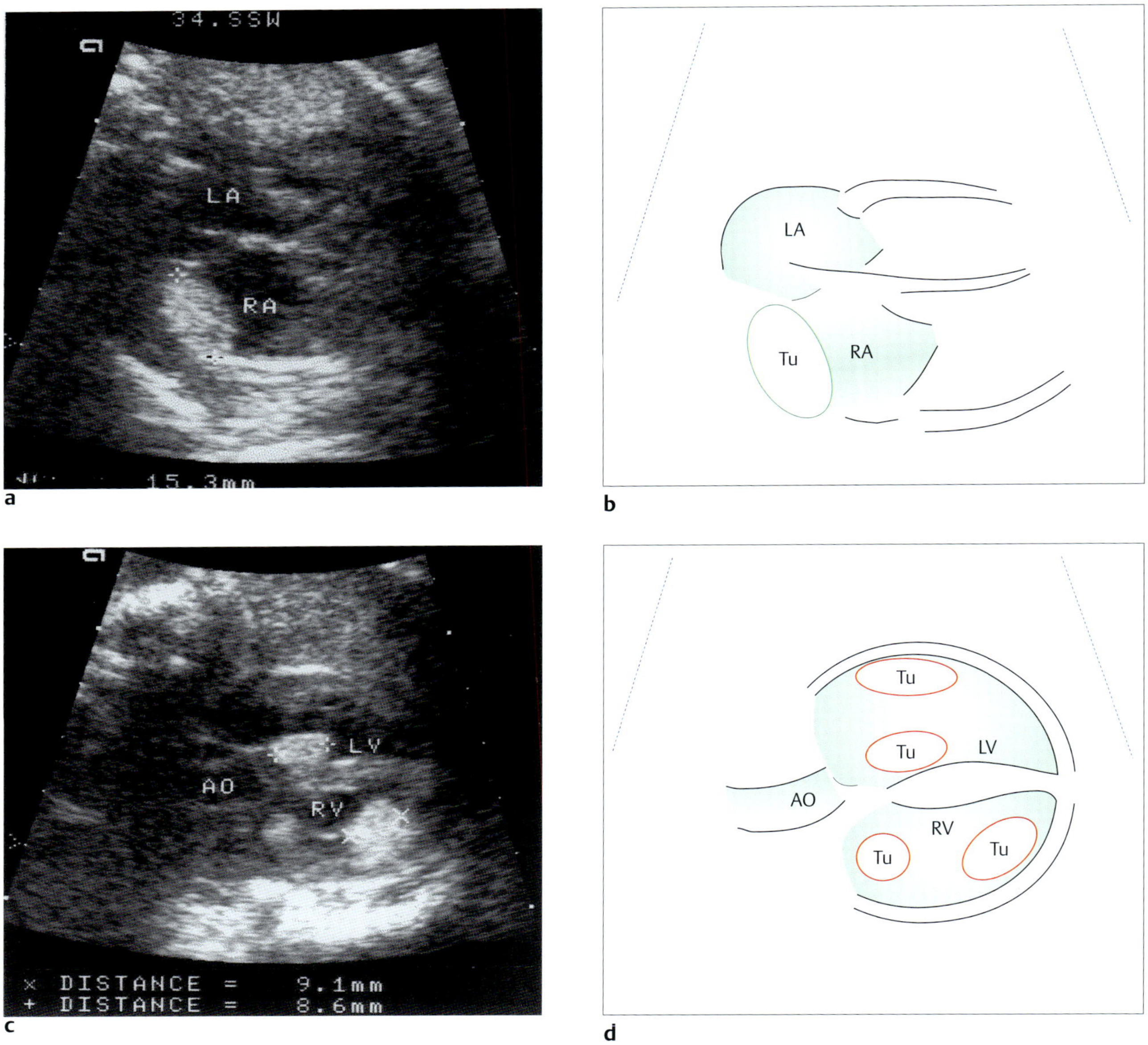

Fig. 22.**26** Cardiac rhabdomyomas in tuberous sclerosis (34th week of gestation).
a, b Scan in the atrial plane demonstrates a tumor (Tu) 15.3 mm in diameter (cursors in **a**) located on the roof of the right atrium (RA). LA = left atrium.
c, d Scan in the ventricular plane demonstrates three tumors: one on the subaortic left ventricular septum (8.6 mm in diameter), one at the right ventricular apex (9.1 mm in diameter), and a smaller tumor located in the right ventricular outflow tract. LV = left ventricle; RV = right ventricle; AO = aorta.

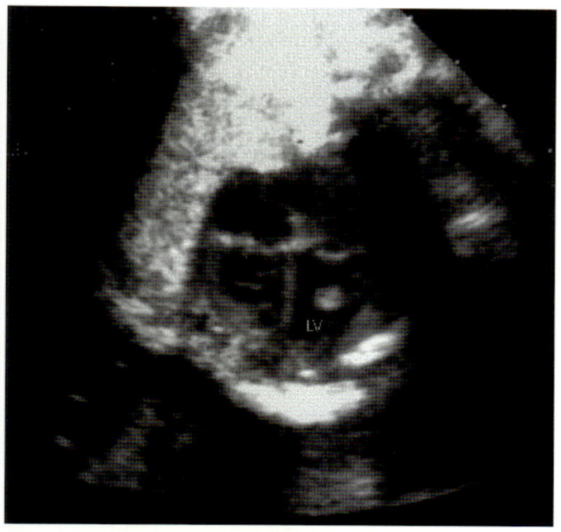

Fig. 22.**27** Four-chamber view (30th week of gestation) shows a rounded, echogenic structure in the papillary muscle of the mitral valve. The feature pulsated with the valve movements and did not cause mitral valve dysfunction. It disappeared during the first few months of life.

Differential diagnosis. Differentiation is required from a harmless mitral valve anomaly, which appears as a rounded, hyperechoic structure in the papillary muscles that moves synchronously with the pulse and has an undetermined cause (described as the "tennis ball sign" or "golf ball sign"). It does not interfere with mitral valve function (Fig. 22.**27**) and resolves after birth. Today it is thought that this feature is associated with a chromosome abnormality in approximately 1–2 % of fetuses.

Anomalies of Cardiac Position

Primary Anomalies of Cardiac Position

- **Situs inversus of the thoracic organs:** dextrocardia with the cardiac apex pointing toward the right side.
- **Complete situs inversus:** situs inversus of the thoracic and abdominal organs.

Both conditions may occur in isolation but are frequently combined with heart defects or with a heterotaxy syndrome.

- **Ectopia cordis.** In ectopia cordis, all or part of the heart is displaced anterior to the sternum. It may be associated with a chromosome abnormality, omphalocele, or an epigastric hernia. A congenital heart defect is usually present.
- **Thoracopagus.** Twins conjoined in the sternal region have very complex cardiac anomalies, often with symmetrical involvement of both hearts.

Secondary Anomalies of Cardiac Position

- **Lung malformations.** Lung hypoplasia, lung tumors, or cystic adenomatoid malformation of one lung leads to displacement of the heart and mediastinum.
- **Diaphragmatic hernia.** A diaphragmatic hernia is often present on the left side and is associated with cardiac and mediastinal displacement into the right side of the chest.
- **Diaphragmatic effusions.** Copious, unilateral diaphragmatic effusions displace the heart toward the opposite side.

Differential Diagnostic Considerations

Differential Diagnosis of Ventricular Hypoplasia

- Hypoplastic right ventricle:
  - Tricuspid atresia
  - Pulmonary atresia with an intact ventricular septum

  Common feature: pulmonary trunk is often hypoplastic.
- Hypoplastic left ventricle:
  - Hypoplastic left heart syndrome
  - Aortic valve atresia, critical aortic stenosis

Common feature: hypoplastic ascending aorta.

Differential Diagnosis of Right Atrial and Right Ventricular Dilatation

- Premature closure of the ductus arteriosus
- Premature closure of the foramen ovale
- Total anomalous pulmonary venous return
- Coarctation of the aorta
- Global heart failure (e.g., supraventricular tachycardia, fetofetal transfusion syndrome)

Differential Diagnosis of an Enlarged Right Atrium

- Tricuspid valve dysplasia
- Ebstein anomaly

Common feature: tricuspid insufficiency.

Differential Diagnosis of an Enlarged Left Atrium

- Severe valvular aortic stenosis
- Myocarditis, dilatative cardiomyopathy

Critical Postnatal Defects

Ductus-dependent defects. The following congenital heart defects can lead to a critical hemodynamic situation in the newborn due to dependence on the ductus arteriosus. These cases require early prostaglandin E_1 therapy or prompt catheter intervention or corrective surgery.

- Critical semilunar valve stenosis (aortic or pulmonary stenosis)
- Pulmonary atresia
- Tricuspid atresia
- Critical coarctation of the aorta, critical aortic valve stenosis
- Hypoplastic left heart syndrome
- Transposition of the great arteries

Decreased blood flow in the descending aorta caused by postnatal closure of the ductus arteriosus can lead to renal failure with oliguria or anuria, necrotizing enterocolitis (decreased intestinal blood flow), and functional liver failure with coagulopathy and protracted shock.

Decreased lung perfusion following ductal closure in the presence of pulmonary or tricuspid valve atresia leads to a hypoxic state that is not correctible by mechanical ventilation. With a D-transposition of the great arteries, persistence of the ductus arteriosus and foramen ovale is necessary for life.

Defects dependent on an atrial septal defect. Infants with ASD-dependent defects must undergo a balloon atrioseptostomy (Rashkind maneuver) after birth. These are defects that involve atresia of an AV or semilunar valve or a D-transposition of the great arteries.

- Pulmonary atresia with an intact ventricular septum
- Tricuspid atresia
- Mitral atresia, hypoplastic left heart syndrome
- D-transposition of the great arteries
- Total anomalous pulmonary venous return

All infants with congenital heart defects that are dependent on a persistent ductus arteriosus or foramen ovale, and thus require immediate therapeutic intervention, should be delivered at a center where the services of a pediatric cardiologist and cardiac surgeon are available.

Management of Suspected Congenital Heart Disease

When a congenital heart defect is suspected, the pregnant woman should be referred to a tertiary perinatal center where (1) the heart defect and any associated extracardiac diseases or genetic defects can be investigated and (2) appropriate family counseling can be offered. Ideally, the patient should present at 20–21 weeks' gestation to allow sufficient time for all necessary diagnostic tests. The same applies to the targeted exclusion of heart defects in high-risk pregnancies.

References

1 Allan LD, Crawford DC, Shita SK et al.: Familial recurrence of congenital heart disease in a prospective series of mothers referred for fetal echocardiography. Amer. J. Cardiol. 58 (1986) 334

2 Bosi G, Scorrano M, Tosato G, Forini E, Chakrokh R and the Working Party of the Italian Society of Ped. Cardiology: The Italian Multicentric Study on Epidemiology of Congenital Heart Disease: First Step of the Analysis. Cardiol. Young 9 (1999) 291

3 Boughman JA, Neill CA, Ferencz C, Loffredo CA: The genetics of congenital heart disease. In Ferencz C, Rubin JD, Loffredo CA, Magee C (eds.): Perspectives in pediatric cardiology. Vol. 4. Epidemioloy of congenital heart disease: the Baltimore-Washington Infant Study 1981–1989. Futura, Mount Kisco NY 1993

4 Buskens E, Grobbert D, Frohn-Muldet I, Wladimiroff J, Hess J: Aspects of the aetiolgy of congenital heart disease. Eur. Heart J. 16 (1995) 584

5 Chinn A, Fitzsimmons J, Shepard TH, Fantel AG: Congenital heart disease among spontaneous abortuses and stillborn fetuses: prevalence and associations. Teratology 40 (1989) 475

6 Copel J, Cullen M, Green J, Mahoney M, Hobbins J, Kleinman C: The frequency of aneuploidy in prenatally diagnosed congenital heart disease: an indication for fetal karyotyping. Amer. J. Obstet. Gynecol. 158 (1988) 409

7 Gerlis LM: Cardiac malformations in spontaneous abortions. Int. J. Cardiol. 7 (1985) 29

8 Hoffman JIE: Incidence of Congenital Heart Disease: I. Postnatal Incidence. Ped. Cardiol. 16 (1995) 103

9 Mennicke K, Schwinger E: Genetische Aspekte kongenitaler fetaler Herzerkrankungen. Gynäkologe 30 (1997) 181

10 Nora JJ, Nora AH: Maternal transmission of congenital heart diseases: new recurrence risk figures and the questions of cytoplasmic inheritance and vulnerability to teratogens. Amer. J. Cardiol. 59 (1987) 459

11 Nora JJ, Nora AH: Update on counseling the family with a first-degree relative with a congenital heart defect. Amer. J. Med. Genet. 29 (1988) 137

12 Paladini D, Calabro R, Palmieri S, Andrea T: Prenatal diagnosis of congenital heart disease and fetal karyotyping. Obstet. Gynecol. 81 (1993) 679

13 Rose V, Gold RJM, Lindsey G, Allen M: A possible increase in the incidence of congenital heart defects among the offspring of affected parents. J. Amer. Coll. Cardiol. 6 (1985) 376

14 Tennstedt C, Chaoui R, Körner H, Dietel M: Spectrum of congenital heart defects and extracardiac malformations associated with chromosomal abnormalities: results of a seven year necropsy study. Heart 82 (1999) 34

15 Schwanitz G, Zerres K, Gembruch U, Bald R, Gamerdinger F, Hansmann M: Prenatal detection of heart defects as an indication for chromosome analysis. Ann. Genet. 33 (1990) 79

16 Ursell PC, Byrne JM, Strobino BA: Significance of cardiac defects in the developing fetus: a study of spontaneous abortuses. Circulation 72 (1985) 1232

17 Vermilion RP: Basic Physical Principles. In Snider AR, Serwer GA, Ritter SB (eds.): Echocardiography in Pediatric Heart Disease, 2nd ed. Mosby-Year Book, St. Louis 1997

18 Whittemore R, Wells JA, Castellsague-Pique X, Holabird NB: Congenital heart defects in the progeny of affected mothers versus fathers. Circulation 78 (Suppl. II) (1988) 396 (abstract)

23 Use of Color Doppler in Echocardiography

U. Gembruch

Importance of Color Doppler Echocardiography in Prenatal Diagnosis

The introduction of color Doppler echocardiography (synonym: two-dimensional Doppler echocardiography) can be considered a milestone in the prenatal diagnosis of heart defects and disturbances of cardiac function. While it is true that most fetal heart defects can be diagnosed in the second and third trimesters by two-dimensional imaging with a modern high-resolution ultrasound scanner (two-dimensional echocardiography), the ability to simultaneously display blood flow can greatly facilitate the diagnosis of complex cardiac anomalies. Color-flow imaging can also supply essential information on the hemodynamics of specific heart defects, which can vary greatly with the degree and severity of the defect. Moreover, prognostic evaluations can be made antenatally by analyzing the changes in intracardiac blood flow.

Some heart defects cannot even be diagnosed without the aid of color Doppler imaging. This is particularly true when fetal echocardiography is performed during the first and early second trimesters, when the flow image is useful and sometimes essential for locating the desired cardiac scan planes.

This chapter deals primarily with the situations and anomalies in which color Doppler echocardiography provides an essential adjunct to two-dimensional echocardiography. Further information on color Doppler echocardiography and its role in fetal echocardiography can be found in many current articles and textbooks[4–6, 8–11, 14, 28, 31].

Examination of the Normal Heart

Equipment Settings

There is no need to explore the physical aspects of color Doppler sonography in this chapter. One essential aspect of fetal echocardiography is using the correct equipment settings.

Pulse repetition frequency. Because color Doppler sonography is itself a pulsed Doppler technique, aliasing will occur when the sampled blood flow exceeds the selected velocity range, i.e., when the detected Doppler shift frequencies exceed the Nyquist limit. Given the high blood flow velocities that occur in the fetal heart, it is best to use a relatively initial high pulse repetition frequency (PRF) setting in fetal echocardiography. On the other hand, some flows in and around the fetal heart will require a low PRF setting, such as pulmonary venous flow, flow across the foramen ovale, and also diastolic ventricular inflow and systolic outflow into the two great arteries when an unfavorable insonation angle is used. The PRF setting is considered optimal when the imaged blood flow completely fills the associated vessel or cardiac chamber with no aliasing.

To achieve the highest possible spatial and temporal resolution of intracardiac blood flow, the ROI (region of interest) box should be set to display color flow in the smallest possible area. This is necessary to ensure a high line density and frame rate. The wall filters in fetal echocardiography are generally set higher than in the Doppler scanning of peripheral vessels. The persistence is set relatively low due to the desired high frame rate.

Variance mode. An essential step in the initial color setup is to select the variance mode. When this mode is selected, increasing variance (bandwidth) of the velocities and Doppler shift frequencies about the mean velocity or Doppler shift frequency in the selected sample volume causes more and more green pixels to be added to the basic flow color. When the flow is directed toward the transducer, red becomes increasingly yellow; when the flow is away from the transducer, blue becomes increasing turquoise. The variance mode setting is a rapid way to detect disturbed or turbulent blood flow in the heart such as that caused by valve stenosis, valve regurgitation, or intracardiac shunts. For this reason, the variance mode is preferred over the velocity mode in fetal echocardiography. The latter mode, which is standard in obstetric Doppler, displays higher Doppler shift frequencies in brighter shades of color. This is less important in fetal echocardiography because, based on the Doppler equation, the encoded colors do not reflect the flow velocities but are always influenced by velocity and insonation angle. For this reason, absolute velocity measurements in the heart are always performed with a spectral Doppler system, generally pulsed Doppler, making an effort to keep the insonation angle in the range 0–10° or 170–180°. Angle correction should not be used for intracardiac flow studies. Based on the cosine function of the insonation angle in the Doppler equation, angle correction at very low angles is unnecessary while at higher angles it leads to gross errors in the estimation of blood flow velocities. These errors result from the fact that blood flow in the central portion of the heart or great arteries may not move parallel to the vessel wall.

Imaging jets. In imaging flow jets through stenotic or regurgitant valves and intracardiac shunts, absolute velocities can be measured only if the beam angle is close to 0° or 180° in relation to the jet. Angle corrections are not used for these measurements because they lead to extreme overestimation of the jet velocity due to the eddies that form in the parajet. Due to the high Doppler shift frequencies, it is often necessary to use continuous-wave Doppler so that flow can be sampled without aliasing. Also, when the velocity mode is used instead of variance mapping, a turbulent jet may not produce an abnormal color Doppler flow pattern because the encoded mean velocities may fall within the normal range due to the tremendous variance of Doppler shift frequencies in the jet area. This is the main reason why variance mapping is always preferred in color Doppler echocardiography over the velocity mapping traditionally used in obstetrics.

Examination Technique

Screening. Color Doppler echocardiography permits rapid screening of the fetal heart for abnormal blood flow patterns. When color-flow abnormalities are found, they should always be confirmed with spectral Doppler, as this is the only technique that can positively distinguish aliasing from turbulence. In cases with unfavorable access angles to the fetal heart and in early echocardiography, color Doppler is helpful in quickly locating the veins and arteries leading to and away from the heart. It provides a natural adjunct to two-dimensional echocardiography, in which a 90° insonation angle is best for defining cardiac structures and vessels, as opposed to the parallel angle (0° or 180°) that is best for visualizing blood flow in Doppler echocardiography.

Procedure. Doppler echocardiography follows a standard segmental protocol that starts by imaging the visceral situs, the descending aorta, and the inferior vena cava. Next the visceroatrial, atrioventricular, and ventriculoarterial connections are defined along with the blood flow patterns in those areas. The PRF should be continually adjusted at this stage to obtain an optimum flow image that is not distorted by aliasing. For example, a low PRF is generally necessary to define pulmonary venous inflow into the left atrium, while a relatively high PRF is better for defining the origin of the great arteries from the two ventricles. A relatively low PRF is needed to demonstrate the brachiocephalic vessels arising from the aortic arch, which generally have an unfavorable angle relative to the transducer.

Structures visualized. Besides the blood flow in the inferior and superior vena cava and pulmonary veins (with practice, patience, and the proper transducer, all four pulmonary veins can be identified as early as the 13th week of gestation), the following are easily defined with color Doppler ultrasound: atrioventricular flow across the tricuspid and mitral valves; ventriculoarterial flow across the pulmonary and aortic valves; the two pulmonary arteries, the ductus arteriosus, and the aortic arch with the origins of the brachiocephalic vessels; interatrial blood flow across the foramen ovale; and the venous flow in the coronary sinus, which travels from left to right along the posterior cardiac wall facing the diaphragm[3]. On the other hand, the coronary arteries arising from the aortic sinus can be visualized in the normal fetus only under extremely favorable examination conditions (advanced gestational age, or a fetus lying dorsoanterior with the aortic root close to the transducer, allowing the use of relatively high frequencies) or in cases of myocardial hypoxia with extreme dilatation and increased perfusion of the coronary vascular bed (Fig. 23.**1**), which may occur in chronic and acute hypoxemic states[2, 19].

Differentiation of abnormal flow patterns. Aliasing can occur even with normal blood flow patterns and velocities, depending on the transducer frequencies that are used (when the transducer frequency is doubled, the same blood flow velocities produce twice the Doppler frequency shift). Aliasing is most pronounced in the great arteries and especially in the ductus arteriosus, where the highest flow velocities normally occur. Thus, when abnormal blood flow patterns are detected in the area of the semilunar valves and ductus arteriosus, spectral Doppler analysis of flow velocity waveforms should be performed to differentiate between aliasing and disturbed blood flow due to pathological obstructions. This points to another advantage of color Doppler echocardiography: areas with abnormal flow patterns can be quickly located with color Doppler and then interrogated by selective positioning of the pulsed Doppler sample volume or the CW (continuous-wave) Doppler sampling beam. This application of color Doppler is also useful in studies of normal and abnormal hemodynamics, as it permits very accurate positioning of the pulsed Doppler sample volume, thereby increasing the precision of absolute velocity measurements compared with conventional duplex Doppler echocardiography.

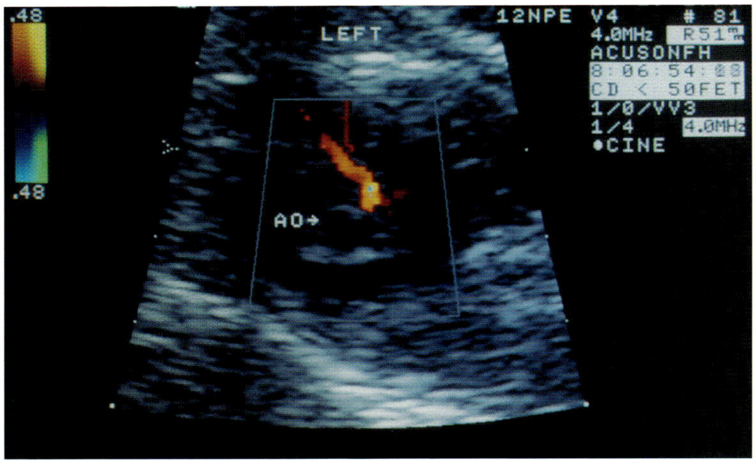

Fig. 23.**1** Origin of the right coronary artery from the aortic sinus (AO) in a severely growth-retarded fetus (29 weeks + 5 days).

Cardiac Valve Regurgitation

Functional Physiological Tricuspid Regurgitation

Owing to the specific features of the fetal circulation, regurgitation through the atrioventricular (AV) valves most commonly involves the tricuspid valve apparatus. The right heart dominance and higher ventricular afterloads, along with the structural peculiarities of the tricuspid valve apparatus, explain why functional tricuspid valve regurgitation occurs in 6–7% of all fetuses[21,27], while functional mitral valve regurgitation is almost unknown. This "physiological" tricuspid regurgitation is transient, is generally confined to early and mid-systole, has a maximum velocity no higher than 2 m/s, and produces a relatively small color-flow jet within the atrium. Rarely, there are also cases in which transient, physiological tricuspid regurgitation is holosystolic, occupies more of the atrial area, and reaches velocities higher than 2 m/s.

Pulmonary valve regurgitation is rare, with an incidence of approximately 0.5%[30], and aortic regurgitation is much rarer still. The observation that both tricuspid regurgitation and pulmonary valve regurgitation occur during periods of fetal breathing indicates that the degree of distention and afterload are factors that affect functional valve regurgitation.

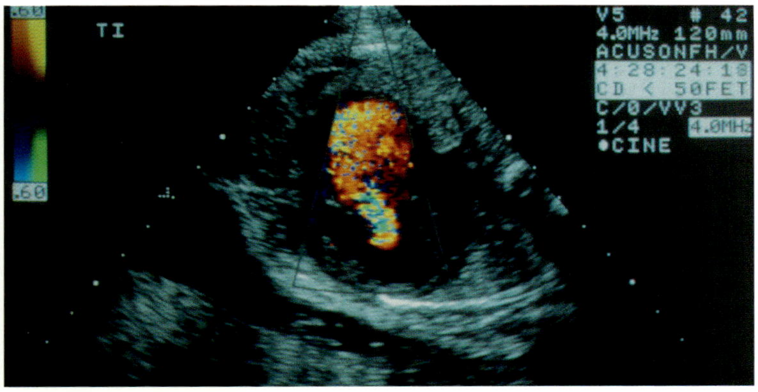

a

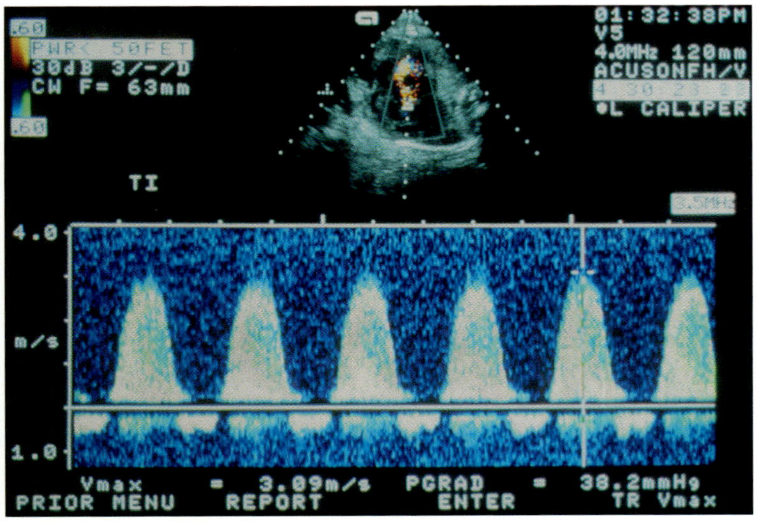

b

Fig. 23.**2** Severe, holosystolic tricuspid regurgitation in a fetus with tricuspid valve dysplasia and pulmonary atresia (33 weeks + 4 days).
a Color Doppler shows a long regurgitant jet covering a large area within the dilated right atrium.
b Spectral analysis with continuous-wave Doppler shows that the regurgitation is holosystolic with a peak velocity of 3.09 m/s.

Pathological Tricuspid Regurgitation

Cardiac valve malformations. Severe tricuspid regurgitation occurs as a result of tricuspid valve malformations due to dysplasia (Fig. 23.**2**) and in the setting of Ebstein anomaly. This regurgitation can lead to massive dilatation of the fetal right atrium and in some cases to venous pressure elevation culminating in fetal hydrops (see Anomalies of Atrioventricular Blood Flow).

Stretching of the valve ring. Stretching of the tricuspid valve ring leads to secondary functional tricuspid regurgitation that is no longer within the normal range. The stretching can result from volume overload of the ventricle and/or pressure loads due to outflow tract obstruction. For example, relatively mild forms of tricuspid regurgitation result from flow obstructions in the left heart that cause volume loading of the right ventricle. They also result from volume overloads due to arteriovenous shunts. Outflow tract obstructions such as constriction of the ductus arteriosus, which is usually drug-induced[27] and rarely occurs spontaneously[23], and severe pulmonary obstruction with an intact interventricular septum or in absent-pulmonary-valve syndrome can lead secondarily to tricuspid valve regurgitation.

Tricuspid and Mitral Valve Regurgitation

Myocardial diseases and tachyarrhythmias. Besides a volume overload that distends the valve ring, AV valve regurgitation may also be caused by myocardial dysfunction due to an infectious disease (myocarditis) or cardiomyopathy or by a rise of ventricular pressure with secondary papillary muscle dysfunction due to local hypoxia. AV valve regurgitation in fetal tachyarrhythmias most commonly results from a tachycardia-induced "cardiomyopathy" that develops above a critical heart rate due to myocardial hypoxemia (Fig. 23.**3**)[16,25]. Most myocardial perfusion takes place in diastole, as the extravascular wall pressure is much lower in this phase than during systole. But the length of diastole is greatly shortened when tachyarrhythmia is present. The final stage of severe myocardial hypoxia is marked by AV valve regurgitation on both the right and left sides of the heart[1].

Fetofetal transfusion syndrome. In this syndrome the elevated preload and afterload in the recipient twin lead initially to tricuspid valve regurgitation and, in the advanced stage, to mitral valve regurgitation.

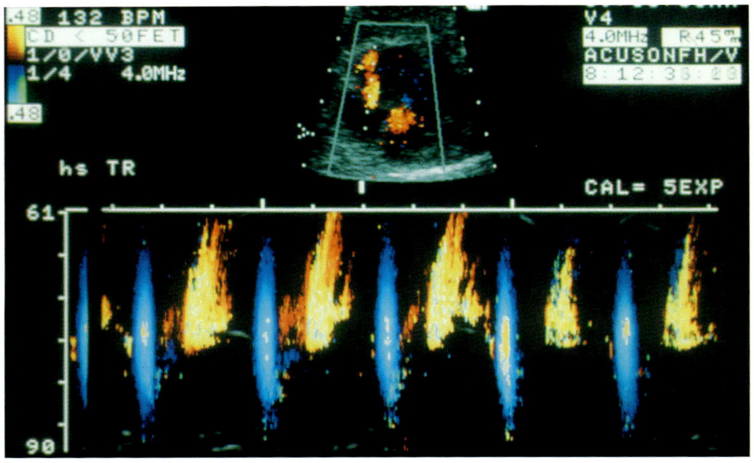

Fig. 23.**3** Holosystolic tricuspid regurgitation (encoded in yellow) following drug-induced conversion of fetal supraventricular tachycardia to a sinus rhythm of 132 bpm, demonstrated by M-mode Doppler echocardiography (29 weeks + 1 day). Diastolic inflow from the right atrium into the right ventricle is encoded in blue.

23

Isolated mitral valve regurgitation. Mitral regurgitation as an isolated phenomenon is more commonly a result of left ventricular outflow tract obstruction in which there is an associated hypoxemia-induced disturbance of left ventricular myocardial and valvular function, whose end stage is endocardial fibroelastosis.

Semiquantification of AV Valve Regurgitation

While the severity of AV valve regurgitation cannot be quantified by Doppler echocardiographic methods, it can be assessed by semiquantitative grading.

Pressure–time gradient. The maximum velocity measurement reflects the pressure gradient that exists between the ventricle and atrium. This gradient is most strongly influenced by the pressure within the ventricle, which in turn depends on the afterload (outflow tract obstruction and/or arterial hypertension) and on the filling and myocardial function of the affected ventricle. It also depends on the pressure at the atrial level. The pressure–time gradient ($\Delta p/\Delta t$) of the regurgitant jet provides information on the systolic function of the ventricle, noting that the pressure rise is slowed when ventricular function is impaired. If the maximum velocity of the regurgitant jet is greater than 4 m/s, this indicates the presence of an outflow tract obstruction and/or arterial hypertension like that occurring in the recipient twin in fetofetal transfusion. Based on the simplified Bernouli equation ($P_{grad} = 4\,V_{max}^2$), a maximum regurgitant jet velocity of 4 m/s reflects an instantaneous pressure difference of 64 mmHg. The systolic blood pressure of the fetus does not exceed this value, however[8].

Spatial extent of the jet. A semiquantitative assessment of the severity of tricuspid and mitral valve regurgitation is based on the temporal duration of the regurgitation and on the spatial extent of the color-flow jet, but the spatial extent of a jet depends more on the velocity of the regurgitated blood than on its volume. Also, the necessary standardization of instrument settings and of the Doppler signal display limits the value of spatial jet parameters in the color Doppler image for the semiquantitative assessment of AV valve incompetence.

Duration of valve regurgitation. For the present, the best parameter appears to be the temporal duration of AV valve regurgitation, which can be determined by pulsed or CW Doppler or accurately measured in the M-mode color Doppler display[12]. As a rule, milder degrees of AV valve regurgitation occur in early to mid-systole while severe, significant regurgitation is holosystolic. This has been demonstrated in AV canal malformations, where holosystolic AV regurgitation was associated with the development of fetal hydrops while regurgitation confined to early and mid-systole did not lead to hydrops[15]. Similar studies in fetuses with tachycardia-induced "cardiomyopathies" allow us to assess the severity of myocardial dysfunction: a steady decline in the temporal and spatial parameters for the semiquantification of AV valve incompetence with increasing time from conversion to sinus rhythm indicates a steady improvement in myocardial function. The severity of AV valve incompetence also correlates with the extent of changes in venous blood flow patterns and with the time required for the complete remission of hydrops[16, 25].

Anomalies of Visceroatrial Blood Flow

Anomalies involving the afferent veins of the heart can be investigated much more easily by the adjunctive use of color Doppler than by two-dimensional echocardiography alone. The smaller the fetus, the more difficult it is to differentiate the pulmonary veins and the veins about the liver, including the ductus venosus and inferior vena cava, in the gray-scale image and to define their sites of entry into the corresponding atrium. These structures are much easier to identify with color Doppler sonography, especially when anomalies are present.

Inferior and superior vena cava. For example, when the upper portion of the inferior vena cava is interrupted, the azygos vein drains much of the venous blood from the lower half of the body. In these cases the azygos vein, which generally opens into the superior vena cava, can be demonstrated more easily with color Doppler (Fig. 23.**4**). This finding is usually seen in conjunction with a visceral situs ambiguus, especially left-sided isomerism. The same applies to anomalies involving the umbilical vein and ductus venosus. Color Doppler imaging also facilitates the diagnosis of a persistent left superior vena cava by demonstrating blood flow toward the heart.

Anomalous pulmonary venous return. Color Doppler is particularly useful in diagnosing the various types of anomalous pulmonary venous return. These veins are not visible in the gray-scale image until the second trimester, but they can be iden-

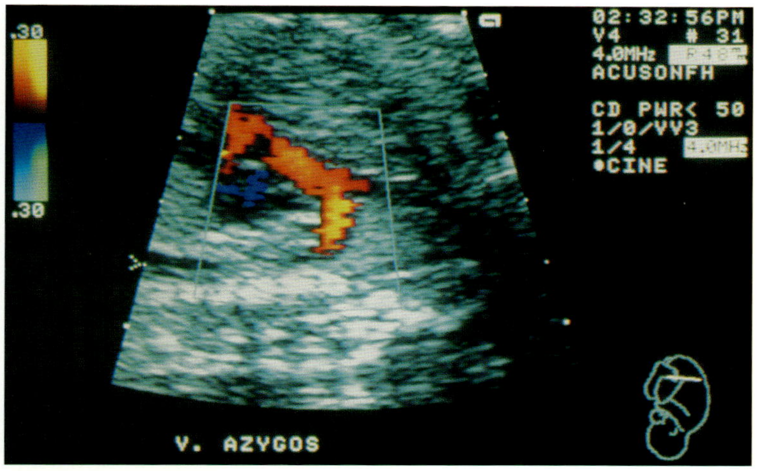

tified as early as the 12th week in the color Doppler image. All types of anomalous pulmonary venous return (supracardiac, cardiac, and infracardiac) can be diagnosed more easily and confidently with the aid of color Doppler.

◁ Fig. 23.**4** Color Doppler image of blood draining from an azygos vein into the superior vena cava (30 weeks + 2 days).

Anomalies of Atrioventricular Blood Flow

Tricuspid and mitral valve atresia. Atresia of the tricuspid and mitral valve apparatus is easily diagnosed in the two-dimensional image. Color Doppler can additionally demonstrate an absence of blood flow across the atretic valves. As an aid to differential diagnosis in fetuses with outflow tract obstructions in the right and left heart, the pulse repetition frequency should be progressively lowered during the examination, since there may still be low-velocity inflow into the hypoplastic right or left ventricle in fetuses with pulmonary atresia and a hypoplastic right heart or in fetuses with aortic atresia and a hypoplastic left heart (Fig. 23.**5**). An interatrial left-to-right shunt across the foramen ovale is typically present in these cases of severe left ventricular obstruction and is much easier to visualize with color Doppler sonography.

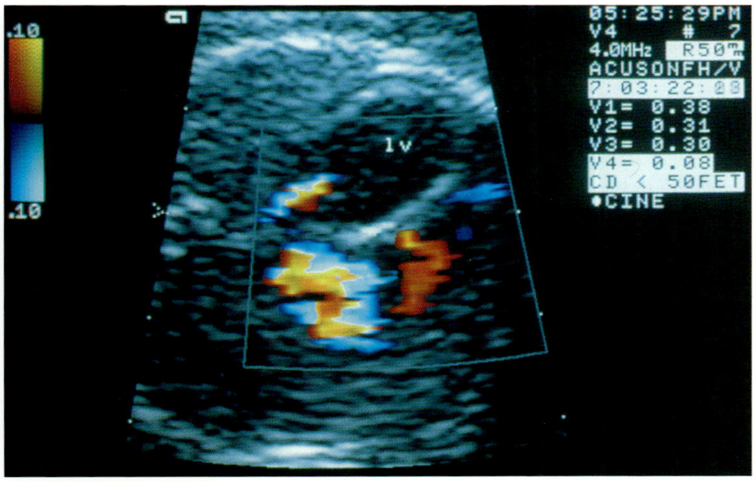

Fig. 23.**5** Hypoplastic left heart with endocardial fibroelastosis resulting from a severe aortic obstruction (21 weeks). Flow across the mitral valve into the left ventricle (lv) can be demonstrated only when a very low pulse repetition frequency is used.

Tricuspid valve dysplasia and Ebstein's anomaly. Both in tricuspid valve dysplasia and in Ebstein's anomaly of the tricuspid valve, severe tricuspid regurgitation can develop during fetal life, leading to massive enlargement of the right atrium. Less commonly there may be venous pressure elevation sufficient to cause fetal hydrops. Here again, color Doppler can aid in the diagnosis and semiquantitative grading of tricuspid valve regurgitation. Color Doppler M-mode echocardiography can be used to measure the duration of the regurgitant jet, and two-dimensional color Doppler echocardiography can be used to determine the length and area of the jet as described above.

Associated outflow tract obstruction. Tricuspid regurgitation is especially pronounced in cases of associated outflow tract obstruction (pulmonary atresia, pulmonary stenosis), with some authors placing tricuspid dysplasia with pulmonary valve atresia under the heading of pulmonary atresia with an intact interventricular septum. However, unlike pulmonary atresia with an intact interventricular septum and without tricuspid dysplasia, hypoplasia of the right ventricle does not occur in pulmonary obstructions with tricuspid dysplasia (Fig. 23.**2**). The right ventricle may even be dilated in some of these cases. It should be noted that even when the pulmonary valve is patent in cases of pronounced tricuspid valve dysplasia with severe tricuspid regurgitation, there will no longer be any detectable forward flow across the pulmonary valve[5, 33]. In these cases the entire pulmonary arterial bed is perfused by retrograde flow through the ductus arteriosus, mimicking the classic features of pulmonary valve atresia in the antenatal period. Once the infant has been delivered, forward flow occurs across the pulmonary valve because the right ventricular afterload becomes substantial due to the fall in pulmonary vascular resistance.

Anomalies of Ventriculoarterial Blood Flow

Pulmonary and aortic stenosis. Stenoses of the pulmonary valve and aortic valve generally produce a turbulent color-flow pattern behind the valve, accompanied by prestenotic flow acceleration in front of the valve (Fig. 23.**6**). If a relatively large ventricular septal defect is present, however, even a very high-grade stenosis may not produce a turbulent pattern because the blood is easily routed through the ventricular septal defect into the other major artery (e.g., into the overriding aorta with pulmonary stenosis). In these cases, retrograde blood flow can be detected in the artery with the stenotic valve, i.e., reverse perfusion of the pulmonary arterial bed through the ductus arteriosus in cases of severe pulmonary stenosis, and reverse flow from the ductus arteriosus into the aortic arch and ascending aorta in cases of severe aortic stenosis with a ventricular septal defect. These blood flow patterns are characteristic of atresias of the great arteries and are an important aid to differential diagnosis. The basic rule still applies, however, that abnormal blood flow patterns like those associated with stenotic jets or reverse blood flow due to severe valvular obstructions should always be confirmed by pulsed Doppler spectral analysis. Even with severe hypoplasia of the aortic arch, an isolated coarctation of the aorta, or an interruption of the aortic arch (types I–III), color Doppler flow imaging is very helpful for accurately defining the anatomical relationships.

Hypoplasia of the associated arteries. When a severe valve obstruction (stenosis or atresia) is present, the associated artery is usually hypoplastic. With two-dimensional echocardiography, it is often extremely difficult to define the hypoplastic vessel and thus differentiate between aortic atresia, pulmonary atresia, and a common truncus arteriosus. Color Doppler echocardiography can be very helpful in these cases by demonstrating reverse blood flow in the hypoplastic artery (Fig. 23.**7**) or, with a common truncus arteriosus, by showing the pulmonary arteries arising from a common trunk. The same is true in a double-outlet right ventricle with or without pulmonary stenosis, which often coexists with a malposition of the great vessels (Fig. 23.**8**), and in the tetralogy of Fallot. In either of these anomalies, a turbulent blood flow pattern is not necessarily found in the pulmonary artery even with severe pulmonary stenosis, because blood can easily drain from the right ventricle into the overriding aorta as a result of intrauterine circulatory dynamics. For this reason, the magnitude of the pressure gradient cannot be appreciated until postnatal circulatory adaptation is completed.

Transposition of the great vessels. Transposition of the great vessels, which more often takes the form of an "uncorrected" D-transposition than a "corrected" L-transposition, is also accessible to diagnosis by two-dimensional echocardiography. Again, however, color Doppler echocardiography makes it eas-

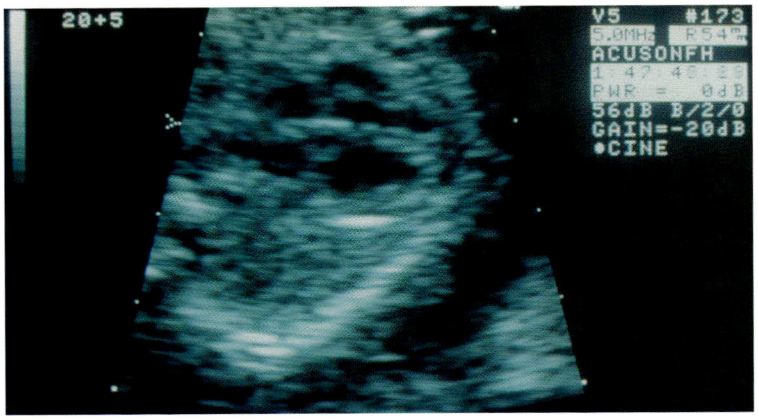

a

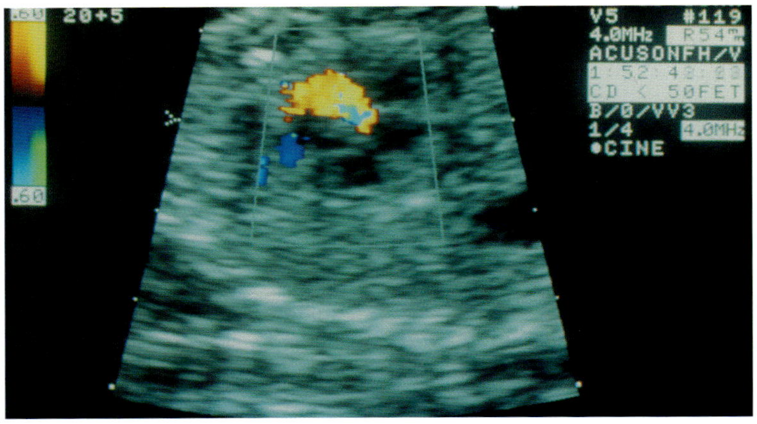

b

Fig. 23.6 Critical aortic stenosis (20 weeks + 5 days).
a Normal-appearing four-chamber view in two-dimensional echocardiography.
b With color Doppler echocardiography, the aortic stenosis can be diagnosed by the stenotic jet in the initial segment of the ascending aorta.

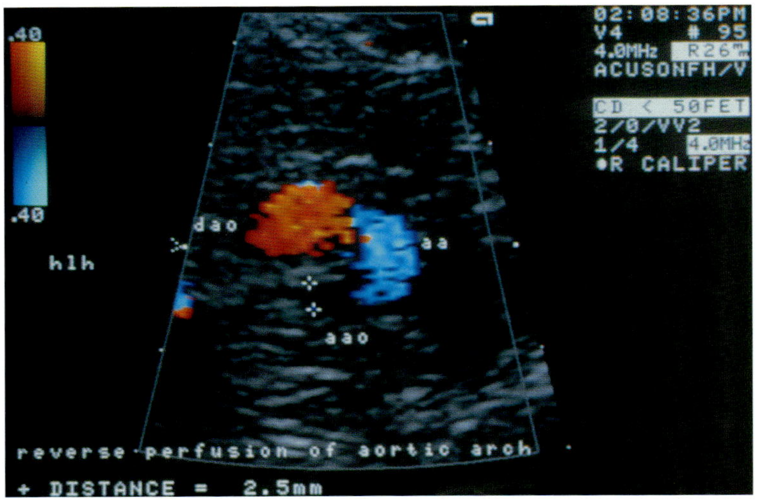

Fig. 23.7 Hypoplastic left heart with aortic atresia (31 weeks). The aortic arch (aa) and hypoplastic ascending aorta (aao, diameter 2.5 mm) are perfused entirely by retrograde blood flow from the ductus arteriosus. The descending aorta (dao) is perfused by antegrade flow.

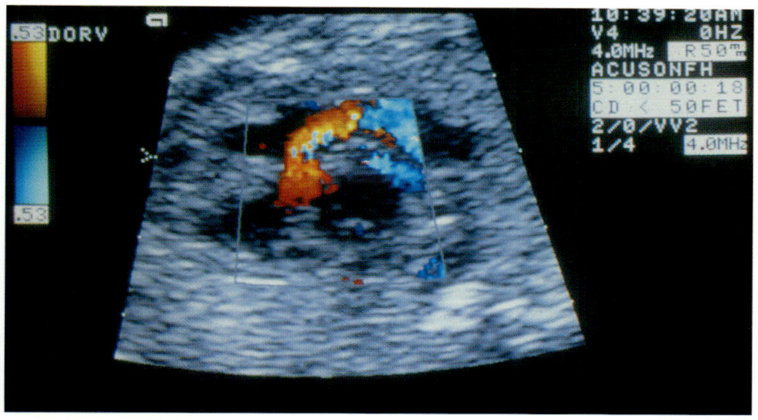

Fig. 23.**8** Double-outlet right ventricle (DORV) with malposition of the great arteries. Blood ejected from the left ventricle passes through a VSD into the aorta, which arises entirely from the right ventricle. The pulmonary artery is hypoplastic but can be identified as a second vessel, also carrying antegrade flow, that runs parallel to the aorta.

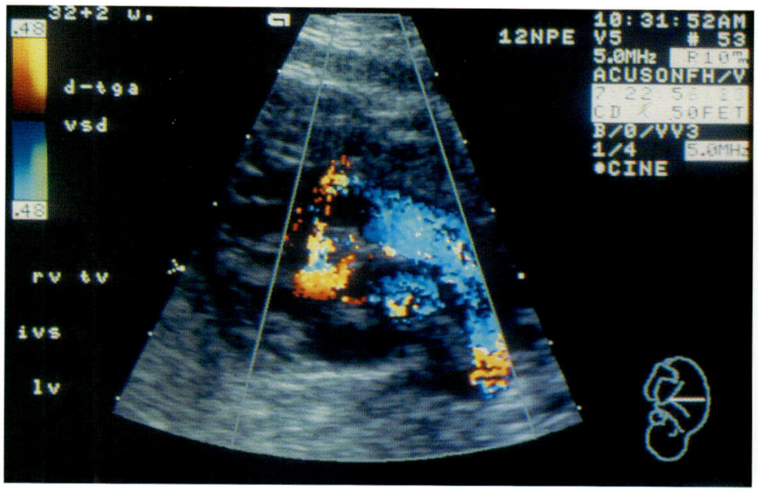

Fig. 23.**9** D-transposition of the great arteries with a VSD (32 weeks + 2 days). The origin and parallel course of the two great arteries can be appreciated. The color Doppler image also shows a systolic left-to-right shunt through a small outlet VSD.

ier to locate the abnormally positioned arteries, especially in cases with associated pulmonary stenosis and in early echocardiography. As in other anomalies, color Doppler echocardiography can increase the accuracy of fetal echocardiography by detecting or excluding associated malformations of the heart with greater certainty than two-dimensional echocardiography alone (Fig. 23.**9**).

Anomalies of Blood Flow through the Cardiac Septa

Interatrial shunt reversal. Color Doppler flow mapping can clearly demonstrate the physiological right-to-left interatrial shunt, both in the four-chamber view and in the basal short-axis view at the level of the origin of the great arteries ("circle and sausage view"). The latter view shows that the blood flow from the ductus venosus is directed across the foramen ovale into the left atrium, while blood from the inferior vena cava passes through the tricuspid valve into the right ventricle. A reversal of the interatrial shunt, with shunting of blood from the left atrium to the right atrium, occurs when the pressure in the left atrium is elevated due, for example, to a severe outflow tract obstruction at the aortic level.

Atrioventricular septal defect. Atrioventricular septal defect (synonyms: AV canal defect, endocardial cushion defect) is a combined defect involving the atrial septum primum and the ventricular septum in the inflow tract. Usually it is associated with a malformation of the AV valve apparatus. This anomaly is easily diagnosed with two-dimensional echocardiography. Color Doppler sonography confirms the diagnosis by demonstrating common blood flow into both ventricles and also showing AV valve regurgitation, which is almost always present in this defect. With very severe AV valve regurgitation in the setting of an atrioventricular septal defect, the rising pressure in the right atrium may lead to venous pressure elevation and fetal hydrops[15]. The semiquantitative evaluation of AV valve regurgitation in fetuses with atrioventricular septal defect was described earlier in the section on valve regurgitation.

Ventricular septal defects. Most isolated ventricular septal defects are perimembranous defects that involve the pars membranacea of the interventricular septum and an adjacent part of the muscular septum. Most of these defects are large and are easily appreciated in the two-dimensional echocardiogram. Muscular ventricular septal defects in the outflow tract, which qualify as a form of conotruncal anomaly, are usually so large that they can be diagnosed in the two-dimensional echocardiogram. They are frequently combined with an overriding aorta, often as part of a tetralogy of Fallot or double-outlet right ventricle.

Bidirectional shunt across a ventricular septal defect. The most common ventricular septal defects are small defects located in the trabecular part of the muscular septum. They are extremely difficult to diagnose prenatally, and most are not detectable in the two-dimensional image. Color Doppler echocardiography is the only technique available for diagnosing these defects. As noted earlier, it is essential that variance mapping be used. Because pressure differences between the right and left ventricles are either absent or very small due to the parallel arrangement of the fetal circulatory system, transient pressure differences arise during the cardiac cycle and increase during late pregnancy. As a result, color Doppler can demonstrate turbulent shunts across the interventricular septum, with maximum blood flow velocities of approximately 40 cm/s in the second trimester and 1.20 m/s at term[13]. Typically, a right-to-left shunt is observed in systole and a left-to-right

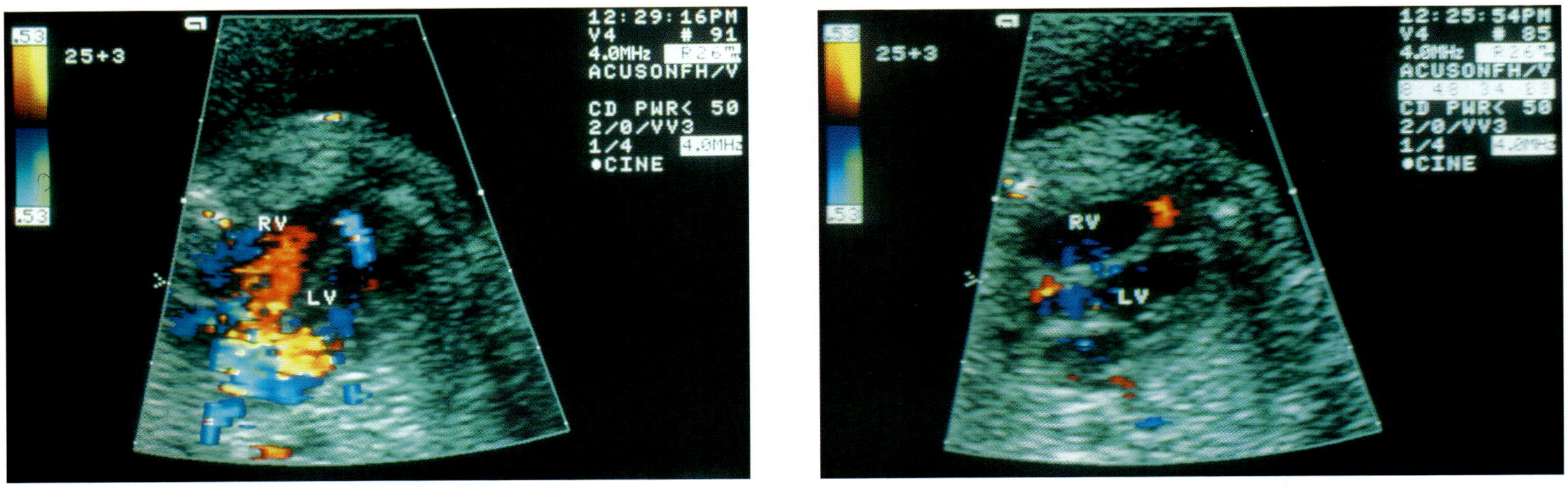

a b

Fig. 23.**10** Small, muscular ventricular septal defect with a bidirectional shunt pattern, not appreciated in the two-dimensional image (25 weeks + 3 days).

a A right-to-left shunt is present during systole.
b A left-to-right shunt appears during diastole.

23

shunt in diastole (Fig. 23.**10**). This observation may be due to differences in the patterns of excitation and relaxation and changes in the afterload of both ventricles (the afterload of the right ventricle is nearly constant in the second and third trimesters, while the afterload of the left ventricle rises steadily until term). The effect of afterload on shunt direction is supported by the observation that, with a D-transposition of the great arteries, a left-to-right shunt occurs across the ventricular septal defect in systole (Fig. 23.**9**) while a right-to-left shunt occurs in diastole[13].

Unidirectional shunts across a ventricular septal defect. If the shunt across a ventricular septal defect is in one direction only, the presence of an inflow tract obstruction and/or an outflow tract obstruction should be excluded. For example, tricuspid valve atresia is associated with the presence of a unidirectional left-to-right shunt across the ventricular septal defect. Other unidirectional shunt patterns consist of a right-to-left shunt due to pulmonary stenosis with a ventricular septal defect or a left-to-right shunt due to aortic stenosis and atresia with a ventricular septal defect.

Color Doppler Sonography in Fetal Arrhythmias

Tachycardia-induced "cardiomyopathy." Color Doppler sonography is necessary in the diagnosis and classification of fetal arrhythmias only when done as part of a complete echocardiographic workup of the fetus to exclude associated cardiac anomalies. Also, functional disturbances that are secondary to fetal arrhythmias are manifested in the form of AV valve regurgitation, which can be diagnosed and semiquantitatively analyzed with color Doppler, as described above. AV valve regurgitation in the setting of fetal arrhythmias most commonly occurs in association with tachyarrhythmias (tachycardia-induced "cardiomyopathy") and complete AV block. In the latter condition, pulmonary valve insufficiency has been observed in the setting of cardiac decompensation.

Differential diagnosis of fetal arrhythmia. The diagnosis and differential diagnosis of fetal arrhythmias are based on the demonstration of wall motion, valve motion, and blood flow patterns with a high temporal resolution. This is best accomplished by the use of M-mode echocardiography and pulsed Doppler echocardiography[18]. Wall and valve movements and blood flow patterns in different segments of the heart are recorded and analyzed for their temporal relationships. By demonstrating the effects of electrical excitation in the heart (atrial and ventricular systoles, opening movements of the AV and semilunar valves, ventricular inflow and outflow), the examiner can obtain indirect information on the spread of the cardiac impulse and diagnose the type of arrhythmia that is present[18].

Color Doppler M-mode echocardiography. The same M-mode beam can be used to simultaneously record a conventional M-mode echocardiogram and a color-flow map, with correspondingly high temporal resolution. With proper placement of the M-mode beam, a precise analysis can be made of the time intervals between hemodynamic events in veins, atria, ventricles, and arteries. The movements of the cardiac valves and walls are simultaneously recorded so that they can be correlated with hemodynamic events and the latter can be correlated with specific phases of the cardiac cycle. With color M-mode echocardiography, then, it is almost always possible to classify the underlying arrhythmia. This cannot always be done with conventional M-mode echocardiography, whether because of an unfavorable fetal lie or insufficient wall motion during atrial contractions[12].

Summary

Information added by color Doppler sonography of the fetal heart. Color Doppler mapping of blood flow patterns in the fetal heart and in the veins and arteries leading to and from the heart is an essential part of the echocardiographic examination of the fetus[32]. While most heart defects can be diagnosed by means of two-dimensional structural analysis, the addition of color Doppler imaging supplies a variety of additional, essential information (Tables 23.**1**, 23.**2**). Some heart defects can be detected only with the aid of color Doppler echocardiography. For others, the rapid screening of the fetal heart for flow abnormalities yields important information that, when combined with subsequent pulsed or CW Doppler spectral analysis, makes it possible to diagnose valvular stenosis, valvular insufficiency, and interventricular shunts. In cases with complex heart defects, color Doppler sonography is an essential tool for making a confident diagnosis that encompasses all cardiac abnormalities. With some defects, moreover, color Doppler supplies additional parameters that provide more individualized information on prognosis and on the further intrauterine and postnatal course of the disease.

Role of fetal color Doppler echocardiography. Various authors have attempted to define the role of color Doppler echocardiography within the framework of fetal echocardiography. For example, Copel et al.[7] found in their retrospective analysis of fetal cardiac abnormalities that Doppler color-flow mapping was essential to the correct anatomical diagnosis in 29% of the fetuses, was helpful but not essential in 47%, was not helpful in 24%, and even led to misinterpretations in several cases. From a critical standpoint, however, it should be noted that this type of study depends strongly on the experience of the examiner. It also depends on the technique that is used in conducting the examination (one examiner makes heavy use of color Doppler in locating cardiac structures, another examiner less so) and also on the technical quality of the two-dimensional image and the color Doppler echocardiogram. Another key factor is gestational age. Cardiac structures are generally easier to define in

Table 23.**1** General advantages of color Doppler sonography

- Rapid screening of the heart for abnormal flow patterns (variance mapping) to detect or exclude jets associated with stenosis, valve regurgitation, or interventricular shunts.
- Rapid and optimum placement of the Doppler sample volume for absolute flow measurements and the qualitative evaluation of blood flow patterns.
- More rapid localization of normal cardiac structures, especially in early examinations for heart defects.
- Particularly advantageous in identifying hypoplastic arteries and demonstrating retrograde blood flow patterns in those vessels (diagnosis and differential diagnosis of severe pulmonary stenosis, pulmonary atresia, severe aortic stenosis, aortic atresia, and common truncus arteriosus).
- In selected cases, the intrauterine blood flow patterns make it possible to determine the severity of the heart defect and offer an intrauterine and postnatal prognosis.

Table 23.**2** Relevant diagnostic information that color Doppler echocardiography adds to two-dimensional echocardiography for various heart defects

Heart defect	Information added by color Doppler sonography
Interrupted inferior vena cava	Demonstrates caudocranial blood flow in the azygos vein parallel to the descending aorta and its junction with the superior vena cava
Anomalous pulmonary venous return	Locates the pulmonary veins and demonstrates their anomalous termination at the supra-, intra- or infracardiac level
Type I atrial septal defect	Associated mitral valve regurgitation
Complete atrioventricular septal defect	Inflow into both ventricles through a common AV valve; demonstration and semiquantitation of AV valve regurgitation, including multiple jets
Tricuspid dysplasia, Ebstein anomaly	Demonstration and semiquantitation of tricuspid valve incompetence; evaluation of right ventricular outflow tract and pulmonary valve (caution: pulmonary artery flow may be retrograde even with an open pulmonary valve)
Tricuspid atresia	No evidence of antegrade flow through the tricuspid valve; rudimentary right ventricle is filled through a VSD; helpful in evaluating the position of the great arteries
Pulmonary atresia with an intact IVS	Shows only retrograde perfusion of the pulmonary arteries via the ductus arteriosus; demonstrates inflow through the dysplastic tricuspid valve into the hypoplastic right ventricle, also tricuspid valve regurgitation
Pulmonary atresia with a VSD	Demonstrates only retrograde flow through the ductus arteriosus into the generally hypoplastic pulmonary arterial system; blood flow from the right ventricle into the broad aorta, which is usually overriding
Pulmonary stenosis without a VSD	Shows a stenotic jet at the pulmonary valve
Pulmonary stenosis with a VSD (also combined with other heart defects)	May not demonstrate a stenotic jet when a large VSD is present; may show additional or exclusive retrograde blood flow in the pulmonary arterial system via the ductus arteriosus
Aortic atresia with a hypoplastic left heart	Shows only retrograde blood flow in the aortic arch and ascending aorta; no inflow (mitral atresia) or scant inflow (mitral valve dysplasia) into the hypoplastic left ventricle and mitral insufficiency; left-to-right shunt across the foramen ovale
Aortic stenosis	Stenotic jet at the aortic valve; a critical aortic stenosis with initial left ventricular dysfunction is marked by mitral valve regurgitation and increasing retrograde flow through the aortic arch, accompanied by an interatrial left-to-right shunt

Continued ▷

Table 23.2 (Continued)

Heart defect	Information added by color Doppler sonography
Coarctation of the aorta, tubular hypoplasia of the aortic arch, interrupted aortic arch	Locates and defines the hypoplastic arch segment and origins of the brachiocephalic arteries; differentiates coarctation from interrupted aortic arch; classifies the interrupted arch as to type; demonstrates blood flow through an associated ventricular septal defect
Double-outlet right ventricle	Defines the usually underdeveloped pulmonary artery; may show accelerated or disturbed flow patterns in the pulmonary valve area due to pulmonary stenosis
Tetralogy of Fallot	Defines the usually underdeveloped pulmonary artery; may show accelerated or disturbed flow patterns in the pulmonary valve area due to pulmonary stenosis
Absent-pulmonary valve syndrome	Demonstrates bidirectional flow in the dilated pulmonary arteries with stenotic and/or regurgitant patterns in the area of the absent pulmonary valve; detects or excludes associated agenesis of the ductus arteriosus
Transposition of the great arteries	Confirms the two-dimensional echocardiographic diagnosis and demonstrates small ventricular septal defects; with a "corrected" transposition, occasionally shows accelerated or disturbed flow patterns in the pulmonary valve area due to associated pulmonary stenosis
Ventricular septal defect (VSD)	Small muscular VSDs can be diagnosed only by the interventricular jet revealed by color Doppler echocardiography; typical bidirectional shunt pattern with systolic right-to-left shunt and diastolic left-to-right shunt through the VSD; unidirectional shunts signify additional obstructions in the inflow tract and/or outflow tract; with transposition of the great arteries, the biphasic shunt pattern is reversed

the B-mode image in late pregnancy than in early pregnancy. For examinations performed in the late first trimester and early second trimester, color Doppler echocardiography is of crucial importance not only in diagnosing cardiac anomalies but also in locating and identifying normal blood vessels[17, 20].

Specific heart defects. There are a number of heart defects for which color Doppler echocardiography provides essential additional information (Table 23.**2**). Color Doppler is the only technique that can detect small muscular ventricular septal defects, which are associated with demonstrable intracardiac shunts starting in the mid-second trimester. Color Doppler also adds essential information in the diagnosis of anomalies of the great arteries, especially when one of the arteries is hypoplastic. Color Doppler in these cases aids in localizing the vessels, which is often difficult in cases of malposition and transposition where one of the arteries is hypoplastic. Semilunar valve stenosis can also be diagnosed more easily with color Doppler but requires confirmation by Doppler spectral analysis. Finally, color Doppler echocardiography appears to be essential in the diagnosis of total anomalous pulmonary venous return because these vessels are difficult to define with two-dimensional echocardiography, even in the second trimester. Another domain of color Doppler echocardiography is in locating sites of AV valve regurgitation, which can be semiquantitatively analyzed with Doppler echocardiography. Some of these sites are physiological, but others occur in association with cardiac and extracardiac fetal anomalies. Color flow can also improve the accuracy of spectral Doppler measurements and volumetry for determining ventricular outflow, since preliminary color Doppler echocardiography can be used to find the most favorable insonation angle for spectral Doppler echocardiography.

References

1 Baschat AA, Gembruch U: Triphasic umbilical venous blood flow with prolonged survival in severe intrauterine growth retardation: a case report. Ultrasound Obstet. Gynecol. 8 (1996) 201–205

2 Baschat AA, Gembruch U, Reiss I, Gortner L, Diedrich K: Demonstration of fetal coronary blood flow by Doppler ultrasound in relation to the arterial and venous flow velocity waveforms and perinatal outcome—the "heart-sparing effect". Ultrasound Obstet. Gynecol. 9 (1997) 162–172

3 Baschat AA, Gembruch U: Examination of fetal coronary sinus by Doppler ultrasound. Ultrasound Obstet. Gynecol. 11 (1998) 410–414

4 Chaoui R, Bollmann R: Die fetale Farbdoppler-Echokardiographie. Teil 1: Allgemeine Grundlagen und normale Befunde. Ultraschall Med. 15 (1994) 100–104

5 Chaoui R, Bollmann R: Die fetale Farbdoppler-Echokardiographie. Teil 2: Fehlbildungen des Herzens und der großen Gefäße. Ultraschall Med. 15 (1994) 105–111

6 Chiba Y, Kanzaki T, Kobayashi H, Murakami M, Yutani C: Evaluation of fetal structural heart disease using color flow mapping. Ultrasound Med. Biol. 16 (1990) 221–229

7 Copel JA, Morotti R, Hobbins JC, Kleinmann CS: The antenatal diagnosis of congenital heart disease using fetal echocardiography: Is color flow mapping necessary? Obstet. Gynecol. 78 (1991) 1–8

8 Copel JA, Kleinman CS: The abnormal fetal heart. In Copel JA, Reed KL (eds.): Doppler ultrasound in obstetrics and gynecology. Raven Press, New York 1995, 2209–2217

9 DeVore GR, Hornstein J, Siassi B, Platt LD: Fetal echocardiography VII: Doppler color flow mapping: A new technique for the diagnosis of congenital heart disease. Amer. J. Obstet. Gynecol. 156 (1988) 1054–1064

10 DeVore GR: The use of color Doppler imaging to examine the fetal heart. Normal and pathologic anatomy. In Jaffe R, Warsof SL (eds.): Color Doppler imaging in obstetrics and gynecology. McGraw-Hill, New York 1992, 121–154

11 Gembruch U, Hansmann M, Redel DA, Bald R: Fetal two-dimensional Doppler echocardiography (colour flow mapping) and its place in prenatal diagnosis. Prenat. Diagn. 9 (1989) 535–547

12 Gembruch U, Bald R, Hansmann M: Die farbkodierte M-mode-Doppler-Echokardiographie bei der Diagnostik fetaler Arrhythmien. Geburtsh. u. Frauenheilk. 50 (1990) 286–290

13 Gembruch U, Bald R, Redel DA, Hansmann M: Shunt patterns of fetal ventricular septal defects. Ultrasound Obstet. Gynecol. 1 (Suppl. 1) (1991) 82

14 Gembruch U, Chatterjee M, Bald R, Redel DA, Hansmann M: Color flow mapping of fetal heart. J. Perinat. Med. 19 (1991) 27–32

15 Gembruch U, Knöpfle G, Chatterjee M et al.: Prenatal diagnosis of atrioventricular canal malformation using up-to-date echocardiographic technology (a report of 14 cases). Amer. Heart J. 121 (1991) 1489–1497

16 Gembruch U, Redel DA, Bald R, Hansmann M: Longitudinal study in 18 cases of fetal supraventricular tachycardia: Doppler-echocardiographic findings and pathophysiological implications. Amer. Heart J. 125 (1993) 1290–1301

17 Gembruch U, Knöpfle G, Bald R, Hansmann M: Early diagnosis of fetal congenital heart diseases by transvaginal echocardiography. Ultrasound Obstet. Gynecol. 3 (1993) 310–317
18 Gembruch U, Somville T: Intrauterine Diagnostik und Therapie fetaler Arrhythmien. Gynäkologe 28 (1995) 329–345
19 Gembruch U, Baschat AA: Demonstration of fetal coronary blood flow by color-coded and pulsed wave Doppler sonography: a possible indicator of severe compromise and impending demise in intrauterine growth retardation. Ultrasound Obstet. Gynecol. 7 (1996) 10–16
20 Gembruch U, Baschat AA, Knöpfle G, Hansmann M: First- and early second-trimester diagnosis of fetal cardiac anomalies. In Wladimiroff JW, Pilu G (eds.): Ultrasound and the fetal heart. Parthenon Publishing Group, New York 1996 39–46
21 Gembruch U, Smrcek J: The prevalence and clinical significance of tricuspid valve regurgitation in normally grown fetuses and those with intrauterine growth retardation. Ultrasound Obstet. Gynecol. 9 (1997) 174–182
22 Gembruch U: Prenatal diagnosis of congenital heart defects. Prenat. Diagn. 17 (1997) 1283–1298
23 Hofstadler G, Tulzer G, Altmann R, Schmitt K, Danford D, Huhta J: Spontaneous closure of the human fetal ductus arteriosus—A cause of fetal congestive heart failure. Amer. J. Obstet. Gynecol. 174 (1996) 879–883
24 Hornberger LK, Sahn DJ, Kleinmann CS, Copel JA, Reed KL: Tricuspid valve disease with significant tricuspid insufficiency in the fetus: Diagnosis and outcome. J. Amer. Coll. Cardiol. 17 (1991) 167–173
25 Krapp M, Gembruch U, Baumann P: Venous blood flow pattern suggesting tachycardia-induced "cardiomyopathy" in the fetus. Ultrasound Obstet. Gynecol. 10 (1997) 32–40
26 Respondek ML, Kammermeier M, Ludomirsky A, Weil SR, Huhta JC: The prevalence and clinical significance of fetal tricuspid valve regurgitation with normal heart anatomy. Amer. J. Obstet. Gynecol. 171 (1994) 1265–1270
27 Respondek ML, Weil SR, Huhta JC: Fetal echocardiography during indomethacin treament. Ultrasound Obstet. Gynecol. 5 (1995) 86–89
28 Sharland GK, Chita SK, Allan LD: The use of colour Doppler in fetal echocardiography. Int. J. Cardiol. 28 (1990) 229–236
29 Sharland GK, Chita SK, Allan LD: Tricuspid valve dysplasia or displacement in intrauterine life. J. Amer. Coll. Cardiol. 17 (1991) 944–949
30 Smrcek J, Germer U, Gembruch U: Functional pulmonary valve regurgitation in the fetus. Ultrasound Obstet. Gynecol. 12 (1998) 254–259
31 Stewart PA, Wladimiroff JW: Fetal echocardiography and color Doppler flow imaging: the Rotterdam experience. Ultrasound Obstet. Gynecol. 3 (1993) 168–175
32 Stümpflen I, Stümpflen A, Wimmer M, Bernaschek G: Effect of detailed fetal echocardiography as part of routine prenatal ultrasonographic screening on detection of congenital heart disease. Lancet 348 (1996) 854–857
33 Yeager SB, Parness I, Sanders S: Severe tricuspid regurgitation simulating pulmonary atresia in the fetus. Amer. Heart J. 115 (1988) 906–908

24 Structure of the Human Placenta and Pathomorphological Changes in Placental Insufficiency

K. R. Reitnauer

Structure of the Human Placenta

Weight and Dimensions

The human placenta is a disk-shaped organ composed of the fetal chorionic plate, which bounds the amniotic cavity along with the three-vessel umbilical cord; the basal plate with the spiral arteries and veins, which bounds the uterine wall; and the intervening villi with the intervillous spaces (Fig. 24.**1**).

The human placenta is of the hemochorionic type, in which the fetal villous trees are immersed in the maternal blood. Because of this arrangement, the fetal chorionic epithelium that lines the villi is bathed directly by maternal blood[15]. At term the placenta has an approximate weight of 500 g, a diameter of 17–19 cm, a basal surface area of 250 cm^2, and a thickness of 2–2.5 cm.

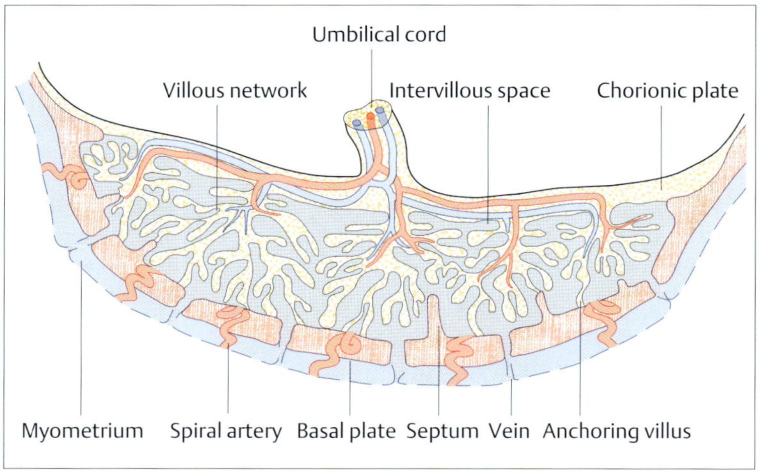

Fig. 24.**1** Schematic diagram of the mature human placenta. (Modified from Schiebler and Kaufmann.)

Early Development of the Human Placenta

Five stages (Fig. 24.**2a–e**) can be recognized in the early development of the human placenta[5,6,30,46].

Preimplantation stage. The preimplantation stage begins with formation of the zygote, which matures to a blastocyst through repeated cell divisions. The blastocyst consists of an inner embryoblast and outer trophoblast (Fig. 24.**2a**).

Stage of the implanted blastocyst. At this stage, which begins on about day 7 after fertilization, the embryonic pole reaches the endometrium and adheres to it. This is followed by a rapid increase in the thickness of the trophoblast, which actively penetrates the endometrium through a histolytic action.

Two types of trophoblastic cell can be distinguished at this stage: an outer syncytiotrophoblast, so named because it is formed by the fusion of adjacent cells, and an inner cytotrophoblast, which functions as a stem cell pool for the syncytiotrophoblast.

Increasing numbers of lacunae form within the syncytiotrophoblast, separated from one another by syncytial trabeculae. By about the 12th day after fertilization, the implanted blastocyst is completely covered by the endometrium (Fig. 24.**2b**).

Primary villi. The "primary villi" are now formed by the ingrowth of cytotrophoblastic cells from the primary chorionic plate into the central portions of the trabeculae (Fig. 24.**2c**).

Secondary villi. Connective tissue cells migrate from the extraembryonic mesoderm of the chorionic cavity along the axis of the primary villi, transforming the latter into secondary villi. Only the basal portions of the villi are spared and remain in the primary villus stage; they form the cell columns. The adjacent, confluent lacunar system becomes the intervillous space. As the trophoblast burrows deeper into the endometrium, it erodes maternal blood vessels, causing maternal blood to bathe the intervillous space (Fig. 24.**2d**).

Tertiary villus stage. The tertiary villus stage begins on about day 18 postconception. It is characterized by the formation of embryonic capillaries and hematopoietic stem cells from the mesenchymal cells located in the villi. This is accompanied by an ingrowth of the "allantoic vessels" from the embryo across the umbilical cord and chorionic plate into the villi. These vessels gain attachment to the capillaries that develop locally in the villi (Fig. 24.**2e**), creating the necessary foundation for the development of an embryoplacental circulation[13,30,34].

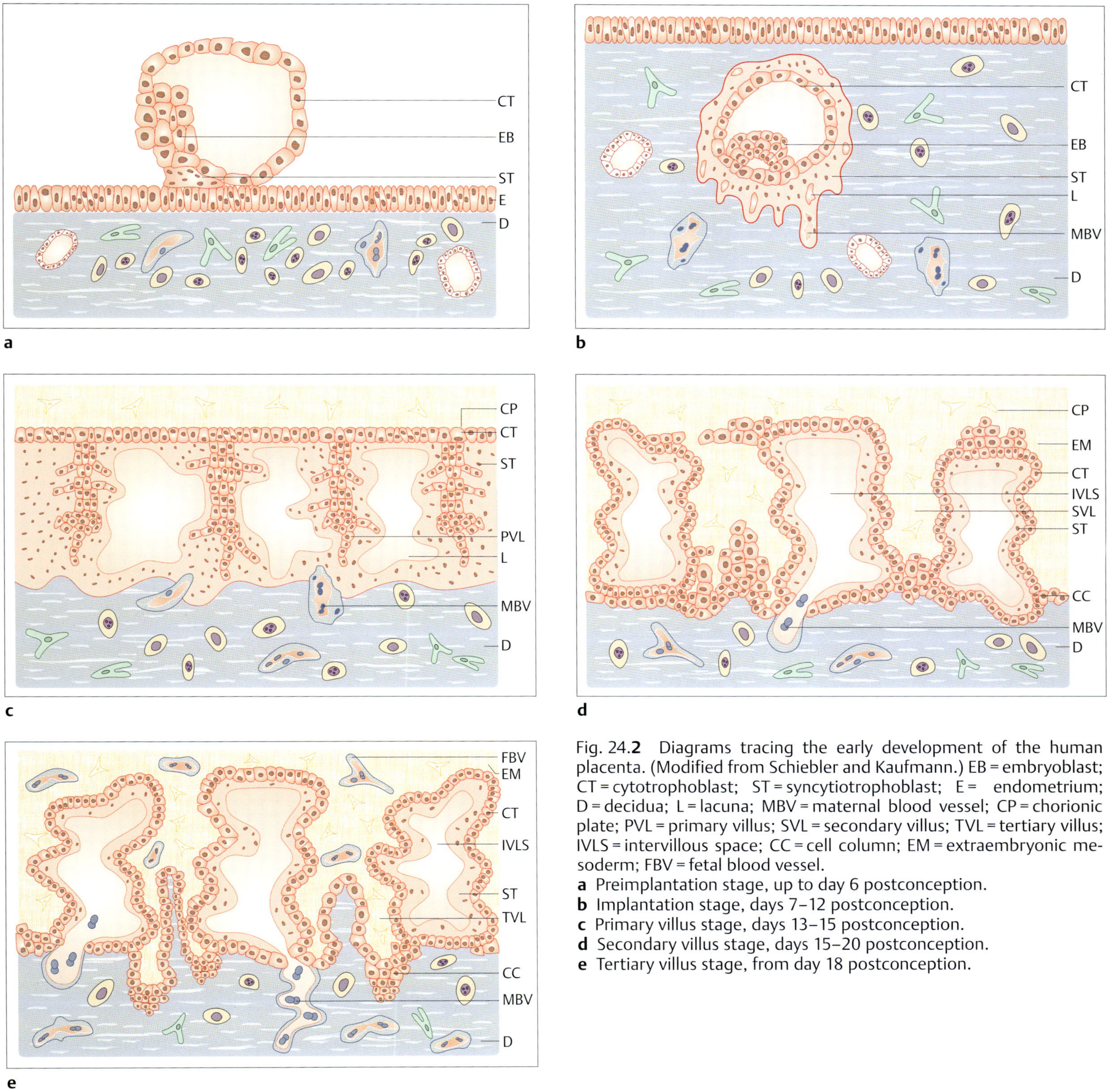

Fig. 24.**2** Diagrams tracing the early development of the human placenta. (Modified from Schiebler and Kaufmann.) EB = embryoblast; CT = cytotrophoblast; ST = syncytiotrophoblast; E = endometrium; D = decidua; L = lacuna; MBV = maternal blood vessel; CP = chorionic plate; PVL = primary villus; SVL = secondary villus; TVL = tertiary villus; IVLS = intervillous space; CC = cell column; EM = extraembryonic mesoderm; FBV = fetal blood vessel.

a Preimplantation stage, up to day 6 postconception.
b Implantation stage, days 7–12 postconception.
c Primary villus stage, days 13–15 postconception.
d Secondary villus stage, days 15–20 postconception.
e Tertiary villus stage, from day 18 postconception.

Chorion frondosum. The embryo in the 8th week of gestation is surrounded, from inside to outside, by the amniotic cavity, the amnion, and the adjacent chorionic cavity. Outside this are the chorionic plate, the trabecular and lacunar system described above, and the basal plate. The latter structures form a uniform, bushy chorion frondosum (Fig. 24.**3**). The outermost structures are the decidua basalis and decidua capsularis. The decidua parietalis lines the uterine cavity (Fig. 24.**4a**).

Chorion laeve. While vascularization of the villi starts at the implantation pole toward the end of the 3rd week of gestation, the villi located at the pole facing the uterine lumen begin to regress. This regression is well advanced by the 10th week, creating an almost villus-free chorion laeve that can be distinguished from the placenta. Starting in about the 14th week, the embryonic membranes become increasingly apposed to the uterine wall (Fig. 24.**4b**). The decidua capsularis becomes fragmentary and partially fuses with the decidua parietalis[5, 6, 15, 34].

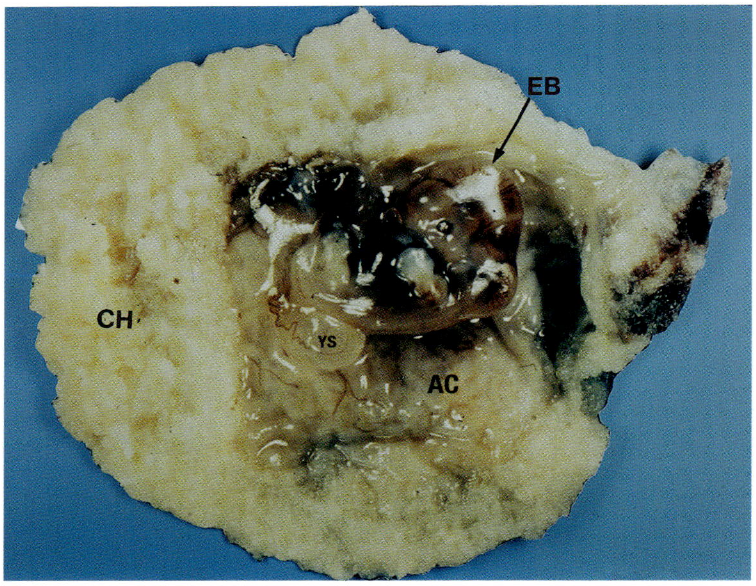

Fig. 24.**3 a** Embryo and chorion frondosum. EB = embryo; AC = amniotic cavity; YS = yolk sac; CH = chorion frondosum.

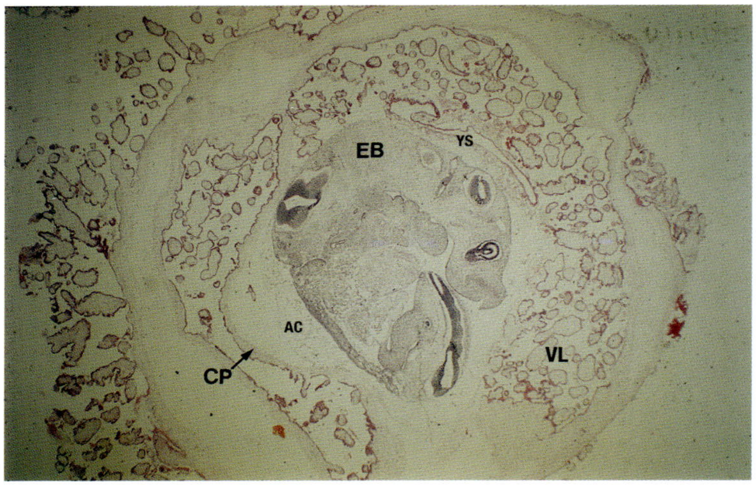

Fig. 24.**3 b** Histological appearance of an embryo, with the amniotic cavity and chorion frondosum (H & E, ×10). EB = embryo AC = amniotic cavity; CP = chorionic plate; VL = villus; YS = yolk sac.

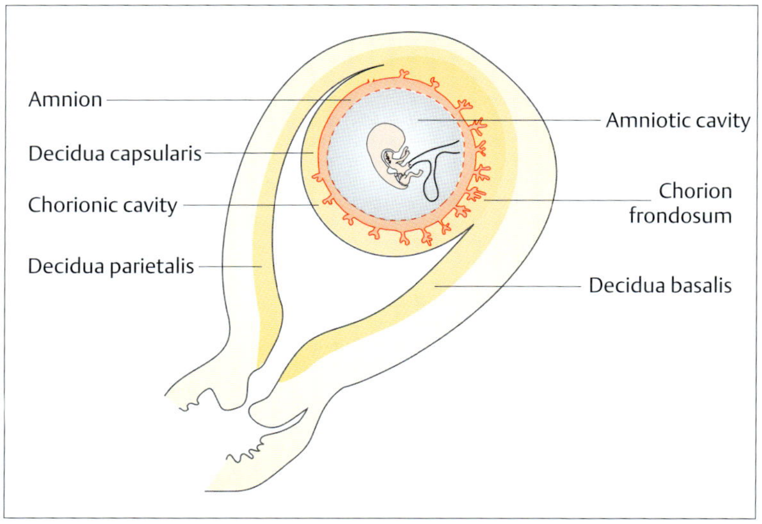

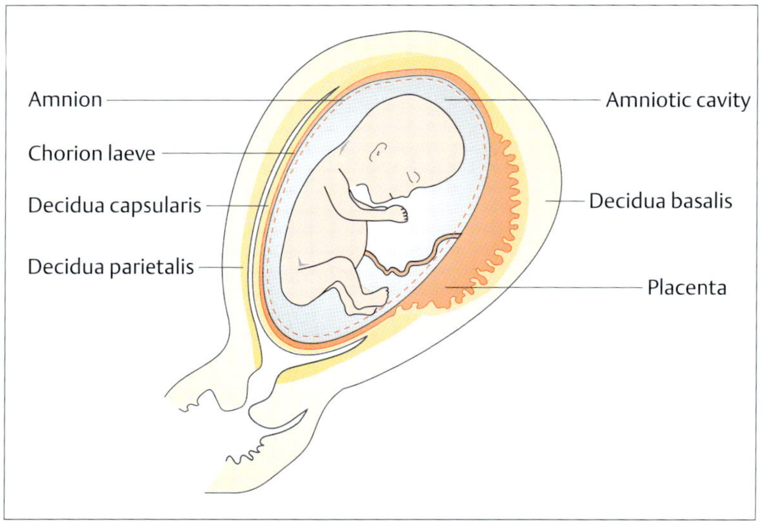

Fig. 24.**4** Diagram showing the relationship between the embryonic membranes and uterine wall.
a Week 8 postconception.

b Week 15 postconception.

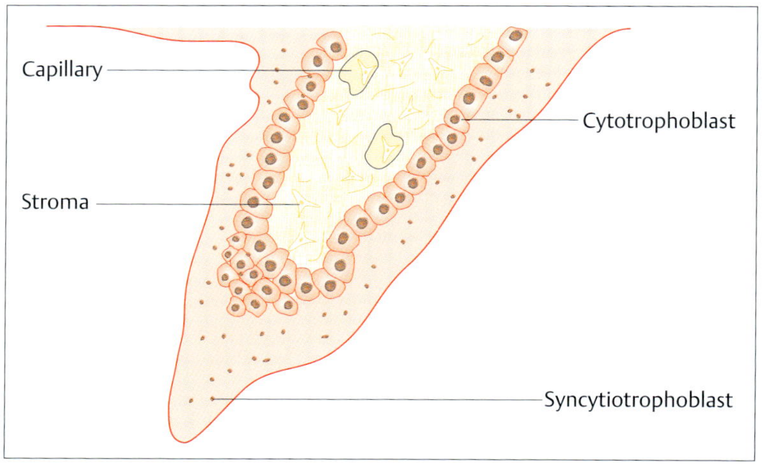

Fig. 24.**5** Mesenchymal villus with a "sprout." (Modified from Kaufmann and Scheffen.).

Structure of the Villous Tree

As the growth of the placenta progresses, different types of villi become increasingly distinct[25,26].

Mesenchymal villi. "Mesenchymal villi" are trophoblastic sprouts, initially fungiform, that are the precursors of all the other villi (Fig. 24.**5**). They are comparable in structure to the primary and secondary villi that form during early development and consist of a centrally located, proliferating cytotrophoblast and an outer syncytiotrophoblast. They are invaded by mesenchymal cells and subsequently vascularized with the formation of new villi[11].

Immature intermediate villi. Large-caliber, immature intermediate villi predominantly develop from the mesenchymal villi during the first and second trimesters. They are distinguished by a coarse-meshed reticular stroma with blood vessels and small, intercommunicating stromal channels (Fig. 24.**6a**). These channels provide transit routes for the Hofbauer cells—fetal tissue macrophages that play an integral role in host defenses and influence the morphogenesis and angiogenesis of the villi [6, 10, 50].

Mainstem villi. The immature intermediate villi are transformed into mainstem villi that connect the chorionic plate with the villous tree. Each villous stem branches into first-order through fourthth-order rami and first-order through tenth-order ramuli. The mainstem villi are characterized by a condensed fibrous stroma, large afferent and efferent blood vessels with a distinct tunica media, and a paravascular fibrous cuff (Fig. 24.**6b**). Their function is to stabilize the villous tree and thus the circulation of blood. The dominant process that occurs during the first and second trimesters is the expansion of the villous tree[11].

Mature intermediate villi and terminal villi. During the third trimester, by contrast, predominantly mature intermediate villi are formed from the mesenchymal villi. They contain many arterioles, capillaries, and venules and are direct precursors of the terminal villi (Fig. 24.**6c**). The terminal villi are equipped with numerous capillaries that have undergone sinusoidal transformation and cause the villous surface to protrude, forming "metabolic membranes" (Fig. 24.**6d**). Thus, the dominant process in the third trimester is maturation of the villi rather than growth of the villous tree[11].

Figure 24.**7** shows a high-power view of a fetal villous tree that includes mainstem, intermediate, and terminal villi.

The development of the terminal villi is dependent on capillary growth in the mature intermediate villi. Once capil-

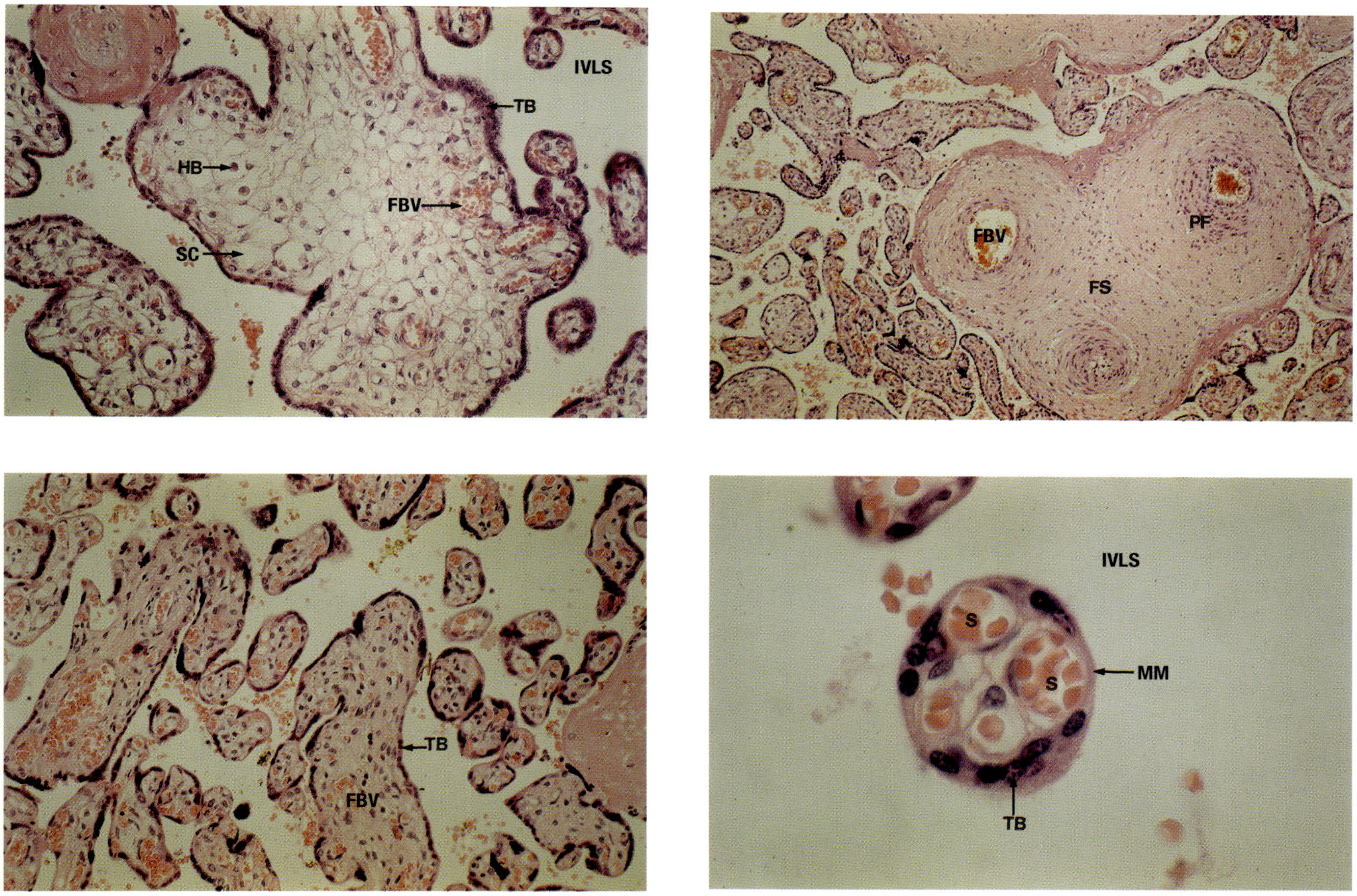

a b c d

Fig. 24.**6** The different types of villi in the human placenta.
a Immature intermediate villus with a reticular stroma, stromal channels, and Hofbauer cells (H & E, ×200). SC = stromal channel HB = Hofbauer cell; FBV = fetal blood vessel; TB = trophoblast; IVLS = intervillous space.
b Mainstem villus with a fibrous stroma and paravascular fibrous cuff (H & E, ×100). FS = fibrous stroma PF = paravascular fibrous cuff; FBV = fetal blood vessel.
c Mature intermediate villus with many arterioles, capillaries, and venules (H & E, ×200). FBV = fetal blood vessel TB = trophoblast.
d Terminal villi with sinusoids and metabolic membranes (H & E, ×1000). S = sinusoid MM = metabolic membrane; TB = trophoblast; IVLS = intervillous space.

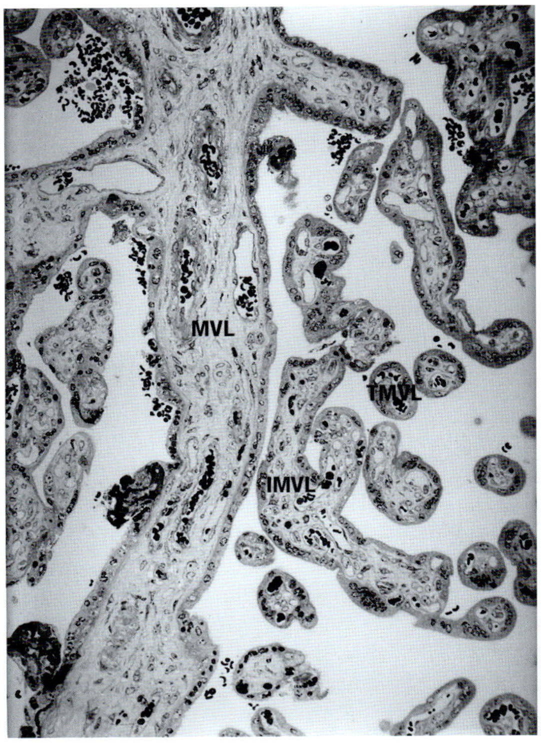

Fig. 24.7 High-power view of a villous tree (semithin section, H & E, ×200). MVL = mainstem villus IMVL = intermediate villus ; TMVL = terminal villus.

lary growth exceeds the length of the villus, the capillary becomes coiled and causes a passive outpouching of the villous surface. This process creates a cluster-of-grapes pattern of terminal branches that represent new terminal villi. Imbalances between capillary and villous growth can lead to various disturbances of villous maturation[6, 27].

Microstructure of the Terminal Villus

Cytotrophoblast. The terminal villus is approximately 30–80 μm in size (Fig. 24.**8**). With advancing gestation, the inner cytotrophoblast becomes more and more fragmentary. It provides a stem cell pool for the outer syncytiotrophoblast, with oxygen apparently serving as an important regulator. Hypoxic damage to the syncytiotrophoblast stimulates proliferation of the cytotrophoblast, followed by fusion of the cells[6].

Syncytiotrophoblast. The outer syncytiotrophoblast is no longer capable of cell division. It shows increasing differentiation and its functions include hormone production.

Villous capillaries. With sinusoidal expansion of the fetal villous capillaries, the syncytial nuclei and organelles are pushed aside. The trophoblastic and capillary basement membranes partially fuse together, forming the metabolic membrane that is so essential for metabolic and gaseous exchange. In the mature placenta, approximately 25–40% of the surface area of the terminal villi consists of these vasculosyncytial membranes, which are 0.5–1 μm in thickness[46].

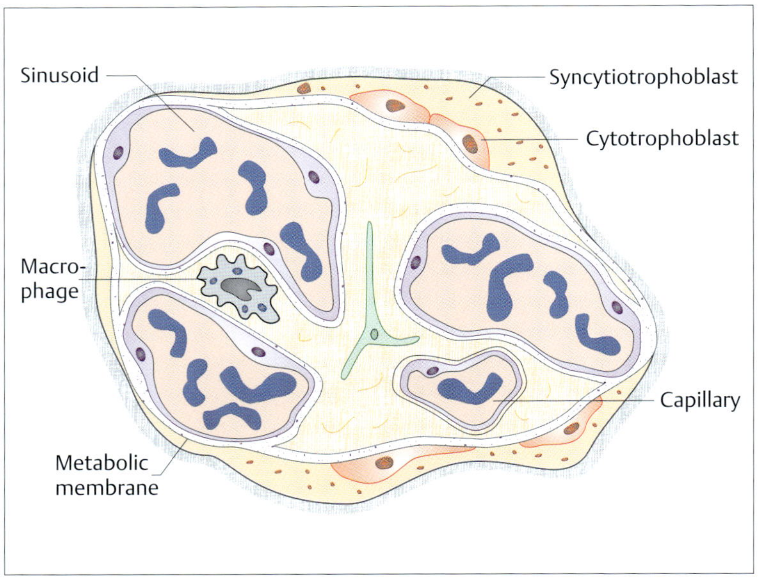

Fig. 24.**8 a** Schematic diagram of the microstructure of a terminal villus.

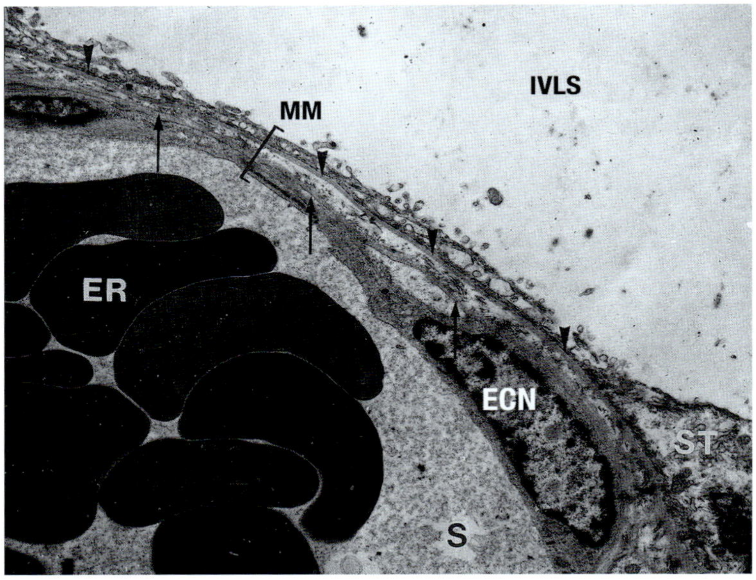

Fig. 24.**8 b** Transmission electron micrograph of the metabolic membrane of a terminal villus. S = sinusoid; ST = syncytiotrophoblast; MM = metabolic membrane; ER = erythrocyte; ECN = endothelial cell nucleus; IVLS = intervillous space; basement membrane of trophoblast (arrowheads); basement membrane of capillary (arrows).

Maturation of the Placenta

The maturation of the placenta serves to shorten the distance between the maternal and fetal blood. Becker[5] defined four signs of placental maturity that can be useful in the qualitative assessment of placental maturation.

First sign of maturity: a steady decrease in villous diameters accompanied by an increase in the total villus surface area.

Second sign of maturity: increasing vascularization of the villi and sinusoidal transformation of the villous capillaries. This process starts at about the end of the 5th month of gestation, increases markedly during the final six weeks, and reaches its high point at term. The villus stroma is drastically reduced, causing a substantial decrease in the fetomaternal diffusion pathway.

Third sign of maturity: decreasing thickness of the villus trophoblast with advancing gestation, accompanied by displacement of the trophoblastic nuclei and the formation of metabolic membranes. The trophoblastic nuclei move to the sides and form nuclear bridges between adjacent villi. This creates a supportive "external cytoskeleton"[5] that interlinks the villi and prevents collapse of the villous network in response to significant changes in maternal blood pressure. The formation of metabolic membranes and trophoblastic bridges is the criterion for the third sign of placental maturity. As vascularization of the villi increases, the "diffusion placenta" is progressively transformed into a "vascularization placenta"[5].

Fourth sign of maturity: luminal narrowing of the large arterial vessels of the mainstem villi and the formation of paravascular fibrous cuffs. This leads to stiffening of the vessels, and the resulting pressure constriction with a slowing of flow in the downstream vascular system serves to decompress the pressure-sensitive peripheral villi and protect the delicate sinusoids in the terminal villi[5].

Vascular Architecture of the Villous Tree

By about the 6th week after fertilization, an embryoplacental circulation is established by the fusion of allantoic vessels from the umbilical cord with the capillaries that develop locally in the villi[6, 13]. Different wall structures result from the different pressure gradients that exist in the afferent and efferent vessels.

Umbilical cord. The umbilical cord transmits a vein carrying arterialized blood along with two arteries, which anastomose near the placenta in 96% of cases in order to equalize pressures and distribute the blood more uniformly (Hyrtl's anastomosis).

Villous tree. Each villous tree is supplied by a centrally located artery and an accompanying vein, which progressively branch into arterioles, capillaries, and venules. The vessel walls are characterized histologically by an almost complete absence of elastic fibers. Numerous processes pass between the endothelial and muscle cells, creating a "musculoendothelial system" that is important in local autonomic vasoregulation, since the placenta is devoid of nerves[36].

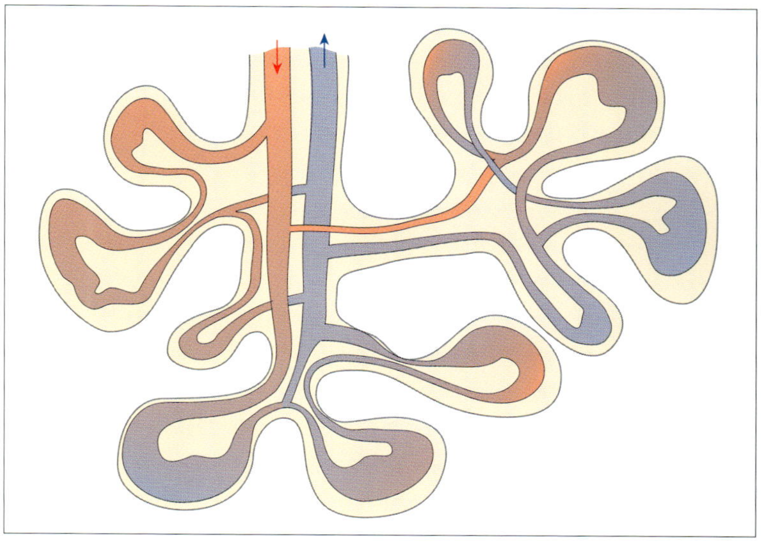

Fig. 24.**9** Arrangement of the fetal blood vessels in a cluster of terminal villi.

Terminal villi. The capillaries of the terminal villi, which lack a tunica media, show local dilatative, sinusoidal transformations that serve to decelerate the blood flow (Fig. 24.**9**). To ensure an optimum transfer of substances and gaseous exchange, the fetal blood passes successively through the capillary loops of 2–5 adjacent terminal villi before it returns to the postcapillary venule[6, 27].

Regulation of Villous Blood Flow

From 40% to 57% of the fetal cardiac output flows through the two umbilical arteries into the placenta. The rate of umbilical blood flow at term is approximately 110 (ml/min)/kg body weight. The arterial fetal blood pressure is approximately 53 mmHg, the intravillous capillary pressure is approximately 35 mmHg, and the umbilical venous pressure is about 20 mmHg[46]. While the fetal heart beat represents an important driving force, it appears that there is a peripheral regulation of villus blood flow. Immunohistochemical studies have demonstrated the presence of contractile actin- and desmin-positive myofibroblasts in the vessels and stroma of the villi (Fig. 24.**10**). It may be that substances such as catecholamines, angiotensin, and prostaglandin exert a hormonal control of villous blood flow. There can be no neural regulation of villous blood flow, as the placenta is devoid of nerves[46].

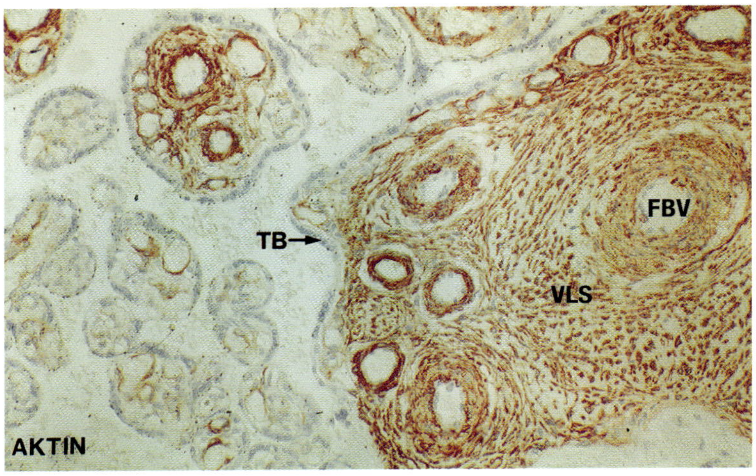

Fig. 24.**10** Immunohistochemical detection of actin-positive myofibroblasts in the fetal vessel walls and villous stroma (×200). TB = trophoblast; FBV = fetal blood vessel; VLS = villous stroma.

Concept of the Placentone

The “placentone” is a fetomaternal functional unit comprised of a fetal villous tree (cotyledon) and the maternal spiral artery that supplies it[42]. The central portions of the placentone serve chiefly as a growth and regeneration area, while the peripheral portions are for the fetomaternal exchange of gas and nutrients[42].

Gas and nutrient transfer. By injecting contrast material into cannulated spiral arteries, Beck[4] showed that the oxygen-rich maternal blood first collects in a centrally located, loosely structured cavity in the placentone (Fig. 24.**11**). From there it flows radially into the intervillous space, which is now of capillary size. These areas are characterized histologically by the presence of numerous terminal villi. With their closely packed arrangement, these villi significantly slow the movement of maternal blood, permitting an optimum fetomaternal exchange of gas and nutrients. The oxygen-depleted venous maternal blood then flows back through the subchorionic space and regions between the villous trees into the veins of the basal plate and into the marginal sinus. As a result, the distribution and direction of maternal blood flow are regulated essentially by the varying density of the villi within the placentones[4,33].

Morphology and Physiological Transformation of the Maternal Basal-Plate Vessels

The maternal blood vessels of the basal plate increasingly undergo regressive changes during the course of pregnancy. Extravillous trophoblastic cells play an essential role in this process[6,9,38,39]. Figure 24.**12** shows a cross section of the basal plate with a maternal spiral artery and a villus anchored in the basal plate, with adjacent cytotrophoblastic and syncytiotrophoblastic cells.

Extravillous trophoblastic cells. Some of the proliferating cytotrophoblasts fuse, as described above, to form the outer syncytiotrophoblasts. Others invade the placental bed and the maternal preplacental vessels. Similarly to tumor cells, these cytotrophoblastic cells show an altered expression of adhesion molecules and matrix molecules and leave the villous basement membrane[6,7,9,43]. These extravillous trophoblastic cells deeply infiltrate the decidua, losing their cellular cohesion. After reaching a depth of more than five cell layers from the villous basement membrane, they stop dividing and start to

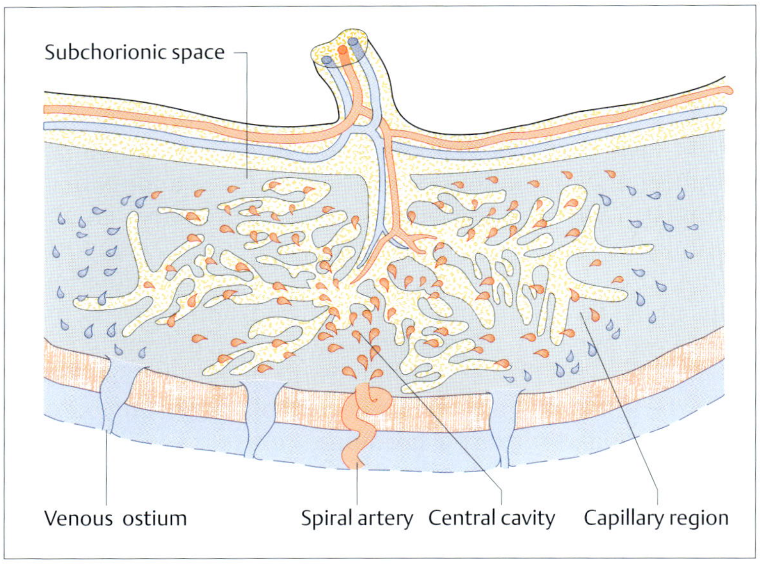

Fig. 24.**11** Schematic diagram of maternal blood flow in a fetomaternal functional unit (placentone; red = arterial blood, blue = venous blood).

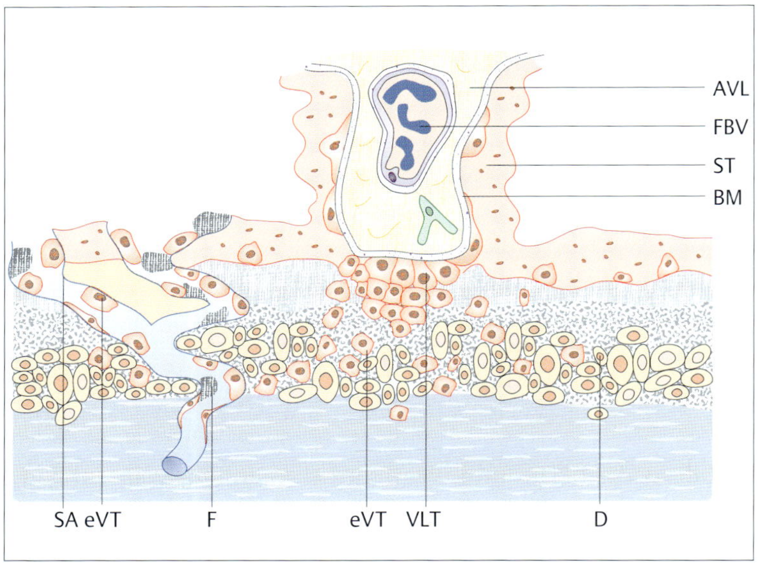

Fig. 24.**12** Schematic diagram showing the histological structure of the basal plate with an adjacent anchoring villus and the arrangement of the villous and extravillous trophoblasts. (Modified from Benirschke and Kaufmann[6].) AVL = anchoring villus; FBV = fetal blood vessel; ST = syncytiotrophoblast; BM = basement membrane; SA = spiral artery; F = fibrinoid; D = decidua; VLT = villous trophoblast; eVT = extravillous trophoblast.

differentiate. In the process, these cells form a "matrix-type fibrinoid"—a modified basement membrane material in nonpolarized form whose functions include creating a barrier to further trophoblast invasion[6]. This "intermediate" trophoblast located outside the villi fulfills at least some of the criteria of a malignant tumor invading the maternal tissue. This infiltration is controlled and self-limiting, however[43].

Intravascular trophoblast invasion. The preplacental maternal arteries are also invaded and remodeled by trophoblastic cells. This process is marked by a decrease of muscular and elastic fibers in the arteries and an increase in intramural fibrinoid deposits (Fig. 24.**13**)[6,14,38].

Two stages can be distinguished in this process. An initial stage of intravascular trophoblast invasion occurs in just the 8th to 10th weeks of gestation and mainly affects arteries located in the central area of the placental bed[32]. A second wave of trophoblast invasion occurs in the 14th to 16th weeks of gestation. In this stage the cells migrate into myometrial segments of the maternal arteries and transform them into broad, relatively stiff channels that are exempt from local vasoregulation. The luminal diameters of the vessels are increased from approximately 200 μm to 1000–2000 μm where they open into the intervillous space, with a corresponding decline of flow resistance[6]. It has been theorized that these physiological vessel wall changes fail to occur in maternal hypertensive disorders of pregnancy[8,39].

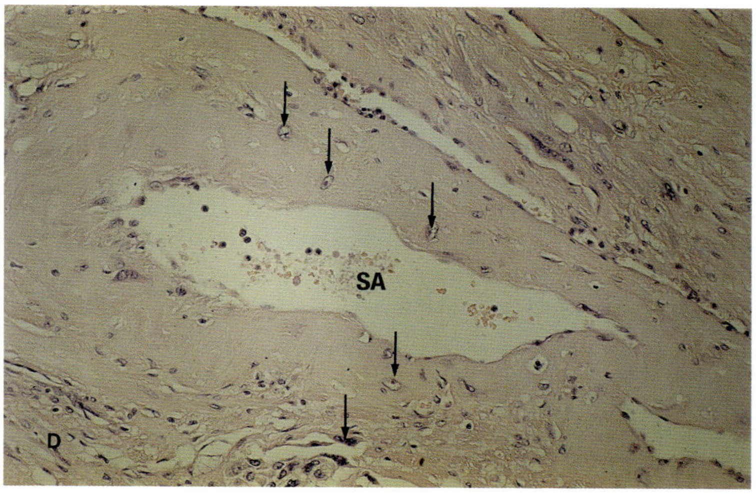

Fig. 24.**13** Physiological fetal trophoblast invasion of a maternal preplacental blood vessel (H & E, ×200). SA = spiral artery D = decidua; extravillous trophoblastic cells (arrows).

Placental Insufficiency

Definition and Etiology of Placental Insufficiency

From a pathoanatomical standpoint, placental insufficiency refers to a decline in the performance of the organ as a result of demonstrable structural abnormalities.

Intraplacental causes. A disturbance of metabolic and exchange processes in the placenta can result from a reduction of perfusion capacity, diffusion capacity, or both[46]. It can have various causes:

- Deficient placental growth (in weight and/or area of attachment)
- Abnormalities of villous maturation
- Decreased blood flow (maternal and/or fetal)
- Parenchymal inflammation
- Tumors

Extraplacental causes. Placental insufficiency may also have extraplacental causes. Preplacental or postplacental abnormalities of the maternal or fetal circulatory system are generally not accessible to pathoanatomical diagnosis.

The placenta normally undergoes a continuous increase in size and weight during the course of pregnancy, accompanied by maturation of the villous trees with a reduction in villous diameters and increasing vascularization. The resulting decrease of impedance can be quantified by Doppler sonography of the umbilical arteries. In several studies in which the histomorphological findings in placentas were correlated with Doppler flow measurements, abnormal Doppler velocity waveforms were found to be associated with significantly smaller placentas (reduced area of attachment and volume) and lower fetal weights[19,20,47]. Moreover, it was common to find markedly reduced numbers of vessels and metabolic membranes per villus in groups with zero or negative flow[21,47,48].

The results in these cases point to the presence of an overriding disorder that is responsible both for the smaller fetoplacental dimensions and for the underdifferentiation of the villous trees, leading to frank placental insufficiency[47]. It may be that an implantation abnormality with reduced fetal trophoblastic invasion of the maternal basal plate and spiral arteries may be the cause of placental dysfunction in these cases[22]. The further clinical course in cases with Doppler abnormalities is critically influenced by the extent of the placental reserve capacity.

Placental Compensatory Mechanisms

The human placenta has a number of compensatory mechanisms at its disposal:

- Initially, placental dysfunction evokes an increase in fetal and/or maternal blood flow.
- More serious disturbances of villous maturation can evoke a compensatory increase in placental growth.
- With a primary small placenta or a placenta whose functional parenchyma has been reduced by extensive obliterative lesions, an acceleration of placental maturation may be observed[5,46].

Classification of Placental Insufficiency by its Progression

Acute and chronic forms. A distinction is drawn between acute and chronic forms of placental insufficiency. The acute form develops in minutes or hours and causes a global impairment of oxygen and other gaseous exchange followed by intrauterine asphyxia (acute respiratory insufficiency). Chronic placental insufficiency develops over a period of weeks or months and leads to intrauterine growth retardation (nutritional insufficiency)[5].

Morphological Counterparts of Latent or Overt Placental Insufficiency

Disorders of Villous Maturation

Timing and consequences. Villous maturation disorders may occur in early pregnancy, when the dominant processes are placental growth and the expansion of the villous trees, or in later pregnancy when the dominant process is placental maturation. A severe villous maturation disorder occurring in early gestation leads to fetal loss. Maturation defects that occur in advanced stages of pregnancy can cause fetal growth retardation but may also lead to intrauterine death. In evaluating villous maturation disorders, it is important to take into account fetal weight and dimensions and compare them with those of the placenta, e.g., by determining the placental-fetal index.

Arrested Villous Maturation

Figure 24.**14** shows an example of arrested villous maturation[46], known also as maturity arrest[5] or persistent immaturity[6]. The histological features are compared with a normally branched villous network in the 28th week of gestation. Villous maturation may become arrested in early or late pregnancy, appearing histologically as large-caliber, poorly branched villi with decreased vascularization, thin vascular tunica media, and the absence of a paravascular fibrous cuff (persistence of immature intermediate villi). This maturation defect may occur in the setting of chromosome abnormalities, diabetes mellitus, or blood group isoimmunization.

Chromosome abnormalities. Chromosome abnormalities such as trisomies, triploidies, and XO monosomies usually lead to pregnancy loss during the first 12 weeks. Other possible effects are abnormal fetal development or intrauterine death in later stages of pregnancy, particularly with numerical chromosome anomalies[46]. A cytogenetic analysis should be done whenever possible, since morphological examination of the placenta rarely identifies criteria that can prove a genetic anomaly. It is common, however, to find underweight placentas with large, immature "cat's paw" villi resulting from abnormal villous branching. These changes are considered part of the phenotype of the underlying chromosome disorder[46].

Diabetes mellitus. Arrest of villous maturation is a common finding in pregnant women with poorly controlled diabetes mellitus. It is characterized by poor villous branching with hypoplastic blood vessels, microcystic edema, and poor differentiation of the chorionic epithelium as a possible response to maternal hyperglycemia[46]. The children of these mothers show hyperinsulinism combined with accelerated longitudinal growth, visceromegaly, obesity, and corresponding weight gain[12]. Appropriate correction of the diabetic metabolism can prevent the development of these changes[6, 46].

Rhesus incompatibility. Another disease that may be associated with arrested villous maturation is rhesus incompatibility (Fig. 24.**14a**). In this condition it is common to find large-caliber, immature villi with coarse villous edema and numerous erythroblasts in the fetal capillaries. Rhesus incompatibility is an immunohemolytic anemia that develops when anti-erythrocyte maternal antibodies cross the placental barrier. The fetal anemia and hypoxia evoke a dramatic increase in the fetal cardiac output. The resulting cardiomegaly with heart failure and hypoxic capillary-wall damage are considered the

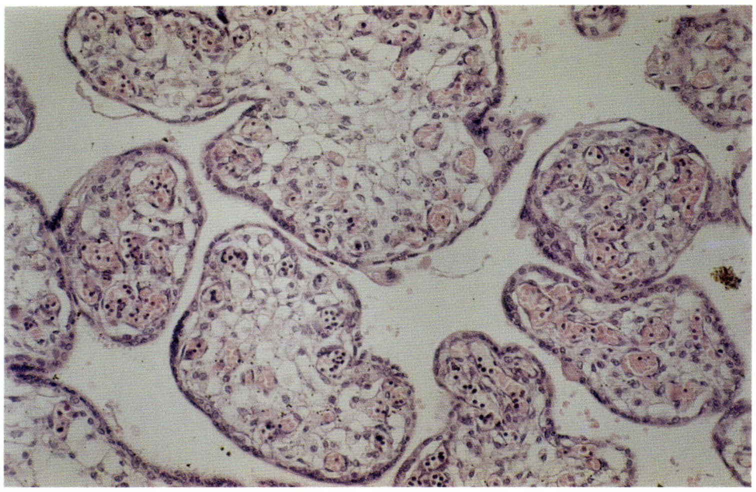

Fig. 24.**14 a** Arrest of villous maturation in rhesus incompatibility (28th week of gestation, H & E, ×200).

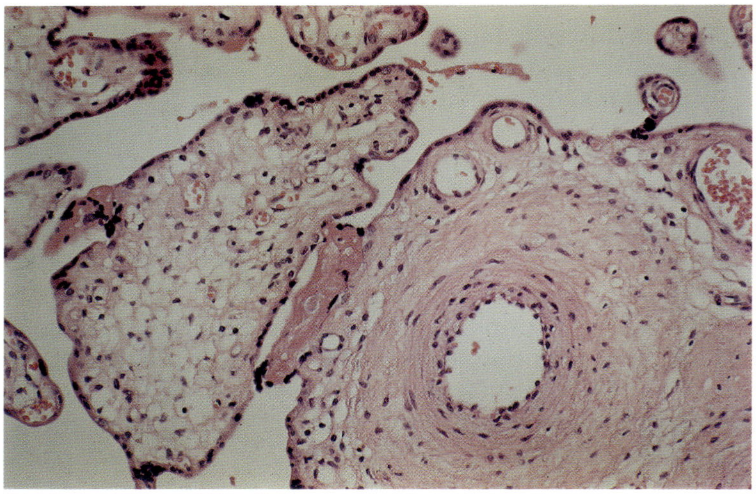

Fig. 24.**14 b** Normally branched villous network (28th week of gestation, H & E, ×200).

major causes of fetal and placental edema, which may progress to hydrops[6,46]. Even villous capillary-wall ruptures with bleeding into the intervillous space are occasionally observed[6]. Fetal erythropoiesis is increased, and there are increased numbers of immature nucleated precursor cells.

Electron microscopy of affected placental villi demonstrates necrotic syncytiotrophoblasts, hyperplasia of the cytotrophoblasts, and thickening of the villous basement membrane[24]. These changes may be caused by direct immune reactions [49].

In some cases, villous immaturity evokes a compensatory increase in placental growth and ineffectual capillary proliferation in the immature intermediate villi[5]. For villous maturation to become arrested, maternal antibodies must cross the placental barrier in early pregnancy, when the villous trees are still immature. The placental changes take some time to develop, and therefore a histological diagnosis cannot be made before the 26th week of gestation[46]. For comparison, the placenta in Fig. 24.**14b** shows a normal degree of development for the 28th week of gestation.

Retarded Villous Maturation

An example of retarded villous maturation[46] is shown in Fig. 24.**15a**. Also known as retarded maturity[5] or terminal villous deficiency[6], this maturation disorder can be diagnosed by the second trimester of pregnancy. Histological sections show insufficient branching of the fetal villous trees with a deficiency of terminal villi and decreased vascularization of the villi. Since the metabolic membranes are reduced in number, the diffusion path is lengthened. The intrauterine fetal supply can be maintained only through compensatory placental hyperplasia. These placentas often show a poor compensatory capacity at birth, however, and this can lead to acute fetal/neonatal hypoxia[5]. Retarded villous maturation can occur in conditions such as postdate pregnancy, poorly controlled diabetes mellitus, and blood group isoimmunization, but often there is no discernible cause. For comparison, Fig. 24.**15b** shows the appearance of a largely mature placenta in the 40th week of gestation.

Dissociated Villous Maturation Disorder with Prevalence of Immaturity

Figure 24.**16a** shows an example of a dissociated villous maturation disorder "with prevalence of immaturity"[46] in the 38th week of gestation. The fetal cotyledons have an approximately normal branching pattern, but they show a focal or extended failure of organization into different types of villi in terms of stromal and vascular development[46]. This villous maturation disorder does not appear before the late second trimester. It is on a continuum with retarded villous maturation.

Histological examination shows a markedly decreased capillary content in the villi, which are still bordered predominantly by two trophoblastic layers. Diffusion capacity is again compromised by a decreased number of metabolic membranes. Figure 24.**16b** illustrates normal placental development for the 38th week of gestation.

Chorangiosis

Chorangiosis[46] is a villous maturation disorder characterized by hyperplasia of the villous capillaries, which often show insufficient sinusoidal transformation (Vogel type I chorangiosis, Fig. 24.**17a**). The diffusion path is correspondingly lengthened. This disorder may be manifested as early as the first trimester and is often viewed as an attempt by the placenta to compensate for primary placental hypotrophy or a secondary decrease in placental size due to extensive infarctions. Chorangiosis may also occur in association with maternal anemia, poorly controlled diabetes mellitus, blood group isoimmunization, pregnancies at high altitudes, or a maternal cyanotic heart defect[6,46]. Studies in experimental animals have shown that chronic hypoxemia leads to endothelial proliferation with increased fetal capillarization of the villi and increased tortuosity of those vessels[3]. The pathogenesis has been related to the production of growth factors by fetal macrophages that have been damaged by hypoxia[37]. Figure 24.**17b** shows a placenta that is normally developed for the 34th week of gestation.

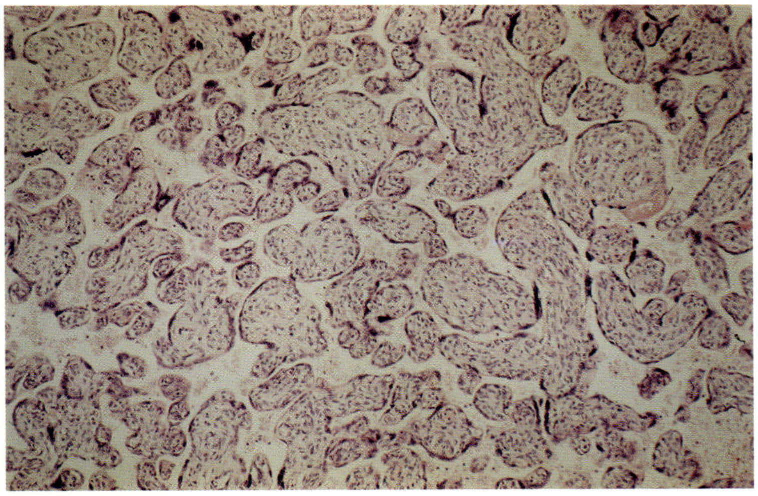

Fig. 24.**15a** Retardation of villous maturation (40th week of gestation, H&E, ×100).

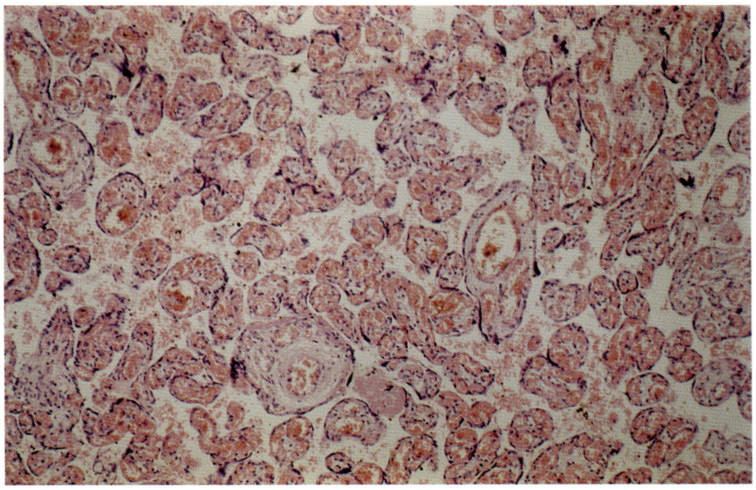

Fig. 24.**15b** Mature placenta (40th week of gestation, H&E, ×100).

Specific Obstetric Problems

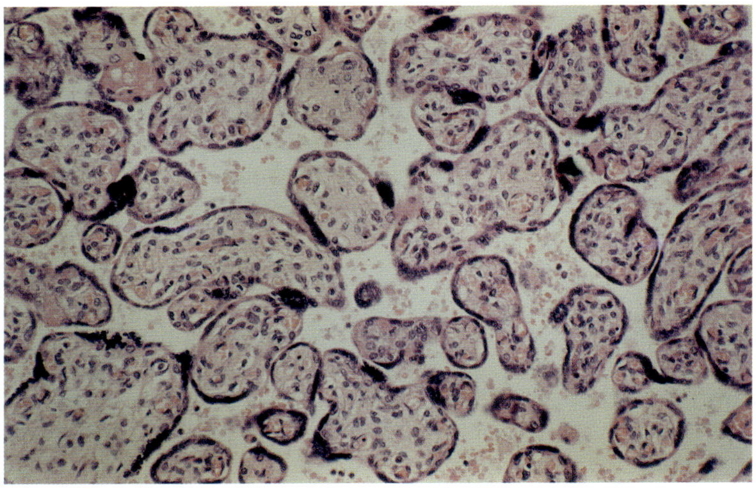

Fig. 24.**16 a** Dissociated maturation defect with prevalence of immaturity (38th week of gestation, H & E, ×200).

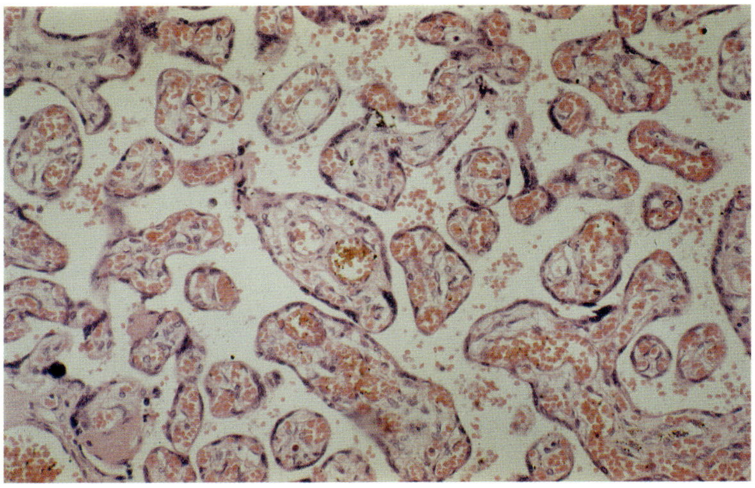

Fig. 24.**16 b** Mature placenta (38th week of gestation, H & E, ×200).

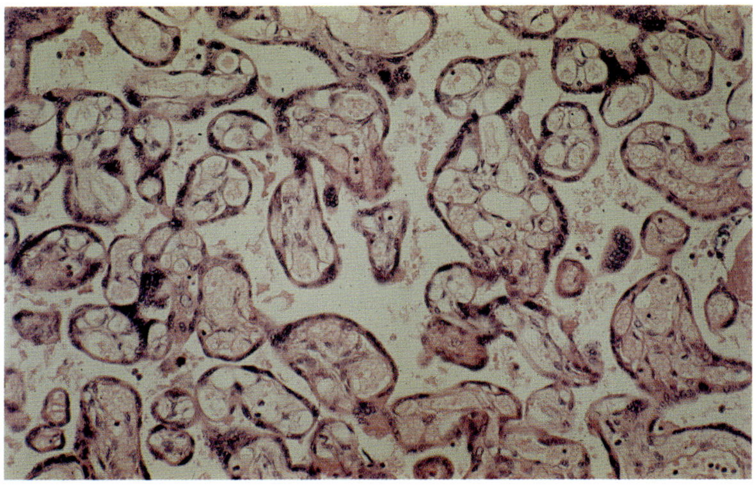

Fig. 24.**17 a** Vogel type I chorangiosis (34th week of gestation, H & E, ×200).

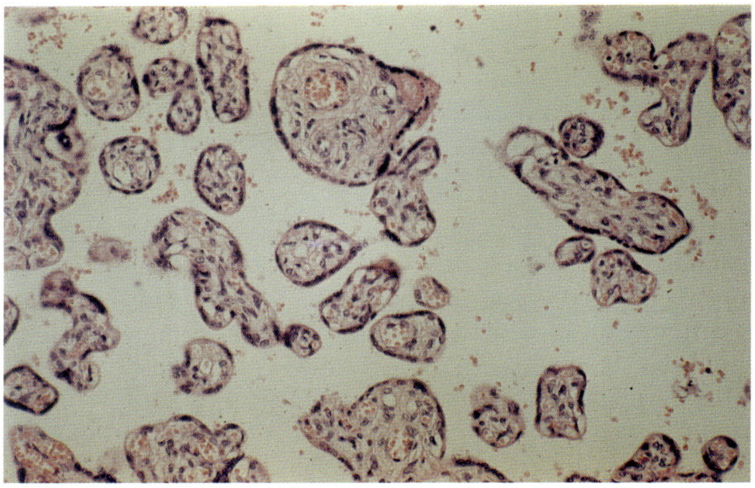

Fig. 24.**17 b** Normally branched villous network (34th week of gestation, H & E, ×200).

Accelerated Villous Maturation

Accelerated villous maturation[6] is commonly seen, for example, in underweight placentas that have a small basal area and extensive adjacent obliterative placental lesions. It is characterized by asynchronous maturation of the fetus and placenta, and many authors interpret it as an attempted compensatory response by the placenta. Known also as premature villous maturation[5], this disorder has the histological appearance of an almost fully mature placenta while the fetus is still immature and appropriate for dates.

Deficiency of Intermediate Villi

Figure 24.**18a** shows the histological appearance of this deficiency[46]—a focal or generalized villous maturation disorder in which the intermediate villi are decreased in number. Normally fibrosed mainstem villi and myriad terminal villi dominate the picture. As a result, the perfusion capacity of the placenta is diminished. Since the placenta has no additional growth reserves, it is pushed to the limit in terms of fetal supply and, according to Becker[5], requires "rescue by birth." For comparison, Fig. 24.**18b** shows the appearance of an essentially normal-for-date placenta in the 29th week of gestation.

Fetal Obliterative Vascular Diseases

Obliterative fetal vascular diseases most commonly occur in the large endplate vessels and mainstem villous vessel, causing fibrous narrowing or even occlusion of the vessel lumina.

Endarteropathy obliterans. A typical representative of these diseases is endarteropathy obliterans, which is defined as reactive fibrous tissue proliferation in the vessel walls in response to intrauterine insults[5,46]. Figure 24.**19** shows an originally large mainstem villous vessel that has been narrowed by fibrous plaque. In most cases the precipitating cause of this process cannot be determined either clinically or morphologi-

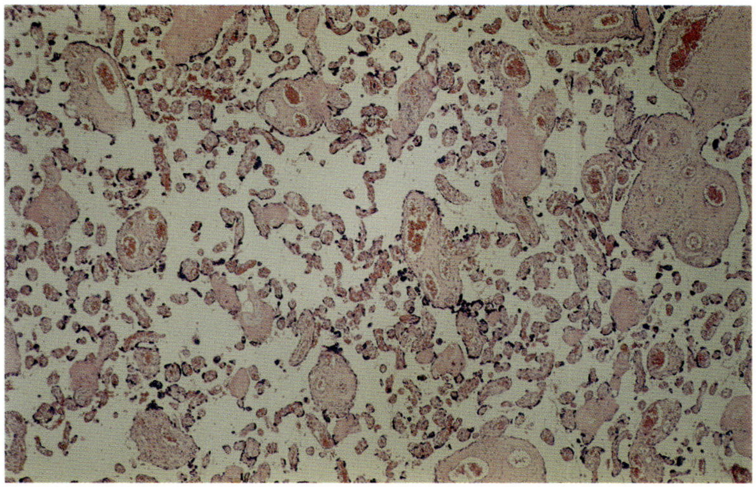

Fig. 24.**18 a** Deficiency of intermediate villi (29th week of gestation, H & E, ×40).

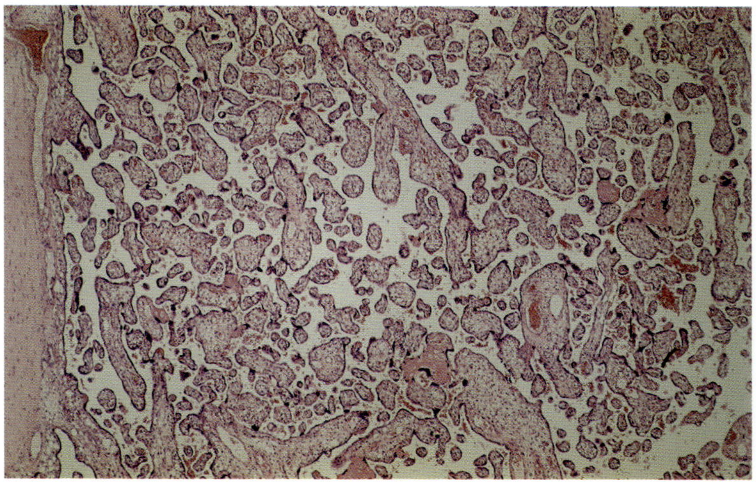

Fig. 24.**18 b** Normally branched villous network (29th week of gestation, H & E, ×40).

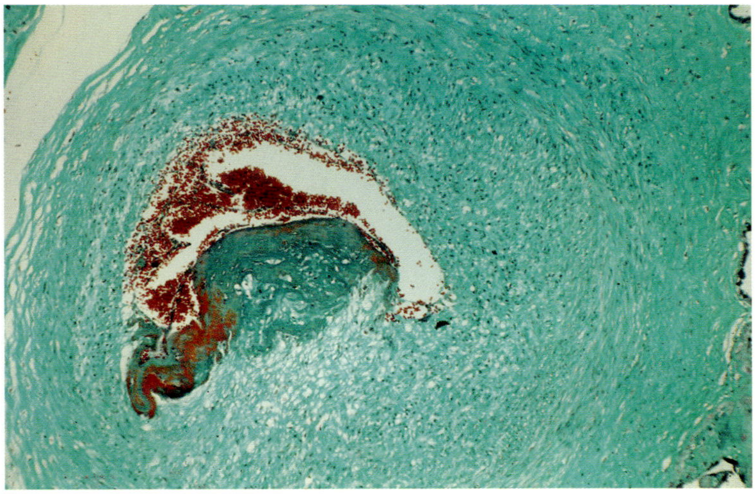

Fig. 24.**19** Endarteropathy obliterans in a mainstem villous vessel (MG, ×100).

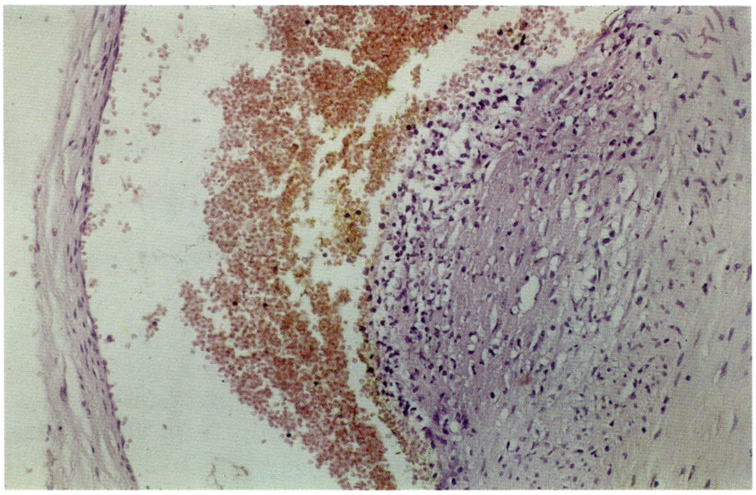

Fig. 24.**20** Endarteritis obliterans in a mainstem villous vessel (H & E, ×200).

cally. Endarteropathy obliterans may be found in the setting of intrauterine infections, diabetes mellitus, blood group isoimmunization, autoimmune disorders, maternal nicotine abuse, and other conditions[5,46].

Endarteritis obliterans. This is an obliterative process in which inflammatory infiltrates or pathogenic organisms can be identified[46]. Endarteritis obliterans most commonly occurs in congenital syphilis or rubella (Fig. 24.**20**).

HIV-associated vasculopathy. Jimenez et al.[23] were able to detect allantoic vasculopathy in 18 of 34 placentas from HIV-positive mothers. Histological sections revealed foci of endothelial cell necrosis, media proliferation, altered collagen synthesis, fibrinoid vessel wall insudation, mononuclear infiltration, and complicating thrombus formation. These vascular changes may be the result of abnormal immune responses. The presence of these changes showed a positive correlation with clinical manifestations of maternal immune suppression, i.e., less than a 1 : 1 ratio of maternal T_4 helper cells to T_8 suppressor cells and a fall in the T_4 lymphocyte count to less than 500/μL.

The same authors observed villous maturation defects with prevalence of immaturity in more than one-third of the placentas examined. The villous immaturity may have resulted from the vasculopathy. Inflammatory changes in the placental parenchyma and/or the fetal membranes are no more common in HIV-infected mothers than in the general population.

The detection of vasculopathy in the allantoic vessels is not specific for HIV infection, nor does it provide evidence of any fetal infection that may be present[23].

Fetal thrombi. Fetal thrombi with luminal narrowing or occlusion are most commonly found in large mainstem villous vessels and endplate vessels. Many of these thrombi are conglutination thrombi that have formed in a setting of endarteritis or endarteropathy. Figure 24.**21** shows endplate vessels occupied by older fetal thrombi that are already becoming organized. Thrombi may be found in small or large villous vessels as a re-

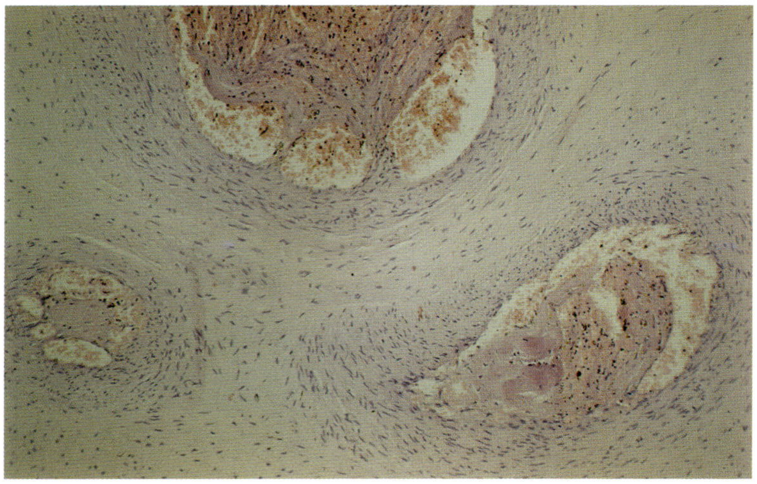

Fig. 24.21 Fetal thrombi in end-plate vessels (H & E, ×100).

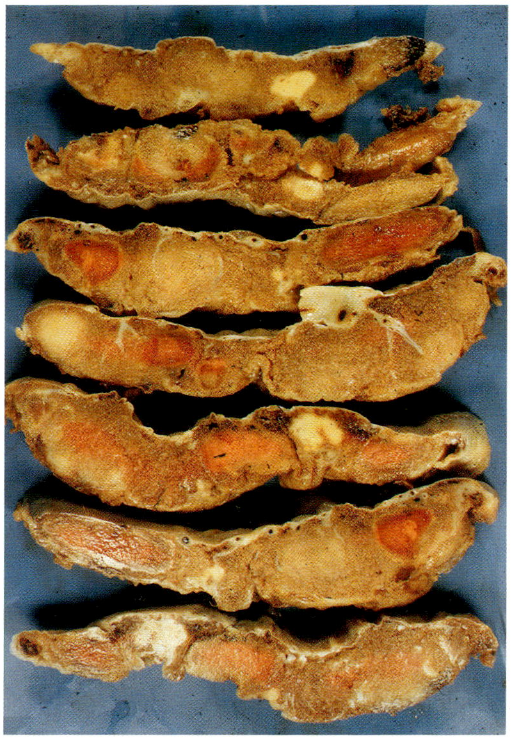

Fig. 24.22 Multiple obliterative foci in a placenta (35th week of gestation).

sult of umbilical cord complications such as strangulation or knotting of the cord, nuchal cord, or prolapse of the cord.

Fetal fibrin-rich microthrombi. Fetal fibrin-rich microthrombi are found chiefly in terminal vascular branches at the periphery of the villi. They are usually a result of generalized intrauterine shock with disseminated intravascular coagulation [46].

Sequelae. The sequelae of chronic obliterative vascular diseases and fetal thrombi include hypovascularity or even avascularity of the villi at the periphery of the placenta. Occurrence of the disease during the first half of pregnancy leads to focal immaturity of the villi in the downstream villous tree. With onset in the second half of pregnancy, histological examination shows obliteration of the villous capillary lumina and fibrosis of the villous stroma. Placental infarction does not occur, however, because the fetal villi are supplied by maternal blood in the intervillous space.

Abnormalities of Maternoplacental Blood Flow

Figure 24.**22** shows a placenta in the 35th week of gestation permeated by multiple obliterative lesions of varying size and color (some are red, others grayish-yellow or white). Their pathogenesis is based on an abnormality of maternoplacental blood flow. Given the large compensatory reserve of the placenta, generally only extensive lesions that occupy more than one-third of the organ lead to fetal growth restriction. If more than half of the placenta is obliterated, a critical placental insufficiency develops that is life-threatening to the fetus[5].

Vogel distinguished among the following types of lesion[46]:

- Placental infarctions
- Reticular infarctions
- Intervillous thrombi
- Hematomas
- Subchorionic and basal pseudoinfarctions

Placental infarctions. Placental infarctions are areas of parenchymal necrosis caused by an interruption of the maternal blood supply, which may involve either a fetomaternal circulatory unit (cotyledon infarction) or several adjacent placentones (massive infarction).

Infarctions are classified by their age as acute, subacute, or chronic. A fresh infarct has a dark red color and friable consistency. Later it turns brownish red and finally grayish-white in the chronic stage.

Figure 24.**23** shows the histological appearance of a chronic infarction with complete villous necrosis and loss of the chorionic epithelium. Some villi are only vaguely identifiable as such ("ghost villi"). The decreased maternal blood supply may result from atherosclerotic changes, spiral artery thrombosis or aneurysms (e.g., in a setting of arterial hypertension), or a disorder of fat metabolism. Other potential causes are arteriopathies and arteriolopathies in diabetes mellitus, autoimmune disorders, and preeclampsia. Exogenous insults can also produce vessel wall lesions that cause luminal narrowing[46].

Reticular infarctions. Reticular infarctions ("gitterinfarcts") are obliterative lesions of variable size, often with maplike borders, that show no consistent relationship to placentones[46]. Figure 24.**24** shows the histological appearance of a reticular infarction, in which multiple groups of villi are encased by homogeneous perivillous and intervillous fibrin masses. The histological spectrum ranges from a loss of chorionic epithelium to complete villous necrosis. Various causes have been suggested besides maternal hypoxemia, including latent herpes simplex infections and abnormal immune reactions between the maternal and fetal tissue[6]. Extensive reticular infarctions can restrict intervillous maternal blood flow and retard further development and maturation of the villi in that region[5].

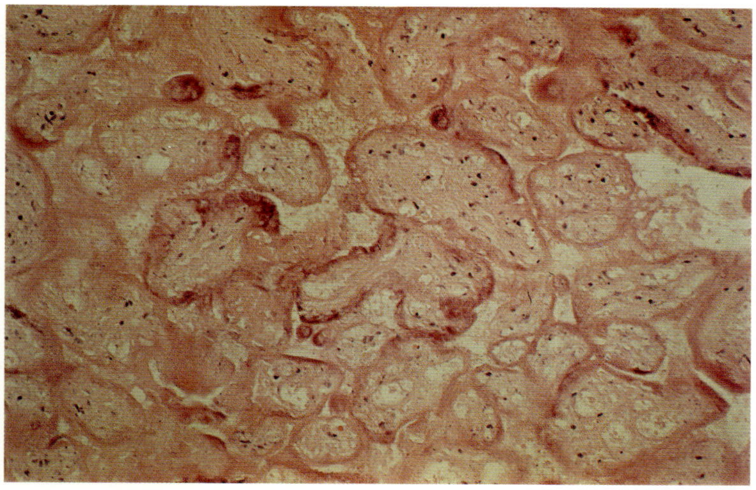

Fig. 24.**23** Histological appearance of a chronic placental infarction (H&E, ×200).

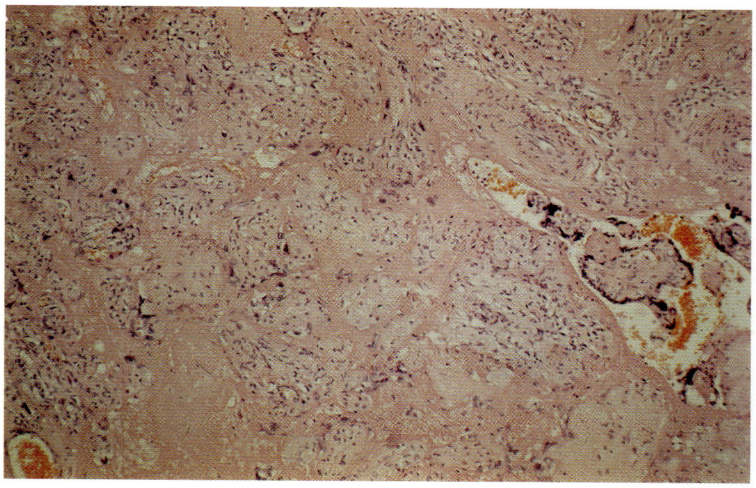

Fig. 24.**24** Microscopic view of a reticular infarction (35th week of gestation, H&E, ×100).

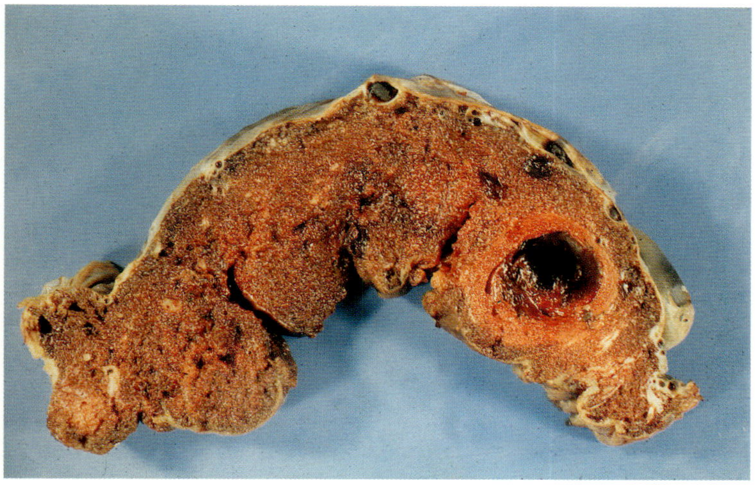

Fig. 24.**25a** Intervillous thrombus (37th week of gestation).

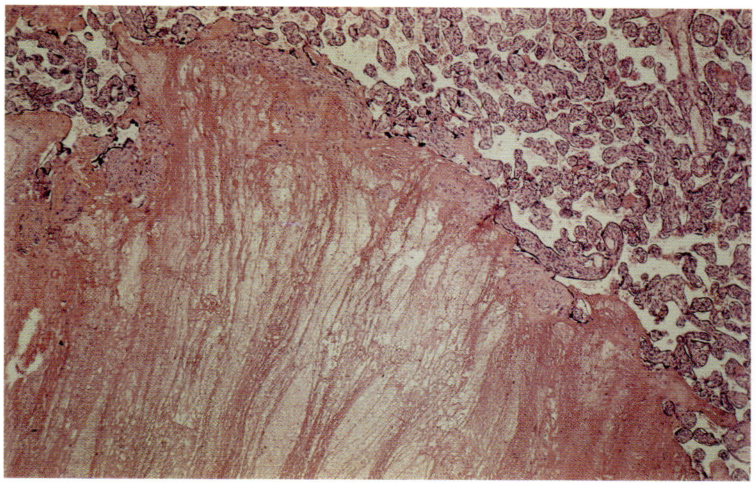

Fig. 24.**25b** Histological appearance of a thrombus in the intervillous space (37th week of gestation, H&E, ×40).

Intervillous thrombus. Figure 24.**25a** shows an example of an intervillous thrombus, appearing as a sharply circumscribed, dark red area in the intervillous space, which corresponds to an intravascular space. Coagulation and conglutination thrombi can be distinguished from each other by histological examination. Figure 24.**25b** shows the histological appearance of a conglutination thrombus in the intervillous space causing villous displacement.

Intervillous thrombi may be caused by maternal venous outflow obstruction and/or by an irregularly structured intervillous space in which alternating zones of narrowing and dilatation lead to turbulence or stasis of maternal blood flow. Another possible cause of intervillous thrombi is damage to the villous chorionic epithelium, due for example to injuries, hypoxemia, or toxic or inflammatory changes and ruptures of villous vessels[46].

Placental hematoma. Placental hematomas are abnormal collections of blood outside the intervillous space. A common example is retroplacental hematoma, which may be detectable in cases of placental abruption.

Subchorionic and basal pseudoinfarcts. Pseudoinfarcts are caused by the deposition of homogeneous fibrin masses[46]. Generally they have no pathological significance but serve to compensate for placental growth discrepancies and apparently increase the stability of the villous tree. Also, by limiting the size of the relatively villus-poor intervillous space below the chorionic plate and in the basal plate region, they can improve blood flow to centrally located placental areas that contain more villi[5].

Placental Changes in Hypertensive Disorders of Pregnancy

Acute atherosis. Extensive true infarctions are a relatively common finding in hypertensive disorders of pregnancy. Most of these infarcts are located in the central areas of the placenta. Characteristic but nonspecific vessel wall changes may be found in maternal preplacental arteries. Histological sections show endothelial cell damage, transmural fibrinoid necrosis, aggregations of foam cells, lymphohistiocytic infiltrates, and thrombi (Fig. 24.**26**). These vessel wall changes can result in

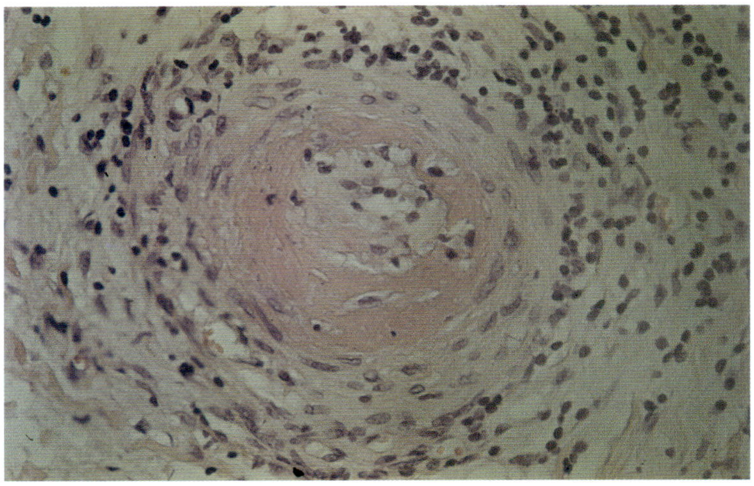

Fig. 24.**26** Acute arterial vasculopathy with transmural fibrinoid necrosis, lymphohistiocytic infiltrates, collections of foam cells, and luminal narrowing (H & E, ×400).

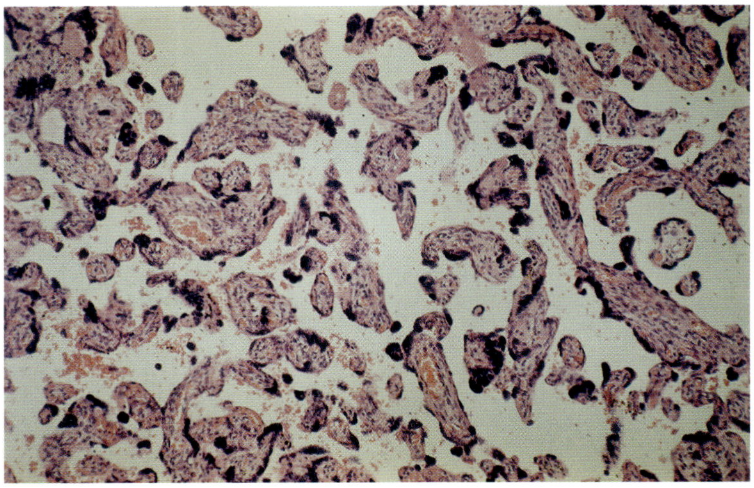

Fig. 24.**27** Numerous syncytial nuclear buds and knots on the villous surfaces ("Tenney-Parker changes," 32nd week of gestation, preeclampsia, H & E, ×100).

luminal narrowing or even occlusion. Known also as "acute atherosis," the vessel wall lesions may occur at intradecidual or even intramyometrial sites[5,6,46].

Lupus erythematosus and antiphospholipid antibody syndrome. Vasculopathies can also occur in other maternal diseases, most notably lupus erythematosus and antiphospholipid antibody syndrome[1,35]. In these conditions it is common to detect antiphospholipid antibodies in the maternal serum, especially a "lupus anticoagulant" and anticardiolipin antibodies. This heterogeneous class of autoantibodies directed against negatively charged phospholipids of proteins, coagulation factors, and platelets can cause an immune coagulopathy with thromboemboli and frequently leads to a decidual vasculopathy[6,16,35].

Typical changes. Other common but nonspecific findings in hypertensive pregnancy disorders are restricted growth of the placenta and a decrease in its basal area. Villous maturation may be appropriate for gestational age. It is not unusual to find either compensatory early villous maturation or, less commonly, retarded villous maturation.

Tenney–Parker changes. Another fairly typical finding is "Tenney–Parker changes," which consist of syncytial epithelial buds and knots on the villous surfaces (Fig. 24.**27**)[45]. The significance of these solid or fragmented nuclear clumps in late pregnancy is a controversial issue. They have been variously interpreted as harmless sectioning artifacts[29], as villous surface irregularities that occur artifactually in flat sections but are still useful diagnostically, or as real structures related to the sprouting of new terminal villi[11,28,41]. These syncytial epithelial buds and knots are seen with some frequency under hypoxemic conditions (e.g., maternal anemia), in hypertensive disorders of pregnancy, and in pregnancies at high altitudes.

Placental abruption. An increased incidence of premature separation of the placenta has been reported in hypertensive disorders of pregnancy—probably a result of the chronic perfusion deficit and extensive fibrin deposition[5].

Causes. This disease most likely traces its pathogenesis to the time of placentation during the first trimester[44]. Morphometric studies of preeclamptic placentas with abnormal Doppler findings have shown that these placentas often have decreased weight and volume and reduced vascular parameters. Based on these findings, abnormal implantation of the placenta has been postulated as the cause[22,47]. A failure of normal preplacental vessel-wall transformation due to deficient fetal intravascular trophoblast invasion is considered to be the main cause of hypertensive pregnancy disorders[8]. As a result of this, the maternal spiral arteries are still susceptible to the effects of vasoconstrictor substances even after 18 weeks' gestation.

Immunological maladaptation between the mother and fetus may have causal significance in hypertensive disorders of pregnancy. Raymond interprets preeclampsia as an exaggerated maternal immune response to immature fetal trophoblastic cells. In many preeclamptic women, he observed small, superficially implanted placentas that had reactively formed an excess of immature trophoblasts. These cells could not properly invade and remodel the spiral arteries due to an abnormal expression of adhesion and matrix molecules. Apparently, large numbers of these functionally incompetent fetal trophoblastic cells trigger the maternal immune system and stimulate antibody formation, leading to secondary vessel wall damage[39].

Inflammatory Placental Changes

Extensive inflammatory changes in the placenta can also compromise the function of the organ. Amniotic-type inflammations with chorioamnionitis, endplate vasculitis, omphalovasculitis, and/or funiculitis are distinguished from parenchymal-type placentitis [46].

Amniotic-Type Inflammations

In amniotic-type inflammations, the causative organisms may reach the amniotic surface by the ascending route through an open gestational sac. If the sac is closed, the organisms may reach the placenta from the internal cervical os by the transmembranous or intramembranous routes. Possible fetal complications are infections of the lung and/or gastrointestinal tract acquired from the contaminated amniotic fluid. An extensive spread of organisms via the fetal bloodstream can lead to generalized fetal sepsis or septicopyemia.

Parenchymal-Type Placentitis

Causes. Placentitis of the parenchymal type is especially likely to cause placental insufficiency. Signs of exudative and/or proliferative inflammation are found in the intervillous space, on villous surfaces, and/or in the villous stroma. The inflammation may be caused by bacteria, viruses, fungi, parasites, or by hypoxic states or toxicity leading to chorionic epithelial lesions. In many cases, however, the etiology of the placentitis cannot be determined clinically or morphologically.

Route of infection. Infectious parenchymal placentitis is generally acquired by the maternal hematogenous route, with primary invasion of the intervillous space causing an initial acute intervillous thromboangiitis[46]. The source of the infection may be a maternal sore throat, pneumonia, or bacterial endocarditis. In other cases an inflammation may spread directly by the transdecidual route from maternal purulent endometritis, or chorioamnionitis may spread to the placental margin or basal plate by the transamniotic route.

Perivillitis. In florid perivillitis, sites of chorionic epithelial necrosis and fibrin deposition are found on the villous surfaces, which are permeated by maternal granulocytic and/or mononuclear cell infiltrates. In later stages, fibroblast proliferation with collagen fiber proliferation often leads to complete obliteration of the intervillous space.

Villitis. Infectious villitis occurs when pathogenic organisms reach the villi by fetal hematogenous spread. It is characterized by necrosis of the villous stroma and chorionic epithelium and by inflammatory cell infiltrates. In some cases the specific nature of the intrauterine infection can be determined from the composition of the inflammatory infiltrate[46].

Histological findings. Figure 24.**28** illustrates the histological appearance of acute intervillous thromboangiitis, perivillitis, and villitis secondary to intrauterine listeriosis. The infection usually spreads by the transplacental route, leading to multiple abscess formation in the placenta. In other cases, listeria present in the stool or vagina may spread by the ascending route and incite a phlegmonous chorioamnionitis with purulent liquefaction. In both modes of infection, a generalized septic granulomatosis may develop in the fetus[12].

Figure 24.**29** shows the histological features of a vaguely granulomatous inflammation of the placenta with the formation of multinucleated Langhans-type giant cells like those seen in tuberculosis or a varizella zoster infection[6].

Figure 24.**30** shows the histological appearance of an extensive fungal infection of the placenta in the 17th week of pregnancy with a retained maternal intrauterine device. Premature rupture of the membranes had led to the development of an extensive, florid chorioamnionitis, perivillitis, and villitis with evidence of multiple *Candida albicans* hyphae.

Effects. Cases of extensive parenchymal placentitis with the spread of organisms into the fetal circulation can lead to fetal sepsis or septicopyemia with the development of multiple inflammatory foci in the fetal organs and intrauterine fetal death. In less severe cases, the compromise of placental function caused by the inflammatory changes can lead to restriction of fetal growth.

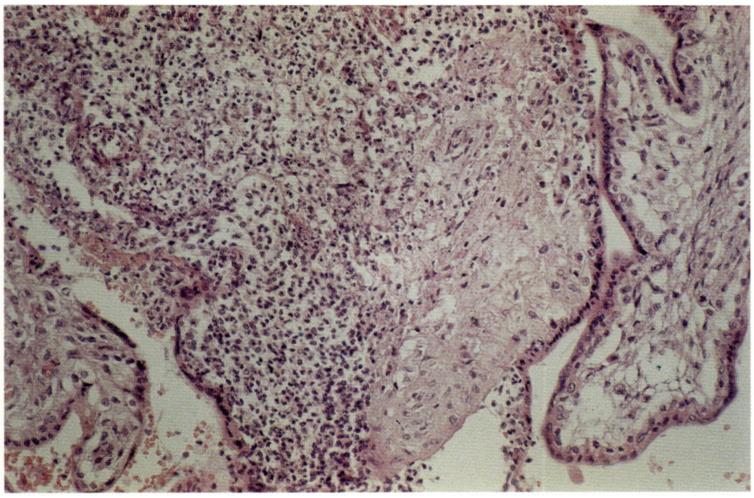

Fig. 24.**28** Suppurative intervillous thromboangiitis, perivillitis, and necrotizing villitis in intrauterine listeriosis (26th week of gestation, H&E, ×200).

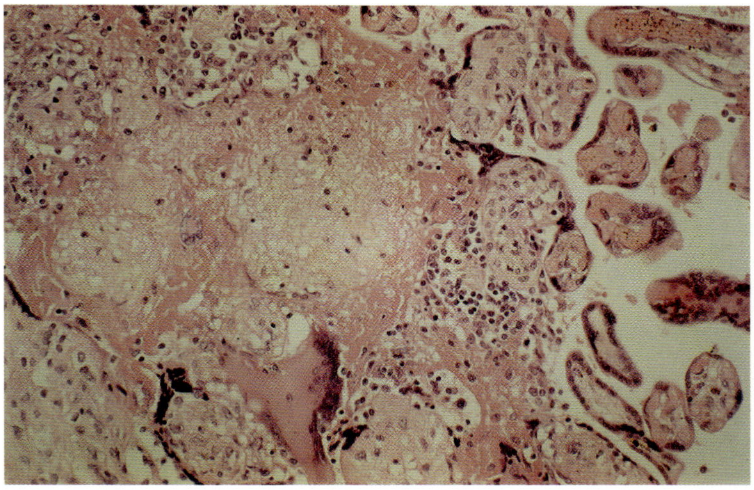

Fig. 24.**29** Vaguely granulomatous villitis and perivillitis with a multinucleated Langhans giant cell (35th week of gestation, H&E, ×200).

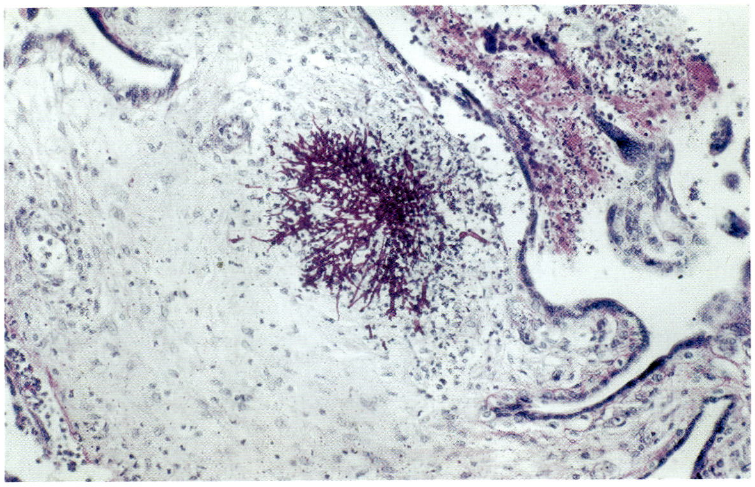

Fig. 24.**30** Parenchymal placentitis with evidence of *Candida albicans* (17th week of gestation, PAS, ×200).

Placental Tumors

Chorangiomas

Chorangiomas are relatively common benign tumors that can cause impairment of placental function. When sufficiently large, they can be diagnosed with ultrasound. Chorangiomas are solitary or multiple tumors that usually have a sharply circumscribed capsule. Most are less than 5 cm in diameter, but in rare cases chorangiomas may occupy a large area of the placenta, with a corresponding reduction of functional parenchyma. These tumors may bulge into the amniotic cavity, compressing the umbilical cord vessels. Chorangiomas have a lobulated cut surface, a reddish-brown to yellow color, and a relatively soft consistency (Fig. 24.**31a**).

Histological findings. Endotheliomatous, capillary, and cavernous forms can be distinguished histologically. Figure 24.**31b** shows a chorangioma with numerous vascular spaces embedded in a stroma of varying cellularity. Large tumors in particular may harbor regressive changes in the form of necrosis, intratumoral hemorrhage, thrombi, and calcifications.

Cause. Chorangiomas are believed to be caused by abnormal fetal angiogenesis of the placenta[6, 46]. These tumors communicate with the fetal circulatory system.

Complications. Besides placental insufficiency caused by the reduced parenchyma, rare complications of chorangiomas include hemorrhage, placental abruption, and Kasabach–Meritt syndrome. This syndrome is characterized by hemolytic anemia due to the mechanical destruction of red cells, and by the presence of disseminated microthrombi[18]. Chorangiomas are frequently accompanied by polyhydramnios, which is attributed to the transudation of fluid from the tumor surface into the amniotic cavity and/or increased fetal urine excretion resulting from the increased placental output[18].

Very vascular chorangiomas can lead to fetal cardiomegaly due to excessive arteriovenous shunting of blood and increased venous return to the fetal heart. In turn, this cardiomegaly can lead to heart failure and generalized hydrops fetalis[17, 18].

An association of chorangiomas with other fetal malformations such as cutaneous angiomas is occasionally described[31].

Trophoblastic Tumors

Trophoblastic tumors are very rare neoplasms caused by the invasive proliferation of trophoblasts, which may seed metastatic deposits. Choriocarcinomas and trophoblastic tumors of the placental bed may occur in association with molar pregnancies and spontaneous abortions, and rare cases may be seen after normal and ectopic pregnancies[6, 46].

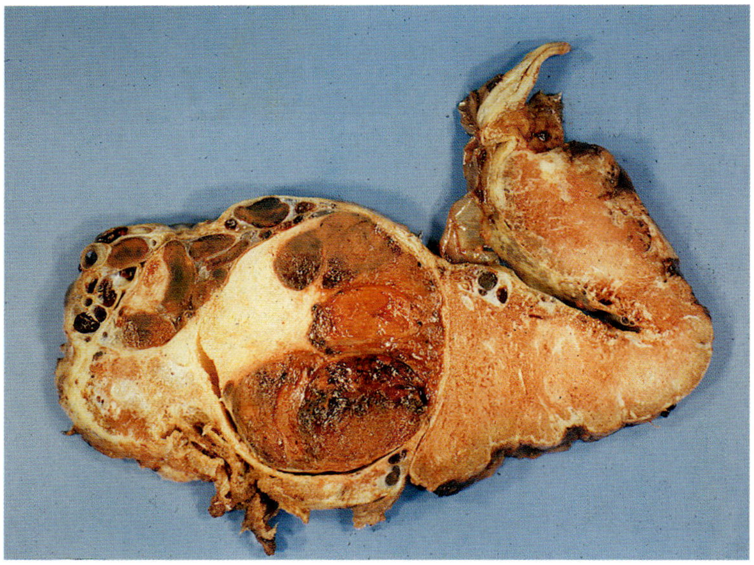

Fig. 24.**31 a** Cut surface of a placenta with a 10 cm chorangioma (37th week of gestation).

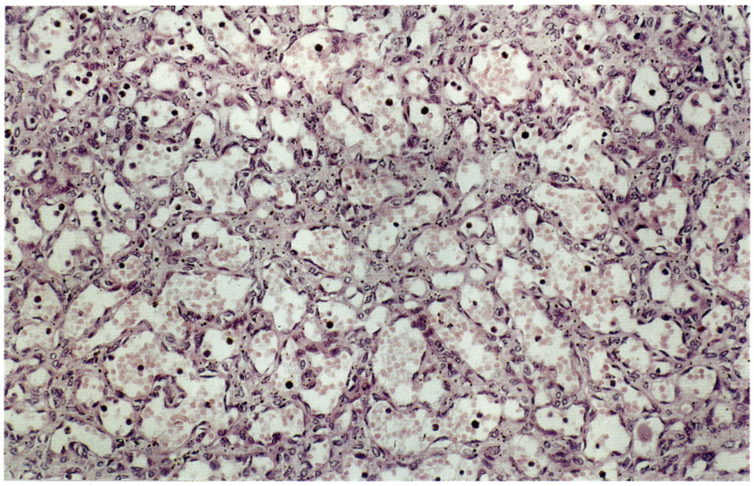

Fig. 24.**31 b** Histological appearance of chorangioma (H & E, ×200).

Secondary Tumors

Secondary tumors of the placenta are rare. Embolized tumor cells from carcinomas in other maternal organs have occasionally been found in the intervillous space, but these cells did not infiltrate the villous stroma[5, 46]. True metastatic deposits in the placenta, with transplacental spread to the fetus, have been described in some patients with malignant tumors[5, 46].

References

1 Abramowski CR, Veges ME, Swinehert G: Decidual vasculopathy of the placenta in lupus erythematodes. New Engl. J. Med. 303 (1980) 668–672

2 Altschuler G: Chorangiosis: an important placental sign of neonatal morbidity and mortality. Arch. Pathol. Lab. Med. 108 (1984) 71–74

3 Bacon BJ, Gilbert RD, Kaufmann P, Smith AD, Trevino FT, Longo LD: Placental anatomy and diffusing capacity in guinea pigs following long-term maternal hypoxia. Placenta 5 (1984) 475–488

4 Beck T: Der materne Blutfluß durch die menschliche Plazenta. Z. Geburtsh. Perinat. 186 (1982) 65–71

5 Becker V, Schiebler Th H, Kubli F: Die Plazenta des Menschen. Thieme, Stuttgart 1981

6 Benirschke K, Kaufmann P: Pathology of the human placenta. Third Edition. Springer, Berlin 1995

7 Blankenship TN, Enders AC, King BF: Trophoblastic invasion and the development of uteroplacental arteries in macaque: Immunhistochemical localisation of cytoceratins, desmin, type IV collagen, laminin and fibronectin. Cell Tissue Res. 272 (1993) 227–236

8 Brosens UA: Morphological changes in the uteroplacental bed in pregnancy hypertension. Clin. Obstet. Gynaec. 4 (1977) 573–583

9 Burrows TD, King A, Loke YW: Expression of integrins by human trophoblast and differential adhesion to laminin or fibronectin. Hum. Reprod. 8 (1993) 475–484

10 Castellucci M, Zaccheo D, Pescetto G: A three-dimensional study of the normal human placental villous core. I. The Hofbauer cells. Cell Tissue Res. 210 (1980) 235–247

11 Castellucci M, Scheper M, Scheffen I, Celona A, Kaufmann P: The development of the human placental villous tree. Anat. Embryol. 181 (1990) 117–128

12 Cottier H, Hess MW, Keller HU, Boos B, Schindler R, Zimmermann A: Pathogenese. Ein Handbuch für die ärztliche Fortbildung. Springer, Berlin 1980

13 Demir R, Kaufmann P, Castellucci M, Erbengi T, Kotowski A: Fetal vasculogenesis and angiogenesis in human placental villi. Acta. Anat. (Basel) 136 (1989) 190–203

14 De Wolf F, De Wolf-Peeters C, Brosens I: Ultrastructure of the spiral arteries in the human placental bed at the end of the normal pregnancy. Amer. J. Obstet. Gynecol. 117 (1973) 833–848

15 Drews U: Taschenatlas der Embryologie. Thieme, Stuttgart 1993

16 Gröne HJ: Systemischer Lupus erythematosus und Antiphospholipid-Syndrom. Pathologe 17 (1996) 405–416

17 Hadi HA, Finley J, Strickland D: Placental chorioangioma: Prenatal diagnosis and clinical significance. Amer. J. Perinat. 10 (1993) 146–149

18 Hirata GI, Damon I, Masaki I, O'Toole M, Medearis AL, Platt LD: Color flow mapping and doppler velocimetry in the diagnosis and management of a placental chorioangioma associated with nonimmune fetal hydrops. Obstet. Gynecol. 81 (1993) 850 –852

19 Hitschold T, Müntefering H, Berle P: Histologische Plazentabefunde bei diastolischem Null- oder Negativflow der Nabelarterien: Eine dprospektive Untersuchung unter Berücksichtigung der Dauer der klinischen Beobachtung. Geburtsh. Frauenheilk. 52 (1992) 219–224

20 Hitschold T, Beck T, Müntefering H, Berle P: Plazentamorphometrie bei diastolischem Null- und Negativflow der Nabelarterien. Geburtsh. u. Frauenheilk. 52 (1992) 270–274

21 Hitschold T, Weiss E, Beck T, Müntefering H, Berle P: Low target birth weight or growth retardation? Umbilical doppler flow velocity waveforms and histometric analysis of fetoplacentar vascular tree. Amer. J. Obstet. Gynecol. 168 (1993) 1269–1274

22 Hitschold T, Ulrich S, Kalder M, Müntefering H, Berle P: Blutströmungsprofile der Arteria Uterina. Korrelation zur Plazentamorphologie und zu klinisch-geburtshilflichen Daten im Rahmen der Präeklampsie. Z. Geburtsh. u. Neonat. 199 (1995) 8–12

23 Jimenez E, Unger M, Vogel M et al.: Morphologische Untersuchungen an Plazenten HIV-positiver Mütter. Pathologe 9 (1988) 228–234

24 Jones CJP, Fox H: An ultrastructurell study of the placenta in materno-fetal rhesus incompatibility. Virchows Arch. Pathol. Anat. Histol. 379 (1978) 229–241

25 Kaufmann P, Sen DK, Schweikhart G: Classification of human placental villi. I. Histology and scanning electron microscopy. Cell tissue Res. 200 (1979) 409–423

26 Kaufmann P, Luckhardt M, Schweickhart G, Cantle SJ: Cross-sectional features and three-dimensional structures of human placental villi. Placenta 8 (1987) 235–247

27 Kaufmann P, Luckhardt M, Leiser R: Three-dimensional representation of the fetal vessel system in the human placenta, Trophoblast. Res. 3 (1988) 113–137

28 King BF, Menton DN: Scanning electron microscopy of human placental villi from early and late gestation. Amer. J. Obstet. Gynecol. (1975) 824–828

29 Küstermann W: Über „Proliferationsknoten“ und „Synzytialbrücken“ der menschlichen Plazenta. Anat. Anz. 150 (1981) 144–157

30 Langman J: Medizinische Embryologie. Die normale menschliche Entwicklung und ihre Fehlbildungen. 6. Auflage. Thieme, Stuttgart 1980

31 Manzke H, Mau G: Korrelation schwangerschaftsanamnestischer und klinischer Befunde mit dem Auftreten von Naevi flammei bei Neugeborenen. Monatsschr. Kinderheilk. 123 (1975) 124–127 und in Vogel, M: Atlas der morphologischen Plazentadiagnostik. 2. ed. Springer, Berlin 1996

32 Matijewic R, Meekins JW, Walkinshaw SA, Neilson JP, McFadyen IR: Spiral artery blood flow in the central and peripheral areas of the placental bed in the second trimester. Obstet. Gynecol. 86 (1995) 289–292

33 Moll V, Künzel W: Der utero-plazentare Kreislauf. Z. Geburtsh. u. Perinat. 178 (1974) 1–11

34 Moore KL: Grundlagen der medizinischen Embryologie. Enke, Stuttgart 1990

35 Nayar R, Lage JM: Placental changes in a first trimester abortion in maternal systemic lupus erythematodes with antiphospholipid syndrome: A case report and review of literature. Hum. Pathol. 27 (1996) 201–206

36 Nicolov SpD, Schiebler TH: Über Endothelzellen in Zottengefäßen der reifen menschlichen Plazenta. Acta. Anat. Basel 110 (1981) 338–344

37 Ogawa S, Leary J, Clauss M et al.: Modulation of endothelial cells (EC) function in hypoxia: alterations in cell growth and the response to monocyte-derived mitogenic factors. J. Cell. Biochem. Suppl. 15 F (1991) 213 und in: Benirschke KP, Kaufmann P: Die Plazenta des Menschen. 3. Auflage. Springer, Berlin 1995

38 Pijnenborg R, Dixon G, Robertson WB, Brosens I: Trophoblastic invasion of human decidua from 8 to 18 week of pregnancy. Placenta 1 (1980) 3–19

39 Raymond WR, Patterson P: Preeclampsia is assosiated with an excess of proliferative immature intermediate trophoblast. Hum. Pathol. 26 (1995) 594–600

40 Reshetnikova OS, Burton GJ, Milovanov AP: Effects of hypobaric hypoxia on the fetoplacental unit: The morphometric diffusing capacity of the villous membrane at high altitude. Amer. J. Obstet Gynecol. 171 (1994) 1560–1565

41 Sala MA, Matheus M, Valeri V: Volume density of syncytial sprouts and its regional variation in the normal human placenta. Gegenbaurs morphol. Jahrb. 129 (1983) 489–493

42 Schuhmann R, Wehler V: Histologische Unterschiede an Plazentazotten innerhalb der materno-fetalen Strömungseinheit: Ein Beitrag zur funktionellen Morphologie der Plazenta. Arch. Gynecol. 210 (1971) 425–439
43 Shih IM, Kurman RJ: Expression of melanoma cell adhesion molecule in intermediate trophoblast. Lab. Invest. 75 (1996) 377–388
44 Steck T, Würfel W: Die Bedeutung immunologischer Faktoren bei der Ätiologie der schwangerschaftshypertensiven Hypertonie. Zentralbl. Gynäkol. 117 (1995) 3–10
45 Tenney B, Parker F: The placenta in toxemia of pregnancy. Amer. J. Obstet. Gynecol. 39 (1940) 1000–1005 and Benirschke K, Kaufmann P: Pathology of the human placenta. 3. ed. Springer, Berlin 1995
46 Vogel M: Atlas der morphologischen Plazentadiagnostik. 2. ed. Springer, Berlin 1996
47 Weickert U, Reitnauer K, He J, Ertan AK, Schmidt W, Remberger K: Histopathologische Plazenta-Morphometrie und Dopplersonographie der Nabelschnurgefäße bei Frühgeborenen. Geburtsh. u. Neonat. 121 (1999) 7–13
48 Weiss E, Hitschold T, Müntefering H, Berle P: Dopplersonographie der Art. umbilicalis: Differenzierte Diagnostik bei der intrauterinen Mangelentwicklung. Geburtsh. Frauenheilk. 49 (1989) 466–471
49 Wentworth P: The placenta in cases of hemolytic disease. Amer. J. Obstet. Gynecol. 98 (1967) 283–289
50 Werb Z: How the macrophage regulates its extracellular environment. Amer. J. Anat. 166 (1983) 237–256

25 Morphological and Morphometric Studies of the Placenta with Doppler Abnormalities of the Fetal Umbilical Arteries*

T. Hitschold

Abnormalities of pregnancy and delivery relating to placental insufficiency syndrome can have a variety of causes. As an "organ between organisms," the placenta plays a central role in this process. A disturbance in the functional capacity of this organ can be described in terms of its clinical, pathophysiological, and pathomorphological aspects.

Clinical Aspects of Placental Insufficiency

The clinical aspects are illustrated by intrauterine growth retardation and prematurity, which frequently occur in association with placental insufficiency. Etiological diagnosis, early detection, and treatment continue to be subjects of comprehensive clinical and theoretical research.

Established diagnostic procedures. Traditionally, ultrasound biometry and biochemical surveillance methods have been the established diagnostic procedures available to the obstetrician. These procedures are essentially retrospective in nature, as they are used to measure an existing condition of restricted fetal growth and monitor its progression. Intervention is warranted if the fetal heart rate (FHR) trace suggests frank or impending fetal asphyxia, but this test is already measuring the effects of placental insufficiency on fetal cardiovascular regulation, which may consist of hypoxemia, hypoxia, or acidosis.

Pathophysiological Aspects of Placental Insufficiency

The pathophysiological aspects of placental insufficiency are generally investigated in experimental animal studies and mainly involve uteroplacental and fetoplacental circulatory mechanisms. The central importance of hemodynamics in all partial functions of the placenta is widely acknowledged. Any reduction in blood flow decreases the exchange capacity of the placenta and reduces metabolic processes due to an insufficient supply of oxygen and nutrients.

Fetoplacental blood flow. Fetoplacental blood flow, unlike other circulatory compartments, is not subject to neural or endocrine regulatory mechanisms. Basically, a fetus grows within the limits imposed by its placenta. These limits are broad enough to provide a safety reserve, but they appear to be genetically determined. A variety of factors during pregnancy can exhaust this safety reserve, causing the system to decompensate. In cases of this kind we can appreciate the inadequacies that exist in current established monitoring techniques, which cannot readily distinguish antenatally between a genetically small fetus and one that is growth-retarded due to placental insufficiency.

New possibilities with Doppler sonography. The advent of noninvasive Doppler sonography has brought significant advantages from a methodological standpoint. Doppler ultrasound is already being utilized in many clinical examinations. It can be used to determine flow resistance and impedance, making it possible to evaluate the perfusion of a selected region. Blood flow measurements with chronically implanted transducers in fetal sheep have shown that the greatest blood pressure drop in the fetoplacental circulation takes place across the villous tree[8]. This suggests that the resistance and impedance in the fetoplacental compartment are critically determined by the vascularity of the placental villi and that the umbilical cord vessels, for example, play only a minor role. Impedance changes in this region, as measured by qualitative resistance indices in the fetal umbilical arteries, appear to precede other parameters such as fetal movement patterns and FHR changes by days or even weeks[54], signaling a need for appropriate surveillance. This is a major diagnostic advance in populations at risk and has been confirmed in prospective randomized studies[39, 53] showing a reduction of obstetric emergencies when the Doppler findings are known. Given the difficulties of accurately determining vascular diameters and angles in the umbilical cord vessels, quantitative measurements have largely been discarded in favor of the qualitative analysis of flow velocity waveforms based on the determination of resistance indices (S/D ratio, resistance index, pulsatility index).

* Dedicated with gratitude to my esteemed teachers, Professor P. Berle and Professor H. Müntefering

Pathomorphological Aspects of Placental Insufficiency

Functional placental morphology[5] has described the morphological correlates of placental insufficiency such as infarction and villous stromal fibrosis while also noting that the problem of placental insufficiency is basically a matter of quantity: it is caused less by typical, isolated placental changes than by a disproportion between supply and demand.

Fetoplacental villous and vascular architecture. Classic studies by Schuhmann[41–43] described the fetal cotyledons, with their central spiral artery terminations, as the smallest functional unit ("placentone") of the fetoplacental villous and vascular tree. Kaufmann et al.[27–31] made additional discoveries in electron-microscopic examinations of serial sections, so that today we have a fairly detailed picture of the fetoplacental villous and vascular architecture. Basically it consists of a parallel arrangement of flow units that are represented by a dichotomous pattern of vascular branches, starting with the endplate vessels and extending to the villous vessels in the individual placentones.

Placental and fetal weight. Vogel[55] explored the relationship between placental growth and maturation, i.e., links between the size of the organ and its internal structural differentiation, identifying placental weight as an important indicator of the functional competence of the placenta. This is expressed, at least to a degree, in the relationship between fetal weight and placental weight. Ultimately it is uncertain what determines this relationship. But the placenta and fetus do have a common genome, and so it is reasonable to assume that the relationship is genetically determined. Increased ratios of placental and fetal weight may signal an imbalance in this relationship and are commonly associated with intrauterine asphyxia. This parameter is clinically relevant, therefore, but it cannot be reliably determined in the antenatal period.

Maturation of the villous tree. Studies by Castelluci and Kaufmann[7] and Schweikhart[44] based on combined scanning electron-microscopic and histological examinations have brought forth a new concept of defective placental maturation. These authors distinguished the "branches" of the fetoplacental villous tree (the mainstem villi) from the older and younger "twigs" (mature and immature intermediate villi) and the "leaves" (terminal villi). Deviations from the normal maturation of the villous tree may be associated with clinically detectable abnormalities (growth restriction, acidosis, asphyxia)[5, 44, 45]. Initial studies in the late 1980 s[14, 15, 26] pointed to a correlation between Doppler sonographic findings and qualitative morphological placental findings.

Morphometric studies of the terminal villi. Stoz et al.[46–48], Noack et al.[37], Teasdale[5], and other authors (review of the literature in Beck[2]) have performed morphometric investigations and described changes in the terminal villi, also called "absorption villi," in cases of maternal diabetes, preeclampsia, and growth restriction. Beck[3, 4], who carried out systematic computer-assisted histometric measurements at the periphery of the placentone, found changes in the absorption villi in connection with various typical pregnancy complications such as prematurity, acidosis, diabetes, and preeclampsia. Beck found compelling evidence that standardized morphometric measurements of the peripheral absorption villi in a placentone are representative of the organ as a whole, using this as a basis not only for measuring individual villous properties but also for calculating functionally relevant internal surface areas. This new ability to evaluate functional aspects of placental morphology that can be quantitatively documented has contributed greatly to our understanding of pathophysiological relationships.

S/D ratio in the umbilical arteries. Giles et al.[11] made the breakthrough discovery that the arterioles in third-order villous stems were greatly reduced—whether by defective angiogenesis or secondary obliteration—in human fetuses that showed an increased S/D ratio in the umbilical arteries. Other authors confirmed these findings[6, 35], proving that resistance changes in the placenta could be detected by the qualitative analysis of Doppler waveforms sampled from the umbilical arteries.

Embolization of the fetoplacental circulation. Morrow et al.[36] were able to induce absent or reverse end-diastolic flow by embolizing the fetoplacental circulation in fetal sheep with 50 μm microspheres. Kaufmann[31] found that the size of these microspheres matched the mean diameter of the vessels in the smaller mainstem villi of the human placenta (fitth-order to eighth-order branches), and Giles[11] further supported these findings. Trudinger et al.[52] and Schmidt et al.[40] occluded the terminal villi with 15 μm microspheres, as this dimension matches the size of the capillaries at that level. While this increased the S/D ratio, it did not induce an absence of end-diastolic flow. Trudinger[52] did measure a fall of the fetal arterial P_{O_2}, however, demonstrating the importance of the terminal villi for gaseous exchange.

Umbilical resistance and placental reserve capacity. Thompson[51] used a mathematical computer model of a resistor and capacitor circuit to simulate the dichotomous branching pattern of the fetoplacental circulation and study the effects of different variables on the flow patterns in the umbilical arteries. For example, the umbilical resistance rises as the number of obliterated small vessels increases. This effect is more pronounced when there are fewer branches in the placental vascular tree. Thus, the smaller the placenta and the more severe the maturation defects in the villous tree, the more pronounced the resistance effect. These studies also showed that approximately 60% of the vessels must be obliterated before any resistance effect is noted in the waveform indices. This is the mathematical expression of a "placental reserve capacity" that guards against fetal compromise due to minor pathological changes.

Functional competence of the placenta. Morphometric techniques[2, 42] can be used to document certain quantitative properties of the villi at a selected site in the fetomaternal flow units (Fig. 25.**1**). Metabolically active villous surface areas can be cal-

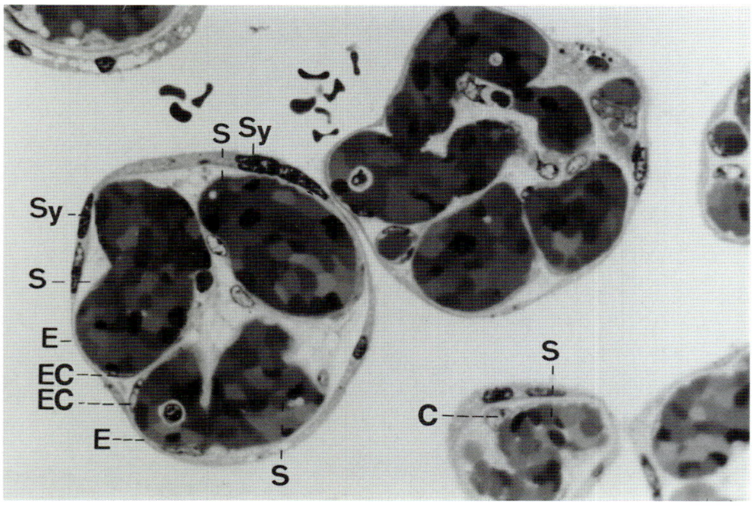

◁ Fig. 25.1 Two terminal villi with vessels showing sinusoidal dilatation (S), epithelial plates (E), endothelial cells (EC), syncytiotrophoblasts (Sy), and cytotrophoblasts (C). Semithin section, toluidine blue, ×2000.

culated to gain an impression of the functional competence of the organ and of any disturbance of these mechanisms that occurs in certain pregnancy complications.

Validation of Doppler Findings by Placental Histology

The size of the vascular tree (the intravillous blood volume) is of special interest in the validation of Doppler ultrasound findings by placental histology.

Resistance Index of the Umbilical Arteries

Figure 25.**2** shows the relationship between the fetal intravillous blood volume and the umbilical artery resistance index (RI) determined during the last week before birth. We found a direct relationship between these variables, with a correlation coefficient of $r = -0.703$. The larger the vascular tree of the placenta, the lower the resistance to blood flow, i.e., the lower the RI in the umbilical artery. Conversely, cases with a particularly low villous blood volume show increased Doppler indices or a loss of antegrade end-diastolic blood flow velocities (absent end-diastolic flow).

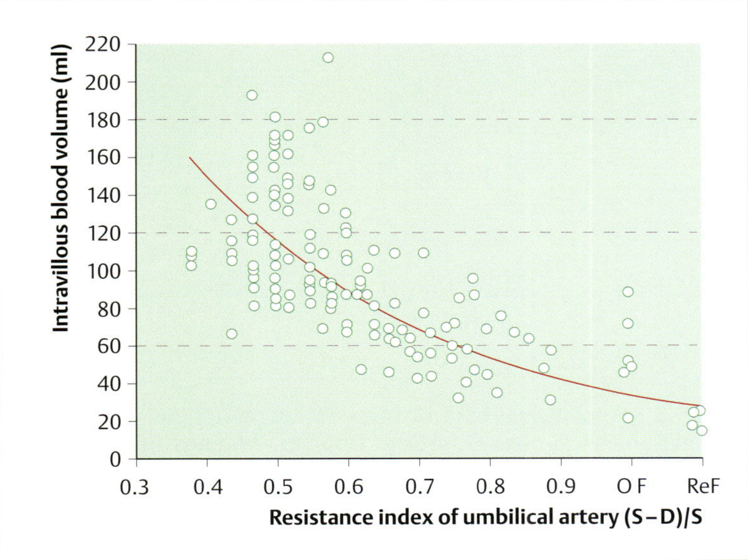

Fig. 25.2 Resistance index in the umbilical arteries plotted against the size of the fetoplacental vascular tree, expressed as the fetal intravillous blood volume (second-degree polynomial regression).
OF = absent end-diastolic flow; ReF = reverse end-diastolic flow.
Curve formula: $y = 283.593 – 407.741x\ 155.50x^2$
$r = -0.703$; $r^2 = 0.508$

Placental reserve capacity. Analogously to the mathematical model of Thompson[51], our results indicate that the RI values have a large range of normal variation, which represents the "placental reserve capacity." An RI of 0.5 (corresponding to an S/D ratio of 2.0) may be associated with an intravillous blood volume of 80–180 ml determined by our method. This is in good agreement with the findings of other authors on volume flow[11,24,40], which must be reduced by more than 50% before a qualitative change in resistance index is seen. This also indicates that Doppler sonography of the umbilical arteries, while not an acute study, is capable of detecting significant disturbances of vascularization and reductions in the fetoplacental blood volume.

Reduced intravillous blood volume. A low fetal intravillous blood volume of less than 85 ml is almost always associated with an increased RI value that exceeds 0.66 (corresponding to an S/D ratio of 3). We have therefore taken that value as a cutoff in determining the relationship of intravillous blood volume to fetal risk. This is shown in Fig. 25.**3**. As the graph indicates, fetal risks have a higher association with a reduced blood volume than with normal values of intravillous blood volume. This parameter is clinically relevant, therefore, and can be evaluated with Doppler ultrasound. In cases with a small intravillous blood volume (yellow bars in the graph), the rates of small-for-gestational-age (SGA) fetuses ($p = 0.000$), the incidence of severe acidosis (pH < 7.15) ($p = 0.004$), the number of fetuses with a low Apgar score (< 8) ($p = 0.029$), and the cesarean section rates for impending intrauterine asphyxia ($p = 0.000$) are significantly increased compared with fetuses whose placentas have a sufficiently large intravillous blood volume (blue bars in the graph).

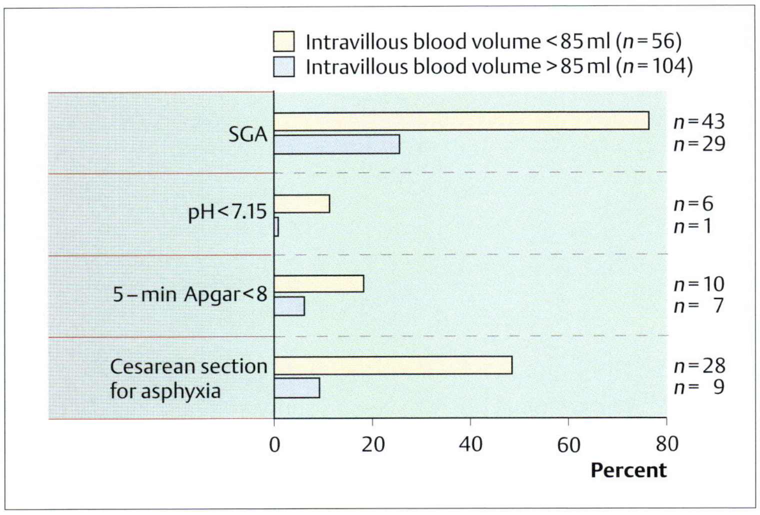

Fig. 25.3 Perinatal risks as a function of the size of the fetoplacental vascular tree ($n = 160$). With a small intravillous blood volume (yellow bars), the rates of SGA fetuses ($p = 0.000$), the incidence of severe acidosis (pH < 7.15) ($p = 0.004$), the number of fetuses with a low 5-min Apgar score (< 8) ($p = 0.029$), and the rate of cesarean sections for impending intrauterine asphyxia ($p = 0.000$) are significantly increased relative to fetuses whose placentas have an adequate intravillous blood volume (blue bars). Fisher's exact test.

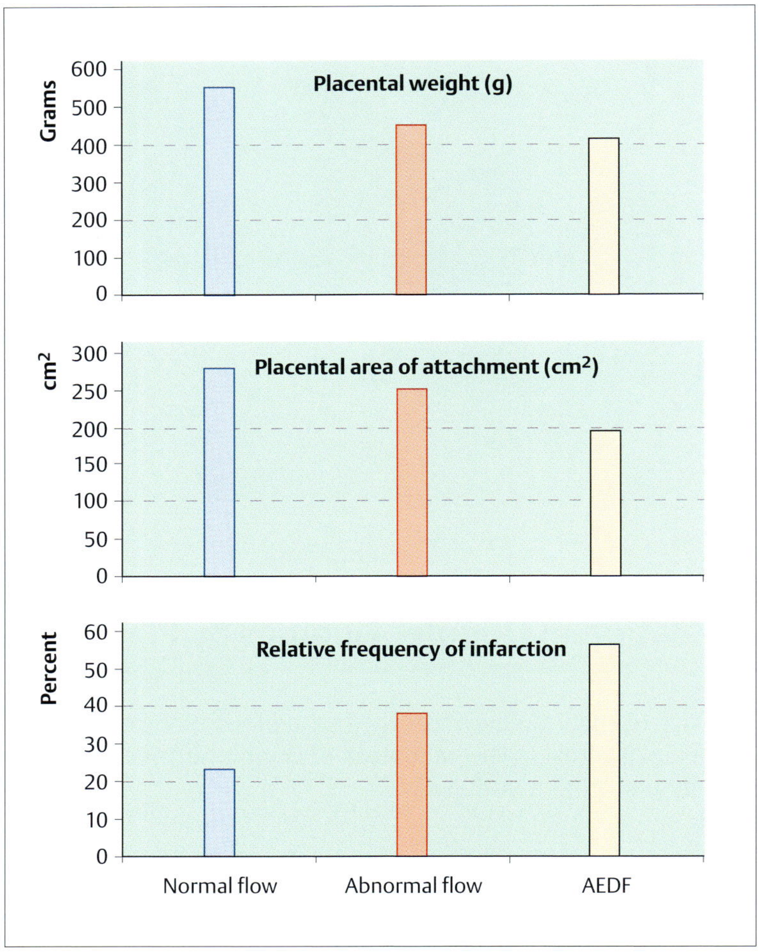

Fig. 25.4 Placental weight, area of placental attachment, and frequency of infarction versus Doppler findings in the umbilical arteries. Normal flow: $n = 193$. Abnormal flow: $n = 31$. Absent end-diastolic flow (AEDF): $n = 9$.

End-Diastolic Blood Flow Velocities in the Umbilical Arteries

A relative reduction or absence of end-diastolic blood flow velocities in the umbilical arteries reflects an increased impedance in the downstream vascular bed and therefore may signify abnormal vascularization of the placenta[12, 13, 15]. On the other hand, a steady rise of end-diastolic blood flow velocities with advancing gestation occurs in normal pregnancies and reflects a decline of impedance due to placental growth as well as maturation of the villous and vascular tree [49]. Our study in a nonselected population showed that as the severity of Doppler abnormalities increased, there was a decrease in placental weight and attachment area and an increase in the frequency of infarctions (Fig. 25.**4**) and defective maturation and development of the villous tree (Fig. 25.**5**)[17, 23].

Absent end-diastolic flow in the umbilical arteries. We did a comparative case–control study in a selected group of 37 patients with absent end-diastolic flow (AEDF) in the umbilical arteries and in matched controls of equal gestational age (± 3 days) with normal umbilical artery flow[16]. This matched-pair study design enabled us to compare placental histology and macromorphometric findings independently of gestational age (Table 25.**1**). For the same gestational ages in both groups, the mean birthweight of cases with AEDF was significantly lower (Wilcoxon test: $p = 0.0001$) than in the matched controls. The rate of intrauterine growth retardation (IUGR) was approximately 10 times higher with this flow pattern (76% vs. 8%, McNemer test: $p = 0.0000$). This supply deficit is reflected in the macromorphometric placental data.

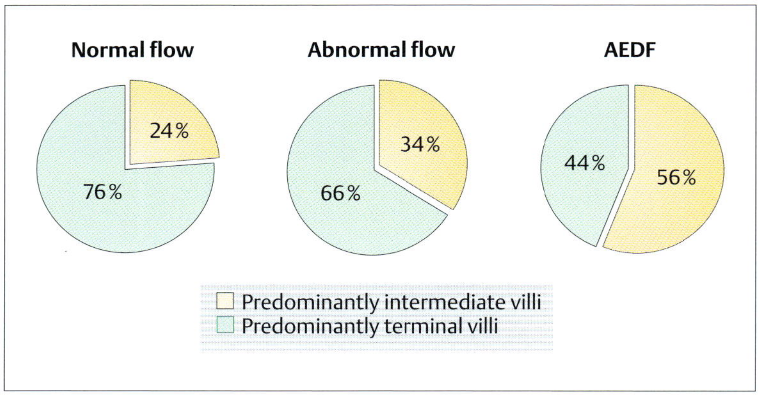

Fig. 25.5 Percentages of villous types in the peripheral villous tree as a function of umbilical artery Doppler findings. Normal flow: $n = 193$. Abnormal flow: $n = 31$. Absent end-diastolic flow (AEDF): $n = 9$.

Table 25.**1** Gestational age, birthweight, and macromorphometric placental findings in absent or reverse end-diastolic umbilical artery flow compared with normal matched controls

	Normal matched controls ($n = 37$)	Absent or reverse end-diastolic flow ($n = 37$)	p (Wilcoxon test)
Delivery (days gestation)	231 ± 25	231 ± 25	–
Birthweight (g)	1801 ± 710	1425 ± 652	0.0001
SGA < 10. percentile	n = 3 (8 %)	n = 26 (76 %)	0.0023
Placental weight (g)	362 ± 117	273 ± 153	0,0001
Area of placental attachment (cm^2)	213 ± 78	151 ± 54	0.0001

SGA = Small for gestational age.

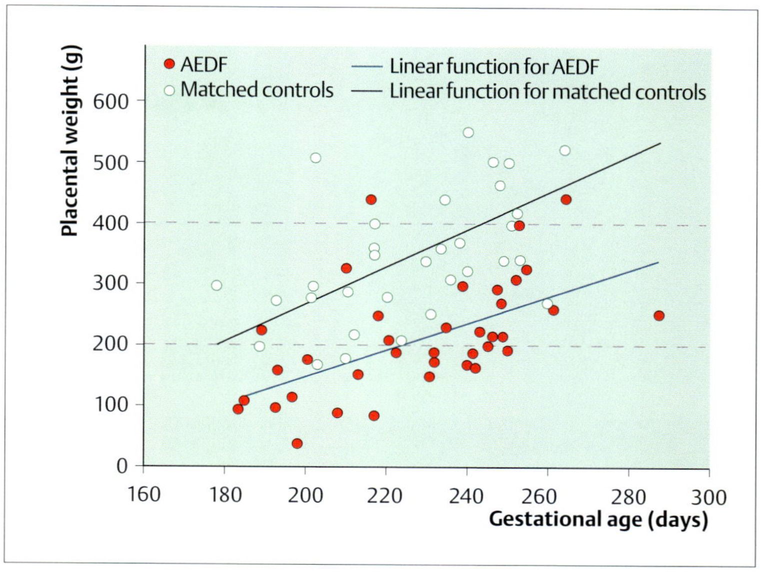

Fig. 25.**6** Placental weight as a function of gestational age in a group with absent or reverse end-diastolic flow (AEDF, $n = 37$) and in matched controls of equal gestational age with normal flow ($n = 37$).
Regression functions for AEDF:
$y = -274.76 + 2.13x$; $r = 0.568$; $r^2 = 0.322$
For matched controls: $y = -339.01 + 3.03x$; $r = 0.637$; $r^2 = 0.405$

Placental weight. With AEDF in the umbilical arteries, we find that the mean placental weight at a given gestational age is reduced by approximately 100 g. The area of placental attachment is also reduced. When placental weight is plotted againstgestational age for both study groups using a linear regression function (Fig. 25.**6**), we find that while the organ weight is decreased in AEDF as described above, growth of these placentas is still observed with advancing gestational age. The curve has a flatter slope than in the matched controls, however. This could indicate an early implantation defect with a consequent delay in placental growth and maturation as a pathogenic mechanism for the development of AEDF in the umbilical artery. Accordingly, a relationship is also found between abnormal findings in the placental vascular bed[56] and morphological placental findings[20] and the flow patterns recorded in the uterine arteries.

Microscopic findings. The microscopic evaluation of villous maturation in umbilical artery AEDF does not show development appropriate for gestational age. This condition is characterized by severe increases of flow resistance with absent or reverse end-diastolic flow in the umbilical arteries in a small placenta, with associated abnormalities of the central and peripheral villous and vascular trees. The dominant finding is chronic placental infarctions followed by delayed villous maturation and endarteropathy obliterans. Perhaps as a result of these changes, the impaired maturation of the peripheral villous tree is often accompanied by accelerated villous maturation and angiosis of the terminal villous vessels reflecting a compensatory mechanism. Figure 25.**7** shows histological examples.

Defective maturation of the peripheral villous tree. The progression of this pathology over time is not yet fully understood. A number of recent studies suggest that changes in the peripheral villous tree are of major significance with regard to increased impedance in the fetoplacental compartment. Jackson et al.[25] and Macara et al.[34] found vascular maldevelopment in the terminal villous compartment of growth-restricted fetuses with abnormal flow patterns in the umbilical arteries. The number and luminal diameters of the stem villous vessels were unchanged relative to the control group, but their wall thickness was significantly decreased in cases with abnormal flow. Detailed electron-microscopic studies by Krebs et al.[33] confirmed abnormal maturation of the peripheral villous tree in pregnancies with AEDF in the umbilical arteries, which validates our histometric studies (Fig. 25.**2**). Together with the results of Nordenvall et al.[38], whose radiographic studies showed abnormal dichotomous branching of endplate vessels and few cotyledons in placentas from fetuses with AEDF, it is reasonable to assume a generalized maldevelopment of the central and peripheral villous tree as the pathogenic principle that underlies abnormal umbilical artery blood flow. Overall, however, the capillary compartment in the terminal villi probably does affect fetoplacental impedance and thus the Doppler velocity waveforms in the umbilical arteries.

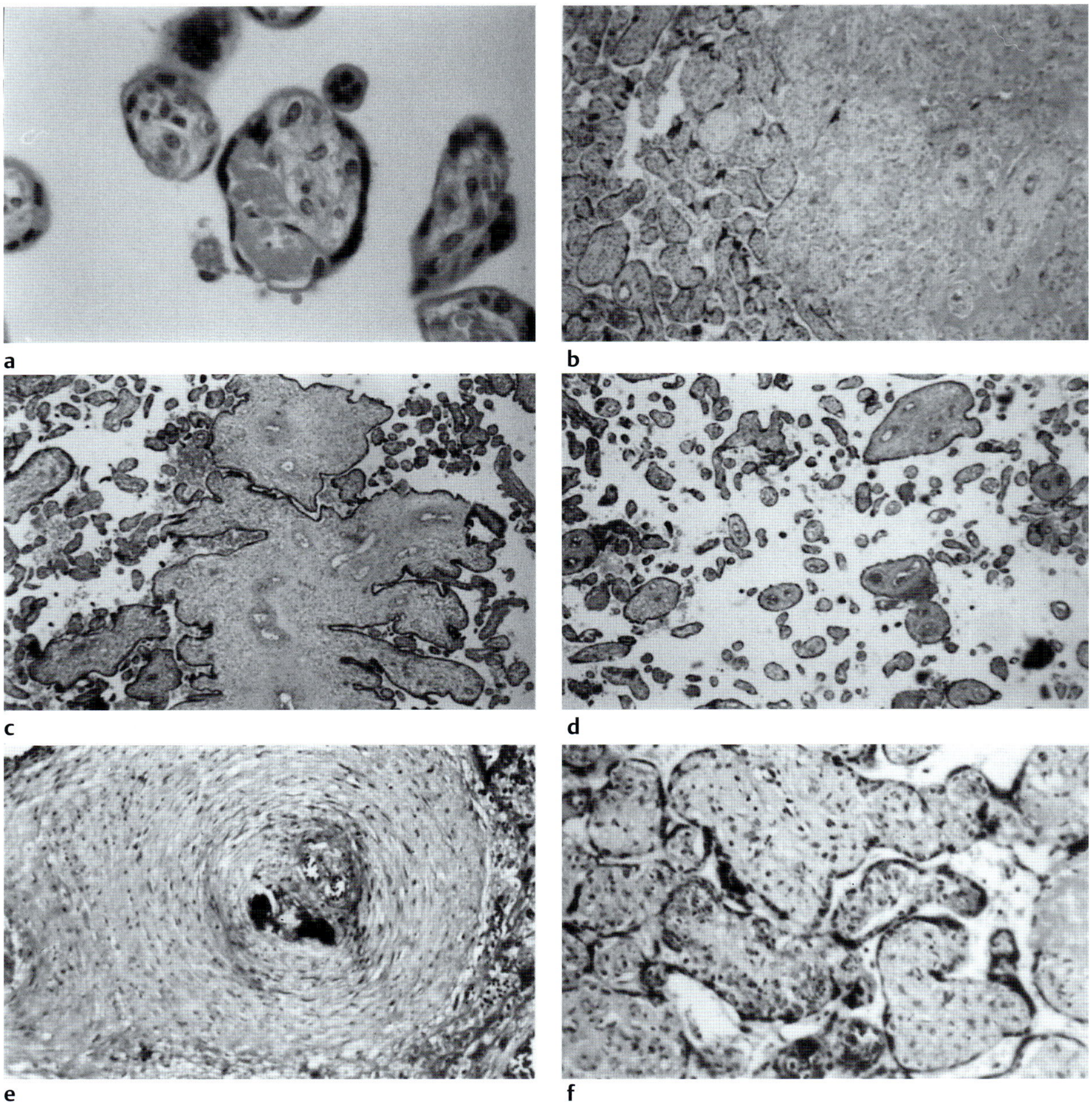

Fig. 25.**7** Placentas from fetuses with AEDF in the umbilical arteries.
a AEDF placenta in the 17th week of gestation. Terminal villus with microthrombosis in the sinusoidal vessels (H & E, ×250).
b AEDF placenta in the 35th week of gestation. Microinfarct with "ghost villi" (H & E, ×100).
c AEDF placenta in the 31st week of gestation. Trisomy 18, arrested ramification and deficient vascularization of the malformed villi (H & E, ×125).
d AEDF placenta in the 35th week of gestation. Intermediate villi are deficient almost none are seen between the mainstem villi and the finely branched terminal villi (H & E, ×100).
e AEDF placenta in the 36th week of gestation. End-arteropathy obliterans of a mainstem villous vessel with intravascular calcium deposition and signs of recanalization (Pearce, ×80).
f AEDF placenta in the 40th week of gestation. Intermediate villi show villositis and a marked disturbance of vascularization (H & E, ×100).

Clinical and Diagnostic Value of Doppler Sonography of the Umbilical Arteries

To investigate the clinical and diagnostic value of umbilical artery Doppler, we examined morphometrically the placentas of SGA fetuses with abnormal umbilical artery flow patterns, SGA fetuses with normal umbilical artery flow, and a control group of eutrophic (AGA, "appropriate for gestational age") fetuses with normal flow[19]. The Doppler waveforms were recorded during the final week before delivery. The data are shown in Table 25.**2**.

Particularly noteworthy is the fact that there were no significant differences in any of the tested histometric criteria between the groups of SGA and AGA fetuses with normal Doppler waveforms. This means that both groups were equally well endowed at the level of the terminal villi and that Doppler sonography of the umbilical artery can demonstrate this antenatally. By contrast, the placentas of SGA fetuses with abnormally increased resistance indices in the umbilical arteries showed a reduction of vascularization and diffusion parameters.

Diagnosis of fetal growth restriction. These findings support the thesis that it is possible to differentiate a genetically small and thus noncompromised fetus with normal umbilical artery waveforms from a growth-restricted fetus with abnormal flow that is threatened by placental insufficiency[9, 57]. The former is not truly growth-restricted but only biometrically small, and these pregnancies can be followed on an outpatient basis as there is no increase in perinatal risk. On the other hand, cases with abnormal umbilical artery waveforms warrant a more thorough diagnostic workup and may require inpatient monitoring and delivery. This was convincingly demonstrated by Weiss[59].

Low resistance indices. Initial studies[22] indicate not only that abnormally high Doppler resistance indices in the fetal umbilical arteries have major diagnostic implications, but that markedly low indices below the 10th percentile may also be associated with increased fetal risk. These cases involve a premature, "accelerated" maturation of the villous and vascular

Table 25.2 Macromorphometric and micromorphometric data in SGA fetuses as a function of Doppler findings compared with a control group of eutrophic fetuses

	Weight < 10th percentile, S/D ratio > 3 ($n = 9$)	**Weight < 10th percentile, S/D ratio < 3 ($n = 7$)**	**Weight appropriate for gestational age (AGA), S/D ratio < 3 ($n = 14$)**
Gestational age (days)	247 ± 23**	280 ± 15 NS	276 ± 11
Birthweight (g)	1680 ± 527*	2617 ± 217***	3503 ± 676
Placental weight (g)	359 ± 126 NS	421 ± 75**	598 ± 164
Decidual attachment area (cm^2)	183 ± 63 NS	223 ± 57*	293 ± 71
Placental-fetal weight index	0.219 ± 0.063*	0.160 ± 0.030 NS	0.170 ± 0.031
Mean villous surface area (μm^2)	1997 ± 279 NS	2116 ± 398 NS	2273 ± 361
Mean villous circumference (μm)	158 ± 10 NS	163 ± 15 NS	169 ± 13
Diffusion path (μm)	4.6 ± 1.6 *	3.3 ± 0.4 NS	3.7 ± 0.7
Mean vascular surface area (μm^2)	476 ± 213 **	924 ± 310 NS	884 ± 229
Number of vessels per villus (n)	4.1 ± 0.7 NS	3.8 ± 0.7 NS	4.2 ± 0.6
Degree of vascularization (%)	29.2 ± 8.8 **	42.8 ± 7.5 NS	38.9 ± 8.7
Mean surface area of a single vessel (μm^2)	139 ± 38 *	252 ± 112 NS	213 ± 81
Mean length of epithelial plates (μm)	11.2 ± 1.5 **	14.7 ± 2.9 NS	13.4 ± 2.2
Number of epithelial plates/villus (n)	1.8 ± 0.3 NS	1.8 ± 0.5 NS	1.8 ± 0.4
Percentage of epithelial plates on villous circumference (%)	7.1 ± 0.9*	9.0 ± 1.5 NS	7.9 ± 1.3

Significance was calculated for the adjacent column, i.e., weight < 10th percentile with normal flow versus weight < 10th percentile with abnormal flow, and weight < 10th percentile with normal flow versus AGA weight with normal flow.
Wilcoxon test: * $p < 0.05$; ** $p < 0.01$; *** $p < 0.001$; NS, not significant.

tree, which accounts for the low resistance. These placentas reach the limits of their performance reserve too early. As a result, we find a high incidence of prematurity, growth retardation, and infants with low Apgar scores and pH values. These cases reaffirm the correlation between Doppler findings and placental morphology.

Comments. It appears that the morphological correlates described above can be detected antenatally by Doppler scanning of the umbilical arteries to assess the impedance of the fetoplacental compartment. Doppler sonography thus provides an important new tool for the evaluation of individual fetal risk. Based on these considerations, it is at least conceivable that this modality can further reduce perinatal mortality and particularly the rate of intrauterine fetal deaths.

References

1 Beck T, Höckel M, Friese K: Plazenta-Reifegrad und histologischer Befund: Klinisch prospektive Untersuchung an einem Kollektiv von Termingeburten und Frühgeburten. Z. Geburtsh. u. Perinat. 192 (1988) 24–32

2 Beck T: Funktionelle Plazentadiagnostik mit halbautomatischer Histometrie. Vergleichende klinische, histometrische und endokrinologische Untersuchungen typischer Risikoschwangerschaften. Habilitationsschrift Mainz 1989

3 Beck T: Placentone architecture as a structural basis for histometric investigation of the human placenta. In Soma H (ed.): Placenta: Basic research for clinical application. Karger, Basel 1991, 46–58

4 Beck T: Placental morphometry using a computer assisted measuring programme: reference values for normal pregnancies at term. Arch. Gynecol. Obstet. 249 (1991) 135–147

5 Becker V: Allgemeine und spezielle Pathologie der Plazenta. In Becker V, Schiebler TH, Kubli F (eds.): Die Plazenta des Menschen. Thieme, Stuttgart 1981, 251–379

6 Bracero LA, Beneck D, Kirshenbaum N, Pfeiffer M, Stalter P, Schulman H: Doppler velocimetry and placental disease. Amer. J. Obstet. Gynecol. 161 (1989) 388–393

7 Castelluci M, Kaufmann P: A three-dimensional study of the normal human placental villous core: II. stromal architecture. Placenta 3 (1982) 269–285

8 Dawes GS: The umbilical circulation. Amer. J. Obstet. Gynecol. 84 (1962) 1634–1648

9 Divon MY, Guidetti DA, Braverman JJ, Oberlander E, Langer O, Merkatz IR: Intrauterine growth retardation – a prospective study of the diagnostic value of real-time sonography combined with umbilical artery flow velocimetry. Obstet. Gynecol. 72 (1988) 611–614

10 Giles WB, Trudinger BJ, Baird PJ: Fetal umbilical artery flow velocity waveforms and placental resistance: pathological correlation. Brit. J. Obstet. Gynaecol. 92 (1985) 31–38

11 Giles WB, Lingman G, Marsál K, Trudinger BJ: Fetal volume blood flow velocity waveform analysis: a comparison. Brit. J. Obstet. Gynaecol. 93 (1986) 462–465

12 Hitschold T, Weiss E, Berle P, Müntefering H: Histologische Placentabefunde bei Terminüberschreitung: Korrelation zwischen placentarer Reifungsretardierung, fetal-outcome und dopplersonographischen Befunden der Nabelarterie. Z. Geburtsh. u. Perinat. 193 (1989) 42–46

13 Hitschold T, Weiss E, Beck T, Berle P, Lehmann S, Müntefering H: Gepulste Dopplersonographie der Nabelarterie und fetoplazentarer Widerstand: Eine histometrische Untersuchung an Gestoseplazenten im Vergleich mit einem Normalkollektiv. Geburtsh. u. Frauenheilk. 49 (1989) 1056–1067

14 Hitschold T, Weiss E, Müntefering H, Berle P: Evaluation of placental risk using pulsed Doppler ultrasound to asses flow velocities in the umbilical arteries: Correlation with histometric data and clinical findings. Placental Comm. 199 (1990) 168–169

15 Hitschold T, Weiss E, Beck T, Müntefering H, Berle P: Beeinflußt die Vaskularisation der Placenta fetalis die enddiastolischen Blutflußgeschwindigkeiten in den Nabelarterien? Geburtsh. u. Frauenheilk. 50 (1990) 623–627

16 Hitschold T, Müntefering H, Berle P: Histologische Plazentabefunde bei diastolischem Null- oder Negativflow der Nabelarterien: Eine prospektive Untersuchung unter Berücksichtigung der Dauer der klinischen Beobachtung. Geburtsh. u. Frauenheilk. 52 (1992) 219–224

17 Hitschold T, Müntefering H, Berle P: Dopplersonographie der Art. umbilicalis zur Beurteilung des fetoplazentaren Gefäßbaumes: Ein prospektiver histologischer Vergleich. Ultraschall in Med. 13 (1992) 162–165

18 Hitschold T, Braun S, Weiss E, Berle P, Müntefering H: A case of discordant flow velocity waveforms in non-anastomosing umbilical arteries: A morphometric analysis. J. Matern. Fetal. Invest. 2 (1992) 215–219

19 Hitschold T, Weiss E, Beck T, Müntefering H, Berle P: Low target birth weight or growth retardation? Umbilical Doppler flow velocity waveforms and histometric analysis of the fetoplacental vascular tree. Amer. J. Obstet. Gynec. 168 (1993) 1260 –1264

20 Hitschold T, Ulrich S, Kalder M, Müntefering H, Berle P: Blutströmungsprofile der Arteria uterina: Korrelation zur Plazentamorphologie und zu klinisch-geburtshilflichen Daten im Rahmen der Präeklampsie. Z. Geburtsh. u. Neonat. 199 (1995) 8 –12

21 Hitschold T, Müntefering H, Ulrich S, Berle P: Does extremely low fetoplacental impedance as estimated by umbilical artery doppler velocimetry also indicate fetuses at risk ? Ultrasound Obst. Gynecol. 8–1 (1996) 36

22 Hitschold T, Müntefering H, Berle P: Die niedrige fetoplazentare Impedanz: Ein neuer Dopplerparameter zur Erkennung gefährdeter Feten. Z. Geburtsh. u. Frauenheilk. 58 (1998) 551–555

23 Hönig, B: Histomorphologische Plazentaveränderungen und fetale Risiken in Abhängigkeit von dopplersonographischen Befunden der Nabelarterien. Inauguraldissertation Universität Mainz 1996

24 Hüneke B, Schröder H: Doppler-sonographische Untersuchungen des fetalen Blutstroms im Tierexperiment. Gynäkologe 25 (1992) 341–351

25 Jackson MR, Walsh AJ, Morrow RJ, Mullen JB, Lye SJ, Ritchie JWC: Reduced placental villous tree elaboration in small-for-gestational-age pregnancies: Relationship with umbilical artery Doppler waveforms. Amer. J. Obstet. Gynecol. 172 (1995) 18–25

26 Jimenez E, Vogel M, Arabin B, Wagner G, Mirsalin P: Correlation of ultrasonic measurement of the utero-placental and fetal blood flow with the morphological diagnosis of the placental function. Trophoblast Res. 3 (1988) 325–334

27 Kaufmann P, Bruns U, Leiser R, Luckhardt M, Winterhager E: The fetal vascularization of term human placental villi. II. Intermediate and terminal villi. Anat. Embryol. 173 (1985) 203–214

28 Kaufmann P, Luckhardt M, Schweikhart G, Cantle SJ: Cross sectional features and three-dimensional structure of human placental villi. Placenta 8 (1987) 235–247

29 Kaufmann P, Luckhardt M, Leiser R: Three-dimensional representation of the fetal vessel system in the human placenta. Trophoblast Res. 3 (1988) 113–137

30 Kaufmann P: Architecture of normal villous tree. In Benirschke K, Kaufmann P (eds.): Pathology of the human placenta. 2nd edition. Springer, Berlin 1990, 81–113

31 Kaufmann P, Kosanke G, Leiser R, Scheffen I, Schweikhart G: Morphologische und morphometrische Grundlagen der Gefäßversorgung der menschlichen Plazenta. In Fendel H, Funk A, Jung H (eds.): Pränatale Dopplerdiagnostik. Steinkopf, Darmstadt 1992, 1–11

32 Kingdom JCP, Bürrell SJ, Kaufmann P: Pathology and clinical implications of abnormal umbilical artery Doppler waveforms. Ultrasound Obstet. Gynecol. 9 (1997) 271–286

33 Krebs C, Macara LM, Leiser R, Bowmann AW, Greer IA, Kingdom JP: Intrauterine growth restriction with absent end-diastolic flow velocity in the umbilical artery is associated with maldevelopment of the placental terminal villous tree. Amer. J. Obstet. Gynecol. 175 (1996) 1534–1542

34 Macara L, Kindgom JP, Kohnen G, Bowman AW, Greer IA, Kaufmann P: Elaboration of stem villous vessels in growth restricted pregnancies with abnormal umbilical artery Doppler waveforms. Brit. J. Obstet. Gynaecol. 102 (1995) 807–812

35 McCowan LM, Mullen BM, Ritchie K: Umbilical artery flow velocity waveforms and the placental vascular bed. Amer. J. Obstet. Gynecol. 157 (1987) 900–902

36 Morrow RJ, Adamson SL, Bull SB, Ritchie JWK: Effect of placental embolization on the umbilical arterial velocity waveform in fetal sheep. Amer. J. Obstet. Gynecol. 161 (1989) 1055–1060

37 Noack EJ, Stoz F, Schuhmann RA: Morphometrische Untersuchungen an Plazentazotten reifer menschlicher Plazenten. Z. Geburtsh. u. Perinat. 185 (1982) 155–160

38 Nordenvall M, Ullberg U, Laurin G, Sandstedt B, Ulmsten U: Placental morphology in relation to umbilical artery blood velocity waveforms. Eur. J. Obstet. Gynecol. Reprod. Biol. 40 (1991) 179–190

39 Reuwer PJ, Sijmons EA, Rietman GW, van Thiel MWM, Bruinse HW: Intrauterine growth retardation: Prediction of perinatal distress by Doppler ultrasound. Lancet ii (1987) 415–418

40 Schmidt KG, di Tommasso M, Silverman NH, Rudolph AM: Evaluation of changes in unilateral blood flow in the fetal lamb by Doppler waveform analysis. Amer. J. Obstet. Gynecol. 164 (1991) 1118–1126

41 Schuhmann R, Wehler V: Histologische Unterschiede an Plazentazotten innerhalb der materno-fetalen Strömungseinheit. Ein Beitrag zur funktionellen Morphologie der Plazenta. Arch. Gynäk. 210 (1971) 425–439

42 Schuhmann R, Borts H, Geier G, Kraus H: Über die Plazentone der reifen menschlichen Plazenta. Z. Geburtsh. u. Perinat. 181 (1977) 13–25

43 Schuhmann R, Stoz F, Maier M: Histometrische Untersuchungen an Plazentonen menschlicher Plazenten. Z. Geburtsh. u. Perinat. 190 (1986) 196–203

44 Schweikhart G: Morphologie des Zottenbaumes der menschlichen Plazenta. Orthologische und pathologische Entwicklung und klinische Relevanz. Habilitationsschrift Mainz 1985

45 Schweikhart G, Kaufmann P, Beck T: Morphology of placental villi after premature delivery and its clinical relevance. Arch. Gynecol. 239 (1986) 101–114

46 Stoz F, Schuhmann RA, Noack EJ: Morphometrische Befunde bei EPH-Gestose. Z. Geburtsh. u. Perinat. 186 (1982) 72–75

47 Stoz F, Schuhmann RA, Noack EJ: Morphometrische Untersuchungen an Plazenten reifer Mangelgeborener. Z. Geburtsh. u. Perinat. 187 (1983) 142–145

48 Stoz F, Schuhmann RA, Schmidt A: Morphometric investigations of terminal villi of diabetic placentas in relation to the White classification of diabetes mellitus. J. Perinat. Med. 15 (1987) 193–198

49 Stoz F, Schuhmann RA, Schebesta B: The development of the placental villus during normal pregnancy: Morphometric data base. Arch. Gyn. Obstet. 244 (1988) 23–32

50 Teasdale F: Histomorphometry of the human placenta in maternal preeclampsia. Amer. J. Obstet. Gynecol. 152 (1985) 25–31

51 Thompson RS, Trudinger BJ: Doppler waveform pulsatility index and resistance, pressure and flow in the umbilical placental circulation: An invesigation using a mathematical model. Ultrasound. Med. Biol. 16–5 (1990) 449–458

52 Trudinger BJ, Stevens D, Connelly A et al.: Umbilical artery flow velocity waveforms and placental resistance: The effects of embolization of the umbilical circulation. Amer. J. Obstet. Gynecol. 157 (1987) 1443–1448

53 Trudinger BJ, Giles WB, Cook CM: Umbilical artery flow velocity waveforms in high-risk pregnancy: randomised controlled trial. Lancet i (1987) 495–497

54 Visser GHA, Sadovsky G, Nicolaides KH: Antepartum heart rate patterns in small-for-gestational-age third-trimester fetuses: Correlations with blood gas values obtained at cordocentesis. Amer. J. Obstet. Gynecol. 162 (1990) 698–703

55 Vogel M: Atlas der morphologischen Plazentadiagnostik. Springer, Berlin 1992

56 Voigt HJ, Becker V: Dopplersonographie und Histomorphologie der uteroplazentaren Zirkulation. In Fendel H, Funk A, Jung H (eds.): Pränatale Dopplerdiagnostik. Steinkopf Darmstadt 1992, 13–19
57 Weiss E, Hitschold T, Berle P: Dopplersonographie der Art. umbilicalis: Differenzierte Diagnostik bei der intrauterinen Mangelentwicklung. Geburtsh. u. Frauenheilk. 49 (1988) 466–471
58 Weiss E, Berle P: Klinisches Management bei Feten mit diastolischem Null- oder Negativflow der Nabelarterien: Dauer der klinischen Beobachtung und Fetal Outcome. Z. Geburtsh. u. Perinat. 195 (1991) 37–42
59 Weiss E: Die Blutflußmessung in der Schwangerschaft. Bedeutung der Dopplersonographie in der klinischen Anwendung. Habilitationsschrift Universität Mainz 1995

Gynecological Ultrasound

26 Sonographic and Doppler Sonographic Examination of Uterine Anomalies

S. Kupesic and A. Kurjak

Classification of Uterine Anomalies

Congenital uterine malformations have a variable incidence that is usually estimated at 3–4%. Less than half of these anomalies have clinical manifestations[1,8,24]. Uterine anomalies can be classified into the following main types:

- Unicornuate uterus
- Uterus didelphys
- Bicornuate uterus
- Septate uterus

Unicornuate uterus. Unicornuate uterus results from a developmental error in one of the müllerian ducts. One rudimentary horn may be present, and implantation in this horn creates a very high likelihood of an abnormal pregnancy or tubal gestation. This anomaly can be diagnosed very accurately with three-dimensional (3 D) ultrasound (Fig. 26.**1**).

Uterus didelphys. The didelphic uterus is based on a failure of fusion of the paired müllerian ducts, resulting in duplication of the uterine corpus and cervix. These patients generally have no difficulties with menstruation or with coitus. Pregnancy, however, is associated with an increased risk of abnormal fetal presentations and prematurity.

Bicornuate uterus. If the müllerian ducts undergo only partial fusion, a bicornuate uterus develops with a single cervix and variable separation of the uterine horns. This anomaly is associated with high rates of early pregnancy loss, prematurity, and breech presentation (Fig. 26.**2**).

Septate uterus. Incomplete absorption of the septum between the müllerian ducts leads to defects ranging from a partial septum to significant separation of the endometrial uterine cavity. A complete failure of absorption leads to a longitudinal vaginal septum, called a "double vagina."

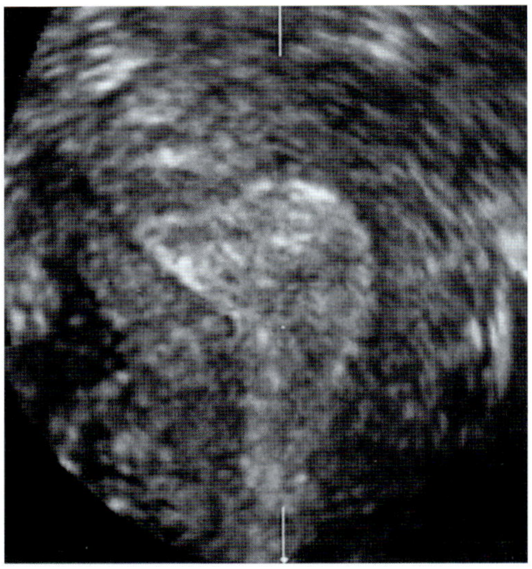

Fig. 26.**1** Three-dimensional ultrasound image of a unicornuate uterus. Absence of the left uterine horn is clearly demonstrated.

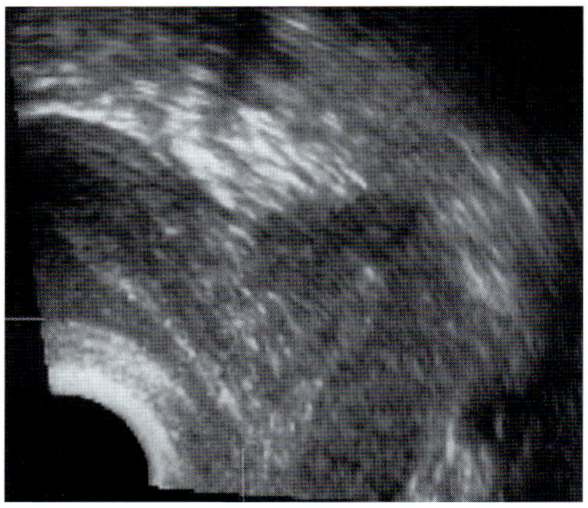

Fig. 26.**2** Three-dimensional ultrasound image of a bicornuate uterus. Note the separation of the two horns and the concave shape of the fundus.

Diagnosis and Complications of Septate Uterus

The septate uterus accounts for up to 50% of symptomatic uterine anomalies[6, 8].

Spontaneous abortion. The risk of spontaneous abortion in this group during the first trimester is between 28% and 45%, while the risk of second-trimester abortion is approximately 5%[8]. Premature births, abnormal fetal presentations, preterm labor, and dystocia are typical complications of septate uterus[10]. Deficient blood flow to the septum is the presumed cause of the pregnancy loss[6]. Electron-microscopic studies by Fedele et al.[6] showed that the endometrium covering the septum of the malformed uterus undergoes few if any preovulatory changes. This may play a role in the pathogenesis of primary infertility in women with a septate uterus.

Surgical intervention. Unfavorable obstetric conditions can be eliminated by surgical correction of the intrauterine septum. Formerly the septum was removed by transabdominal metroplasty[6], but hysteroscopic treatment has now become the method of choice. The advantage of this simple and effective treatment is that it does not leave a scar in the myometrium. Cararach et al.[2] and Goldenberg et al.[9] report pregnancy rates of 75% and 88.7% following hysteroscopic surgery.

Transvaginal sonography. The technical simplicity and effectiveness of hysteroscopy have compelled the clinician to make an early, reliable diagnosis of uterine anomalies. Transvaginal ultrasound has a sensitivity of almost 100% in screening for congenital uterine anomalies[19, 26], although it cannot positively distinguish between different types of anomalies[20, 21].

Hysterosalpingography. Hysterosalpingography (HSG) is an invasive diagnostic procedure that involves the use of contrast medium and x-rays. Although the uterine cavity is clearly delineated by HSG, it is not always possible to detect smaller anomalies or distinguish among different types of lateral fusion defects.

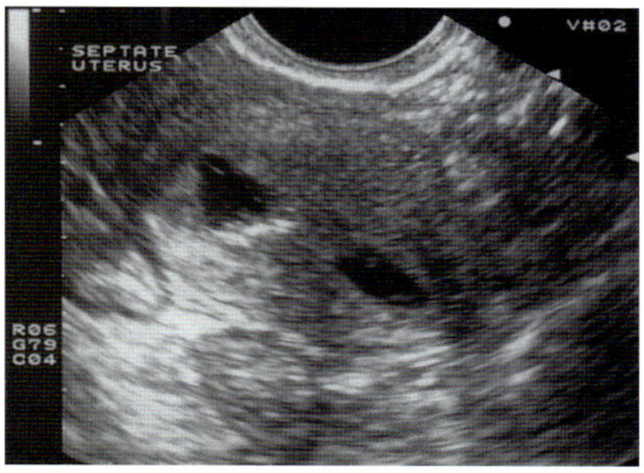

Fig. 26.3 Hysterosonographic image of a septate uterus demonstrates two endometrial echoes separated by a thick septum.

Hysterosonography. Hysterosonography[22] is a technique in which the uterine cavity is distended with saline solution and then examined with transvaginal ultrasound. This simple and minimally invasive procedure supplies anatomical images of the endometrium and myometrium, clearly defines the septate uterus, and even allows the thickness and length of the septum to be measured[23] (Fig. 26.3).

Other diagnostic procedures. Although several reports note the high accuracy of magnetic resonance imaging[3, 17] and three-dimensional ultrasound[11] in the diagnosis of uterine anomalies, so far these techniques have rarely been used in the routine investigation of these conditions. In patients scheduled for corrective surgery, the examination is usually supplemented by another invasive procedure, CO_2 hysteroscopy[25].

Ultrasound in the Diagnosis and Treatment of Septate Uterus—Authors' Results

Our study is an attempt to evaluate the combined use of transvaginal ultrasound, transvaginal color and pulsed Doppler sonography, hysterosonography, and three-dimensional ultrasound in the preoperative evaluation of the septate uterus[15]. In the second part of the study, we analyzed the obstetric and perinatal complications of septate uterus and assessed the reproductive outcome after hysteroscopic treatment.

Patients and Methods

Study patients. A total of 420 infertile women undergoing hysteroscopic surgery were included in the study. Table 26.1 summarizes the intraoperative findings in the 420 patients. An intrauterine septum was present in 278 patients and was surgically corrected. Forty-three of the women with a septate uterus had a prior history of recurrent spontaneous abortions, 71 women had a history of one spontaneous abortion (56 in the first trimester, 15 in the second trimester), 82 had primary infertility, and 20 had had premature deliveries, including six breech presentations and two transverse presentations. Seventy-six patients had a prior history of ectopic pregnancy.

B-mode imaging and Doppler sonography. All of the patients underwent transvaginal B-mode and Doppler ultrasound examinations during the luteal phase of their cycle. First the uterus was systematically examined to assess its position, size, and morphology. Uterine morphology was carefully defined in the B-mode image, giving special attention to the endometrium in the sagittal and transverse planes. The septum could be recognized as an echogenic structure dividing the uterus into two cavities. Next, transvaginal color Doppler scanning was performed by a different examiner who did not know the results of the B-mode examination.

Table 26.**1** Intraoperative findings in 420 infertile women undergoing hysteroscopy

Findings	Number of patients
Submucous leiomyoma	46
Endometrial polyp	35[a]
Intrauterine adhesionsn	19*
Septate uterus	278
Arcuate uterus	28
Bicornuate uterus	16[b]
Total	422

[a] One patient with an endometrial polyp and one patient with intrauterine adhesions had a septate uterus.

[b] Diagnosed by the combined use of laparoscopy and hysteroscopy.

With permission from Kupesic S, Kurjak A.: Septate uterus: detection and prediction of obstetrical complications by different forms of ultrasonography. J Ultrasound Med. 17 (1998) 631–636.

Blood flow in the septum and myometrium was assessed by color and pulsed Doppler examination. Flow velocity waveforms were sampled from all blood vessels that could be visualized, and the resistance index (RI) was calculated. Starting with the maximum frequency, the RI was calculated by using the formula: (maximum systolic velocity–end-diastolic velocity) ÷ maximum systolic velocity.

Hysterosonography. Each patient was placed in a gynecological examination chair for the intrauterine instillation of isotonic saline solution. In a total of 76 patients, the cervix was visualized with a speculum and swabbed with iodine solution. A catheter with an outer diameter of 1.6 mm and inner diameter of 1.1 mm was carefully inserted into the cervical canal. The catheter cuff was inflated with 1.5–2 ml of sterile saline to prevent leakage of the instilled fluid. From 10 to 20 ml of isotonic saline solution was slowly injected with a syringe to distend the uterine cavity. The speculum was then removed, and the vaginal transducer was inserted. Scans in the transverse and sagittal planes defined the septum as an echogenic structure dividing the uterine cavity into two parts.

Three-dimensional ultrasound. Eighty-six women undergoing hysteroscopy were also examined by 3 D ultrasound. This was preceded by a transvaginal B-mode, color Doppler and pulsed Doppler examination. Hysterosonography was additionally performed in 12 of these women. The examiner did not know the results of the previous examinations. Three mutually perpendicular sectional planes were displayed simultaneously on the monitor, permitting a detailed analysis of uterine morphology. Coronal scans were particularly helpful in diagnosing the uterine anomaly.

Results

Table 26.**2** reviews the sensitivity, specificity, and positive and negative predictive values of transvaginal sonography, transvaginal color and pulsed Doppler, hysterosonography, and three-dimensional ultrasound in the diagnosis of septate uterus.

Transvaginal ultrasound. In 264 cases a presumptive diagnosis of septate uterus was made on the basis of transvaginal B-mode findings. Fourteen patients had a false-negative diagnosis. Thus, the sensitivity of transvaginal sonography in the diagnosis of septate uterus was 94.96%.

Transvaginal color and pulsed Doppler ultrasound. A septate uterus could be diagnosed in 276 cases by Doppler examination, corresponding to a sensitivity of 99.28%. A septate uterus was incorrectly diagnosed in one patient with an endometrial polyp and in another patient with intrauterine adhesions. Thus, the reliability of Doppler sonography was decreased when other intracavitary lesions such as an endometrial polyp or submucous leiomyomas were present.

Color and pulsed Doppler examinations demonstrated a septal blood supply in 198 of the patients (71.22%) (Fig. 26.**4**).

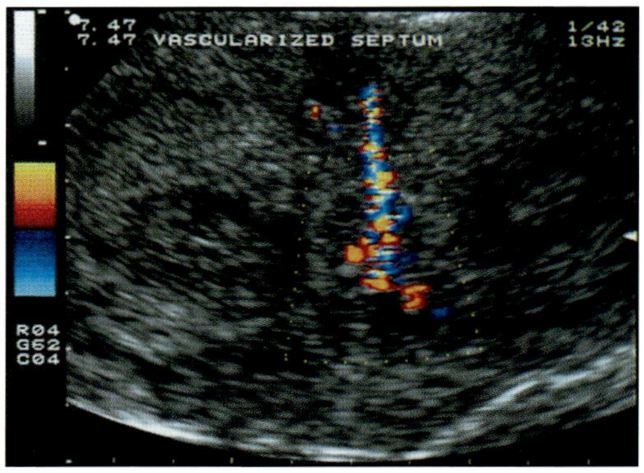

Fig. 26.**4** Transvaginal ultrasound image of a septate uterus. Two separate endometrial compartments are visible during the proliferative phase of the cycle. Color Doppler demonstrates small myometrial vessels in the septum.

Table 26.**2** Sensitivity, specificity, and positive (PPV) and negative predictive values (NPV) of various imaging procedures in the diagnosis of septate uterus in 420 patients with a history of infertility or spontaneous abortion

Imaging procedure	Sensitivity (%)	Specificity (%)	PPV (%)	NPV (%)
Transvaginal B-mode ultrasound	94.96	92.86	95.65	91.77
Transvaginal color and pulsed Doppler ultrasound	99.28	99.30	99.64	98.60
Hysterosonography	100.00	95.65	98.18	100.00
Three-dimensional ultrasound	93.55	96.55	98.31	87.50

With permission from Kupesic S, Kurjak A.: Septate uterus: detection and prediction of obstetrical complications by different forms of ultrasonography. J Ultrasound Med. 17 (1998) 631–636.

The RI values in the septal vessels ranged from 0.68 to 1.0 (mean RI = 0.84 ± 0.16) (Fig. 26.**5**). No diastolic blood flow was detected in 18 patients, while continuous diastolic flow was present in all the rest.

Hysterosonography. In 76 women, the intrauterine instillation of isotonic saline solution was recommended prior to hysteroscopy. A clear-cut diagnosis of septate uterus was made in 54 (71.5 %) of the patients. The sensitivity and negative predictive value of hysterosonography following transvaginal color Doppler scanning were 100 %. In one woman with dense intrauterine adhesions, however, the uterine septum could not be detected even by hysterosonography.

Three-dimensional ultrasound. Three-dimensional images of good quality were obtained in 86 patients (Fig. 26.**6**). The 3 D ultrasound findings agreed with hysteroscopic findings in 58 patients with a septate uterus. However, an arcuate uterus was diagnosed by 3 D ultrasound in four women who had a septate uterus. The uterine cavity in these patients appeared to be deformed by a fibroid in the fundal area. A false-positive diagnosis of septate uterus was made in one patient with intrauterine adhesions.

Radiographic hysterosalpingography. Radiographic HSG was performed in 188 women during a 12-month period preceding our study. A septate uterus was diagnosed in 49 of these patients based on the Reuter criteria[23]. A septate uterus was assumed to be present when the angle between the two cavities was < 75°, a bicornuate uterus when the angle was > 105°. In 15 cases (7.98 %), hysterosalpingography showed a deformed uterine cavity but no evidence of a congenital uterine anomaly. The overall sensitivity of radiographic HSG in the diagnosis of septate uterus was only 26.6 %.

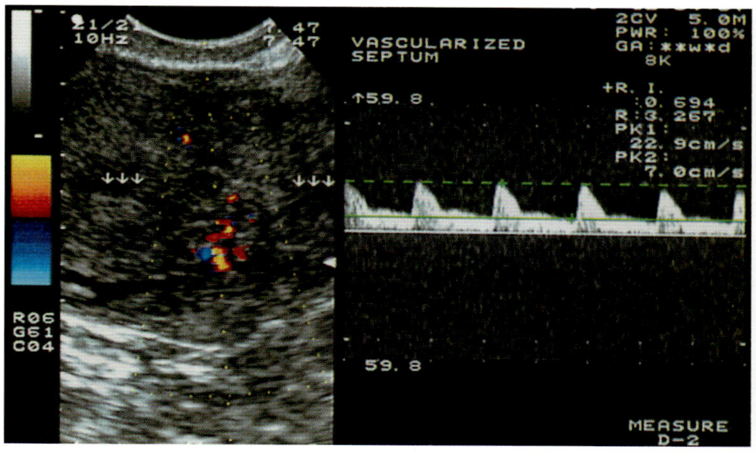

Fig. 26.5 Same patient as in Fig. 26.**4**. Pulsed Doppler waveform analysis shows a moderately high RI (0.69), which is typical of the radial arteries.

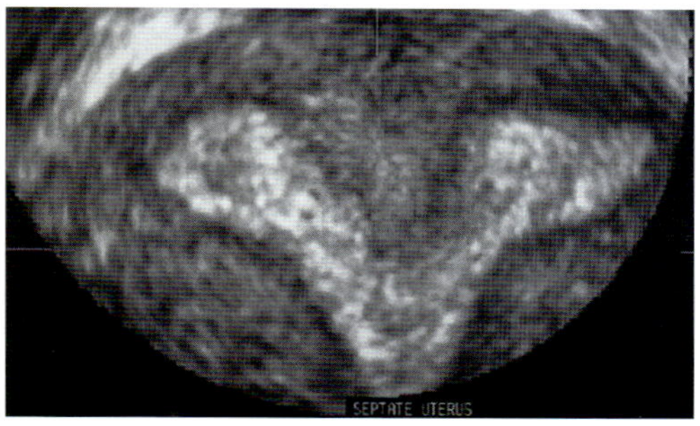

Fig. 26.**6** Three-dimensional ultrasound image of a septate uterus, which shows a normal outer uterine contour and a thick septum extending into the uterine cavity.

Obstetric complications. In the second part of our five-year study (1992–1996), we compared the obstetric complications in 278 patients with a septate uterus with an average control group. Early pregnancy loss occurred in 114 of the 278 patients (41.1 %), compared with a 15 % incidence in the control group. Late abortions and preterm deliveries occurred in 35 of the 278 patients (12.59 %), compared with 7 % in the controls. Intrauterine growth retardation occurred in two pregnancies with a septate uterus (8.7 %), versus 6 % in the controls. The incidence of intrauterine fetal death was 4.35 % in our patients and 0.5 % in our control group. Placental abruption was found in one patient with a septate uterus (4.35 %), and placenta previa was diagnosed in another patient (4.35 %). Breech presentation occurred in six women with a septate uterus (26.9 %) and transverse presentation in two (8.70 %). Since septate uterus has a significantly higher statistical association with abnormal fetal presentations, the rate of cesarean sections was also significantly higher (34.78 %). Cervical insufficiency during pregnancy was documented in nine patients (25.71 %) with a septate uterus.

Ectopic pregnancy was diagnosed in 76 of the patients (27.34 %), which was twice the incidence found in the control group (13.3 %). Seven patients with a septate uterus were found to have bilateral ectopic pregnancies.

Follow-up after hysteroscopic surgery. We followed the reproductive outcomes in 116 patients (32 with primary infertility, 16 after one spontaneous abortion, 12 after preterm deliveries, and 26 after recurrent abortions) who had undergone hysteroscopic surgery for an intrauterine septum. The prospective follow-up period was scheduled for 24 months in all patients. The pregnancy rate in the study group was 50.86 %: 44 patients (74.58 %) with term deliveries, 11 (18.64 %) with first-trimester abortion, and 4 (6.78 %) with early pregnancy loss. The other patients (162) were followed in the same way but for periods less than 24 months, and so they are not reported in this study.

New Thoughts on Old Problems

In the past, at least two diagnostic procedures have been routinely used in the diagnosis of congenital uterine anomalies. The attending gynecologist should be aware, however, that an extended diagnostic workup delays treatment, increases costs, and leads to greater risks and discomfort for the patient[23]. Patients with a septate uterus require a prompt, accurate diagnosis as a prelude to surgical correction of their disorder.

Histology of the septal endometrium. Fedele et al.[7] showed that an intrauterine septum can be a cause of primary infertility. When they examined the endometrial surface morphology of biopsy samples taken in the preovulatory phase of the cycle, they found significant structural differences between the septal endometrium and the endometrium of the lateral uterine wall. These histological changes included a reduced number of irregularly distributed glandular ostia, incomplete ciliogenesis on ciliated cells, and a decreased ratio of ciliated to nonciliated cells. These changes were indications that the differentiation and estrogen-stimulated maturation of the septal endometrium were irregular. Since the hormone levels in the study patients were normal for cycle phase, the most plausible hypothesis is that the mucosa covering the septum is poorly responsive to estrogens, perhaps because of deficient blood flow to the septal connective tissue.

Muscle fiber content of the septum. March[16] believed that the septum was composed of fibroelastic tissue, while Fayez[5] thought that it consisted mostly of connective tissue with a small amount of muscle fibers. Our study results could not confirm this. Color and pulsed Doppler ultrasound demonstrated septal blood flow in 71.22 % of the patients, indicating that most of the septa were supplied by myometrial vessels.

Dabirashrafi et al.[4] performed histological examinations of uterine septa from 16 patients who underwent abdominal metroplasty. Four uterine biopsies were taken from each patient: one from the septum near the serosal layer, one from the middle of the septum, one from the tip of the septum, and one from the left posterior uterine wall. The authors found less connective tissue in the septum (confirmed by the Bonferroni criteria for multiple comparisons) and a greater average amount of muscle tissue, muscle interlacing, and vessels with a muscle wall. These findings were contrary to the classic view on the histological makeup of the intrauterine septum. The small amount of connective tissue in the septum could be responsible for the poor decidual transformation and placentation that occur following implantation at this site[4,5]. Also, the greater amount of muscle tissue and interlaced muscle fibers in the septum could lead to pregnancy loss as a result of increased or uncoordinated contractions.

Septal height, septal blood flow, and obstetric complications. In a more recent study at our department[14], no correlation was found between the height of the intrauterine septum and the occurrence of obstetric complications ($p > 0.05$). Abortions and complications during late pregnancy had the same frequency in patients with small septa less than one-third the height of the uterine cavity and in patients whose septum occupied more than two-thirds of the uterine cavity. The same correlation was found for septal thickness: obstetric complications were equally common in patients with thin and thick septa ($p > 0.05$). The spontaneous abortion rate did show a significant correlation with septal blood flow, however. The patients with a vascularized septum had a significantly higher incidence of spontaneous abortions in early pregnancy and complications in late pregnancy than the patients with a nonvascularized septum ($p < 0.05$).

Information gained with color Doppler. Uterine morphology, including the endometrial layer and uterine musculature, can be accurately evaluated with transvaginal ultrasound. Color flow can simultaneously display the morphology and vascular system of the uterus, supplying valuable information on the nature and extent of uterine anomalies. Color Doppler contributes to evaluation of the myometrium by demonstrating the myometrial vessels. Doppler can also be used to diagnose a scant septal blood supply and/or inadequate development of the endometrium in patients with a septate uterus[13,14].

Uses of three-dimensional ultrasound. Three-dimensional ultrasound can provide uterine scans that accurately demonstrate the notch in the fundal region and the length of the intrauterine septum (Fig. 26.**7**). In our experience, however, this technique can give the false impression of an arcuate uterus in patients who have a leiomyoma in the fundal region. This lesion imparts a concave shape to the uterine cavity, while the dimple in the fundus appears indistinct. The value of 3 D ultrasound is also limited by acoustic shadows that are cast by uterine fibroids, an irregular endometrial layer, and a small uterine cavity (due to intrauterine adhesions).

Comments. Our study[15] showed conclusively that obstetric complications are more common in patients with a septate

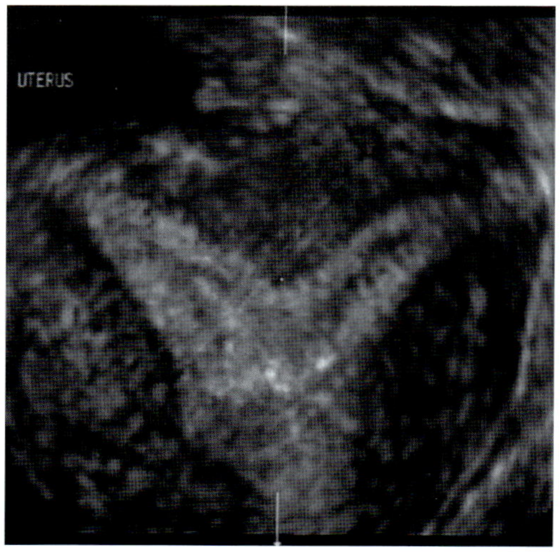

Fig. 26.**7** Another three-dimensional ultrasound image of a septate uterus. A thick septum subdivides the uterine cavity. The extent of the anomaly is clearly defined.

uterus than in other women. In particular, ectopic pregnancy was twice as common in these patients (27.34%) as in the control group (13.3%). A possible cause might be menstrual reflux, which is common in women with uterine anomalies and may impede the transport of the fertilized oocyte into the uterine cavity. Our study clearly showed that patients with primary infertility or recurrent pregnancy problems benefited from surgical removal of the intrauterine septum.

In the conventional management of a septate uterus, intervention is delayed until the first obstetric problem arises, since a large percentage of patients have no complications[8]. But given the higher infertility rates in women with a septate uterus and the good results that can be achieved with endoscopic surgery, we are obliged to recommend hysteroscopy as soon as we diagnose this condition, if possible even before the patient conceives[12,14,15]. It appears that incision of the septum can prevent implantation at an unfavorable site, either by revascularization of the uterine connective tissue in the fundal area or by suppressing uterine contractions that might expel the pregnancy[7]. Since both events can be demonstrated with color or pulsed Doppler ultrasound, this technique can be efficiently utilized both for the diagnosis of congenital anomalies and for the follow-up of hysteroscopic surgery.

References

1 Ashton D, Amin HK, Richart RM, Neuwirth RS: The incidence of asymptomatic uterine anomalies in women undergoing transcervical tubal sterilization. Obstet. Gynecol. 72 (1988) 28–30

2 Cararach M, Penella J, Ubeda J, Iabastida R: Hysteroscopic incision of the septate uterus: scissors versus resectoscope. Hum. Reprod. 9 (1994) 87–89

3 Carrington BM, Hricak M, Naruddin RN: Mullerian duct anomalies: MR evaluation. Radiology 170 (1990) 715–720

4 Dabrashrafi H, Bahadori M, Mohammad K, Alavi M, Moghadami-Tabrizi N, Zandinejad R: Septate uterus: New idea on the histologic features of the septum in this abnormal uterus. Amer. J. Obstet. Gynecol. 172 (1995) 105–107

5 Fayez JA: Comparison between abdominal and hysteroscopic metroplasty. Obstet. Gynecol. 68 (1986) 399–403

6 Fedele L, Arcaini L, Parazzini F, Vercellini P, Nola GD: Metroplastic hysteroscopy and fertility. Fertil. Steril. 59 (1993) 768–770

7 Fedele L, Bianchi S, Marchini M, Franchi D, Tozzi L, Dorta M: Ultrastructural aspects of endometrium in infertile women with septate uterus. Fertil. Steril. 65 (1996) 750–752

8 Gaucherand P, Awada A, Rudigoz RC, Dargent D: Obstetrical prognosis of septate uterus: a plea for treatment of the septum. Eur. J. Obstet. Gynecol. Reprod. Biol. 54 (1994) 109–112

9 Goldenberg M, Sivan E, Sharabi Z: Reproductive outcome following hysteroscopic management of intrauterine septum and adhesions. Hum. Reprod. 10 (1995) 2663–2665

10 Heinonen PK, Saarikoski S, Pystynen P: Reproductive performance of women with uterine anomalies. An evaluation of 182 cases. Acta Obstet. Gynecol. Scand. 61 (1982) 157–162

11 Jurkovic D, Giepel A, Gurboeck K, Jauniaux E, Natucci M, Campbell S: Three dimensional ultrasound for the assessment of uterine anatomy and detection of congenital anomalies: a comparison with hysterosalpingography and two-dimensional sonography. Ultrasound Obstet. Gynecol. 5 (1995) 233–237

12 Keltz MD; Olive DL, Kim AH, Arici A: Sonohysterography for screening in recurrent pregnancy loss. Fertil. Steril. 67 (1997) 670–674

13 Kupesic S, Kurjak A: Uterine and ovarian perfusion during the periovulatory period assessed by transvaginal color Doppler. Fertil. Steril. 3 (1993) 439–443

14 Kupesic S, Kurjak A: Comparison of B-mode, color Doppler, threedimensional ultrasound and hysterosonography in detection of septate uteri. Am. J. Obstet. Gynecol. (1998)

15 Kupesic S, Kurjak A: Pregnancy after diagnosis and treatment of uterine anomalies. Croat Med. J. (1998)

16 March CM: Hysteroscopy as an aid to diagnosis in female infertility. Clin. Obstet. Gynecol. 26 (1983) 302–312

17 Marshall C, Mintz DI, Thickman D, Gussman H, Kressel Y.: MR evaluation of uterine anomalies. Radiology 148 (1987) 287–289

18 McShane PM, Reilly RJ, Schiff L: Pregnancy outcome following Tompkins metroplasty. Fertil. Steril. 40 (1983) 190–194

19 Nicolini U, Bellotti B, Bonazzi D, Zamberleti G, Battista C: Can ultrasound be used to screen uterine malformation? Fertil. Steril. 47 (1987) 89–93

20 Randolph J, Ying Y, Maier D, Schmidt C, Riddick D: Comparison of real time ultrasonography, hysterosalpingography, and laparoscopy/hysteroscopy in the evaluation of uterine abnormalities and tubal patency. Fertil. Steril. 5 (1986) 828–832

21 Reuter KL, Daly DC, Cohen SM: Septate versus bicornuate uteri: errors in imaging diagnosis. Radiology 172 (1989) 749–752

22 Richman TS, Viscomi GN, Cherney AD, Polan A: Fallopian tubal patency assessment by ultrasound following fluid injection. Radiology 152 (1984) 507–510

23 Salle B, Sergeant P, Galcherand P, Guimont I, De Saint Hilaire P, Rudigoz RC: Transvaginal hysterosonographic evaluation of septate uteri: a preliminary report. Hum. Reprod. 11 (1996) 1004–1007

24 Sorensen S: Estimated prevalence of mulerian anomalies. Acta Obstet. Gynecol. Scand. 67 (1988) 441–445

25 Taylor PJ, Cumming DC: Hysteroscopy in 100 patients. Fertil. Steril. 31 (1979) 301–304

26 Valdes C, Malini S, Malinak LR: Ultrasound evaluation of female genital tract anomalies: a review of 64 cases. Am. J. Obstet. Gynecol. 149 (1984) 285–290

27 Doppler Examination of the Normal Endometrium and Benign Endometrial Changes

S. Kupesic and A. Kurjak

Changes in the Normal Endometrium during the Menstrual Cycle

The endometrium appears sonographically as a central, echogenic layer in the uterine wall. Its structural details are clearly defined by transvaginal imaging[6]. The thickness and structure of the endometrium depend upon the plasma levels of circulating estrogens and progestins.

Menstrual phase. During menstruation in the healthy, fertile woman, the two functional layers of the endometrium (the stratum compactum and spongiosum) are shed[36], leaving only the stratum basale, from which the endometrium is regenerated. Hormone withdrawal and the changes in the spiral arteries are key components of this mechanism. Increased coiling of the spiral arteries leads to an arrest of circulation, which in turn causes progressive tissue ischemia. Vasoconstriction of the spiral arteries and necrosis of their vessel walls finally lead to menstrual bleeding[24]. The hypoechoic areas that are sometimes visible with ultrasound reflect this structural breakdown of the endometrium. As the menstrual phase progresses, both hypoechoic areas (blood) and hyperechoic areas (sloughed endometrium) can be seen. As menstruation ends, the endometrium appears sonographically as a thin, almost linear, slightly irregular echogenic layer.

Early follicular phase. The endometrium is normally less than 5 mm thick in the early follicular phase. The endometrial glands, lined by relatively flat cells, are now almost tubular. Mitoses become more numerous, and blood vessels grow from the stratum basale toward the endometrial surface, where a capillary network is formed. At this time the endometrium appears sonographically as a hyperechoic layer. In some cases it cannot be clearly distinguished from the myometrium.

Ovulation. The endometrial glands are even more numerous around the time of ovulation, and the endometrial thickness averages 10 mm. A three-layered endometrial structure is typical of the follicular phase. The hyperechoic structure of the endomyometrial junction is most conspicuous at this time.

Secretory phase. This phase is characterized by a marked increase in glycogens, acid phosphatases, and lipids in the endometrium. The endometrium appears uniformly hyperechoic, losing its three-layered structure and hypoechoic border. During this phase of the cycle, the endometrium appears markedly hyperechoic to the myometrium.

Midluteal phase. The endometrium shows its greatest sonodensity in the midluteal phase, when it appears uniformly hyperechoic. Posterior acoustic enhancement is typical of this phase of the cycle.

Changes in Endometrial Blood Flow during the Menstrual Cycle

Transvaginal color and pulsed Doppler ultrasound can be used to examine endometrial perfusion under normal and pathological conditions.

Comparison of normal and stimulated cycles. The increase in endometrial blood flow during the course of the menstrual cycle is based on the blood flow changes that occur in the uterine, arcuate and radial arteries (Fig. 27.**1**). The changes that occur in the flow velocity waveforms of the spiral arteries during the normal ovulatory cycle have been tracked with Doppler ultrasound[26]. On the day before ovulation, the resistance index (RI) is equal to 0.54 ± 0.03 (Fig. 27.**2**) and the blood flow velocity begins to rise. The RI reaches its lowest point (0.49 ± 0.05) between days 16 and 18 of the cycle. In the stimulated cycle, by contrast, the resistance rises on the last day before ovulation. It may be that the induction of ovulation evokes this uterine response, which should be evaluated with

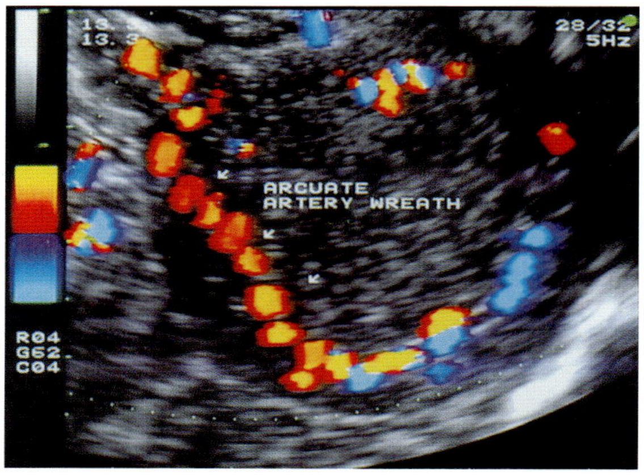

Fig. 27.**1** Transvaginal scan of the uterine vascular supply: uterine arteries at the level of the corpus–cervix junction, arcuate arteries encircling the uterus, and radial arteries within the myometrium.

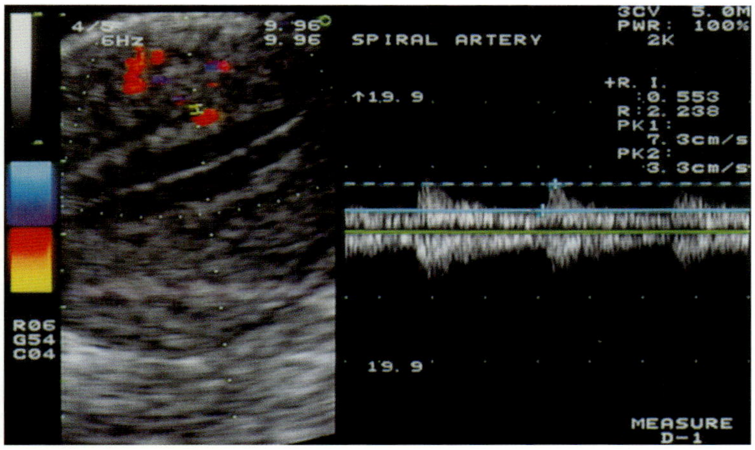

Fig. 27.**2** Color Doppler signals from the periphery of the multilayer endometrium. Increased blood flow velocity and a decreased resistance index (RI = 0.55) are noted on the day of ovulation.

Doppler scans prior to embryo transfer. Endometrial blood flow appears to provide a noninvasive parameter that is better for predicting uterine receptivity than the flow velocity in the uterine artery. Analysis of the blood flow changes in the spiral arteries should be used in predicting successful implantation, investigating unexplained fertility problems, and referring patients with abnormal endometrial blood flow for appropriate treatment[25] (Table 27.**1**).

Zaidi et al.[46] recently published interesting data on this subject. They examined 96 women in an IVF program on the day of human chorionic gonadotropin (hCG) administration by transvaginal color Doppler sonography. They assessed endometrial thickness and morphology, the presence or absence of intra- and subendometrial blood flow, intraendometrial vascular penetration, and subendometrial blood flow velocity on the day of hCG administration and related the results to pregnancy rates. The overall pregnancy rate was 32.3 %. There was no significant difference between the pregnant and nonpregnant groups in terms of endometrial thickness, subendometrial peak systolic flow velocity V_{max}, or subendometrial pulsatility index (PI). The pregnancy rates based on endometrial morphology were not significantly different: 17.6 % for type A (hyperechoic) endometrium, 33.3 % for type B (isoechoic), and 35.6 % for type C (three-layered). In eight patients, subendometrial blood flow and intraendometrial vascularization were not detected. When endometrial blood flow was not detected, implantation did not occur. The pregnancy rates related to the depth of vascular penetration into the endometrial and subendometrial regions were 26.7 % for the subendometrial region (zone 1), 36.4 % for the outer hyperechoic region (zone 2), and 37.9 % for the inner hyperechoic region (zone 3), but the differences were not statistically significant. Of the cycles with type A endometrium, 23.5 % showed no subendometrial color flow. This was greater than the frequency of absent color flow noted in the type C endometrium.

Comparison of the normal cycle and luteal insufficiency. Our most recent study[27] has clearly shown that flow resistance in the uterine, spiral, and radial vessels decreases at the transition from the follicular to the luteal phase in women with normal endometrial development. But in patients with delayed endometrial development and luteal insufficiency, rising uterine vascular resistance is measured during the course of the luteal phase. In women with a luteal phase defect, the impedance values in the spiral arteries were increased in the preovulatory phase (RI = 0.70 ± 0.06, $p < 0.001$), midluteal phase (RI = 0.72 ± 0.01, $p < 0.001$), and late luteal phase (RI = 0.72 ± 0.04, $p < 0.001$). Because the most significant deviations from normal impedance values in the intraovarian and subendometrial vessels were seen in patients with luteal insufficiency, we may conclude that color and pulsed Doppler ultrasound is helpful in the assessment of luteal phase adequacy. Moreover, Doppler ultrasound can be used along with, or even in place of, hormonal and histological markers in the evaluation of uterine receptivity.

Table 27.**1** Blood flow in the spiral arteries during the preovulatory phase

Time from ovulation (days)	Maximum systolic blood flow velocity (cm/s)	Resistance index	Pulsatility index
–3	6.21 ± 0.04	0.55 ± 0.02	0.86 ± 0.05
–2	6.02 ± 0.09	0.54 ± 0.03	0.85 ± 0.09
–1	6.32 ± 0.12	0.48 ± 0.04	0.83 ± 0.12
0	6.68 ± 0.68	0.48 ± 0.06	0.84 ± 0.14
+ 1	7.46 ±1.31	0.49 ± 0.05	0.72 ± 0.12

With permission from Kupesic S, Kurjak A: Uterine and ovarian perfusion during the periovulatory period assessed by transvaginal color Doppler. Fertil. Steril. 60 (1993) 439–443.

Gynecological Ultrasound

Submucous Leiomyomas

The uterine fibroid is one of the most common tumors that occur in the lesser pelvis of women during their reproductive years[45]. Leiomyomas may be intramural, submucous, or subserous, and some are pedunculated. The clinical hallmarks of submucous leiomyomas are metrorrhagia, pain in the lesser pelvis, or infertility. The presence and severity of the symptoms depend largely on the number, size, and location of the tumors. The ultrasound diagnosis of leiomyoma is based on a change in uterine contour, uterine enlargement, and/or a change in echo texture.

Detection by transvaginal sonography and hysteroscopy. Because the proportions of smooth muscle and connective tissue in leiomyomas are highly variable, these benign tumors vary greatly in their sonographic appearance from hypoechoic to hyperechoic. As a result, they are occasionally misinterpreted as endometrial polyps, blood, or mucus. Fedele et al.[14] tested the accuracy of transvaginal sonography in the detection of small submucous myomas in patients who underwent transvaginal sonography and hysteroscopy before a scheduled hysterectomy. The sensitivity of transvaginal sonography was 100%, its specificity was 94%, the positive predictive value of an abnormal finding was 81%, and the positive predictive value of an abnormal examination was 100%. These figures are comparable to the values reported for hysteroscopy (100%, 96%, 87%, and 100%, respectively). Transvaginal ultrasound is more accurate than hysteroscopy in the localization of leiomyomas but is less accurate in distinguishing between a leiomyoma and an endometrial polyp.

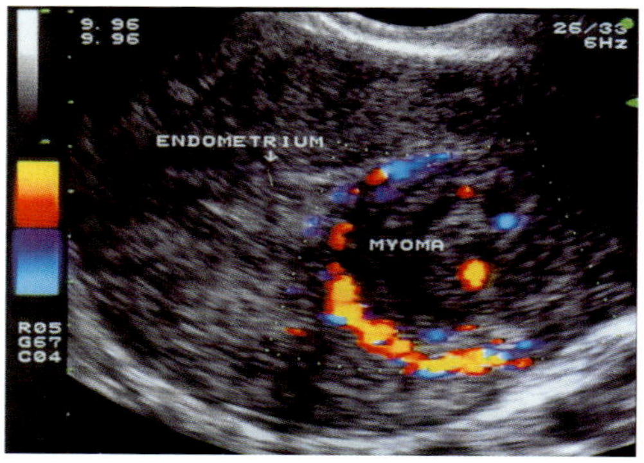

Fig. 27.**3** Submucous leiomyoma, delineated by its increased vascularity. The color flow signals map the rich vascular supply at the tumor base.

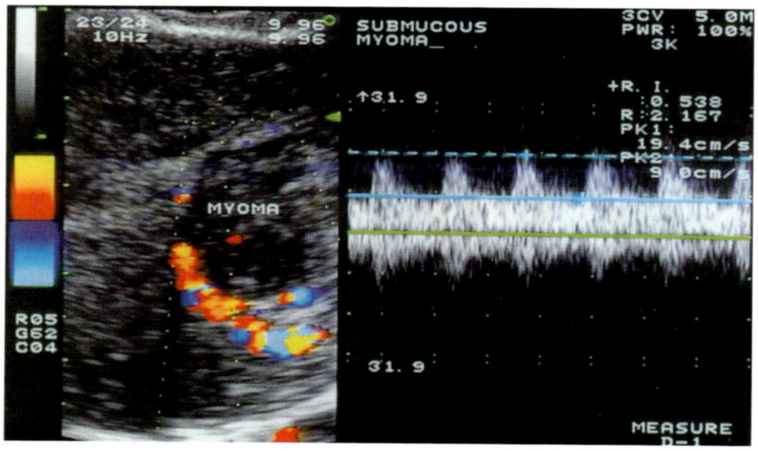

Fig. 27.**4** Same patient as in Fig. 27.**3**. Analysis of the flow velocity waveform (right) indicates moderate vascular resistance (RI = 0.54).

Leiomyomas and infertility. The relationship between leiomyomas and infertility or habitual abortion is still poorly understood. Many women with leiomyomas can conceive without difficulty and carry their pregnancy to term. On the other hand, pregnancy rates of 10–89% have been reported in previously infertile patients who had their fibroids removed[43]. The cause of fibroid-associated infertility is still unknown, however. It has been theorized that the tumors may decrease the area available for implantation, decrease the contractility of the uterine musculature, cause venous changes in the endometrium, or obstruct the cervix and fallopian tubes, causing interference with sperm transport.

Vascularization of uterine leiomyomas. With transvaginal and color Doppler sonography, it is possible to measure blood flow in small vessels and also to obtain more reproducible measurements[31] (Table 27.**2**).

Kurjak et al.[31] examined 161 women: 101 patients with palpable uterine fibroids and 60 healthy volunteers. Flow velocity waveforms were used to analyze the vascular resistance in the uterine arteries and in the main arteries supplying the definable fibroids. Color flow signals were detected at the periphery of the tumors and also at their center (Fig. 27.**3**). Diastolic flow was always detectable in these vessels and was usually greater than in the uterine artery. The mean RI of myometrial blood flow was 0.54, and the mean pulsatility index (PI) was 0.89 (Fig. 27.**4**). All of the tumors were benign on histological examination, even when the RI was very low. Low RI values were found in necrotic tumors and in tumors with secondary degenerative or inflammatory changes. Examination of the uterine arteries in the control group showed a mean RI of 0.84 and a PI of 2.525. Significantly lower values were measured in the fibroid group: 0.74 for RI and 1.65 for PI. The flow velocity, RI, and PI were each measured between the 5th and 8th days of the menstrual cycle.

Table 27.**2** Uterine artery blood flow parameters in patients with palpable (vascularized) uterine fibroids and in healthy volunteers

Blood flow parameter	Control group ($n = 60$)	Patients with vascularized fibroids ($n = 81$)
Velocity (cm/s)	34.4 ± 12.25	47.08 ± 18.46
Resistance index (RI)	0.84 ± 0.09	0.74 ± 0.09
Pulsatility index (PI)	2.52 ± 0.87	1.65 ± 0.49

With permission from Kurjak A, Kupesic S, Miric D: The assessment of benign uterine tumor vascularization by transvaginal color Doppler. Ultrasound Med. Biol. 18 (1992) 645–649.

27

The results of this study show that the vascularization of a tumor depends largely on its size and location and on the extent of secondary degenerative changes. Large, laterally situated myomas, especially those with necrotic, degenerative, and inflammatory changes, usually show increased diastolic flow and a decreased resistance index ($RI_{min} = 0.35$).

Myomas during pregnancy. Other factors in addition to size should be considered in the sonographic evaluation of myomas: their location, their spatial relationship to the placenta, their echogenicity[12], and their Doppler values[23]. A significantly increased incidence of threatened abortion, premature labor, placental abruption, and lower abdominal pain was observed in patients with myomas ($p < 0.001$)[12]. Placental abruption was most common with myomas larger than 200 cm^3, submucous myomas, and myomas located under the placenta.

Kessler et al.[23] showed that conventional ultrasound was unable to distinguish between a myoma and local thickening of the uterine wall due to a transient contraction. In such cases, repeat scanning approximately 30 min later was able to differentiate contractions, which usually resolved, from myomas. The authors examined 10 patients using B-mode and color Doppler sonography. In five patients with myomas, they observed splaying of the vessels around the tumor, whereas in five patients with contractions, there was no vascular displacement in the area of local myometrial thickening. Thus, color Doppler sonography can help eliminate the need for a prolonged ultrasound examination in questionable cases.

RU 486 and leuprolide acetate. Reinisch et al.[37] studied the effect of RU 486 and leuprolide acetate on uterine artery blood flow and uterine volume in patients with myomas. Uterine artery blood flow declined steadily in the patients receiving RU 486, showing a 40% decrease. The patients who received leuprolide acetate showed a 21% decrease. Both groups also showed a significant decrease in uterine volume at three months. It was assumed that the reduction in uterine artery blood flow provided the mechanism for the decrease in uterine size and for the decreased uterine blood loss at the time of surgery or hysteroscopy.

Endometrial Polyps

Because flow resistance is markedly decreased in and around endometrial polyps, an inexperienced examiner might mistake these lesions for a uterine malignancy. Endometrial polyps are best demonstrated in the early proliferative phase or following the injection of a "negative contrast medium" into the uterine cavity.

Histology and blood supply. Endometrial polyps may be single or multiple, occur in pedunculated and various other forms, and are often composed of hyperplastic basalis tissue[16,32]. Almost two-thirds contain no functional endometrium, and many show the histological features of cystic hyperplasia. They derive their blood supply from preexisting terminal branches of the uterine arteries. Blood flow can be demonstrated in multiple separate vessels, and the blood flow velocity can be analyzed (Table 27.3, Fig. 27.5). Diastolic flow can always be detected, and the RI is usually higher than 0.45[30,32] (Fig. 27.6). The RI tends to be lower in polyps with necrotic and inflammatory changes ($RI_{min} = 0.37$). Polypoid structures may appear in infertile women on GnRH (gonadotropin-releasing hormone) therapy but usually disappear in the next cycle if in-vitro fertilization was unsuccessful.

Table 27.3 Vascularization of benign uterine changes

Type of uterine change	Number	RI
Submucous leiomyoma	38	0.54 ± 0.06
Adenomyosis	62	0.57 ± 0.08
Endometritis	28	0.50 ± 0.06
Incomplete abortion	31	0.41 ± 0.02
Endometrial polyp	46	>0.45

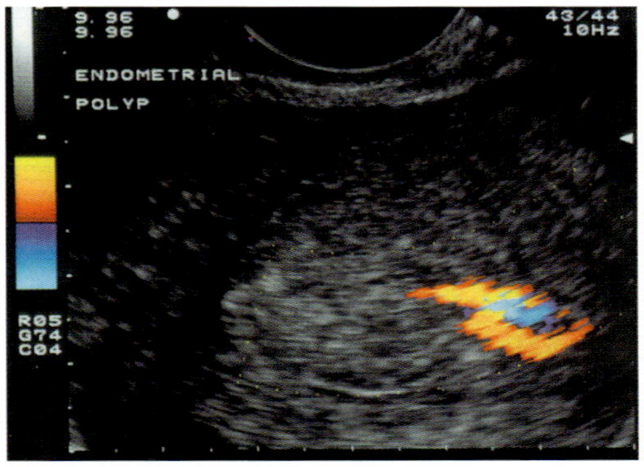

Fig. 27.5 An oblique scan in a premenopausal patient shows a demarcated zone of increased echogenicity, typical of an endometrial polyp. The normally branching vessels are well defined.

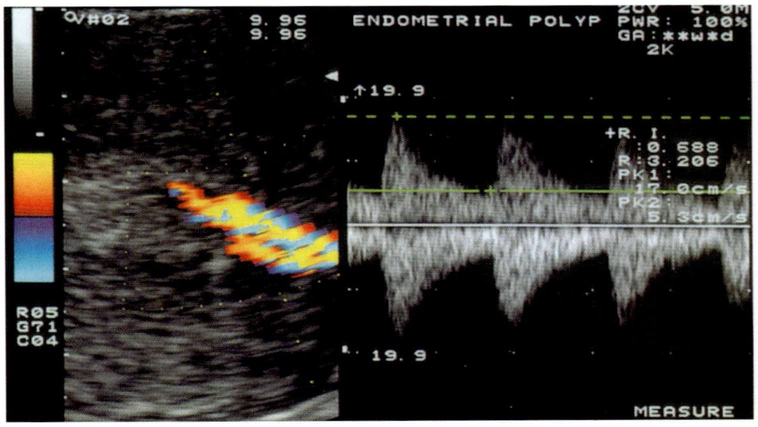

Fig. 27.6 Same patient as in Fig. 27.5. Analysis of the flow velocity waveform (right) shows a high resistance index (0.69), which is typical of an endometrial polyp.

Tamoxifen and incidence of endometrial polyps. Tamoxifen is a nonsteroidal antiestrogen widely used in the hormonal treatment of breast cancer. Studies are currently under way to determine whether it could also be used in healthy women at increased risk for breast cancer. The weak estrogen-like action of tamoxifen on the endometrium calls for vigilance in the management of these cases. Since several studies have reported endometrial carcinomas during tamoxifen use, patients receiving tamoxifen should be regularly examined. A number of pathological changes have been described during long-term tamoxifen use (20 mg/day)[20], including epithelial metaplasia, simple and atypical hyperplasia, endometrial polyps, and endometrial carcinoma[1]. Endometrial changes present sonographically as abnormal endometrial thickening and nonhomogeneous increased echogenicity with numerous small cystic lesions. At least three studies have linked tamoxifen therapy in postmenopausal breast cancer patients to a high incidence of endometrial polyps[3, 13, 35]. Achiron et al.[3] found that 44% of these patients had a conspicuous "honeycomb-like" endometrial structure on B-mode ultrasound, which was associated with a correspondingly high occurrence (40%) of endometrial polyps.

On the other hand, two large studies in postmenopausal women on tamoxifen therapy showed no correlation between an endometrial thickness >5 mm and abnormal endometrial findings[8, 44].

Tamoxifen and endometrial blood flow. So far, little research has been done on the effect of tamoxifen on endometrial blood flow. Achiron et al.[2] described blood flow changes in the endometrium and subendometrial regions. On examining asymptomatic, postmenopausal women with an endometrial thickness <5 mm who were taking tamoxifen, these authors found increased endometrial blood flow with a significant fall of RI compared with an untreated control group. Another study by the same authors[1] showed a markedly lower RI in women with endometrial thickening, especially in the presence of endometrial polyps, than in women with a normal endometrium (mean value of 0.39 vs. 0.79). After the endometrial polyps were removed, the RI values returned to normal, supporting the notion of a benign, transitory effect of long-term tamoxifen therapy on the endometrium.

Endometrial adenofibromas. Huang et al.[19] described two cases of endometrial adenofibroma that occurred during prolonged tamoxifen use. Adenofibromas are rare, benign, mixed mesodermal tumors. Their sonographic appearance differs from that of endometrial polyps, endometrial hyperplasia, and endometrial carcinoma and more closely resembles the "snowstorm" appearance of trophoblastic disease (hydatidiform mole). They can be differentiated with color and pulsed Doppler ultrasound: lacunae with turbulent blood flow are characteristic of trophoblastic diseases, whereas low-resistance blood flow is suggestive of endometrial adenofibroma.

Endometrial Hyperplasia

The endometrial echo in postmenopausal women generally appears simply as a linear echo 1–3 mm thick. The thickness of the echo is increased in association with certain benign uterine changes such as endometrial hyperplasia. An endometrial thickness >14 mm in premenopausal women and >5 mm in postmenopausal women warrants further investigation[32]. The peak incidence of adenomatous hyperplasia is between 40 and 50 years of age.

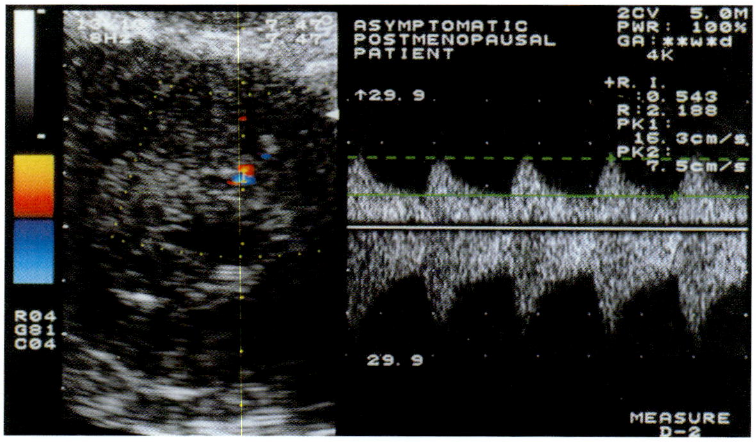

Fig. 27.**7** Thickened endometrium (10 mm) in an asymptomatic postmenopausal woman. Note the peripheral arrangement of the vessels and the moderate RI of 0.54, which is typical of endometrial hyperplasia.

Differentiation from endometrial carcinoma. Benign and malignant changes cannot be differentiated by their morphology alone, and so transvaginal B-mode scanning is not sufficient to distinguish hyperplasia from carcinoma. Both hyperplasia and carcinoma display typical vascular structures, however, and therefore they can be detected with sensitive Doppler instruments[16, 30]. Branches from normal vessels extending into the peripheral part of the endometrium form a typical pattern seen with hyperplasia (Fig. 27.**7**). On Doppler examination, blood flow signals are usually recorded from the periphery of the hyperplastic endometrium (Table 27.**3**). Significant differences in RI values have been discovered between endometrial carcinoma (mean RI = 0.42) and endometrial hyperplasia (mean RI = 0.50)[28] (Figs. 27.**8**, 27.**9**). Some clinicians have already begun to consider sonographic and Doppler parameters when selecting patients for biopsy or observation. This appears to be particularly advantageous in older patients in poor general health and in patients with cervical stenosis.

Gredmark et al.[17] found that 15% of women with postmenopausal bleeding had adenomatous hyperplasia or endometrial carcinoma, while 50% were found to have an atrophic endometrium. This suggests that ultrasound and Doppler sonography should be included in the workup of postmenopausal bleeding, both to avoid the unnecessary biopsy of atrophic endometrium and to disclose any ovarian abnormalities that may be present.

Kurjak et al.[33] reported that endometrial carcinomas could be detected even in asymptomatic women by noninvasive

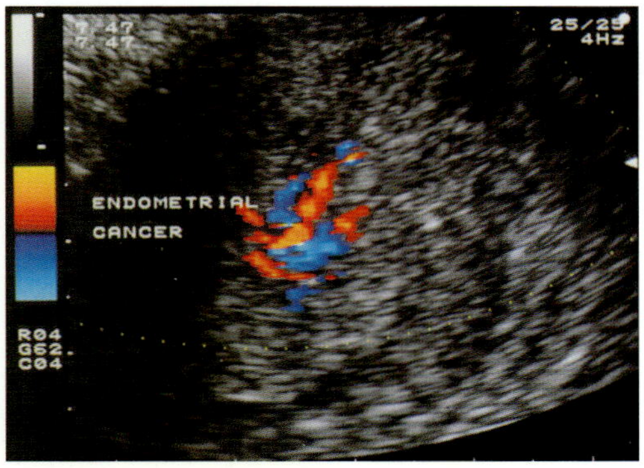

Fig. 27.**8** Peripheral neovascularization of endometrial carcinoma, demonstrated by color Doppler.

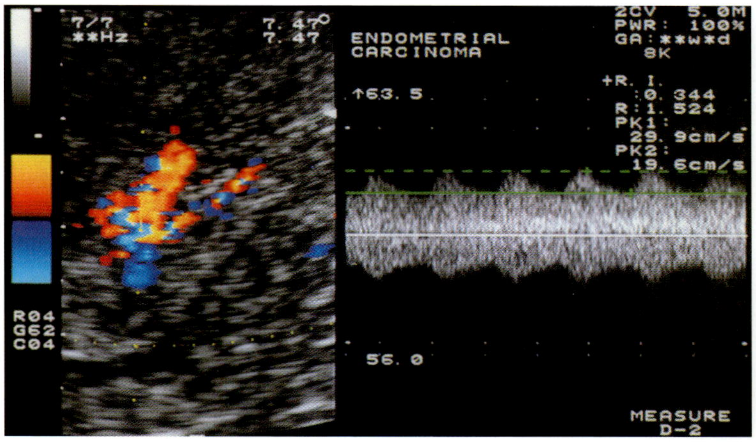

Fig. 27.**9** The Doppler waveform indicates a low RI (0.34), which is typical of endometrial carcinoma.

transvaginal color and pulsed Doppler sonography and that this method could even be used for screening.

Sheth, however, found an overlap of PI and RI values between benign and malignant causes of endometrial thickening[39]. Endometrial arterial flow was seen in 23 of 36 (64%) benign endometrial lesions. The mean RI was 0.48 ± 0.13 (range 0.27–0.84) while the mean PI was 0.72 ± 0.33 (0.31–1.77). Abnormal vascularity was found in 56% of endometrial carcinomas. The mean PI in this group was 0.71 ± 0.32 (0.42–1.17), while the mean RI values were the same as for the benign lesions: 0.48 ± 0.15 (0.34–0.69).

Chan et al.[7] found that transvaginal sonography was superior to color Doppler imaging in the detection of endometrial pathology, but neither method could distinguish benign from malignant lesions. By contrast, Bonilla-Musoles et al. [5] found a positive correlation between uterine artery resistance values and the likelihood of carcinoma in postmenopausal women. Based on the positive correlation between arterial flow resistance and time in years since menopause[29], it should be possible to estimate the cancer risk in postmenopausal women who show decreased vascular resistance.

Adenomyosis

Adenomyosis has been found in 10–50% of autopsy series and in 5.6–61.5% of surgical specimens[34, 38]. Uterine adenomyosis is characterized by an ingrowth of endometrial tissue into the myometrium. This tissue may directly adjoin the endometrium or it may deeply penetrate the myometrium and even reach the serosa.

Sonographic and clinical characteristics. Most patients with adenomyosis have either a normal-sized uterus or nonspecific uterine enlargement[4]. Ultrasound typically shows a diffusely enlarged uterus with a thickened myometrium that has a "Swiss cheese" appearance due to blood and tissue collections in the muscle[40] (Fig. 29.**10**). Severe cases generally show an unsettled echo pattern in the central endometrial layer. Occasionally the entire uterus is hypoechoic, but large cysts are rarely found. The typical symptoms of adenomyosis include dysmenorrhea, lower abdominal pain, and menometrorrhagia. Dysmenorrhea is caused by bleeding from the endometrial tissue within the myometrium. Hysterosalpingography occasionally shows the passage of contrast material into the myometrium.

Role of color Doppler. Fedele et al.[15] tested the value of endovaginal ultrasound in the diagnosis of diffuse adenomyosis. Sensitivity and specificity were 80% and 74%. With color Doppler sonography and spectral analysis, it is possible to examine the uterine blood flow in adenomyosis and compare it with uterine blood flow in other benign conditions[31, 34]. The authors found that the mean RI of blood flow in the myometrium reached 0.56, while the uterine artery RI was decreased (0.75) compared with healthy volunteers (0.87). The differences between leiomyomas and adenomyosis are explained partly by the fact that leiomyomas have a higher estrogen receptor density than the surrounding myometrium. As a result, leiomyomas are responsive to changes in the luteal hormone level, whereas adenomyosis is marked by a deficiency of estrogen and progesterone receptors[10].

Differentiation from uterine malignancy. Hirai et al.[18] evaluated 44 benign uterine masses and seven uterine malignancies to determine whether Doppler sonography was useful in differentiating adenomyosis from uterine malignancies. For this purpose they used an adenomyosis score based on the parameters of myometrial thickness, structure, contour, and also Doppler findings. The RI values tended to be lower for malignant lesions (mean RI = 0.40 ± 0.07) than for adenomyosis (mean RI = 0.57 ± 0.08). The latter value also corresponded to the mean RI for leiomyomas (0.57). The maximum blood flow velocity V_{max} was higher for malignant lesions than for adenomyosis. The V_{max} in leiomyomas was only slightly higher than in adenomyosis. Thus, statistically significant differences were found between malignant neoplasms and adenomyosis for RI and for V_{max}. A slight difference in V_{max} was found between adenomyosis and leiomyomas, but there was no difference in RI.

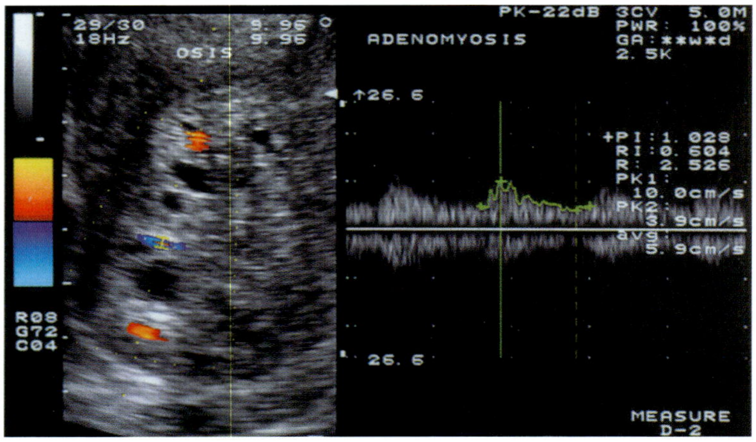

Fig. 27.**10** Color Doppler reveals diffuse blood flow in the thickened myometrium ("Swiss cheese" appearance) in a dysmenorrheic patient. A moderate to high RI (0.60) was recorded from the endometrial tissue that had invaded the myometrium. Histology confirmed adenomyosis.

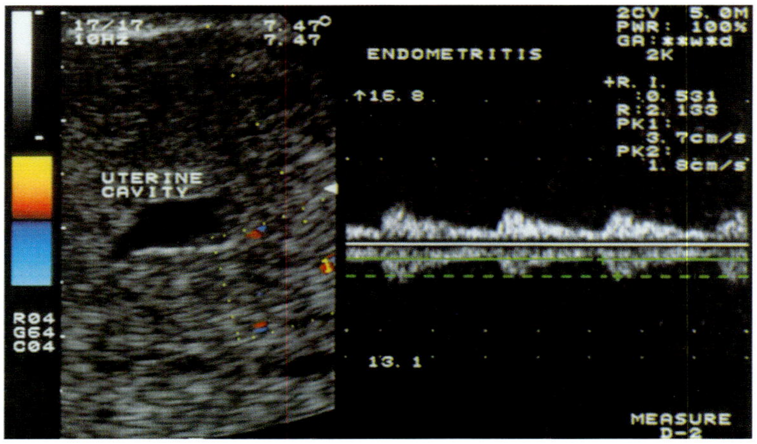

Fig. 27.**11** Transvaginal scan of the uterus shows increased thickness, echogenicity, and vascularity of the endometrium and inner third of the myometrium in a patient with endometritis. Moderate to high impedance values (RI = 0.53) are measured in the corresponding vessels.

Endometritis

Endometritis is characterized by a thickened endometrium that shows increased echogenicity and vascularization[34]. The RI is moderately increased to $>0.50 \pm 0.06$ (Table 27.**3**, Fig. 27.**11**). The clinical symptoms and vascularity tend to regress over time, while the RI values remain elevated.

Incomplete Abortion

While the value of Doppler sonography in predicting pregnancy outcomes is still uncertain, it is clear that the use of Doppler sonography can significantly reduce uncertainties in the diagnosis of early pregnancy loss.

The manifestations of incomplete abortion vary depending on gestational age and the amount of tissue that has already been expelled at the time of examination. Ultrasound generally shows fluid and echogenic tissue fragments of varying size within the uterine cavity. Color Doppler demonstrates good perfusion of gestational products that have not been expelled. The good perfusion is based on the dilatation of the spiral arteries and venous vessels caused by the active trophoblast.

A low vascular resistance ($RI = 0.41 \pm 0.01$) can be detected in the endometrium even if there is no identifiable conceptus (Table 27.**3**). However, the combined presence of hyperechoic and hypoechoic areas in the endometrium is still the most important criterion for diagnosing incomplete abortion.

Decidua

The differentiation of endometrial structures, decidual tissue, normal early pregnancy, and molar pregnancy is critically important in the differential diagnosis of amenorrhea[9]. Ectopic pregnancy should be considered in all patients who have a positive pregnancy test and an empty uterus with a decidual reaction or a small, central fluid collection. Color Doppler sonography is helpful in patients with nonspecific adnexal findings, as it can detect a decrease in vascular resistance on the affected side. Unlike the two concentric decidual circles of early intrauterine pregnancy, the decidual ring in an early ectopic pregnancy consists of only one layer.

Doppler parameters in intrauterine and ectopic pregnancies. When Jurkovic et al.[21] compared the Doppler parameters of intrauterine and ectopic pregnancies, they found no differences in the blood flow impedance of the uterine arteries, spiral arteries, and luteal vessels. However, the peak systolic velocity in the uterine arteries was identified as a parameter that very accurately reflects the increased blood flow in an ectopic pregnancy. Thus, the use of transvaginal color Doppler sonography appears to be useful as an aid to therapeutic decision-making (noninterventional conservative therapy, minimally invasive surgery, or laparotomy).

References

1 Achiron R, Grisaru D, Golan-Porat N, Lipitz S: Tamoxifen and the uterus: an old drug tested by new modalities. Ultrasound Obstet. Gynecol. 7 (1996) 374–378

2 Achiron R, Lipitz S, Frenkel Y, Mashiach S: Endometrial blood flow response to estrogen replacement therapy and tamoxifen in asymptomatic postmenopausal women: a transvaginal Doppler study. Ultrasound Obstet. Gynecol. 5 (1995) 411–414

3 Achiron R, Lipitz S, Sivan E, Goldenberg M, Horovitz A, Frenkel Y, Maschiah S: Changes mimicking endometrial neoplasia in postmenopausal, tamoxifen-treated women with breast cancer: a transvaginal Doppler study. Ultrasound Obstet. Gynecol. 6 (1995) 116–120

4 Bohlman ME, Ensor RE, Sanders RC: Sonographic findings in adenomyosis of the uterus. Amer. J. Radiol. 184 (1987) 765–766

5 Bonilla Musoles F, Marti MC, Ballester MJ, Raga F, Osborne NG: Normal uterine arterial blood flow in postmenopausal women assessed by transvaginal color Doppler ultrasonography. J. Ultrasound Med. 14 (1995) 491–494

6 Callen PW, De Martini WJ, Filly RA: The central uterine cavity echo: a useful sign on the ultrasonographic evaluation of the female pelvis. Radiology 131 (1979) 187–190

7 Chan FY, Chau MT, Pun TC et al.: Limitation of transvaginal sonography and color Doppler imaging in the differentiation of endometrial carcinoma from benign lesions. J. Ultrasound Med. 13 (1994) 623–628

8 Cohen I, Rosen D, Trepper R et al.: Ultrasonographic evaluation of the endometrium and correlation with endometrial sampling in postmenopausal patients treated with tamoxifen. J. Ultrasound Med. 5 (1993) 275–278

9 Dodson MG, Gast M: Early pregnancy. In Dodson MG (ed.): Transvaginal ultrasound. Churchill Livingstone, New York 1995, 187–217

10 Droegmueller E: Endometriosis and adenomyosis. In Manning S, Steinborn E, Salway J (eds.): Comprehensive gynecology. Mosby Year Book, St. Louis 1992, 545–576

12 Exacoustos C, Rosati P: Ultrasound diagnosis of uterine myomas and complications in pregnancy. Obstet. Gynecol. 82 (1993) 97–101

13 Exacoustos E, Zupi E, Cangi B, Chiaretti M, Arduini D, Romaninis C: Endometrial evaluation in postmenopausal breast cancer patients receiving tamoxifen: an ultrasound, color flow Doppler hysteroscopic and histological study. Ultrasound Obstet. Gynecol. 6 (1995) 435–442

14 Fedele L, Bianchi S, Dorta M et al.: Transvaginal ultrasonography versus hysteroscopy in the diagnosis of uterine submucous myomas. Obstet. Gynecol. 77 (1991) 745–748

15 Fedele I, Bianchi S, Dorta M et al.: Transvaginal ultrasonography in the diagnosis of diffuse adenomyosis. Fertil. Steril. 58 (1992) 94

16 Fleischer AC, Kepple DM, Entman SS: Transvaginal sonography of uterine disorders. In Timor-Tritsch IE, Rottem S (eds.): Transvaginal sonography. 2nd edn. Elsevier, New York 1991, 109–130

17 Gredmark T, Huint S, Havel G, Mattsson LA: Histopathological findings in women with postmenopausal bleeding. Brit. J. Obstet. Gynecol. 102 (1995) 133–136

18 Hirai M, Shibata K, Sagai H, Sekiya S, Goldberg BB: Transvaginal pulsed on color Doppler sonography for the evaluation of adenomyosis. J. Ultrasound Med. 14 (1995) 529–532

19 Huang KT, Chen CA, Cheng WF et al.: Sonographic characteristics of adenofibroma of the endometrium following tamoxifen therapy for breast cancer: two case reports. Ultrasound Obstet. Gynecol. 7 (1996) 363–366

20 Ismail SM: Pathology of the endometrium treated with tamoxifen. J. Clin. Pathol. 47 (1994) 827–833

21 Jurkovic D, Bourne T, Jauniaux E, Campbell S, Collins PW: Transvaginal color Doppler study of blood flow in ectopic pregnancy. Fertil. Steril. 57 (1992) 68–73

22 Jauniaux E, Jurkovic D, Campbell S: In vivo investigation of the anatomy and the physiology of early human placental circulations. Ultrasound Obstet. Gynecol. 1 (1991) 435–445

23 Kessler A, Mitchell X, Goldberg BB: Myoma vs. contraction in pregnancy: differentiation with color Doppler imaging. J. Clin. Ultrasound 21 (1993) 241–244

24 Keye WR, Yuen BH, Jaffe RB: New concepts in the physiology of the menstrual cycle. Clin. Endocrinol. Metab. 2 (1973) 251–258

25 Kupesic S: The first three weeks assessed by transvaginal color Doppler. J. Parinat. Med. 24 (1996) 301–317

26 Kupesic S, Kurjak A: Uterine and ovarian perfusion during the periovulatory period assessed by transvaginal color Doppler. Fertil. Steril. 60 (1993) 439–443

27 Kupesic S, Kurjak A, Vujisic S, Petrovic Z: Luteal phase defect: comparison between Doppler velocimetry, histological and hormonal markers. Ultrasound Obstet. Gynecol. 9 (1997) 1–8

28 Kupesic-Urek S, Shalan H, Kurjak A: Early detection of endometrial cancer by transvaginal color Doppler. EUROBS, 49 (1993) 46–49

29 Kurjak A, Kupesic S: Ovarian senescence and ist significance on uterine and ovarian perfusion. Fertil. Steril. 3 (1995) 532–537

30 Kurjak A, Kupesic S: Transvaginal color Doppler and pelvic tumor vascularity: lessons learned and future challenges. Ultrasound Obstet. Gynecol. 6 (1995) 1–15

31 Kurjak A, Kupesic-Urek S, Miric D: The assessment of benign uterine tumor vascularization by transvaginal color Doppler. Ultrasound Med. Biol. 18 (1992) 645–649

32 Kurjak A, Kupesic S, Zalud I, Predanic M: Transvaginal color Doppler. In Dodson MG (ed.): Transvaginal ultrasound. Churchill Livingstone, New York 1995, 325–339

33 Kurjak A, Shalan H. Kupesic S et al.: An attempt to screen asymptomatic women for ovarian and endometrial cancer with transvaginal color and pulsed Doppler sonography. J. Ultrasound Med. 13 (1994) 295–301

34 Kurjak A, Zalud I: The characterization of uterine tumors by transvaginal color Doppler. Ultrasound Obstet. Gynecol. 1 (1991) 50–52

35 Lahti E, Blanco G, Kaupilla A, Apaja-Sarkkinen M, Taskinen PJ, Laatikainen T: Endometrial changes in postmenopausal breast cancer patients receiving tamoxifen. Obstet. Gynecol. 81 (1993) 660–664

36 Lawn AM: The ultrastructure of the endometrium. J. Reprod. Fertil. 37 (1974) 239–242

37 Reinsch RC, Murphy AA, Morales, Yen SS: The effects of RU 486 and leuprolide acetate on uterine artery blood flow in the fibroid uterus: a prospective randomized study. Amer. J. Obstet. Gynecol. 170 (1994) 1623–1627

38 Seidler D, Laing FC, Jeffrey RB Jr, Wing VW: Uterine adenomyosis – a difficult sonographic diagnosis. J. Ultrasound Med. 6 (1987) 345–348

39 Sheth S, Hamper VM, McCollum ME, Caskey CI, Rosenshein NB, Khurman RJ: Endometrial blood flow analysis in postmenopausal women: can it help differentiate benign from malignant causes of endometrial thickening. Radiology 195 (1995) 661–665

40 Siedler D, Lang FC, Jeffrey RB, Wing VW: Uterine adenomyosis—a difficult sonographic diagnosis. J. Ultrasound Med. 6 (1987) 345–349

41 Soules MR, McCarty KS Jr: Leiomyoma: steroid receptor content. Variation within normal menstrual cycles. Amer. J. Obstet. Gynecol. 143 (1982) 6–11

42 Stabile I, Grudzinkas J, Campbell S: Doppler ultrasonographic evaluation of abnormal pregnancies in the first trimester. J. Clin. Ultrasound 18 (1990) 497–501

43 Thureck RW: Uterine leiomyomata. In: Garcia CR, Mastroianni L, Amelar RD, Dubin L (eds.): Current therapy of infertility. BC Decker Inc. Toronto 1988, 80–82

44 Uziely B, Lewin A, Brufman G. Dorembus D, Mor-Yosef S: The effect of tamoxifen on the endometrium. Breast Cancer Res. Treat. 26 (1993) 101–105

45 Vollenhoven BJ, Lawrence AS, Healy D: Uterine fibroids: a clinical review. Brit. J. Obstet. Gynecol. 97 (1990) 285–298

46 Zaidi J, Campbell S, Pittrof R, Tan SL: Endometrial thickness, morphology, vascular penetration and velocimetry in predicting implantation in an in vitro fertilization program. Ultrasound Obstet. Gynecol. 6(3) (1995) 191–198

28 Transvaginal Color Doppler and Pulsed Doppler Diagnosis of Benign Changes in the Uterine Myometrium

A. Kurjak and S. Kupesic

Examination Technique, Anatomy, and Physiology

Transvaginal color Doppler sonography can provide detailed images of the entire uterus, myometrial and endometrial changes, and the uterine vessels[9].

Scan planes. The uterus can be imaged in three cardinal planes: long axis, semichorionic or semiaxial, and short axis[8,9]. The long axis view is obtained by inserting the ultrasound probe into the vagina and imaging the uterus in its greatest longitudinal dimension. A semiaxial scan is obtained by rotating the probe 90° to the long axis, delineating the full width of the uterus. For the short axis view, the probe is withdrawn to the middle third of the vagina and directed toward the anterior fornix wall.

With an anteflexed uterus, the probe handle is angled posteriorly to direct the beam forward. This technique is reversed for a retroflexed uterus. In a semiaxial scan of the anteflexed uterus, the cervix is the first structure to appear on the monitor. In a retroflexed uterus, the first structure to be visualized is the uterine fundus.

Uterine size. The normal size of the uterus varies as a function of parity and menopausal status. In nulliparae, the uterus measures approximately 6 cm in length and 3–4 cm in its anteroposterior and transverse diameters[8]. The uterine long axis may measure up to 8 cm in parous women, decreasing to approximately 4–6 cm after menopause.

Blood supply. The uterus derives its blood supply from a complex vascular network arising from the uterine arteries. The main branches penetrate to about the outer third of the myometrium and form the network of arcuate arteries that encircle the uterus[4]. Smaller branches called the radial arteries run toward the uterine lumen[18]. On reaching the inner third of the myometrium and crossing the myometrial–endometrial junction, they become the spiral arteries.

Contractions. The myometrial muscle fibers undergo contractions during a normal menstrual cycle. Focal contractions occur in the middle or outer muscle layer during pregnancy[8], while subendometrial contractions are common during menstruation and ovulation[1,30]. These contractions are directed toward the cervix during menstrual bleeding and away from the cervix during ovulation. Presumably, contractions during the periovulatory period play an important role in sperm transport, and a deficiency of these contractions can lead to fertility problems.

Leiomyomas (Fibroids)

Forms and symptoms. Uterine leiomyomas, also known as fibroids or myomas, are by far the most common pelvic tumors in women of childbearing age[7,20,39]. They arise from the smooth muscle and soft tissues of the uterine corpus and fundus. Only 3% are of cervical origin[26]. Leiomyomas are more often multiple than solitary and vary greatly in size. They may be confined to the uterine musculature (intramural) or may extend toward the uterine cavity, becoming submucosal. Centripetal growth leads to subserous or pedunculated forms[7]. The intramural type is the most common, the subserous type the least common[20]. Clinical symptoms such as metrorrhagia, pelvic pain, and infertility usually occur in patients with submucous leiomyomas, while subserous leiomyomas tend to be asymptomatic. The presence and severity of symptoms correlate closely with the number, size, and location of the tumors.

Sonographic features. The ultrasound diagnosis of leiomyomas is based on uterine contour changes, uterine enlargement, and changes in echo texture. The sonographic structure ranges from hypoechoic to hyperechoic, depending on the proportions of muscle and connective tissue. If secondary changes have developed (necrosis, intratumoral hemorrhage, degeneration, calcification), they present a broad spectrum of sonographic appearances (Fig. 28.**1**).

Leiomyomas that enlarge and outstrip their blood supply develop central ischemia, usually followed by varying degrees of degenerative change. There are reports[14] that almost 25% of fibroids are associated with calcifications ranging from small calcific foci to large deposits or complete calcification. Because of their variable appearance, leiomyomas may be misinterpreted as solid ovarian tumors, endometrial polyps, blood, or mucus[8]. Fedele et al.[5] investigated the accuracy of transvaginal ultrasound in the detection of small submucous myomas in women who had undergone both transvaginal sonography and hysteroscopy before a scheduled hysterectomy. The sensitivity and specificity of transvaginal sonography and hysteroscopy were comparable in this study.

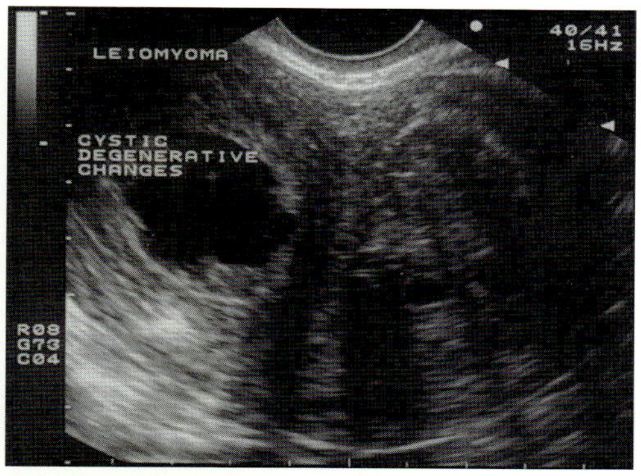

Fig. 28.**1** Transvaginal ultrasound scan of a leiomyoma with central ischemia and necrosis.

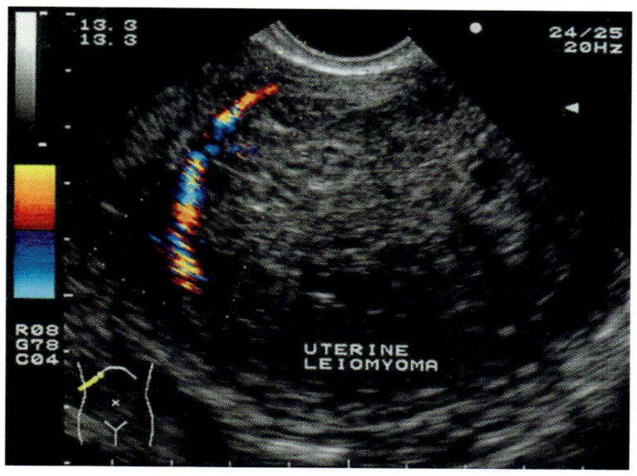

Fig. 28.**2** Transvaginal scan of a patient with a posterior, isoechoic leiomyoma. Color Doppler shows a large-caliber vessel at the tumor periphery.

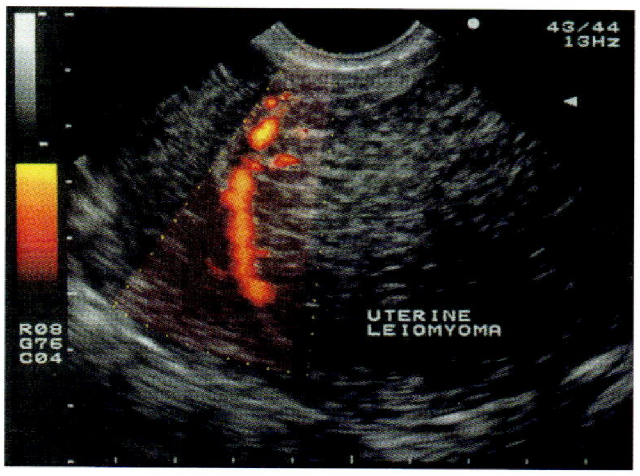

Fig. 28.**3** Same patient as in Fig. 28.**2**. The vessel at the border of the leiomyoma is defined more clearly by power Doppler.

Leiomyomas and infertility. The clinical and/or surgical management of leiomyomas are directed by the symptoms and size of the palpable mass and the desire of the patient to preserve her fertility[39]. Because leiomyomas tend to occur later in a woman's reproductive life, the trend toward delaying parenthood could increase the association of fibroids with infertility. Nevertheless, the relationship between these benign pelvic tumors and infertility/pregnancy failure is not yet fully understood. A change in the shape of the uterine cavity or an obstruction of the cervix and fallopian tubes could interfere with sperm transport. Leiomyomas can also lead to venous dilatation, with consequent swelling of the myometrium and endometrium. This would impede the implantation of a fertilized ovum and might also restrict the blood supply to the developing embryo. Myoma-related mehorrhagia could result from congestion and dilatation of the adjacent endometrial venous plexus, with an associated adverse effect on the myometrial, radial, and arcuate veins.

Vascularization of Leiomyomas

The blood flow impedance in small vascular branches can be reproducibly measured with transvaginal color Doppler ultrasound[16,27]. Small tumor-feeding vessels are very easy to detect with this method. Transvaginal color Doppler has been used to investigate the blood supply to leiomyomas[23,37] and the physiological and pathophysiological characteristics of blood flow in the uterine artery[13,21,24,28,33,38]. Transvaginal color Doppler has proved particularly helpful for the in-vivo differentiation of benign and malignant uterine tumors[23,27]. It has been shown that benign uterine masses derive their blood supply from preexisting normal vessels (Figs. 28. **2**, 28.**3**). Analysis of the color Doppler waveforms showed a slight decrease in blood flow impedance or similar perfusion patterns as in the terminal branches of the uterine arteries in the normal myometrium.

Premenopausal patients. Examinations in premenopausal women consistently demonstrated diastolic blood flow, which was usually increased relative to the blood flow in the uterine artery. Color Doppler also demonstrated flow at the periphery and/or center of the leiomyoma. The mean RI of blood flow in leiomyomas was 0.54, and the mean PI was 0.89 (Fig. 28.**4**). Low resistance indices were found in cases with necrosis and secondary degenerative or inflammatory changes within the myoma (Fig. 28.**5**).

In our own studies of patients with leiomyomas and healthy volunteers, the same technique was used to investigate blood flow in the uterine arteries. In patients with leiomyomas, we found increased blood flow velocities and decreased RI and PI values in both uterine arteries (Fig. 28.**6**). The different vascularization patterns observed in this study could be useful in predicting the growth rates of these benign uterine tumors.

Postmenopausal patients. Several interesting observations were made in women with shrunken fibroids after menopause[33]. The blood flow at the periphery of the tumors showed a higher RI than in premenopausal women, and some cases showed a complete absence of end-diastolic blood flow. Transvaginal color Doppler can also be used to evaluate the effect of hormone replacement therapy on the thickness of the endometrium and myometrium. Moreover, Doppler examinations can supply additional information on the quality of blood flow in the uterine, myometrial, and endometrial vessels. It can be used to assess the effect of treatment measures on uterine morphology and perfusion and even answer the question of whether hormone replacement therapy increases the risk of a myometrial neoplasm.

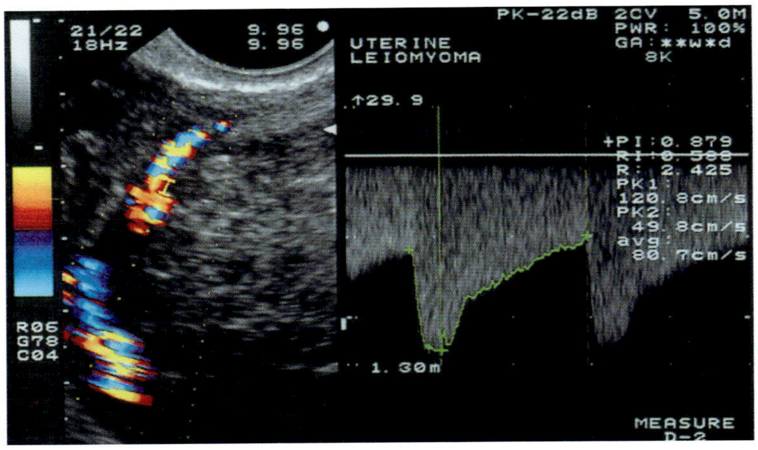

Fig. 28.**4** Same patient as in Figs. 28.**2** and 28.**3**. The Doppler spectrum indicates a moderate RI (0.59). Histopathology confirmed the benign nature of the tumor.

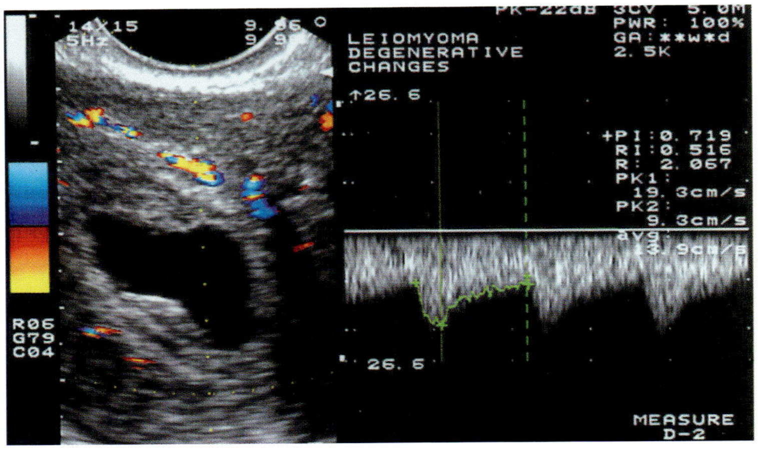

Fig. 28.**5** Pulsed Doppler signals from a small vessel at the border of the leiomyoma indicate a small variation between systolic and end-diastolic blood flow (RI = 0.52).

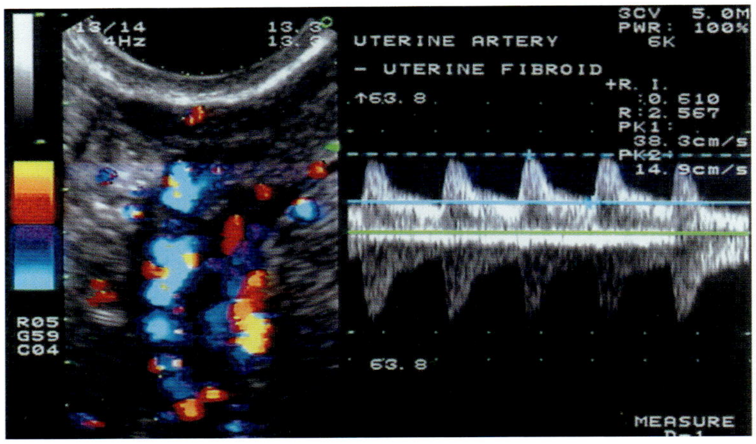

Fig. 28.**6** A giant intramural leiomyoma has caused a significant fall of vascular resistance in the uterine arteries (RI = 0.61).

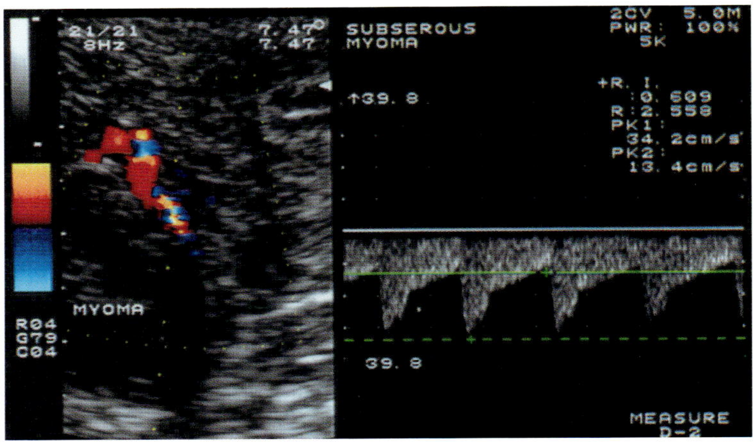

Fig. 28.**7** Transvaginal scan of a subserous leiomyoma. The color Doppler image shows normal branching of the vessels arising at the periphery of the uterus. The spectral trace indicates moderate to high impedance in the tumor vessels (RI = 0.61).

Differentiation from ovarian tumors. Color Doppler sonography is useful in differentiating intraligamentous fibroids from solid ovarian tumors (Fig. 28.**7**). Even if the uterine tumor is located outside the uterine corpus, Doppler can still reveal waveforms that are typical of the uterine vascular network. The flow pattern in these cases generally resembles the pattern of normal myometrial blood flow arising from the terminal branches of the uterine arteries. By contrast, the vessels that supply adnexal tumors are derived from the ovarian vessels.

Leiomyomas during pregnancy. Since leiomyomas are responsive to female sex hormones, they frequently enlarge during early pregnancy[29,36]. Transvaginal color Doppler sonography is the method of choice for observing hemodynamic changes in the maternal and fetal vascular system of the placenta, both in normal early pregnancies and in pregnancies with uterine fibroids[19,22,25]. No significant difference in blood flow impedance ($p > 0.05$) was observed between the two groups. Leiomyomas affect neither the hemodynamics of the utero-placental circulation nor the maternal or fetal components of the placental circulation. The marked rise of blood flow velocity measured in the radial arteries from the 10th to 13th weeks of gestation ($p < 0.001$) is probably due to the increased metabolism of the growing fibroid cells.

Management of Uterine Leiomyomas and the Importance of Color Doppler Sonography

Current treatment options for uterine fibroids include nonsurgical therapies (especially the use of gonadotropin-releasing hormones, GnRH agonists) as well as various surgical options such as abdominal, laparoscopic, or hysteroscopic myomectomy and abdominal or vaginal hysterectomy (which may be laparoscopy-assisted).

Medical Treatment with GnRH Agonists

Mechanism of action. Hormonal therapy is best for reducing the size of submucous leiomyomas to relieve postoperative symptoms, facilitate hysteroscopic resection, or temporize until menopause in perimenopausal women. Synthetic GnRH agonists inhibit estrogen production, leading to a decrease in uterine size. Filicori et al.[6] were the first to report on a patient with uterine fibroids and anemia who was treated with GnRH agonists. Since then, many studies have been published confirming the efficacy of GnRH agonists in reducing the size of the uterus and leiomyomas and suppressing menstrual bleeding.

Reduction in uterine volume. Friedman[10] documented an average 40–50% reduction of uterine volume relative to pretreatment size in women who were treated with GnRH agonists for 3–6 months. Generally the decrease in uterine volume was greatest during the first 12 weeks of therapy. More than a 50% reduction in uterine volume was achieved in 20% of the women, and a moderate reduction of 26–50% was achieved in half of the women. A slight reduction of up to one-fourth the pretreatment size was obtained in 25% of the women, and in 5% the uterus enlarged by up to 25% of the pretreatment size. GnRH agonists did not continue to afford protection after they were discontinued. The myomas generally returned to their original size within four months after the end of treatment.

Measurement of RI. The first Doppler examination of the uterine arteries in fibroid patients treated with GnRH agonists was performed by Matta et al. in 1988[33]. Their study, which involved only eight patients, showed a significant rise of resistance in the uterine vessels after four months' treatment with GnRH agonists. During that time the RIs were measured in the uterine arteries and in the larger fibroid vessels. After treatment, the RI rose significantly from an average of 0.52 in the uterine arteries and 0.48 in the fibroid arteries to respective values of 0.92 and 0.91. The authors concluded that a decrease in uterine vasculature was responsible for the reduction in uterine volume after treatment with GnRH agonists.

Effect of RU 486 and leuprolide acetate. Reinsch et al.[35] recently studied the effect of RU 486 (mifepristone, a synthetic steroid with both antiprogesterone and antiglucocorticoid properties) compared with leuprolide acetate on uterine blood flow in 14 women. Both groups showed a rise of RI values in the uterine arteries and a reduction in uterine volume following three months' treatment. The RI rose in 2–51% of the patients who had received RU 486 and in 8–39% of the patients treated with leuprolide acetate. The authors interpreted the rise of RI in response to GnRH agonist therapy as a hypoestrogenic effect, while the decreased blood flow in response to RU 486 was apparently caused by other, previously unknown mechanisms.

Our studies. In one study we observed 25 women with symptomatic uterine leiomyomas, 88% of whom complained of metrorrhagia, 52% of lower abdominal pain or pressure, and 40% of dysmenorrhea. We performed transabdominal and/or transvaginal sonography in these patients to examine the uterus and adnexa. The volume of the myomas was calculated using the formula for an ellipsoid (0.521 × length × width × depth). Pulsed and color Doppler ultrasound were used to identify and quantitate blood flow in the main arteries supplying the myomas. The initial tumor volumes were between 25 and 390 mm^3 (mean value 85 mm^3). After this examination, 1200 µg/day of buserelin (Suprefact, Höchst) was administered intranasally. Transvaginal color Doppler examinations were performed for follow-up during and at the end of GnRH analogue therapy. The mean RI in the principal myoma-feeding vessels was 0.54 ± 0.08. The resistance in these vessels rose markedly during the first three months of treatment to a mean RI of 0.78, while the myoma volume decreased to 76% of the initial value. After six months' treatment, the mean myoma volume had been reduced to 56% of the pretreatment volume, with color Doppler showing a rise of impedance to RI = 0.88 in the uterine arteries and RI = 0.82 in the myoma vessels in 80% of the patients treated.

We also observed that three myomas disappeared completely and had not recurred by six months after the end of treatment. The remaining myomas increased their volume by 24% of the initial posttreatment size. Nine of the 25 patients were referred for surgical treatment with improved hemoglobin levels. The reduced uterine and myoma volumes, which now enabled operative treatment, resulted in less intraoperative blood loss, facilitated enucleation of the myomas, and reduced the number of postoperative complications [31,32]. Four perimenopausal patients with severe metrorrhagia experienced menopause during the therapy.

Surgical Treatment

Methods. The surgical treatment options for uterine leiomyomas include abdominal, laparoscopic, and hysteroscopic myomectomy as well as abdominal and vaginal hysterectomy. The main advantage of myoma enucleation is that it preserves the uterus and thus the patient's fertility. No benefit has been demonstrated for laparoscopic removal, however. Several published cases document possible complications including uterine suture dehiscence[15]. Hysterectomy is the definitive treatment for symptomatic uterine fibroids in women who have concluded their childbearing. The mortality risk of hysterectomy for a benign indication is approximately 1 : 1000[2,40].

Laser therapy. A small study[34] described a very promising alternative to the above surgical treatments for uterine fibroids: laparoscopic myolysis with the Nd:YAG laser. The laser applicator was introduced laparoscopically into the myoma. During treatment, the applicator was repeatedly inserted into the leiomyoma to reduce the stroma, denature tumor proteins, destroy tumor vessels, and reduce the tumor volume. The insertion sites were spaced 5–7 mm apart, and the blood vessels at a 3–5 mm radius around each insertion site were coagulated. Multiple punctures were effective in obliterating the myomas. Since the uterus remains fully intact, there is no impairment of hormonal or sexual function. At six-month follow-up, the echo structure of the coagulated myomas so closely resembled normal myometrium that only an experienced examiner could tell the difference. A long-term study of seven patients showed no evidence of tumor regrowth. Histological examination showed complete devascularization of the myoma tissue with subsequent necrosis[34].

Electrocautery. Currently we are completing a study to evaluate the size, vascularity, and degenerative changes in myomas whose feeding vessels were electrocoagulated by the laparoscopic route. Color Doppler ultrasound was used to demonstrate the myoma vessels during the electrocautery. Accurate spatial placement of the bipolar needles in the vascularized plane of the myoma was compared with "blind" electrocautery. We believe that the treatment can be made more successful and the procedure shortened with the aid of color Doppler.

We therefore investigated key variables such as the amount of bleeding and the reduction in myoma size following laparoscopic electrocautery in one group of patients in which color Doppler had been used and another group in which the entire myoma had been "blindly" coagulated. We found that the size and vascular supply of the myomas showed a much greater regression in the color Doppler group than in the group without color Doppler guidance. These initial results underscore the value of color Doppler sonography in the management of leiomyomas, not only to better document lesion size and structure but also as an aid to preoperative planning and intraoperative guidance for resection or electrocautery.

Vascular Diseases in the Lesser Pelvis (Varicose Veins or Arteriovenous Malformations)

These lesions are easily demonstrated by duplex sonography or color Doppler imaging[17] (Figs. 28.**8**, 28.**9**). Doppler sonography is the modality of choice for distinguishing these vascular lesions from cystic structures without internal echoes such as hydrosalpinx and from cysts diagnosed by B-mode imaging (Figs. 28.**10**– 28.**12**).

Uterine adenomyosis is discussed in Chapter 27 (Doppler Examination of the Normal Endometrium and Benign Endometrial Changes).

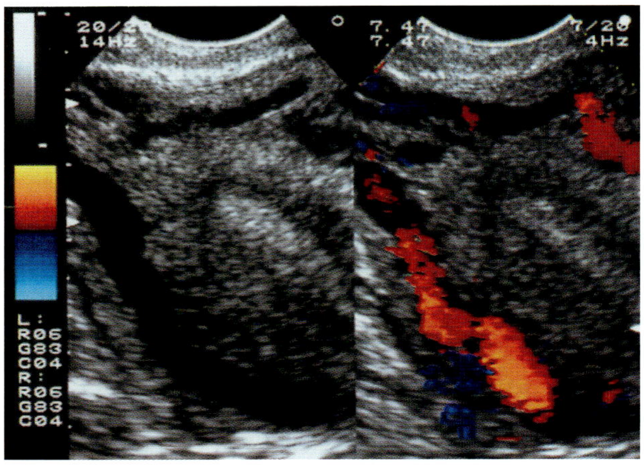

Fig. 28.**8** Color Doppler image (right) reveals a dilated uterine vein in the outer third of the myometrium.

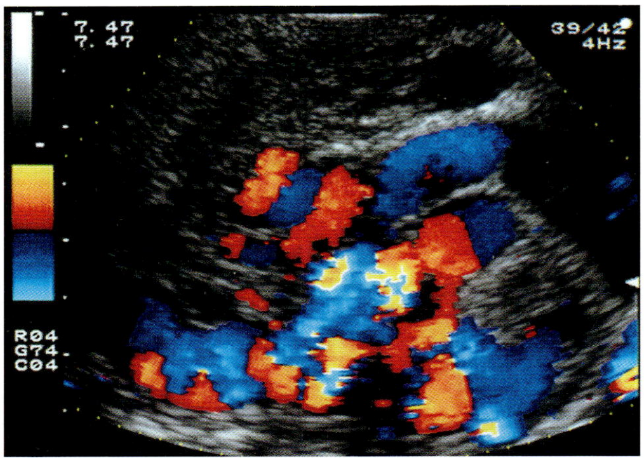

Fig. 28.**9** Color Doppler is the method of choice for diagnosing pelvic congestion syndrome.

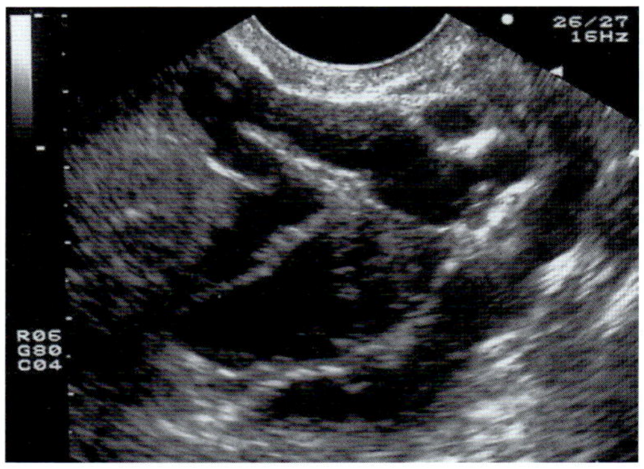

Fig. 28.**10** Transvaginal sonogram shows irregular structures near the uterus of a woman with chronic lower abdominal pain. A definitive diagnosis could not be made from the B-mode image.

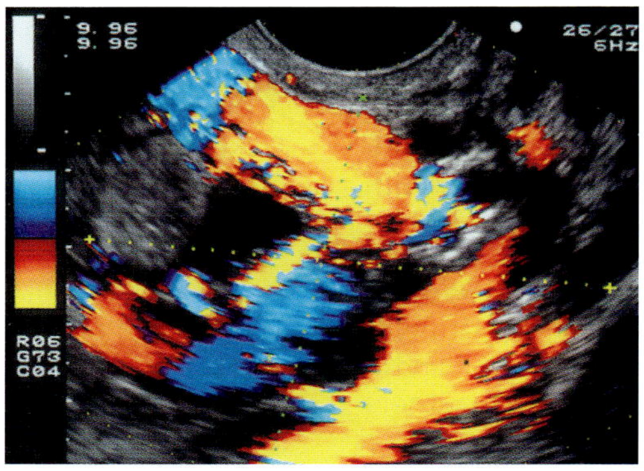

Fig. 28.**11** Same patient as in Fig. 28.**10**. The color Doppler image shows dilated pelvic veins consistent with pelvic congestion syndrome.

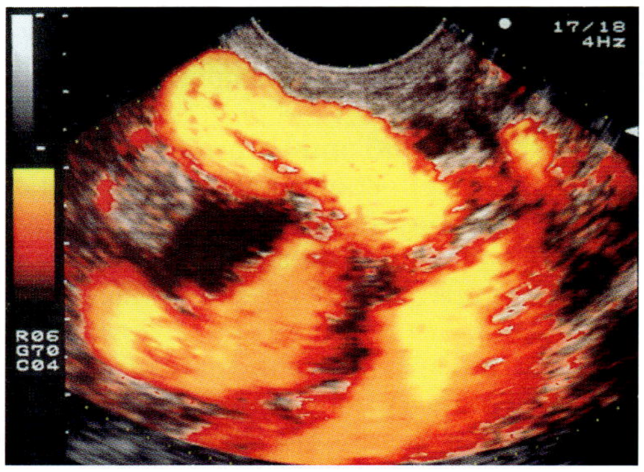

Fig. 28.**12** Same patient as in Figs. 28.**10** and 28.**11**. Appearance of the dilated pelvic veins in the power Doppler mode.

References

1 De Vries K, Lyons EA, Ballard G, Levi CS, Lindsay DJ: Contractions of the inner third of the myometrium. Am. J. Obstet. Gynecol. 162 (1990) 679–682

2 Dicker RC, Greenspan JR, Strauss LT, Cowart MR, Scally MJ, Peterson HB: Complications of abdominal and vaginal hysterectomy among women of reproductive age in the United States. Am. J. Obstet. Gynecol. 144 (1982) 841–848

3 Donnez J, Nisolle M: Hysteroscopic surgery. Curr. Opin. Obstet. Gynecol. 4 (1992) 439–455

4 Du Bose TJ, Hill LW, Heningan HW Jr et al.: Sonography of arcuate uterine blood vessels. J. Ultrasound Med. 4 (1985) 229–233

5 Fedele L, Bianchi S, Dorta M et al.: Transvaginal ultrasonography versus hysteroscopy in the diagnosis of uterine submucous myomas. Obstet. Gynecol. 77 (1991) 745–748

6 Filicori M, Hall DA, Loughlin JS, Rivier J, Vale W, Crowley WV Jr: A conservative approach to the management of uterine leiomyoma: pituitary desensitization by a luteinizing hormone-releasing hormone analogue. Am. J. Obstet. Gynecol. 147 (1983) 726–727

7 Fleischer AC, Entman SS, Porrath SA, James AE: Sonographic evaluation of uterine malformations and disorders. In Sanders RC (ed.): The principles and practice of ultrasonography in obstetrics and gynecology. Appleton Century Crofts, Norwalk 1985, 531

8 Fleischer AC, Kepple DM: Benign conditions of the uterus, cervix and endometrium. In Nyberg DA, Hill LM, Bohm-Velez M, Mendelson EB (eds.): Transvaginal ultrasound. Mosby Year Book, St. Louis 1992, 21–43

9 Fleischer AC, Kepple DM, Entman SS: Transvaginal sonography of uterine disorders. In Timor-Tritsch IE, Rottem S (eds.). Transvaginal sonography, 2nd ed. Elsevier Publishing, New York 1991, 119–130

10 Fiedman AJ: Treatment of uterine myomas with GnRH agonists. Semin. Reprod. Endocrinol. 11 (1993) 154–161

11 Friedman AJ, Daly M, Juneau-Norcross M, Rein MS: Predictors of uterine volume reduction in women with myomas treated with a gonadotropin releasing-hormone agonist. Fertil. Steril. 58 (1992) 413–415

12 Garry R, Erian J, Grochmal SA: A multi-centre collaborative study into the treatment of menorrhagia by Nd-YAG laser ablation of the endometrium Brit. J. Obstet. Gynaecol. 98 (1991) 357–362

13 Goswamy RK, Streptoe PC: Doppler ultrasound studies of the uterine artery in spontaneous ovarian cycles. Hum. Reprod. 3 (1988) 721–726

14 Gross BH, Silver TM, Jaffe NH: Sonographic features of uterine leiomyomas: analysis of 41 proven cases. J. Ultrasound Med. 2 (1983) 401

15 Harris WJ: Uterine dehiscence following laparoscopic myomectomy. Obstet. Gynecol. 80 (1992) 545–546

16 Hata T, Hata K, Senoch D et al.: Transvaginal Doppler flow mapping. Gynecol. Obstet. Invest. 27 (1989) 217–218

17 Jain KA, Jeffrey BR, Sommer FG: Gynecologic vascular abnormalities: diagnosis with Doppler ultrasound. Radiology 178 (1991) 549–551

18 Jurkoric D, Jauniaux E, Kurjak A et al.: Transvaginal color Doppler assessment of the uteroplacental circulation in early pregnancy. Obstet. Gynecol. 77 (1991) 365–369

19 Kupesic-Urek S, Kurjak A: Transvaginal color Doppler in the assessment of uterine artery blood flow in early pregnancy complicated by uterine mass. J. Matern. Fetal. Invest. 1 (1991) 163

20 Kurjak A: Sonography of the uterus. In Kurjak A (ed.): Atlas of ultrasonography in obstetrics and gynecology. Mladost, Zagreb 1986, pp. 245–247

21 Kurjak A, Jurkovic D, Alfirevic Z, Zalud I: Transvaginal color Doppler imaging. J. Clin. Ultrasound 18 (1990) 227

22 Kurjak A, Kupesic-Urek S, Miric D: Transvaginal color Doppler in the assessment of uterine blood flow changes during the pregnancy in patients with fibroids. J. Perinat. Med. 19 (Suppl. 2) (1991) 81

23 Kurjak A, Kupesic-Urek S, Miric D: The assessment of benign uterine tumor vascularization by transvaginal color Doppler. Ultrasound Med. Biol. 18 (1992) 645–649

24 Kurjak A, Kupesic-Urek S, Schulman H, Zalud I: Transvaginal color Doppler in the assessment of ovarian and uterine perfusion in infertile women. Fertil. Steril. 56 (1991) 870–873

25 Kurjak A, Predanic M, Kupesic-Urek S, Zudenigo D, Matijevic R, Salihagic A: Transvaginal color Doppler in the study of early normal pregnancies and pregnancies associated with uterine fibroids. J. Matern. Fetal. Invest. 2 (1992) 81–85

26 Kurjak A, Zalud I: Uterine masses. In Kurjak A (ed.): Transvaginal color Doppler. Parthenon Publishing, Lancet 1991, 123–135

27 Kurjak A, Zalud I: The characterization of uterine tumors by transvaginal color Doppler. Ultrasound Obstet. Gynecol.1 (1991) 50–52

28 Kurjak A, Zalud I, Jurkovic D, Alfirevic Z, Miljan M: Transvaginal color Doppler for the assessment of pelvic circulation. Acta Obstet. Gynecol. Scand. 68 (1989) 131

29 Lev-Toaff AS, Coleman BG, Arger PH et al.: Leiomyomas in pregnancy: sonographic study. Radiology 164 (1987) 375–380

30 Lyons EA, Taylor PI, Zheng XH et al.: Characterization of subendometrial myometrial contractions throughout the menstrual cycle in normal fertile women. Fertil. Steril. 55 (1991) 771–774

31 Maheux R, Guilloteau C, Lemay A, Bastide A, Fazecas AR: Regression of leiomyomata uteri following hypoestrogenism induced by repetitive luteinizing hormone agonist treatment: preliminary report. Fertil. Steril. 42 (1984) 644–646

32 Maheux R, Lemay A, Merat P: Use of intranasal luteinizing hormone-releasing hormone agonist in uterine leiomyomas. Fertil. Steril. 47 (1987) 229–233

33 Matta WHM, Stabile I, Shaw RW, Campbell S: Doppler assessment of uterine blood flow changes in patients with fibroids receiving the gonadotropin-releasing hormone agonist Buserelin. Fertil. Steril. 49 (1988) 1083–1085

34 Nisolle M, Smets M. Malvaux V, Anaf V, Donnez J: Laparoscopic myolysis with the Nd: YAG laser. J. Gynecol. Surg. 9 (1993) 95–99

35 Reinsch RC, Murphy AA, Morales AJ, Yen SC: The effects of RU 486 and leuprolide acetate on uterine artery blood flow in the fibroid uterus: a prospective, randomized study. Amer. J. Obstet. Gynecol. 170 (1994) 1623–1628

36 Rosati P, Bellatti U. Exacoustos C et al.: Uterine myoma in pregnancy: ultrasound study. Int. J. Gynaecol. Obstet. 28 (1989) 109–117

37 Shimamoto K, Sakuma S, Ishigaki T, Makino N: Intratumoral blood flow: evaluation with color Doppler echography. Radiology 165 (1987) 683–685

38 Sholtes MCW, Wladimiroff JW, van Rijen HJM, Hop WCJ: Uterine and ovarian flow velocity waveforms in the normal menstrual cycle: a transvaginal color Doppler study. Fertil. Steril. 53 (1989) 981–985

39 Tureck RW: Uterine leiomyomata. In Garcia CR, Mastroianni L, Amelar RD, Dubin L (eds.): Current therapy of infertility. BD Decker Inc., Toronto 1988, 80–83

40 Wingo PA, Huezo CM, Rubin GL, Ory HW, Peterson HB: The mortality risk associated with hysterectomy. Amer. J. Obstet. Gynecol. 152 (1985) 803–808

29 Use of Color Doppler in the Evaluation of Suspicious Endometrial Findings

M. Holländer, C. Villena-Heinsen, A. K. Ertan, and W. Schmidt

Incidence of Endometrial Carcinoma

Endometrial carcinoma has become the most common tumor of the female genital tract. While the incidence of cervical cancer is definitely on the decline, owing largely to effective Pap screening programs, the incidence of endometrial cancer is slightly on the rise. The rate of new cases in Saarland, Germany, during the 1970 s and 1980 s was between 20 and 30 per 100,000 population, or approximately 15 cases per 100,000 based on the World Standard Population[24]. The risk is highest in obese postmenopausal women, with a peak age incidence of 60–65 years (normally distributed)[24]. Additional risk factors are diabetes mellitus, arterial hypertension, infertility, unopposed estrogen use, tamoxifen therapy, and genetic predisposition[11, 16, 19, 40].

Diagnostic Investigation of Suspicious Endometrial Findings

Transvaginal sonography. Malignancy should be suspected in women with postmenopausal bleeding or other bleeding irregularities and when thickened or irregular endometrium is noted at ultrasound. There is no question that transvaginal sonography has added much new information to endometrial examinations compared with transabdominal scanning. On closer scrutiny, however, we find that vaginal sonography is inaccurate in a considerable number of the cases examined and that different authors have cited very different cutoff levels for endometrial thickness at which curettage is urgently indicated, ranging from 4 mm[10] to as much as 10 mm[23]. At present, sensitivity and specificity data are not sufficient to eliminate the need for histological evaluation[6, 22, 31, 33, 34, 43]. As a result, fractional curettage is still the "gold standard" for excluding cancer in women with postmenopausal bleeding or sonographic abnormalities, even though endometrial cancer is actually found in only a small percentage of cases[9, 32]. To avoid unnecessary biopsies in this predominantly older population with an increased surgical risk, noninvasive methods are being sought that can differentiate benign endometrial changes from carcinoma.

Color Doppler Sonography

Neovascularization of tumors. The hope that color Doppler sonography will yield additional diagnostic information is based on the following theoretical considerations. Solid tumors have the capacity to induce angiogenesis, and they require these newly formed vessels in order to grow[13, 14]. These new vessels have a distinctive histological morphology: they are dilated, saccular, and tortuous with arteriovenous shunts and intervenous connections. The vessel walls contain very little smooth muscle, giving the vessels a lower impedance to flow than "normal" arterioles. This results in a markedly lower flow velocity[15, 20, 28, 37].

Doppler parameters. Color Doppler is being used in an attempt to detect this altered flow pattern. Waveforms are recorded to derive objective parameters that may be useful for noninvasive benign–malignant discrimination. Because absolute flow velocity depends on the incidence angle of the color Doppler beam and an accurate angle correction often cannot be made, it is better to use standard angle-independent parameters that indicate flow resistance: the S/D ratio, the resistance index $(RI) = (S-D)/S$, or the pulsatility index $(PI) = (S-D)/mean$, where S = maximum frequency shift in systole, D = minimum frequency shift in diastole, and mean = mean value.

Breast and ovarian cancers. The Doppler examination of blood vessels has already been applied to the differential diagnosis of other gynecological tumors, particularly breast and ovarian cancers. With regard to ovarian cancer, initial studies reporting extremely positive results[17, 26] gave way to increasingly critical studies giving reports of poor accuracy and specificity[38, 39]. Contradictory studies have also been published on breast cancer diagnosis. While Sohn in 1992 found a statistically significant decline of maximum resistance indices in breast cancers compared with benign tumors and Madjar in 1994 noted higher velocity sums in breast malignancies, the workers at our center were unable to confirm these positive results either as discriminatory parameters or as a means of assessing breast cancer prognosis[41].

Endometrial findings. In the early 1990 s, the first studies appeared on the use of color Doppler sonography in the differential diagnosis of endometrial pathology. The authors initially limited their work to large-vessel examinations such as that of the uterine artery[4, 5, 21], but soon they also studied smaller vessels in the endometrium and also in tumors. Most of the studies analyzed the angle-independent parameters RI and/or PI, since the sampled vascular segments are often very short and precise angle correction is difficult to achieve.

Examination of the Uterine Artery

Several authors found significant differences in flow resistance between benign and malignant endometrial lesions. Bourne et al.[4] stated a cutoff value of PI = 2.0 for the uterine artery. Chan et al.[8] also found significant PI differences, but only between normal and abnormal endometrium; benign lesions could not be distinguished from their malignant counterparts. The authors used a higher cutoff value than Bourne et al., with PI_{mean} = 2.17 for endometrial carcinoma. Similarly, Merce[30] could distinguish only between normal endometrium and endometrial disease (i.e., benign and malignant changes). Kupesic-Urek examined both the uterine artery and endometrial vessels[25] and found significant differences. The uterine artery resistance indices were considerably less than those reported by Merce (Table 29.**1**).

Table 29.**1** Review of the literature on uterine artery examinations, with significant results

Author	Number	Vessel	Indices for malignant lesions	Indices for benign lesions	Analysis
Bourne 1990[4]	54	Uterine artery/ (endometrium)	PI_{mean} = 0.89	PI_{mean} = 3.82	Cutoff value: PI = 2.0
Bourne 1991[5]	138	Uterine artery	PI_{mean} = 0.91	$PI_{mean} \geq 2.53$[a]	
Mercé 1991[30]	64	Uterine artery/ myometrial	RI_{mean} = 0.79 RI = 0.66	$RI_{mean} > 0.87$ $RI > 0.78$	([b]) Statistically significant
Kupesic-Urek 1993[25]	276	Uterine artery endometrial	RI = 0.53 RI = 0.37	RI = 0.76 RI = 0.54	Statistically significant
Chan 1994[8]	67	Uterine artery	PI_{mean} = 2.17	A: PI_{mean} = 3.41 B: PI_{mean} = 2.28	A: normal, B: benign pathology; significant differences between A and B and between A and cancer

[a] Includes different groups with benign pathology.
[b] No distinction made between malignant and benign pathology, only between normal and abnormal.

Experience at the Department of Obstetrics and Gynecology, Homburg University Hospital, Saar

Patients and Methods

We studied 95 women from 17 to 88 years of age (average 57.8 years) with clinically suspicious (menorrhagia, metrorrhagia, postmenopausal bleeding) or sonographically suspicious (structure, diameter) endometrial findings who had been referred to our department for further evaluation.

Sonographic and color Doppler sonographic examinations. All patients underwent hysteroscopic biopsy or fractional curettage to establish a histological diagnosis. The surgery was preceded by a transvaginal B-mode ultrasound examination of the internal genitalia with assessment of endometrial diameter and structure. This was immediately followed by a color Doppler examination to determine whether endometrial and subendometrial blood vessels could be visualized and, if so, how many. We also attempted to record a frequency-shift waveform from each endometrial vessel that could be visualized. These waveforms were then used to determine the angle-independent parameters of S/D ratio, RI, and PI along with the smallest value (min), greatest value (max), and mean value (mean) for each patient. We also documented menopausal status, the use of hormonal products, and the presence of risk factors for endometrial cancer (obesity, diabetes mellitus, arterial hypertension, positive family history). All examinations were performed with an Acuson 128 XP/10 scanner with integrated color Doppler using a 5 MHz endovaginal transducer that could be switched to 5 MHz pulsed color Doppler mode. The scanner was operated over a power range of 50–500 mW/m^2, and color sensitivity was set to the lowest velocity range. For maximum sensitivity in locating vessels, the color Doppler gain was set just below the noise threshold.

Fractional curettage. Nineteen of the women (20%) were premenopausal, 62 (65.3%) were postmenopausal, and 14 (14.7%) were perimenopausal. The indication for curettage was a bleeding disorder (postmenopausal bleeding, menorrhagia, metrorrhagia) in 74.8% of the patients and suspicious endometrial findings in 16.8%. Other causes (lower abdominal pain, search for a primary tumor, etc.) were present in 8.4%. Examination of the curettage material revealed a malignant condition in 23 cases (24.2%) and a benign condition in 72 cases (75.8%). As expected, the great majority of cancer patients (87%) were in menopause. A histological breakdown of the malignant lesions revealed 21 adenocarcinomas, one müllerian mixed tumor, and one high-sited cervical carcinoma (Tables 29.**2** and 29.**3**).

Table 29.2 Patients grouped by menopausal status and benign/malignant histology

Menopausal status	Malignant lesions		Benign lesions	
	Number	Percent	Number	Percent
Total	23	24.2%	72	75.8%
Premenopausal	2	8.7%	17	23.6%
Perimenopausal	1	4.3%	13	18.1%
Postmenopausal	20	87.0%	42	58.3%

Table 29.3 Histological diagnoses

Histological diagnosis	Number
Endometrium in desquamation phase	1
Endometrium in secretory phase	4
Autolytic endometrium	1
No endometrium detected	3
Residual proliferation or transitional menopausal endometrium	14
Quiescent, atrophic, or fibrotic endometrium (appropriate for age)	11
Polypous or cystic-glandular hyperplasia	34
Adenomatous hyperplasia (diffuse/polypous)	2
Atypical adenomatous hyperplasia	1
Hydatidiform mole	1

a
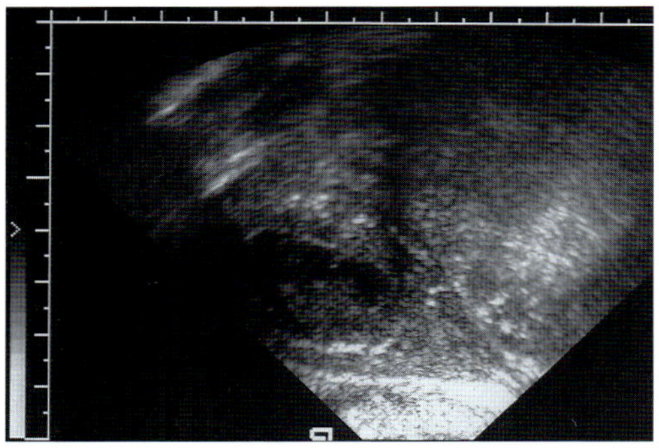

b
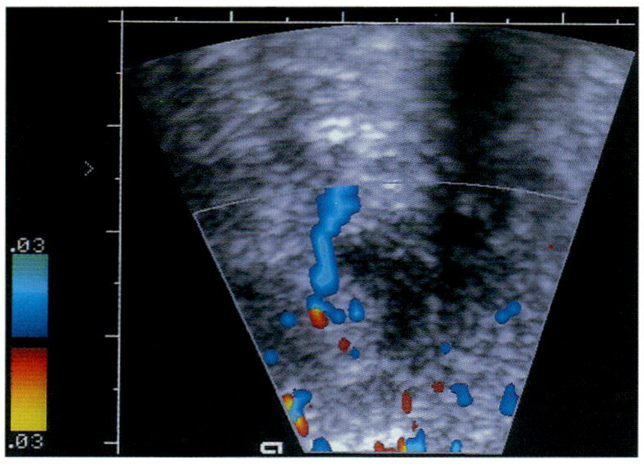

Fig. 29.1 Endometrial carcinoma (pT1 c pN0 G2 adenocarcinoma) in a 72-year-old woman.
a B-mode image.
b Color flow image of the endometrial vessels, showing heavy vascularization.

Visualization and Morphology of the Vessels

Visualization of endometrial vessels. We were able to visualize endometrial vessels in 7% of all the patients in our study. Vessels could be defined in 100% of the patients with malignant disease (Fig. 29.**1**) but in only 66.6% of the patients with a benign endometrial condition. This difference is statistically significant ($p = 0.0034$). It is interesting to note the relatively higher percentage of patients with benign findings after menopause in whom endometrial vessels could not be visualized: 43% versus 24% of the premenopausal patients (not statistically significant at $p = 0.27431$ (Fig. 29.**2**). Other authors also described differences in the visualization of endometrial vessels between benign and malignant tumors. Different authors have reported visualization rates of 33.3%[12], 56%[35], and 100%[3,17,25]. The differences are even greater in benign endometrial conditions, however, ranging from no endometrial vessels[12,17,21,27] to finding vessels in 100% of the women examined[1].

Readable waveforms. Readable waveform indices could be recorded in approximately 83% of the cancer patients (Fig. 29.**3**) but in only 38 women with benign endometria (= 52.8%). This difference is also statistically significant. When we consider only women with definable vessels, we were able to record readable waveforms in approximately 80% of cases, regardless of whether the lesions were malignant or benign (malignant 83%, benign 79%), enabling us to calculate waveform parameters.

Comparison with other studies. Aleem et al.[2] also found significant differences in the ability to visualize and sample benign and malignant endometrial lesions, but in a substantially smaller percentage of cases. They could demonstrate endometrial vessels in 43% of carcinoma cases (and myometrial vessels in 93%) but in only 12% of cases with benign hyperplasia and in no cases with leiomyomas (although myometrial vessels were visualized in 63% and 55% of these cases, respectively). Thus, the nonvisualization of blood vessels cannot be

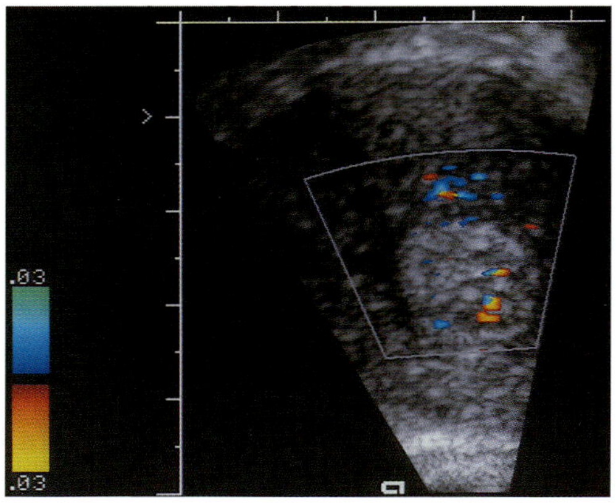

Fig. 29.2 Scant vascularity is observed in postmenopausal bleeding with a benign etiology (glandular-cystic polyp of the uterine mucosa).

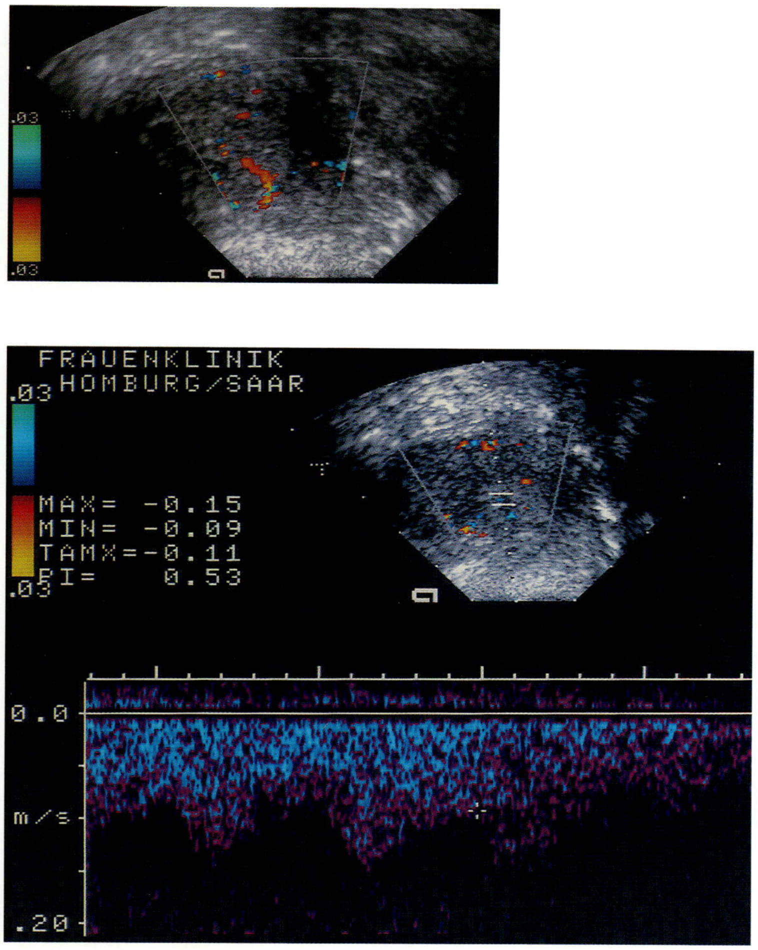

Fig. 29.**3** Endometrial carcinoma (pT1 b pN0 G1–2 adenocarcinoma) in a 64-year-old woman.
a Color flow image.
b Resistance index = 0.40.

used as a criterion for excluding malignancy—especially since we found that as we gained experience with the method we were able to define endometrial vessels more frequently even in cases with a benign histology (Table 29.**4**).

Number of vessels sampled. The number of vessels that could be sampled by Doppler sonography in our examinations was six or fewer. This number was similar for benign and malignant lesions: 2.6 ± 1.2 vessels for malignant endometrium versus 2.5 ± 1.3 for benign lesions. The only marked difference was seen in the premenopausal women: 3.5 vessels in patients with a malignant disease versus 2.1 vessels in patients with a benign condition. A larger population of premenopausal cancer patients would have to be studied in order to draw further conclusions.

Vascular morphology. With regard to vascular morphology, Aleem[2] noted a dense vascular distribution in 80% of malignant lesions, contrasting with the more isolated vessels supplying benign lesions, although he looked only at myometrial vessels. Carter et al.[7] described the endometrial blood supply of benign lesions as fine vascular branches with no "hot spots," differing from the increased intratumoral flow seen in carcinomas. We noted a different vascular distribution in our study as well, finding that long segments of endometrial vessels could be visualized only in association with endometrial cancers (Figs. 29.**1**, 29.**2**). This has been cited as a possible criterion for benign–malignant differentiation, but it must be viewed very cautiously because of its subjectivity. In patients examined with an advanced ultrasound scanner, there were cases in which long vascular segments could be defined in the endometrium with no histological evidence of malignancy. It appears, then, that technological advances in ultrasound instruments are basically providing improved sensitivity in visualizing blood vessels and defining their morphology.

Table 29.**4** Visualization and sampling rates of endometrial vessels reported by various authors under various conditions; large differences in visualization rates are reported in different studies

Author	Visualization rates with malignant lesions	Visualization rates with benign lesions	Influences/special conditions
Achiron 1995		100% (45/45)	on tamoxifen
Aleem 1995[2]	43%	12% Hyperplasia 9% Control	no hormones, postmenopause
Alge 1996	not specified (n = 8)	46% without hormones 79% with hormones	With/without hormone replacement; rates not reported for cancer cases
Bonilla-Musoles[3]	100% (n = 2)	45.5% Hyperplasia 3.6% Proliferative endometrium	With hormone replacement
Flam 1995[12]	33.3% (9/27)	0	
Hata 1991[17]	100% (n = 10)	0 (n = 31)	
Hata 1992	77.7% (7/9)		
Juhasz 1990[21]	89% (8/9)	21% Leiomyomas 0% Adenomyosis	Myometrial vessels in benign cases
Kupesic-Urek 1993[25]	100% (n = 26)	67% (167/250)	
Sheth 1995[35]	56% (5/9)	64% (23/36)	No hormone replacement therapy
Our Study 1997	100% Sampled: 82.6%	66.6% Sampled: 52.8%; 79.2% in patients with definable vessels	

Resistance Indices of Endometrial Vessels

RI. In the study by Kupesic-Urek et al.[25], the resistance indices in the endometrial vessels were considerably lower than in the uterine arteries, with an RI of 0.37 for carcinoma and RI = 0.54 for benign endometrium. Juhasz et al.[21] reported similar values, whereas other studies with considerably higher values found significant differences, although the indices for carcinoma in those studies were already as high as those for benign endometrium in the previous studies [2,37]. In our study population, the resistance indices tended to show lower values in malignant endometrium, especially when minimum values were compared: 0.54 versus 0.57. This difference is not significant, however ($p = 0.2849$). The mean and median values showed very slight differences: $RI_{mean} = 0.62$ for carcinoma versus $RI_{mean} = 0.61$ for benign endometrium (Table 29.**5**). The range of RI values was higher in malignant endometrium than in benign endometrium, but there was an almost complete overlap of values between the two groups. This large overlap does not permit a benign-malignant differentiation in any given case.

PI and S/D ratio. We obtained similar results for the other waveform parameters—the pulsatility index and S/D ratio. Other authors also reported nonsignificant results. Sheth and Sladkevius are the only authors who published values for endometrial vessels. Sheth examined only endometrium thicker than 8 mm. Sladkevius did not find significant differences in the endometrial vessels but did in the uterine artery. The minimum resistance indices reported by Sheth[35] were lower in both groups than in our study. Sladkevius, working with the pulsatility index, also found lower values (Tables 29.**6** and 29.**7**).

Table 29.**5** Mean (and median) values of RI_{min} (smallest resistance index) and RI_{mean} (mean resistance index) with standard deviation, shown separately for benign and malignant lesions

Lesion category	RI_{min}	RI_{mean}
Benign	0.57 ± 0.11 (0.56) 0.30 – 0.83	0.61 ± 0.10 (0.61) 0.43 – 0.83
Malignant	0.54 ± 0.18 (0.50) 0.29 – 0.83	0.62 ± 0.14 (0.60) 0.39 – 0.86

Effect of Menopausal Status and Hormone Use

RI in pre- and postmenopausal cancers. We did not find a correlation between waveform indices and menopausal status. The only notable finding was a slightly decreased minimum and mean resistance index in carcinomas after menopause compared with premenopausal cases (0.53 versus 0.58 for RI_{min} and 0.61 versus 0.67 for RI_{mean}). Statistical significance is indeterminate, however, given the small case numbers of premenopausal endometrial cancer. Sohn et al. likewise found no differences in blood flow between cancerous and healthy endometria before menopause, but they did observe significant differences in Doppler parameters after menopause ($RI_{min} = 0.54$ with malignant findings and $RI_{min} = 0.65$ with benign findings).

Tamoxifen therapy. It is interesting to consider the effects of hormones that influence endometrial metabolism. Patients on tamoxifen therapy are of particular interest, because B-mode ultrasound in these patients often reveals a thickened, irregular endometrium that does not correlate with histological abnormalities. An average of 2.5 vessels could be sampled in cancer patients receiving tamoxifen therapy (two vessels in one case, three in another). This is consistent with the average number of 2.13 generally reported for this group. The average number of sampled vessels in women with benign endometrium on tamoxifen was 0.43, which is well below the general average of 1.33 vessels (Fig. 29.**4**). The waveform pa-

Table 29.**6** Reference list for endometrial vessels with significant results

Author	Number	Vessels	Indices for malignant lesions	Indices for benign lesions	Analysis
Juhasz 1990[21]	168	Endometrial vessels	$RI_{min} = 0.38$	$RI_{min} > 0.50$	Cutoff value: RI = 0.50
Hata 1991[17]	68	Subendometrial vessels (arcuate arteries)	$RI_{min} = 0.54$	$RI_{min} > 0.68$ ([a])	Statistically significant
Kupesic-Urek 1993[25]	276	Endometrial uterine artery	RI = 0.53 RI = 0.37	RI = 0.76 RI = 0.54	Statistically significant
Kurjak 1993[27]	750	Endometrial vessels	$RI_{mean} = 0.42$	$RI_{mean} = 0.65$	Statistically significant
Kurjak 1994[28]	5013	Endometrial/myometrial vessels			Screening: 6 cancers detected at $RI_{min} \leq 0.42$
Aleem 1995[2]	42	Endometrial/myometrial Gefäße	$RI_{mean} = 0.53$ $PI_{mean} = 0.88$	$RI_{mean} \geq 0.64$ $PI \geq 1.11$ (*)	Statistically significant for myometrial vessels
Sohn 1993	35	Endometrial vessels?	$RI_{min} = 0.54$	Premenopausal RI = 0.55	Statistically significant in post-menopausal patients

[a] Includes different groups with benign pathology

Table 29.**7** Studies reporting nonsignificant waveform differences in benign-malignant differentiation

Author	Number	Vessels	Indices for malignant lesions	Indices for benign lesions	Remarks
Sheth 1995[35]	45 (9 cancers)	Endometrial	RI_{min} = 0.48 PI = 0.71	RI_{min} = 0.48 PI = 0.72	Endometrium ≥ 8 mm, flow in 5/9 cancer patients
Sladkevius 1994[36]	138 (24 cancers)	Uterine artery, subendometrial and endometrial	Endometrial: PI = 0.7 Subendometrial: PI = 0.8 Uterine artery: PI = 1.4	Endometrial: PI = 0.7 Subendometrial: PI = 0.8 Uterine artery: PI = 1.8	Significant in uterine artery
Chan 1994[8]	67 (17 cancers)	Uterine artery	PI_{mean} = 2.17	PI_{mean} = 2.28	No significant differences between cancer and benign lesions
Flam 1995[12]	39 (27 cancers)	Uterine artery	PI_{mean} = 1.70 after radiotherapy: PI = 1.22	PI_{mean} = 1.94	Only significant differences in cancer cases before and after radiotherapy
Our Data	95 (23 cancers)	Endometrial	RI_{min} = 0.54 RI_{mean} = 0.62 PI_{mean} = 1.12	RI_{min} = 0.57 RI_{mean} = 0.61 PI_{mean} = 1.10	No significant differences

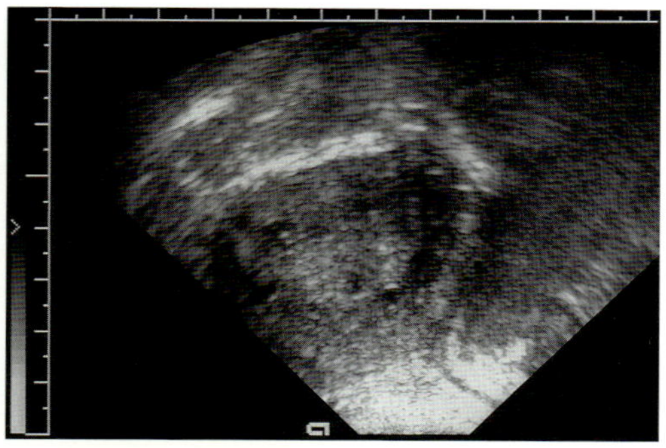

a

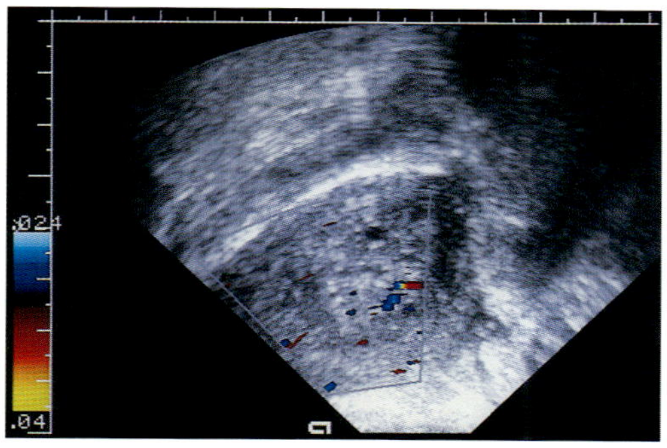

b

Fig. 29.**4** Fibrocystic endometrial polyp with atrophic mucosa in a patient on tamoxifen therapy.
a B-mode image.
b Long vascular segments are not visualized.

rameters of the sampled vessels showed marked differences in these cases, with $RI_{min} = 0.40$ and $RI_{mean} = 0.51 \pm 0.01$ for cancer compared with the markedly high values of $RI_{min} = 0.72 \pm 0.16$ and $RI_{mean} = 0.72 \pm 0.15$ for benign findings. As far as we can tell from the small case numbers, these differences are not statistically significant ($p = 0.1025$ and 0.1213).

Estrogen/progestin therapy. The differences were smaller and not significant in patients taking other hormones such as estrogens and/or progestins. The resistance indices in patients with benign histology who were taking estrogens were in the average range for all patients. The RIs were slightly below average in patients taking progestins and slightly above average in patients taking an estrogen–progestin combination. Patients with cancer who were taking estrogens had slightly higher-than-average resistance indices ($RI_{min} = 0.60 \pm 0.18$ and $RI_{mean} = 0.64 \pm 0.13$), and patients without cancer who were taking progestins had slightly lower indices, with $RI_{min} = 0.43 \pm 0.18$ and $RI_{mean} = 0.52 \pm 0.10$. The differences were not statistically significant. Zalud could likewise find no significant effect of hormone replacement therapy on resistance indices. On the other hand, Achiron et al.[1] found significantly lower resistance indices when the estrogen level reached its peak at midcycle. When the progesterone level subsequently rises, the resistance indices return to higher values.

Effect of Histopathological Parameters, with Reference to Prognostic Factors

When we looked at the histological diagnoses separately, we were struck by the particularly low resistance indices in a patient with a müllerian mixed tumor ($RI_{min} = 0.40$, $RI_{mean} = 0.51$) and by the high resistance indices in a patient with atypical adenomatous hyperplasia ($RI_{min} = RI_{mean} = 0.67$). Also, tumors at a higher stage or grade tended to show lower impedance values, although the differences were not significant: $RI_{mean} = 0.67 \pm 0.20$ for stage Ia tumors, $RI_{mean} = 0.61 \pm 0.14$ for stage Ib tumors, and $RI_{mean} = 0.57 \pm 0.11$ for stage Ic tumors. As for tumor grades, the resistance indices ranged from $RI_{mean} = 0.65$ in grade 1 tumors and $RI_{mean} = 0.62 \pm 0.16$ in grade 2 cancers to $RI_{mean} = 0.57 \pm 0.09$ in grade 3 cancers (not significant). Hata et al. also noted a tendency toward finding lower resistance indices in tumors of a higher stage (though less pronounced) or grade. Again, the differences were not significant.

Subendometrial and Myometrial Vessels

In 24 patients, we additionally defined subendometrial and myometrial vessels and sampled their waveforms. We found that the differences between the vessels in patients with cancerous and benign endometria were even smaller: the minimum and mean resistance indices in cancer patients were slightly (not significantly) higher than in patients with benign histology. In all cases the resistance indices found in the endometrial vessels were markedly lower than in waveforms sampled from subendometrial and myometrial vessels, with an abrupt change of impedance occurring at the endometrial–subendometrial junction (Table 29.**8**).

Table 29.**8** Resistance indices for subendometrial and myometrial vessels

	RI_{min}	RI_{mean}	Number
Malignant	0.66 ± 0.10	0.71 ± 0.09	4
Benign	0.65 ± 0.15	0.69 ± 0.15	20

Summary

On the whole, the color Doppler examination of endometrial vessels and the analysis of waveform indices do not appear to be sufficiently accurate for the differential diagnosis of endometrial changes, nor can they provide a reliable screening method for endometrial carcinoma. Despite the fact that the patients in our study whose endometrial vessels could not be visualized with color Doppler consistently had benign histological findings, this is not sufficient to justify the omission of histological evaluation, especially for less experienced examiners. Nevertheless, when resistance indices are combined with the evaluation of vascular structural abnormalities and the visualization of long, branched endometrial vascular segments, which are suggestive of carcinoma, the color Doppler examination provides clues that can advance the differential diagnosis of endometrial disease.

References

1 Achiron R, Levran D, Lipitz S, Dor J, Sivan E, Mashiach S: Endometrial blood flow response to hormone replacement therapy in women with premature ovarian failure: a transvaginal Doppler study. Fertil. and Steril. 63 (1995) 550–554
2 Aleem F, Predanic M, Calame R, Moukhtar M, Pennisi J: Transvaginal Color and Pulsed Doppler Sonography of the Endometrium: A Possible Role in Reducing the Number of Dilatation and Curettage Procedures. J. Ultrasound Med. 14 (1995) 139–145
3 Bonilla-Musoles F, Ballester MJ, Marti MC, Raga F, Osborne NG: Transvaginal color doppler assessment of endometrial status in normal postmenopausal women. The Effect of Hormone Replacement Therapy. J. Ultrasound Med. 14 (1995) 503–507
4 Bourne TH, Campbell S, Whitehead MI, Royston P, Steer CV, Collins WP: Detection of endometrial cancer in postmenopausal women by transvaginal ultrasonography and colour flow imaging. Brit. Med. J. 301 (1990) 369
5 Bourne TH, Campbell S, Steer CV, Royston P, Whitehead MI, Collins WP: Detection of Endometrial Cancer by Transvaginal Ultrasonography with Color Flow Imaging and Blood Flow Analysis: A Preliminary Report. Gynecol. Oncol. 40 (1991) 253–259
6 Brandner P, Gnirs J, Neis KJ, Hettenbach A, Schmidt W: Der Stellenwert der Vaginosonographie in der non-invasiven Beurteilung des Endometriums am postmenopausalen Uterus. Geburtshilfe. Frauenheilkd. 51 (1991) 734–740
7 Carter JR, Fowler JM, Carlson JW, Carson LF, Adcock LL, Twiggs LB: Prediction of malignancy using transvaginal color flow Doppler in patients with gynecologic tumors. Int. J. Gynecol. Cancer 3 (1993) 279–284
8 Chan FY, Chau MT, Pun TC et al.: Limitations of transvaginal sonography and color Doppler imaging in the differentiation of endometrial carcinoma from benign lesions. J. Ultrasound Med. 13 (1994) 623–628
9 Choo YC, Mak KC, Hsu C, Wong TS, Ma HK: Postmenopausal Uterine Bleeding of Nonorganic Cause. Obstet. and Gynecol. 66 (1985) 225–228
10 Conoscenti G, Meir YJ, Fischer-Tamaro L et al.: Endometrial assessment by transvaginal sonography and histological findings after D & C in women with postmenopausal bleeding. Ultrasound Obstet. Gynecol. 6 (1995) 108–115
11 Fisher B, Costantino JP, Redmond CK and other NSABP contributors: Endometrial cancer in Tamoxifen-treated Breast cancer patients: Findings from the National Surgical Adjuvant Breast and Bowel Project (NSABP) B-14. Journal nat. Cancer Inst. 86 (1994) 527–537
12 Flam F, Almström H, Hellström AC, Moberger B: Value of Uterine Artery Doppler in Endometrial Cancer. Acta Oncol. 34 (1995) 779–782
13 Folkman J: Tumour Angiogenesis: Therapeutic Implications. New Engl. J. Med. 18 (1971) 1182–1186
14 Folkman J: How is Blood Vessel Growth in Normal and Neoplastic Tissue?—G. H. A. Clowes Memorial Award Lecture Cancer Research 46 (1986) 467–473
15 Gamill SL, Shipkey FH, Himmelfarb EH, Parvey LS, Rabinowits JG: Roentgenology—pathology correlation study of neovascularization. Amer. J. Radiol. 126 (1981) 376
16 Gusberg SB: Diagnosis and Principles of Treatment of Cancer of the Endometrium. In: Female genital cancer. New York 1988, 337–360
17 Hata K, Makihara K, Hata T, Takahashi K, Kitao M: Transvaginal color Doppler imaging for hemodynamic assessment of reproductive tract tumors. Int. J. Gynecol. Obstet. 36 (1991) 301–308
18 Hata K, Hata T, Kitao M: Intratumoral Blood Flow Analysis in Endometrial Cancer: Does it Differ among Individual Tumor Characteristics? Gynecol. Oncol. 61 (1996) 341–344
19 Hulka BS: Links between hormone replacement therapy and neoplasia. Fertil. and Steril. 62(Suppl.2) (1994) 168–175
20 Jain RK, Ward-Harley K: Tumor blood flow—characterization, modifications and role in hyperthermia. Trans. Sonics Ultrasonics 31 (1984) 504–509
21 Juhasz B, Kurjak A, Lampe L, Zalud I, Crvenkovic G, Hernadi Z: Tissue characterization by transvaginal colour Doppler for the evaluation of gynecological tumors. 2. Clinical experiences. Acta Med. Hung. 47 (1990) 149–156
22 Karlsson B, Granberg S, Wickland M et al.: Transvaginal ultrasonography of the endometrium in women with postmenopausal bleeding—a Nordic trial. Amer. J. Obstet. Gynecol. 172 (1995) 1488–1494
23 Klug PW, Leitner G: Die Gegenüberstellung vaginalsonographischer und histologischer Befunde am Endometrium. Geburtshilfe u. Frauenheilk. 49 (1989) 797–802
24 Kolles H, Stegmaier C, Seebach HB, Ziegler H: Deskriptive Epidemiologie und Prognose maligner gynäkologischer Tumoren. Geburtshilfe u. Frauenheilk. 49 (1989) 573–578

25 Kupesic-Urek S, Shalan H, Kurjak A: Early detection of endometrial cancer by transvaginal color Doppler. Eur. J. Obstet. Gynecol. Reprod. Biol. 49 (1993) 46–49

26 Kurjak A, Schulman H, Sosic A, Zalud I, Shalan H: Transvaginal ultrasound, color flow, and Doppler waveform of the postmenopausal adnexal mass. Obstet. Gynecol. 80 (1992) 917–921

27 Kurjak A, Shalan H, Sosic A et al.: Endometrial carcinoma in postmenopausal women: Evaluation by transvaginal color Doppler ultrasonography. Amer. J. Obstet. Gynecol. 169 (1993) 1597–1603

28 Kurjak A, Shalan H, Kupesic S et al.: An attempt to screen asymptomatic women for ovarian and endometrial cancer with transvaginal color and pulsed Doppler sonography. J. Ultrasound Med. 13 (1994) 295–301

29 Kurjak A: An Atlas of Transvaginal Color Doppler: Current State of the Art (Encyclopedia of visual medicine series). Limited 2. Rev. ed. II Series. Parthenon 1994

30 Merce LT, Lopez Garcia G, De La Fuente F: Doppler ultrasound assessment of endometrial pathology. Acta Obstet. Gynecol. Scand. 70 (1991) 525–530

31 Osmers R, Völksen M, Rath W, Teichmann A, Kuhn W: Vaginosonographische Messungen des postmenopausalen Endometriums zur Früherkennung des Endometriumkarzinoms. Geburtsh. u. Frauenheilk. 49 (1989) 262–265

32 Procope BJ: Aetiology of Postmenopausal Bleeding. Acta Obstet. Gynecol. Scand. 50 (1971) 311–313

33 Schramm T, Kürzl R, Schweighart C, Stuckert-Klein AC: Endometriumkarzinom und Vaginalsonographie: Untersuchungen zur diagnostischen Validität. Geburtsh. u. Frauenheilk. 55 (1995) 65–72

34 Seelbach-Göbel B, Rempen A, Kristen P: Transvaginaler Ultraschall am Endometrium in der Postmenopause. Geburtsh. u. Frauenheilk. 55 (1995) 59–64

35 Sheth S, Hamper UM, McCollum ME, Caskey CI, Rosenshein NB, Kurman RJ: Endometrial blood Flow Analysis in Postmenopausal Women: Can it help differentiate benign from Malignant Causes of Endometrial Thickening? Radiology 195 (1995) 661–665

36 Sladkevius P, Valentin L, Marsál K: Endometrial thickness and Doppler velocimetry of the uterine arteries as discriminators of endometrial status in women with postmenopausal bleeding: A comparative study. Amer. J. Obstet. Gynecol. 171 (1994) 722–728

37 Sohn C, Holzgreve W: Ultraschall in Gynäkologie und Geburtshilfe. Thieme, Stuttgart 1995

38 Tekay A, Jouppila P: Validity of pulsatility and resistance indices in classification of adnexal tumors with transvaginal color Doppler ultrasound. Ultrasound Obstet. Gynecol. 2 (1992) 338–344

39 Tekay A, Jouppila P: Controversies in assessment of ovarian tumors with transvaginal color Doppler ultrasound. Acta Obstet. Gynecol. Scand. 75 (1996) 316–329

40 van Leeuwen FE, Benraadt J, Coebergh JWW et al.: Risk of endometrial cancer after tamoxifen treatment of breast cancer. Lancet 343 (1994) 448–452

41 Villena-Heinsen C, Ertan AK, Tossounidis I, Holländer M, König J, Schmidt W: Diagnostische Aussagekraft der Farb-Doppler-Sonographie bei Mammatumoren. Geburtsh. u. Frauenheilk. 55 (1995) 541–547

42 Villena-Heinsen C, Mink D, Tossounidis I et al.: Ist eine Prognoseeinschätzung beim Mammakarzinom mittels farbkodierter und Spektral-Doppler-Sonographie möglich? Tumordiagn. u. Ther. 16 (1995) 187–193

43 Weigel M, Friese K, Strittmatter HJ, Melchert F: Sonographische Beurteilung des postmenopausalen Endometriums: Ist die Messung der Dicke ausreichend? Ultraschall in Med. 15 (1994) 117–121

30 Malignant Uterine Tumors

A. Kurjak and S. Kupesic

Because ultrasonography can detect even subtle changes in the endometrium and myometrium, it is the modality of choice in the evaluation of many uterine tumors. This chapter explores the diagnostic capabilities of transvaginal color and pulsed Doppler ultrasound as a noninvasive method of investigating malignant uterine tumors.

Endometrial Carcinoma

Incidence

The interest of oncologists in endometrial carcinoma has grown considerably in recent years, due primarily to the increased prevalence of this disease[6]. The rising incidence of endometrial cancer is due in part to better socioeconomic conditions, enabling more women to reach the advanced age at which the disease is most common. Other factors are the growing popularity of estrogen replacement therapies and also nutritional factors, especially higher cholesterol levels.

Endometrial carcinoma has become the leading cancer of the female genital tract in many developed countries. Except for Japan, the incidence is highest in industrialized nations. More than 39,000 new cases are reported annually in the United States alone[14].

More than 70% of endometrial carcinomas are stage I tumors at diagnosis, which currently have a five-year survival rate of 80%. Five-year survival rates fall dramatically as the tumor stage increases[34]. The greater the extent of myometrial invasion, the more swiftly the disease progresses[23].

Risk Factors

Age and genetic factors. Age is the greatest risk factor for the development of endometrial carcinoma. Less than 25% of all cases occur before menopause. The incidence of endometrial cancer is highest between 55 and 65 years of age[31]. Genetic factors also appear to play an important role, as the disease is more prevalent among white women than black women.

Estrogen. Nearly 40–50% of all endometrial cancers develop from atrophic endometrium, independent of hormonal influences. Nevertheless, a link is believed to exist between hormonal therapy and endometrial carcinoma. Unopposed high-dose estrogen therapy administered for a prolonged period of time increases the relative risk compared with short-term, low-dose therapy. The incidence in women who are treated with estrogen *and* progestin during menopause is even lower than in the general female population[14,37]. Even cancers that develop during unopposed estrogen use tend to show a lower histological grade and clinical stage compared with the general population. If myometrial invasion is present, the foci tend to be superficial and generally have a good prognosis.

Although it is possible for a complex or atypical hyperplasia to progress to endometrial cancer, endometrial hyperplasia is considered more of an indicator of carcinoma than an actual precursor.

Obesity and diabetes. The link between endometrial cancer and other risk factors has been analyzed in various epidemiological studies. Obesity increases the risk of endometrial cancer as a result of increased endogenous estrogen production and greater bioavailability of the estrogen that is produced[5,37]. Diabetes is associated with a 2.8 relative risk for endometrial carcinoma and thus ranks behind age, weight, and socioeconomic status. Hypertension is common in women with endometrial cancer but does not appear to be an independent risk factor[5].

Target Group for Screening

The American Cancer Society has recommended selective screening for all women who are considered to be at high risk for endometrial cancer. Unfortunately, even the screening of women with risk factors such as obesity, nulliparity, late menopause, diabetes, unopposed estrogen use, or a history of colon cancer, breast cancer, or pelvis radiation will detect only around 50% of all endometrial cancers. Moreover, the cancers in women who do not have these risk factors tend to be more aggressive and poorly differentiated.

Because the principal risk factor is age, screening should be extended to all postmenopausal women. One extra screening examination per year is also advisable for women who are at increased risk.

Screening: Dream or Reality?

Endometrial carcinoma has traditionally been viewed as a very treatable form of cancer. Generally it takes a less fulminating course than other gynecological malignancies, with the result that, in many cases, a lower priority is placed upon endometrial cancer in terms of screening and early detection.

Increased incidence. Three decades ago, cervical carcinoma had a four times higher incidence than endometrial carcinoma. Today, however, the incidence of endometrial carcinoma is twice that of cervical carcinoma[5]. On the one hand, these changes document the success of early screening programs for cervical cancer; but they also reflect an absolute increase in endometrial carcinoma and underscore the need for a reliable screening method.

Uterine bleeding. The earliest sign of endometrial cancer is uterine bleeding[7]. The most frequent cause of postmenopausal bleeding, however, is sclerotic vascular changes leading to venous or arterial rupture[4]; fewer than 10% of women with postmenopausal bleeding have endometrial cancer. The established diagnostic protocol is based upon fractional dilation and curettage, which is performed mainly in selected patients at risk (e.g., dysfunctional bleeding during menopause). The costs and risks that are associated with inpatient fractional curettage, plus the low diagnostic yield of endometrial cancers with this procedure, have prompted the development of numerous new devices for obtaining endometrial material for cytological analysis or tissue samples for histological evaluation in the ambulatory setting.

Cytological Methods

Improved screening by cervical cytology has resulted in the early detection of cervical premalignant lesions. This method has little importance in the early diagnosis of endometrial cancer, however, and it is unreliable as a screening test for that disease[38,40].

Smears taken directly from the uterine cavity could solve this problem. It is hoped that this cytological method will be easier to perform, cause less discomfort, reduce costs, and provide interpretable samples of atrophic endometrium. Endometrial smears are difficult to interpret, however, and there are no standard morphological criteria for diagnosing hyperplasia. Moreover, a positive cytological sample still requires histological confirmation[8,17].

Histological Methods

Intracervical curettage is not a practical screening method for endometrial cancer, as it has been estimated that up to 14% of postmenopausal women have an obliterated cervical canal[28]. Endometrial biopsy is generally diagnostic in women who have postmenopausal bleeding and endometrial cancer. The main advantages of endometrial biopsy over cytology are its significantly greater sensitivity and higher specimen quality. It is an invasive procedure, however, and the cost–benefit ratio is an important factor to be considered.

In summary, it would be very beneficial to have a procedure that is less invasive than diagnostic biopsy, provides a comparably high detection rate, and has an acceptably low false-positive rate. If this method could detect endometrial cancer at an early stage in asymptomatic women, it could substantially reduce the morbidity and mortality of this disease.

Ultrasound

In recent years there have been remarkable advances in imaging procedures that allow clinicians to detect physiological and pathological changes in the fetal genital tract. The various sonographic procedures in particular have evolved into an important modality for the evaluation of uterine tumors.

Transabdominal Ultrasound

Since ultrasound has become widely utilized in gynecology, many reports have been published describing the sonographic appearance of the normal uterus and of uterine abnormalities. The typical appearance of the postmenopausal endometrium is a single, echogenic line that is markedly thinner than the premenopausal endometrium. The inner layer of the myometrium appears as a thin hypoechoic ring surrounding the endometrium. B-mode ultrasound has also proved to be a very accurate method for determining endometrial thickness, yielding results that show very good agreement with actual measurements in surgical specimens[24].

Endometrial thickness. Endometrial carcinoma typically produces a thickened endometrium of varying echogenicity. There are other lesions that closely mimic this appearance, however, including endometrial hyperplasia, endometrial polyps, ovarian carcinoma that has spread to the endometrium, hematometra, mucometra, and pyometra[25]. A thin endometrium (< 5 mm) is a significant finding in postmenopausal women with vaginal bleeding, because a thin endometrium is usually associated with atrophic or benign changes. This finding, then, can obviate the need for endometrial biopsy or curettage. A major disadvantage of transabdominal ultrasound is the fact that endometrial thickness cannot be accurately measured in a retroflexed uterus. The endometrium is also difficult to evaluate in patients with uterine prolapse[30].

Echogenicity. Investigators have looked for a possible connection between endometrial echogenicity and tissue differentiation (tumor grade). High-intensity echoes were a more common finding in well differentiated or moderately differentiated carcinomas, probably because these tumors contain large numbers of glands[10]. Heterogeneous echo patterns were found in poorly differentiated carcinomas[10]. However, these endometrial echo patterns may also be seen in other conditions, such as endometrial hyperplasia, hematometra, pyometra, and adenomyosis.

Depth of invasion. The most appropriate treatment for endometrial carcinoma depends on the depth of myometrial invasion. Deep invasion means a considerably higher risk of metastasis, and preoperative irradiation or a more extensive surgical procedure is indicated[3]. The five-year survival rate also correlates closely with the depth of myometrial invasion: 93.7% with no invasion, 88.1% with the invasion of superficial myometrial layers, and 36.2% with deep invasion[24]. In most cases the degree of myometrial invasion can be determined very accurately with ultrasound[24]. There are cases in which the accuracy of the measurement is compromised by intracavitary exophytic tumor growth, leiomyomas, or other lesions that alter the structure of the myometrium and distort the uterine wall. Obesity and a retroflexed uterus can also make it difficult to accurately determine the depth of myometrial invasion by endometrial carcinoma with transabdominal ultrasound.

Transvaginal Ultrasound

Transvaginal ultrasound has several advantages over transabdominal scanning: there is no need for a full urinary bladder, and higher transducer frequencies can be used for the more detailed visualization and assessment of genital tract morphology. Transvaginal sonography can define the extent of myometrial or cervical invasion by endometrial carcinoma and can thus make an important contribution to preoperative staging.

Endometrial thickness. The thickness of the endometrium, as measured by transvaginal ultrasound, correlates closely with the presence of endometrial carcinoma[25]. Schoenfeld et al.[35] found that the normal postmenopausal endometrium is thin, showing a total (double layer) thickness of 6–8 mm. Fleischer et al.[11] even stated that an endometrial thickness of 2–3 mm (single layer) should be considered a normal finding in postmenopausal women. In any case, a double-layer endometrial thickness of more than 10 mm in postmenopausal women is suggestive of hyperplasia or carcinoma[15] (Fig. 30.**1**).

Cutoff value. In a prospective study, 539 asymptomatic postmenopausal women were examined by transvaginal sonography. Every fifth woman whose endometrial thickness (single layer) measured more than 4 mm at ultrasound was found to have endometrial carcinoma. The authors concluded that a cutoff value of >4 mm (single layer) appeared to be a better indicator of endometrial carcinoma than postmenopausal bleeding[33]. In a multicenter study that included eight Scandinavian gynecology departments, Wikland et al.[39] examined 1000 women with postmenopausal bleeding who had been admitted for fractional curettage. They did not find endometrial carcinoma when the endometrial thickness was ≤ 4 mm (single layer). The average endometrial thickness in patients with endometrial carcinoma was 18 mm (range of 5–55 mm).

Nasri et al.[29] performed transvaginal ultrasound scanning on 111 postmenopausal women. Of the 103 women with bleeding, 93 underwent fractional curettage and 10 were managed conservatively with a repeat scan at six months. The authors found that 29 patients (31 %) had abnormal endometrial histology when endometrial thickness was >5 mm. They recommended an endometrial thickness of 5 mm as an appropriate cutoff level for the conservative management of patients with postmenopausal bleeding or in screening programs for endometrial cancer.

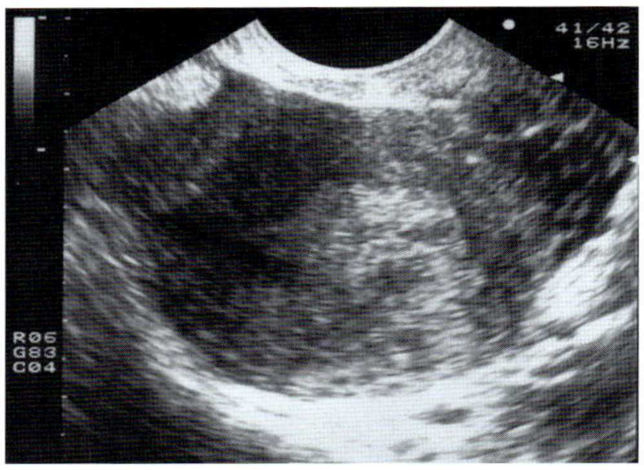

Fig. 30.**1** Anteflexed uterus with an endometrial thickness of 11 mm in a postmenopausal woman. Histological examination revealed a well-differentiated stage Ia adenocarcinoma by the FIGO classification.

Practical recommendations. This concept can be applied clinically in postmenopausal women. Finding an endometrial thickness less than 5 mm at ultrasound could avoid unnecessary surgery in these women, which is important in this age group. An endometrial thickness of 5 mm (single layer) or 10 mm (double layer) may also be a useful screening criterion for endometrial carcinoma.

The studies cited above are sufficient to justify a recommendation that all postmenopausal women undergo transvaginal sonography. If the maximum endometrial thickness is only 1 mm, the next follow-up should be performed one year later. In women with an endometrial thickness of 2–3 mm, repeat scans should be obtained every 2–3 months. Women with an endometrial thickness of 4 mm or more should undergo dilation and curettage, and a repeat scan should be obtained three months later. It should be noted in this regard that thickening of the endometrium can occur in women who are receiving estrogen therapy. Transvaginal ultrasound scanning is not beneficial in this subset of patients.

The surrounding hypoechoic ring should not be included in the measurement of endometrial thickness, as it represents the inner layer of the myometrium. This layer is symmetrical and intact in the normal postmenopausal uterus and also in women with endometrial polyps and simple or atypical hyperplasia. With invasive endometrial carcinoma, however, ultrasound demonstrates absence or irregularity of this layer[30].

At present, transvaginal ultrasound appears to be the preferred study for evaluating the endometrium. It is a simple, noninvasive procedure that is well tolerated by most patients and can be used for concomitant screening of the ovaries.

Transvaginal Color and Pulsed Doppler Sonography

The introduction of transvaginal color and pulsed Doppler ultrasound has made it possible to investigate the vascularization of benign and malignant pelvic tumors. When color Doppler is applied to the endometrium, it can increase the sensitivity of the transvaginal scan by displaying blood vessels as color-encoded areas superimposed on the B-mode image[20].

Uterine arteries. When the uterine arteries occupy a normal anatomical location, they can usually be seen lateral to the cervix in a longitudinal scan. A characteristic flow velocity waveform can be recorded from the uterine arteries[21]. A study done in postmenopausal women showed a marked difference in uterine artery flow resistance between patients with endometrial carcinoma and healthy subjects[1c]. This could be a result of tumor angiogenesis. It should be noted in this regard that a marked decrease of uterine artery flow resistance occurs during estrogen replacement therapy and that this effect is only partially reversed with progestin[1c].

Tumor vessels. When endometrial carcinoma is present, areas of neovascularization can be seen within the tumor itself (Fig. 30.**2**). These areas typically appear as thin, irregular ves-

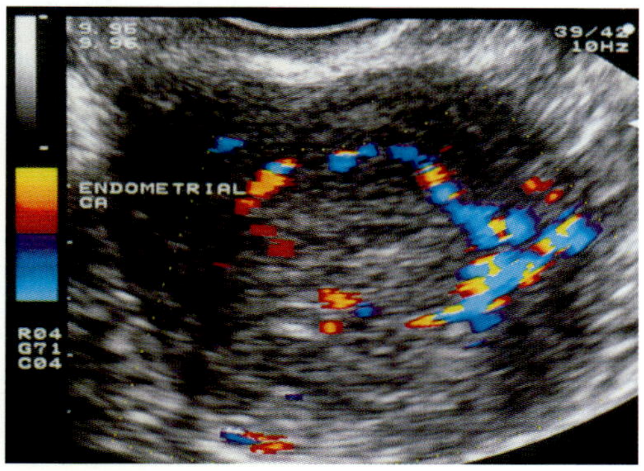

Fig. 30.**2** Color Doppler appearance of peripheral and intratumoral neovascularity in a patient with endometrial carcinoma.

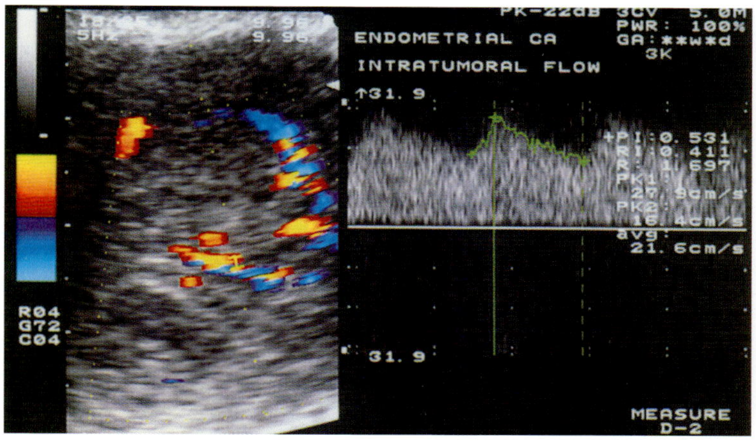

Fig. 30.**3** Analysis of intratumoral blood flow in endometrial carcinoma indicates a low resistance index (RI = 0.41).

sels that are irregularly distributed in the tumor and exhibit abnormal blood flow patterns[12,18,22,36]. Analysis of the waveforms sampled from these areas indicates a very low flow resistance (Fig. 30.**3**). This shows that flow assessment can be used to decrease the false-positive rates in screening examinations based on endometrial thickness[2,22].

Protocol for Transvaginal Sonography

Transabdominal sonography was used for many years to detect endometrial carcinoma and assess the depth of myometrial invasion[24]. Better results can be achieved by scanning with a high-frequency endovaginal probe, as the transvaginal route permits a better evaluation of the endomyometrial junction, which is very important in terms of myometrial invasion. With the advent of color and pulsed Doppler ultrasound, we developed a protocol for transvaginal sonography in order to standardize both the examination and its interpretation.

We evaluated the entire genital tract[22], giving particular attention to the uterus, which was examined in both longitudinal and transverse scans so that all of the endometrium could be evaluated.

B-mode ultrasound. B-mode imaging was used to evaluate the following:

- Endometrial thickness, which is measured in longitudinal section from the anterior subendometrial halo to the opposite side (double-layer measurement, Fig. 30.**4**).
- Endometrial echogenicity, which is classified as hypoechoic or hyperechoic to the myometrium or as nonhomogeneous.
- Presence or absence of intracavitary fluid (Fig. 30.**5**).
- Integrity of the subendometrial halo or presumed depth of myometrial invasion (less than half the myometrium = superficial; more than half = deep).

Color Doppler sonography. Color Doppler sonography was used to evaluate the following:

- Signs indicative of irregular vascular structures (Fig. 30.**6**).
- Localization of tumor vessels (intratumoral = within the endometrial echo; peritumoral = around the outside of the endometrial echo).

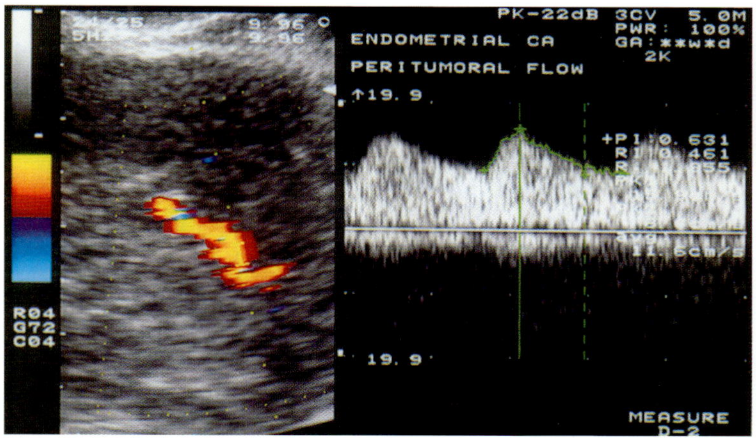

Fig. 30.**4** Analysis of peritumoral blood flow in the same patient indicates a slightly higher resistance index (RI = 0.46).

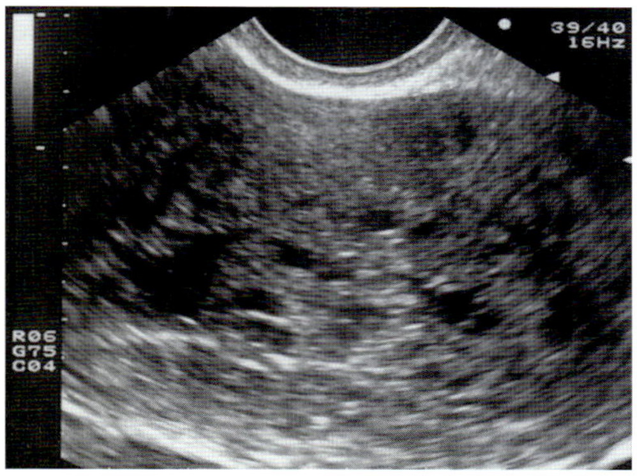

Fig. 30.**5** Postmenopausal uterus with thickened endometrium containing small cysts. The patient is receiving tamoxifen therapy. A small intracavitary fluid collection is visible on the left side.

- Measurements of peak systolic flow velocities and impedance in these vessels using the resistance index (RI). In women on hormone replacement therapy, the effect of the therapy on endometrial thickness and blood flow velocities should be taken into account (Fig. 30.**7**).

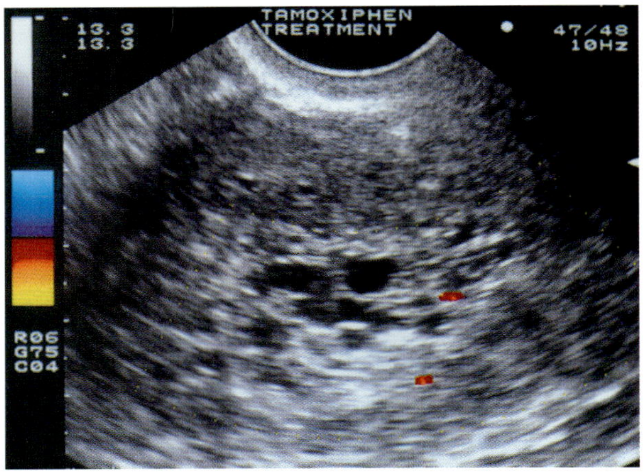

Fig. 30.6 Same patient as in Fig. 30.5. Color Doppler sonography demonstrates peripheral blood vessels.

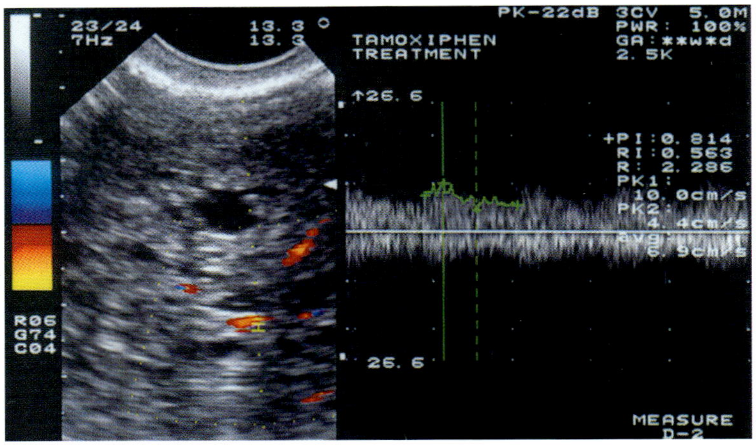

Fig. 30.7 Same patient as in Figs. 30.5 and 30.6. A moderately high RI (0.56) is recorded from the vessels in Fig. 30.6, consistent with a benign process. An endometrial polyp was diagnosed at histological examination.

Authors' Experience

We conducted a study[22] to determine whether the protocol outlined above would be useful in the early detection of endometrial carcinoma and, in positive cases, assessing the extent of myometrial invasion. A group of 750 postmenopausal women were examined by transvaginal sonography one day before a scheduled hysterectomy, which was indicated for various gynecological conditions. Thirty-five of the women had endometrial carcinoma.

Endometrial thickness. Ninety percent of the carcinomas showed an endometrial (tumor) thickness greater than 10 mm (usually > 20 mm). The endometrium in the remaining 10% was between 5 and 10 mm thick. Endometrial thickness was less than 5 mm in 100% of the patients with an atrophic endometrium and in 73% with a normal endometrium. In approximately half of the carcinoma cases, the endometrium had either a hyperechoic or nonhomogeneous appearance at ultrasound.

Intrauterine fluid. While many studies have dealt with endometrial thickness, little attention has been given to the presence of intrauterine fluid. It has been found, however, that the presence of intrauterine fluid correlates with a malignant process in the genital tract arising from the uterine fundus, cervix, ovaries, or fallopian tube. Whenever intrauterine fluid is detected, a detailed tumor search should be initiated. We found that the presence of intrauterine fluid was associated with endometrial carcinoma in most cases, although the absence of intracavitary fluid did not exclude a malignant process.

Myometrial invasion. A subendometrial ring (halo) was visible in 87% of endometrial carcinomas, and interruption of the halo was a sign of myometrial invasion. This invasion was classified as superficial in three cases and deep in 15 cases. The presence and depth of myometrial invasion were accurately determined with ultrasound (relative to histological findings) in 92% of the carcinoma cases.

Tumor vessels. Areas of neovascularity could be detected in the endometrial carcinomas. The newly formed vessels were classified as intratumoral (color pixels within the endometrium) or peritumoral (color pixels bordering the endometrium). The blood flow velocity in the intratumoral vessels was lower than in the peritumoral vessels (Figs. 30.8 and 30.9). Overall, 91% of the endometrial carcinomas displayed abnormal intratumoral and/or peritumoral blood flow with low impedance values (RI = 0.42 ± 0.02). The average RI in the carcinoma vessels was significantly lower than in the endometrial vessels of patients with hyperplasia (Table 30.1).

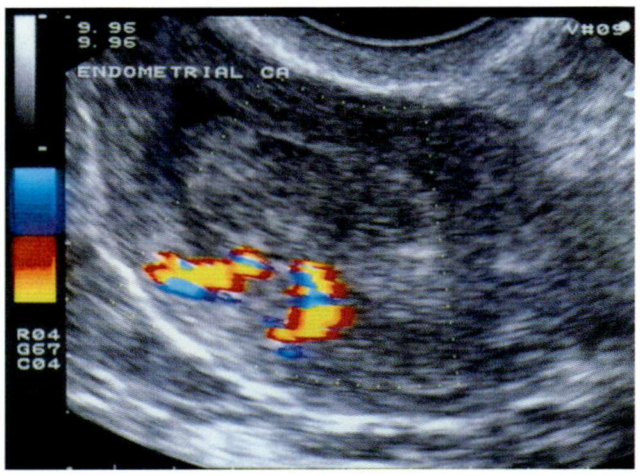

Fig. 30.8 Color Doppler shows areas of intense neovascularization within the myometrium, consistent with myometrial invasion by endometrial carcinoma.

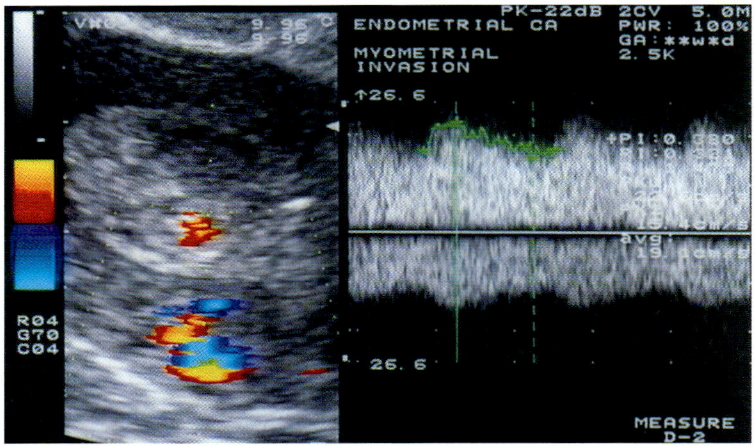

Fig. 30.9 Same patient as in Fig. 30.8. Low impedance values (RI = 0.32, PI = 0.38) are detected within the myometrium. Histology confirmed invasive endometrial carcinoma.

Table 30.**1** Flow detection, resistance index, and peak systolic velocity in relation to histological findings

Histology	Number of patients	Flow detection		Resistance index		Peak systolic velocity (cm/s)	
		Number	%	Mean	SD	Mean	SD
Atrophic	10	0	0	–	–	–	–
Normal	643	0	0	–	–	–	–
Hyperplastic	62	5	8	0,65*	0,05	7,0*	2,1
Endometrial carcinoma	35	32	91	0.42*	0.02	17.1**	2.7

[a] Statistically significant ($p < 0.05$)

[b] Peak systolic velocity only in peritumoral vessels.

With permission from Kurjak A et al: Endometrial carcinoma in postmenopausal women: evaluation by transvaginal color Doppler ultrasonography. Am. J. Obstet. Gynecol. 169 (1993) 1597 – 603.

Summary of the results. In this study the following features were most strongly suggestive of endometrial carcinoma:

- Endometrial thickness greater than 5 mm
- Nonhomogeneous or hyperechoic endometrial structure
- Presence of intrauterine fluid

Myometrial invasion was usually present when the subendometrial halo was interrupted. The intratumoral blood vessels showed a higher flow velocity than the peritumoral blood vessels. The average RI in endometrial carcinoma was significantly lower than the RI in endometrial hyperplasia. This could be an important criterion for differentiating the two conditions in diagnostic practice.

Three cases could not be evaluated because the endometrium had been curetted a week before the ultrasound examination and could not be assessed sonographically. This shows that the sequence of diagnostic procedures is an important aspect that should always be considered if optimum sonographic results are to be achieved.

Three stage I endometrial carcinomas were asymptomatic and were detected by following the above protocol, which is based on increased endometrial thickness (>20 mm) and the detection of intratumoral and/or peritumoral blood flow with a low RI.

Follow-up study. In a more recent study at our department, Illjas et al.[16a] examined 288 postmenopausal women who underwent hysterectomy for various indications (incontinence, prolapse, uterine descent, leiomyomas). Fourteen endometrial carcinomas were detected in postoperative histological examinations.

The diagnostic accuracy of transvaginal Doppler sonography in this study was 93%, as 13 of the 14 endometrial carcinomas were detected with ultrasound prior to hysterectomy. Endometrial thickness was more than 10 mm in 36% of the patients with endometrial carcinoma; it was between 6 and 10 mm in 57%. Only one patient with endometrial carcinoma had an endometrial thickness equal to 5 mm. Areas of neovascularity were found in 93% of the patients. The newly formed vessels were classified as intratumoral (color pixels within the endometrium) or peritumoral (color pixels bordering the endometrium). The average resistance index was 0.39 ± 0.03 in the intratumoral vessels and 0.43 ± 0.03 in the peritumoral vessels. The RI values in the endometrial carcinomas were significantly lower than in the nine cases with endometrial hyperplasia (RI = 0.64 ± 0.05).

Review of the Literature

Transvaginal B-Mode and Doppler Sonography

The appearance of endometrial carcinoma at transvaginal ultrasound has been described in the literature[21] as a thickened, nonhomogeneous endometrial band. In advanced cases the endometrium appears thickened, has an irregular structure, shows inhomogeneous echogenicity, and is poorly delineated from the myometrium (absence of the subendometrial halo). The depth of myometrial invasion, which typically appears as a hypoechoic area, can be determined accurately by transvaginal sonography. However, endometrial hyperplasia cannot be distinguished from carcinoma by B-mode imaging alone, since both conditions present with endometrial thickening[2,12,36]. It appears that diagnostic accuracy can be substantially improved by adding color and pulsed Doppler ultrasound, because most endometrial carcinomas exhibit abnormal blood flow (tumor angiogenesis) with low impedance values[2,16].

Angiogenesis and tumor grade. Abulafia et al.[1] evaluated angiogenesis in endometrial hyperplasia and in stage I endometrial carcinoma and investigated the relationship between neoangiogenesis, tumor grade, and the depth of tumor invasion. Three groups of patients were examined: a control group of patients who underwent hysterectomy for benign uterine changes ($n = 19$), patients with endometrial hyperplasia ($n = 24$), and patients with stage I endometrial carcinoma ($n = 34$). All hysterectomy specimens were stained immunohistochemically for factor VIII-related antigen, which is a sensitive and specific marker for vascular endothelium. Areas of endometrium with the deepest myometrial invasion or with the highest grade of endometrial hyperplasia and highest angiogenic intensity were selected for examination.

Increased angiogenic intensity (based on microvessel counts) was found in complex endometrial hyperplasia compared with simple hyperplasia. Angiogenesis in stage Ia endometrial carcinoma was comparable to that found in complex endometrial hyperplasia. Angiogenesis in invasive endometrial carcinoma (stages Ib and Ic) was more pronounced than in complex endometrial hyperplasia or stage Ia endometrial carcinoma. Higher tumor grades were found to be directly correlated with angiogenic intensity. Due to the limited number of patients with stage Ib and Ic endometrial carcinoma, the depth of tumor invasion could not be analyzed independently of tumor stage. Despite this limitation, it was found that angiogenesis increased with the depth of invasion and tumor grade. This suggests that the extent of neovascularization in endometrial carcinoma is linked to the grade of tumor differentiation.

The key result of this study is that both endometrial hyperplasia and endometrial carcinoma have a rich vascular supply and can therefore be detected with sensitive Doppler instruments. Also, the intensity of angiogenesis in stage I endometrial carcinomas shows a direct correlation with the depth of tumor invasion and tumor grade.

Uterine artery PI. Bourne et al.[2] studied flow impedance in the uterine arteries of women with endometrial carcinoma ($n = 17$), women with normal endometrium ($n = 85$), and women receiving hormonal therapy ($n = 35$). The authors defined an arbitrary cutoff value of <1.5 for the PI as a positive test result (Fig. 30.**10**). Using this cutoff, they achieved a positive predictive value of 94% and a negative predictive value of 91% in women with postmenopausal bleeding. Better predictive values could probably have been achieved by sampling more specific vessels than the uterine artery, such as the intratumoral and peritumoral vessels[18,36]. Moreover, the PI appears to be a less sensitive and specific impedance parameter than the RI in the evaluation of tumor vessels.

RI in the arcuate and spiral arteries. Hata et al.[16] found a significantly lower resistance index ($RI = 0.535 \pm 0.158$) in the arcuate arteries in 10 endometrial carcinomas than in the normal uterus ($RI = 0.767 \pm 0.75$). The RI in this study was higher than the RI calculated in our studies because Hata et al. measured values in the arcuate arteries whereas we sampled the spiral arteries. Also, the study by Hata et al. was performed in only 10 cancer patients.

Merce et al.[26] examined 45 patients with metrorrhagia and compared their findings with 19 healthy volunteers. The authors used duplex sonography and sampled Doppler signals from both uterine arteries and from the intramyometrial (radial and arcuate) arteries. A significantly lower RI was found in the uterine and intramyometrial arteries of women with abnormal endometrial findings, including two carcinomas, than in women with normal findings. The authors concluded that the RI of the intramyometrial (radial and arcuate) arteries was a considerably more accurate and specific indicator than the RI of the uterine arteries and was therefore a better predictor of positive findings.

Criteria for evaluating uterine vascularity. Aleem et al.[1a] attempted to establish color and pulsed Doppler sonographic criteria for evaluating uterine vascularity in postmenopausal women in order to reduce the number of unnecessary dilation and curettage procedures. The prospective study involved 42 postmenopausal patients who were examined before undergoing fractional curettage. Twenty patients had symptoms such as vaginal bleeding or an enlarged uterus, and 22 patients were asymptomatic. The criteria used to evaluate the endometrium were endometrial thickness (cutoff value of 8 mm), rates of visualization, and the distribution pattern of myometrial (peritumoral) and endometrial (intratumoral) vessels, along with pulsatility and resistance indices.

Endometrial thickness was greater than 8 mm in all cases of endometrial carcinoma (14 of 14 cases), endometrial hyperplasia (8 of 8 cases), and one endometrial polyp. Endometrial thickness was less than 8 mm in the nine patients with leiomyomas and in the asymptomatic controls. The visualization rates of myometrial and endometrial vessels in cases of endometrial carcinoma were 93% and 43%, respectively, which were significantly higher than for cases with benign endometrial changes. Dense vascularity was found in 80% of the endometrial carcinomas, contrasting with the "loose" vascularity that is more typical of endometrial hyperplasia. Intratumoral vessels were seen in 43% of the carcinoma patients but in only 12% of the patients with endometrial hyperplasia. The average PI and RI values for these vessels in cases of endometrial carcinoma were 0.73 ± 0.11 and 0.50 ± 0.05, respectively.

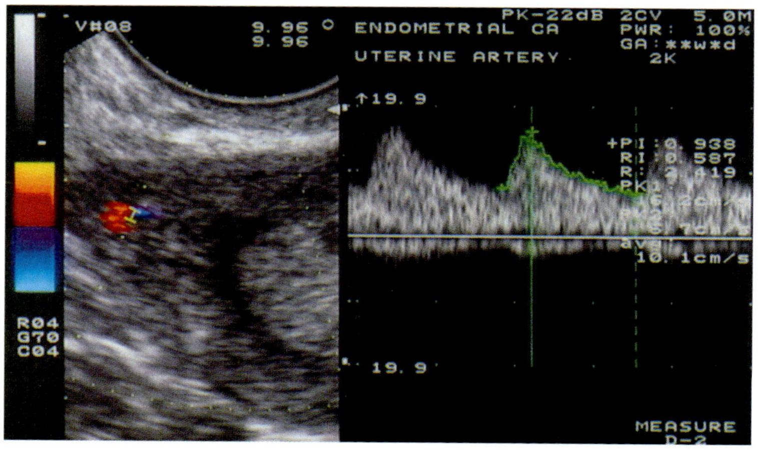

Fig. 30.**10** Same patient as in Figs. 30.**8** and 30.**9**. A low RI (0.59) is also found in the uterine arteries of this patient with invasive endometrial carcinoma. This finding indicates myometrial invasion.

Comments. Transvaginal color Doppler sonography appears to be an effective method of screening for endometrial carcinoma and ovarian carcinoma in the same sitting. Hence, it can provide an important addition to oncological preventive medicine in women. The use of this technique could also reduce the number of dilation and curettage procedures, thereby reducing the costs to payers and lowering the risks to patients.

Intra-arterial chemotherapy. Hata et al.[15a] used transvaginal Doppler ultrasound and MRI to assess the efficacy of intra-arterial chemotherapy for endometrial carcinoma. The RI in the intratumoral vessels showed no correlation with the tumor volume, but significant differences were measured between the RI values before ($mean = 0.58 \pm 0.15$) and after ($mean = 0.77 \pm 0.13$) intra-arterial chemotherapy. The authors' conclusion is that transvaginal color Doppler is a useful method for evaluating hemodynamic changes after intra-arterial chemotherapy and can demonstrate the response of the endometrium to the therapy.

Three-dimensional Sonography—New Discoveries with a New Method?

Significance of endometrial volume. In a study by Hata et al.[15], 3D ultrasound was used to assess the significance of endometrial volume in women with postmenopausal bleeding[17a]. The endometrial volume in cases of endometrial carcinoma differed significantly from the volume of normal or hyperplastic endometria. A cutoff volume of 13 ml was 100% sensitive in the diagnosis of carcinoma. There was only one false-positive result in a patient with endometrial hyperplasia, resulting in a specificity of 98.8% and a positive predictive value of 91.7%. Endometrial volume correlated with tumor grade: both the thickness and volume of the endometrium were smaller in well-differentiated tumors than in moderately or poorly differentiated tumors. Endometrial thickness and volume in patients who subsequently had to undergo surgery were compared with the depth of myometrial invasion and tumor stage. It was found that the sonographically determined values for endometrial volume and thickness rose with increasing depth of invasion. In addition, a relatively large primary tumor volume was found in patients with advanced disease. Only patients with tumor volumes greater than 25 ml were found to have positive pelvic lymph nodes at operation. These findings agree with a previous study by Schink et al.[34a] in which tumors smaller than 2.0 cm in diameter were associated with only a 4% risk of lymph node involvement and a 98% five-year survival rate. Thus, tumor size has proved to be a significant prognostic factor that is independent of tumor grade and the depth of myometrial invasion.

Visualization of the uterine cavity. In a study published by Bonilla-Musoles et al.[1b], endometrial thickness and homogeneity in 36 women with postmenopausal bleeding were evaluated with 3 D ultrasound following distention of the uterine cavity with sterile saline solution. The results of 3 D hysterosonography were compared with transvaginal sonography, transvaginal sonohysterography, transvaginal color Doppler, and hysteroscopy. The results with transvaginal 3 D sonography agreed with the findings obtained by hysteroscopy. Thus, 3 D hysterosonography with a negative contrast medium appears to improve the visualization of the uterine cavity and the evaluation of endometrial thickness and of myometrial and cervical invasion in patients with endometrial carcinoma.

In our experience, the 3 D power Doppler mode is excellent for evaluating the tortuous, thin-walled vessels that grow into the uterine cavity and myometrium when an invasive cancer is present (Fig. 30.**11**). This is facilitated by the three-dimensional and sector image display options that are available (Fig. 30.**12**).

Indications. Three-dimensional sonography is currently used for the following applications:

- Defining the complex vascular structures, which often appear fragmentary, in patients with endometrial carcinoma (this may be facilitated by contrast enhancement)
- Determining the depth of invasion of endometrial carcinoma (staging)
- Monitoring tumor response to chemotherapy and/or radiotherapy[17b].

Interpretation of 3 D information. Problems still exist in the interpretation of 3 D power Doppler data. The examiner must be familiar with the complexities of data processing for proper digital manipulation of the 3 D display. It should also be realized that this data processing is very time-consuming and that optimal, standardized software must be used to obtain reproducible results. Various parameters such as the pulse repetition frequency (PRF), wall filter setting, receiver gain, color priority, and frame rate must be optimized for the 3 D display. By merely changing the color setting, the operator can radically alter the quantitative results and the 3 D rendering of vascular structures. We have found that information loss does not occur with power Doppler imaging, even with a perpendicular beam–vessel angle, and that very little decrease in color intensity occurs during diastole.

With the technology available today, the assessment of vascular density with 3 D power Doppler ultrasound is a very subjective matter. It is reasonable to expect that this problem will be solved by integrated software that can perform 3 D blood-flow measurements in tumors. The use of contrast media may aid in the detection of very small tumor vessels and thus improve the sensitivity and specificity of the method.

Comments. Three-dimensional ultrasound with a power Doppler facility could play an important role in assessing the depth of invasion of endometrial carcinoma. The ability to simultaneously display areas of neovascularization in the endometrium and myometrium increases diagnostic accuracy. Because the method had a negative predictive value of 100% in our studies[17b], it could justify the election of less radical surgery in patients who do not have deep myometrial invasion, helping to reduce morbidity and lower costs in the health care industry[13a]. Meanwhile, patients with a tumor stage that warrants a primary aggressive approach can be referred directly for the appropriate treatment.

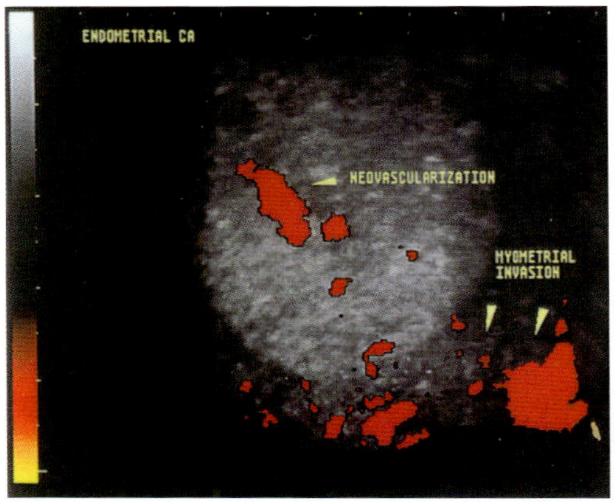

Fig. 30.**11** Three-dimensional image of abnormal endometrial proliferation in a postmenopausal woman who had undergone fractional curettage. Histopathological examination revealed endometrial carcinoma. Power Doppler demonstrates both peripheral and central neovascularity. Careful examination of the posterior uterine wall showed abnormal vessels in the inner portions of the myometrium, consistent with superficial myometrial invasion.

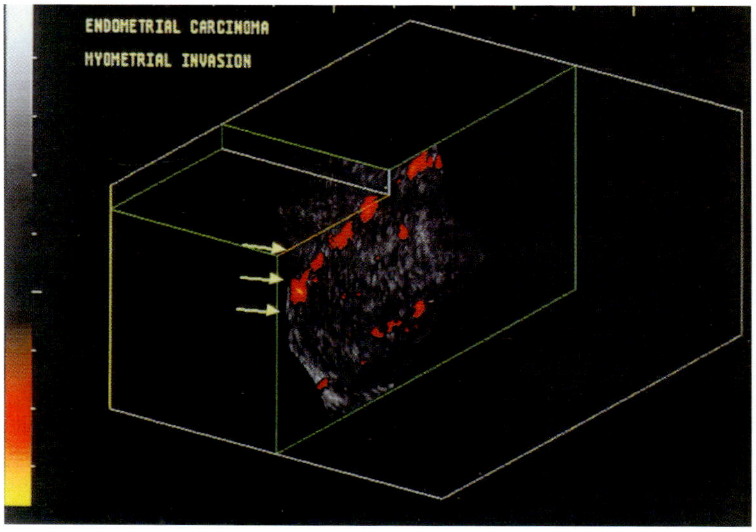

Fig. 30.**12** The "niche display mode" portrays a slice from a three-dimensional image. Rotating and shifting the coordinate system is helpful in detecting irregular myometrial vessels that would indicate myometrial invasion. The stored image volume is subdivided into a series of parallel slices, and the examiner can scroll through them to locate a myometrial area with signs of neovascularization. This type of study can accurately determine both the location and depth of myometrial infiltration.

Uterine Sarcoma

Uterine sarcoma is a rare tumor that accounts for only 1–3% of all genital tract malignancies in women and from 3% to 7.4% of malignant tumors of the uterine corpus[32] (Fig. 30.**13**). It is a rare but extremely aggressive tumor that tends to metastasize early, leading to death[9]. There are still many unanswered questions regarding this disease, and there is still no satisfactory procedure for making an early and accurate diagnosis. Moreover, it is expected that the incidence of uterine sarcomatous disease will rise in the near future as more gynecologists opt for the conservative management of uterine leiomyomas[27].

Authors' Experience

In patients who were treated at our center for uterine sarcoma, all documented findings including surgery reports and histological results were analyzed for this study. The RI in both uterine arteries, the RI in the newly formed tumor vessels, and peak systolic velocity were determined. To our knowledge, this is the largest clinical series of uterine sarcomas to be evaluated with Doppler ultrasound. The data from eight patients with uterine sarcoma could be analyzed; 75% of the women were postmenopausal, and the most frequent complaint was vaginal bleeding.

RI and peak systolic velocity. Irregular blood vessels were detected in all sarcomas (100%), while tumor blood flow was detected in only 30% of the myomas (Fig. 30.**14**). The RI values decreased from the normal uterus to the myomatous uterus to the sarcomatous uterus. The mean RI in the sarcomas was 0.37 ± 0.03 (Fig. 30.**15**). The peak systolic velocity also decreased from normal to myomatous to sarcomatous uterus (16.8 ± 6.4 cm/s in sarcomas).

We studied tumor blood flow in both benign (myomas) and malignant uterine tumors (sarcomas) in order to discover criteria that might indicate a sarcoma and to improve the accuracy of ultrasound in differentiating between these tumors. Ultrasound in patients with uterine sarcoma typically showed irregular, scattered, small-caliber vessels with low impedance and low intratumoral and peritumoral blood flow velocities. Additionally, both uterine arteries in sarcoma patients showed a low RI and a low peak systolic velocity compared with women who had a normal or myomatous uterus (Fig. 30.**16**).

Comments. There has been an urgent need for a diagnostic method that can ensure the precise differentiation in vivo of benign and malignant uterine tumors. Transvaginal color Doppler sonography appears to have great potential in this regard.

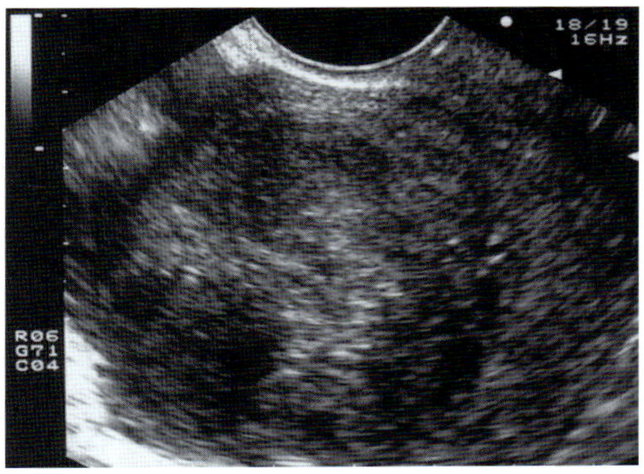

Fig. 30.**13** Nonhomogeneous uterine tumor in a postmenopausal woman. The nonhomogeneity of the mass is caused by degenerative changes and intratumoral hemorrhage.

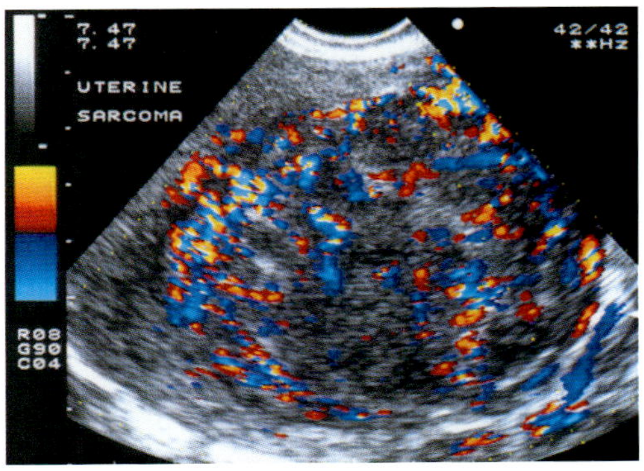

Fig. 30.**14** Same patient as in Fig. 30.**13**. Color Doppler image shows numerous, irregularly distributed vessels consistent with a malignant tumor.

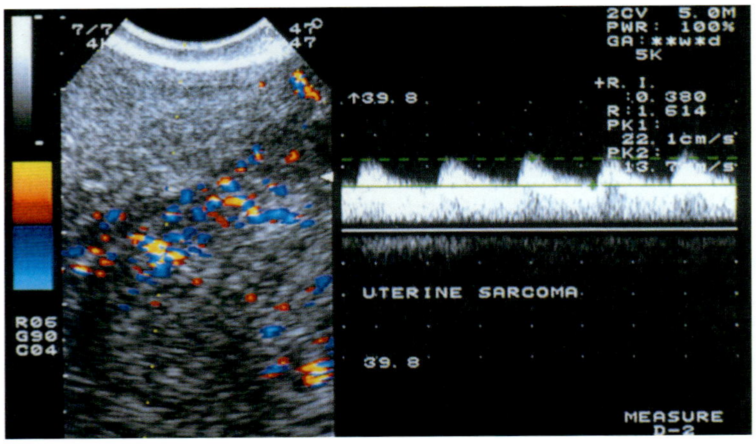

Fig. 30.**15** Same patient as in Figs. 30.**13** and 30.**14**. Analysis of the power Doppler waveform indicates a low RI (0.34).

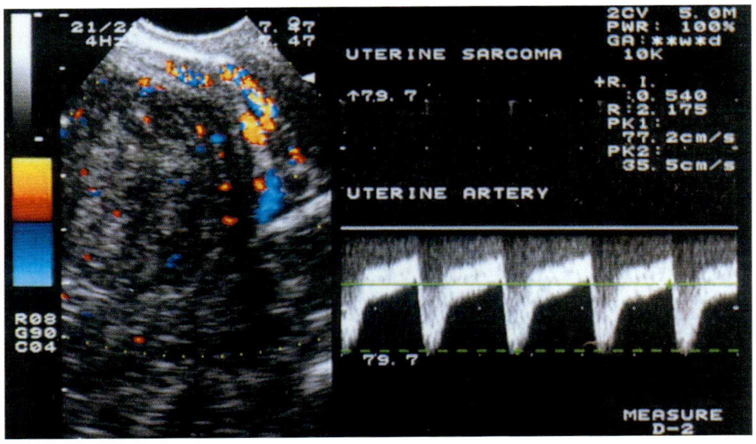

Fig. 30.**16** Same patient as in Figs. 30.**13**– 30.**15**. The low RI (0.54) in the uterine artery is a sign of advanced uterine sarcoma.

Cervical Carcinoma

The uterine cervix varies considerably in size and thickness. It occupies a larger portion of the uterus before menarche and after menopause than in women of childbearing age. Visualization of the cervix with transvaginal ultrasound may prove difficult if the probe is placed too close to the cervix. This problem is solved by withdrawing the probe slightly and by angling it slightly backward in patients with an anteflexed uterus or slightly forward in patients with a retroflexed uterus[13].

Positioning the transducer. The success of color Doppler sonography in the diagnosis of uterine malignancies does not extend to cervical carcinoma. Hata et al.[16] were unable to detect abnormal blood flow signals in half of their cases, even in patients with advanced cervical cancers. It is assumed that the new blood vessels that form during the early stage of cervical carcinoma are too small to be detected with current ultrasound instruments and that the blood flow velocity is below the resolution limit of the scanners. These problems arise even in the examination of advanced cervical cancers.

Uterine artery. We examined blood flow in the descending branch of the uterine artery in 89 patients with histologically proven cervical carcinoma and in a control group of 24 healthy women. We were unable to visualize the descending branch of the uterine artery in 18% (16/89) of the patients with cervical carcinoma and in 17% (4/24) of the controls. The RI was markedly lower in the carcinoma patients than in the healthy women. The PI in the carcinoma patients was also lower than in the control group, but statistical significance was borderline.

Nevertheless, we hope that color Doppler sonography will prove to be a useful modality in the follow-up of patients who have been treated for cervical carcinoma.

Conclusion

Transvaginal color Doppler sonography is a noninvasive procedure for evaluating the vascularity of uterine tumors. It can be used in any given patient as often as desired. The ability to display and analyze specific flow velocity waveforms will undoubtedly yield new data on normal and abnormal vascular structures in and around tumors. With color Doppler ultrasound, we have a technology that may enable us to reliably differentiate benign from malignant tumors in the foreseeable future.

References

1 Abulafia O, Triest WE, Sherer DM, Hansen CC, Ghezzi F: Angiogenesis in endometrial hyperplasia and stage I endometrial carcinoma. Obstet. Gynecol. 86 (1995) 479–485

1a Allem F, Predanic M, Calame R, Moukhtar M, Pennisi J: Transvaginal color and pulsed Doppler sonography of the endometrium: a possible role in reducing the number of dilatation and curettage procedures. J. Ultrasound Med. 14 (1995) 139 –145

1b Bonilla-Musoles F, Raga F, Osborne NG, Blanes J, Coelho F: Three-dimensional hysterosonography for the study of endometrial tumors: comparison with conventional transvaginal sonography, hysterosalpingography and hysteroscopy. Gynecol. Oncol. 65 (1997) 245–252

1c Bourne TH, Campbell S, Steer CV, Royston P, Whitehead MI, Collins WP: Detection of endometrial cancer by transvaginal ultrasonography with color flow imaging and blood flow analysis: A preliminary report. Gynecol. Oncol. 40 (1991) 253 –259

2 Bourne TH, Reynold KMM, Campbell S: Screening for ovarian and uterine carcinoma. In Nyberg DA, Hill LM, Bohm-Velez M, Mendelson EB (eds.): Transvaginal Ultrasound. Mosby Year Book, St. Louis 1992, 267–283

3 Cacciatore B, Lehtovirta P, Wahlstrom: Preoperative sonographic evaluation of endometrial cancer. Amer. J. Obstet. Gynecol. 160 (1989) 133–137

4 Choo Y, Mak K, Hsu C, Wong T: Postmenopausal uterine bleeding of monorganic cause. Obstet. Gynecol. 66 (1985) 225–231

5 Cramer DW, Knapp RC: Review of epidemiologic studies of endometrial cancer and exogenous estrogen. Obstet. Gynecol. 54 (1979) 521–526

6 Creasman WE, Soper JT: Endometrial carcinoma. In Garcia CR, Mikuta JJ, Rosenblum NG (eds.): Current therapy in surgical gynecology. Decker Inc., Toronto 1987, 200

7 Creasman WT, Boronow RC, Morrow CP, DiSaia PJ, Blessing J: Adenocarcinoma of the endometrium and its metastatic lymph node potential. Gynecol. Oncol. 4 (1976) 239–243

8 Creasman WT, Weed JC: Carcinoma of endometrium (FIGO stages I & II). Clinical features and management. In Copplesand M (ed.): Gynecologic Oncology. Churchil Livingstone, Edinburgh 1981, 562–577

9 El-Naggar AK, Abdul-Karim FW, Silva EG, McLemore D, Garnesy L: Uterine stromal neoplasms: A clinicopathologic and DNA fow cytometric correlation. Hum. Pathol. 22 (1991) 897–903

10 Fleischer AC, Dudley BS, Entman SS et al.: Myometrial invasion by endometrial carcinoma: Sonographic assessment. Radiology 162 (1987) 307–310

11 Fleischer AC, Gordon AN, Entman SS: Transabdominal and transvaginal sonography of pelvic masses. Ultrasound Med. Biol. 15 (1989) 529–533

12 Fleischer AC, Gordon AN, Entman SS: Transvaginal scanning of the endometrium. J. Clin. Ultrasound 18 (1990) 337–349

13 Fleischer AC, Kepple DM: Benign conditions of the uterus, cervix and endometrium. In Nyberg DA, Hill LM, Bohm-Velez M, Mendelson EB (eds.): Transvaginal ultrasound. Mosby Year Book, St Louis 1992, 38

13a Gabrielli S, Marabini A, Bevini M et al.: Transvaginal sonography vs. hysteroscopy in the preoperative staging of endometrial carcinoma. Ultrasound Obstet. Gynecol. 7 (1996) 443–446

14 Gelfand MM, Forenczy A: A prospective 1-year study of estrogen and progestin in postmenopausal women: effects on the endometrium. Obstet. Gynecol. 74 (1989) 398–403

15 Goldstein SR, Nachtigall M, Sayder JK, Nachtigall L: Endometrial assessment by vaginal ultrasonography before endometrial sampling in patients with postmenopausal bleeding. Amer. J. Obstet. Gynecol. 163 (1990) 119–123

15a Hata T, Hata K, Fujiwaki R, Manabe A, Kitao M: Hypertensive intraarterial chemotherapy for endometrial carcinoma by transvaginal Doppler ultrasound and magnetic resonance imaging. J. Clin. Ultrasound 23 (1995) 407–411

16 Hata T, Hata K, Senoh D et al.: Doppler ultrasound assessment of tumor vascularity in gynecological disorders. J. Ultrasound Med. 8 (1989) 309–314

16a Illjas M, Marton U, Hanzevacki M: Color Doppler in the assessment of endometrial carcinoma. In Kurjak A, Kupesic S (eds.): Doppler in gynecology and infertility. Edizioni Internazionali, Rom 1996, 134–137

17 Ivorsen DE, Sagdal E: The value of endometrial cytology: A comparative study of the Gravlee Jet-Washer, Isaacs cell Sampcher and Endoscan versus curretage in 600 patients. Obstet. Gynecol. Surg. 40 (1985) 14–20

17a Jurkovic D, Geipiel A, Gruboeck K, Jauniaux E, Natucci M, Campbell S: Three dimensional ultrasound for the assessment of uterine morphology. Ultrasound Obstet. Gynecol. 67 (1991) 2791–2794

17b Kupesic S, Kurjak A: Three dimensional power Doppler ultrasound in the staging of endometrial adenocarcinoma. Ultrasound Obstet. Gynecol. (1998)

18 Kupesic-Urek S, Shalan H, Kurjak A: Early detection of endometrial cancer by transvaginal color Doppler. Eur. J. Obstet. Gynecol. Reprod. Biol. 49 (1993) 46–49

19 Kurjak A, Kupesic-Urek S, Miric D: The assessment of benign uterine tumors by transvaginal color Doppler. Ultrasound Med. Biol. 18 (1992) 645–649

20 Kurjak A, Zalud I: The characterization of uterine tumors by transvaginal color Doppler. Ultrasound Obstet. Gynecol. 1 (1991) 50–52

21 Kurjak A, Zalud I: Uterine masses. In Kurjak A (ed.): Transvaginal color Doppler. Parthenon, Lancs New Jersey 1991, 123–135

22 Kurjak A, Shalan H, Sosic A, Benic S, Zudenigo D, Kupesic S: Endometrial carcinoma in postmenopausal women: Evaluation by transvaginal color Doppler sonography. Amer. J. Obstet. Gynecol. 169(6) (1993) 1597–1603

23 Lehtovirta P, Cacciatore B, Wahlstrom T et al.: Ultrasonic assessment of endometrial cancer invasion. J. Clin. Ultrasound 15 (1987) 519–524

24 Lehtovirta P, Cacciatore B, Wahlstrom T, Ylostalo P: Ultrasonic assessment of endometrial cancer invasion. J. Clin. Ultrasound 15 (1987) 519–524

25 Mendelson EB, Bohm-Velez M, Joseph N, Neiman HL: Endometrial abnormalities: Evaluation with transvaginal sonography. Amer. J. Radiology 150 (1988) 139–142

26 Merce LT, Garcia L, De La Fuente F: Doppler ultrasound assessment of endometrial pathology. Acta Obstet. Gynecol. Scand. 70 (1991) 525–530

27 Meyer WR, Mayer AR, Daimond MP, Carcangiu ML, Schwartz PE, DeChernay AH: Unsuspected leiomyosarcoma: Treatment with a gonadotropin-releasing hormone analogue. Obstet. Gynecol. 75 (1990) 529–531

28 Morrow CHP, DiSaia PH, Townsend DE: Current management of endometrial carcinoma. Obstet. Gynecol. 42 (1973) 399–406

29 Nasri MA, Shepherd JH, Setchwell ME, Lowe DG, Chard T: The role of vaginal scan in measurement of endometrial thickness in postmenopausal women. Brit. J. Obstet. Gynecol. 98 (1991) 740–475

30 Nasri MN, Cuast GJ: Correlations of ultrasound findings and endometrial histopathology in postmenopausal women. Brit. J. Obstet. Gynecol. 96 (1989) 1333–1338

31 Office of population censuses and survey: Mortality statistics 1989. England and Wales. Her Majesty‹s Stationary Office, London 1990

32 Olah KS, Gee H, Blunt S, Dunn JA, Chan KK: Retrospective analysis of 318 cases of uterine sarcoma. Eur. J. Cancer 27 (1991) 1095–1099

33 Osmers R: Transvaginal sonography in endometrial cancer. Ultrasound Obstet. Gynecol. (1992) 2–3

34 Requard CK, Wicks JD, Mettler FA Jr: Ultrasonography in the staging of endometrial adenocarcinoma. Radiology 140 (1981) 781–785

34a Schink J, Rademaker A, Miller DS, Larain J: Tumor size in endometrial cancer. Cancer 67 (1991) 2791–2794

35 Schoenfeld A, Levmvi H, Hirsch M, Pardo J, Ovadia J: Transvaginal sonography in postmenopausal women. J. Clin. Ultrasound 18 (1991) 350–358

36 Shalan H., Kurjak S, Sosic A: Endometrial carcinoma in postmenopausal women and transvaginal color and pulsed Doppler sonography. Ultrasound Obstet. Gynecol. 2(Suppl.1) (1992) 160

37 Voigt LF, Weiss NS, Daling JR, McKnight B, Van Belle G: Progestagen supplementation of exogenous estrogens and risk of endometrial cancer. Lancet 338 (1991) 274–276

38 Weid G, Bartels PH, Bobbo M et al.: Frequency and reliability of diagnostic cytology of the female genital tract. Acta Cytol. 25 (1981) 543–549

39 Wikland M, Granberg S, Karlsson B: Assessment of the endometrium in the postmenopausal women by vaginal sonography. Ultrasound Quarterly 10(1) (1992) 15–27

40 Zucker PK, Kadson EJ, Feldstein ML: The validity of Pap smear parameters as predictors of endometrial pathology in menopausal women. Cancer 58 (1986) 2258–2263

31 Blood Flow Changes in Cervical Carcinoma Treated by Primary Chemotherapy

C. Villena-Heinsen, A. K. Ertan, M. Holländer, and W. Schmidt

Treatment of Cervical Carcinoma

Tumor volume. In patients with cervical carcinoma, the tumor volume has a critical impact on local tumor control, the incidence of distant metastases, the disease-free interval, and total survival time[3]. Accordingly, patients with a "bulky" tumor have a poorer prognosis than women with smaller tumors at the same FIGO (International Federation of Gynecologists and Obstetricians) stage[6]. Both radiotherapy and surgical excision are of only limited value as curative treatments [15].

Cisplatin-based combination chemotherapy. Because stage-specific survival rates have not improved with traditional treatment modalities during the past 40 years[16], clinical experimentation is justified in an effort to improve treatment outcomes in this population. Primary, cisplatin-based systemic combination chemotherapy has shown high response rates between 60% and 100%[1,17,24]. Sardi et al. reported a significant prolongation of disease-free interval ($p = 0.009$) and total survival ($p = 0.05$) in patients with a bulky tumor more than 4 cm in diameter[17]. Good local control is achieved with primary intra-arterial chemotherapy[9,18,21].

Intra-arterial chemotherapy. Bilateral, superselective intra-arterial chemotherapy with cisplatin is done to achieve a maximum reduction in tumor size with minimum systemic toxicity, the goal being to improve the operability of the tumor. Response to the therapy, as measured by a reduction in tumor volume, was objectively and sequentially documented by manual palpation, B-mode ultrasound, and MR imaging. The progression of tumor markers was also documented.

Assessing Treatment Response with Pulsed Color Doppler Sonography

We also studied pulsed color Doppler sonography in our search for an additional indicator of treatment response. The goal of our study was to evaluate the accuracy of this procedure in monitoring the response of large cervical carcinomas to neoadjuvant intra-arterial chemotherapy.

Authors' Studies

Patients and Methods

Eight patients (7 premenopausal) between 32 and 54 years of age (average 43 years) with a bulky tumor at clinical stage Ib2 to IIb were examined by color Doppler sonography immediately before their first and second chemotherapy cycles and prior to operative treatment. A clinical and MRI assessment of tumor size was also performed at these times. The examination protocol, equipment technology, and instrument settings were standardized in this prospective study. The examinations were performed with an Acuson 128 XP10 using a 5 MHz endovaginal probe and color Doppler system.

Parameters. The primary tumor was examined closely and systematically for blood vessels using color Doppler ultrasound. The following parameters were documented in the tumor area:

- Number of sampled vascular segments
- Peak systolic blood flow velocity within the sampled vascular segments
- Mean resistance index (RI_{mean}) within the sampled vascular segments

The RI_{mean} in the tumor-supplying vessels in the area of both uterine arteries was also determined:

$$RI = \frac{\text{peak systolic velocity} - \text{end-diastolic velocity}}{\text{peak systolic velocity}}$$

Treatment Regimen

Both uterine arteries were punctured and catheterized by the transfemoral route using the Seldinger technique, and inflow for the pending chemotherapy was confirmed by spiral CT. Next, 25 mg of cisplatin was infused into the uterine artery for two hours. This was repeated 24 hours later.

Three weeks later, a second treatment cycle was performed according to the same regimen.

After another three-week interval, a radical Wertheim–Meigs operation was performed with para-aortic and paracaval lymph node sampling. Adjuvant radiotherapy was added in some cases, depending on the definitive tumor histology.

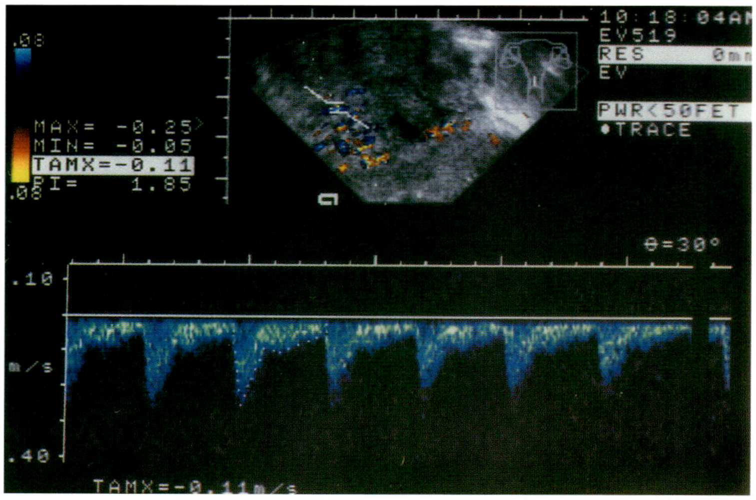

Fig. 31.**1** Pretherapeutic examination (before the first chemotherapy cycle). Color Doppler image shows dense intratumoral vascularity in the transverse scan.

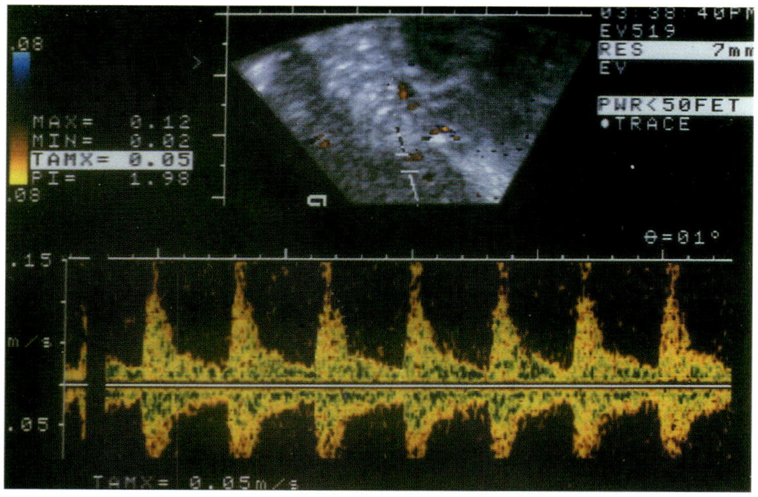

Fig. 31.**2** A reduction of intratumoral vascularity is noted before the second cycle of intra-arterial chemotherapy.

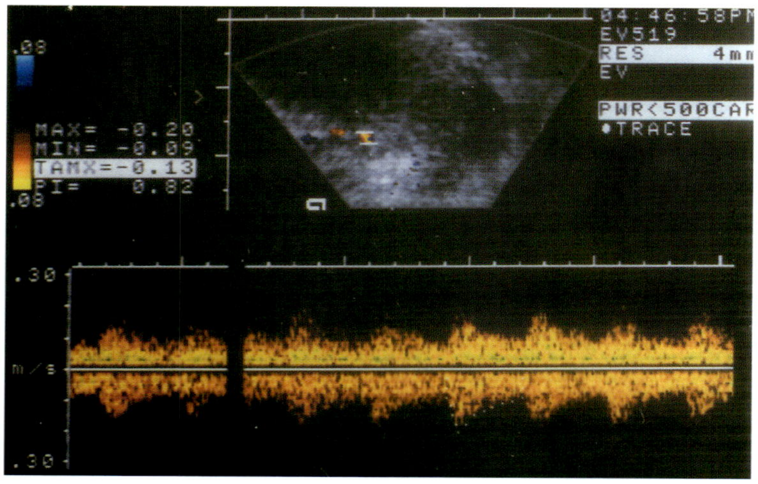

Fig. 31.**3** A further decrease in vascularity is noted immediately before surgery (3 weeks after the second chemotherapy cycle).

Results

Tumor volume. The mean tumor volume was 98 cm^3 before treatment began (the smallest tumor volume was 36 cm^3, the largest 203 cm^3). All of the tumors responded to intra-arterial chemotherapy. The preoperative tumor volume after two cycles of intra-arterial chemotherapy was 46 cm^3 (15–83 cm^3). The average reduction in tumor volume was 51% (28–80%) ($p = 0.01$). Three patients showed a partial remission (>50% reduction in tumor volume), and five showed a minor response (<50% reduction in tumor volume). The reduction in tumor volume was also clinically apparent in the eight patients.

The goal of improving operability was achieved in seven of the eight patients. One patient, who had a stage T2b tumor with an initial volume of 203 cm^3, showed positive local response but developed fulminating, unilateral parametrial spread to the pelvic wall. Surgical resection was withheld in this case, therefore, and radiotherapy was administered.

SCC tumor marker. Four of eight patients showed an abnormally high level of serum SCC tumor marker before treatment. The mean level was 5.1 ng/ml, the lowest level was 3.6 ng/ml, and the highest was 6.4 ng/ml (normal serum level ≤ 2.5 ng/ml). All of the patients showed a fall of tumor marker level in response to chemotherapy ($p = 0.06$). Only one of four patients still had an abnormally high SCC level prior to surgery. After surgery it returned to normal and remained there. Following two cycles of intra-arterial chemotherapy, a positive correlation was found between the reduction in tumor volume and the fall in tumor marker level (Spearman correlation coefficient $r = 0.65$; $p = 0.08$).

Intratumoral vascularity. A steady decrease was observed in the number of vascular segments that could be defined and sampled with Doppler ultrasound, the peak systolic flow velocity, and the RI_{mean}. An average of 3.5 intratumoral vascular segments were sampled before treatment, compared with 3.2 before the second treatment cycle and 2.7 just before surgery. This decrease was not statistically significant. A statistically significant correlation was found between the reduction in vessel count and the reduction in tumor volume following two cycles of intra-arterial chemotherapy ($p = 0.02$). Figures 31.**1**–31.**3** document the decline of vascularity in response to therapy.

Doppler parameters. The peak systolic flow velocity was 0.16, 0.14, and 0.13 m/s at the successive points in the study.

A corresponding pattern was seen for RI_{mean}, which was 0.73 before treatment and 0.71 and 0.67 preoperatively. These trends were not statistically significant.

The RI_{mean} in both uterine arteries showed very little change in response to therapy. The mean value for both arteries was initially 0.79. It was still 0.79 before the second chemotherapy cycle and 0.76 before surgery.

These Doppler parameters did not correlate with the therapeutic response measured by a reduction in tumor volume and a fall of tumor marker levels. Table 31.**1** shows the progression of the mean values of the Doppler parameters during the study.

Table 31.1 Mean vessel counts and Doppler parameters found at different examination times

Intratumoral vascularity				Uterine arteries	
	Vessel count	V_{max} (m/s)	RI_{mean}	RI_{mean}, right artery	RI_{mean}, left artery
Before 1st chemotherapy	3.5	0.16	0.73	0.79	0.79
Before 2nd chemotherapy	3.2	0.14	0.71	0.78	0.79
Preoperative	2.7	0.13	0.67	0.75	0.78

Discussion

Value of Color Doppler Sonography in Monitoring Treatment Response

Relatively little has been published on the use of color Doppler sonography in evaluating treatment response. On the whole, however, the procedure appears to be useful for this purpose.

Breast carcinoma. Kedar et al.[8] found a strong correlation between changes in tumor vascularity and changes in tumor size. When a specially designed quantitative scoring system was applied in women with breast carcinoma[4,5], it was concluded that color Doppler sonography was of value in predicting the response of the cancer to primary (neoadjuvant) medical therapy.

Fibrocystic breast changes and nonpuerperal mastitis. Madjar et al.[12] observed a measurable reduction of blood flow in hormonally treated fibrocystic breasts. The decrease in the mean frequency shift detectable by CW Doppler and the sum of the frequency shifts correlated with an improvement in subjective complaints and palpable findings.

In another study, pulsed color Doppler sonography was used in seven patients with nonpuerperal mastitis to evaluate their response to treatment with antibiotics and bromocriptine. The peak systolic and end-diastolic flow velocities, which initially were abnormally high compared with the contralateral breast, returned to normal in response to the medical therapy[2].

Trophoblastic tumors. Several groups of authors[7,14,20] investigated the value of color Doppler sonography in evaluating the response of trophoblastic tumors to therapy. Hsieh et al.[7] compared 23 patients with trophoblastic tumors with a control group of 55 nonpregnant women and 15 patients with a treated trophoblastic tumor. The group with trophoblastic tumors showed a significantly higher peak systolic flow velocity and significantly lower RI_{mean} in both uterine arteries ($p < 0.0001$). The peak systolic velocity fell significantly during chemotherapy ($p < 0.001$). This decrease correlated with the treatment response as measured by serum β-hCG and B-mode ultrasound.

Park et al.[14] compared the therapeutic response in 16 women with trophoblastic tumors with the S/D ratio in the uterine arteries. Starting with similar initial values, the responders showed a significantly lower S/D ratio compared with the nonresponders ($p < 0.05$).

Tepper et al.[20] documented a strong inverse correlation between serum β-hCG and the intratumoral RI in three patients.

On the whole, these results support the usefulness, accuracy, and sensitivity of color Doppler sonography in evaluating treatment response.

Currently there are no reports in the international literature on similar studies in patients with cervical carcinoma. Certainly there are no data on the use of color Doppler in evaluating response to intra-arterial chemotherapy.

Benign–Malignant Differentiation of Cervical Lesions with Color Doppler

Some results have been published on the use of pulsed color Doppler sonography in the benign–malignant discrimination of cervical lesions[11,19].

RI and PI in the uterine artery. Kurjak and his group[11] examined 89 patients with histologically proven cervical carcinoma and compared them with a control group of 24 healthy women. The descending branch of the uterine artery was examined with Doppler ultrasound. The uterine artery could not be visualized in 18% of the carcinoma patients and 17% of the controls. The RI was significantly lower in the carcinoma patients than in the healthy women. The PI in the carcinoma group was also lower than in the controls, but statistical significance was borderline. The authors believe that this new technique is less effective in evaluating cervical lesions than ovarian tumors.

RI_{min}. Sohn et al.[19] examined 33 women with cervical lesions (19 malignant and 14 benign; four of the malignant lesions developed before menopause). The authors did not specify the exact site at which the RI was determined (intratumoral, intracervical, in the uterine artery, etc.). The malignant cervical lesions showed a minimum RI (RI_{min}) of 55% ± 8 SD, while the RI_{min} for the benign lesions was 79% ± 11 SD. The premenopausal patients had a mean RI_{min} of 58% ± 9 SD. The RI values were approximately the same in the group of premenopausal women with benign lesions and in the group with malignant lesions. Marked differences were found, however, between the group of postmenopausal women with benign lesions and the group with malignant lesions. Based on these results, the authors advocate the separate examination of premenopausal and postmenopausal patients and state that RI_{min} has potential value in the benign–malignant differentiation of cervical lesions.

RI_{mean}. The RI_{mean} in our study was significantly higher in intratumoral vessels and in both uterine arteries at all examination times. The pretherapeutic intratumoral RI_{mean} was 0.73, and the mean pretherapeutic RI_{mean} for both uterine arteries was 0.79. The preoperative intratumoral RI_{mean} was 0.67, and the preoperative RI_{mean} for both uterine arteries was 0.765.

We cannot directly compare the study results because the objective Doppler analyses were based on different parameters such as RI_{min} and RI_{mean}. By analogy with the results in other gynecological tumors, however, it appears that the RI_{min} in the group of patients studied here is not significantly lower than the RI_{mean} (0.73). Thus it is considerably higher than would be expected in the age group of premenopausal women.

Blood Flow Parameters and Menopausal Status

The dependence of color Doppler-determined blood flow parameters on menopausal status is an important discovery. Our own results confirm this fact in breast tumors[22,23]. In patients evaluated for infertility, characteristic cycle-dependent variations in these parameters have been documented in the uterine arteries and also in the ovaries. Kupesic found a higher RI in the uterine artery during the proliferative phase of the cycle (0.85–0.90) than in the secretory phase (0.83–0.85), measuring the lowest values immediately after ovulation and during the midsecretory phase[10].

It may be that cervical blood flow is subject to similar variations, but we do not yet know the form or extent to which a cervical tumor might be subject to this dynamic.

Blood Flow Parameters and Chemotherapy

Number of vessels. Quantitative changes in tumor vascularity have been documented with color Doppler ultrasound. A progressive decline in the number of vascular segments that can be visualized with color Doppler reflects the continuous destruction of tumor vessels—probably a result of tumor necrosis induced by chemotherapy. Park[14], for example, describes the disappearance of tumor blood flow by color Doppler in response to chemotherapy. The very strong correlation that has been found between a decreased number of intratumoral vessels and a reduction in tumor volume demonstrates the high sensitivity of this parameter in monitoring treatment response.

V_{max} and RI_{mean}. The decrease in peak systolic velocity and RI_{mean} that has been found in tumors responsive to chemotherapy is consistent with the observation that proliferative malignant lesions are associated with a rise in the peak systolic velocity and resistance index[13,22]. The net result of tumor necrosis and partial destruction of the tumor blood supply is a fall of the elevated intratumoral and peritumoral pressures and flow velocities. The observed reduction in peak systolic velocity correlates with study results in patients with trophoblastic tumors[7,14,20] and breast cancer[2,12]. On the other hand, the pattern of changes seen in the intratumoral RI_{mean} differs from the patterns seen in other tumor entities[2,7,14,20]. Neither of the two parameters shows a statistically significant correlation with treatment response, however, as measured by a reduction in tumor volume or falling tumor marker levels.

Contrary to expectations, cervical tumors that are initially bulky and later undergo significant shrinkage in response to therapy do not show demonstrable changes in blood flow. These results contrast with those of Hsieh[7] on trophoblastic tumors, in which the RI_{mean} in the uterine arteries increased during chemotherapy as a function of treatment response. This effect might depend on the extent of tumor regression and thus may become measurable at a later time after a further reduction in tumor size.

Sensitivity of color Doppler sonography. On the whole, the results document the sensitivity of color Doppler sonography in evaluating treatment response based on small intratumoral vessels. But despite the general reduction in vascular density, peak systolic velocity, and RI_{mean}, a strict correlation with a decrease in tumor volume and tumor markers does not appear to exist in any given case. It appears that the blood flow patterns in large cervical carcinomas detected by pulsed color Doppler sonography do correlate with response to intra-arterial chemotherapy, indicating that color Doppler can provide a new, noninvasive modality for evaluating treatment response.

Summary

Bulky cervical carcinomas have a particularly poor prognosis among cervical malignancies. Radical surgery and curative radiotherapy often fail to meet expectations. Preoperative reduction of tumor volume is a good way to improve operability. Imaging procedures (ultrasound, CT, and MRI) can objectively demonstrate changes in tumor volume in response to therapy.

Study. Eight women with a FIGO stage Ib2 –IIb bulky tumor were examined in a prospective study. Conventional B-mode and color Doppler examinations were performed immediately before each of two chemotherapy cycles and before surgical treatment, using an Acuson model 128 XP10 system. The scan technique, equipment, and instrument settings were standardized. A probe frequency of 5 MHz was used for transvaginal sonography and color Doppler examinations.

Changes in intratumoral vascularity were observed during the course of chemotherapy. The average number of sampled vascular segments per examination decreased from 3.5 to 2.7. The peak systolic velocity fell from 0.16 to 0.13 m/s. The mean resistance index (RI_{mean}) declined from 0.73 to 0.67. Despite a significant reduction in tumor volume, no change of RI_{mean} was measured in the uterine arteries. The decrease in the number of intratumoral vessels correlated with the reduction in tumor volume ($p = 0.002$).

Conclusions. The number of intratumoral vessels is a sensitive parameter for detecting therapy-induced changes in the vascularization of large cervical carcinomas. Thus, color Doppler sonography can provide a new, noninvasive method for evaluating treatment response. However, the parameters of intratumoral peak flow velocity and RI_{mean} and the RI_{mean} in the uterine arteries do not correlate with treatment response as measured by a reduction in tumor volume and elevated tumor marker levels.

References

1 Benedetti Panici P, Greggi S, Scambia G et al.: High-dose cisplatin and bleomycin neoadjuvant chemotherapy plus radical surgery in locally advanced cervical carcinoma: a preliminary report. Gynecol. Oncol. 41 (1991) 212–216

2 Blohmer J U, Bollmann R, Chaoui R, Kurten A, Lau H U: Die Mastitis nonpuerperalis in der Realtime- und Farbdoppler-Sonographie. Geburtshilfe Frauenheilkd. 54 (1994) 161–166

3 Burghardt E et al.: Results of surgical treatment of 1028 cervical cancers studied with volumetry. Cancer 70 (1992) 648–655

4 Cosgrove D O, Bamber J C, Davey J B, McKinna J A, Sinnet H D: Color Doppler Signals from Breast Tumours. Radiology 176 (1990) 175–180

5 Cosgrove D O, Kedar R P, Bamber J C et al.: Breast Diseases: Color Doppler US in Differential Diagnosis. Radiology 189 (1993) 99–104

6 Durrance F, Fletcher G, Rutledge F: Analysis of central recurrent disease in stage I and II squamous cell carcinomas of the cervix on intact uterus. Am. J. Roentgenol. 106 (1969) 831–838

7 Hsieh F J, Wu C C, Chen C A, Chen T M, Hsieh C Y, Chen H Y: Correlation of uterine hemodynamics with chemotherapy response in gestational trophoblastic tumors. Obstet. Gynecol. 83 (1994) 1021–1025

8 Kedar R P, Cosgrove D O, Smith I E, Mansi J L, Bamber J C: Breast Carcinoma: Measurement of tumor response to primary medical therapy with color Doppler flow imaging. Radiology 190 (1994) 825–830

9 Knapstein P G, Kreienberg R, Beck T H, Mahlke M, Mitze M, Düber C: Intraarterielle präoperative Chemotherapie fortgeschrittener Zervixkarzinome. Geburtsh. u. Frauenheilk. 51 (1991) 156–160

10 Kupesic S, 1994, persönliche Mitteilung.

11 Kurjak A, Shalan H, Kupesic S et al.: Transvaginal color Doppler sonography in the assessment of pelvic tumor vascularity. Ultrasound Obstet. Gynecol. 3 (1993) 137–154

12 Madjar H: Benign disease assessment with Doppler. In Ioannidou-Mouzaka L, Agantis N J, Karydas J (eds.): Senology. Excerpta Medica, Amsterdam, 1992, 121–123

13 Madjar H, Prömpeler H, Wolfahrt R, Bauknecht T, Pfleiderer A: Farbdopplerflußdaten von Mammatumoren. Ultraschall in Med. 15 (1994) 69–76

14 Park Y W, Kim D K, Cho J S et al.: The utilization of Doppler ultrasonography with color flow mapping in the diagnosis and evaluation of malignant trophoblastic tumors. Yonsei Med. J. 35 (1994) 329–335

15 Perez C A et al.: Effect of tumor size on the prognosis of carcinoma of the uterine cervix treated with irradiation alone. Cancer 69 (1992) 2796–2806

16 Petterson F (ed.): 21 st Annual report on the results of treatment in gynecological cancer. Int. J. Gyn. Obstet. 36 (Suppl.) (1991) 27–130

17 Sardi J, Sananes C, Giaroli A et al.: Results of a prospective randomized trial with neoadjuvant chemotherapy in stage IB, bulky, squamous carcinoma of the cervix. Gynecol. Oncol. 49 (1993) 156–165

18 Scarabelli C, Tumolo S, De Paoli A et al.: Intermittent pelvic arterial infusion with peptichemio, doxorubicin and cisplatin for locally advanced and recurrent carcinoma of the uterine cervix. Cancer 60 (1987) 25–30

19 Sohn Ch, Meyberg G, v. Fournier D, Bastert G: Die Durchblutung maligner und benigner Tumoren des inneren Genitale. Geburtsh. u. Frauenheilk. 53 (1993) 395–399

20 Tepper R, Shulman A, Altaras M et al.: The role of color Doppler flow in the management of nonmetastatic gestational trophoblastic disease. Gynecol. Obstet. Invest. 38 (1994) 14–17

21 Villena-Heinsen C, Mink D, Lung-Kurt S et al.: Preoperative intraarterial chemotherapy for bulky cervical carcinoma in stage IB–IIB. Reg. Cancer Treat. 1 (1994) 17–21

22 Villena-Heinsen C, Ertan A K, Tossounidis I, Holländer M, König J, Schmidt W: Diagnostische Aussagekraft der Farbdopplersonographie bei Mammatumoren. Geburtsh. u. Frauenheilk. 55 (1995) 541–547

23 Villena-Heinsen C, Alexander C, Tossounidis I, Holländer M, Ertan A K, König J, Schmidt W: Influence of Menopausal State on Colour Doppler Flow Parameters of Breast Tumours and healthy mammary Tissue. European Journal of Ultrasound 6 (1997) 49–52

24 Weiner S A, Aristizabal S, Alberts D, Survit E A, Deatherage-Deuser R N: A phase II trial of mitomycin, vincristine, bleomycin and cisplatin (MOBP) as neoadjuvant therapy high-risk cervical carcinoma. Gynecol. Oncol. 30 (1988) 1–6

32 Color Doppler Imaging of Benign Adnexal Masses—A Spectrum of Findings

A. Kurjak, S. Kupesic, and A. K. Ertan

Adnexal masses cause a great deal of concern because of their malignant potential and the limited options for reliable preoperative benign–malignant discrimination. This is particularly difficult when bizarre structures such as dermoid cysts, large endometriomas, complex corpus luteum cysts, and cystadenomas are seen.

Appearance of Normal Ovaries by B-Mode and Color Doppler Ultrasound

The normal ovary can be defined relatively clearly by transvaginal sonography. The first ultrasound images of normal ovaries were published by Kratochwil[16]. It is usually possible to identify follicles or the corpus luteum. The ovary is normally located posteromedial to the hypogastric vein. A normal-size ovary is mobile and may change its position during the examination. The position of the ovary may be permanently altered due to postinflammatory adhesions.

Color Doppler sonography can demonstrate the vascular supply of normally functioning ovaries. Monitoring follicular growth with ultrasound is an established procedure in the investigation of infertility. Blood flow can be clearly visualized at the margin of the developing follicle. The development of the corpus luteum can be plainly observed, and it is easier to record color-flow signals from the ovary during the luteal phase. A dense color pattern characterizes an active corpus luteum, even if this cannot be appreciated in the B-mode image.

Specific Adnexal Masses

Enlarged ovaries can be subdivided into three categories: cystic changes, cystic-solid ovarian masses, and solid ovarian tumors.

Cystic and Cystic-Solid Ovarian Masses

Polycystic Ovaries

Polycystic ovary syndrome (PCO syndrome) is characterized by enlarged, somewhat globular ovaries whose greatest diameter exceeds the anteroposterior diameter of the uterine fundus. Polycystic ovaries are generally twice as large as normal ovaries, but approximately one-third of patients have ovaries of normal size. Polycystic ovaries contain small cystic structures (<10 mm), and the volume of the ovarian stroma is increased[3, 18]. The cysts may be located chiefly at the periphery of the ovary or may permeate the stroma diffusely.

Differential diagnosis. Increased stroma is the most important sign differentiating polycystic ovaries from multifollicular ovaries. The latter may be a transient feature of normal development during puberty, or may be seen in patients with secondary amenorrhea due to severe weight loss[29]. Ovaries with a polycystic appearance may be seen in women who take oral contraceptives. These changes also occur in association with certain endocrine disorders, pituitary adenomas, or virilizing ovarian and adrenal tumors. The intraovarian vessels in polycystic ovaries are located within the stroma (Fig. 32.**1a**). The mean resistance index is 0.54 and does not vary with the phases of the menstrual cycle (Fig. 32.**1b**).

Functional Ovarian Cysts

Functional ovarian cysts are the most common type of cystic adnexal mass. They are easily identified at ultrasound, generally appearing as unilateral cystic structures with a smooth, thin wall and clear fluid contents[1]. These cysts arise from unruptured follicles and are usually less than 10 cm in diameter. The cyst wall is partially bordered by normal ovarian tissue. Pericystic blood flow shows a moderate mean flow velocity with an RI = 0.52 ± 0.06.

Corpus Luteum Cysts

Corpus luteum cysts have varying appearances on transvaginal sonography. Their internal echoes, caused by a retracting blood clot, make them difficult to distinguish from other benign and malignant ovarian masses. A persistent corpus luteum cyst may exceed 10 cm in diameter. Its contents may be liquid, solid, or mixed, and internal septa or even papillary structures may be seen (Fig. 32.**2a**). Unfortunately, the cysts typically show high angiogenic intensity marked by numerous blood vessels with low impedance values (RI = 0.46 ± 0.08) (Fig. 32.**2b**). To avoid a false-positive diagnosis from this "great imitator" of malignancy, it is very important to examine premenopausal patients at the start of their menstrual cycle[21].

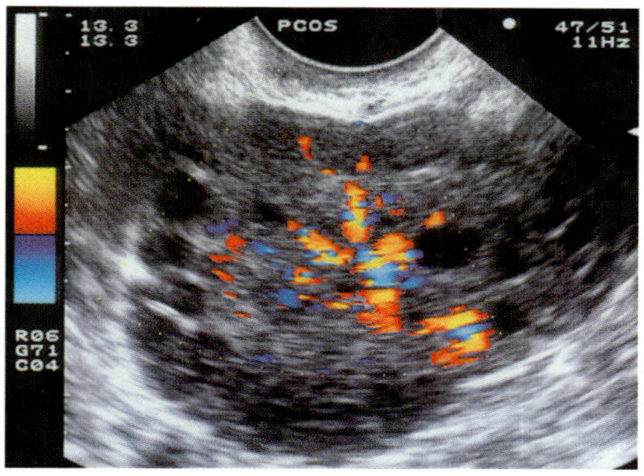

Fig. 32.**1 a** Transvaginal scan of a polycystic ovary. The ovarian stroma is enlarged and is pushed outward by numerous small, crowded cystic structures. The vascularity of the stroma is markedly increased.

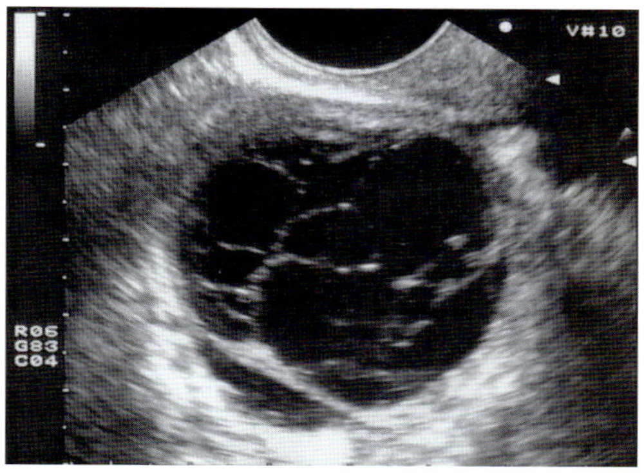

Fig. 32.**2 a** Transvaginal scan of a corpus luteum cyst. Color Doppler demonstrates pericystic blood flow.

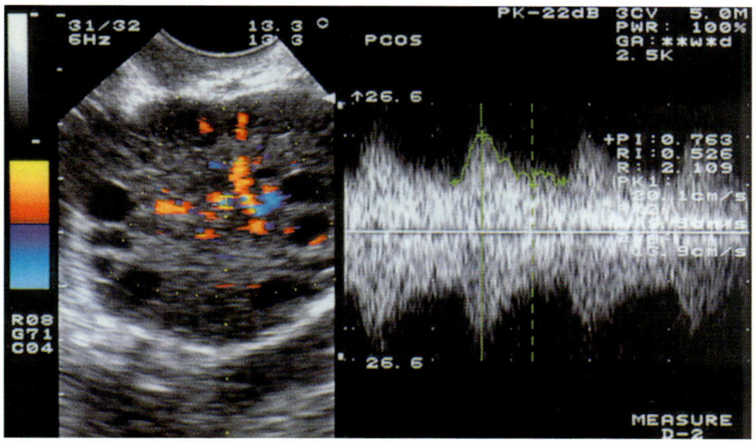

Fig. 32.**1 b** Pulsed Doppler scan of the stromal vessels indicates a moderate impedance to blood flow (RI = 0.53).

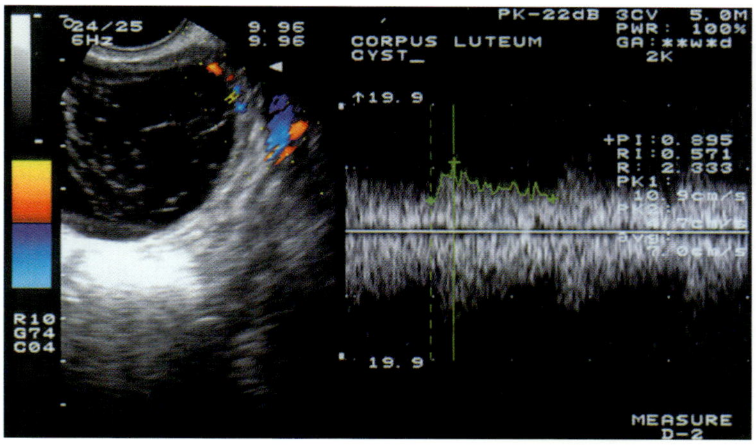

Fig. 32.**2 b** A high blood flow velocity and moderate to low resistance index (0.57) characterize the typical blood flow pattern of a corpus luteum cyst.

Serous and Mucinous Cystadenomas

The most common ovarian epithelial tumors are serous and mucinous cystadenomas. They typically appear as multilocular cysts at ultrasound. These cysts are usually large and contain clear, hypoechoic fluid and linear septa, which are more conspicuous in the mucinous type. The most important feature is the presence of thin septations less than 3 mm thick. Papillary structures may be observed in both serous and mucinous cystadenomas (Fig. 32.**3a**).

The location of the vessels and the type of angiogenesis are important differentiating features from malignant tumors. The hallmark of a benign lesion is a moderate flow resistance (RI = 0.50 ± 0.08) measured in peripherally located vessels that are clearly delineated from one another (Fig. 32.**3b**). The ves-

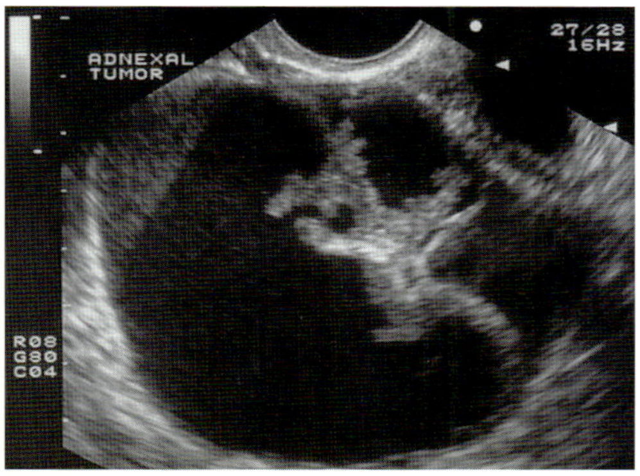

Fig. 32.**3 a** Transvaginal sonogram of a complex tumor. The mass contains several thick septa, papillary outgrowths, and areas containing a clear, viscous fluid.

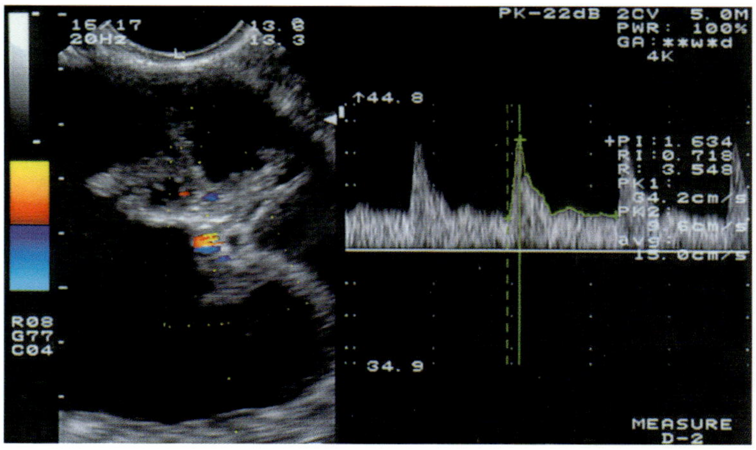

Fig. 32.**3 b** Same patient as in Fig. 32.**3 a**. Doppler measurements indicate a high vascular impedance (RI = 0.72), suggestive of a benign tumor. Histopathology revealed a serous cystadenoma.

sels located in the septa generally have a slightly lower RI (0.48 ± 0.04).

Paraovarian Cysts

Paraovarian cysts develop from the Gartner duct and rarely differ in appearance from functional cysts. Paraovarian cysts may measure only 2–3 cm, but most are considerably larger[2,23]. A thin, smooth wall, the absence of internal septa, echo-free fluid, and normal-appearing ovarian tissue are consistent with a paraovarian cyst. Usually these lesions are not associated with an increase in vascularity.

Endometrioma

Hyperechoic cyst contents are most commonly found in mucinous cystadenomas and endometriomas. A homogeneous, moderately intense internal echo pattern is a common finding in ovarian endometriomas. Kupfer et al.[19] observed this pattern in 82% of the cases they examined. It may be limited to one or more cystic structures within a multilocular mass or may be visible throughout the mass. The echo pattern may result from blood degradation products and changes during the menstrual cycle. The irregular hyperechoic areas within the mass may correlate with more recent hemorrhage. In most cases the endometriotic cyst wall is clearly demarcated from normal adjacent ovarian stroma.

a

b

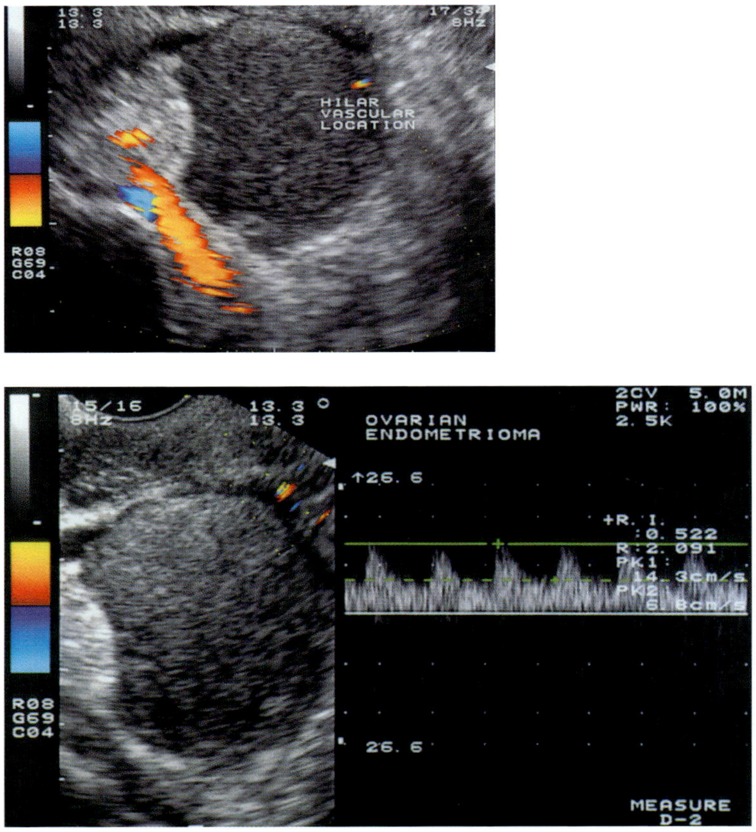

Fig. 32.**4** **a** Ovarian endometrioma with a homogeneous, hyperechoic internal structure. A copious blood supply is visible at the ovarian hilum.
b Same patient as in Fig. 32.**4 a**. Doppler waveform analysis indicates a moderate impedance to blood flow (RI = 0.52).

Vascularity. The most intense vascularity is found in the area of the ovarian hilum (Fig. 32.**4a**) and is detected in 78.6% of endometriomas[20]. The RI value recorded from this region is usually greater than 0.45. Recent Doppler studies[20] showed different vascularization patterns in endometriomas during the proliferative and secretory phases of the menstrual cycle. In the early stage of endometrioma formation, which is marked by intense angiogenic activity, a low to moderate impedance is measured (RI = 0.44 ± 0.06). It is believed that the thickness of the collagen layer, fibrotic changes, and hemorrhagic foci affect the tumor blood supply and the diffusion of nutrients into the endometrioma. High impedance values (RI = 0.51 ± 0.09) are typical of advanced stages (Fig. 32.**4b**). Additionally, the rise of intratumoral pressure caused by the accumulation of "chocolate fluid" alters the vascular supply, eventually blunting the response to endogenous and exogenously administered hormones.

Cystic Teratomas

Approximately 15% of all ovarian masses are germ cell tumors, and 96% of these are benign cystic teratomas[7,12,31]. Most of these lesions develop as asymptomatic adnexal masses, but there is up to 16% incidence of dermoid cyst torsion[7,12,36], and occasionally such a lesion may rupture and incite peritonitis[7,36]. From 1% to 3% of all ovarian teratomas are malignant[7,12,311,36]. Caruso et al.[7] reported on 305 patients with ovarian teratomas. The average age of patients with a malignant tumor component was 60.8 years.

Sonographic features. Several authors did retrospective studies on the various echo patterns of cystic teratomas. Generally these patterns enabled them to distinguish the tumors from other ovarian lesions. Quinn et al.[32] described a specific sonographic appearance that they called the "Rokitansky protuberance." Other sonographic signs of ovarian teratoma are a "dermoid snowstorm" or "tip of the iceberg"[13], a "fat–fluid level"[30], and a "dermoid mesh"[28] . Additionally, a dense acoustic shadow has been found behind hyperechoic components in all echo patterns. Bizarre structures and the absence of pathognomonic patterns have occasionally made it difficult to diagnose dermoid tumors. This led Cohen and Sabbagha [8] to propose additional sonographic criteria such as an echogenic protuberance accompanied by a cystic echo pattern, thin bandlike echogenic structures, and/or increased echogenicity with or without cystic components (Fig. 32.**5**).

Differential diagnosis. Unfortunately, malignant tumors occasionally have the same or similar appearance, and even an experienced examiner using a high-resolution scanner may make an erroneous diagnosis by looking at morphological criteria alone. The complex texture, thick walls, and solid echogenic clots may cause confusion with ovarian endometriosis. Pelvic inflammatory diseases can also mimic a number of other lesions such as dermoid tumors, endometriomas, and even malignant neoplasms. An accurate and reliable pretherapeutic differential diagnosis is necessary to ensure an appropriate treatment strategy, since malignancies require an aggressive approach that includes major surgery while endometriomas, dermoid tumors, and inflammatory conditions can be managed conservatively or by minimally invasive surgery.

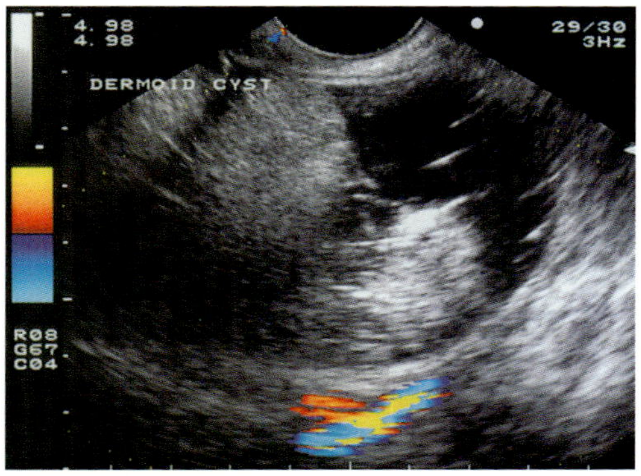

Fig. 32.**5** Transvaginal image of a dermoid cyst with solid echogenic components and bizarre intracystic structures. Color Doppler sonography did not demonstrate intratumoral blood flow.

Vascularity. Another parameter that can be used for benign–malignant discrimination is vascularity. Malignant ovarian tumors generally show a diffuse pattern of vascularity that includes dilated, tortuous vascular segments with a paucity of smooth muscle in the tunica media[21]. Tumor angiogenesis is characterized by numerous arteriovenous shunts. These vessels have a low impedance to blood flow, which is manifested by a high flow velocity and low resistance index.

Increased blood flow is also seen in association with benign lesions such as tubo-ovarian abscesses[17], ovarian endometriomas[20], hemorrhagic corpus luteum cysts, and dermoid cysts with an inflammatory component. All of these lesions are characterized by an enlarged intravascular space that can mimic a malignant-type low-resistance blood flow pattern.

Use of color Doppler sonography. Numerous studies have been published during the past seven years in which transvaginal B-mode and color Doppler ultrasound were used in an attempt to reduce the rate of false-positive findings in malignant tumor screening[4, 5, 11, 17, 20–22, 24–26, 33, 34].

Jain et al.[15] examined 50 adnexal masses that included five dermoid cysts. One false-negative finding by transvaginal sonography was found at pathological examination to be a borderline malignant dermoid tumor. This tumor had a high flow impedance, suggesting a benign diagnosis. Also, one dermoid tumor showed a low flow impedance (RI < 0.40) and was incorrectly classified as an ovarian carcinoma.

In a study by Hata et al.[14], 63 women with ovarian masses were examined with transvaginal ultrasound. Fourteen of the masses were dermoid tumors, and seven of these were suspected to be malignant on the basis of transvaginal sonographic findings. Four cystic teratomas showed signs of malignancy on color Doppler examination. An unusually high cutoff value of 0.72 was used for the RI in this study.

Weiner et al.[35] examined 62 women with adnexal masses. A dermoid cyst was diagnosed in 10 of these patients by histopathological examination. In three women with cystic teratomas, B-mode ultrasound showed signs suspicious for malignancy. Two of these patients had elevated serum CA 125 levels above 35 IU/ml. Transvaginal color and pulsed Doppler ultrasound analysis showed a high impedance value (PI > 1.0) in all these tumors, however, and therefore color Doppler sonography was useful in avoiding false-positive findings.

Fleischer et al.[10] combined transvaginal sonography with color Doppler imaging in the analysis of 96 adnexal masses. When conventional transvaginal sonography was used by itself, four of six dermoid cysts were diagnosed correctly. When color Doppler sonography was added, all six of the lesions were correctly identified.

Campbell et al.[6] evaluated the usefulness of transvaginal sonography with color Doppler imaging as a level-two diagnostic study in screening programs for women with a strong family history of ovarian cancer. The most frequent causes of false-positive findings were endometriosis (4 of 9 cases) and cystic teratomas (2 of 9 cases).

Timor-Tritsch et al.[34] correlated the sonographic and histopathological results for 115 adnexal masses. Nine dermoid cysts were correctly diagnosed by using a morphological scoring system and color Doppler measurements. The RI was greater than 0.46 and the PI was greater than 0.62.

Our own results[22] show that increased vascularity can be detected in a small portion of cystic teratomas (27%). The flow impedance in these cases was above the cutoff value that we proposed for ovarian cancer screening. In cystic teratomas, low- to moderate-impedance flow signals (RI = 0.42–0.72) were recorded in areas where high cell proliferation or inflammation was found on histopathological examination. By contrast, we found no increase of vascularity in inactive tumor masses.

Solid Ovarian Masses

Fibromas

The most common solid benign ovarian tumors are fibromas (and some dermoids). These tumors usually appear as an echogenic mass with smooth, rounded borders (Fig. 32.**6**).

Intratumoral vascularity (central or peripheral) is rarely seen. If it is present, the vessels generally show a high impedance to blood flow.

Pelvic Inflammatory Disease

Pelvic inflammatory disease is a serious complication of sexually transmitted microbial infections, which can permanently damage the organs of the upper genital tract. Approximately 30% of all infertility cases and 50% of ectopic pregnancies are causally related to prior inflammatory processes. Adnexal findings such as enlarged ovaries, tubular anechoic structures about the adnexa, or complex adnexal masses are the most common manifestations. Complex adnexal masses with inter-

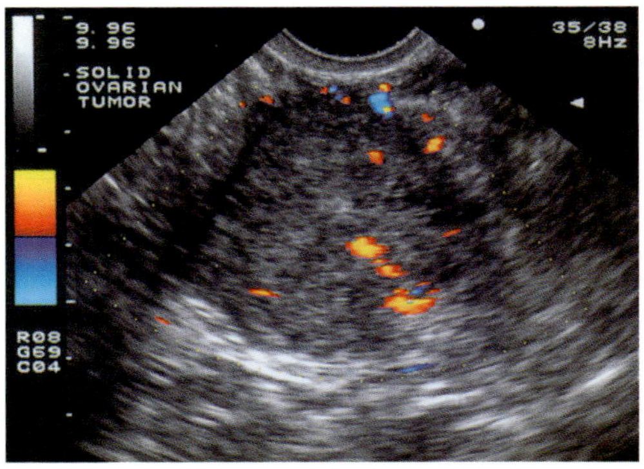

Fig. 32.**6** Transvaginal scan of a solid ovarian tumor.

nal septa and irregular outer contours, scattered echogenic structures, and fluid levels may be observed. These findings suggest the presence of a tubo-ovarian abscess.

Use of color Doppler sonography. Because the tubo-ovarian abscess can mimic a number of benign and malignant adnexal conditions (tubal abortion, hematosalpinx, ovarian tumors), an analysis of clinical and biochemical parameters should be supplemented by B-mode and Doppler ultrasound examinations in order to make a correct diagnosis. The local inflammatory mediators that arise in an acute pelvic inflammation cause intense vasodilation, leading to a fall in the resistance index (RI = 0.53 ± 0.09). The ensuing fluid collection within the fallopian tubes alters blood flow characteristics by compressing the vessels that run in the tube wall. With further progression of the inflammatory processes, fibroblasts proliferate and scar tissue is formed. This causes a decrease in regional blood flow, manifested by a rising resistance index (RI = 0.71 ± 0.07). Doppler examination can be very helpful in these cases for the differentiation of chronic inflammatory processes, which occasionally feature pseudopapillary structures that are morphologically similar to a malignant tumor (Fig. 32.**7**). Thus, the typical absence of vascular structures detectable with Doppler ultrasound in chronic inflammatory processes is a helpful sign in differentiating these conditions from ovarian carcinoma.

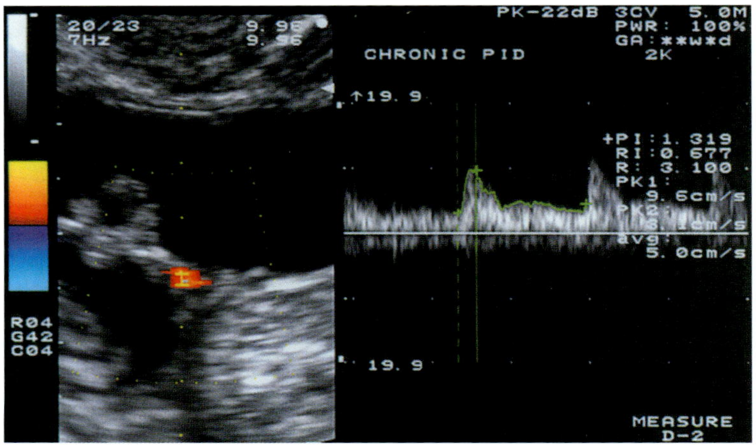

Fig. 32.**7** Chronic pelvic inflammatory disease. Color flow signals are visible in the tubal wall and pseudopapillary projections. Pulsed Doppler shows a moderate flow impedance (RI = 0.67), suggestive of a benign lesion.

Conclusions

Predictive value. Before the advent of transvaginal color and pulsed Doppler ultrasound, morphological criteria were the only parameters available for the evaluation of adnexal masses. A careful review of the Doppler literature suggests that transvaginal color Doppler sonography, with its high predictive value, can help in the evaluation of adnexal masses and the selection of an appropriate treatment (follow-up, minimally invasive surgery, or laparotomy) (Table 32.**1**).

Differentiation of ovarian masses. The rates of false-positive and false-negative findings in the differentiation of ovarian masses can be effectively reduced by using the scoring system proposed by our group[17,20,22]. When the scoring systems for endometriosis and dermoid cysts are compared, several common features are seen: childbearing age, multiple positive sonographic findings, and a mass with thick walls and high-level internal echoes. Some important differences are also noted, such as location (retrouterine with endometriosis, lateral with dermoid cysts), bilaterality (with endometriosis) versus unilaterality (with dermoid tumors), as well as the visualization of blood vessels and the type of vascularity that is seen. Ovarian endometriomas were vascularized in 88.3% of cases, usually in the area of the ovarian hilum, whereas cystic teratomas were predominantly avascular (72.6%).

It is still possible, of course, to confuse endometriomas and dermoid cysts. But an experienced examiner can differentiate these entities from ovarian carcinoma with a high degree of confidence.

Table 32.**1** Transvaginal color Doppler sonography in the evaluation of benign adnexal masses

Histopathology	Detection of blood flow			Resistance index	SD
	n	n	%		
Functional cysts					
– Follicular cyst	92	84	91.3	0.52	0.06
– Corpus luteum cyst	104	104	100	0.46	0.08
Dermoid cysts	32	9	28.1	0.48	0.10
Cystadenoma	56	50	89.3	0.50	0.08
Fibroma	7	5	71.4	0.46	0.04
Theca-granulosa cell tumor	2	2	100	0.60	0.04
Brenner tumor	4	3	75	0.50	0.08
Endometriosis	152	137	90.1	0.49	0.11
Inflammatory processes	184	134	72.8	0.54	0.12
All cases	633	528	83.4	0.50	0.08

Adapted from Kurjak A, Kupesic S: Scoring system for prediction of ovarian endometriosis based on transvaginal color and pulsed Doppler sonography. Fertil. Steril. 62 (1994) 81–88.

Minimally invasive surgical procedure. We have found transvaginal color and pulsed Doppler sonography to be a reliable noninvasive preoperative diagnostic procedure. By doing this examination routinely, we have been able to increase the rate of laparoscopic procedures at our department in recent years without misinterpreting a single malignant lesion.

Since most of our patients with benign masses are of childbearing age, they benefit from a minimally invasive surgical procedure that preserves their fertility. Given the fact that the incidence of malignancies is very low in this age group[7, 12, 31], we believe that, in selected cases, a technically proficient laparoscopic procedure that minimizes the risk of intra-abdominal spillage of the cyst contents provides a safe, effective, and advantageous alternative to laparotomy.

References

1 Auslender R, Atlas I, Lissak A, Bornstein J, Atad J, Abramovici H: Follow-up of small, postmenopausal ovarian cysts using vaginal ultrasound and CA-125 antigen. J. Clin. Ultrasound 24 (1996) 175–178

2 Barloon TJ, Brown BP, Abu-Yousef MM, Warnock NG: Paraovarian and paratubal cysts: preoperative diagnosis using transabdominal and transvaginal sonography. J. Clin. Ultrasound 24 (1996) 117–122

3 Battaglia C, Artini PG, Genazzani AD et al.: Color Doppler analysis in lean and obese women with polycystic ovary syndrome. Ultrasound Obstet. Gynecol. 7 (1996) 342–346

4 Bourne T, Campbell S, Steer C et al.: Transvaginal color flow imaging: a possible new screening technique for ovarian cancer. Brit. Med. J. 299 (1989) 1367–1370

5 Brown DL, Frates MC, Laing FC et al.: Ovarian masses: can benign and malignant lesions be differentiated with color and pulsed Doppler US. Radiology 190 (1994) 333–336

6 Campbell S, Bourne TH, Reynolds K et al.: Role of color Doppler in an ultrasound-based screening programme. In Sharp F, Mason WP, Creasman W (eds.): Ovarian cancer 2. Chapman & Hall Medical, London 1992, 237–247

7 Caruso PA, Marsh MR, Minicowitz S et al.: An intense clinicopathologic study of 305 teratomas of the ovary. Cancer 27 (1971) 348

8 Cohen L, Sabbagha R: Echo patterns of benign cystic teratomas by transvaginal ultrasound. Ultrasound Obstet. Gynecol. 3 (1993) 120–123

9 Di Meglio A, Di Meglio G, Esposito A et al.: Echo patterns of ovarian dermoid tumor. Eur. J. Gynecol. Oncol. 9 (1988) 242–245

10 Fleischer AC, Cullinan JA, Kepple DM et al.: Conventional and color Doppler transvaginal sonography of pelvic masses: a comparison of relative histologic specificities. J. Ultrasound Med. 12 (1993) 705–712

11 Fleischer AC, Cullinan JA, Peery CV, Jones HW: Early detection of ovarian carcinoma with transvaginal color Doppler ultrasonography. Amer. J. Obstet. Gynecol. 174 (1996) 101–106

12 Gallion H, Van Nagell JR, Donaldson ES et al.: Immature teratoma of the ovary. Amer. J. Obstet. Gynecol. 146 (1983) 361–365

13 Guttman PH Jr: In search of the elusive benign cystic ovarian teratoma: application of the ultrasound „tip of the iceberg“ sign. JCU 5 (1977) 403–406

14 Hata K, Hata T, Manabe A et al.: A critical evaluation of transvaginal Doppler studies, transvaginal sonography, magnetic resonance imaging, and CA 125 in detecting ovarian cancer. Obstet. Gynecol. 80 (1992) 922–926

15 Jain KA: Prospective evaluation of adnexal masses with endovaginal gray-scale and dupley and color Doppler US: correlation with pathologic findings. Radiology 191 (1994) 63–67

16 Kratochwill A, Urban G, Friedrich F: Ultrasonic tomography of the ovaries. Ann. Chir. Gynecol. 61 (1972) 211–214

17 Kupesic S, Kurjak A, Pasalic L et al.: The value of transvaginal color Doppler in the assessment of pelvic inflammatory disease. Ultrasound Med. Biol. 21 (1995) 733–738

18 Kupesic S, Kurjak A, Stilinovic K: The assessment of female infertility. In Kurjak A (ed.:): An atlas of transvaginal color Doppler. Parthenon Publishing, London 1994, 171–197

19 Kupfer MC, Schwimer SR, Lebovic J: Transvaginal sonographic appearance of endometriomata: spectrum of findings. J. Ultrasound Med. 11 (1992) 129–133

20 Kurjak A, Kupesic S: Scoring system for prediction of ovarian endometriosis based on transvaginal color and pulsed Doppler sonography. Fertil. Steril. 62 (1994) 81–88

21 Kurjak A, Kupesic S: Transvaginal color Doppler and pelvic tumor vascularity: lessons learned and future challenges. Ultrasound Obstet. Gynecol. 6 (1995) 1–15

22 Kurjak A, Predanic M: New scoring system for prediction of ovarian malignancy based on transvaginal color Doppler sonography. J. Ultrasound Med. 11 (1992) 631–638

23 Kurjak A, Predanic M, Kupesic S, Zalud I: Adnexal masses malignant ovarian tumors. In Kurjak A (ed.): An atlas of transvaginal color Doppler. Parthenon Publishing, London 1994, 291–316

24 Kurjak A, Predanic M. Kupesic-Urek S et al.: Transvaginal color and pulsed Doppler assessment of adnexal tumor vascularity. Gynecol. Oncol. 50 (1993) 3–9

25 Kurjak A, Shalan H, Kupesic S et al.: Transvaginal color Doppler sonography in the assessment of pelvic tumor vascularity. Ultrasound Obstet. Gynecol. 3 (1993) 137–154

26 Lerner JP, Timor-Tritsch IE, Federman A et al.: Transvaginal ultrasonographic characterization of ovarian masses with an improved, weighted scoring system. Amer. J. Obstet. Gynecol. 170 (1994) 81–85

27 Mais V, Guerriero S, Ajossa S et al.: Transvaginal ultrasonography in the diagnosis of cystic teratoma. Obstet. Gynecol. 85 (1995) 48–52

28 Malde HM, Kedar RP, Chadha D et al.: Dermoid mesh: A sonographic sign of ovarian teratoma. AJR 159 (1992) 1349–1350

29 Obbrai M, Lyrich SS, Holder G, Jackson R, Tang L, Butt WE: Hormonal studies on women with polycystic ovaries diagnosed by ultrasound. Clin. Endocrinol. 32 (1990) 467–474

30 Owre A, Pedersen JF: Characteristic fat-fluid level at ultrasonography of ovarian dermoid cyst. Acta Radiol. 23 (1991) 317–319

31 Peterson WF, Prevost EC, Edmunds FT et al.: Benign cystic teratomas of the ovary; a clinicostatistical study of 100 cases with a review of the literature. Am. J. Obstet. Gynecol. 70 (1955) 368–382

32 Quinn SF, Erickson S, Black WC: Cystic ovarian teratomas: The sonographic appearance of the dermoid plug. Radiology 155 (1985) 477–478

33 Schulman H, Conway C, Zalud I et al.: Prevalence in a volounteer population of pelvic cancer detected with transvaginal ultrasound and color flow Doppler. Ultrasound Obstet. Gynecol. 4 (1994) 414–420

34 Timor-Tritsch IE, Lerner JP, Monteagudo A et al.: Transvaginal ultrasonographic characterization of ovarian masses by means of color flow-directed Doppler measurements and a morphologic scoring system. Amer. J. Obstet. Gynecol. 168 (1993) 909 –913

35 Weiner Z, Thaler I, Beck D et al.: Differentiating malignant from benign ovarian tumors with transvaginal color flow imaging. Obstet. Gynecol. 79 (1992) 159–162

36 Woodruff JD, Protos P, Peterson WF: Ovarian teratoma. Amer. J. Obstet. Gynecol. 102 (1968) 702–715

33 Malignant Adnexal Tumors

A. Kurjak, S. Kupesic, and A. K. Ertan

Color Doppler Sonography of Adnexal Malignancies

Since transvaginal color Doppler sonography (TVCD) was first used to assess the vascularity of the ovaries[2,45], there has been a diversity of opinion as to its value in diagnosing adnexal malignancies. Most of the studies published on this topic have shown that ovarian carcinomas display characteristic blood flow patterns compared with benign ovarian masses. The overlap in the blood flow parameters of malignant and benign adnexal masses has become the main discussion point in efforts to accurately differentiate malignant and benign lesions based on their vascular characteristics.

Publications on this topic (since 1989) have reported marked differences between malignant and benign adnexal lesions. While TVCD portrays malignant adnexal tumors as extremely vascular masses with a very low resistance index in the tumor vessels, benign lesions display a small number of blood vessels with a very high resistance index. However, as more experience has been gained with this method, growing numbers of studies have shown a significant overlap in the blood flow parameters of benign and malignant lesions. Moreover, there are some benign lesions that exhibit the same blood flow parameters as malignant neoplasms, and vice versa. Ultimately, however, there has been general agreement that transvaginal color Doppler sonography is a very useful method that can be of significant help in making diagnostic and therapeutic decisions in routine clinical situations.

Review of the Literature

Selection of publications. For this chapter we analyzed 33 articles published between 1989 and 1996 by 27 different hospitals and institutions based in 14 countries. Because some centers published more than one article on this topic during the seven-year period, for most centers we selected a single article with the most representative results.

Evaluation of TVCD results. Table 33.**1** reviews the publications that we selected according to these criteria. The majority are from departments of obstetrics and gynecology (20 in all), and the remaining seven are from radiology departments. Most of the studies[14] were published in 1994. Based on their conclusions, the studies were classified according to whether the authors felt that, in assessing vascularity and differentiating between malignant and benign adnexal masses, transvaginal Doppler scanning was definitely useful, was somewhat useful, or was not helpful at all. We discovered that 51% of the studies concluded that transvaginal Doppler was definitely useful, while 19% found it to be of no value. Thirty percent found that transvaginal Doppler was useful to a limited degree.

Sensitivity and specificity. Table 33.**2** shows the results of the 33 selected studies based on sensitivity, specificity, positive predictive value, and negative predictive value. A close look at these parameters shows that the sensitivity and specificity of TVCD have declined since its introduction. Several factors have contributed to this significant decline. First, the many departments that practice the procedure have followed different examination protocols. Second, different scanners and different equipment settings have been used (e.g., different resolutions). Other important factor are examiner experience and the personal attitude of the examiner toward the new method. For example, of the 27 facilities that were included in the review, only one radiology department rated TVCD as a valuable diagnostic procedure. Three of the other six radiology departments found that TVCD was only somewhat better than traditional procedures, and the remaining three found that it was definitely inferior to conventional ultrasound (Table 33.**3**). By contrast, 55% of the OB/GYN departments rated TVCD as very helpful, 30% rated it as somewhat helpful, and only 15% found it to be of no value. We can explain this discrepancy by noting that color Doppler is widely considered to be more of an adjunct to conventional ultrasound—which is widely used by gynecologists—than a separate procedure in itself.

Expectations of a new diagnostic method. While OB/GYN departments interpret a new diagnostic technique within the context of clinical information, radiologists are more apt to view the pathological finding in isolation, with little awareness of the clinical data. Accordingly, gynecologists tend to evaluate a new method on the basis of how the new information (even when minimal or not statistically significant) can add to the overall clinical impression. Hence they greet the method with much more enthusiasm and optimism than do radiologists.

The most frequent criticism is that the Doppler examination contributes very little to the clinical management of patients with adnexal tumors. We shall therefore discuss the capabilities of color Doppler sonography in detail, presenting

Table 33.**1** How color Doppler sonography has been assessed in various studies published since 1989

Authors	Year	Department	Number of tumors evaluated	Assessment
Hata et al.[30]	1989	Gynecology	21	Positive
Fleischer et al.[21]	1991	Radiology	43	Positive
Kurjak et al.[44]	1991	Gynecology	680	Positive
Weiner et al.[71]	1992	Gynecology	53	Positive
Kawai et al.[36]	1992	Gynecology	24	Positive
Tekay et al.[68]	1992	Gynecology	72	Negative
Hata et al.[29]	1992	Gynecology	64	Negative
Kurjak et al.[43]	1992	Gynecology	83	Positive
Hamper et al.[59]	1993	Radiology	31	Indifferent
Schneider et al.[27]	1993	Gynecology	55	Positive
Timor-Tritsch et al.[69]	1993	Gynecology	115	Positive
Jain[34]	1994	Radiology	50	Indifferent
Weiner et al.[17]	1994	Gynecology	18	Positive
Levine et al.[46]	1994	Radiology	35	Negative
Brown et al.[10]	1994	Radiology	44	Indifferent
Valentin et al.[70]	1994	Gynecology	149	Negative
Bromley et al.[9]	1994	Gynecology	33	Indifferent
Carter et al.[14]	1994	Gynecology	30	Indifferent
Prompeler et al.[53]	1994	Gynecology	83	Indifferent
Chou et al.[15]	1994	Gynecology	108	Positive
Wu et al.[73]	1994	Gynecology	410	Positive
Zaneta et al.[75]	1994	Gynecology	76	Positive
Salem et al.[57]	1994	Radiology	102	Negative
Sawicki et al.[60]	1994	Gynecology	65	Positive
Sengoku et al.[58]	1994	Gynecology	28	Positive
Franchi et al.[26]	1995	Gynecology	129	Indifferent
Maly et al.[47]	1995	Gynecology	102	Positive
Stein et al.[65]	1995	Radiology	169	Negative
Carter et al.[13]	1995	Gynecology	89	Indifferent
Fleischer et al.[19]	1995	Radiology	126	Positive
Buy et al.[12]	1996	Gynecology	132	Indifferent
Rehn et al.[56]	1996	Gynecology	259	Indifferent
Predanic et al.[52]	1996	Gynecology	106	Positive

its advantages as well as disadvantages and pointing out ways in which the results might be overinterpreted.

We hope that by the end of the chapter we will have refuted the arguments of the "Doppler opponents" as we are convinced that color Doppler blood flow measurements provide very useful additional information and help to distinguish patients who require early, immediate intervention from those in whom a wait-and-see approach is justified.

Table 33.**2** Review of the literature on the sensitivity, specificity, positive predictive value, and negative predictive value of Doppler sonography

Authors	Sensitivity	Specificity	Positive predictive value	Negative predictive value
Hata et al.[30]	100	100	100	100
Fleischer et al.[21]	100	83	73	100
Kurjak et al.[44]	96	99	98	99
Weiner et al.[71]	94	97	94	94
Kawai et al.[36]	88	100		
Tekay et al.[68]	82	72	35	96
Hata et al.[29]	92	53	59	90
Kurjak et al.[43]	96	95	96	95
Hamper et al.[59]	66	76	40	90
Schneider et al.[27]	94	56	47	96
Timor-Tritsch et al.[69]	94	99	94	99
Jain[34]	70	82		
Weiner et al.[17]	86	100	92	100
Levine et al.[46]	25	89		
Brown et al.[10]	100	79		
Valentin et al.[70]	100	53		
Bromley et al.[9]	66	81		
Carter et al.[14]	57	78	68	69
Prompeler et al.[53]	95	86		
Chou et al.[15]	88	92	85	94
Wu et al.[73]	68	97		
Zaneta et al.[75]	91	85		
Salem et al.[57]	79	77	37	96
Sawicki et al.[60]	100	94	95	100
Sengoku et al.[58]	82	92	93	79
Franchi et al.[26]	76	72	68	93
Maly et al.[47]	100			
Stein et al.[65]	43	56	56	
Carter et al.[13]				
Fleischer et al.[19]	92	86	86	98
Buy et al.[12]	71	67	43	87
Rehn et al.[56]	67	53	22	89
Predanic et al.[52]	86	83	32	98

Table 33.**3** Comparison of the assessment of transvaginal color Doppler sonography (TVCD) by radiologists and gynecologists

Assessment of TVCD	Radiology	Gynecology	Total
Useful	1 (14%)	11 (50%)	12 (51%)
Somewhat useful	3 (43%)	6 (30%)	9 (30%)
Not useful	3 (43%)	3 (15%)	6 (19%)

Neoangiogenesis

The use of color Doppler sonography in tumor diagnosis is based on the hypothesis that the unrestricted growth of tumors requires angiogenesis[22,23]. The formation of new blood vessels and the further development of preexisting vessels are influenced by specific angiogenic factors[24].

Physiological and pathological angiogenesis. Angiogenesis is a physiological phenomenon that occurs in the endometrium during implantation[32] and in the ovary during follicular maturation[8]. During oncogenesis, however, angiogenesis also occurs as a pathological process[22]. Most tumors that have reached more than 2–3 mm in size cannot continue to grow without augmenting their blood supply. Thus, the growth of carcinomas and their metastases requires an adequate vascular system that can supply sufficient nutrients[25]. As studies on tumor angiogenesis have shown, tumors derive their blood supply from preexisting, previously organ-supplying blood vessels as well as from new vessels that form in response to tumor-produced angiogenic factors. Tumor vessels more commonly develop from veins than from arteries supplying the affected organ. Consequently, the walls of tumor vessels usually contain much fewer smooth muscle cells than normal vessels. The location, structure, flow velocities, and impedance values of the tumor vessels can be analyzed by TVCD.

Detecting Blood Vessels and Defining their Location

Peripheral and central tumor vessels. Tumor vessels can be grossly categorized as central or peripheral[21]. Although this classification is somewhat misleading anatomically, it is helpful in describing the location of tumor vessels that are detectable with ultrasound. It is reasonable to assume that peripheral tumor vessels are derived from preexisting vessels in the affected region, whereas central vessels are newly formed in response to tumor-elaborated angiogenic factors or intratumoral necrotic processes. If we compare the blood supply of malignant and benign adnexal lesions, we find that benign lesions are usually supplied by peripheral and pericystic vessels, while malignant tumors more often have a central type of vascular supply[41, 47] (Fig. 33.**1**). The presence of color-flow signals in regular septa is rarely detectable and, in positive cases, does not signify malignancy[12].

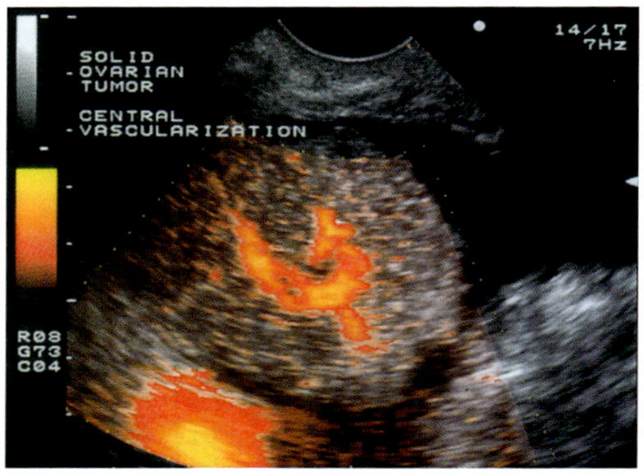

Fig. 33.**1** Copious central blood flow in a malignant tumor, demonstrated by power Doppler imaging.

Tumor size. Color Doppler sonography can be very helpful for benign–malignant discrimination in cases with equivocal morphological findings. The usefulness of the method has been particularly emphasized in the evaluation of semisolid and solid-cystic masses[44,70]. If there are morphological criteria suggestive of malignancy, TVCD can define the vascular supply in tumors that are at least 1 cm in size[12]. Tumors that appear quite small with transvaginal B-mode ultrasound will appear considerably larger when subsequently examined with TVCD, and a tumor previously considered benign may be classified as malignant on the basis of its vascular features[65].

Vascular diameter and "vascular density." Most studies have consistently detected blood flow in malignant tumors[7,21,27,28,34,44,57,58,73]. A few investigators found an absence of color flow in a malignant tumor[10,34,36,41,68]. This discrepancy may be due to the fact that the minimum diameter that a vessel must have to be detected sonographically varies with the quality of the ultrasound system. It has been found that vascular diameters in benign lesions were between 0.01 and 0.03 mm and that approximately 9–12 vessels could be counted per 10-power field. The vascular diameters in borderline tumors ranged from 0.01 to 0.1 mm, and 10–20 vessels could be counted per field. In malignant tumors, the vascular diameters ranged from 0.01 to 0.1 mm and 20–30 vessels could be counted per field[12].

Vascular Patterns

Diffuse and isolated vascular patterns. Only a few studies have attempted to evaluate or quantitate tumor vascularity. In one of the first series of studies, we tried to classify tumor vascular patterns into a "diffuse" type and an "isolated" type. A diffuse pattern was one in which multiple color-flow signals could be detected, while an isolated pattern was defined as the presence of only one color signal within the definable mass[28]. We found that a diffuse vascular pattern was three times more common in the central, solid portion of malignant tumors (80%) than in benign masses (33%). It is reasonable to assume, moreover, that high angiogenic activity is present in areas with a diffuse vascular pattern (Fig. 33.**2**). These results are consistent with the above observation that the number of blood vessels that can be counted at 10-power magnification is significantly higher in malignant tumors (20–30) than in benign lesions (8–12).

Quantification of vascular density. Based on the observations described above, we believe that an objective analytical method based on computer quantification is needed to express the likelihood of malignancy as indicated by vascular density. One such system for the quantification of tumor vascularity, described in an experimental study of nine transplanted murine tumors, showed excellent agreement with histopathological results[50]. This system appears to provide an accurate depiction of tumor vascularity. The time–activity

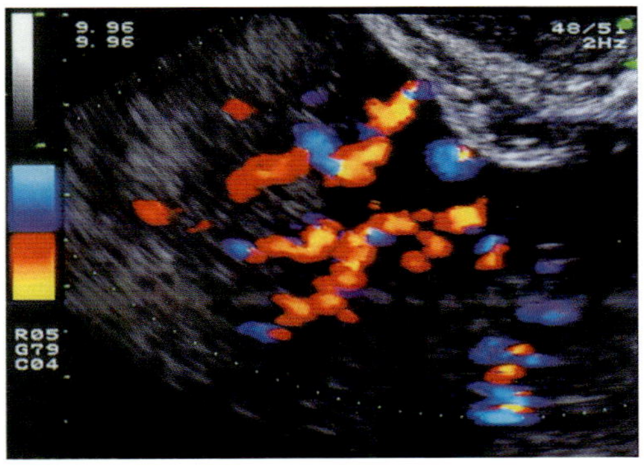

Fig. 33.**2** Numerous blood vessels distributed diffusely in an ovarian mass suggest intense angiogenesis, which is typical of malignant tumors.

curves showed markedly higher blood flow in an experimental group injected with an exotoxin than in a control group injected with saline solution. The quantification of vascular density by TVCD was also more accurate following the injection of an intravascular agent. One hypothesis to explain the resistance of tumors to chemotherapeutic agents is that only very small amounts of these agents are delivered to poorly perfused areas of the tumor. It is generally agreed that when a chemotherapeutic agent is administered together with a drug that increases blood flow, the agent can penetrate more deeply into the tumor tissue, including hypoxic or ischemic areas, and that this can achieve far better results. The quantification scheme devised by Meyerowitz et al.[50] may make it possible to assess the likelihood of malignancy and to monitor the tumor response to chemotherapy on the basis of blood flow and vascular density. This technique could also be used to determine whether the vascular density of a tumor correlates with the likelihood of metastasis. It is expected that the adjunctive use of 3D sonography with Doppler ultrasound will permit a more accurate evaluation of the blood flow within a tumor.

Pulsed Doppler Waveforms

Early diastolic notch. Besides the distribution of blood vessels in adnexal masses, the shape of the flow velocity waveform has established itself as a critically important hemodynamic criterion (Fig. 33.**3**). The presence of a notch in the early diastolic portion of the waveform, called the "early diastolic notch," has been found more frequently in benign tumors than in malignant tumors[20,47]. Experimental studies with tumors inoculated in rabbit flanks showed that vessels in an area of active tumor growth displayed only a small number of smooth-muscle cells in the muscular coat and more closely resembled sinusoids than normal arterioles. These sinusoidal vascular spaces have also been observed in human hepatic and adnexal tumors in areas of active tumor growth. This type of vascularity can account for the low flow velocity and low impedance found in tumor vessels[62]. In blood vessels with a normally developed muscular coat, systole is followed by a brief relaxation phase of the vessel wall during which antegrade blood flow occurs[62]. This blood flow pattern with an early diastolic notch is observed in normal ovarian vessels during the proliferative phase of the menstrual cycle[30].

Absence of the diastolic notch. The absence of a diastolic notch in the Doppler waveform is presumably due to an absence or relative deficiency of smooth-muscle cells in the vessel wall, as the muscle cells are responsible for the initial flow resistance during the first half of diastole and subsequent relaxation of the vessel wall[18]. It is important to note, however, that a diastolic notch can also be found at vascular bifurcations. Absence of the diastolic notch has been observed in newly formed vessels in the wall of the corpus luteum, again presumably due to the paucity of muscle cells in the wall of these "young" vessels (Fig. 33.**4**). Although the early diastolic notch is most commonly observed in benign cystic masses, it is also found in 7% of multilocular solid malignant tumors[70]. Additionally, Parsons reports that the vascular plexus in the wall of the corpus luteum is supplied by larger vessels in which, interestingly, the blood flow has a higher impedance and flow velocity than in the more distal branches[51].

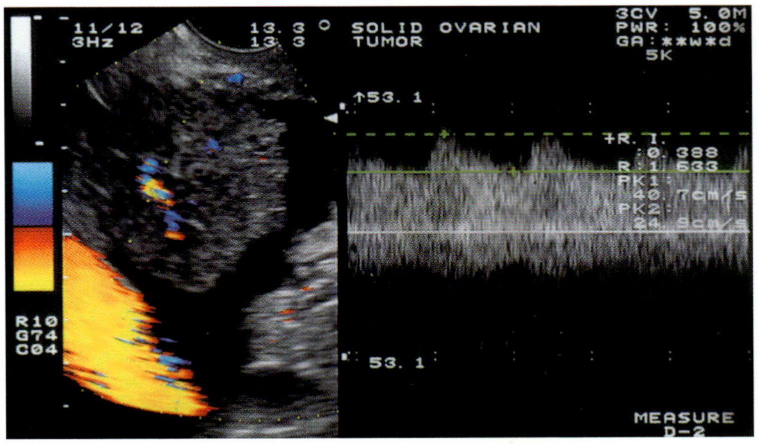

Fig. 33.**3** The Doppler waveform indicates high diastolic flow and a low resistance index (RI = 0.38). Histopathological examination confirmed that the tumor was malignant.

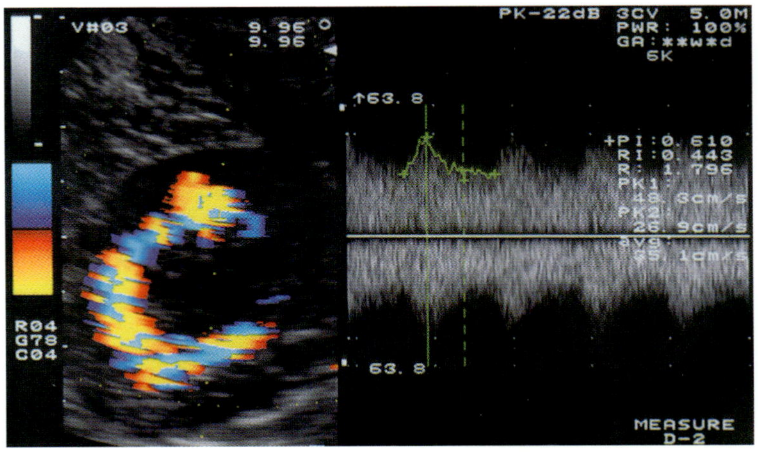

Fig. 33.**4** Increasing vascularization of the corpus luteum. The low resistance index (RI = 0.44) is typical of blood flow in the corpus luteum.

Vascular Impedance

We presume that the diversity of opinions on Doppler ultrasound for evaluating the vascular characteristics of malignant adnexal tumors is based on the great diversity of results that have been reported in different studies in recent years.

Three different assessments of TVCD. It should be emphasized at this point that Doppler waveform analysis and vascular impedance continue to be the most important criteria in the evaluation of tumor vessels. As a result, most studies are concerned chiefly with differences in vascular impedance between benign and malignant adnexal masses. The authors of these studies can be divided into three main groups. The first group presents Doppler results of high sensitivity and specificity[15, 19, 21, 36, 44, 52, 58, 60, 71, 73, 75] and claims that Doppler sonography significantly improves upon the accuracy of B-mode ultrasound and that its potential value as a screening method warrants further study. The second group of investigators is less optimistic in interpreting their results[10, 14, 27, 29, 34, 53, 59], claiming that while Doppler sonography has definite potential, it does not significantly facilitate the process of routine clinical decision-making at the present time. Finally, the third group[9, 46, 56, 68] believes that Doppler cannot add to the conventional diagnostic workup, failing to consider that the problem may be less the method itself than the inexperience of the examiner and an unfavorable patient selection. It is remarkable that the first group, who used the method from the very beginning, achieved the highest sensitivity and specificity in the diagnosis of malignant tumors even when using less advanced instruments with relatively poor resolution. It may be objected, of course, that the uncritical enthusiasm of these examiners biased their appraisal of the method. But it should also be considered that the first group had more experience with ultrasound, and this explains why their subsequently published studies yielded similarly good results[6, 17–20, 31, 40, 41, 72].

Areas of overlap. More recent studies have shown smaller impedance differences and larger areas of overlap between the flow parameters of benign and malignant tumors. The fact remains, however, that differences in vascularity do exist between these tumors and that lower impedance values are found in the vessels of malignant lesions. The resistance indices (RI) and pulsatility indices (PI) in benign and malignant ovarian tumors are compared in Table 33.**4**. We believe that the areas of overlap may be caused by the range of variation of PI and RI values in any given tumor. Various authors have noted this range of variation in adnexal tumors[20, 36, 41, 68, 70, 71] and have emphasized the importance of sampling indices in vessels that accurately represent the tumor. It is also important to have some basic knowledge of Doppler physics and adequate experience in the use of Doppler instrumentation. Additionally, the examiner should know the potential sources of errors and artifacts before practicing the method in a patient.

Table 33.**4** Resistance indices (RI and PI) of malignant and benign ovarian tumors

Authors	Index	Malignant adnexal tumors	Benign adnexal tumors
Hata et al.[30]	RI	0.469 ± 0.11	0.96 ± 0.17
Fleischer et al.[21]	PI	0.3 – 1.5	0.5 – 4.0
Kawai et al.[36]	PI	0.53 ± 0.65	1.44 ± 0.05
Tekay et al.[68]	PI	0.5 (0.5 – 0.9)	0.6 (0.5 – 3.5)
Hata et al.[29]	RI	0.50 ± 0.11	0.69 ± 0.18
Kurjak et al.[43]	RI	0.38 (0.27 – 0.61)	0.52 (0.46 – 1.0)
Hamper et al.[59]	RI	0.5 ± 0.17 (0.27 – 0.67)	0.77 ± 0.33 (0.2 – 1.0)
Schneider et al.[27]	RI	0.52 (0.2 – 1.0)	0.84 (0.24 – 1.0)
Timor-Tritsch et al.[69]	RI	0.39 (0.2 – 0.53)	0.63 (0.23 – 0.98)
Levine et al.[46]	RI	0.47 ± 0.11	0.57 ± 0.17
Brown et al.[10]	RI	0.39 ± 0.09 (0.25 – 0.50)	0.62 ± 0.16 (0.34 – 0.90)
Valentin et al.[70]	PI	0.9 – 0.94	0.18 – 0.96
Carter et al.[14]	RI	0.6 ± 0.1	0.7 ± 0.2
Prompeler et al.[53]	RI	0.40 (0.22 – 0.66)	0.68 (0.26 – 1.0)
Chou et al.[15]	RI	0.41 (0.18 – 0.68)	0.68 (0.36 – 0.89)
Zaneta et al.[75]	RI	0.46 ± 0.10 (0.27 – 0.99)	0.72 ± 0.14 (0.43 – 0.90)
Salem et al.[57]	PI	0.82 ± 0.38 (0.3 – 1.89)	1.44 ± 0.65 (0.3 – 3.5)
Sengoku et al.[58]	PI	0.57 ± 0.14	2.42 ± 0.67
Franchi et al.[26]	RI	0.49 (0.28 – 0.78)	0.72 (0.48 – 0.98)
Maly et al.[47]	RI	0.5 (0.3 – 0.6)	0.7 (0.5 – 1.0)
Stein et al.[65]	RI	0.53 ± 0.16 (0.27 – 0.83)	0.65 ± 0.18 (0.27 – 0.98)
Buy et al.[12]	RI	0.54 ± 0.11 (0.28 – 0.77)	0.59 ± 0.14 (0.34 – 1.0)
Predanic et al.[52]	RI	0.33 ± 0.03 (0.23 – 0.45)	0.57 ± 0.02 (0.35 – 1.0)

Table 33.**5** Blood flow velocities in malignant and benign ovarian tumors

Authors	Blood flow velocities (cm/s)	
	Malignant adnexal tumors	Benign adnexal tumors
Fleischer et al. (21)	7–61 cm/s	16–37 cm/s
Kurjak et al. (41)	14.4–26.2 cm/s	20.2–27.3 cm/s
Carter et al. (14)	13.8 ± 10.7 cm/s	14.4 ± 9.9 cm/s
Prompeler et al. (53)	47.1 (14.6–105) cm/s	17.5 (5.2–61.5) cm/s

Blood Flow Velocities

Several authors have documented abnormal Doppler spectra with high flow velocities in the periphery of malignant tumors, suggesting that these signals may be caused by arteriovenous anastomoses[11,72]. Other authors support this hypothesis[35,48,61,66,67]. It was suggested that a cutoff value of 40 cm/s might be useful in differentiating malignant from benign tumors based on blood flow velocity[16]. However, while these high flow velocities and cutoff values have been successfully used in breast cancer diagnosis, they have not been confirmed for adnexal tumors[21,41]. Table 33.**5** reviews the flow velocities that have been detected by various groups of investigators. It is noteworthy that only one study group described a significant difference in flow velocities between benign and malignant tumors and found that flow velocity was superior to resistance index as an indicator of tumor behavior[53].

Stages of Malignant Tumors

Tumor grade and size. Neoangiogenesis is a common phenomenon in malignant ovarian tumors, but the intensity of neovascularization depends on individual tumor characteristics[5]. Thus, a marked decrease in the resistance indices of adnexal tumors may reflect the grade of tumor malignancy[74]. Animal studies have shown that Doppler sonography can detect angiogenesis even in tumors that have a small volume (25 mg)[54]. This demonstrates that angiogenesis is detectable even in carcinomas that are still within the ovarian capsule and in low-grade malignancies. Indeed, several groups of authors have shown that a stage I carcinoma can be diagnosed with TVCD ultrasound[7,19,42] (Fig. 33. **5**). One group described 2 of 18 stage I ovarian carcinomas based entirely on abnormal blood flow patterns detected in normal-sized ovaries[42]. Another group detected 3 of 17 stage I carcinomas based on flow parameters[19]. In another study, two stage I tumors did not show abnormal blood flow even though they were already larger than 15 cm. Most likely these undetected carcinomas were too low-grade to induce significant angiogenesis, but it is also possible that the vessels were too small to be detected with available equipment.

Decreasing impedance with tumor progression. The new power or energy-mode Doppler instruments can define even the smallest blood vessels. Paradoxically, it has been found that even tiny intraparenchymal arterioles in normal or benign tissues may display a low impedance and low flow velocities, which can lead to false-positive results. Nevertheless, there is a definite tendency to find an incremental decrease in impedance from benign tumors toward borderline tumors, early-stage carcinoma, and finally to advanced-stage malignancies[74]. This observation is supported by in-vivo and histopathological studies showing an incrementation of angiogenesis in melanocytes with tumor progression. These data are consistent with the fact that neoplasms require an increased blood supply in order to undergo rapid, aggressive growth.

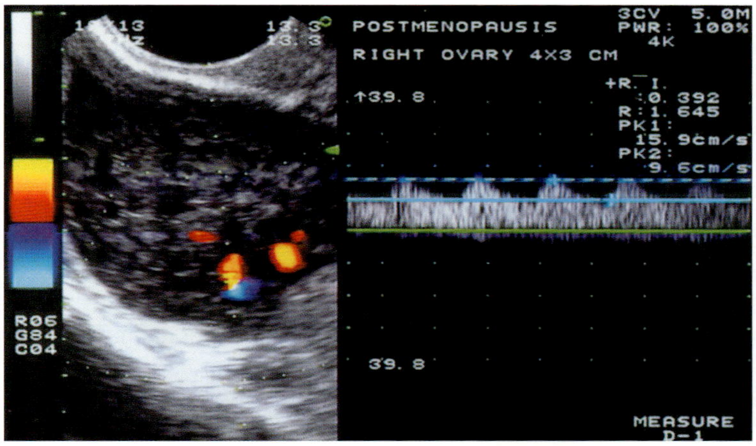

Fig. 33.**5** Color and pulsed Doppler signals recorded from a slightly enlarged ovary in a postmenopausal patient. The low resistance index is suggestive of malignancy. This was confirmed by histopathological examination.

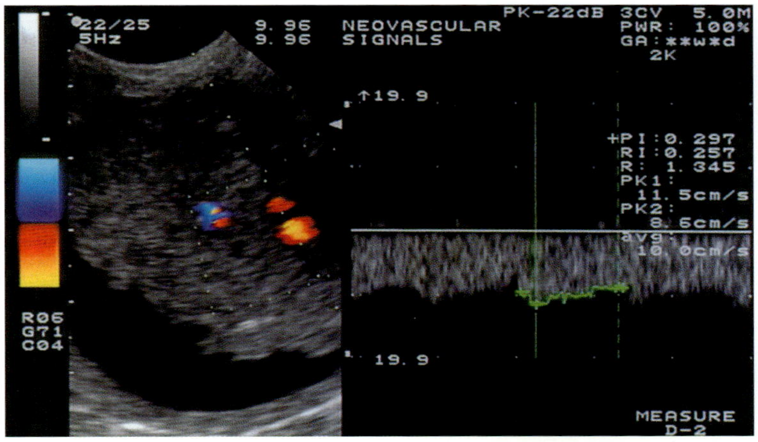

Fig. 33.**6** New vessels that have formed within the solid portion of a complex adnexal tumor enable further tumor growth. These vessels have very few smooth muscle cells in the tunica muscularis, accounting for the very low vascular impedance (RI = 0.26).

Representativeness of individual tumor vessels. As noted earlier, the impedance values sampled from one area of a tumor are not necessarily representative of the tumor as a whole[64]. For example, Doppler signals may be sampled from a main drainage area or from an area in which the vessel walls are very permeable, leading to stasis and the formation of arteriovenous shunts with very low impedance[4] (Fig. 33.**6**). Conversely, a high interstitial pressure may exist only in certain tumor areas, leading to high local impedance values.

False-Positive Results

It has been shown that while significant differences exist in the vascular characteristics of malignant and benign tumors, some degree of overlap is bound to occur. This overlap can be minimized by knowing the physiological changes that occur during the menstrual cycle and the morphological characteristics of adnexal masses. The examiner should also realize that overlaps will occur in the examination of any biological material. Nevertheless, even an experienced sonographer using good equipment is not immune to occasional misinterpretations.

Increased blood flow. A major source of diagnostic errors is increased blood flow occurring under physiological conditions. For example, increased blood flow and significantly decreased impedance may be seen in the preovulatory follicle and in the corpus luteum, which is why the cycle phase should always be considered when interpreting blood flow measurements in the premenopausal ovary.

Corpus luteum. The "angiogenic ring" around the dominant follicle is most prominent just before ovulation. The perifollicular blood flow velocity tends to increase, while vascular impedance falls[39]. Angiogenesis continues to increase after follicular rupture and corpus luteum formation, and there is further dilatation of the ovarian stromal vessels. As a result, the normal corpus luteum and benign lesions of the corpus luteum are a frequent source of false-positive results in differentiating between benign and malignant lesions. The blood flow to the corpus luteum, occasionally described as the "ring of fire," most likely results from circumscribed dilatation of the ovarian stromal vessels or from "luteal conversion"[49] caused by locally rising levels of E_3 prostaglandins. It is known that these prostaglandins have a potent vasodilatory action[2,55]. The correct evaluation of luteal blood flow is also hampered by the sonographic appearance of the corpus luteum itself, which appears as a cystic structure with irregular margins and a nonhomogeneous internal echo pattern (Fig. 33.**7**). Thus, cystic structures with echogenic contents, high flow velocity, and low impedance can lead to misinterpretation and a false-positive diagnosis (Fig. 33.**8**). For this reason, Doppler ultrasound examinations should be performed during the early proliferative phase of the cycle. It should be noted, however, that increased ovarian blood flow may be found even during the first five days of the cycle due to the persistence of corpus luteum activity[55].

Tubo-ovarian abscess and endometriosis. Both of these benign adnexal lesions can simulate malignancy, as both are associated with marked vascularization that generally occurs in response to inflammation or an inflammatory component[37,63]. Additionally, a hormonal imbalance in overweight patients can give rise to blood flow patterns with a low resistance index[38].

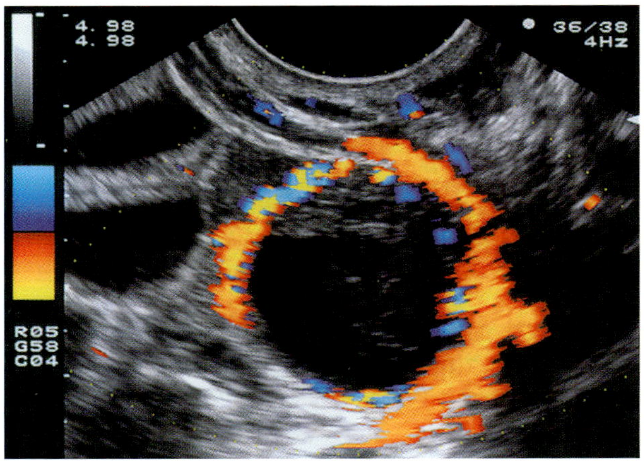

Fig. 33.**7** Transvaginal color Doppler scan of a nonspecific corpus luteum cyst. Strongly dilated vascular channels penetrate into the hemorrhagic cavity of the ruptured follicle.

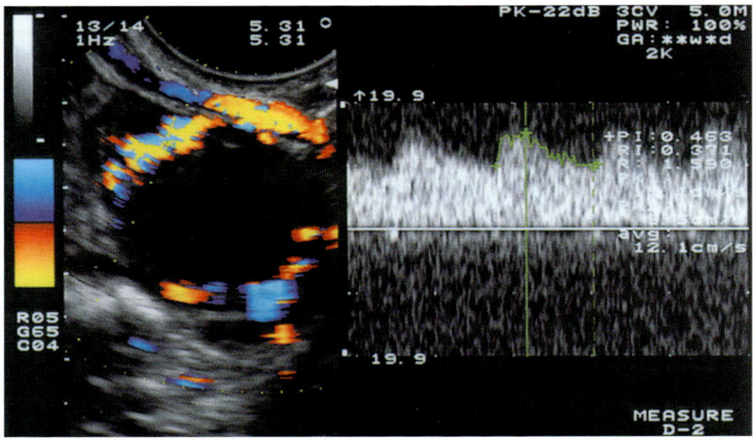

Fig. 33.**8** Increased blood flow indicating an active corpus luteum. The low resistance index in the angiogenic area (RI = 0.37) could result in a false-positive diagnosis of ovarian carcinoma.

Conclusions

TVCD defines the vascularity of adnexal masses and thus provides information on the histology and metabolism of the masses. Accordingly, blood flow data should be viewed as an indicator of tumor angiogenic intensity but not as an indicator of malignancy itself. Today it seems clear that initial attempts to classify ovarian masses based entirely on impedance measurements were too simplistic. This problem was partially solved by the introduction of additional "vascular parameters" such as the location and arrangement of the vessels, their waveform patterns, the presence or absence of an early diastolic notch, and the measurement of blood flow velocities. Even so, it is not always possible to differentiate between benign and malignant tumors based solely on the analysis of blood flow parameters. A frequent criticism of the Doppler method is that the examiner is always influenced by the B-mode image when using Doppler, i.e., an examiner who finds malignant-type morphological criteria in the B-mode image will look more carefully for abnormal blood flow patterns than in cases where B-mode findings suggest a simple cyst.

Nevertheless, when Doppler sonography is used correctly by an experienced examiner, it provides significant additional diagnostic information that can confirm a tentative diagnosis based on morphological criteria.

References

1 Aleem F, Predanic M: Transvaginal color Doppler determination of the ovarian and uterine blood flow characteristics in polycstic ovarian disease. Fertil. Steril. 65 (1996) 510–516

2 Alila HW, Corradino RA, Hansel W: A comparison of the effects of cyclooxygenase prostanoids on progesterone production by small and large bovine luteal cells. Prostaglandins 36 (1988) 259–270

3 Barnhill RL, Fandrey K, Levy MA, Mihm MC Jr, Human B: Angiogenesis and tumor progression of melanoma: quantification of vascularity in melanocytic nevi and cutaneous malignant melanoma. Lab. Invest. 67 (1992) 57–62

4 Blood CH, Zetter BR: Tumor interactions with the vasculature: angiogenesis and tumor metastasis. Biochem. Biophys. Acta 1032 (1990) 89–118

5 Bourne TH: Should clinical decisions be made about ovarian masses using transvaginal color Doppler? Ultrasound Obstet. Gynecol. 4 (1994) 257–260

6 Bourne TH, Campbell S, Reynolds KM et al.: Screening for early familial ovarian cancer with transvaginal ultrasonography and color flow imaging. Brit. Med. J. 306 (1993) 1025–1029

7 Bourne TH, Campbell S, Steers CV, Whitehead MI, Collins WP: Transvaginal colour flow imaging: a possible new screening technique for ovarian cancer. Brit. Med. J. 299 (1989) 1367–1370

8 Bourne TH, Jurkovic D, Waterstone J, Campbell S, Collings WP: Intrafollicular blood flow during human ovulation. Ultrasound Obstet. Gynecol. 1 (1991) 215–219

9 Bromley B, Goodman H, Benacerraf BR: Comparison between sonographic morphology and Doppler waveform for the diagnosis of ovarian malignancy. Obstet. Gynecol. 83 (1994) 434–437

10 Brown DL, Frates MC, Laing FC et al.: Ovarian masses: Can benign and malignant lesions be differentiated with color and pulsed Doppler US? Radiology 190 (1994) 333–336

11 Burns PN, Halliwell M, Webb AJ, Wells PNT: Ultrasonics Doppler studies of the breast. Ultrasound Med. Biol. (1982) 127–143

12 Buy JN, Ghossain MA, Hugol D et al.: Characterization of adnexal masses: Combination of color Doppler and conventional sonography compared with spectral Doppler analysis alone and conventional sonography alone. Amer. J. Roentgenol. 166 (1996) 385–393

13 Carter JR, Lau M, Fowler JM, Carlson JW, Carson LF, Twiggs LB: Blood flow characteristics of ovarian tumors: Implications for ovarian cancer screening. Amer. J. Obstet. Gynecol. 172 (1995) 901–907

14 Carter J, Saltzman A, Hartenbach E, Fowler J, Carson L, Twiggs LB: Flow characteristics in benign and malignant gynecologic tumors using transvaginal color flow Doppler. Obstet. Gynecol. 83 (1994) 125–130

15 Chou CY, Chang CH, Yao BL, Kuo HC: Color Doppler ultrasonography and serum CA 125 in the differentiation of benign and malignant ovarian tumors. J. Clin. Ultrasound 22 (1994) 491–496

16 Dock W, Grabanwoger F, Metz V, Elbenberger K, Farres M: Tumor vascularization: assessment with duplex sonography. Radiology 181 (1991) 241–244

17 Einer Z, Beck D, Brandes JM: Transvaginal sonography, color flow imaging, computed tomography scanning, and CA 125 as a routine follow-up examination in women with pelvic tumor: Detection of recurrent disease. J. Ultrasound Med. 13 (1994) 37 –41

18 Fleischer AC, Cullinan JA, Jones III HW, Peery CV, Bluth RF: Serial assessment of adnexal masses with transvaginal color Doppler sonography. Ultrasound Med. Biol. 21 (1995) 435–443

19 Fleischer AC, Cullinan JA, Peery CV, Jones III HW: Early detection of ovarian carcinoma with transvaginal color Doppler ultrasonography. Amer. J. Obstet. Gynecol. 174 (1996) 101–106

20 Fleischer AC, Rodgers WH, Kepple DM, Williams LL, Jones III HW: Color Doppler sonography of ovarian masses: a multiparameter analysis. J. Ultrasound Med. 12 (1993) 41–48

21 Fleischer AC, Rodgers WH, Rao BJ et al.: Assessment of ovarian tumor vascularity with transvaginal color Doppler sonography. J. Ultrasound Med. 10 (1991) 563–568

22 Folkman J: Tumor angiogenesis. Adv. Cancer Res. 48 (1985) 2641–2645

23 Folkman J, Cotran RS: Relation of vascular proliferation to tumor growth. Int. Rev. Exp. Pathol. 16 (1976) 207–212

24 Folkman J, Klaysburn M: Angiogenic factors. Science 235 (1987) 442–447

25 Folkman J, Watson K, Ingber D, Hanahan D: Induction of angiogenesis during the transition from hyperplasia to neoplasia. Nature 339 (1989) 58–61

26 Franchi M, Beretta P, Ghezzi F, Zanaboni F, Goddi A, Salvator S: Diagnosis of pelvic masses with transabdominal color Doppler, CA 125 and ultrasonography. Acta Obstet. Gynecol. Scand. 75 (1995) 734–739

27 Hamper UM, Sheth S, Abbas FM, Rosenshein BN, Aronson D, Kurman JR: Transvaginal color Doppler sonography of adnexal masses: Differences in blood flow impedance in benign and malignant lesions. Am. J. Roentgenol. 160 (1993) 1225–1228

28 Hata K, Hata T, Manage A, Kitao M: Ovarian tumors of low malignant potential: transvaginal Doppler ultrasound features. Gynecol. Oncol. 45 (1992) 259–264

29 Hata H, Hata T, Manabe A, Sugimura K, Kitao M: A critical evaluation of transvaginal Doppler studies, transvaginal sonography, magnetic resonance imaging, and CA 125 in detecting ovarian cancer. Obstet. Gynecol. 80 (1992) 922–926

30 Hata T, Hata K, Senoh D et al.: Doppler ultrasound assessment of tumor vascularity in gynecologic disorders. J. Ultrasound Med. 8 (1989) 309–314

31 Hata K, Makihara K, Hata T, Takahashi K, Kitao M: Transvaginal color Doppler imaging for hemodynamic assessment of r eproductive tract tumors. Int. J. Gynecol. Obstet. 36 (1991) 301–308

32 Jaffe R, Warsof SL: Transvaginal color Doppler imaging in the assessment of uteroplacental blood flow in the normal first-trimester pregnancy. Am. J. Obstet. Gynecol. 164 (1991) 781–785

33 Jain RK: Transport of molecules across tumor vasculature. Cancer Met. Rev. 6 (1987) 559–561
34 Jain KA: Prospective evaluation of adnexal masses with endovaginal grayscale and duplex and color Doppler US: Correlation with pathologic findings. Radiology 191 (1994) 63–67
35 Jellins J, Kossoff G, Boyd J, Reeve TS: The complementary role of Doppler to the B-mode examination of the breast. J. Ultrasound Med. 10 (1983) 29–35
36 Kawai M, Kano T, Kikkawa F, Maeda O. Oguchi H, Tomoda Y: Transvaginal Doppler ultrasound with color flow imaging in the diagnosis of ovarian cancer. Obstet. Gynecol. 79 (1992) 163–167
37 Kupesic S, Kurjak A, Pasalic L. Benic S: Transvaginal color Doppler in the assessment of pelvic inflammatory disease. J. Ultrasound Med. Biol. 6 (1995) 733–738
38 Kupesic S, Kurjak A: Scoring system for prediction of ovarian endometriosis based on transvaginal color and pulsed Doppler sonography. Fertil. Steril. 62 (1994) 81–88
39 Kurjak A, Kupesic S, Schulman H, Zalud I: Transvaginal color flow Doppler in the assessment of ovarian and uterine blood flow in infertile women. Fertil. Steril. 56 (1991) 870–873
40 Kurjak A, Predanic M: New scoring system for prediction of ovarian malignancy based on transvaginal color Doppler sonography. J. Ultrasound Med. 11 (1992) 631–638
41 Kurjak A, Predanic M, Kupesic-Urek S, Jukic S: Transvaginal color and pulsed Doppler assessment of adnexal tumor vascularity. Gynecol. Oncol. 50 (1993) 3–9
42 Kurjak A, Shalan H, Matijevic R, Predanic M, Kupesic-Urek S: Stage I ovarian cancer by transvaginal color Doppler sonography: a report of 18 cases. Ultrasound Obstet. Gynecol. 3 (1993) 195–198
43 Kurjak A, Schulman H, Sosic A, Zalud I, Shalan H: Transvaginal ultrasound, color flow, and Doppler waveform of the postmenopausal adnexal mass. Obstet. Gynecol. 80 (1992) 917–921
44 Kurjak A, Zalud I, Alfirevic Z: Evaluation of adnexal masses with transvaginal color ultrasound. J. Ultrasound Med. 10 (1991) 295–297
45 Kurjak A, Zalud I, Jurkovic D, Alfirevic Z, Miljan M: Transvaginal color Doppler of the assessment of pelvic circulation. Acta Obstet. Gynecol. Scand. 68 (1989) 131–136
46 Levine D, Feldstein VA, Babcook CJ, Filly RA: Sonography of ovarian masses: Poor sensitivity of resistive index for indentifying malignant lesions. Am. J. Roentgenol. 162 (1994) 1355–1359
47 Maly Z, Riss P, Deutinger J: Localization of blood vessels and qualitative assessment of blood flow in ovarian tumors. Obstet. Gynecol. 85 (1995) 33–36
48 Maniasan M, Bamber JCA: A preliminary assessment of an ultrasonic Doppler method for the study of blood flow in human breast cancer. Ultrasound Med. Biol. 8 (1983) 257–261
49 Merce LT, Garces D, Barco MJ, de la Fuente F: Intraovarian Doppler velocimetry in ovulatory, dysovulatory and anovulatory cycles. Ultrasound Obstet. Gynecol. 2 (1992) 197–202
50 Meyerowitz CB, Fleischer AC, Pickens DR et al.: Quantification of tumor vascularity and flow with amplitude color Doppler sonography in an experimental model: Preliminary results. J. Ultrasound Med. 15 (1996) 827–833
51 Parsons AK: Ultrasound of the human corpus luteum. Ultrasound Quarterly 12 (1994) 127–166
52 Predanic M, Vlahos N, Pennisi J, Moukhtar M. Aleem FA: Color and pulsed Doppler sonography, gray-scale imaging, and serum CA 125 in the assessment of adnexal disease. Obstet. Gynecol. 88 (1996) 283–288
53 Prompeler HJ, Sauerbrei WM, Latternann U, Pfleiderer A: Quantitative flow measurements for classification of ovarian tumors by transvaginal color Doppler sonography in postmenopausal patients. Ultrasound Obstet. Gynecol. 4 (1994) 406–413
54 Ramos I, Fernandez LA Morse SS, Fotune KL, Taylor KJW: Detection of neovascular signal in a 3-day Walker 256 rat carcinosarcoma by CW Doppler ultrasound. Ultrasound Med. Biol. 14 (1988) 123–126
55 Raud J: Vasodilatation and inhibition of mediator release represent two distinct mechanisms of prostaglandin modulation of acute mast cell-dependent inflammation. Brit. J. Pharmacol. 99 (1990) 449–454
56 Rehn M, Lohmann K, Rempen A: Transvaginal ultrasonography of pelvic masses: Evaluation of B-mode technique and Doppler ultrasonography. Am. J. Obstet. gynecol. 175 (1996) 97–104
57 Salem S, White LM, Lai J: Doppler sonography of adnexal masses: The predictive value of the Pulsatility index in benign and malignant disease. Am. J. Roentgenol. 163 (1994) 1147–1150
58 Savicki E, Spiewankiewicz B, Cendrowski K, Stelmachow J: Transvaginal Doppler ultrasound with colour flow imaging in benign and malignant ovarian lesions. Clin. Exp. Obst. Gynecol. 22 (1995) 137–142
59 Schneider VL, Schneider A, Reed KL, Hatch KD: Comparison of Doppler with two-dimensional sonography and CA 125 for prediction of malignancy of pelvic masses. Obstet. Gynecol. 81 (1993) 893–988
60 Sengoku K, Satoh T, Saitoh S, Abe M, Ishikawa M: Evaluation of transvaginal color Doppler sonography, transvaginal sonography and CA 125 for prediction of ovarian malignancy. Int. J. Gynecol. Obstet. 46 (1994) 39–43
61 Shimamoto K, Sakuma S, Ishigaki T, Makino N: Intratumoral blood flow: evaluation with color Doppler echography. Radiology 165 (1987) 683–685
62 Shubik P, Vascularization of tumors: a review. J. Cancer Res. Clin. Oncol. 103 (1982) 211–219
63 Sladkevicius P, Valentin L, Marsál K: Blood flow velocity in the uterine and ovarian arteries during menstruation. Ultrasound Obstet. Gynecol. 4 (1994) 421–427
64 Srivastava A, Laidler P, Davies RP, Horgan K, Hughes LE: The prognostic significance of tumor vascularity in intermediate thickness (0.76–4.0 mm thick) melanoma: a quantitative histologic study. Am. J. Pathol. 133 (1988) 419–423
65 Stein SM, Laifer-Narin S, Johnson MB et al.: Differentiation of benign and malignant adnexal masses. Relative value of gray-scale, color Doppler, and spectral Doppler sonography. Am. J. Roentgenol. 164 (1995) 381–386
66 Taylor KJW, Morse SS: Doppler detects vascularity of some malignant tumors. Diagn. Imaging 10 (1988) 132–136
67 Taylor KJW, Ramos I, Carter D, Morse SS, Snower D, Fortune K: Correlation of Doppler US tumor signals with neovascular morphologic features. Radiology 166 (1991) 57–62
68 Tekay A, Jouppila P: Validity of pulsatility and resistance indices in classification of adnexal tumors with transvaginal color Doppler ultrasound. Ultrasound Obstet. Gynecol. 2 (1992) 338–344
69 Timor-Tritsch IE, Lerner JP, Monteagudo A, Santos R: Transvaginal ultrasonographic charaterization of masses by means of color flow-directed Doppler measurements and a morphologic scoring system. Am. J. Obstet. Gynecol. 168 (1993) 909–913
70 Valentin L, Sladkevicius P, Marsál K: Limited contribution of Doppler velocimetry to the differential diagnosis of extrauterine pelvic tumors. Obstet. Gynecol. 83 (1994) 425–433
71 Weiner Z, Thaler I, Beck D, Rottem S, Deutsch M, Brandes JM: Differentiation malignant from benign ovarian tumors with transvaginal color flow imaging. Obstet. gynecol. 79 (1992) 159–162
72 Wells PNT, Halliwell M, Skidmore R, Webb AJ, Woodcock JP: Doppler studies of the breast. Ultrasound Med. Biol. 15 (1977) 231–235
73 Wu CC, Lee CN, Chen TM, Lai JI, Hsieh CY, Hsieh FJ: Factors contributing to the accuracy in diagnosing ovarian malignancy by color Doppler ultrasound. Obstet. Gynecol. 84 (1994) 605–608
74 Wu CC, Lee CN, Chen TM et al.: Incremental angiogenesis assessed by color Doppler ultrasound in the tumorigenesis of ovarian neoplasms. Cancer 73 (1994) 1251–1256
75 Zaneta G, Vergani P, Lissoni A: Color Doppler ultrasound in the preoperative assessment of adnexal masses. Acta Obstet. Gynecol. Scand. 73 (1994) 637–641

34 Three-Dimensional Power Doppler Sonography in Evaluating the Angiogenesis of Ovarian Tumors

A. Kurjak, S. Kupesic, and B. Breyer

It has been known for more than 25 years that the development of new blood vessels is necessary to sustain the growth, invasion, and metastasis of malignant tumors[27,69,81]. Angiogenesis is fundamental to tumor growth, as it supplies the tumor with essential oxygen and nutrients while eliminating degradation products. As angiogenesis progresses, moreover, increasing numbers of tumor cells appear at peripheral circulatory sites, promoting the formation of metastases[57,91]. Cancer cells, even in premalignant stages of tumor development, frequently activate an angiogenic "switch" that induces resting vascular cells to form new vessels. Various observations suggest that the regulation of angiogenesis is independent of tumor cell proliferation. This suggests that it may be possible to develop medications that can selectively inhibit angiogenesis, providing an effective adjunct to traditional chemotherapy, which attacks the tumor cells directly[27,33,69,81]. As a result, tumor angiogenesis research has become a important topic[22,46–48,85].

Contribution of Transvaginal Color Doppler Sonography

Tumor size. Angiogenesis is a familiar phenomenon in malignant ovarian tumors, but the intensity of neovascularization depends on individual tumor characteristics[8]. The incremental decline of the resistance index in adnexal tumors reflects the increase in angiogenesis and can be considered an indicator of malignant potential[96]. Studies in experimental animals have shown that neoangiogenesis is detectable with Doppler ultrasound even in tumors that have a small volume (25 mg)[70]. This shows that angiogenesis is detectable even in carcinomas that are still within the ovarian capsule and in low-grade malignancies. Several groups of investigators have shown that a stage I carcinoma can be diagnosed with Doppler ultrasound[9,25,50] (see Chapter 33, p. 310 ff).

Power Doppler. The new power or energy-mode Doppler instruments can define even the smallest blood vessels. Paradoxically, it has been found that even tiny intraparenchymal arterioles in normal or benign tissues may exhibit low impedance and low flow velocities, leading to false-positive results. Nevertheless, there is a definite tendency to find decreasing vascular impedance from benign and borderline tumors to early-stage carcinoma and finally to advanced malignancies[96] (see Chapter 33, p. 310 ff).

Overlaps. Since transvaginal color Doppler sonography was first used to assess the vascularity of the ovaries[9,54], there have been varying opinions as to its value in the diagnosis of malignant adnexal lesions. Most studies published on this topic agree that ovarian malignancies display characteristic blood flow patterns compared with benign lesions. But the overlap in the blood flow parameters of malignant and benign ovarian tumors has become the main stumbling block in efforts to improve the differentiation of ovarian tumors based on vascular characteristics. What can three-dimensional (3D) power Doppler imaging contribute to finding an optimum solution?

Power Doppler imaging. Power Doppler imaging, known also as power Doppler angiography, has been in clinical use for several years[3,14,74,75,93]. In this technique the shade and brightness of the color signal represent the total energy of the Doppler signal. The advantages of power Doppler imaging are that it is more sensitive to low-velocity flows than standard color Doppler, it is unaffected by the beam–vessel angle, and it is free of aliasing effects. All flow signals from a designated region are recorded, producing an image that resembles a conventional angiogram. Experience with power Doppler in obstetrics and gynecology is still limited, however[28,29,64,66].

Three-Dimensional Imaging

Rendering. The ultrasound image displayed on the monitor is actually two-dimensional, whereas the data acquired from the insonated object are three-dimensional. New 3D ultrasound scanners register the data that they display on the two-dimensional monitor in three dimensions. “Rendering” refers to the data-processing methods that are used to produce a three-dimensional impression on the monitor by means of shading and image rotation.

Analog and digital data acquisition. Data acquisition in three-dimensional ultrasound may be analog or digital. In the analog method, the ultrasound image is generated by a broad acoustic beam. This produces “fuzzy” images that simulate the appearance of a three-dimensional object, especially when the object contrasts sharply with its surroundings (e.g., fetal body parts surrounded by amniotic fluid). This method is relatively economical but has limited value because the three-dimensional data cannot actually be stored as such. In the digital method of data acquisition, the 3D data are stored digitally in a computer and can thus be retrieved later to reconstruct the whole 3D image or portions of the image viewed from different angles. The analog method will not be discussed further, as it does not furnish 3D data that can be mathematically processed and manipulated.

Three-Dimensional Imaging of Vascular Patterns

Organ relations. The 3D display can supply additional information in various diagnostic situations[2,20,23,90]. It gives the examiner a fast, simple means of displaying multiple, superimposed vessels in one image and defining their relationship to one another and to tumors or surrounding tissues. With a 3D display, the examiner can visibly portray these organ relations on the monitor instead of recording them in individual sectional images and then piecing them together mentally into a three-dimensional image. The 3D power Doppler vascular display enables the physician to define questionable structures in much greater detail, which can expedite the conduct of the examination and the clinical decision-making process.

Detection of infarcted areas. In renal transplantation patients, for example, the conventional sonographic test for allograft rejection is to evaluate blood flow based on the resistance index of the interlobular vessels[42]. A newer approach is to calculate power Doppler signal intensities in follow-up examinations in order to detect chronic rejection[1,60]. Today we can also construct 3D vascular images that demonstrate the morphology and branching pattern of the intrarenal vessels so that we can detect any nonvascularized areas in the allograft that may represent infarctions.

Types of Image Acquisition and Equipment for Three-Dimensional Ultrasound

Data acquisition in two planes. At the present time, two main types of equipment are available for the 3D display of vascular structures. The first type is a computer program that does not require the use of extra hardware (e.g., Color Power Angio in the HDI-3000 or HDI-5000 from ATL, or SSD-1700 with Volume Mode from Aloka). These systems use freehand data acquisition without a position sensor. The *x*–*y* plane is fixed (corresponding to the width of each individual 2D image), and the sweep range and speed of the transducer in the *z* axis are not defined. Since positional information in the *z* axis does not enter into calculations, often the vascular architecture is not defined with very high precision. In this method, differences in the transducer patterns that are manually traced over the same vascularized area can result in different vascular displays. Because the image acquisition process does not include the calculation of spatial information, 3 D reconstruction time is very short—usually less than 30 seconds (depending on the size of the image lines being reconstructed).

Data acquisition in three planes. The second type of 3D system uses fully digitized processing of the power Doppler data, all of which are stored in computer memory. At present there are two commercially available systems that employ this method. Examples are the Kretz 530-D 3D ultrasound system (Medison) and the 3D FreeScan system with an off-line workstation (Echo-Tech Inc., Germany). In this type of image acquisition, accurate positional information is generated and processed for each two-dimensional image slice. This makes the 3D reconstruction much more precise but also more time-consuming. The data are acquired either with a special mechanical 3D probe (Kretz 530-D) or with a magnetic-field position sensor attached to a 2D ultrasound probe (3D FreeScan). The mechanical 3D ultrasound probe consists of an ordinary 2D probe that is mounted on an axle and is moved in the third dimension by a motor. The 3D probe must be held stationary during 3D data acquisition. The relationships between the individual two-dimensional image data remain constant.

The other system, which uses a magnetic-field position sensor linked to an off-line workstation, is attached to a conventional transducer and yields spatial information for each 2D image based on magnetic field deflections. The probe can be freely moved in any direction.

Display Modes for Three-Dimensional Vascular Images

The HDI-3000 or HDI-5000 system (Color Power Angio from ATL, or SSD-1700 with Volume Mode from Aloka) has built-in programs for viewing three-dimensional images from various angles and demonstrating the vascular architecture in various planes. The 3D objects can be rotated around a point in the horizontal plane, creating a three-dimensional impression. By contrast, the Kretz 530-D and Echo-Technology 3D FreeScan systems use the "ray-casting" technique to interpolate the fully digitized three-dimensional data. With the freehand 3D compound scan technique used in the HDI-3000 or Aloka 1700 systems, the examiner must be very careful to move the transducer at a constant speed. The resulting image is fairly useful for demonstrating the 3D architecture of vessels but is not useful for length or volume measurements. Fully digitized image processing with integrated spatial information (e.g., the Kretz 530-D) can generate a more precise 3D display. The corresponding 3D gray-scale information can also be integrated into the display with this technique. However, data acquisition and processing take considerably more time than in the freehand technique. Also, the examiner must have greater technical knowledge and experience with complex data processing.

Problems in the Interpretation of 3D Power Doppler Data

Although power Doppler has various advantages over conventional Doppler, it is basically a color Doppler display with improved software and not a separate imaging technology. For this reason, various standard color Doppler parameters such as the pulse repetition frequency (PRF), wall filter, color priority, power output, receiver gain, and frame rate must be optimized in 3D imaging for both the qualitative and quantitative analysis of the power Doppler data. Another problem is that the greater data processing requirements of color and power Doppler lead to a marked slowing of the frame rate (generally 2–3 Hz) when the power mode is switched on. One result of this is that very small vessels may not be visualized, especially when a mechanical 3D probe (as in the Kretz 530-D) is used. Moreover, attenuation of the ultrasound beam can sometimes lead to different power intensities in the near and far fields of the ultrasound image.

Ultrasound Technology in Tumor Diagnosis

Morphology of tumors. Various sonographic techniques have been used in an effort to find new approaches to early cancer detection. The first approach to diagnosing malignant tumors with ultrasound was the attempt to differentiate tumors by their morphology. Ultrasound is currently being used to detect malignant tumors at an early stage on the basis of morphological criteria[45], but with very poor specificity[15]. Since we cannot rely on tumor morphology alone, blood flow characteristics are now being used as an additional source of information. A number of scoring systems have been devised for this purpose, but no single system has become widely implemented.

Morphology of the vascular system. To date, relatively little research has been done on vascular morphology as a possible indicator of malignancy. Many clinicians have the impression that blood vessels that supply fast-growing neoplasms have a distribution and branching pattern that distinguishes them from the vessels that supply healthy organs. If this were the case, the morphology of the blood vessels could supply additional information that could not be obtained with conventional studies. However, the description of arborizing vascular patterns is a complex task that can be solved only by complicated mathematical computer operations, which is why we must explore this issue more fully.

Mathematical models. Mathematical models play an important role in the quantification of diagnostic findings. In our case, we analyzed the branching pattern of a vascular tree. This pattern is the result of a mathematical principle that acts repetitively on arborizing vessels in such a way that they consistently branch in a similar pattern at different levels. This process is comparable to the growth that occurs in a real tree. If the underlying principle is altered, the branching pattern will also change. This change is the diagnostic criterion that we are attempting to quantify so that it can be used in early cancer detection. The innovative part of our approach is to apply new mathematical analyses and concepts to data that are acquired with 3D ultrasound technology.

Brief Review of the Mathematical Basis of New Technologies

In many cases, mathematical relationships are known for some time before they find broad practical application[11,61,59]. It is not surprising, then, that nonlinear processes, which have been developed over the past 36 years, have not yet had a significant impact on science and technology and have not been implemented in most practical applications. At the same time, these discoveries have major implications for research in population dynamics, ecology, meteorology, and fluid dynamics. Major industries, most notably the oil companies and the airline and aerospace industries, are investing large sums in projects involving the practical application of this methodology.

Linear and nonlinear equations. Processes in time or space can be described by either linear or nonlinear equations and formulas. Linear mathematical equations can often be solved completely. On the other hand, nonlinear equations, like those occurring in biological systems, raise serious mathematical problems and often can only be partially solved by assuming that a small portion of the nonlinear processes are linear within a narrow range. These linear assumptions afford only a limited look at the underlying processes. The main difference between the two computational processes is that, in linear systems, a small change in various parameters will produce a small change in the final result, whereas a small change in one parameter in a nonlinear system can cause immense, virtually unpredictable changes in the final result.

Fractals. Many processes in nature are recursive, meaning that they repeat according to a consistent pattern (mathematical principle). An example is the division of replicating cells. The starting point of one generation is the end point of the previous generation based on the constantly recurring division process. The end result of such processes, in which various underlying parameters undergo changes, may at first appear fairly chaotic. But if we look at several of the possible outcomes of these processes (e.g., the branching of blood vessels, animal populations, climate changes, etc.), we will eventually notice a pattern, even within the processes themselves, that is repeated at different levels of observation and scale. The objects that display this recurring pattern at different levels are called fractals, and the sum of the various possible states of a system is called the attractor[59, 61].

Current Methods for Evaluating Vascular Geometry and Function

Vascular density. While research with fractals is a very new field, some results have already been published dealing with simple descriptions of vascular geometry. They include reports on correlations between the density of microvessels in ovarian and endometrial carcinomas and the likelihood that these tumors will recur[36, 44]. In these studies vascular density was evaluated independently of vascular distribution as the only meaningful parameter. A similar approach has been taken in color Doppler studies by counting the number of color signals visible within a tumor area. This concept is easily applied to our two-dimensional images.

"Geometric" feature. Our group of investigators is working on ways in which Doppler sonography can be used to characterize tumors. While there still appear to be ways of refining this diagnostic approach, we feel that it is necessary to add a "geometric" feature to the criteria that have previously been used. The analysis of flow indices once appeared to be a very promising approach, but over time it has been found that these indices lack the specificity needed for an effective screening method.

Technique for Evaluating Vascular Geometry

Fractal geometry. Our working hypothesis was that the branching pattern of blood vessels in 1 cm^3 of tissue was like the branching pattern that occurs in 50 cm^3 of the same tissue. Structures that are alike at different levels of scale are called fractals[11, 59, 79]. They occur in natural geometric forms (tree branches, bird feathers, etc.) and in the final states of seemingly chaotic processes that have nothing to do with geometry (e.g., live populations). The growth of blood vessels also appears to exhibit these properties. We make the assumption that vascular branches are an example of fractal geometry[11, 79], i.e., that they follow the same geometric pattern at all levels. The normal vascular tree (arteries and veins) consists of progressively arborizing structures with increasingly smaller branches and calibers. This arrangement can supply the entire body with blood while occupying only 6% of the available space.

Our research is based on the assumption that a change in the fractal dimension is present when the branching pattern becomes irregular, i.e., when normal, regular growth is replaced by disordered growth.

Scalability of vascular branches. Working with our colleagues from pathology, we investigated the differences in the scalability of vascular branches in normal and malignant tissues. Some of the results that we obtained have an important bearing on basic research; but we also discovered that too few blood vessels were present in any given tissue sample for us to draw accurate conclusions on consistent patterns of vascular branching.

Comparison of sonography and histology. The construction of a three-dimensional model of large volume is a great mathematical and technological challenge. In our case, an important aspect is comparing the reconstruction of vascular geometry from histological sections with the results of available imaging procedures, especially 3D ultrasound. Although image resolution has improved considerably in recent years, it is not physically possible to achieve resolutions on the order of 10 μm with any known sonographic technique. Three-dimensional vascular imaging with ultrasound is possible and brings us one step closer to mathematical analysis, but image resolution in 3D ultrasound continues to be a problem. Since we have been working on the quantitative three-dimensional reconstruction and computation of the properties of vascular branching, quantitative 3D data acquisition (especially with the Kretz Voluson system) has improved so much that it appears to be markedly superior to other, previously used imaging procedures.

Combined approach. Methods are apparently available that enable us to measure the fractal properties of vascular growth. From this standpoint, it is best to regard vascular growth as a nonlinear process whose behavior is "chaotic." Nonlinear functions are a powerful tool for describing alternating normal and chaotic behavior—as opposed to ordinary statistical methods, which fail when applied to phenomena in vascular structures or populations, for example. The main problem is that a great deal of vascular growth takes place in the tiniest blood vessels. The diameter of erythrocytes sets a lower limit on how small vascular diameters can be, and this scale marks the end point of vascular proliferation according to a reiterative pattern. We therefore suggest the use of a combined study approach in which the 3D ultrasound imaging of larger-diameter blood vessels is combined with histological findings in vessels that are too small to be defined with ultrasound. This approach still requires further development, however, and necessary allowances must be made in the budgeting of time, equipment, and personnel.

Example of 3D Power-Mode Imaging of Benign and Malignant Gynecological Tumors

Angiogenesis in physiological and benign processes. To date, various types of angiogenesis have been described under physiological and pathological conditions. Physiological angiogenesis has been observed in folliculogenesis, embryogenesis and implantation, chronic inflammations, and some benign tumors[51]. We were able to demonstrate the vessels of the mesovarium using 3D power Doppler angiography. During the preovulatory phase, these blood vessels grow slowly from the hilar area into the stroma with a steadily increasing number of fine, branching vessels (Fig. 34.**1**). After ovulation, we were able to detect luteal cyst formation in some cases (Fig. 34.**2**). Normally the impedance index of the luteal vessels is low. Generally we have observed only a small number of luteal vessels (often only one), which rarely show complicated branching patterns or tortuosity around the cyst (Fig. 34.**3**). This contrasts with the findings that are seen with malignant tumors. With chocolate cysts (Fig. 34.**4**), the vessels often take a linear course, arise from a vessel at the hilum, and branch regularly on the surface of the mass (Fig. 34.**5**). Similar vascular structures are found in association with dermoid cysts (Fig. 34.**6**).

Neoangiogenesis in malignancies. We were able to observe neovascularity in a number of malignant tumors. The tumor vessels in these cases are usually distributed irregularly in the stroma and periphery, and tortuous vessels are sometimes definable on the tumor surface. The main tumor vessel usually

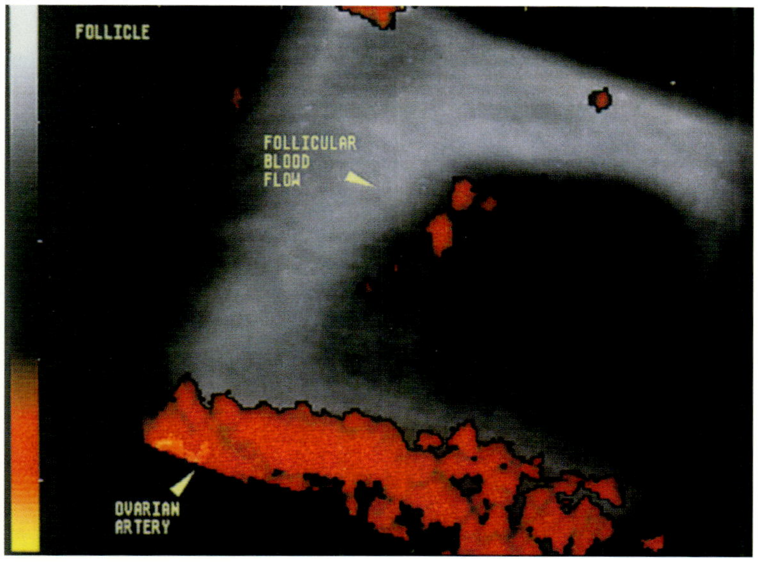

Fig. 34.**1** Three-dimensional display of the vessels surrounding the follicle. The power Doppler image shows the ovarian artery and follicular capillaries. The three-dimensional view affords a vivid image of the preovulatory follicle and its vascular supply.

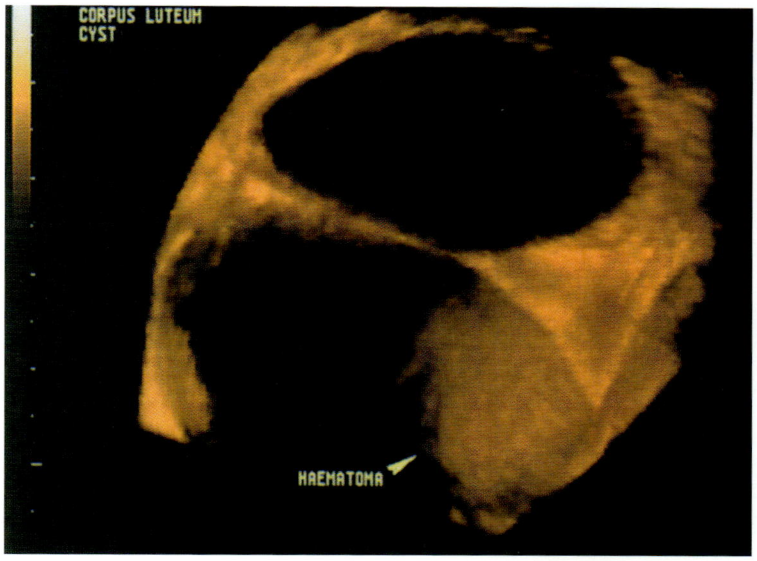

Fig. 34.**2** Three-dimensional display of the corpus luteum. The image clearly demonstrates the blood-filled cavity of the ruptured follicle with echogenic clots.

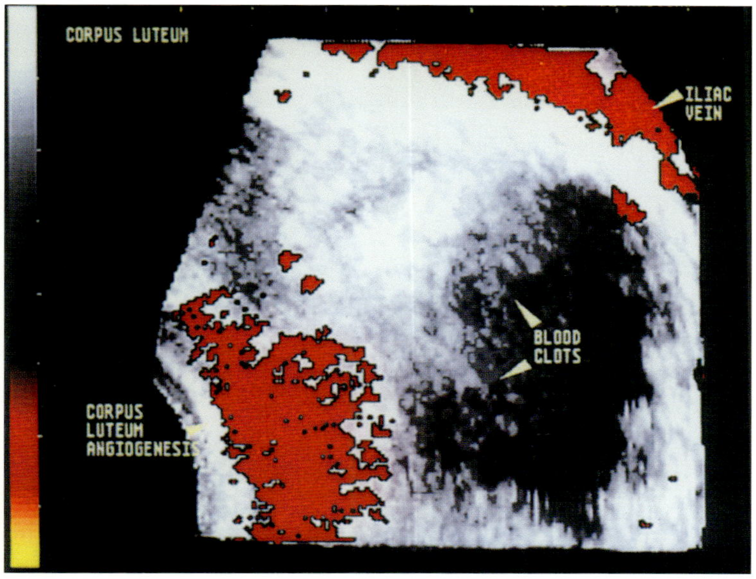

Fig. 34.**3** Three-dimensional display of the early corpus luteum. The nonhomogeneous areas within the corpus luteum represent blood clots. Power Doppler demonstrates the ingrowth of capillary vessels into the cyst lumen.

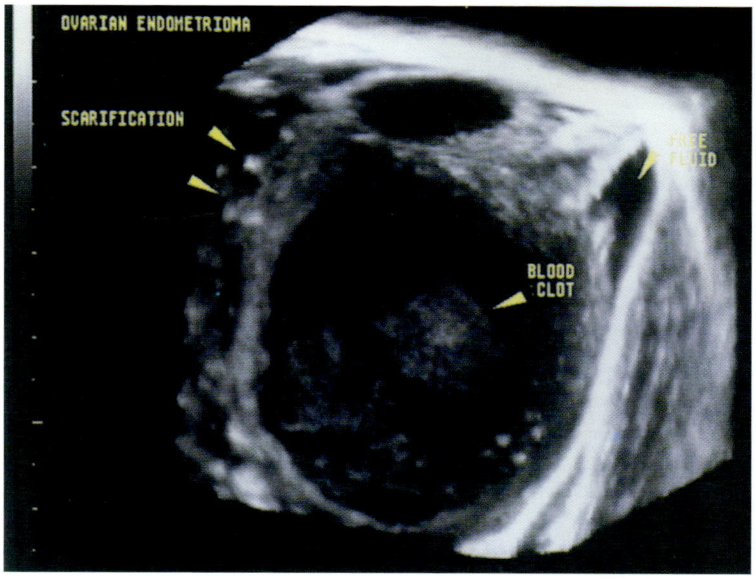

Fig. 34.**4** Three-dimensional display of an ovarian endometrioma. The image shows chocolate-like fluid containing blood clots with normal peripheral echogenicity as a sign of early organization.

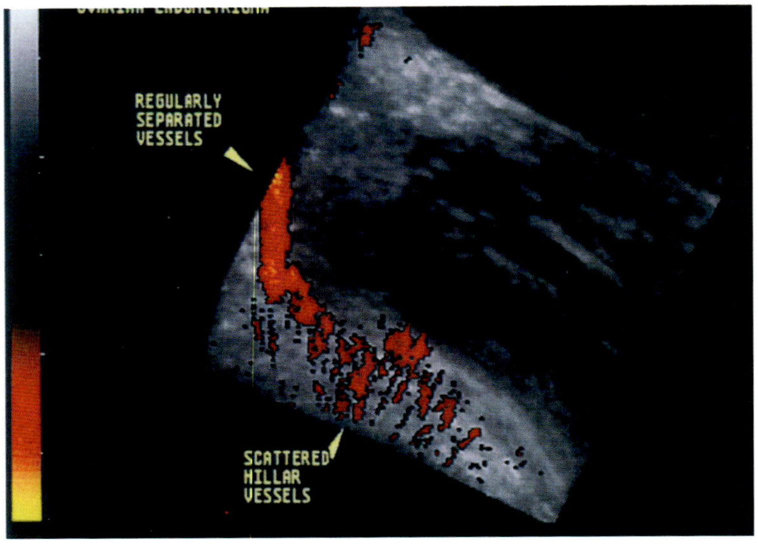

Fig. 34.**5** Same patient as in Fig. 34.**4**. This image shows increased, disseminated vascularization at the ovarian hilum and regularly branched peripheral vessels. Both vascular patterns are typical of ovarian endometriosis and are easily demonstrated by 3D power Doppler imaging.

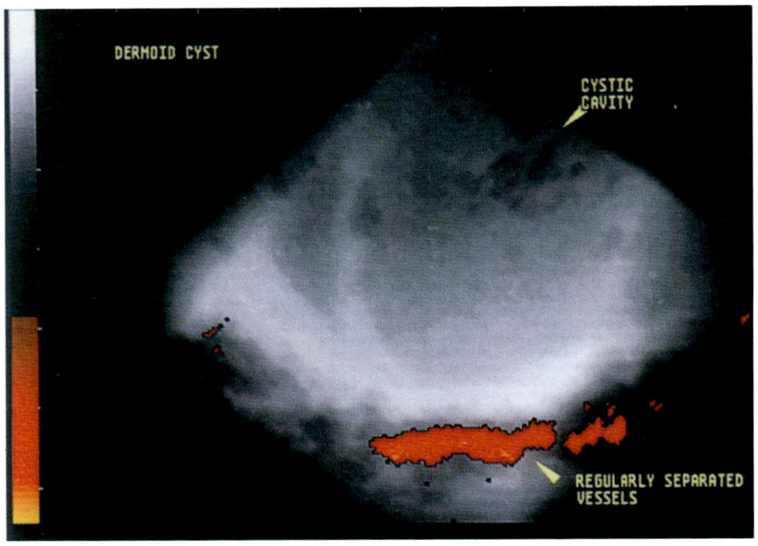

Fig. 34.**6** Three-dimensional display of a dermoid cyst. Bony structures within the lesion cast a typical posterior acoustic shadow. Regularly branched vessels are clearly identified at the periphery of the dermoid cyst by 3D power Doppler imaging.

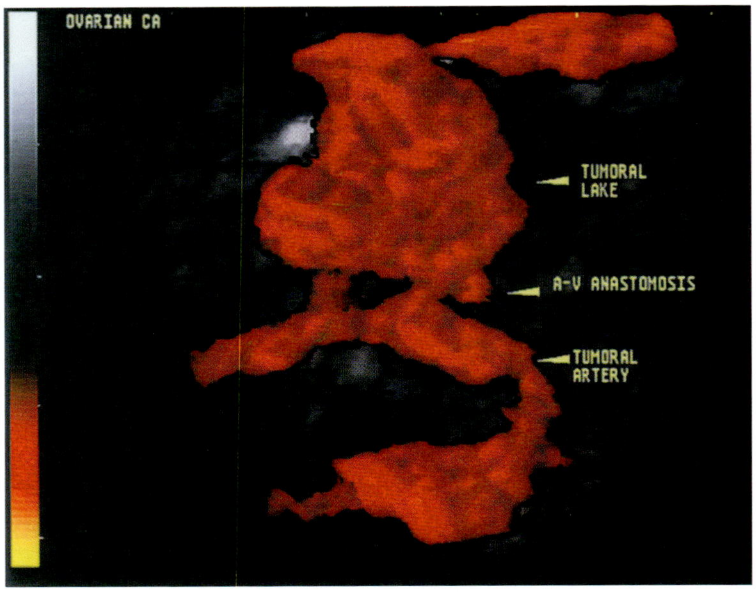

Fig. 34.**7** Malignant neovascularity is characterized by the presence of arteriovenous shunts, stenoses, microaneurysms, and blindly terminating "vascular lakes." These characteristics of tumor angiogenesis can all be demonstrated with 3D power Doppler.

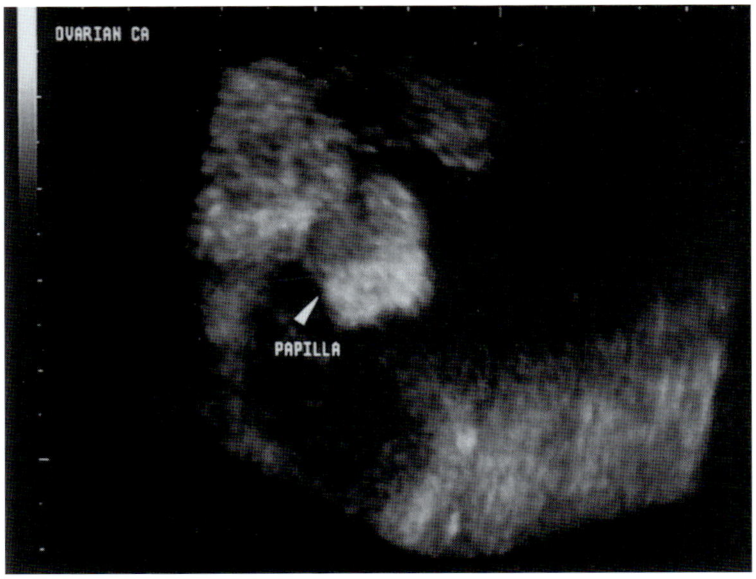

Fig. 34.**8** Three-dimensional display of a complex ovarian mass. The papillary structure projecting into the interior of the cyst is clearly visible.

takes an irregular course and shows a complicated branching pattern. These vessels also show irregular calibers with "spur-like" protuberances (Fig. 34.**7**). These findings are consistent with the results of numerous studies done with conventional color Doppler ultrasound[51]. Three-dimensional ultrasound provides better visualization (Fig. 34.**8**) and gives the examiner a better appreciation of the three-dimensional architecture of the microcirculation (Fig. 34.**9**). At the same time, the resolution of power Doppler scanners has become so good that vessels 1 mm in diameter can be defined[20]. We believe, therefore, that 3D power Doppler ultrasound is a very promising tool for evaluating the angiogenesis of tumors in the lesser pelvis, especially when malignancy is suspected (Figs. 34.**10**– 34.**13**).

Perfusion regions and patterns. In an effort to systematize the description of perfusion, we can often identify four different regions of varying blood flow in malignant tumors: the necrotic region (central zone), a seminecrotic (ischemic) region, a region of stable and adequate microcirculation, and a hyperemic region at the tumor periphery. Tumors with different histologies or different growth rates as well as primary tumors and metastases may exhibit different blood flow pat-

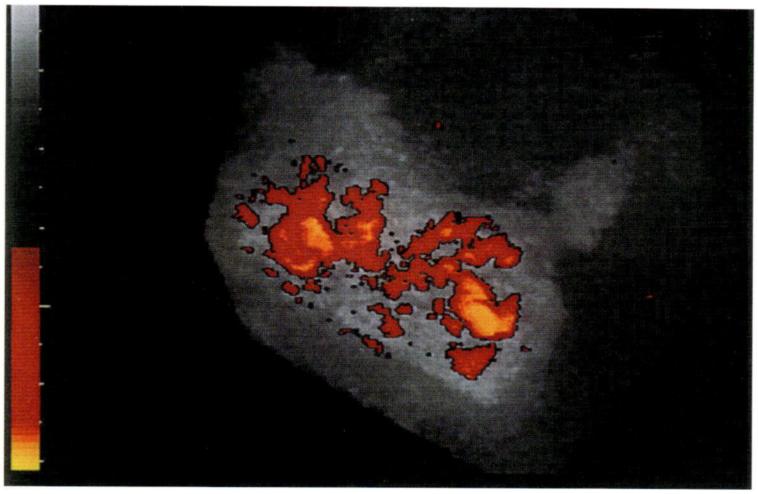

Fig. 34.**9** Same patient as in Fig. 34.**8**. Numerous irregularly distributed vessels are demonstrated within the papillary structure, pointing to the malignant nature of this ovarian tumor. Histopathological examination confirmed adenocarcinoma of the ovary.

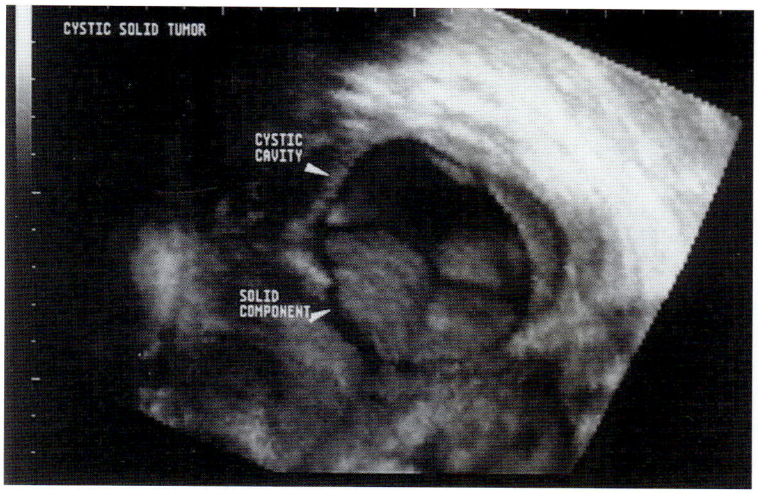

Fig. 34.**10** 3 D ultrasound display of a complex ovarian mass. The surface of the solid component can be precisely analyzed in vivo. The morphology is suspicious for an ovarian malignancy, which was confirmed by histopathology.

Fig. 34.**11** Same patient as in Fig. 34.**10**. Numerous arteriovenous shunts, microaneurysms, and dilated tumor vessels are demonstrated.

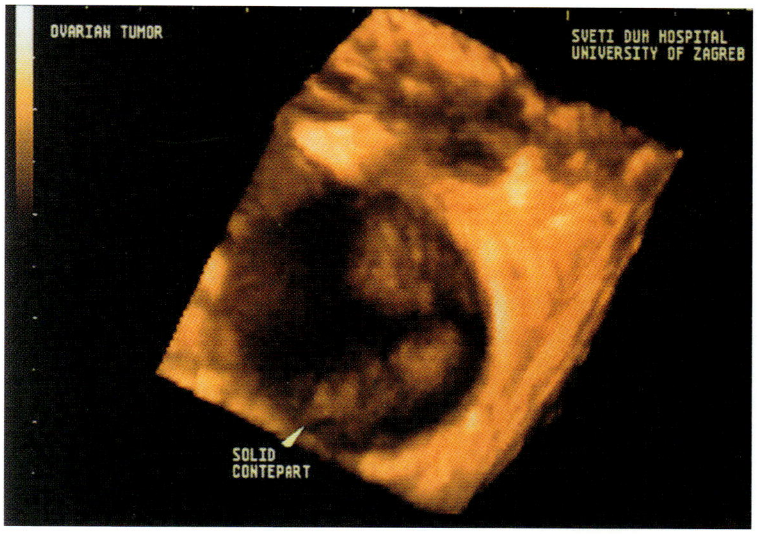

Fig. 34.**12** 3 D ultrasound display of a cystic-solid ovarian mass 3 cm in diameter. The solid component is visible on the right side of the lesion.

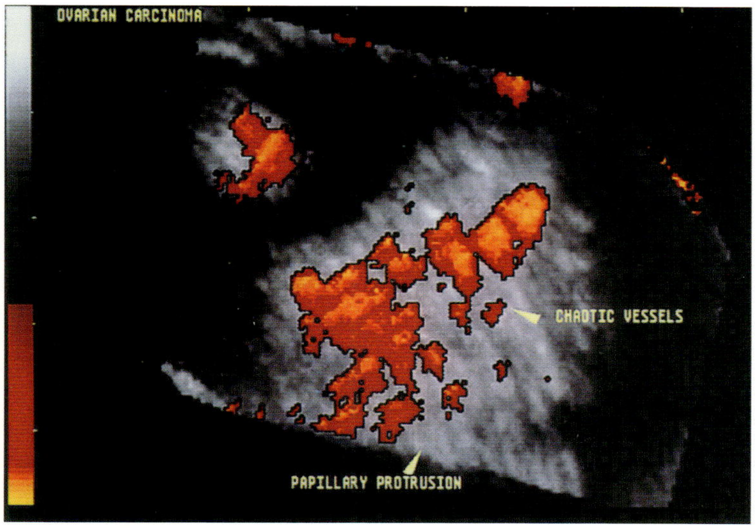

◁ Fig. 34.**13** Same patient as in Fig. 34.**12**. The volume can be rotated in all three planes, clearly demonstrating the irregular course and complex branching pattern of the tumor vessels.

terns. The compact type of trophoblastic tumor (as classified by Hsieh et al.[38,39], which often is a choriocarcinoma, displays the typical pattern of tumor vascularity described above. Three-dimensional power Doppler can thus be used for the in-vivo measurement of tumor vascularity.

Advances in Tumor Therapy

Our discussion is based on more than 10 years' experience with transvaginal color Doppler sonography and its correlation with the results of macroscopic and microscopic pathological studies. We believe that as further technical advances are made, the new mathematical models described above (fractals) and 3D imaging can be used to evaluate tumor-induced angiogenesis with greater precision than is possible by the customary analysis of flow indices.

Angiogenic "switch." Color Doppler sonography thus appears to have a valid clinical role in evaluating the architecture of tumor vessels, justifying the expansion of research in this area. This is further supported by the discovery and characterization of a group of angiogenesis regulators that include both stimulators and inhibitors[33]. In the case of solid tumors, a shift in the balance between stimulators and inhibitors may provide the angiogenic "switch" that enables the tumor to induce the formation of blood vessels to ensure its survival. This process appears to be absolutely necessary for tumor growth to progress beyond the microscopic stage. The goal of most research initiatives in tumor angiogenesis is to find a way to disable this angiogenic "switch," thus providing a new form of cancer therapy.

Angiogenesis inhibitors. Results from the Folkman laboratory[7] have been highly encouraging in terms of pursuing this approach. When Boehm and his team[7] administered a newly discovered angiogenesis inhibitor (endostatin) to mice bearing three different types of tumor, it was found that drug resistance did not develop after multiple treatment cycles and that prolonged tumor dormancy was achieved. This treatment strategy can help to circumvent certain problems that arise in current chemotherapy regimens, such as acquired drug resistance based on the genetic instability of tumor cells or an intrinsic resistance of the tumor parenchyma to drug penetration[43]. Influencing angiogenesis and tumor vascularization by endostatin therapy may provide another effective treatment strategy and perhaps even a preventive measure in the fight against human cancers. Like any new treatment, antiangiogenic therapy raises a number of important questions for the future, such as whether the agent will inhibit the growth of new blood vessels without "disturbing" quiescent blood vessels. Ideally the agent should stop neovascularization in the tumor and cause an arrest of the growth process. The tumor should neither grow nor shrink but should be "frozen" in a state of equilibrium between proliferation and apoptosis while retaining its existing vasculature[37].

Another important question is whether there are tissue-specific differences in vascular structures, and thus in the anatomy of tumor vessels, that could influence the response of a tumor to antiangiogenic therapy. Can 3D power Doppler sonography help to answer some of these questions? There is no doubt that further research on these topics is a challenge for diagnostic ultrasound.

Future Outlook

For the present, the use of 3D power Doppler ultrasound in tumor diagnosis is a qualitative or semiquantitative study, meaning that it can supply information on whether or not vascularization is present[5,55]. Various efforts have been made to use power Doppler imaging to quantify arterial stenoses as an alternative to conventional angiography[31,84]. In another quantitative application, an attempt has been made to calculate vascular density with power Doppler [62]. We hope that this approach can be significantly advanced by the geometric evaluation described above. This can be facilitated by investigating the differences between malignant and benign tumors with Doppler ultrasound and comparing the results with other diagnostic methods such as immunohistochemistry and microvessel density[22]. Contrast agents are another means of improving the results of 3D power Doppler examinations. The use of a contrast agent can increase the detection rate of small vessels. Future refinements in 3D power Doppler programs should include the simultaneous display of 3D gray-scale images (anatomical information) and Doppler vascular images.

Summary

The results of 3D Doppler ultrasound studies described in the literature pose a new challenge and raise new questions concerning the regulation of tumor angiogenesis, vascular density, and the differences in the vascular architecture of benign and malignant tumors. The 3D power Doppler display of tumor vessels appears to have a number of possible clinical applications including the early detection of ovarian and endometrial cancers. The more reliable sonographic assessment of tumor architecture could perhaps improve the diagnostic value of this modality and help to reduce the morbidity and mortality of both types of cancer.

References

1 Akiyama T, Ikegami M, Hara Y et al.: Hemodynamic study of renal transplant chronic rejection using power Doppler sonography. Transplant. Proceed. 28 (1996) 1458–1460
2 Baba K, Satoh K, Sakamoto S, Okai T, Ishii S: Development of an ultrasonic system for three-dimensional reconstructions of the fetus. J. Perinatl. Med. 17 (1989) 19–24
3 Babcock DS, Patricuin H, LaFortune M, Dauzat M: Power Doppler sonography: basic principles and clinical applications in children. Pediatr. Radiol. 26 (1996) 109–115
4 Barnhill RL, Fandrey K, Levy MA, Mihm MC Jr, Human B: Angiogenesis and tumor progression of melanoma: quantification of vascularity in melanocytic nevi and cutaneous malignant melanoma. Lab. Invest. 67 (1992) 57–62
5 Birdwell RL, Ikeda DM, Jeffrey SS, Jeffrey RB: Preliminary experience with power Doppler imaging of solid breast masses. AJR 169 (1997) 703–707
6 Blood CH, Zetter BR: Tumor interactions with the vasculature: angiogenesis and tumor metastastis. Biochem. Biophys. Acta 1032 (1990) 89–118
7 Boehm T, Folkman J, Browder T, O'Reilly MS:Antiangiogenic therapy of experimental cancer does not induce acquired drug resistance. Nature 390 (1997) 404–407
8 Bourne TH: Should clinical decisions be made about ovarian masses using transvaginal color Doppler? Ultrasound Obstet. Gynecol. 4 (1994) 257–260
9 Bourne TH, Campbell S, Steers CV, Whitehead MI, Collins WP: Transvaginal colour flow imaging: a possible new screening technique for ovarian cancer. Brit. Med. J. 299 (1989) 1367–1370
10 Breyer B, Ferek-Petric B, Cikes I: Properties of ultrasonically marked leads. PACE 12 (1989) 1369–1380
11 Breyer B, Kurjak A: Tumor vascularization Doppler measurements and chaos: what to do? Ultrasound Obstet. Gynecol. 5 (1995) 209
12 Bromley B, Goodman H, Benacerraf BR: Comparison between sonographic morphology and Doppler waveform for the diagnosis of ovarian malignancy. Obstet. Gynecol. 83 (1994) 434–437
13 Brown DL, Frates MC, Laing FC: Ovarian masses: Can benign and malignant lesions be differentiated with color and pulsed Doppler US? Radiology 190 (1994) 333–336
14 Bude RO, Rubin JM, Adler RS: Power versus conventional color Doppler sonography in the depiction of normal intrarenal vasculature. Radiology 192 (1994) 777–780
15 Buy JN, Ghossain MA, Hugol D et al.: Characterization of adnexal masses: Combination of color Doppler and conventional sonography compared with spectral Doppler analysis alone and conventional sonography alone. Amer. J. Roentgenol. 166 (1996) 385–393
16 Campbell S, Royston P, Bhan V: Novel screening strategies for early ovarian cancer by transabdominal ultrasonography. Brit. J. Obstet. Gynecol. 96 (1990) 304–311
17 Carter JR, Lau M, Fowler JM, Carlson JW, Carson LF, Twiggs LB: Blood flow characteristics of ovarian tumors: Implications for ovarian cancer screening. Am. J. Obstet. Gynecol. 172 (1995) 901–907
18 Carter J, Saltzman A, Hartenbach E, Fowler J, Carson L, Twiggs LB: Flow characteristics in benign and malignant gynecologic tumors using transvaginal color flow Doppler. Obstet. Gynecol. 83 (1994) 125–130
19 Chou CY, Chang CH, Yao BL, Kuo HC: Color Doppler ultrasonography and serum CA 125 in the differentiation of benign and malignant ovarian tumors. J. Clin. Ultrasound 22 (1994) 491–496
20 Downey DB, Fenster A: Vascular imaging with a three-dimensional power Doppler system. AJR 165 (1995) 665–668
21 Einer Z, Beck D, Brandes JM: Transvaginal sonography, color flow imaging, computed tomography scanning, and CA 125 as a routine follow-up examination in women with pelvic tumor: Detection of recurrent disease. J. Ultrasound Med. 13 (1994) 37–41
22 Emoto M, Iwasaki H, Mimura K, Kawarabayashi T, Kikuchi M: Differences in the angiogenesis of benign and malignant ovarian tumors, demonstrated by analyses of color Doppler ultrasound, immunohistochemistry, and microvessel density. Cancer 80 (1997) 899–907
23 Fine D, Perring S, Herbetko J, Hacking CN, Fleming JS, Dewburz KC: Three-dimensional (3 D) ultrasound imaging of the gallbladder and dilated biliary tree: Reconstruction from real time B scans. Brit. J. Radiology 64 (1991) 1956–1957
24 Fleischer AC: Color Doppler sonography of benign and malignant adnexal masses: a spectrum of findings. In Kurjak A, Fleischer AC (eds.): Doppler Ultrasound in Gynecology. Parthenon, London 1998, 27–36
25 Fleischer AC, Cullinan JA, Peery CV, Jones III JW: Early detection of ovarian carcinoma with transvaginal color Doppler ultrasound. Am. J. Obstet. Gynecol. 174 (1996) 101–106
26 Fleischer AC, Rodgers WH, Rao BJ et al.: Assessment of ovarian tumor vascularity with transvaginal color Doppler sonography. J. Ultrasound Med. 10 (1991) 563–568
27 Folkman J: What is the evidence that tumors are angiogenesis dependent? J. Nat. Cancer Inst. 82 (1989) 4–6
28 Fortunato SJ: The use of power Doppler and color power angiography in fetal imaging. Am. J. Obstet. Gynecol. 174 (1996) 1828–1833
29 Fox DB, Bruner JP, Fleischer AC: Amplitude-based color Doppler sonography of fetus with sacrococcygeal teratoma. J. Ultrasound Med. 15 (1996) 785–787
30 Franchi M, Beretta P, Ghezzi F, Zanaboni F, Goddi A, Salvator S: Diagnosis of pelvic masses with transabdominal color Doppler, CA 125 and ultrasonography. Acta Obstet. Gynecol. Scand. 75 (1995) 734–739
31 Guo Z, Fenster A: Three-dimensional power Doppler imaging: A phantom study to quantify vessel stenosis. Ultrasound Med. Biol. 22 (1996) 1059–1069
32 Hamper UM, Sheth S, Abbas FM, Rosenshein BN, Aronson D, Kurman JR: Transvaginal color Doppler sonography of adnexal masses: Differences in blood flow impedance in benign and malignant lesions. Am. J. Roentgenol. 160 (1993) 1225 –1228
33 Hanahan D, Folkman J: Parameters and emerging mechanisms of the angiogenic switch during tumorigenesis. Cell 86 (1996) 353–354
34 Hata H, Hata T, Manabe A, Sugimura K, Kitao M: A critical evaluation of transvaginal Doppler studies, transvaginal sonography, magnetic resonance imaging, and CA 125 in detecting ovarian cancer. Obstet. Gynecol. 80 (1992) 922–926
35 Hata T, Hata K, Senoh D et al.: Doppler ultrasound assessment of tumor vascularity in gynecologic disorders. J. Ultrasound Med. 8 (1989) 309–314
36 Hollingworth H, Kohn E, Steinberg S, Rothenberg ML, Merino MJ: Tumor angiogenesis in advanced stage ovarian carcinoma. Am. J. Pathol. 47 (1995) 33–41
37 Holmgren L, O'Reilly MS, Folkman J: Dormancy of micrometastases: Balanced proliferation and apoptosis in the presence of angiogenesis suppression. Nature Med. 1 (1995) 149–153
38 Hsieh FJ, Liu CH, Chang FM et al.: Ultrasonography in the diagnosis and management of invasive gestational trophoblastic disease. J. Formosa Med. Assoc. 87 (1988) 139–145
39 Hsieh FJ, Wu CC, Lee CN et al.: Vascular pattern of gestational trophoblastic tumors by color Doppler ultrasound. Cancer 74 (1994) 2361–2365
40 Jain KA: Prospective evaluation of adnexal masses with endovaginal gray-scale and duplex and color Doppler US: Correlation with pathologic findings. Radiology 191 (1994) 63–67
41 Kawai M, Kano T, Kikkawa F, Maeda O, Oguchi H, Tomoda Y: Transvaginal Doppler ultrasound with color flow imaging in the diagnosis of ovarian cancer. Obstet. Gynecol. 79 (1992) 163–167
42 Kelcz F, Pozniak MA, Pirsch JD, Oberly TD: Pyramidal appearance and RI: intensitive and non-specific sonographic indicators of renal transplant rejections. AJR. 155 (1990) 531–535
43 Kerbel RS: Inhibition of tumor angiogenesis as a strategy to circumvent acquired resistance to anticancer therapeutic agents. BioEssays 13 (1991) 31–36
44 Kirschmer CV, Alamis-Amezcus JM, Martin VG et al.: Angiogenesis factor in endometrial carcinoma: a new prognostic indicator? Am. J. Obstet. Gynecol. 174 (1996) 1879–1884
45 Kratochwill A: Ultraschalldiagnostik in Geburtshilfe und Gynaekologie. Thieme, Stuttgart 1968, 84

46 Kupesic S, Kurjak A: Three-dimensional power Doppler ultrasound in the staging of endometrial adenocarcinoma. Ultrasound Obstet. Gynecol. (1998)

47 Kurjak A, Jukic S, Kupesic S, Babic D: A combined Doppler and morphopathological study of ovarian tumors. Eur. J. Obstet. Gynecol. Reprod. Biol. 71 (1997) 147–150

48 Kurjak A, Kupesic S, Ilijas M, Sparac V, Kosuta D: Preoperative diagnosis of primary Fallopian tube carcinoma: Gynecol. Oncol. 68 (1998) 29–34

49 Kurjak A, Shalan H, Kupesic S et al.: Transvaginal color Doppler sonography in the assessment of pelvic tumor vascularity. Ultrasound Obstet. Gynecol. 3 (1993) 137–154

50 Kurjak A, Shalan H, Matijevic R, Predanic M, Kupesic-Urek S: Stage I ovarian cancer by transvaginal color Doppler sonography: a report of 18 cases. Ultrasound Obstet. Gynecol. 3 (1993) 195–198

51 Kurjak A, Schulman H, Predanic M: Pelvic tumor neovascularity. In Kurjak A (ed.): An atlas of transvaginal color Doppler: current state of the art (2nd ed). Parthenon, Carnforth 1993, 231–246

52 Kurjak A, Schulman H, Sosic A, Zalud I, Shalan H: Transvaginal ultrasound, color flow, and Doppler waveform of the postmenopausal adnexal mass. Obstet.Gynecol. 80 (1992) 917–921

53 Kurjak A, Zalud I, Alfirevic Z: Evaluation of adnexal masses with transvaginal color ultrasound. J. Ultrasound Med. 10 (1991) 295–297

54 Kurjak A, Zalud I, Jurkovic D, Alfirevic Z, Miljan M: Transvaginal color Doppler of the assessment of pelvic circulation. Acta Obstet. Gynecol. Scand. 68 (1989) 131–136

55 Lencioni R, Pinto F, Armillotta N, Bartolozzi C: Assessment of tumor vascularity in hepatocellular carcinoma: comparison of power Doppler US and color Doppler US. Radiology 201 (1996) 353–358

56 Levine D, Feldstein VA, Babcook CJ, Filly RA: Sonography of ovarian masses: Poor sensitivity of resistive index for identifying malignant lesions. Am. J. Roentgenol. 162 (1994) 1355–1359

57 Liotta L, Kleinerman J, Saidel F: Quantitative relationships of intravascular tumor cells, tumor vessels, and pulmonary metastases following tumor implantation. Cancer Res. 34 (1974) 997–1004

58 Maly Z, Riss P, Deutinger J: Localization of blood and qualitative assessment of blood flow in ovarian tumors. Obstet. Gynecol. 85 (1995) 33–36

59 Mandelbrot B: Fractals: form, chance and dimension. Freeman 1977

60 Martinoli C, Crespi G, Bertolotto M et al.: Interlobular vasculature in renal transplants: a power Doppler US study with MR correlation. Radiology 200 (1996) 111–117

61 May R: Biological populations with nonoverlaping generations: stable points, stable cycles and chaos. Science 186 (1974) 645–647

62 Meyerowitz CB, Fleischer AC, Picken DR et al.: Quantification of tumor vascularity and flow with amplitude color Doppler sonography in an experimental model: Preliminary results. J. Ultrasound Med. 15 (1996) 827–833

63 Nissen SE, Gurley JC: Application of intravascular ultrasound for detection and quantitation of coronary atherosclerosis. Int. J. Card. Imag. 6 (1991) 165–167

64 Papadimitriou A, Kalogirou D, Antonio G, Petridis N, Kalogirou O: Power Doppler ultrasound: a potential useful alternative in diagnosing pelvic pathological conditions. Clin. Exp. Obstet. Gynecol. 23 (1996) 229–232

65 Parkes C, Wald NJ: Screening for ovarian cancer. In Kurjak A (ed.): An atlas of transvaginal color Doppler. Parthenon, London 1994, 317–328

66 Pooh RK, Aono T: Transvaginal power Doppler angiography of the fetal brain. Ultrasound Obstet. Gynecol. 8 (1996) 417–421

67 Predanic M, Vlahos N, Pennisi J, Moukhtar M, Aleem FA: Color and pulsed Doppler sonography, gray-scale imaging, and serum CA 125 in the assessment of adnexal disease. Obstet. Gynecol. 88 (1996) 283–288

68 Prompeler HJ, Sauerbrei WM, Latternann U, Pfleiderer A: Quantitative flow measurements for classification of ovarian tumors by transvaginal color Doppler sonography in postmenopausal patients. Ultrasound Obstet. Gynecol. 4 (1994) 406–413

69 Rak JW, St. Croix DB, Kerbel RS: Consequences of angiogenesis for tumor progression, metastasis and cancer therapy. Anti-Cancer Drugs 6 (1995) 3–18

70 Ramos I, Fernandez LA, Morse SS, Fotune KL, Taylor KJW: Detection of neovascular signal in a 3-day Walker 256 rat carcinosarcoma by CW Doppler ultrasound. Ultrasound Med. Biol. 14 (1988) 123–126

71 Rankin RN, Fenster A, Downey DB, Munk PL, Levin MF, Vellet AD: Three-dimensional sonographic reconstruction: techniques and diagnostic applications. AJR 161 (1993) 695–702

72 Rehn M, Lohmann K. Rempen A: Transvaginal ultrasonography of pelvic masses: Evaluation of B-mode technique and Doppler ultrasonography. Am. J. Obstet. Gynecol. 175 (1996) 97–104

73 Robb RA: Three-dimensional biomedical imaging: principles and practice. VCH, New York 1995

74 Rubin JM, Adler RS, Fowlkes JB et al.: Fractional moving blood volume: estimation with power Doppler US. Radiology 197 (1995) 183–190

75 Rubin JM, Bude RO, Carson PL, Bree RL, Adler RS: Power Doppler US: a potentially useful alternative to mean frequency-based color Doppler US. Radiology 190 (1994) 853–856

76 Salem S, White LM, Lai J: Doppler sonography of adnexal masses: The predictive value of the Pulsatility index in benign and malignant disease. Am. J. Roentgenol. 163 (1994) 1147–1150

77 Savicki E, Spiewankiewicz B, Cendrowski K, Stelmachow J: Transvaginal Doppler ultrasound with colour flow imaging in benign and malignant ovarian lesions. Clin. Exp. Obstet. Gynecol. 22 (1995) 137–142

78 Schneider VL, Schneider A, Reed KL, Hatch KD: Comparison of Doppler with two-dimensional sonography and CA 125 for prediction of malignancy of pelvic masses. Obstet. Gynecol. 81 (1993) 983–988

79 Schoenfeld A, Levavi H, Tepper R, Breslavski D, Amir R, Ovadia J: Assessment of tumor induced angiogenesis by three dimensional display: confusing Doppler signals in ovarian cancer screening? Ultrasound Obstet. Gynecol. 4 (1994) 516–518

80 Sengoku K, Satoh T, Saitoh S, Abe M, Ishikawa M: Evaluation of transvaginal color Doppler sonography, transvaginal sonography and CA 125 for prediction of ovarian malignancy. Int. J. Gynecol. Obstet. 46 (1994) 39–43

81 Skobe M, Rockwell P, Vosseler S, Fusenig NE: Halting angiogenesis suppresses carcinoma cell invasion. Nature Med. 3 (1997) 1222–1227

82 Srivastava A, Laidler P, Davies RP, Horgan K, Hughes LE: The prognostic significance of tumor vascularity in intermediate thickness (0.76–4,0 mm thick) melanoma: a quantitative histologic study. Amer. J. Pathol. 133 (1988) 419–423

83 Stein SM, Laifer-Narin S, Johnson MB et al.: Differentiation of benign and malignant adnexal masses: Relative value of gray-scale, color Doppler, and spectral Doppler sonography. Amer. J. Roentgenol. 164 (1995) 381–386

84 Steinke W, Meairs S, Ries S, Hennerici M: Sonographic assessment of carotid artery stenosis: comparison of power Doppler imaging and color Doppler imaging. Stroke 27 (1996) 91–94

85 Suren A, Osmers R, Kuhn W: 3D Color Power Angio^{TM} imaging: a new method to assess intracervical vascularization in benign and pathological conditions. Ultrasound Obstet. Gynecol. 2 (1998) 133–138

86 Tekay A, Jouppila P: Validity of pulsatility and resistance indices in classification of adnexal tumors with transvaginal color Doppler ultrasound. Ultrasound Obstet. Gynecol. 2 (1992) 338–344

87 Tong S, Downey DB, Cardinal HN, Fenster A.: A three-dimensional ultrasound prostate imaging system. Ultrasound Med. Biol. 22 (1996) 735–746

88 Timor-Tritsch IE, Lerner JP, Monteagudo A, Santos R: Transvaginal ultrasonographic characteriziation of masses by means of color flow-directed Doppler measurements and a morphologic scoring system. Amer. J. Obstet. Gynecol. 168 (1993) 909–913

89 Valentin L, Sladkevicius P, Marsál K: Limited contribution of Doppler velocimetry to the differential diagnosis of extrauterine pelvic tumors. Obstet. Gynecol. 83 (1994) 425–433

90 Wagner S, Gebel M, Bleck JS, Magnus MP: Clinical application of three-dimensional sonography in hepatobiliary disease. Bildgebung 61 (1994) 104–109

91 Weidner N: Intratumor microvessel density as a prognostic factor in cancer. Amer. J. Pathol. 147 (1995) 9–15

92 Weiner Z, Thaler I, Beck D, Rottem S, Deutsch M, Brandes JM: Differentiation malignant from benign ovarian tumors with transvaginal color flow imaging. Obstet. Gynecol. 79 (1992) 159–162

93 Winsberg F: Power Doppler sonography. J. Ultrasound Med. 15 (1996) 164

94 Woolf SH, Battista RN, Anderson GM, Logan AG, Wang E: Force on the Periodic Health Examinations: Assessing the clinical effectiveness of preventive maneuvers: Analytic principles and systematic methods in reviewing evidence and developing clinical practice recommendations. J. Clin. Epidemiol. 43 (1990) 891–905

95 Wu CC, Lee CN, Chen TM, Lai JI, Hsieh CY, Hsieh FJ: Factors contributing to the accuracy in diagnosing ovarian malignancy by color Doppler ultrasound. Obstet. Gynecol. 84 (1994) 605–608

96 Wu CC, Lee CN, Chen TM et al.: Incremental angiogenesis assessed by color Doppler ultrasound in the tumorigenesis of ovarian neoplasms. Cancer 73 (1994) 1251–1256

97 Zaneta G, Vergani P, Lissoni A: Color Doppler ultrasound in the preoperative assessment of adnexal masses. Acta Obstet. Gynecol. Scand. 73 (1994) 637–641

35 Ovarian Cancer Screening

A. Kurjak, W. Schmidt, A. K. Ertan, and S. Kupesic

35

Incidence and Five-Year Survival Rates of Ovarian Cancer

Approximately 20 000 women in the United States are diagnosed with ovarian cancer each year[11], and almost 60% of these patients die from their disease. Ovarian cancer causes 12 000 deaths per year, making it the fourth leading cause of cancer deaths among women. Because very few women have early or specific symptoms, 70–80% of patients already have metastases at the time of diagnosis[25,51]. Approximately two-thirds of patients have stage III or IV disease when diagnosed. The five-year survival rate for all stages is 36%[66]. This high mortality rate results from the lack of early symptoms and the consequent delay in diagnosis.

Diagnosis at an early stage. Ovarian carcinoma is a very aggressive malignancy. Realistically speaking, management strategies based on radical excision, chemotherapy, and radiotherapy have had very little impact on survival rates. Since patients who are diagnosed with stage I ovarian cancer have a much better prognosis with a five-year survival rate of approximately 75–85%, all current efforts are focusing on early detection. The prospective study data published by Young et al.[81] even showed a five-year survival rate of 96% for patients with stage I disease. These results support the view that treatment outcomes and survival rates can be substantially improved by diagnosing the cancer before it has penetrated the ovarian capsule. It has been estimated that ovarian cancer screening that increased the early detection rate of stage I or II disease from 20% to 80% could reduce the mortality rate by one-half[73]. For many years, the detection of ovarian cancer at an early stage has been considered fortuitous. During the past decade, various methods of examination have been proposed with the goal of allowing a very early diagnosis of ovarian cancer in normal-appearing or visibly altered ovarian tissue. While these methods have had varying degrees of success, there are still many problems in differentiating between malignant and benign ovarian lesions in the living patient. Known differentiating criteria such as the mitotic index or pleomorphism cannot be evaluated with currently established diagnostic tests. Unfortunately, there is no detectable premalignant lesion that can be interpreted as a "preinvasive form" of ovarian carcinoma. As a result, the prognosis of this disease is about the same today as it was 30 years ago. The early diagnosis of ovarian cancer is still more a matter of chance than a triumph of the scientific method.

Requirements of a Screening Program

Definition

Screening programs are based on two assumptions: that prevention is better than cure; and that early diagnosis can allow successful treatment, provided the disease is still localized. Screening, then, is defined as "the identification of healthy-appearing individuals whose risk of suffering from a particular disease now or in the future is large enough to justify diagnostic measures and, in some circumstances, direct preventive measures"[21]. This definition covers two important aspects of screening tests. First, screening tests are not done for the purpose of making a diagnosis or, in positive cases, identifying possible therapeutic interventions. Second, screening tests are intended more for "healthy-appearing individuals" than for patients who seek medical attention for a specific complaint.

WHO criteria. The WHO has established a list of criteria that prospective screening programs should meet[80]. These recommendations form the basis for the publications by the UK Coordinating Committee on Cancer Research on ovarian cancer screening[72]. The WHO criteria are as follows:

- The disease under study should be an important health problem.
- The natural history of the disease must be sufficiently known.
- There must be an identifiable early stage of the disease.
- Treatment at an early stage should be of greater benefit than at a later stage.
- There should be a suitable test for the early stage.
- The test must be acceptable to the screened population.
- There should be adequate provisions for the further diagnosis and treatment of detected abnormalities
- For diseases with an insidious onset, intervals should be determined for repeating the test.
- The risk of suffering physical or psychological harm from the test should be less than the possible benefit.
- The costs of the screening program should be weighed against its benefits.

Screening includes the obligation to take no further actions until the consequences of those actions can be fully assessed.

The screening process for ovarian carcinoma is subject to some inherent limitations based on the nature of the disease.

Screening Methods

Ovarian cancer screening should employ methods that can detect either a change in ovarian structure (e.g., size, morphology) or a change in ovarian function (e.g., the secretion of metabolites into the bloodstream). Additionally, the observed changes should be specific for malignancy, since the incidence of ovarian cancer is relatively low compared with benign ovarian lesions and a positive screen will require a surgical diagnosis. An "ideal" screening test should detect the cancer in its premalignant stage so that it can be treated before it becomes invasive. For example, Pap smears are an effective screening test for cervical carcinoma because they detect not only invasive carcinomas but also precancerous lesions. Unfortunately, there is still no well-defined premalignant form of ovarian carcinoma analogous to intraepithelial neoplasia of the cervix (CIN) or atypical endometrial hyperplasia of the uterine corpus. On the other hand, we know that a benign ovarian cyst may eventually become a nidus for malignant transformation and the development of ovarian carcinoma.

Screening Parameters

Sensitivity and specificity requirements. Numerous methods have been used in attempts to permit the early diagnosis of ovarian cancer. Although many of these methods were impractical or unsuitable, important advances in early detection have been made during recent years. Unfortunately, there is still no evidence that any of the currently available screening tests or combinations of tests have the necessary sensitivity or specificity for detecting early forms of ovarian cancer. It is extremely difficult to establish such a test, because a positive predictive value less than 10% is unacceptable in ovarian cancer screening, and the only subsequent diagnostic option would be laparoscopy or laparotomy[41]. Age is still the best selection criterion for screening. A screening test for ovarian cancer should have a specificity of at least 99.6%. We cannot yet estimate the sensitivity requirement of a screening test based on available data, but it has been suggested that a sensitivity higher than 80 % and a specificity higher than 98% should be adequate for effective screening[79].

Since the ultimate quality standard for a screening test is the survival time of the patients, and this depends in turn on the FIGO tumor stage at the time of diagnosis, it is reasonable to require that a screening test be sensitive enough to detect a carcinoma in FIGO stage I or at least in stage II. Additionally, the screening test must meet the criteria of patient acceptance and a favorable cost–benefit ratio, which are difficult to define.

Definitions and derivations. Several key parameters must be defined in order to assess the effectiveness of a potential screening test. The definitions and derivations of these screening parameters are reviewed in Table 35.**1**.

- The sensitivity or detection rate measures the ability of a test to correctly identify women with ovarian cancer. It represents the number of women who have a positive test result and actually have ovarian cancer.
- The specificity represents the number of women who do not have the disease and have a true-negative test result.
- The rate of false-positive test results can be stated instead of specificity. The false-positive rate represents the number of women who do not have ovarian cancer but still test positive.
- The overall predictive ratio (OAPR) is often used to express the malignant-to-benign ratio. Within all the women who test positive, the OAPR is the ratio of the women who actually have cancer to the women who do not.

Table 35.**1** Definition of screening parameters

Test result	**Actual disease status** **Carcinoma present**	**Carcinoma not present**	
Positive	**A** (true-positive)	**B** (false-positive)	**A + B** (all positive test results)
Negative	**C** (fale-negative)	**D** (true-negative)	**C + D** (all negative test results)
	A + C (all patients with carcinoma)	**B + D** (all patients without carcinoma)	**A + B + C + D** (all test results)

1. Sensitivity = **A** [true-positive] ÷ **(A + C)** [all patients with carcinoma]
2. Specificity = **D** [true-negative] ÷ **(B + D)** [all patients without carcinoma]
3. False-positive rate = **B** [false-positive] ÷ **(B + D)** [all patients without carcinoma]
4. Positive predictive value = **A** [true-positive] ÷ **(A + B)** [all positive test results]
5. Negative predictive value = **D** [true-negative] ÷ **(C + D)** [all negative test results]
6. Prevalence of carcinoma = **(A + C)** [all patients with carcinoma] ÷ **(A + B + C + D)** [all test results]
7. Likelihood of having the disease when testing positive = **A** [true-positive] ÷ **B** [false-positive]

Possible Screening Tests

Bimanual Pelvic Examination

Typical findings. The chance of detecting ovarian cancer at an early stage by bimanual pelvic examination is fairly low. Although palpable masses in the lesser pelvis are very difficult to interpret, the pelvic examination has been the most widely used method for diagnosing ovarian cancer. The examiner should give particular attention to the following:

- A palpable mass in the adnexal region
- Relative immobility of the adnexa due to adhesions and fixations
- An irregular surface structure of the mass, an indolent mass, or a unilateral mass (70% of ovarian carcinomas are unilateral vs. only 5% of benign adnexal masses)

Because tumors located at different sites in the ovary often have different rates of growth, it is not unusual for cystic areas to alternate with soft, rubbery, or indurated components because of differences in blood supply.

Sensitivity. The bimanual pelvic examination is not considered sensitive enough to detect early forms of ovarian cancer. McFarlane et al.[55] reported that only six ovarian malignancies were detected in a total of 18 753 examinations performed in 1319 women over a 15-year period. Andolf et al.[3] found that 16 benign cysts, two borderline tumors, and one ovarian cancer had been missed by manual pelvic examination. Another study by the same authors documented 37 false-negative results in 194 patients[1]. Lundberg et al.[53] also reported a low detection rate of ovarian tumors. These results support the argument that the bimanual pelvic examination is inadequate as a screening test because of its low sensitivity.

Postmenopausal palpable ovaries. The most significant finding in the bimanual pelvic examination is the postmenopausal palpable ovaries (PPO) syndrome described by Barber and Graber[6]. Whenever an ovary that would be normal-sized for a premenopausal woman is palpated after menopause, it should be assumed that a tumor is present. Observation and follow-up are not appropriate for women with PPO syndrome, who should immediately be referred for further testing. It has even been suggested that PPO syndrome is an indication for adnexectomy[5].

Cul-de-sac Washings and Radiological Studies

Instillation of fluid into the cul-de-sac and aspiration of a sample for cytological analysis has been proposed for investigating a suspected ovarian malignancy. This method is useless for early diagnosis, however, because it can detect only disseminated cancers[43, 67].

MRI can also be used to visualize the adnexa. Abnormal ovaries and masses that are 0.5 cm or larger can be demonstrated by MRI[27]. CT is particularly helpful in the evaluation of paravaginal, peritoneal, and retroperitoneal lymph nodes [30]. On the other hand, MRI and CT are costly procedures and are too time-consuming to be used efficiently in ovarian cancer screening.

Tumor Marker

Since the anatomical location and structure of the ovaries make them inaccessible to direct examination, it would be advantageous to have a suitable serum marker for ovarian cancer screening. An antigen that is formed in the early stage of ovarian carcinoma could enter the bloodstream via blood vessels and lymphatics from the well-perfused ovarian stroma or via the free abdominal cavity and peritoneal lymphatics.

CA 125. The most widely studied tumor-associated antigen for ovarian carcinoma is CA 125. This antigen is recognized by a monoclonal antibody that was developed from the ovarian carcinoma cell line[7]. The mean serum level of CA 125 determined by radioimmunoassay is elevated above 30 U/ml in more than 80% of ovarian cancers, but it is similarly elevated in a small percentage of patients with lung cancer, bowel cancer, breast cancer, or pancreatic cancer[8, 35].

In a review of 15 studies[22] in a total of 128 women with stage I ovarian cancer, CA 125 levels above 30 U/ml were measured in 44% of the patients (Table 35.**2**). This does not mean that the CA 125 level is an effective screening parameter, however, because in many cases it does not become elevated until the disease has reached an advanced stage. While it is true that this antigen is detectable months or years before the cancer is diagnosed, presumably it does not become meas-

Table 35.**2** Detection rates (%) based on perioperative CA 125 levels, for various tumor stages

Study	Stage I	II	III	IV
Fuith et al. (1987)	60	80	89	80
Li-juan et al. (1986)	50	100	96	–
Heinone et al. (1985)	0	100	100	75
Halila et al. (1986)	–	0	57	100
Kivinen et al. (1986)	25	100	90	100
Crombach et al. (1985)	60	83	68	100
Schilthuis et al. (1987)	75	100	100	100
Zanaboni et al. (1987)	53	75	85	75
Vergote et al. (1987)	100	–	89	100
Brioschi et al. (1987)	31	100	97	96
Cruickshang et al. (1987)	25	67	94	100
Bast et al. (1983)	100	100	94	100
Patsner und Mann (1988)	40	100	93	100
Malkasian et al. (1988)	29	100	100	67
Zurkawski et al. (1988)	50	83	–	–
Total	44	88	91	94

urable until the tumor has already breached the ovarian capsule. When 35 U/ml is used as the cutoff level, as was done by Zurawski et al.[82], the false-positive rate is 5%, and one-third (4/12) of women with ovarian cancer (all stages) who develop clinical manifestations during the next 18 months can be identified at the time of the CA 125 screen. Unfortunately, elevated CA 125 levels are also found in certain other diseases such as endometriosis, extensive pelvic inflammatory disease, pancreatitis, severe hepatic cirrhosis, and during the first trimester of a normal pregnancy[24]. Consequently, CA 125 is not a specific marker for ovarian cancer. Moreover, a London Hospital study in which only 4 of 11 stage I cancers could be diagnosed found that CA 125 screening did not detect cancers early enough to achieve an improvement in prognosis[59].

Ultrasound

Transabdominal Ultrasound

Ultrasonography of the pelvis and abdomen has established itself as the standard modality for the evaluation of adnexal lesions. With ultrasound, the examiner can detect a pelvic mass, identify its site of origin (ovary, fallopian tube, or uterus), define its internal structure, and exclude or detect concomitant intra-abdominal pathology.

Differentiation of benign and malignant lesions. The effectiveness of ultrasound in differentiating between benign and malignant lesions has been investigated in many studies. Meire et al.[56] were the first authors to investigate whether an ovarian mass could be reliably classified as malignant based on its size and appearance at ultrasound. The presence of fixed septa, a tumor size greater than 5 cm, and multifocality were used in this study as suggestive signs of malignancy. Only 16 of 27 patients with these tumor characteristics were actually found to have an ovarian carcinoma. Ultrasound results have also been disappointing in evaluating masses located outside the pelvis. Lawson[52] reported a 31% accuracy rate in detecting the location of masses and evaluating their size and consistency in 251 cases. Deland et al. could correctly identify ovarian carcinoma in 13 of 14 cases[23]. Their data showed that ovarian tissue with a complex or solid internal structure was associated with malignancy in 70% of cases. Requard et al.[61] found that only 20% of metastases from ovarian cancer in the rectosigmoid, small bowel, and retroperitoneal lymph nodes could be correctly diagnosed by ultrasound examination. Thus, surgical intervention remains the only option for establishing whether a mass is malignant or benign.

Systematic protocol. The reliability of the diagnostic methods is naturally of great importance for the attending gynecologist. The modalities that are most widely used in the preoperative evaluation of ovarian masses are physical examination and sonography. A study by Finkler et al.[26] showed that both procedures have a low sensitivity in relation to all women examined and an extremely low sensitivity in premenopausal women alone. Even in postmenopausal women, who are more likely to have a malignant tumor, ultrasound has a sensitivity of only 47%. Campbell et al.[19] devised the first systematic ultrasound protocol for the early detection of ovarian cancer. They were the first to show that ovarian size and morphology determined by transabdominal ultrasound scanning correlated very well with direct findings at operation.

By contrast, O'Brien et al.[58] rated transabdominal sonography as inferior to physical examination. They found that transabdominal ultrasound had significantly less influence on management than the physical findings noted by the gynecologist.

London Times study. This prospective study was performed in 5479 self-selected asymptomatic women to evaluate the benefit of long-term screening and the development of new screening strategies based on specific changes in ovarian volume. Five women had primary stage Ia or Ib ovarian cancers, and the follow-up study, in which the patients were screened again no later than one year after the initial screen, ultimately showed a detection rate of 100% within this study design. The screening protocol, in which the first screen focused on documenting abnormal ovarian morphology while the second screen focused more on a change in ovarian volume, achieved a false-positive rate of 1.6% and a positive predictive value of 2%, corresponding to a 1 : 50 chance of error in cases with a positive test result. This chance of error results mainly from the difficulty of distinguishing between benign and malignant lesions (benign tumors, hydrosalpinx, etc.).

Routine ultrasound in symptomatic patients. Andolf et al.[4] investigated the value of routine ultrasound in 805 women attending a gynecological outpatient clinic in Sweden. Thirty-nine of the women underwent surgery based on ultrasound findings. One ovarian carcinoma was found at operation, two borderline tumors, and one cancer of the cecum. Because the patients were symptomatic, it is difficult to apply these results to general screening. None of the four tumors had been found by manual pelvic examination. Since there were no reports of follow-up scans, the exact disease status and detection rate cannot be determined. In a later study by Andolf et al.[2], six primary ovarian carcinomas were detected in 801 women who belonged to a high-risk category for ovarian cancer. No additional cancer cases were reported during the next three years, corresponding to a detection rate of 100%. The false-positive rate was 20%.

Studies in asymptomatic women. In ultrasound examinations of asymptomatic women performed by Campbell et al.[17], there was a false-positive rate of 3.5% in the first examination. Four primary ovarian cancers (all stage I) were diagnosed at first screening, and there were no reports of symptomatic new cases during the follow-up period. Three additional stage I cancers were detected at the second screening after periods of 16, 18, and 22 months[17]. Although the rate of false-positive results at first screening appeared to be acceptable, the overall predictive ratio (OAPR) was 1 : 97, i.e., 1 of 97 women with a positive result on screening actually had ovarian cancer. This clearly shows that a second screening method is necessary to reduce the rate of false-positive results and increase the OAPR. This method must also have a very high detection rate, however, so that true-positive results at the first screening do not become false-negative because of the second test.

Comments. The best transabdominal screening examinations for the early detection of ovarian cancer are based either on abnormal ovarian morphology or an ovarian volume above the 97th percentile at the first screening and a specified volume change or abnormal morphology at the second screening, or on abnormal morphology at the first screening and a specified volume change at the second screening[20]. Regardless of which scheme is used, repeat scans should be obtained every 12–18 months. Use of the second scheme results in a positive predictive value of 2% for primary ovarian cancers and 3.8% for all ovarian cancers. The false-positive rate and probability of error in patients with a positive test result are 1.6% (1 : 59) and 1.6% (1 : 26). Although some examiners still consider this false-positive rate too high, the transabdominal ultrasound scan can be classified as a practical screening method.

Transvaginal Ultrasound

Benign–malignant differentiation. Transvaginal ultrasound can reduce the time needed to scan and evaluate the pelvis. The image resolution is markedly better than with transabdominal ultrasound. On the other hand, examination with a transabdominal probe provides a better general view of the pelvis, and even an inexperienced examiner can quickly locate the ovaries with little difficulty, whereas endovaginal scanning requires a step-by-step search for the ovaries. Once the ovaries have been identified, however, the excellent resolution of transvaginal ultrasound will permit even small anatomical structures to be evaluated in detail.

A large number of independent studies have documented the occurrence of changes in the anatomical structure of "healthy" functioning ovaries. These changes include septation, papillary structures, solid and liquid cystic elements, daughter cysts, and solid lesions[62]. The goal of these studies was to learn to differentiate between benign and malignant lesions based upon reproducible criteria.

Scoring system. Many authors have tried to devise a generally accepted scoring system for the transvaginal sonographic characterization of ovarian lesions. Unfortunately, most of these systems had high sensitivity but lacked specificity, or vice versa. Bournet et al.[15] published a prospective study in which a high-risk group of women was screened for the presence of ovarian cancer. The authors used a scoring system based on the morphological appearance of the ovarian lesion (solid, monocystic or multicystic, unilocular or multilocular, regular or irregular cyst margins). The results appeared very promising, but the specificity was inadequate. Other authors tried to develop better scoring systems, but unfortunately without success[34, 40, 63]. In a comparative study of eight series of examinations by Sassone et al.[63], the sensitivity was 62–100%, the specificity was 73–95%, and the positive predictive value was between 31% and 88%. The authors state that one advantage of a numerical scoring system is that the test can be modified by selecting different threshold values until precise cutoff values can be determined. However, many series of examinations must be performed in order to establish an accurate point distribution for different disease processes.

False-positive findings. The main problem in the evaluation of tumor morphology is that some benign diseases can display malignant features. For example, endometriosis and dermoid cysts consistently produce false-positive findings. Van Nagel et al.[74, 75] published studies in which the rates of false-positive results at first screening were 3.1% and 2.3%, respectively. The further course yielded promising results as there were no reported new cases of ovarian cancer after the initial scan.

In another study at King's College Hospital[13], 1601 asymptomatic women with a close relative who had ovarian cancer were screened by transvaginal sonography. Six of the women in the group with a positive screening result had primary ovarian cancer (five stage Ia, one stage III). Three interval ovarian cancers were diagnosed at 24-, 41- and 44-month follow-ups (one stage Ib and two stage III). Including these three cancers, the total detection rate for the 44-month period was 67%.

Comments. It is clear that transvaginal sonography, with its high resolution, can improve the detection rate of early carcinomas. We are still faced with a significant rate of false-positive results, however. Even when screening is limited to women who are at highest risk for ovarian cancer, and thus have a higher prevalence of the disease, only one cancer will be found at operation in every 14 women with a positive test result[18].

Transvaginal Color and Pulsed Doppler Ultrasound

Blood flow characteristics in tumor vessels. Tumors derive their blood supply from normal preexisting vessels and also from vessels that form in response to angiogenic stimulation by the tumor cells[32]. These newly formed vessels differ from normal vessels in several respects: the vessel wall lacks a muscular coat, the vessels are composed mainly of endothelial cell lines, and they may contain tumor cells[68]. The paucity of smooth-muscle cells in the wall of the neovasculature leads to decreased peripheral resistance. Wells et al.[78] and Burns et al.[16] documented an abnormal flow spectrum in the periphery of breast carcinomas. These results were later confirmed by several groups of workers[38, 42, 57, 69]. These blood flow characteristics in tumor vessels are used in color and pulsed Doppler examinations as criteria for distinguishing between malignant and benign tumors (Table 35.**3**). This method is based on the thesis advanced by Judah Folkman in 1972 that tumors cannot grow unless they are able to acquire an additional vascular supply by inducing neoangiogenesis[31].

Resistance indices. Kurjak et al.[50] demonstrated the advantages of this new method. They studied the blood flow patterns of ovarian malignancies and other pelvic tumors and observed a low resistance index in the tumor vessels, with $RI < 0.41$. One granulosa cell tumor led to a false-positive result. In the study by Bourne et al.[14], eight of 18 ovarian tumors were malignant. The pulsatility index (PI) in seven of the eight malignant tumors was less than 1.0 (0.3–0.9). One false-positive result was caused by bilateral dermoid cysts with PI values of 0.4 and 0.8, and one false-negative result by a borderline tumor (serous cystadenoma) with a PI of 5.5. Both groups agree that FIGO stage Ia ovarian cancers can be detected by Doppler sonography and that this method is suitable as a screening test for ovarian carcinoma. Another conclusion was that a high vascular resistance can be used as an exclusion criterion for invasive primary ovarian cancer.

Table 35.3 Comparison of the screening parameters for gray-scale and color Doppler sonography

Authors	n	M/B[a]	Sensitivity (%)	False-positive rate (%)	Method[b]
Kurjak et al. (1989)	20	5/15	– 100	– 3	TVS TVS-CD
Kurjak et al. (1991)	680	56/624	– 96	– 1	TVS TVS-CD
Fleischer et al. (1991)	43	11/32	45 100	18 16	TVS TVS-CD
Fleischer et al. (1991)	26	5/21	100 100	15 17	TVS TVS-CD
Campbell et al. (1992)	7	7/0	– 95	– 6	TVS TVS-CD
Weiner et al. (1992)	53	17/36	94 94	31 6	TVS TVS-CD
Kawai et al. (1992)	24	9/15	57 88	10 0	TVS TVS-CD
Hata et al. (1992)	63	27/36	85 93	31 47	TVS TVS-CD
Timor-Tritsch et al. (1993)	115	16/99	94 94	13 1	TVS TVS-CD
Tekay u. Jouppila (1992)	72	11/61	– 82	– 23	TVS TVS-CD
Kurjak et al. (1993)	83	29/54	– 90	– 5	TVS TVS-CD
Kurjak u. Predanic (1993)	174	38/136	92 97	5 0	TVS TVS-CD
Insgesamt			81 94	20 10	TVS TVS-CD

[a] B = benign ovarian tumors; M = malignant ovarian tumors.

[b] TVS = transvaginal sonography; TVS-CD = transvaginal color and pulsed Doppler sonography

Overlaps and threshold values. Other groups of authors reported higher PI or RI values in malignant tumor vessels or a significant overlap between the PI and RI values of malignant and benign tumors[12,28,29,36,39,44–49,70,71,77]. Smooth transitions occur everywhere in nature, and it is unlikely that exact cutoff points in blood flow parameters could be found to differentiate between malignant and benign lesions. Moreover, blood flow parameters can change along with the stage of a tumor and its metastatic potential. Nevertheless, it is still necessary to define threshold values for PI and RI so that statistical analyses can be performed. The sensitivity of these screening parameters ranges from 46 % to 100% and their specificity from 72% to 99%. It is noteworthy that several studies achieved a sensitivity better than 80% and a specificity of 98%. This again underscores the fact that the measurement of intratumoral blood flow parameters by transvaginal color Doppler is the method of choice in ovarian cancer screening.

Secondary screening test. All of the above studies were performed in patients who were symptomatic or had abnormal physical findings. When color Doppler imaging is added to the transvaginal B-mode examination, the sensitivity of the examination can be improved and the rate of false-positive results can be reduced. Total sensitivity has been increased from 81% to 94%, while the false-positive rate has decreased from 20% to 10%. This demonstrates that a screening program that includes color Doppler ultrasound as a secondary test can meet the above criteria for a combined screening program. The next step is to test this method in a randomized study in asymptomatic women. The study would have to include at least 100 000 women and be continued over at least a 10-year period. Although color Doppler sonography makes a very promising impression, its definitive value cannot be determined until a study of this kind has been completed.

Who Should be Screened?

The false-positive rate can be reduced, and the specificity of a screening test increased, by examining a population that is at increased risk. Owing to the higher prevalence of the disease among those with a positive screening test, a higher percentage of true-positives will be detected. At the same time, the selection of a high-risk group can also help to identify certain factors that have causal significance, resulting in better early detection of the disease and an improved survival rate[9].

Age Distribution

Given the age distribution of ovarian cancer patients, it would be reasonable to limit screening to women over 40 years of age. In 1990, more than 68 000 years of life expectancy were lost in England and Wales due to ovarian cancer (Fig. 35.1). The mortality rate is highest between 60 and 70 years of age and rises significantly after age 40.

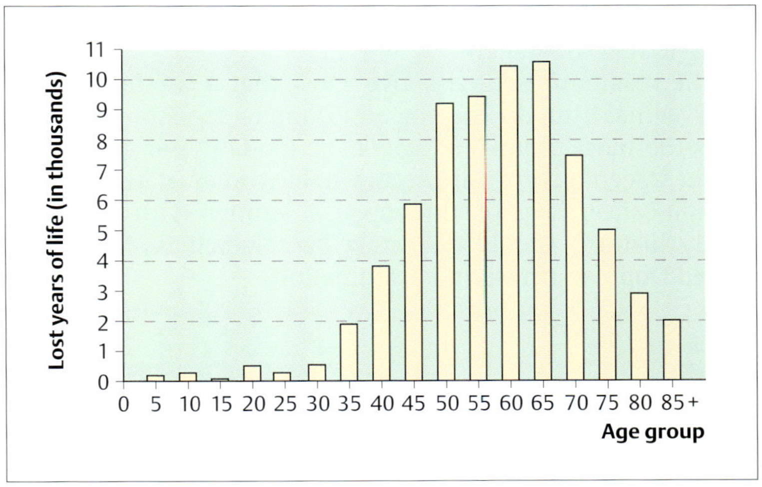

Fig. 35.1 Decreased life expectancy in years due to ovarian cancer in 1990 (England and Wales).

Family History

There have been many reports citing a positive family history as a significant risk factor for the development of ovarian cancer[54,65]. The risk is markedly increased in families where there appears to be a dominant mode of inheritance. A woman with an affected first-degree relative has an estimated 50% lifetime risk of developing ovarian cancer[60]. When two or more close relatives are affected, especially if they are young women or one of them also has breast cancer, a genetic predisposition should be strongly suspected. There is a 50% chance that the daughters or sisters of an affected woman will have the predisposing gene, although not all carriers of the gene will develop the disease. Their lifetime risk is estimated at 40%. The likelihood that a woman with a positive family history of ovarian cancer will develop the disease is 1 : 6, as opposed to 1 : 70 in the average female population[33]. These families, which represent a very interesting group for scientific studies, are the source of approximately 10% of all women with ovarian malignancies.

The genetic links between ovarian, endometrial, and mammary carcinoma have also been investigated[64]. The results prove the existence of a familial combined ovarian-and-breast cancer syndrome, whereas endometrial cancer, which can also run in families, has shown no relationship to this disease. These observations support the thesis that a genetic factor is one cause of ovarian malignancies. Women who have a suspected genetic risk, therefore, should be referred for genetic counseling and regular examinations as part of a special screening program.

Other Risk Factors

Infertility is suspected of increasing the individual risk of ovarian cancer, while oral contraceptives may have a protective effect[76]. The risk is higher for unmarried women and higher still for married women who are childless. Multiple pregnancies and childbearing at an early age appear to decrease the risk of ovarian cancer. Several studies have also ascribed a protective effect to nursing. Women who have had multiple pregnancies or use oral contraceptives tend to have fewer ovulations in their lifetime than other women, and ovulation trauma is considered an important pathogenic factor in ovarian cancer. Tearing of the ovarian capsule during ovulation can allow superficial epithelial cells as well as potential carcinogens to enter the ovarian stroma.

Conclusion

Ovarian cancer is a disease with a poor survival rate once clinical symptoms have appeared. The survival rate can be markedly improved by detecting the disease at an early stage. Although the primary goal of a screening program is to increase the number of cancers that are diagnosed early, the detection of larger but slow-growing "less malignant" tumors also contributes to the improvement of survival rates.

Before any of the screening tests described above can be considered effective, randomized controlled studies must demonstrate a lower mortality rate within the screened population group. In the case of ultrasound screening, this type of conclusion would require a study duration of at least 10–15 years. A positive cost–benefit analysis is also necessary to ensure that payers will make the test available to the population group that it will benefit.

The results presented above show that a combination of transvaginal B-mode imaging and Doppler sonography represents the most promising diagnostic modality for an ovarian cancer screening program. Accordingly, women at high risk for ovarian cancer (e.g., postmenopausal women with a positive family history) should definitely be selected for color and pulsed Doppler ultrasound examination.

References

1 Andolf E, Jorgensen C: A prospective comparison of clinical ultrasound and operative examination of the female pelvis. J. Ultrasound Med. 7 (1988) 617–620
2 Andolf E, Jorgensen C, Astedt B: Ultrasound examination for detection of ovarian carcinoma in risk groups: Obstet. Gynecol. 75 (1990) 106–109
3 Andolf E, Svalenius E, Astedt B: Ultrasonography for early detection of ovarian carcinoma. Brit. J. Obstet. Gynaecol. 38 (1986) 921–925
4 Andolf E, Svalenius E, Asted B: Ultrasonography for early detection of ovarian carcinoma. Brit. J. Obstet. Gynecol. 93 (1986) 1285–1291
5 Barber HRK: Ovarian cancer: diagnosis and management. Am. J. Obstet. Gynecol. 150 (1984) 910–916
6 Barber HRK, Graber EA: The PMPO syndrome (post menopausal palpable ovary syndrom). Obstet. Gynecol. 38 (1971) 921–923
7 Bast RC, Feeney M, Lazarus H et al.: Reactivity of a monoclonal antibody with human ovarian carcinoma. J. Clin. Invest. 38 (1981) 1331–1335
8 Bast RC, Klung TL, St. John E et al.: A radioimmunoassay using a monoclonal antibody to monitor the course of epithelial ovarian cancer. New Engl. J. Med. 309 (1983) 169–174
9 Beral V: The epidemiology of ovarian cancer. In Sharp F, Soutter WP (eds.): Ovarian cancer: the way ahead. Royal College of Obstet. Gynecol., London 1987, 21–23
10 Bhan V, Amso N, Whitehead MI et al.: Characteristics of persistent ovarian masses in asymptomatic women. Brit. J. Obstet. Gynecol. 96 (1989) 1384–1388
11 Boring C, Squires T, Tang T: Cancer statistics 1991. Cancer J. Clinic. 41 (1991) 19–36
12 Bourne TH: Transvaginal color Doppler in gynecology. Ultrasound Obstet. Gynecol. 1 (1991) 359–373
13 Bourne TH, Campbell S, Reynolds KM et al.: Screening for early familial ovarian cancer with transvaginal ultrasonography and color flow imaging. Brit. Med. J.

14 Bourne T, Campbell S, Steer C, Whitehead MI, Collins WP: Transvaginal color flow imaging: A possible new screening technique for ovarian cancer. Brit. Med. J. 299 (1989) 1367–1370
15 Bourne T, Campbell S, Whitehead MI, Collins WP: Transvaginal ultrasound screening for ovarian cancer in a high risk group. Presente at The Third World Congress on Vaginosonography in Gynecology, June 14–17, San Antonio, Texas 1990
16 Burns PN, Halliwell M, Wells PNT, Webb AJ: Ultrasonic Doppler studies of the breast. Ultrasound Med. Biol. 8 (1982) 127–132
17 Campbell S, Bhan V, Royston P et al.: Transabdominal ultrasound screening for early ovarian cancer. Brit. Med. J. 299 (1989) 1363–1367
18 Campbell S, Bourne TH, Whitehead MI, Steer CV, Colling WP: New developments in ultrasound screening for ovarian cancer. J. Ultrasound Med. 9 (Suppl. 1) (1990) 93
19 Campbell S, Goessens L, Goswamy R, Whitehead MI: Real-time ultrasonography for the determination of ovarian morphology and volume. A possible early screening test for ovarian cancer. Lancet 1 (1982) 425–429
20 Campbell S, Royston P, Bhan V, Whithead MI, Collins WP: Novel screening strategies for early ovarian cancer by transabdominal ultrasonography. Brit. J. Obstet. Gynecol. 97 (1990) 304–311
21 Cuckle HS, Wald NJ: Principles of screening. In Wald NJ (ed.): Antenatal and neonatal screening. Oxford University Press, Oxford 1984, 125–139
22 Cuckle HS, Wald NJ: Screening for ovarian cancer. In Miller AB (ed.): Evaluation of screening for cancer. UICC Press, Cambridge 1991, 234–249
23 Deland M, Field A, van Nagell J et al.: Ultrasonography in the diagnosis of tumors of the ovary. Surg. Gynecol. J. Obstet. 18 (1979) 346–348
24 Dordani D, Fleischer AC, Colvin-Huff B, Buckley SL, Jones III HW: Transvaginal sonography and CA 125 for evaluation of the postmenopausal ovary: A prospective pilot study. Med. Rev. 40 (1992) 39–44
25 Einhorn N, Nilsson B, Sjovall K: Factors influencing survival in carcinomas of the ovary. Cancer 55 (1985) 2019–2025
26 Finkler NJ, Benacerraf B, Lavin PT et al.: Comparison of serum CA 125, clinical impression and ultrasound in the preoperative evaluation of ovarian masses. Obstet. Gynecol. 72 (1988) 659–663
27 Fishman-Javit MC, Lovecchio JL, Stein HL: Imaging strategies for MRI of the pelvis. Radiol. Clin. North Amer. 26 (1988) 633–651
28 Fleischer AC, Rodgers WH, Rao BK, Keppler DM, Jones HW: Transvaginal color Doppler sonography of ovarian masses with pathological correlation. Ultrasound Obstet. Gynecol. 1 (1991) 275–278
29 Fleischer AC, Rodgers WH, Rao BK et al.: Assessment of ovarian tumor vascularity with transvaginal color Doppler sonography. J. Ultrasound. Med. 10 (1991) 563–568
30 Fleischer AC, Walsh JW, Jones HW, Shaff MI, James AE: Sonographic evaluation of pelvic masses. Method of examination and role of sonography relative to other imaging modalities. Radiol. Clin. North Amer. 20 (1982) 397–412
31 Folkman J: Anti-angiogenesis: new concept for therapy of solid tumors. Ann. Surg. 133 (1972) 275–279
32 Folkman J: Tumor angiogenesis. Adv. Cancer Res. 48 (1985) 2641–2645
33 Gilda Render Familial Ovarian Cancer Registry. Newsletter 1990
34 Granberg S, Nosrom A, Wikland M: Tumors in the lower pelvis as imaged by vaginal sonography. Gynecol. Oncol. 37 (1990) 224–229
35 Haga Y, Sakamoto K, Egami H et al.: Clinical significance of serum CA 125 values in patients with cancers of the digestive system. Amer. J. Med. Sci. 292 (1986) 30–35
36 Hata K, Hata T, Manabe A, Sugimura K, Kitao M: A critical evaluation of transvaginal Doppler studies, transvaginal sonography, magnetic resonance Imaging, and CA125 in detecting ovarian cancer. Obstet. Gynecol. 80 (1992) 92–926
37 Hata T, Hata K, Senoh D et al.: Doppler ultrasound assessment of tumor vascularity in gynecologic disorders. J. Ultrasound. Med. 8 (1989) 309–314
38 Hata H, Hata K, Yamane Y, Kitao M: Real-time two-dimensional and pulsed Doppler ultrasound detection of intrapelvic neoplastic tumor and abnormal pathogenic changes: preliminary report. J. Cardiovasc. Ultrasonog. 7 (1988) 135–139
39 Hata K, Makihara K, Hata T et al.: Transvaginal color Doppler imaging for hemodynamic assessment of reproductive tract tumors. Int. J. Gynecol. Obstet. 36 (1991) 301–308
40 Herrman UJ Jr, Locher GW, Goldhirsch A: A sonographic patterns of ovarian tumors: prediction of malignancy. Obstet. Gynecol. 69 (1987) 777–781
41 Jacobs IJ, Orham DA: Potential screening tests for ovarian cancer. In Sharp F, Mason WP, Leake RE (eds.): Ovarian cancer. Chapman and Hall, 1989, London 197
42 Jellins J, Kossoff G, Boyd J, Reeve TS: The complementary role of Doppler to the B-mode examination of the breast. J. Ultrasound Med. 59 (1983) 89–93
43 Joppi M, Pezzini MB, Tapparelli E: Washing puncture of puch in early diagnosis and follow-up of ovarian tumors. Eur. J. Gynecol. Oncol. 3 (1983) 60–63
44 Kawai M, Kano T, Kikkawa F, Maeda O. Oguchi H, Tomoda Y: Transvaginal Doppler ultrasound with color flow imaging in the diagnosis of ovarian cancer. Obstet. Gynecol. 79 (1992) 163–167
45 Kurjak A, Predanic M: New scoring system for prediction of ovarian malignancy based on transvaginal color Doppler. J. Ultrasound Med. 11 (1992) 631–638
46 Kurjak A, Predanic M, Shala H: Resistance versus pulsatility index (Letter to the Editor). J. Ultrasound Med. 12 (1993) 16
47 Kurjak A, Schulman H, Sosic A, Zalud I, Shalan H: Transvaginal ultrasound, color flow, and Doppler waveform of the postmenopausal andexal mass. Obstet. Gynecol. 80 (1992) 917–921
48 Kurjak A, Shalan H, Matijevic R, Predanic M: Stage I ovarian cancer by transvaginal color Doppler sonography: A report of 18 cases. Ultrasound Obstet. Gynecol. 3 (1993) 195–198
49 Kurjak A, Zalud I, Alfirevic A: Evaluation of adnexal masses with transvaginal color ultrasound. J. Ultrasound Med. 10 (1991) 295–297
50 Kurjak A, Zalud I, Jurkovic D, Alfirevic Z, Miljan M: Transvaginal color Doppler for the assessment of pelvic circulation. Acta Obstet. Gynecol. Scand. 68 (1989) 131–135
51 Kykes PW, Bradwell AR, Champan CE, Vaugham ATM: Radioimmunotherapy of cancer: Clinical studies and limiting vactors. Cancer Treat. Rev. 14 (1987) 87–96
52 Lawson T, Albarelli J: Diagnosis of gynecologic pelvic masses by gray scale ultrasonography: analysis of specificity and accuracy. Am. J. Roentgenol. 128 (1977) 1003–1006
53 Lundberg WI, Wall JE, Mathers JE: Laparoscopy in evaluation of pelvic pain. Obstet. Gynecol. 42 (1973) 812–816
54 Lynch HT, Watson P, Bewtra C et al.: Hereditary ovarian cancer: Heterogenity in age at diagnosis. Cancer 67 (1991) 1460–1466
55 McFarlane C, Strugis MC, Fetterman FS: Results of an experiment in the control of cancer of the female pelvis organ and report of a fifteen year research. Am. J. Obstet. Gynecol. 69 (1955) 294–299
56 Meire HB, Farrant P, Guha T: Distinction of benign from malignant ovarian cysts by ultrasound. Brit. J. Obstet. Gynecol. 85 (1978) 893–897
57 Minasian M, Bamber JC: A Preliminary assessment of an ultrasonic Doppler method for the study of blood flow in human breast cancer. Ultrasound Med. Biol. 8 (1982) 357–362
58 O'Brain WF, Buck DR, Nash FD: Evaluation of sonography in the initial assessment of the gynecologic patient. Am. J. Obstet. Gynecol. 149 (1984) 598–602
59 Parkes CM, Wald N: Screening for ovarian cancer. In Kurjak A (ed.): Atlas of transvaginal color Doppler. Parthenon, Lanks
60 Ponder BAJ, Peto J, Easton DF: Familial ovarian cancer. In Sharp F, Mason WP, Creasman W (eds.): Ovarian cancer: biology, diagnosis and management. Chapman and Hall, London 1992, 3–8
61 Requard K, Mettler F, Wicks J: Preoperative sonography of malignant ovarian neoplasms. Am. J. Radiol. 137 (1981) 79–82
62 Rottem S, Levit N, Thaler I, Yoffe N et al.: Classification of ovarian lesions by high frequency transvaginal sonography. J. Clin.Ultrasound 18 (1990) 359–363
63 Sassone AM, Timor-Tritsch IE, Artner A, Carolyn W, Warren WB: Transvaginal sonographic characterization of ovarian disease: Evaluation of a new scoring system to predict ovarian malignancy. Obstet. Gynecol. 78 (1991) 70–76

64 Schildkraut JM, Risch N, Thompson W: Evaluating genetic association among ovarian, breast-ovarian, and endometrial cancer: evidence for a breast-ovarian cancer relationship. Amer. J. Hum. Genet. 45 (1989) 521–529
65 Schildkraut JM, Thompson WD: Familial ovarian cancer: a population based case-control study. Amer. J. Epidemiol. 128 (1988) 456–466
66 Silveberg E, Boring CC, Squires TS: Cancer Statistics 1990. Cancer J. Clinic. 40 (1990) 10–23
67 Smith L: Detection of malignant ovarian neoplasms: a review of the literature. 1. Detection of the patient at risk; clinical, radiological and cytological detection. Obstet. Gynecol. Surg. 39 (1984) 313–328
68 Strickland B: The value of arteriography in the diagnosis of bone tumors. Brit. J. Radiol. 32 (1959) 705–712
69 Taylor KJW, Ramos I, Carter D, Morse SS, Snower D, Fortune K: Correlation of Doppler ultrasound tumor signals with neovascular morphologic features. Radiology 166 (1988) 57–61
70 Tekay A, Jouppila P: Validity of pulsatility and resistance indices in classification of adnexal tumors with transvaginal color Doppler ultrasound. Ultrasound Obstet. Gynecol. 2 (1992) 338–344
71 Timor-Tritsch IE, Lerner J, Monteagudo A, Santos R: Transvaginal sonographic characterization of ovarian masses using color-flow directed Doppler measurements (abstract). Ultrasound Obstet. Gynecol. 2 (Suppl. 1) (1992) 171
72 UK Co-ordinating Committee on Cancer Research. Ovarian cancer screening. London 1989
73 van Nagell JR Jr: Ovarian cancer screening (editorial). Cancer 68 (1991) 679–680
74 van Nagell JR, DePriest PD, Puls LE et al.: Ovarian cancer screening in asymptomatic postmenopausal women by transvaginal sonography. Cancer 68 (1991) 458–462
75 van Nagell JR Jr, Higgins RV, Donaldson ES et al.: Transvaginal sonography as a screening method for ovarian cancer. Cancer 65 (1990) 573–577
76 Villard-Mackintosh L, Vessey MP, Jones L: The effects of oral contraceptives and parity on ovarian cancer trends in women under 55 years of age. Brit. J. Obstet. Gynecol. 96 (1989) 783–788
77 Weiner Z, Thaler I, Beck D, Rottem S, Deutchs M, Brandes JM: Differentiating malignant from benign ovarian tumors with transvaginal color flow imaging. Obstet. Gynecol. 79 (1992) 159–162
78 Wells PNT, Halliwell M, Skidmore R, Webb AJ, Woodcock JP: Tumor detection by ultrasonic Doppler blood flow signals. Ultrasonics 15 (1977) 231–234
79 Westhoff C, Randall MC: Ovarian cancer screening: Potential effect on mortality. Am. J. Obstet. Gynecol. 165 (1991) 502–507
80 Wilson JMG, Jungner G: Principles and practice of screening for disease. Public health paper 34., World Health Organization, Geneva 1968
81 Young RC, Walton LA, Ellenberg SS et al.: Adjuvant therapy in stage I and II epithelial ovarian cancer. New Engl. J. Med. 322 (1990) 1021–1027
82 Zurawski VR Jr, Orjaseter H, Andersen A et al.: Elevated serum CA125 levels prior to diagnosis of ovarian neoplasia: Relevance for early detection of ovarian cancer. Int. J. Cancer 42 (1988) 677–680

36 General Aspects of the Ultrasound Investigation of Blood Flow in Breast Tumors

C. Villena-Heinsen, M. Holländer, and W. Schmidt

Historical Development

Systolic modulation. Doppler sonography was first used for the benign–malignant differentiation of breast tumors in 1977 by Wells et al.[21] in a series of nine palpable breast masses using a CW Doppler pencil probe. In the three carcinomas examined, the authors described a characteristic Doppler frequency spectrum with a slow systolic downstroke and high diastolic blood flow. This "systolic modulation" was absent or extremely weak in the six benign tumors. Halliwell[9] interpreted this qualitative difference between benign and malignant tumors as the effect of a low-impedance vascular network. Various groups of authors made similar observations[8,12,22], although they studied small groups using different Doppler systems and different examination techniques.

Unidirectional and bidirectional blood flow. Additionally, Lypacewicz et al.[12] and White and Cledgett[22] described unidirectional blood flow in benign lesions and bidirectional flow in malignant lesions. It was assumed that a causal relationship existed between these different blood flow patterns, detectable by Doppler sonography, and the different histological structures of the neoangiogenic capillary networks in benign and malignant tumors. This assumption was not refuted for years—not even by Burns et al.[4] who, in 1982 qualified the above color Doppler observations and identified the difference in peak systolic frequencies between the tumor and the healthy contralateral breast as the most important parameter for benign–malignant discrimination.

Asymmetrical vascular foci. Madjar et al.[13] observed this characteristic Doppler frequency spectrum in only 54% of carcinomas. They claimed that the presence of an asymmetrical vascular focus relative to the contralateral breast was the most important criterion for malignancy.

Resistance index and pulsatility index. Technical advances led to the use of duplex Doppler scanning and then color Doppler sonography in further studies on the validity of Doppler ultrasound in the benign–malignant differentiation of tumors. Some authors already accepted as fact the hypothesis that tumor neoangiogenesis could be evaluated by Doppler sonography, and they used the lowest measured vascular impedance (minimum resistance index or lowest pulsatility index) as an expression of flow resistance in the capillary bed. The resistance index (RI) was believed to be the most accurate parameter for benign–malignant discrimination.

Parameters and blood flow characteristics. The following parameters and blood flow characteristics have been investigated and evaluated to date:

- Positive or negative blood flow detection
- Number of tumor vessels
- Blood flow parameters independent of beam–vessel angle that are calculated from the arterial waveforms, such as resistance index, pulsatility index, and S/D ratio
- Absolute maximum (systolic) and minimum (end-diastolic) flow velocities
- Number and intensity of single or clustered color pixels in a tumor
- Characteristic modulation of the Doppler waveform
- Comparison of blood flow in lesions and in contralateral healthy ("mirror image") areas in the same patient

Specific Parameters in the Doppler Examination of Breast Tumors

Blood Flow Detection

The first publications on color Doppler imaging in the breast described a simple technique for differential diagnosis: the detection of blood vessels was suspicious for carcinoma, while the absence of detectable flow signals suggested a benign lesion [3]. The value of this parameter was very quickly eroded by the development of more sensitive Doppler systems, which were able to detect blood flow in virtually all healthy breast areas, benign lesions, and breast malignancies. Figure 36.**1** shows a carcinoma in which only the most sensitive Doppler system could detect blood flow, while Fig. 36.**2** shows a fibroadenoma whose intense flow signals raised preoperative suspicion of carcinoma.

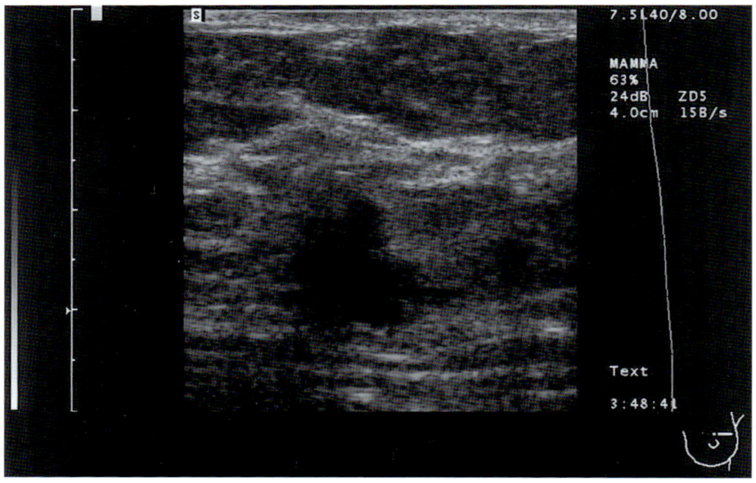

a

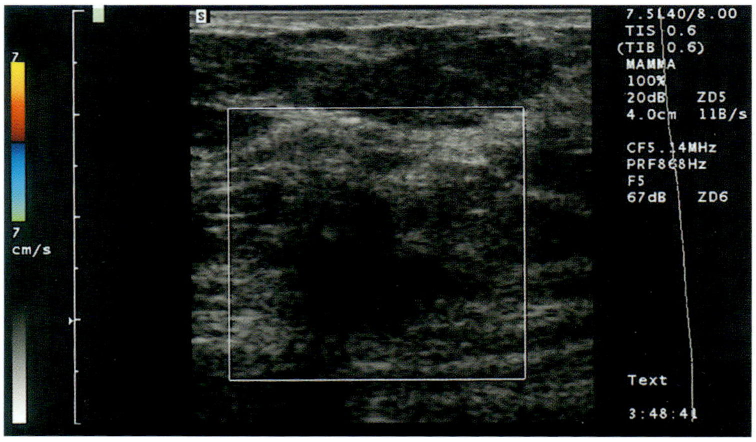

b

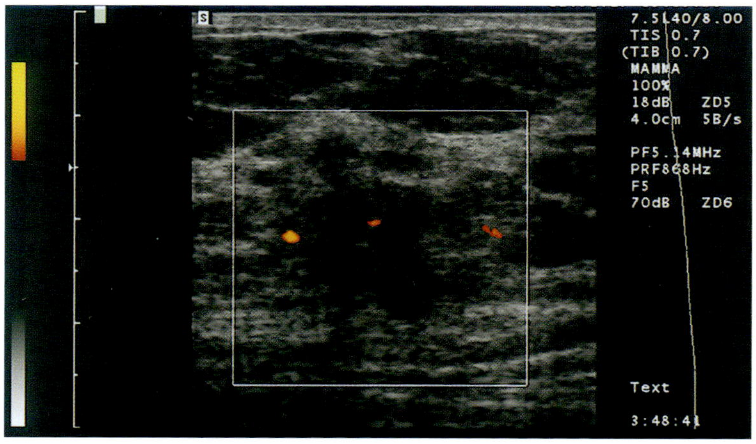

c

Fig. 36.1 Multifocal breast cancer.
a B-mode ultrasound demonstrates a larger, medially situated lesion with a maximum diameter of 15 mm and a smaller, laterally situated lesion with a maximum diameter of 4 mm.
b Contrary to expectations, color Doppler scanning of these lesions does not detect blood flow.
c Only power Doppler, with its higher sensitivity, can detect blood flow within and peripheral to the tumor.

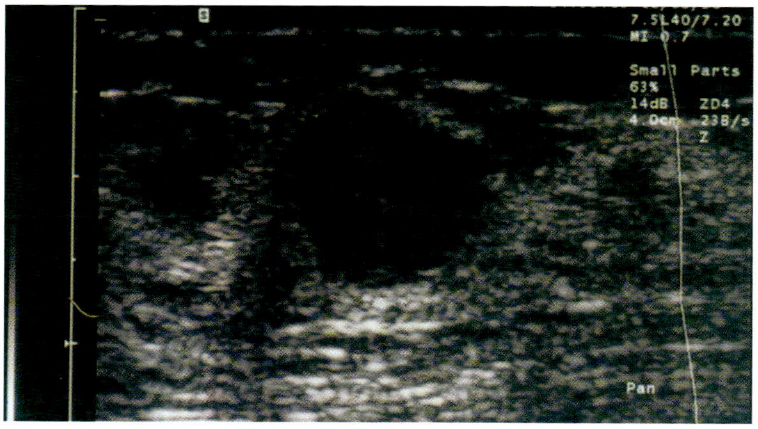

a

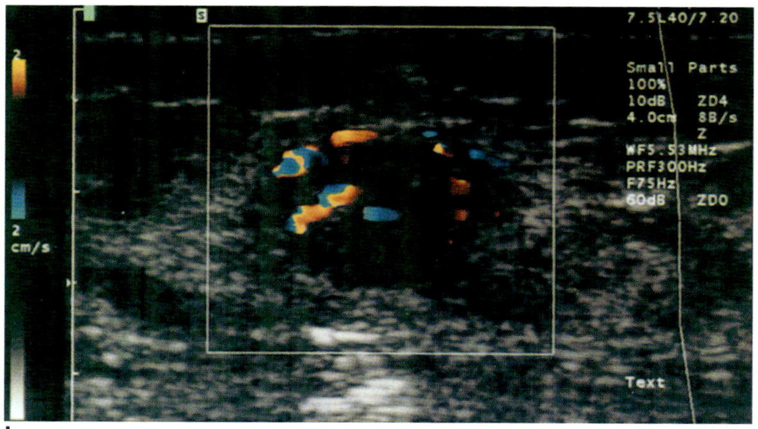

b

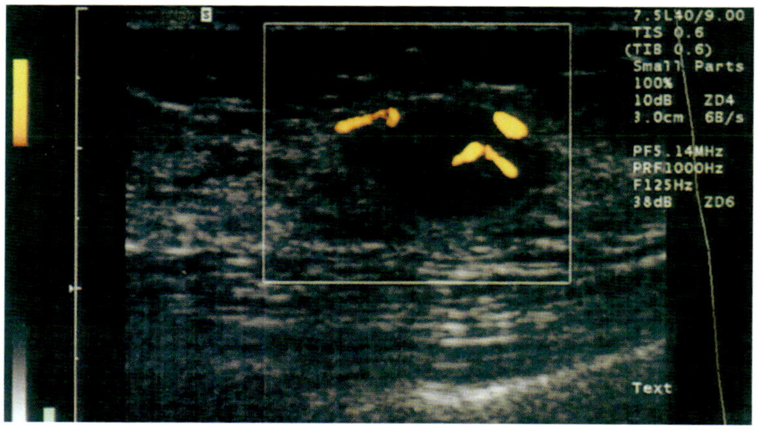

c

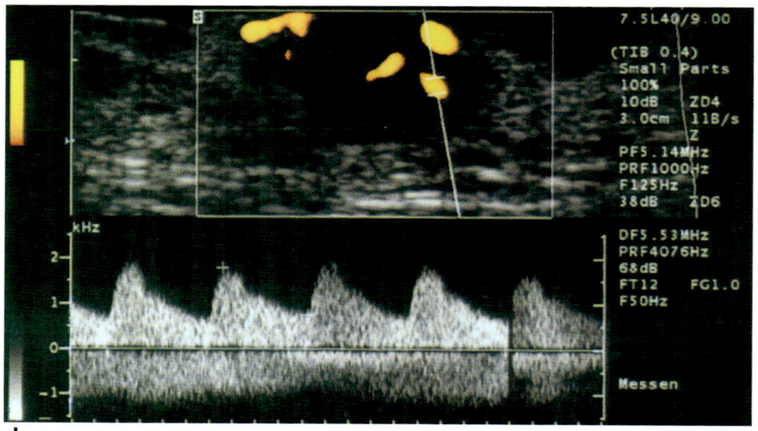

d

Number of Tumor Vessels

Evaluation based on the number of tumor vessels is problematic. The number of vessels in a color-flow image can be determined only by the color pixels or areas that can be visualized. Besides the sensitivity of the color Doppler system, vessel detection also depends on the accuracy, patience, and experience of the examiner. It is also unclear how many vessels the beam will cut, or how often. With CW or pulsed Doppler, the number of vessels can be determined only from the number of vascular signals that are found, and this depends on how systematically the examiner is scanning the area of interest. As Madjar et al.[14] and Villena-Heinsen et al.[17, 19] have reported, malignant tumors are supplied by a significantly greater number of vessels. This number is highly variable, however, and the areas of overlap are so large that a diagnosis based on this parameter will be very uncertain in any given case. It remains to be determined whether three-dimensional vascular imaging can provide better discrimination in this regard (Fig. 36.**3**).

a

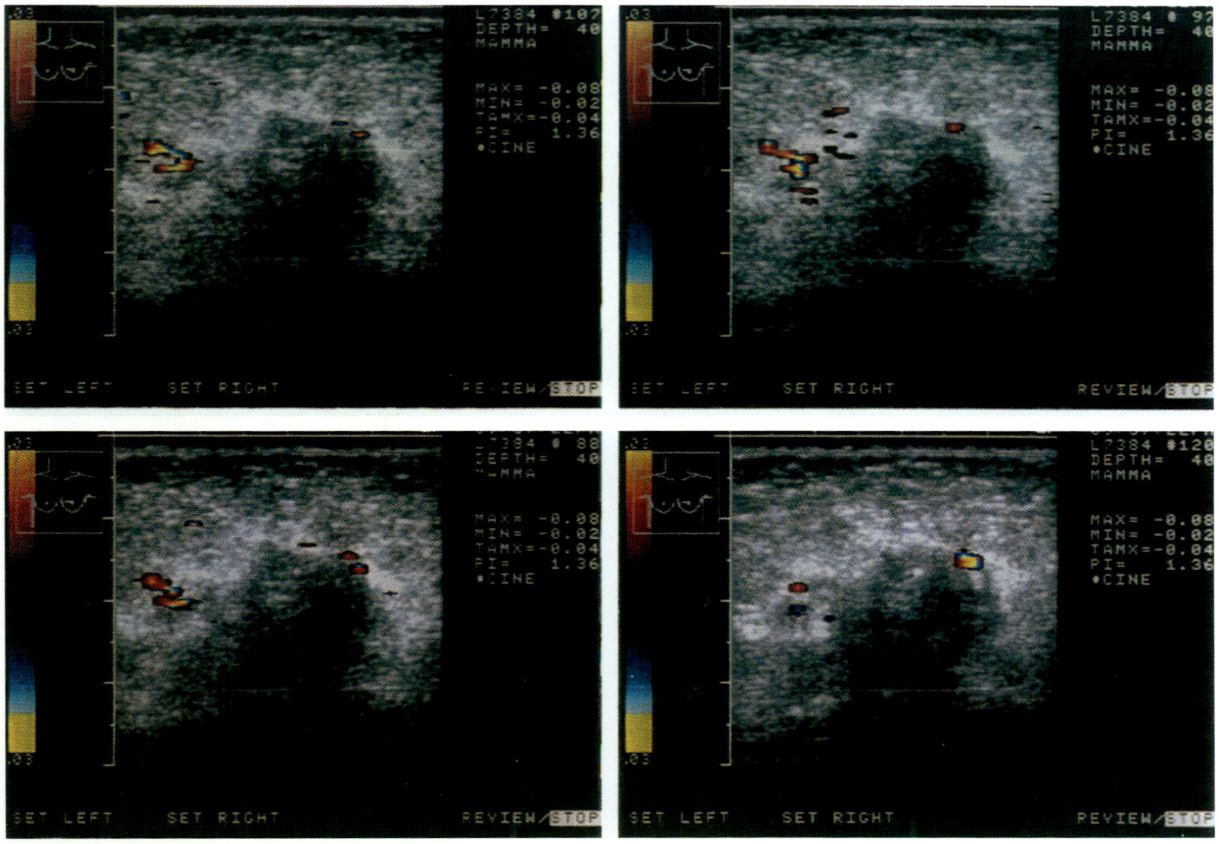

Fig. 36.**3** Counting tumor vessels.
a An accurate tumor vessel count is problematic. In the example shown, multiple scan planes through the carcinoma appear to demonstrate two blood vessels. But when smaller flow signals are also considered, it becomes very difficult to make an exact count. Doppler spectra have to be sampled from each of the color flow signals, and this is an extremely time-consuming process.

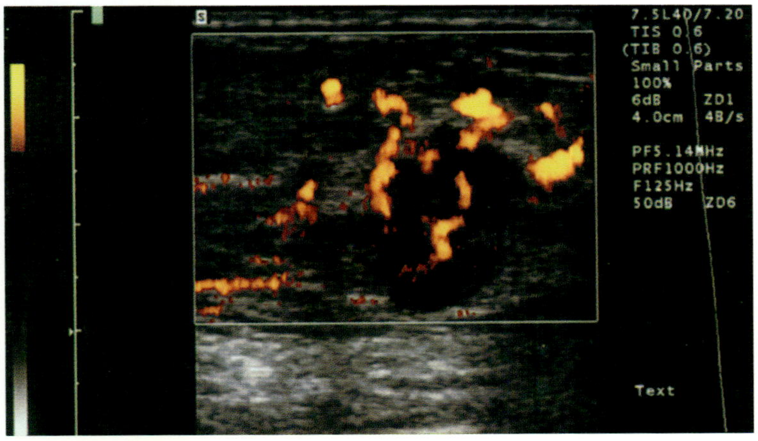

Fig. 36.**3 b** The number of blood vessels in this very vascular breast carcinoma cannot be accurately determined.

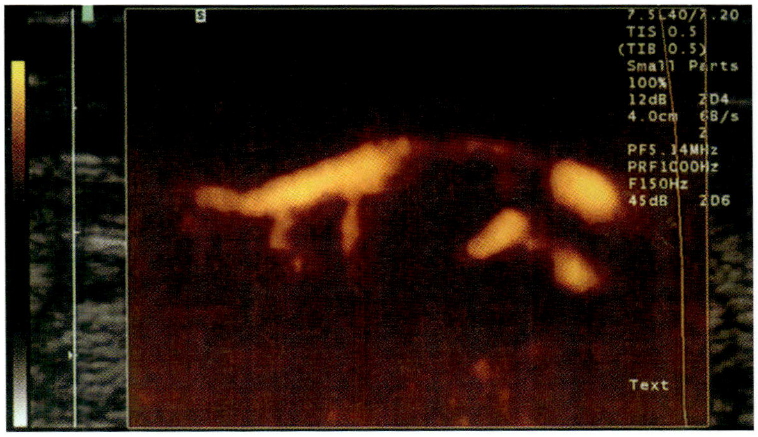

c The vessels may be somewhat easier to count in a three-dimensional flow image with the parenchyma subtracted.

◁ Fig. 36.**2** Fibroadenoma.
a B-mode ultrasound demonstrates a mass 12 mm in diameter.
b Color Doppler reveals intense blood flow.
c Power Doppler (power mode) also demonstrates a number of intratumoral and peripheral vessels.
d Doppler spectrum sampled from an intratumoral vessel in the power mode image indicates a relatively low RI of 0.59.

Resistance Index

The blood flow parameters that reflect vascular impedance appear to be the most reproducible. The resistance index (RI) is most commonly used. Since it ranges from 0 to 1, the RI can also express vascular impedance as a percentage. Most investigators [5,13,15,16] characterize a tumor by the lowest RI that can be recorded in the tumor or its immediate surroundings. Based on the hypothesis that lower flow resistances can be measured in malignant tumors due to morphological differences in the capillary network of malignant and benign lesions, the lowest measured RI was considered a valid discriminatory parameter and was used for tumor characterization. However, clinical experience contradicted the validity of this parameter and this method. Most groups of researchers found a significantly higher minimum RI in malignant tumors. This did not improve the discrimination of benign and malignant tumors, however, and the hypothesis was invalidated. Moreover, the mathematical calculation of minimum RI is subject to irregularities as our own studies illustrate[18] (Fig. 36.**4**).

Absolute Velocities

Several groups of investigators have worked with absolute velocities, especially the maximum systolic velocity[1,2,10,13,15].

Madjar and his group[14] even defined a new blood flow parameter, the flow velocity sum, as a correlate to total blood flow. When the flow velocities in all the tumor vessels were added together, the accuracy of benign–malignant discrimination increased to 90%. The use of flow velocities is still problematic, however, due to the angle dependence of the measured velocities. The need to optimize or correct for the beam–vessel angle is time-consuming and is possible only when the flow can be demonstrated along the vessel axis. This is extremely difficult in the case of small vessels (Fig. 36.**5**).

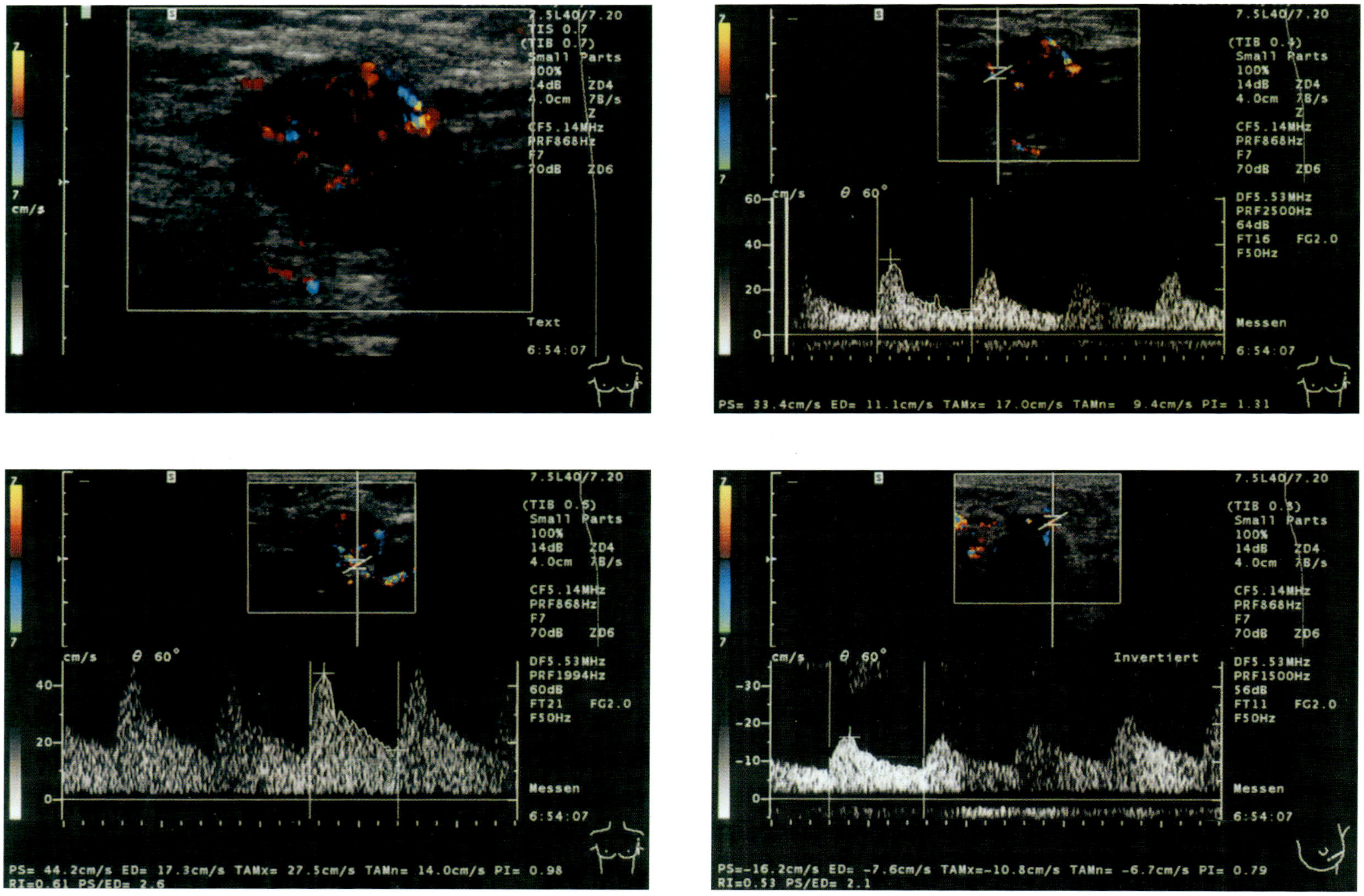

Fig. 36.**4** Breast carcinoma with a rich vascular supply.
a Numerous vessels are visible in the color Doppler image.
b–d The RI is relatively high but shows marked variation within the tumor (0.67, 0.61, 0.53). Is there a representative RI?

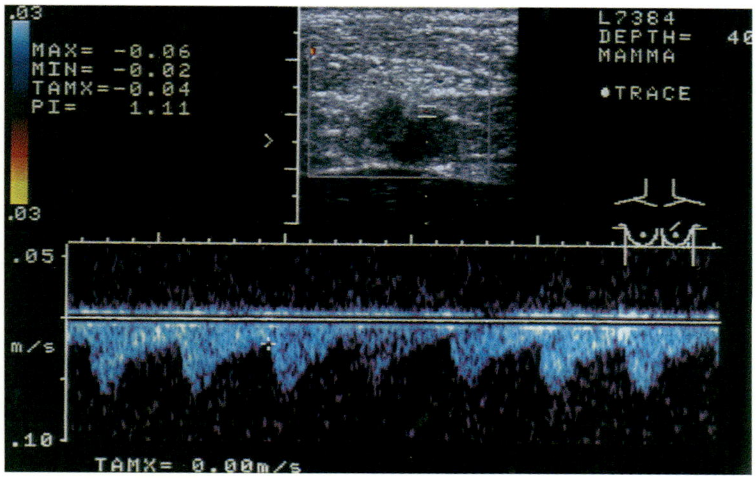

Fig. 36.5 Blood flow detection along the vessel axis.
a Small vessels (slow blood flow) are often difficult to visualize in longitudinal section.

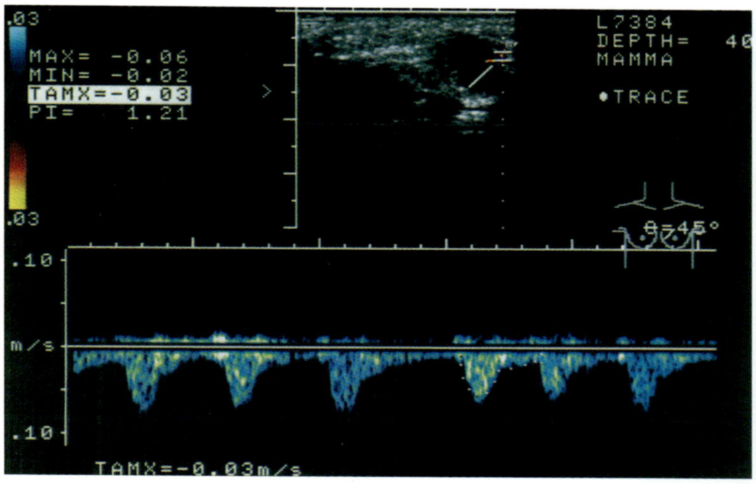

b Blood flow in a peripheral, S-shaped vascular segment. A long-axis vascular scan can be difficult to acquire. The absolute velocity changes with the angle correction, which is almost arbitrary. Ideally, blood flow should be sampled in a straight vascular segment.

Number and Intensity of Color Pixels and Color Areas

Today the number and intensity of individual color pixels and color areas can be evaluated with angle-independent Doppler and non-Doppler methods. The sensitivity of this newer technology is higher than that of conventional color Doppler ultrasound. Disadvantages are a greater susceptibility to errors, problems in comparing findings with different technologies, and the inability to measure flow velocities. There are study results with conventional Doppler in which increased, high-velocity blood flow was found in breast carcinomas using a semiquantitative grading method[6]. One disadvantage of this method is that the color pixels displayed are strongly dependent on the technology used, the equipment settings, and equipment handling by the examiner, and therefore the method has not become established in routine clinical use (Fig. 36.**6**).

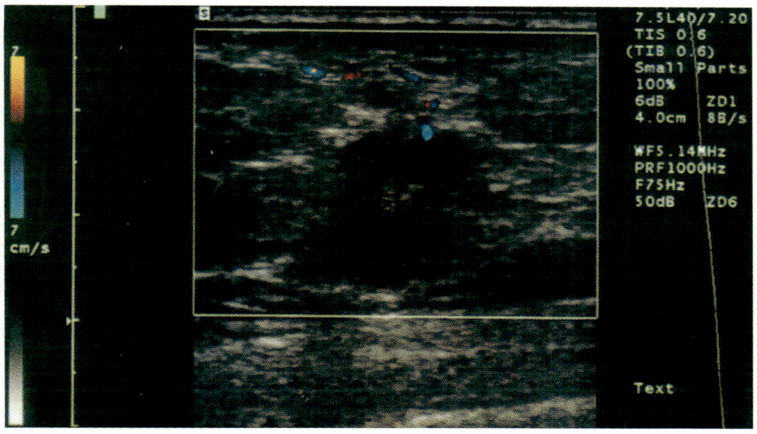

a

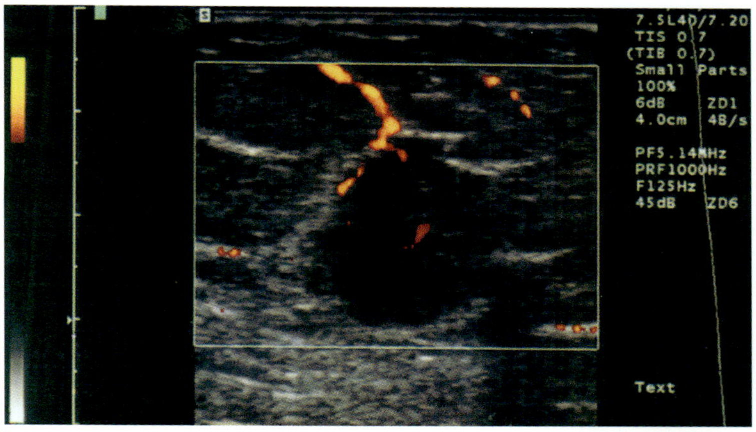

b

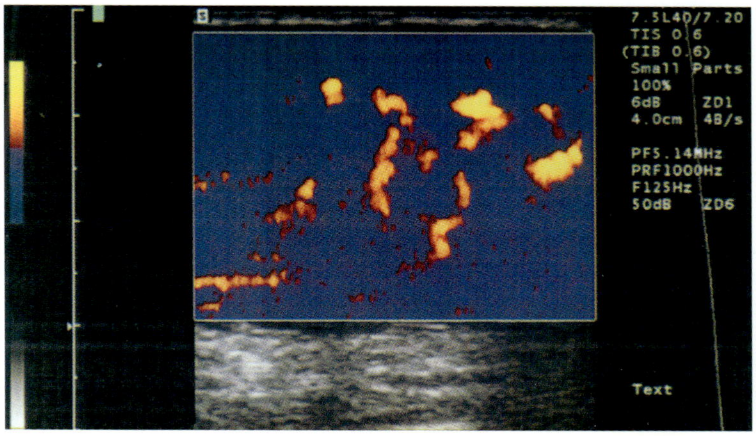

c

Fig. 36.**6** Benign–malignant discrimination based on the number of color pixels and the image area that they occupy has proved to be imprecise. Images **a–c** illustrate the lack of comparability of different technologies.
a S-shaped vascular segment defined by color Doppler.
b Power Doppler demonstrates additional vessels.
c Even more intense vascularity is seen in the subtraction view.

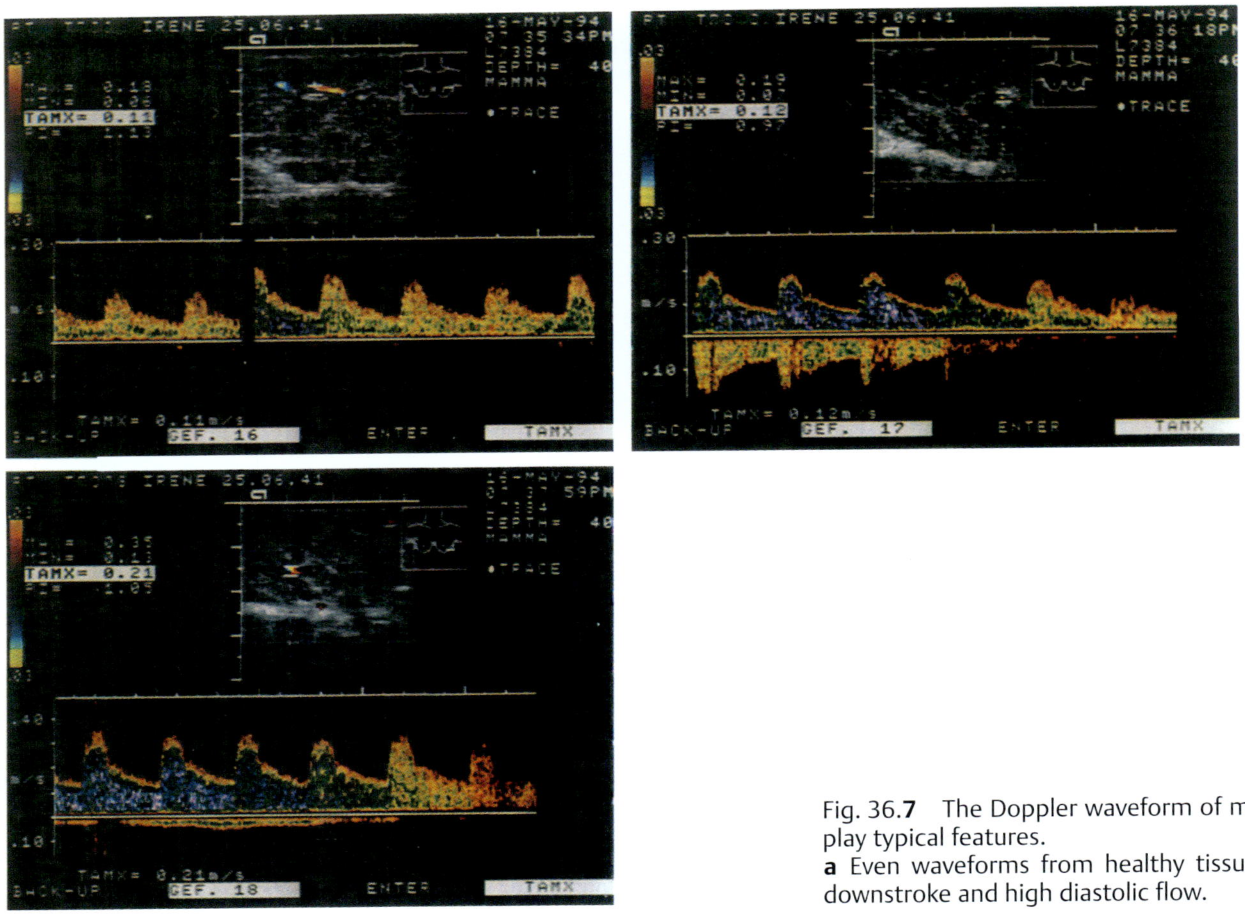

a

Fig. 36.**7** The Doppler waveform of malignant tumors does not display typical features.
a Even waveforms from healthy tissue may exhibit a slow systolic downstroke and high diastolic flow.

b

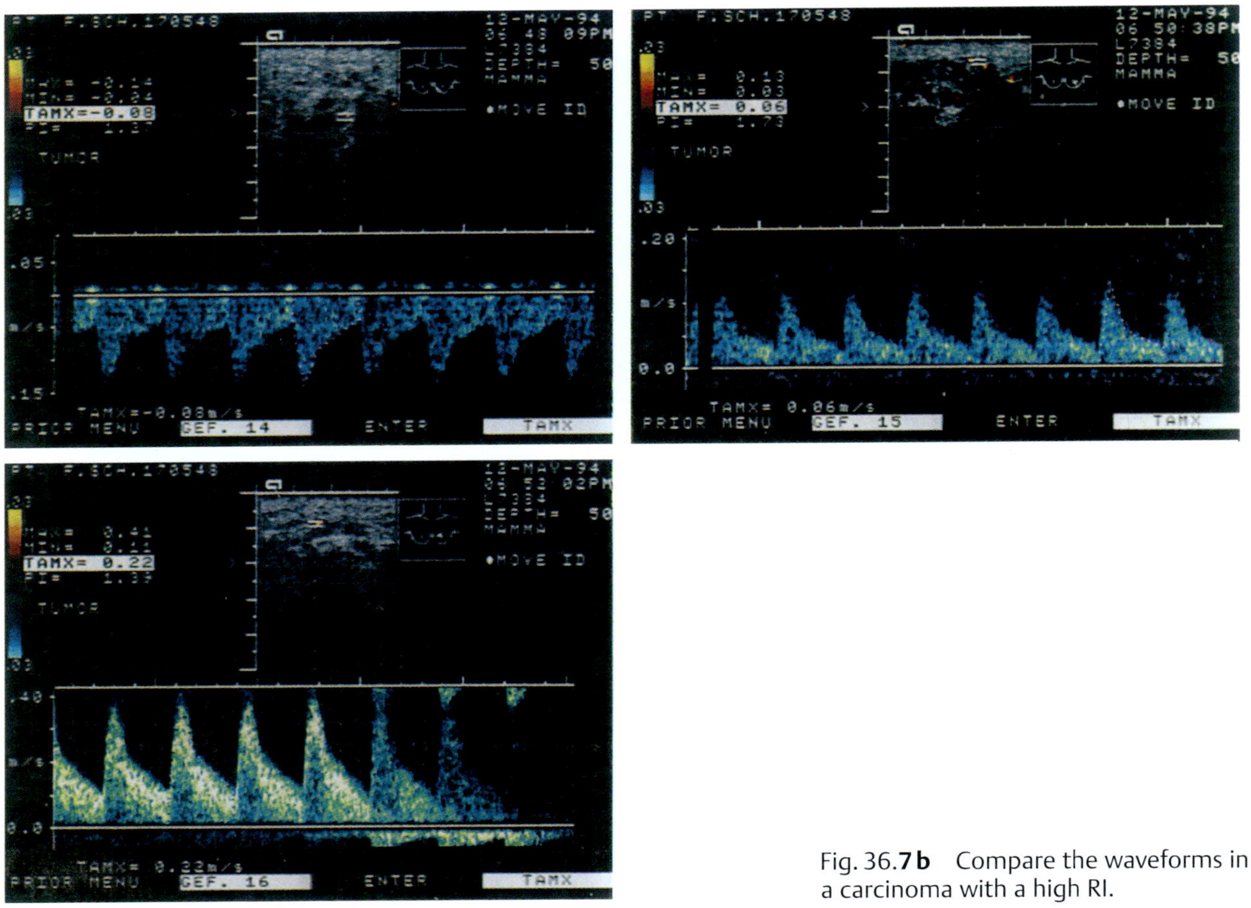

Fig. 36.**7 b** Compare the waveforms in **a** with spectra sampled from a carcinoma with a high RI.

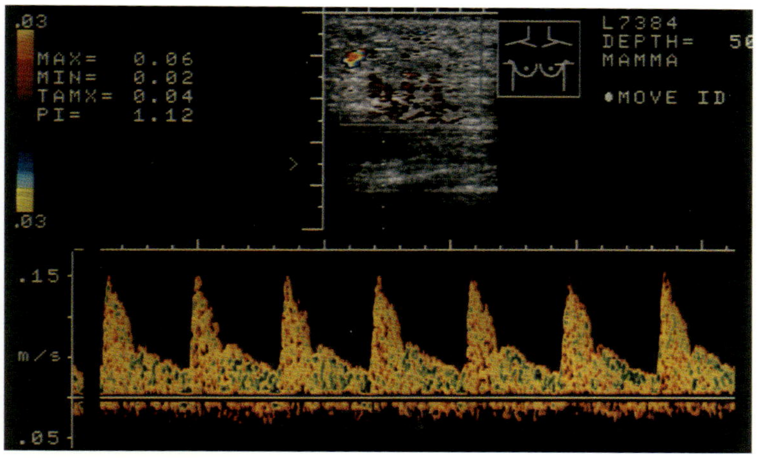

Fig. 36.**7 c** Early diastolic notch in a parenchymal vessel in the axillary tail of the left breast.

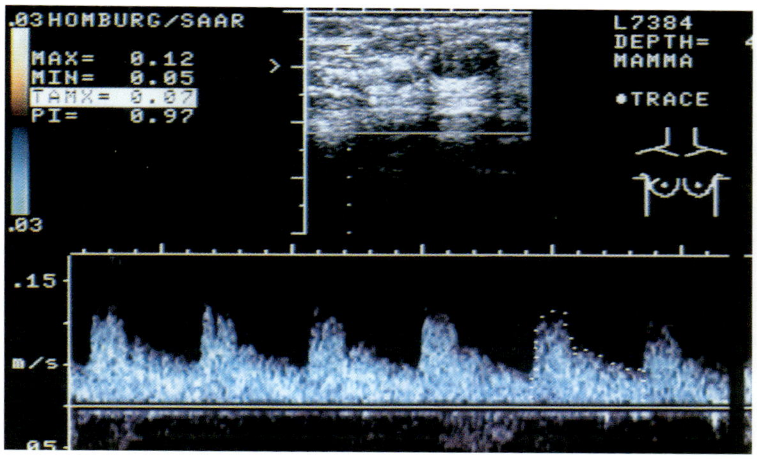

Fig. 36.**7 d** Diastolic notch recorded in close proximity to a fibroadenoma.

Doppler Waveform

Typical features cannot be found in the Doppler waveforms sampled from malignant tumors (Fig. 36.**7a, b**). An early diastolic notch in the Doppler waveform, which signifies high vascular impedance in obstetric ultrasound, is not a differentiating criterion because it is seen only occasionally and also because it may be found in benign lesions and in normal breast tissue (Fig. 36.**7c, d**).

Comparison of "Mirror Image Areas"

Various groups of authors[2, 13, 15, 19] have noted the significance of asymmetry between the tumor and the contralateral healthy breast. Several groups found a statistically significant difference between the tumor area and the healthy parenchymal breast tissue. Madjar et al.[13] found differences in vessel counts and maximum velocity, Blohmer[2] in maximum velocity, and Sohn[15] in the resistance index. Our group[19] found a statistically significant elevation of RI in premenopausal women with a malignant tumor compared with the healthy glandular tissue in the contralateral breast. On the whole, parameters that reflect vascular impedance, particularly the RI, have the best reproducibility. As a result, this parameter has become a focus of international research.

It is our opinion that menopausal status and hormone replacement are important as fundamental factors, regardless of which parameter is studied. But even when these factors are considered, we feel that the comparative approach to breast cancer diagnosis is not sufficiently reliable in individual cases due to the large scatter of the results.

Conceptual Misunderstandings in the Interpretation of Doppler Measurements

Problem areas. The sonographic imaging of blood flow in breast tumors is still a controversial issue. Despite years of testing, this method is still experimental in nature and has not become established in routine clinical use. Despite rapid technological advances in this area, it has not been possible to establish criteria or define blood flow parameters that would improve the discrimination between benign and malignant lesions. Basic problems lie in the unscientific interpretation of relationships between the histopathology and vascular physiology of tumors, the anatomical localization of detected flow signals, and the validity of blood flow characterization by ultrasound measurements.

Unproven claims. Conceptual misunderstandings based on an unscientific interpretation of presumed relationships have colored the debate in recent years. It was often claimed, erroneously, that blood flow could be sampled from the capillary network formed by tumor angiogenesis, yielding characteristic waveforms and impedance values that could be used to evaluate the histomorphology of the angiogenic capillaries of benign and malignant lesions.

In another unproven claim that was accepted more or less as fact, malignant tumors were defined as a flow system in which the resistance to flow was lower than in benign tumors. The results of studies conducted by our group have cast serious doubt on both of these claims[16, 17, 19].

Various authors have speculated whether the resistance indices determined by Doppler sonography or the density of blood vessels in the color Doppler image might correlate with the vascular density determined by microscopy. So far there is no scientific proof that the capillaries formed by tumor angiogenesis can be detected with ultrasound. It is generally acknowledged that color Doppler imaging can demonstrate larger-caliber portions of the vascular network that feed and drain the tumor and can occasionally detect intratumoral vessels as well. Color Doppler is capable only of assessing the blood flow conditions that are associated with tumor neoangiogenesis.

Neoangiogenic potential and color Doppler signals. Lagalla and his colleagues[11] proved that color Doppler blood flow signals do not originate from the vascular network produced by tumor angiogenesis. These authors correlated the detection or non-detection of color Doppler signals in 22 breast carcinomas with the angiogenic potential of the lesion evaluated in histological sections using a scoring system (MAGS = microscopic angiogenesis grading system) based on the number of blood vessels in the histological section, hyperplasia, and the mitosis rate of endothelial cells. Since the histological discoveries of Weidner et al.[20], who proved that the density of tumor angiogenesis in 49 breast carcinomas correlated with a poorer prognosis and an increased rate of distant metastases, it was reasonable to expect that neoangiogenic potential would correlate with the presence and intensity of detectable flow signals. However, the highest neoangiogenic potential (score > 30) was observed in the four tumors that had no detectable color Doppler signals. Meanwhile, a score < 30 was found in 17 of 18 cases with positive color Doppler flow detection. Microscopically, signal detection was found to correlate with vessels larger than 1 mm in diameter! This led the authors to conclude that color Doppler flow detection depends on the caliber of the tumor-feeding vessels and that the absence of flow signals is not a reliable criterion for excluding a malignant tumor.

No objective proof of tumor neovascularity. These results appear to have greater significance that the authors ascribe to them, for, as our own observations have shown[17,19], color Doppler sonography is incapable of recording blood flow signals from neoangiogenic tumor vessels at the present time. We must conclude, then, that the biological principles that have been widely advanced over the years for the color Doppler assessment of tumor angiogenesis are untrue and should be discarded. Of course, angiogenesis is essential for the rapid growth of solid malignant tumors, as Folkman explained in 1971[7]. It is also clear that differences in the prominence of arteriovenous shunts and luminal irregularities (due to the lack of a smooth-muscle layer) are observed in tumor vessels at the histopathological level. There is no proof, however, that the sonographic evaluation of blood flow can detect these characteristic histopathological features or the presumed associated decrease of flow resistance in the capillary bed of breast malignancies. In particular, this type of examination has been unable to objectify the extremely slow blood flows that are claimed to characterize the perfusion of malignant tumors.

Comments. These results also mean that the measurements of resistance indices that have previously been used for benign–malignant discrimination were not obtained in neoangiogenic tumor vessels but in larger tumor-feeding vessels. The wall structure of these vessels does not differ between benign and malignant lesions, and therefore the blood flow is subject to different principles.

When we take into account these conceptual misunderstandings, which run like a thread throughout the literature, we can appreciate why the flows measured in malignant breast tumors may show a higher impedance than in benign lesions.

Chapters 38–40 deal with various factors that can fundamentally influence blood flow parameters. First, the usefulness of the minimum RI as a suitable representative parameter for tumor characterization is investigated. The next chapter explores the effects of menopausal status, hormone replacement therapy, and age on quantifiable blood flow parameters. The final chapter examines the usefulness of the RI in the benign–malignant discrimination and prognostic evaluation of breast tumors.

References

1 Backe J, Mai R, Rempen A: Farbdopplersonographie bei tastbaren Mammatumoren im Vergleich zur kontralateralen Brustdrüse. Ultraschall Klin. Prax. 10 (1995) 1–7

2 Blohmer JU, Bollmann R, Schmalisch A, Chaoui R, Lau HU: Die Differentialdiagnose von Mammatumoren durch den Vergleich der Durchblutung des Tumors mit der kontralateralen Brust mittels farbkodierter, gepulster Dopplersonographie. Geburtsh. u. Frauenheilk. 55 (1995) 1–6

3 Britton PD, Coulden RA: The use of Doppler ultrasound in the diagnosis of breast cancer. Clin. Radiol. 42 (1990) 399–401

4 Burns PN, Halliwell M, Wells PNT, Webb AJ: Ultrasonic Doppler studies of the breast. Ultrasound Med. Biol. 8 (1982) 127–143

5 Campbell S, Bourne TH, Reynolds K et al.: Role of Colour Doppler in an ultrasound-based screening programme. In Sharp F, Mason WP, Creasman W (eds.): Ovarian Cancer 2: Biology, Diagnosis and Management. Chapman and Hall, London 1992, 237 –247

6 Cosgrove DO, Bamber JC, Davey JB, McKinna JA, Sinnett HD: Color Doppler Signals from Breast Tumours. Radiology 176 (1990) 175–180

7 Folkmann J, Merler E, Abernathy C, Williams G: Isolation of a tumor factor responsible for angiogenesis. J. Exp. Med. 133 (1971) 275–288

8 Gros ChM, Dale G, Gairand B: Breast echography: criteria of malignancy and results. In Kurjak A (ed.): Recent Advances in Ultrasound Diagnosis. Excerpta Medica, Amsterdam 1978, 292–298

9 Halliwell M, Atkinson P, Webb AJ, Wells PNT: Breast tumour detection by ultrasound blood flow signals. Proc. Amer. Inst. Ultrasound Med. 1 (1978) 96

10 Heilenkötter U, Jagella P: Farbdopplersonographie exstirpationsbedürftiger Mammatumoren. Geburtsh. Frauenheilk. 53 (1993) 247–252

11 Lagalla R, Caruso G, Marasa L, D'Angelo I, Cardinale AE: Capacità angiogenetica delle neoplasie mammarie e correlazione con le semeiotica color Doppler. Radiol. Med. 88 (1994) 392–395

12 Lypacewicz G, Powalowski T, Lukawska K: Ultrasonic examination of breast tumours with Doppler method. In Filipczynski L, Zieniuk JK (eds.): Proc. 2 nd Congress of the Federation of Acoustical Societies of Europe 1978. Polish Academy of Sciences, Warsaw 1978 Vol. II, 153–156

13 Madjar H, Sauerbrei W, Münch S, Prömpeler H, Schillinger H: Methodenanalyse zur Doppleruntersuchung der weiblichen Brust. Ultraschall in Med. 11 (1990) 196–201

14 Madjar H, Prömpeler H, Wolfahrt R, Bauknecht T, Pfleiderer A: Farbdopplerflußdaten von Mammatumoren. Ultraschall in Med. 15 (1994) 69–76

15 Sohn Ch, Grischke EM, Wallwiener D, Kaufmann M, von Fournier D, Bastert G: Die sonographische Durchblutungsdiagnostik gut- und bösartiger Brusttumoren. Geburtsh. Frauenheilk. 52 (1992) 397–403

16 Villena-Heinsen C, Ertan AK, Tossounidis I, Holländer M, König J, Schmidt W: Diagnostische Aussagekraft der Farbdoppler-Sonographie bei Mammatumoren. Geburtsh. Frauenheilk. 55 (1995) 541–547

17 Villena-Heinsen C, Mink D, Ertan AK, Holländer M, Schmidt W: Bewertung der Aussagekraft der Farb- und Spektraldopplersonographie bei Brusttumoren. In Schmidt W (ed.): Jahrbuch der Gynäkologie und Geburtshilfe 1995/1996. Biermann, Zülpich 1996, S. 121–136

18 Villena-Heinsen C, König J, von Tongelen B et al.: Validity of the minimal Resistance Index for discrimination between benign and malignant Breast Tumours. Eur. J. Ultrasound 7 (1998) 189–193

19 Villena-Heinsen C, Ertan AK, Holländer M, König J, Tossounidis I, Schmidt W: Diagnostische und prognostische Wertigkeit des Gefäßwiderstandes bei Brusttumoren. Ultraschall in Med. 19 (1998) 10–15

20 Weidner NR, Semple JP, Welch WR: Tumour angiogenesis and metastasis: Correlation in invasive breast carcinoma. New Engl. J. Med. 324 (1991) 1–8

21 Wells PNT, Halliwell M, Skidmore B, Webb AJ Woodcock JP: Tumour detection by ultrasonic Doppler blood flow signals. Ultrasonics 15 (1977) 231–232

22 White DN, Cledgett PR: Breast carcinoma detection by ultrasonic Doppler signals. Ultrasound Med. Biol. 4 (1978) 329–335

37 Color Doppler Sonography in the Diagnosis of Breast Cancer

H. Madjar

Evolution of Breast Cancer Diagnosis

Improvements in ultrasonography and mammography. Ultrasonography of the breast has improved dramatically in recent years owing to the development of high-resolution ultrasound systems. Until a few years ago, the investigation of palpable breast masses and the cystic/solid differentiation of mammographic densities were still considered the only goals of breast ultrasound. Higher contrast resolution and spatial resolution led to better soft-tissue discrimination, significantly improving the detection and differential diagnosis of subclinical disease. Mammography, too, has steadily improved through the development of the grid technique along with modern film–screen combinations and tube anodes. Meanwhile, numerous screening studies have documented the value of early diagnosis in reducing the mortality rates from breast cancer.

Problems in mammography. Since breast cancer is the most frequent cancer in women, the reduction of mortality from this disease has major epidemiological significance. While mammography is very sensitive in its ability to detect breast tumors, it cannot provide a definitive diagnosis, and women with abnormal screening results require further testing. In the United States, approximately 70–90% of women with abnormal screening results undergo unnecessary breast surgery for benign conditions. These unnecessary operations are distressful for the patients and can cause scarring that will interfere with further mammographic follow-ups. Negative biopsies have economic ramifications as well. Looking at clinical data in the United States, if we assume that there is a 10% annual positive rate of screening mammograms in a country with 25 million women of screening age and that only 50% of this population will present for screening, we find that approximately 1.25 million women will have suspicious breast findings that warrant further testing. If 50% of these women are referred for breast biopsy, the annual costs amount to billions of dollars, assuming a basic cost of $1000 to $5000 per operation. So from an economic standpoint as well, the specificity of the diagnosis must be substantially improved.

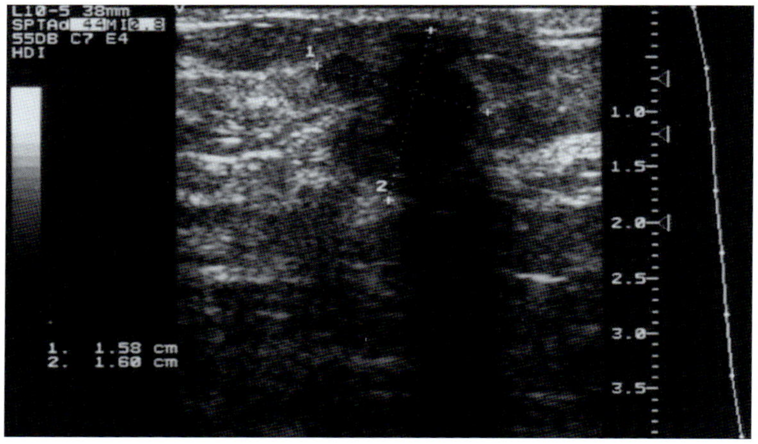

Fig. 37.**1** Typical ultrasound appearance of an invasive ductal breast carcinoma 16 mm in diameter. The image shows a nonhomogeneous, hypoechoic mass with irregular, ill-defined margins and a posterior acoustic shadow.

Problems in ultrasonography. Ultrasound can provide better soft-tissue discrimination than mammography, but there is a considerable overlap of diagnostic criteria between benign and malignant lesions. While carcinomas are typically nonhomogeneous and hypoechoic, with irregular, ill-defined margins and a posterior acoustic shadow (Fig. 37.**1**), there are fibrocystic breast changes and especially scars that mimic these findings and often prompt unnecessary surgery. On the other hand, benign breast lesions such as fibroadenomas typically appear as elliptical, uniformly hypoechoic masses with sharp, well-defined margins and good through-transmission of sound. But there are some fibroadenomas, especially the intracanalicular and proliferative forms, that are nonhomogeneous and have irregular margins (Fig. 37.**2**). Also, there are numerous types of breast cancer that mimic the typical sonographic features of fibroadenoma, such as medullary, mucinous, and solid invasive ductal carcinomas.

The differentiation between cystic and solid masses can also be difficult to accomplish with ultrasound if the contents of the mass are cellular or inspissated (Fig. 37.**3**). Diffuse breast changes are another problem. With severe, proliferative forms of fibrocystic change, disseminated microcalcifications are frequently seen on mammograms. At ultrasound, these breasts often show significant fibrosis with strong sound attenuation (Fig. 37.**4**). Sonograms are difficult to interpret in these cases, making it extremely difficult to rule out cancer or localize a lesion that has been targeted for biopsy (Figs. 37.**5**, 37.**6**). Such diffuse changes pose a major problem, especially in high-risk patients with a strong family history. Also, the sensitivity of mammography is often compromised by the radiographically dense breast parenchyma that exists in younger women. These problems underscore the need for improvements in preoperative breast diagnosis.

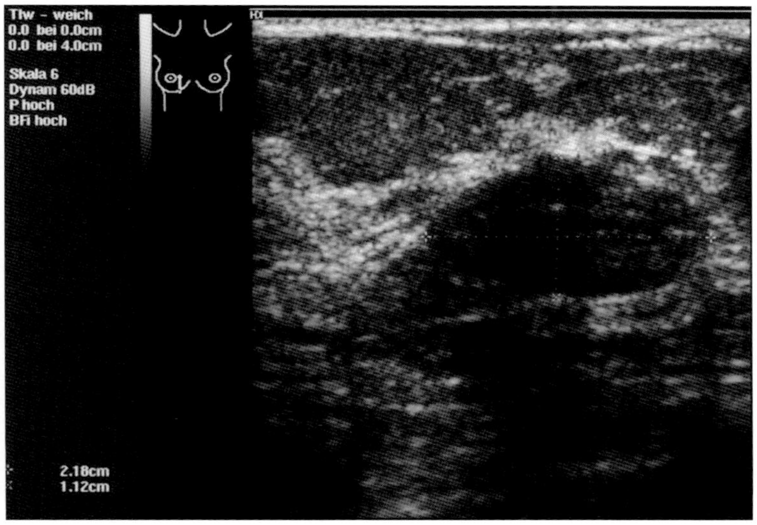

Fig. 37.**2** Intracanalicular fibroadenoma. This benign tumor has atypical sonographic features, appearing as a nonhomogeneous mass with irregular margins.

Assessment of tumor vascularity. Numerous histopathological and molecular biological studies have shown that the development of a malignant tumor from the in-situ stage requires neoangiogenesis[2, 12]. It has also been shown that the degree of neoangiogenesis very likely correlates with the metastatic potential and prognosis of the malignant tumor[13]. These facts have been applied in various diagnostic procedures such as angiography, thermography, and dynamic MRI. Such studies have also shown that the blood flow in malignant tumors is increased in comparison with healthy breast tissue.

Thermography. Thermography can measure temperature differences in the skin caused by local changes in blood flow and metabolism. Deep breast lesions often do not produce measurable changes, however, and many physiological conditions can cause changes in heat distribution. As a result, thermography has proved to be a relatively nonspecific test.

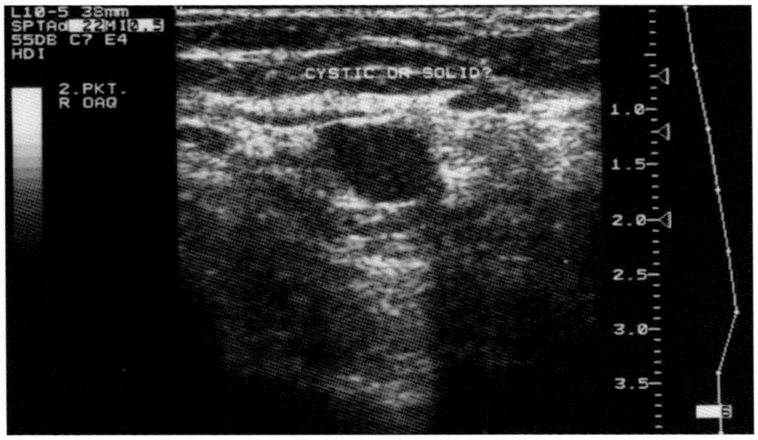

Fig. 37.**3** Inspissated cyst containing hemosiderin. FNA biopsy showed no evidence of proliferative changes. The mass cannot be positively distinguished from a solid tumor by its B-mode appearance.

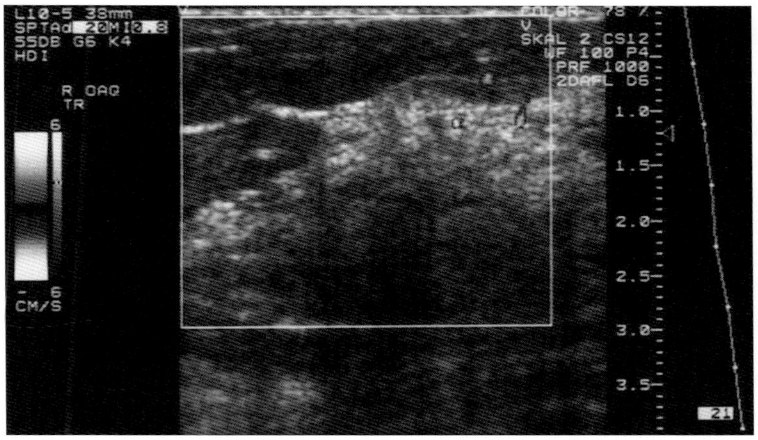

Fig. 37.**4** Severe fibrotic changes with diffuse sound absorption. Both the sonograms and mammograms of this patient were very difficult to read.

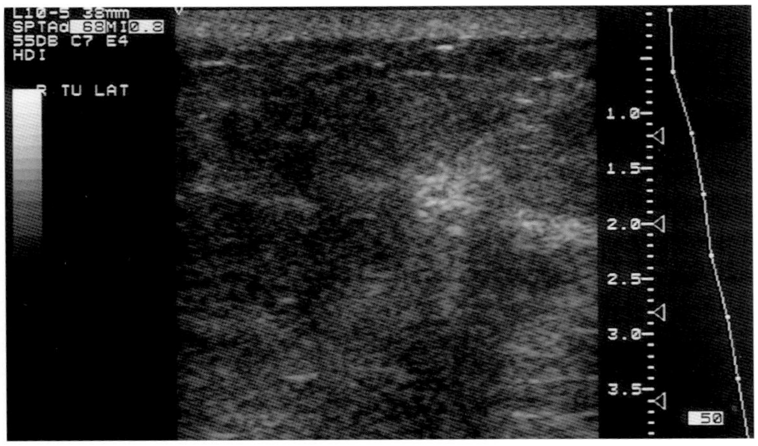

Fig. 37.**5** Diffusely infiltrating ductal carcinoma. The tumor margins are difficult to identify.

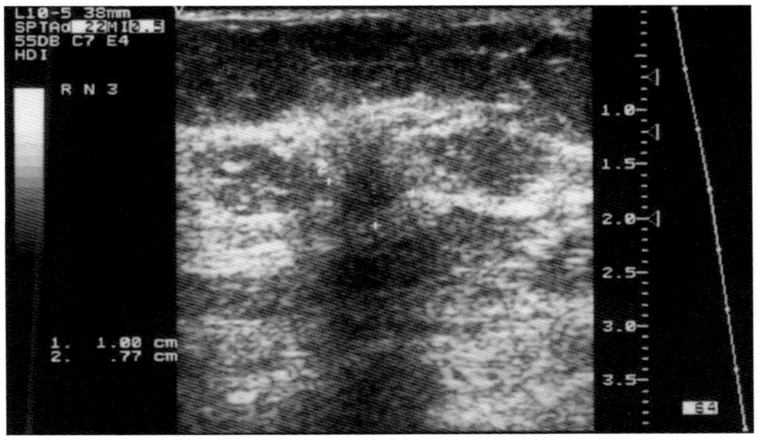

Fig. 37.**6** Diffusely infiltrating lobular carcinoma. Diffuse growth is particularly common with this type of carcinoma, often delaying the diagnosis.

Angiography and MRI. Angiography is too invasive for routine evaluations. Dynamic MRI with contrast administration has demonstrated its reliability, but it is a technologically complex study. Another problem is its limited temporal and spatial resolution, resulting in the frequent detection of blood flow changes that have no clinical, mammographic, or sonographic correlates and often are not referable to specific soft-tissue changes. Also, there has been growing evidence in recent years that physiological blood flow variations can greatly affect the results of MRI. Motion artifacts are another problem that can lead to equivocal or false-positive findings. It would be desirable, then, to have a method for measuring tumor blood flow and simultaneously imaging the associated tissue structures with high temporal and spatial resolution.

Continuous-Wave Doppler

Applications. Continuous-wave (CW) Doppler has been used for many years in neurology and angiology for the diagnosis of vascular stenosis and venous insufficiency. CW Doppler has not been used in abdominal or gynecological examinations because it does not simultaneously display the anatomical region of interest, making it extremely difficult to diagnose deeply situated tumors. CW Doppler has been widely used in pregnant patients for acoustically locating the uterine arteries and recording pregnancy-associated flow changes. This can be done with little difficulty, since the anatomical course of these vessels is known and can easily be located.

Breast tumors. CW Doppler has been used for approximately 20 years in the analysis of tissue and tumor blood flow in the breast. This is possible because the mammary gland is a superficial organ that is easily accessible to blind interrogation with CW Doppler. In particular, it is an easy matter to trace around a palpable breast mass with a hand-held Doppler probe and record vascular signals[3, 5, 11, 14]. Studies showed that malignancies are distinguished from benign lesions by their increased blood flow. As a result, an asymmetry of vascularity is noted when the two sides are compared. The initial studies by Wells et al.[14] were based on the purely acoustic evaluation of an amplified, high-frequency flow signal and did not yet permit a frequency spectrum analysis. Since the broad CW Doppler beam often interrogated multiple tumor vessels at one time, the flow signals were turbulent and rough, and a relatively loud signal was audible even during diastole. Several years passed before frequency spectrum analysis was performed. When detailed comparisons were made between the spectra sampled from malignant and benign lesions, a relatively high diastolic flow component was found even in normal parenchymal vessels of the breast. In contrast to peripheral vascular regions such as the radial artery or other resistance vessels, the systolic waveform is broad and rounded, and generally an early diastolic notch is not seen because practically no muscular vascular resistance occurs in this region (Fig. 37.**7**). When these flow signals are compared with those from malignant tumors, the latter are generally found to have higher systolic and diastolic flow, although the relationship between systole and diastole is usually unchanged[6].

Cyclic blood flow variations. High-frequency CW Doppler has also been used for various other indications. The broad Doppler beam is relatively favorable for evaluating diffuse parenchymal blood flow. Serial studies of physiological variations in blood flow during the normal menstrual cycle have shown that the cyclic flow variations are relatively small for any given individual, ranging from 50 Hz to a maximum of 200 Hz. But when different individuals are compared, the mean frequency shifts of parenchymal blood flow in the breast show differences as large as 500 Hz. Interestingly, blood flow is lowest during the first half of the menstrual cycle and increases markedly from ovulation to menstruation[7]. A pronounced increase equal to several times the baseline level is observed during the initial weeks of pregnancy.

Benign breast diseases. As the severity of benign breast lesions increases, we find an associated increase in blood flow compared with normal breast tissue. Since benign breast diseases are often accompanied by severe symptoms, they are treated with numerous medications, some of which can have serious side-effects. In the past, the only way to monitor treatment response in most patients was by noting improvement in their complaints over time. Doppler sonography, on the other hand, allows us to measure therapeutic response in cases where treatment has caused a reduction in blood flow. Since it is reasonable to assume that the increased blood flow in proliferative forms of fibrocystic change is caused by an increase in metabolism, we may conclude that a reduction in blood flow correlates with a regression of proliferative changes[9].

Chemotherapy for breast cancer. Another problem involves the neoadjuvant chemotherapy of extensive breast cancers and the chemotherapy of recurrent tumors. In the past, there has been no effective method of evaluating therapeutic response in the living patient. The only parameter that can be used for this purpose is tumor size, which is monitored by clinical examination or various imaging procedures. This is an extremely crude method, however, compared with the complicated processes that take place within the tumor. Still, as in benign conditions, the assessment of blood flow reflects the proliferation of the tumor tissue. It has been shown that the measurement of blood flow with Doppler ultrasound provides a sensitive parameter for assessing tumor response to chemotherapy[4].

Advantages and disadvantages. The advantage of CW Doppler is that the signal is not pulsed and high frequencies in the range of 8–10 MHz can be used. Both of these factors contribute to the high sensitivity of CW Doppler, which can detect even small blood flows[6]. One disadvantage of CW Doppler is that, without simultaneous tissue imaging, nonpalpable lesions are difficult to diagnose.

In the past, the combination of CW Doppler and ultrasound imaging required a complex setup in which a CW Doppler probe was incorporated into the central rotary axis of the im-

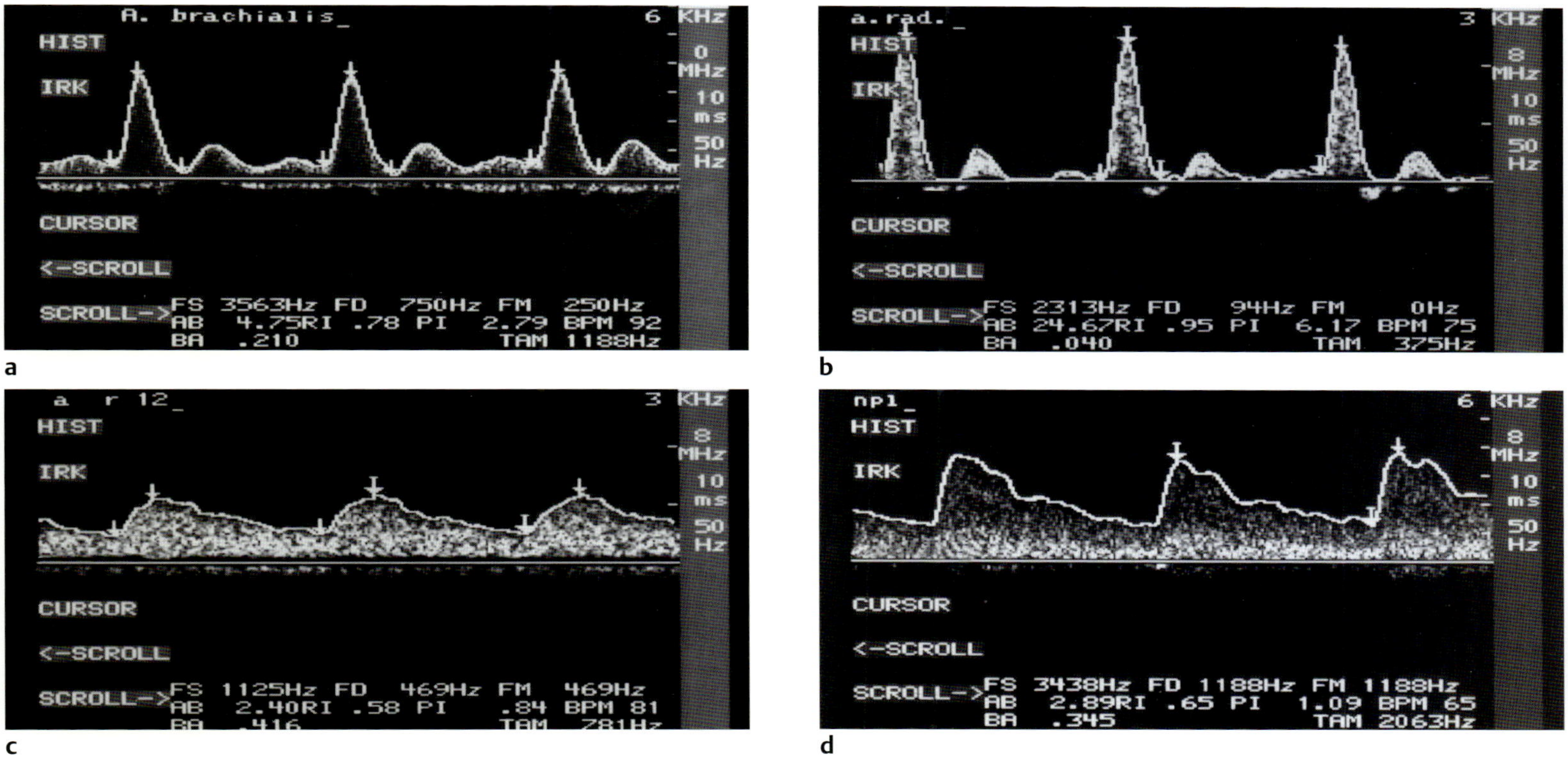

Fig. 37.**7** CW Doppler spectra.
a Brachial artery.
b Radial artery.
c Artery in normal breast parenchyma.
d Artery in a carcinoma.

aging transducers in a water-bath scanner[3]. While this increased the sensitivity of the CW Doppler, water-bath scanning did not become an established method because of its relatively poor spatial resolution compared with modern, high-resolution real-time scanners.

Pulsed Doppler Techniques

Vascular imaging. In principle, duplex scanning fulfills the desire for a method that can assess the vascularity of nonpalpable tumors. The lesion is visualized in the B-mode image, the sample volume is positioned in the vessel of interest, and flow spectra are recorded. However, duplex scanning was designed for use in anatomical regions with grossly visible vessels.

Tumor diagnosis. The vascularity that is associated with a tumor is not completely accessible to diagnostic imaging. It would be necessary to explore the full extent of the tissue volume using a one-dimensional, pulsed, range-limited sample volume, making the examination far too costly and time-consuming. Another problem is that the sensitivity of Doppler equipment is determined chiefly by the transmission frequency, as mentioned above. In most cases the Doppler frequency is well below the frequency of the imaging system. With a 5 MHz imaging transducer, for example, the Doppler frequency is generally in the range of 2–3 MHz. Usually this means that the sensitivity is too low to detect the microvessels in a tumor. We tested this by comparing CW Doppler and duplex scanning in a large series of patients. We found that duplex scanning was extremely difficult to use for this application, for methodological reasons, and that the sensitivity was too low for this type of examination, at least when a 3 MHz Doppler frequency was used[6].

Color Doppler

Tumor vascularity. Since the late 1980s, color Doppler instruments have been available that provide a color-encoded flow image superimposed over a real-time B-mode image (Fig. 37.**8**). This combination satisfies the demand for a combined gray-scale and Doppler technique for evaluating nonpalpable focal breast lesions. As in other techniques, however, the low sensitivity of many color Doppler systems is a serious limiting factor. In the assessment of tumor vascularity, it is important to consider that the type of equipment used has a major bearing on the quality of the examination. Several studies have

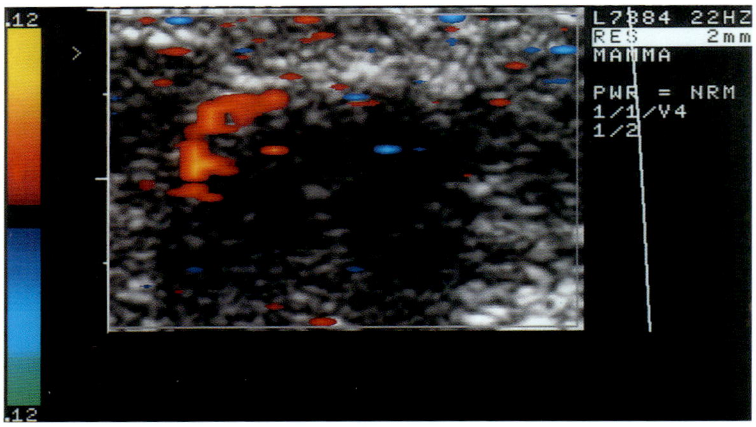

Fig. 37.8 Color Doppler image of an invasive ductal carcinoma 15 mm in diameter (Acuson 128, 7 MHz).

been published in recent years on the color Doppler evaluation of breast lesions[1,10]. These studies have yielded some highly discrepant results. These discrepancies are due largely to the use of different type of equipment[8], but other important factors are the examination technique and especially the equipment settings and transducer selection.

Comparison of different types of equipment. We have tested a number of different color Doppler systems since the early 1990s. In many of the systems used at that time, only 5 MHz linear transducers were available for breast ultrasound, and their operating frequency for color Doppler imaging was only 3 MHz. To provide a standard for comparison, we examined 70 breast carcinomas with high-frequency CW Doppler (10 MHz). We found that in some of the tumors that showed definite blood flow by CW Doppler, no color flow signals were detected. Since malignant tumors can vary greatly in their degree of vascularity, we classified them into different groups. In the cancers that showed particularly high vascularity when examined by CW Doppler, more or less conspicuous flow signals could be detected with all the color Doppler instruments. Often this was not possible in tumors that had a moderate to low degree of vascularity.

Comparison of different transducers. When the same instruments are operated with different transducers, such as the Acuson 128, we were unable to detect blood flow in some tumors using a 5 MHz linear transducer, but we detected definite flow when using a 7 MHz transducer with a 5 MHz Doppler frequency. This illustrates the frequency-dependence of Doppler sensitivity in duplex systems.

Equipment characteristics. The individual characteristics of an ultrasound machine are often difficult to define objectively and can be tested only in examinations on patients. There are, of course, Doppler flow phantoms for measuring the velocity calibration of a Doppler machine, and the technical specifications of the unit provide information on the minimum detectable flow velocity. But there are no technical specifications on the lower size limit of blood vessels that the unit can detect. Consequently, the only way to decide whether a unit is suitable for a particular application is by practical experience. For color Doppler studies, it is important that all patients be examined with the same machine using standardized projections, because differences in these parameters will lead to different results.

Equipment Settings

Frequency, power output, gain, and wall filter. Maximum sensitivity is crucial in Doppler systems used for tumor diagnosis. This depends on various equipment settings. With some transducers, the operating frequency is adjustable over a broad range. The highest possible frequency setting should be used in the Doppler mode. The power output and gain should be set as high as possible, i.e., to levels that are just below the noise threshold. The wall filter, which suppresses vascular pulsations and other motion artifacts, should be set as low as possible. A filter setting of 200 Hz or higher would cause too much sensitivity loss, while a low filter setting (e.g., 25 Hz) would not give adequate suppression of motion artifacts. A filter setting of 50–100 Hz is ideal.

Pulse repetition frequency and aliasing. The pulse repetition frequency (PRF) also has a major effect on sensitivity. While a very high PRF setting, like that used in echocardiography, allows the precise detection of high-velocity flows, it greatly reduces the sensitivity of the Doppler system. On the other hand, a very low PRF leads to a very slow frame rate in the Doppler mode and makes real-time scanning difficult. A PRF of 800–1000 MHz has proved to be an ideal trade-off between a high frame rate and high sensitivity. It should be noted that the flow velocities in malignant tumors can reach values of more than 1 m/s, despite small vessel lumina. When a low PRF is used, this high-velocity flow leads to aliasing—a color reversal that distorts direction and velocity information in the color-flow image. This artifact must be tolerated, because the small vessel lumina make it necessary to use the highest possible sensitivity setting. The artifact itself does not compromise the examination, since the unit must be switched to duplex mode anyway for the quantification of blood flow.

Angle correction. Before duplex is switched on, the direction of the vessel should be checked in the color-flow image. Because the breast is very mobile, it is easy to angle and rotate the transducer to obtain an optimum beam–vessel angle and optimum Doppler signal. Since the probe can be moved about freely on the breast surface, angle corrections are easily accomplished. This should be done before performing spectral analysis in the duplex mode, as it increases the accuracy of flow velocity measurements.

Duplex mode. The duplex mode is used for the quantification of blood flow (Fig. 37.**9**). No fixed recommendations on equipment settings can be made. The optimum PRF setting will depend on the flow velocity that is encountered. If the PRF is set too low, high-velocity flow will cause aliasing and the flow cannot be quantified. As a rule, the power output and gain should be set as high as possible, making certain that background noise does not occur. Generally a low wall filter setting can be used, since the transducer is held stationary during

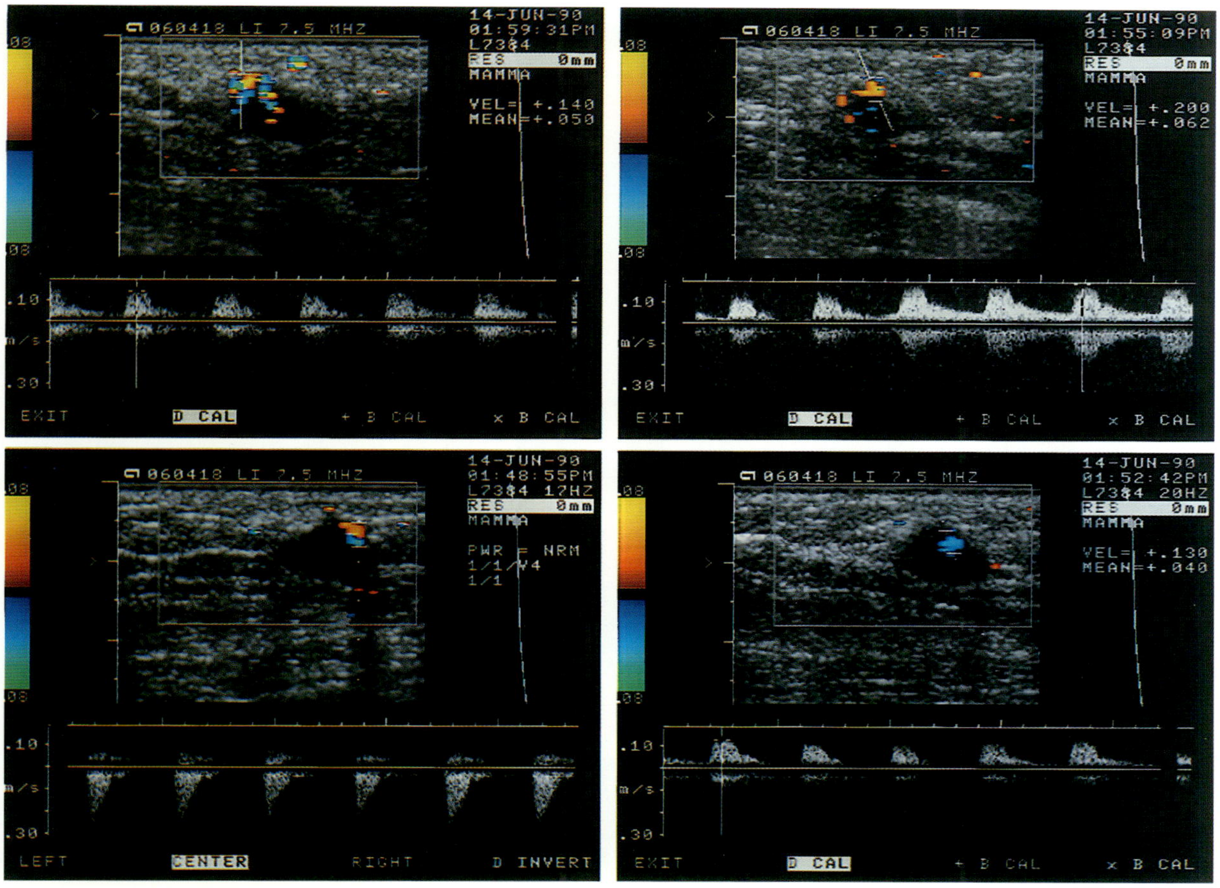

Fig. 37.**9** Color Doppler image of an invasive ductal carcinoma 12 mm in diameter with flow quantitation in the duplex mode (Acuson 128, 7 MHz).

spectral sampling and there should be very little wall motion in the parenchymal and tumor vessels. The ideal filter setting is between 25 and 50 Hz. The flow spectrum should be sampled at the site that yields the best color Doppler flow signal. This is generally the case when the most favorable beam–vessel angle has been achieved. The smaller the beam–vessel angle, the greater the accuracy of the measurement. When the angle exceeds approximately 40°, small inaccuracies begin to cause large changes in the cosine value, and the flow measurement becomes less precise. Generally, however, the examiner can adjust the transducer position to obtain a more favorable insonation angle and a very precise measurement.

Examination Technique

Patient position. The supine position is standard for breast ultrasound examinations, as this position flattens the breast tissue against the chest wall. Raising the arms and clasping the hands behind the neck tightens the pectoralis major muscle, further flattening the breast and holding it in place. Pressure from the transducer additionally flattens the glandular tissue. This is helpful because a thinner tissue layer significantly improves the sound conduction properties of the glandular tissue and permits the use of high ultrasound frequencies. A thicker tissue layer is more resistant to high-frequency ultrasound penetration, causing the image to appear darker. One difficulty is that the attenuation of sound energy cannot be seen in the color Doppler mode. Thus, when a tumor is located deep within the breast, the beam may be so strongly attenuated by the intervening tissues that blood vessels are not detected even though the tumor is heavily vascularized.

B-mode and color Doppler. Before a breast is examined with color Doppler, it should be systematically surveyed in overlapping planes by B-mode scanning. Color Doppler itself is not useful for whole-breast screening because the motion of the transducer causes color artifacts to appear throughout the tissues. After the breast has been completely surveyed by B-mode scanning, any suspicious foci should again be located and selectively scanned for abnormal vessels using color Doppler ultrasound. At this time the transducer should be moved very slowly so that color artifacts are minimized and can be distinguished from small, fluctuating vascular signals. When vessels are detected, it should be determined whether they are normal anatomical vessels. A main supply vessel runs from the axillary artery through the upper outer quadrant and increasingly branches as it approaches the center of the breast (Fig. 37.**10**). Another large supply vessel arises from the internal mammary artery and passes through the upper inner quadrant to the nipple region. Measurements in these vessels indicate high flow velocities. This can be confusing if there is a tumor nearby and the vessels are misinterpreted as tumor vessels. They should be positively identified by tracing the course of the vessels with ultrasound, rotating the transducer if necessary, before an actual detailed flow analysis is performed.

Tumor vessels. All vessels that are seen entering a tumor are classified as tumor vessels. It should be noted that the tissue in breast cancer is often very firm, and many vessels become oc-

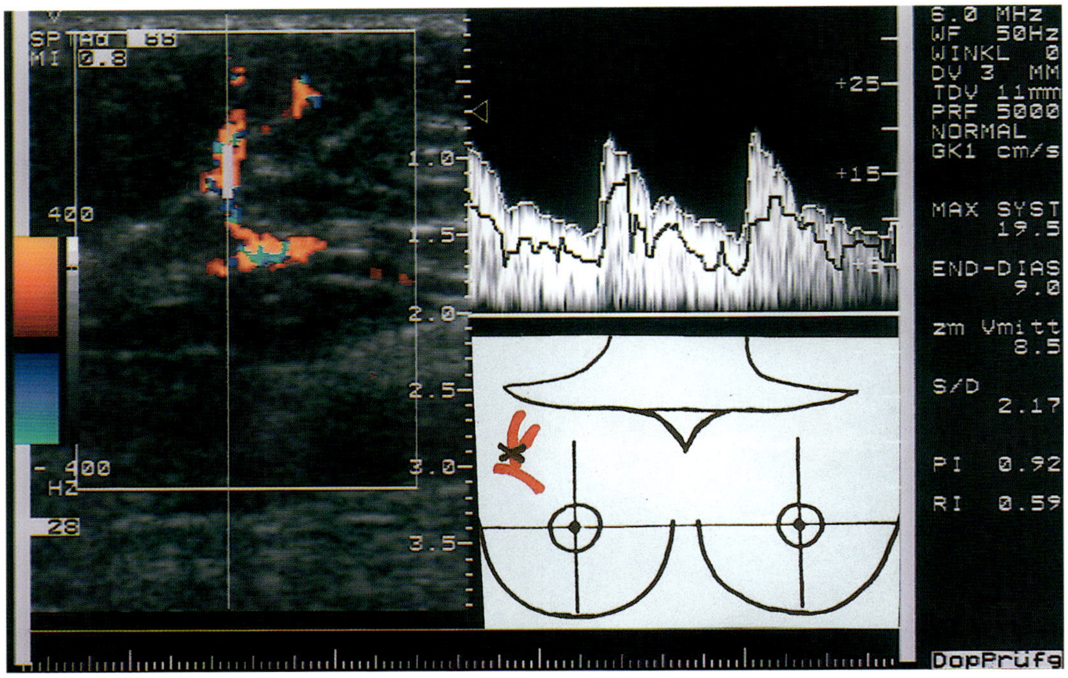

Fig. 37.**10** Color Doppler and duplex image of a main supply vessel in the breast parenchyma. It is important to recognize these normal anatomical breast vessels and distinguish them from tumor vessels to avoid false-positive findings (ATL UM9-HDI, 10 MHz).

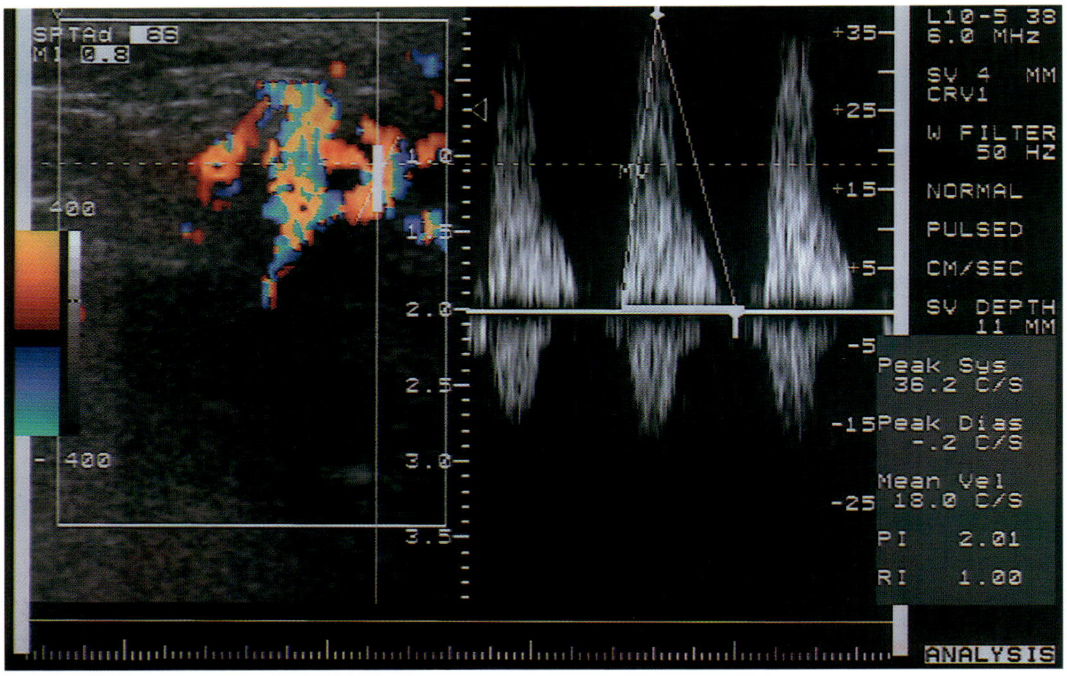

Fig. 37.**11** Color Doppler and duplex image of an invasive carcinoma 25 mm in diameter. The vessels are most conspicuous at the periphery of the tumor and radiate into the tumor mass. Inside the tumor, they collapse due to the surrounding tissue pressure. As a result, the diastolic flow is frequently low or interrupted.

cluded shortly after entering the tumor. Most tumor vessels are found directly adjacent to the tumor and in the peripheral part of the lesion. With malignancies, the vessels enter the tumor in a radial pattern (Fig. 37.**11**). Benign tumors such as fibroadenomas usually exhibit no vessels or, at most, one or two. It is common to find vessels in the normal parenchyma surrounding the lesion. Because fibroadenomas grow by expansion and compress the surrounding tissue, the vessels in that tissue tend to curve around the periphery of the tumor (Fig. 37.**12**). They are not classified as tumor-associated vessels.

Blood Flow Analysis

Tumor vessel count. The color Doppler mode permits a visual assessment of whether there is an abnormal increase of blood flow in the focal lesion. This qualitative assessment is subject to large errors, however, and it does not allow us to compare the results of different examiners. A more objective evaluation is needed. The easiest way to do this is by counting the tumor vessels in the color Doppler mode. In our prospective studies of color Doppler breast imaging, we have found the tumor vessel count to be a useful parameter for benign–malignant discrimination[10].

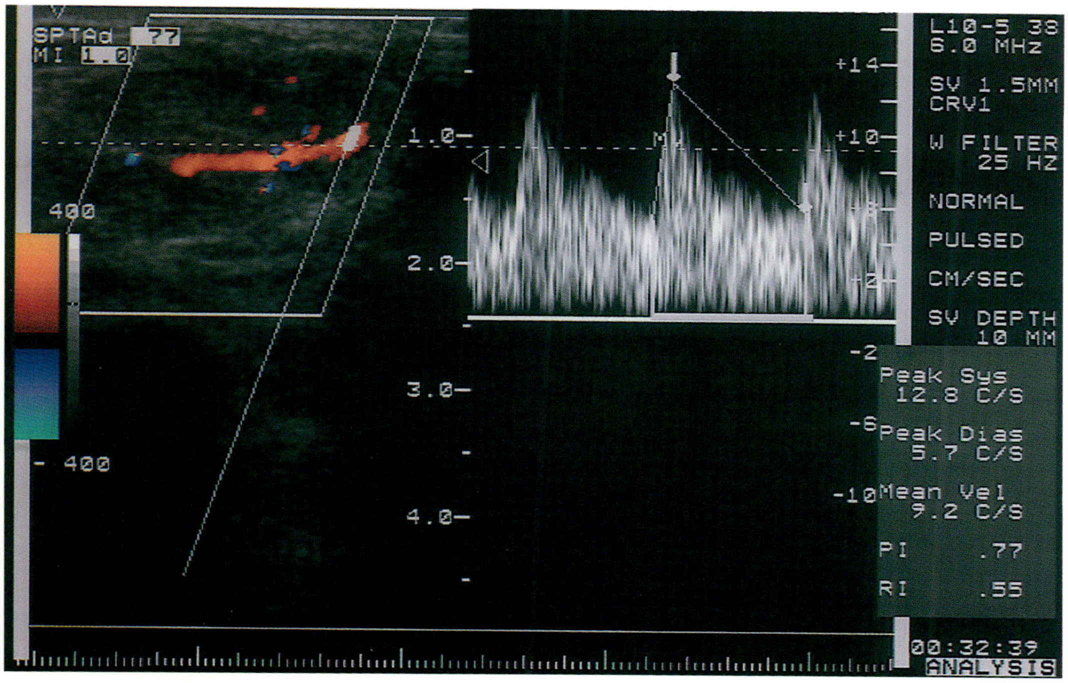

Fig. 37.**12** Color Doppler and duplex image of a fibroadenoma. Intratumoral vessels usually appear sparse or absent. Vessels often skirt the periphery of the tumor, as shown here.

Resistance index. Vascular findings can also be quantified by analyzing the Doppler spectrum. It is known from gynecological Doppler ultrasound that ovarian and endometrial carcinomas have low flow resistance, which is manifested by low RI values in the tumor vessels. High RI values in gynecological masses are suggestive of benignancy. This principle has not been confirmed in CW Doppler or color Doppler studies, however. On visual inspection of the image, the blood vessels in the peripheral part of the tumor show an apparent "cutoff" because they are usually constricted by the firm surrounding tumor (Fig. 37.**11**). Since the tumor vessels lack a muscular coat, the increased flow impedance is caused by pressure from the growing tumor tissue. As a result of this, breast cancers tend to show elevated RI values. We performed RI measurements for comparison with other study results. Because the blood vessels in a tumor are heterogeneous, it is necessary to measure all the vessels. Only a complete survey of tumor blood flow can reveal distinctive impedance features. Also, sampling many different spectra provides a database for analyzing extreme values, such as the minimum resistance index as an extreme value for all the tumor vessels.

Flow velocity. The measurement of flow velocity provides better information on tumor blood flow. The mean blood flow can be determined from the multiplicity of tumor vessels. The problem with this method is that the meticulous measurement of all tumor vessels will include many small vessels that are perfused by slow flow. When these vessels are averaged along with the few large mainstem vessels, they result in a relatively low mean value. Since different examiners differ in how carefully and completely they sample small tumor vessels, the calculated mean value is subject to large variations, which is why this parameter must be handled very carefully. An easier method is to identify the vessel with the highest flow velocity in the color Doppler image. Generally this is the vessel that is seen most clearly in the flow image and has the brightest color. The measurement of this extreme value in the duplex mode provides better discrimination between benign and malignant lesions than determining the mean value.

Flow velocity sum. Total blood flow is an even better benign-malignant discriminator. This cannot be determined by a precise volume flow calculation, but it is possible to add together all the blood flow velocities in all tumor vessels. This "flow velocity sum" is a relatively good indicator of the total blood flow in a tumor, and its calculation is not subject to measurement errors. Even when some small vessels with lower velocities are measured, this will cause little percentage change in the velocity sum if the sum is already high due to the presence of heavily perfused vessels. As a result, this measured value provides a good parameter for differential diagnosis. Its main disadvantage is that the measurement of all vessels can be very time-consuming.

Total blood flow is particularly useful in monitoring the effect of chemotherapy on tumor vascularity. Mean tumor blood flow is an unsatisfactory parameter for this purpose because when tumor blood flow initially decreases in response to chemotherapy, the smaller vessels with low flow velocities will disappear first while the mainstem vessels are still perfused. The paradoxical result of this process is that the mean blood flow may appear to increase even though the total blood flow is decreased.

Study Results

Table 37.**1** shows the principal flow data measured in our study. We counted an average of 12 tumor vessels in each of 82 breast malignancies, as compared with only two vessels in 176 benign breast lesions. The mean and maximum flow velocities and flow velocity sum were significantly higher in carcinomas than in benign lesions. The mean RI was also significantly higher: 0.74 in carcinomas versus

Table 37.**1** Color Doppler and duplex flow data for 82 breast carcinomas and 176 benign breast lesions

Parameter	Carcinomas mean value/SD	Benign lesions mean value/SD	Statistical significance (*p*)
Vessel count	12/11	2/2	< 0.0001
Mean flow	20/9	11/7	< 0.0001
Maximum flow	37/21	13/10	< 0.0001
Flow velocity sum	285/361	29/39	< 0.0001
Mean RI	0.74/0.10	0.68/0.10	< 0.0001
Minimum RI	0.62/0.11	0.63/0.10	> 0.1

0.68 in benign lesions, although the overlap of RI values between benign and malignant lesions is too large for accurate differentiation. Determination of the minimum RI for all the tumor vessels shows a slightly lower value for carcinomas (0.62) than for benign lesions (0.63), but there is an almost complete overlap of minimum RI values between benign and malignant lesions, so the difference is not significant. This demonstrates the variability of the flow data in tumor vessels and shows how important it is to include all the vessels in a flow data analysis.

As Table 37.**1** indicates, the vessel count by color Doppler and the flow velocity sum, as relative measures of total tumor blood flow, reflect the greatest differences between benign and malignant tumors. With the other parameters it will be noted that, while most of the differences are highly significant, the standard deviations show considerable overlap.

Discussion

Malignant tumors. In agreement with all previous study results, we found a marked increase of blood flow in malignant tumors compared with benign lesions. Color Doppler sonography combined with frequency spectrum analysis in the duplex mode provides an effective method for the quantitative differentiation of these flow patterns. Malignant tumors are characterized by many blood vessels with high flow velocities (Fig. 37.**11**). This feature aids in the differential diagnosis of circumscribed masses. It can also improve the detection of diffuse carcinomas, which are easy to miss in the B-mode image (Fig. 37.**13**).

Benign tumors and scars. Benign tumors usually exhibit little or no measurable blood flow on Doppler ultrasound. Of course, this depends strongly on the sensitivity of the equipment used (Figs. 37.**12**, 37.**14**). Doppler sonography can be particularly useful after breast-conserving operations (Fig. 37.**15**). Scars frequently appear as spiculated densities on mammograms. They also have suspicious ultrasound features, appearing as hypoechoic focal lesions with irregular, ill-defined margins and a posterior acoustic shadow. These cases demonstrate the significant value of a noninvasive modality that can improve the differentiation of scars and local recurrences in the previously operated, irradiated breast.

Proliferative and inflammatory lesions. Proliferative lesions are still a problem. It is difficult to distinguish among proliferative fibroadenomas, atypical proliferative fibrocystic change, mural proliferation in cysts, and mastitis (Figs. 37.**16**– 37.**18**). With their increased blood flow, proliferative changes often appear suspicious on color Doppler imaging. Dynamic MRI can be problematic for the same reason. Inflammatory changes can be quickly identified in most cases by treating the patient with antibiotics and dopamine agonists. Regression of increased blood flow should occur within a few days after treatment is begun, thereby distinguishing the condition from inflammatory carcinoma.

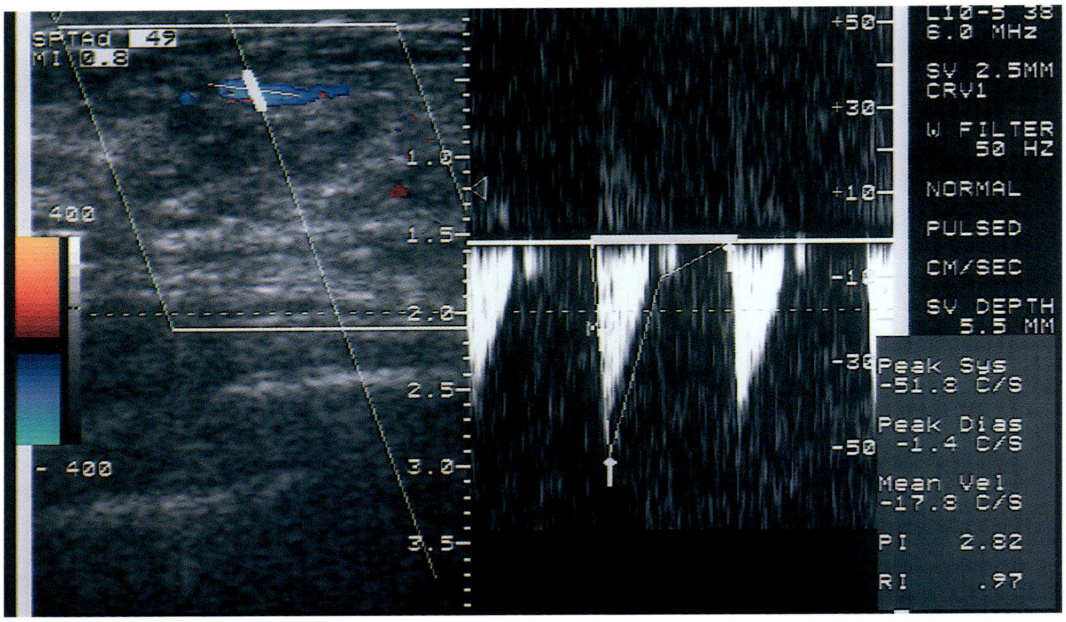

Fig. 37.**13** Diffusely infiltrating lobular carcinoma. The tumor is not visible in the color Doppler image. Dynamic examination with compression revealed an indurated area within the breast. This area contained several small vessels with high flow velocities, which suggested the correct diagnosis.

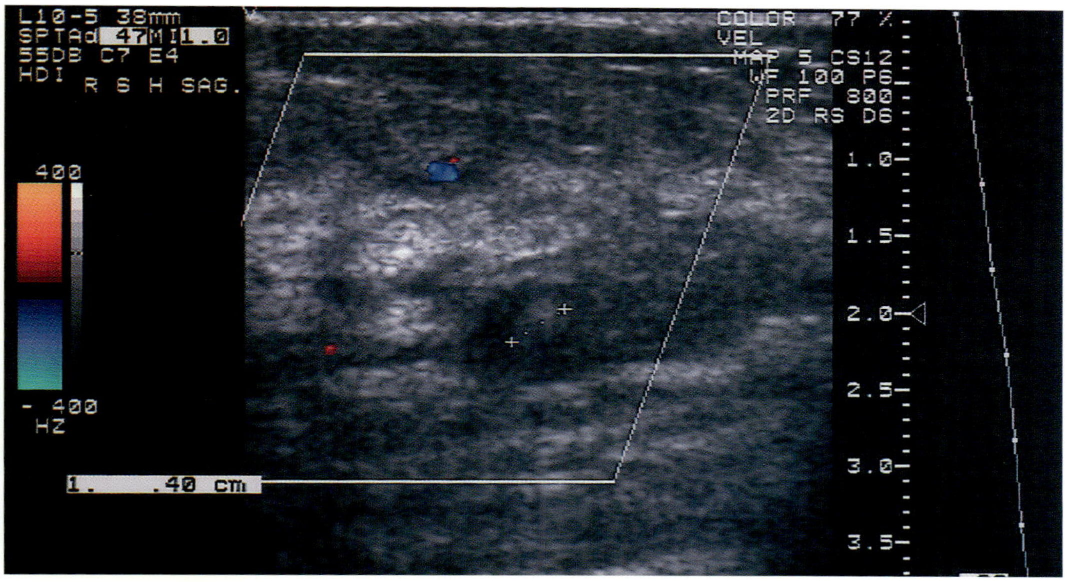

Fig. 37.**14** Color Doppler image of an intracystic papilloma shows no apparent abnormalities.

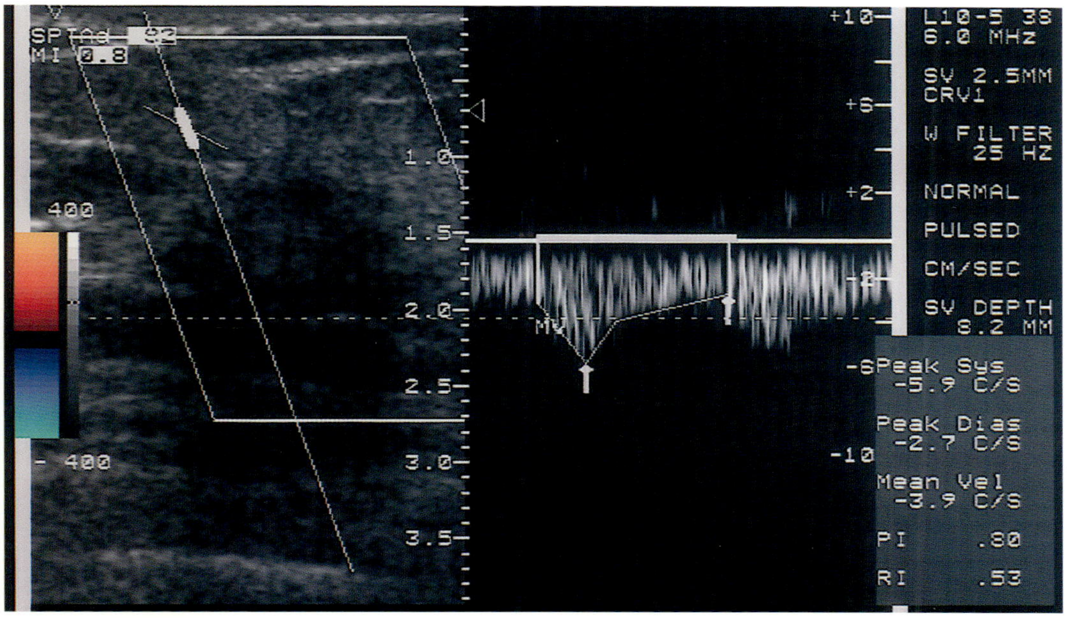

Fig. 37.**15** Hypoechoic scar after breast-conserving surgery for carcinoma. The scar appears avascular in the color Doppler image. A small vessel in the surrounding parenchyma has a typically low flow velocity of 5.9 cm/s.

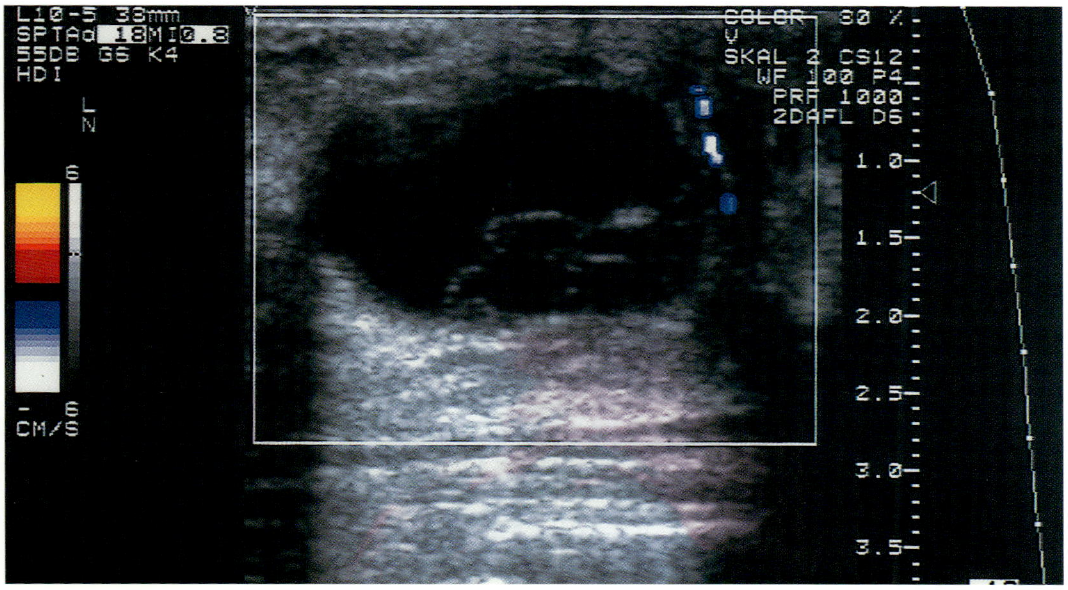

Fig. 37.**16** Mural proliferation in a septated cyst. Color Doppler demonstrates a circumscribed increase in blood flow.

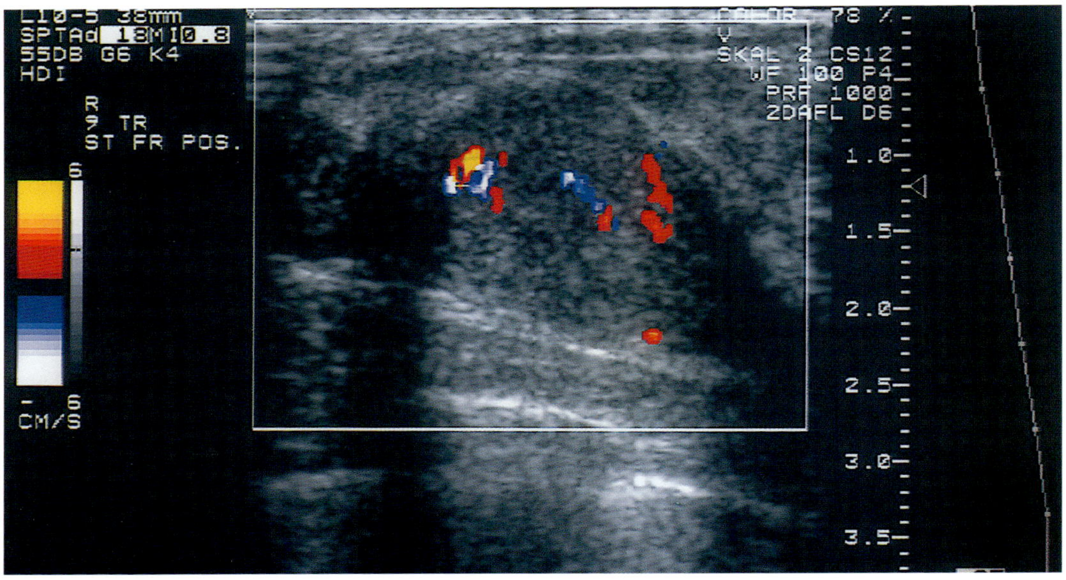

Fig. 37.**17** Color Doppler appearance of a proliferative, hypervascular fibroadenoma. FNA biopsy yielded proliferative cells. Because of the benign findings and for other medical reasons, an expectant approach was taken. Examination four months later showed a marked increase in tumor size, and local excision was performed. Histology confirmed a proliferative fibroadenoma.

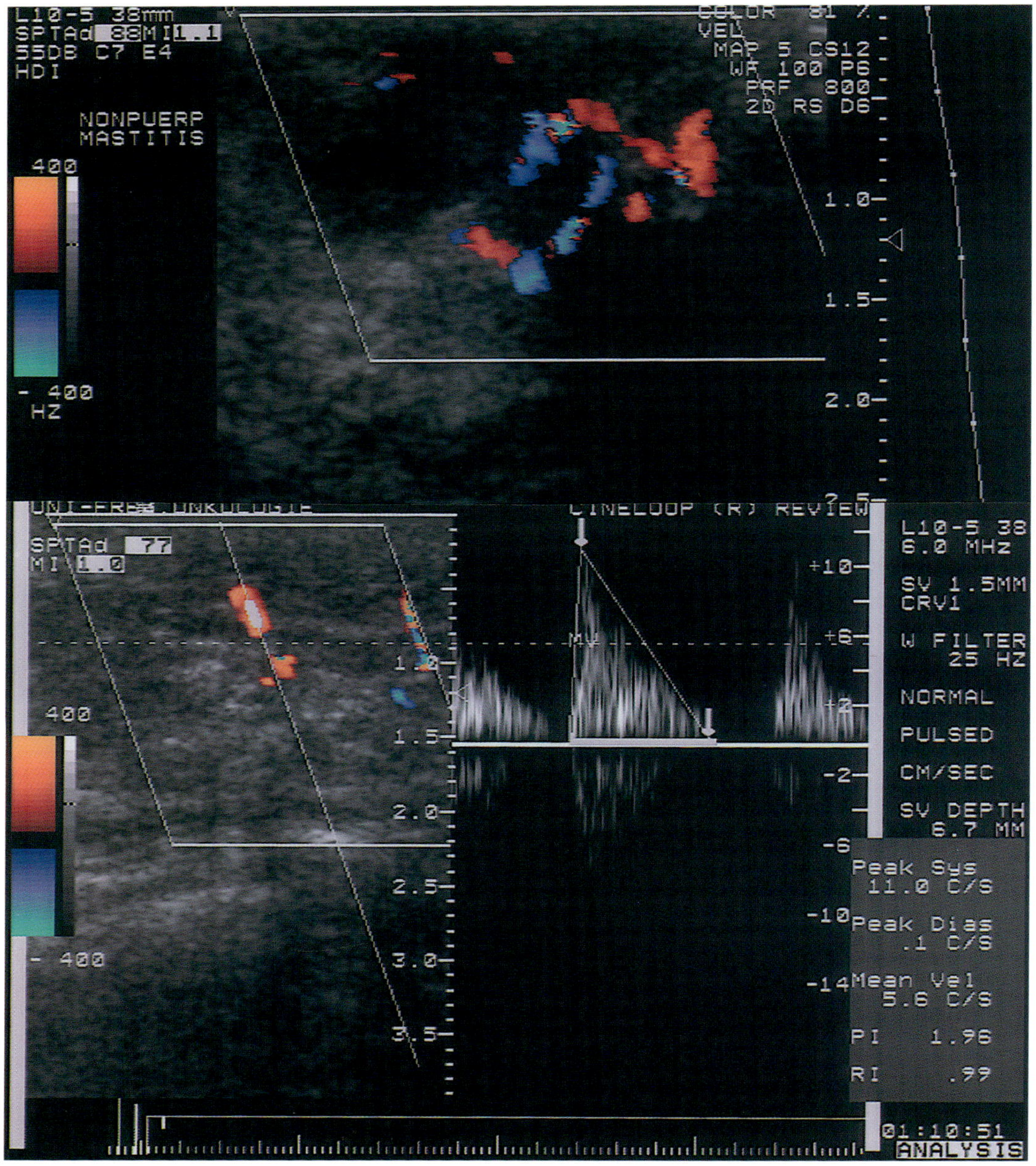

Fig. 37.**18** Nonpuerperal mastitis. The high vessel count by color Doppler suggested an inflammatory carcinoma, but the duplex scan showed relatively low arterial flow velocities with a predominance of venous vessels. Follow-up after anti-inflammatory therapy confirmed the benign nature of the disease.

Conclusions

Although most breast masses can be differentiated by Doppler sonography, this method should be used only by experienced examiners. Due to a lack of standardization of measuring techniques and equipment parameters, not all scanners are equally suitable. In addition, the measurements obtained with one machine cannot be applied to other machines. Even switching transducers can significantly alter the results. Moreover, the biological differences that characterize tumor tissue are not easily defined. In our experience, approximately 10% of malignant tumors exhibit low blood flow, and an examiner who does not properly consider all available diagnostic information may draw a false-negative conclusion. However, it is reasonable to expect that technical improvements, standardized protocols, and better training will help to establish Doppler flow assessment as a routine diagnostic test in the foreseeable future.

References

1 Cosgrove D, Bamber JC, Davey JB, McKinna JA, Sinnet HD: Color Doppler Signals from Breast Tumors. Radiology 176 (1990) 175
2 Folkman J: Tumour angiogenesis. Advan. Cancer Res. 19 (1974) 331
3 Jellins J: Combining imaging and vascularity assessment of breast lesions. Ultrasound Med. Biol. 14 (1988) 121
4 Kedar RP, Cosgrove DO, Smith IE, Mansi JL, Bamber JC: Breast carcinoma: measurement of tumor response to primary medical therapy with color Doppler flow imaging. Radiology 190 (1994) 825–830
5 Madjar H, Sauerbrei W, Schillinger H: Aktueller Stand der Dopplertechniken. In Gebhardt et al. (eds.): Ultraschalldiagnostik 89. Springer, Berlin 1989
6 Madjar H, Sauerbrei W, Münch S, Schillinger H: Continuous-Wave And Pulsed Doppler Studies Of The Breast: Clinical Results And Effect Of Transducer Frequency. Ultrasound Med. Biol. 17 (1991) 31
7 Madjar H, Vetter M, Prömpeler HJ, Wieacker P, Schillinger H: Untersuchungen zur normalen Vaskularisation der weiblichen Brust durch Doppler-Ultraschall. Ultraschall Med. 13 (1992) 171–177
8 Madjar H: Breast examinations with continuous wave and color Doppler. Ultrasound Obstet. Gynecol. 2 (1992) 215
9 Madjar H, Vetter M, Prömpeler HJ, Breckwoldt M, Pfleiderer A: Doppler measurement of breast vascularity under pharmacological treatment of benign breast disease. J. Reproduct. Med. 38(12) (1993) 935–940
10 Madjar H, Prömpeler HJ, Sauerbrei W, Wolfarth R, Pfleiderer A: Color Doppler flow criteria of breast lesions. Ultrasound Med. Biol. 20(9) (1994) 849–858
11 Minasian H, Bamber JC: A preliminary assessment of an ultrasonic Doppler method for the study of blood flow in human breast cancer. Ultrasound Med. Biol. 8 (1982) 357
12 Natsuki S, Kosuke Y: A study on the vascular proliferation in tissues around the tumor in breast cancer. Jpn. J. Surg. 18 (1988) 235
13 Weidner R, Semple JP, Welch WR: Tumor angiogenesis and metastasis-correlation in invasive breast carcinoma. New Engl. J. Med. 324 (1991) 1
14 Wells PNT, Halliwell M, Skidmore R, Webb AJ, Woodcock JP: Tumour detection by ultrasonic Doppler blood-flow signals. Ultrasonics 15 (1977) 231

38 Usefulness of the Minimum Resistance Index in the Benign–Malignant Discrimination of Breast Tumors

C. Villena-Heinsen, M. Friedrich, J. König, and W. Schmidt

Flow Resistance in Malignant Breast Tumors

Number of sampled vessels. Several Doppler flow studies[1,4,5,7] have shown that malignant breast tumors have greater blood flow on average than benign tumors and normal breast tissue. Significantly more vessels could be sampled from malignant tumors, first using a high-frequency continuous-wave (CW) Doppler pencil probe and later using duplex equipment and color Doppler sonography. Despite the consistent results obtained with different technologies, all of these methods involve an indirect assessment of the number of blood vessels actually present. Because the vessels are not imaged in three dimensions, some vessels are not sampled at all while others are inadvertently sampled more than once due to their course. Factors such as the sensitivity of the Doppler equipment, examination time, transducer pressure on the breast, and the patience and experience of the examiner have a systematic influence in these examinations.

Vascular structure and flow resistance. As we know today, the blood flow signals that are sampled with Doppler ultrasound originate from relatively large-caliber supply vessels. Several histopathological studies[2,3] have described typical features of the neovascularity of malignant tumors. It is common, for example, to find arteriovenous shunts and sites where the capillaries lack a muscular coat. These structural differences from the neoangiogenesis of benign tumors result in a lower impedance to blood flow. Because of this, the vessels supplying a malignant tumor should also exhibit low flow resistance, and all of the resistance indices—the minimum, mean, and maximum resistance index (RI)—should be decreased in proportion to the number of tumor vessels that are present.

Number of tumor vessels. Our own study results[6,7] indicate, however, that as the number of tumor vessels increases, the minimum RI progressively declines while the maximum RI increases. The mean RI remains unchanged. Two conclusions can be drawn from this:

1 Neither the minimum RI nor the maximum RI is a satisfactory parameter for characterizing the actual flow resistance of a tumor. They are the end products of a minimization and maximization effect that is proportional to the number of tumor vessels. This means that the likelihood of determining the lowest minimum RI and the highest maximum RI increases when more vessels are present.
2 If a resistance index is to be used in characterizing breast tumors, the best choice would be the mean RI.

That the minimum (and maximum) RI is influenced by the number of vessels sampled is a phenomenon observed not just in breast tumors but also in healthy breast parenchyma.

Authors' Studies

Patients and Methods

Study design. A total of 114 women with a focal breast lesion detectable by preoperative ultrasound underwent a conventional B-mode examination with assessment of tumor size followed by detailed color Doppler scanning of both breasts. First the tumor area and the healthy breast tissue were divided into four quadrants per breast and examined with color Doppler ultrasound. The tumor area was scanned thoroughly and systematically with the goal of sampling as many tumor vessels as possible. To save time, a maximum of two vessels were sampled in each quadrant. Flow signals located more than 2 cm from the tumor margin were assigned to the corresponding quadrant. Figure 38.**1** explains the method used to calculate the resistance indices for the tumor area, for each quadrant, and for the healthy breast tissue. Two different systems were used to calculate the different resistance indices in healthy breast tissue.

Equipment and settings. The examinations were performed with the model 128 XP10 ultrasound system from Acuson (Mountain View, CA, USA). A 7 MHz linear transducer with a 38 mm acoustic window was used. For color Doppler, the transducer automatically switched its operating frequency to 5 MHz. The color Doppler settings were adjusted to optimize the detection of slow flows and weak color signals in the vascular network. Color Doppler sensitivity was set just below the noise threshold to minimize artifacts, and transducer movements were limited to slow angulation. For maximum signal yield, care was taken to maintain a constant transducer pressure throughout the examination.

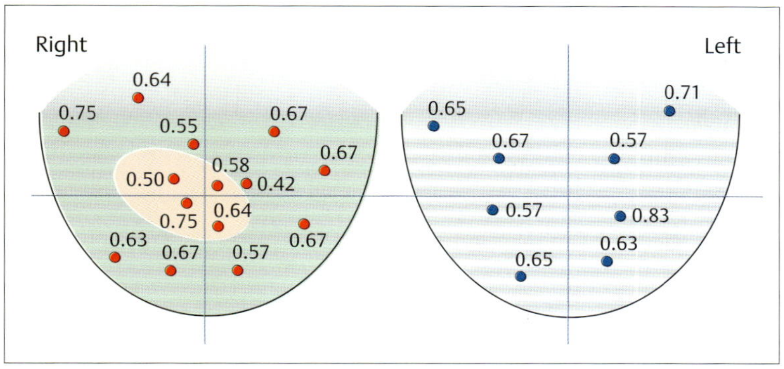

Fig. 38.**1** Methods used to calculate the various resistance indices.

Right breast. Retroareolar mass in a 56-year-old woman, extending predominantly into the upper outer quadrant, classified histologically as invasive ductal carcinoma stage pT3, pN12/18, M0, G3, ER+, PR+, 20% Ki-67-positive cells, aneuploid. Six vascular segments were sampled in the tumor area. The $RI_{min} = 0.42$, the $RI_{mean} = 0.57$, and the $RI_{max} = 0.75$. In the upper inner quadrant, the RI_{min}, RI_{mean}, and RI_{max} were equal to 0.67. RI_{min}, RI_{mean}, and RI_{max} were determined for each quadrant. For example, the values for the right upper outer quadrant were: $RI_{min} = 0.64$, $RI_{mean} = 0.70$, and $RI_{max} = 0.75$. The values for the upper inner quadrant were: $RI_{min} = RI_{mean} = RI_{max} = 0.67$.

Left breast. The resistance indices RI_{min}, RI_{mean}, and RI_{max} in healthy breast tissue were calculated in two ways. (1) Mean values for the RI_{min}, RI_{mean}, and RI_{max} determined for each quadrant were calculated for all eight quadrants (in this case: $RI_{min} = 0.62$, $RI_{mean} = 0.66$, and $RI_{max} = 0.70$). (2) Of all the resistance indices determined (in this case 16), the lowest RI was defined as RI_{min} and the highest RI as RI_{max}, and the RI_{mean} was calculated from all 16 values ($RI_{min} = 0.57$, $RI_{mean} = 0.66$, $RI_{max} = 0.83$).

Results and Discussion

Number of vascular signals. One to nine vascular segments were sampled from the area of malignant tumors and one to six vascular segments from benign tumors. Figure 38.**2** shows the relative frequency of malignant and benign tumors plotted against the number of vessels in the tumor area.

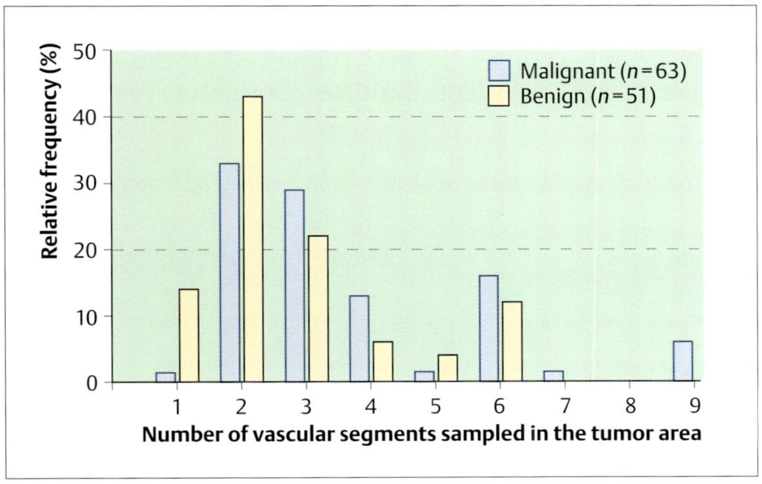

Fig. 38.**2** Percentage frequency of benign and malignant tumors as a function of the number of vascular segments sampled in the tumor area (61 malignant tumors = 100%, 53 benign lesions = 100%).

Table 38.**1** Mean-value comparison of the number of vascular segments sampled in the tumor area, stratified by menopausal status, benignancy, and malignancy

Nature and number of lesions	Vascular segments	*p*-value
All ($n = 114$)	3.3 (± 1.8)	
Benign ($n = 51$)	2.8 (± 0.2)	$p = 0.008$
Malignant ($n = 63$)	3.6 (± 0.2)	
Premenopausal, benign ($n = 36$)	2.9 (± 0.3)	$p = 0.14$
Premenopausal, malignant ($n = 23$)	3.4 (± 0.4)	
Postmenopausal, benign ($n = 15$)	2.5 (± 0.3)	$p = 0.03$
Postmenopausal, malignant ($n = 40$)	3.7 (± 0.3)	

Standard deviation in parentheses

On average, 3.3 vascular signals were detected in each tumor area. The vessels were more numerous in malignant tumors than in benign tumors. This difference was statistically significant only in postmenopausal patients, however (Table 38.**1**).

Dependence of RI_{min} and RI_{max} on the Number of Vessels in the Tumor Region

Presumably, an increasing number of blood flow signals would correlate with an increase in blood flow and a decrease in vascular resistance (Figs. 38.**3**, 38.**4**). It is reasonable to expect that all the resistance indices—minimum, mean, and maximum—would show a more or less parallel decline. Paradoxically, an increase in the number of sampled tumor vessels was associated with a statistically significant decrease in RI_{min} ($p = 0.004$) and a statistically significant increase in RI_{max} ($p = 0.03$). Meanwhile, the RI_{mean} remained constant and was the only blood flow parameter that showed no dependence on the number of vessels sampled.

Figure 38.**5** shows the difference between the resistance indices in the tumor area and the mean indices for the eight breast quadrants for RI_{min} (left), RI_{mean} (center) and RI_{max} (right) as a function of the number of vascular segments sampled in the tumor area. Tumor areas with more vessels display the highest RI_{max} and lowest RI_{min}. These paradoxical results force us to conclude that neither RI_{min} nor RI_{max} reflects the resistance values that are actually present. They are, rather, the products of a minimization and maximization process in which the values are critically determined by the number of potential measurements (vessels).

Dependence of RI_{min} and RI_{max} on the Number of Vessels in Healthy Breast Tissue

This conclusion is supported by a subsequent analysis in which a significantly lower RI_{min} was found in the tumor area—regardless of menopausal status or whether the tumors were malignant or benign—compared with the mean value of RI_{min} calculated from the eight RI_{min} values in the eight healthy breast quadrants ($p = 0.0001$). On average, 3.3 vessels were sampled in the tumor area and an RI_{min} of 0.62 ± 0.11 SD was

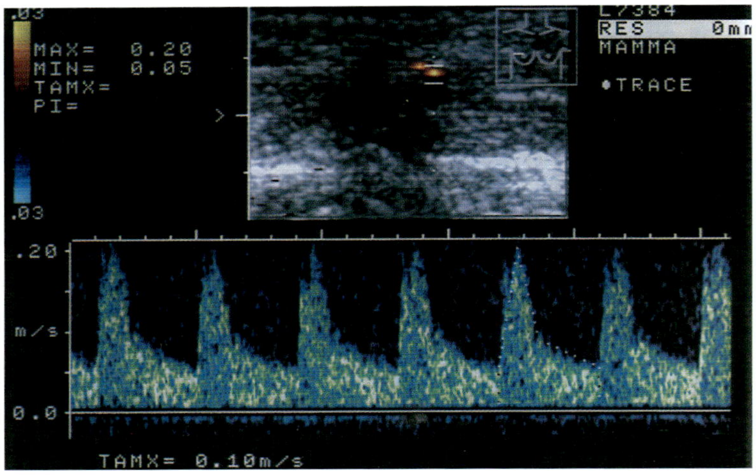

Fig. 38.**3** Hypovascular breast carcinoma with a RI of 0.75.

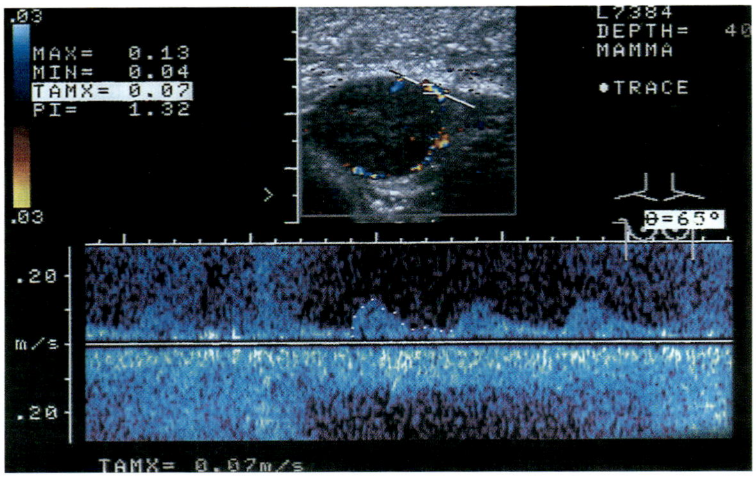

Fig. 38.**4** Hypervascular breast carcinoma with a RI of 0.69. Comparing the two postmenopausal women (who were not receiving HRT) in Figs. 38.**3** and 38.**4**, we note that increasing blood flow is associated with a proportional decrease in RI.

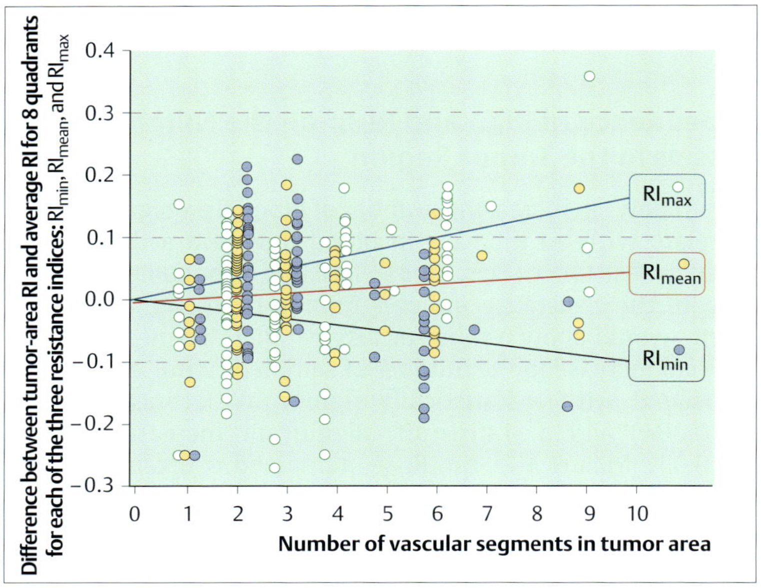

Fig. 38.**5** Differences between the resistance indices in the tumor area and the mean indices for the eight healthy quadrants (positive difference above the zero axis, negative difference below the zero axis) for the parameters RI_{min} (left), RI_{mean} (center) and RI_{max} (right), plotted against the number of vascular segments sampled in the tumor area.

measured in that area. An average of 1.5 vessels were sampled in each of the eight healthy quadrants. The mean value of the RI_{min} of the eight quadrants was 0.65 ± 0.08 SD.

When we compare the tumor region with the lowest RI_{min} of all 8–16 sampled vessels ($RI_{min} = 0.54 \pm 0.1$ SD), we find a highly significant but opposite difference. This means that—analogous to our analysis in the tumor area—the calculation of the lowest RI_{min} and the highest RI_{max} in healthy breast tissue again depends on the number of vessels that are sampled.

This also means that the RI_{min} is significantly lower than in the tumor area when 8–16 vessels (12 on average) can be sampled in the healthy breast tissue. This weakens the basic hypothesis that tumors constitute a low-resistance system, and it implies that the RI_{min} in the tumor area would also reach a value of 0.54 if an average of 12 tumor vessels could be sampled there. This analysis is outlined in Table 38.**2**, which includes a similar analysis for RI_{mean} and RI_{max}.

No Benign–Malignant Discrimination with RI_{min} and RI_{max}

We postulate that a similar number of vessels result in similar values for RI_{min}. This is further supported by the following analysis: of the 114 patients evaluated, one or two vessels were

Table 38.**2** Mean-value comparison of the parameters RI_{min}, RI_{mean}, and RI_{max} in the tumor area and in healthy breast tissue, calculated using the two methods described

Parameters	Mean value for 8 quadrants		Tumor area		all resistance indices (8 quadrants)
Mean number of vessels	1.5		3.3		12
RI_{min} ($n = 114$)	0.65 (± 0.08)		0.62 (± 0.11)		0.54 (± 0.1)
		$p = 0.0001$		$p < 0.0001$	
RI_{mean} ($n = 114$)	0.68 (± 0.07)		0.69 (± 0.09)		0.68 (± 0.07)
		$p = 0.13$			
RI_{max} ($n = 114$)	0.70 (± 0.07)		0.75 (± 0.1)		0.80 (± 0.07)
		$p < 0.0001$		$p < 0.0001$	

Standard deviation in parentheses

sampled from the tumor area in 51 cases. The average RI_{min} in these cases, which included both benign and malignant lesions, was 0.65 (± 0.11 SD). An average of 1.6 vessels were sampled. The mean RI_{min} for the eight quadrants in these cases was 0.66 (± 0.08 SD). On average, 1.5 vessels were sampled per healthy quadrant. With an almost equal number of sampled vessels, no statistically significant differences were found between the calculated values of RI_{min} ($p = 0.67$). Consequently, RI_{min} appears to depend chiefly on the number of vessels examined and not on the type of tissue examined (tumor or healthy tissue). Meanwhile, calculation of the RI_{min} from all 8–16 sampled vessels yielded a significantly lower value of 0.55 (± 0.01 SD).

Since it could be supposed that tumors with a small number of vessels (one or two) may have a higher RI_{min} than tumors with more than two vessels, the first two resistance indices measured in all 114 of the tumors were analyzed in an analogous way. The results were essentially equal. This proves conclusively that RI_{min} is dependent on the number of vessels sampled (Table 38.**3**).

This dependence on the number of measurements also holds for the parameter RI_{max}. It is clear, then, that RI_{min} and RI_{max} are not suitable blood flow parameters for a representative characterization of the vascular resistance in breast tumors. Accordingly, these parameters should not be used for the benign–malignant discrimination of breast lesions.

Table 38.**3** Mean values of RI_{min} in healthy breast tissue (calculated by different methods) compared with mean values in tumor areas having one or two vessels and with the first two vessels sampled in all 114 tumors

1 or 2 vessels in the tumor area ($n = 51$)	**Mean value for 8 quadrants**	**Tumor area with 1 or 2 vessels**	**All RI (8 quadrants)**
Mean number of vessels	1.5	1.6	12
RI_{min}	0.66 (± 0.8)	0.65 (± 0.11)	0.55 (± 0.01)
	$p = 0.67$	$p < 0.00001$	
All tumors ($n = 114$)	**Mean value for 8 quadrants**	**First 2 RIs measured in tumor area**	**All RI (8 quadrants)**
Mean number of vessels	1.5	2.0	12
RI_{min}	0.65 (± 0.01)	0.64 (± 0.01)	0.54 (± 0.01)
	$p = 0.63$	$p < 0.00001$	

Standard deviation in parentheses

Summary

The results of this study prove the existence of a purely mathematical minimization and maximization effect that substantially influences the calculation of RI_{min} and RI_{max}. The likelihood of determining the very lowest RI_{min} (and the very highest RI_{max}) is critically determined by the number of vessels that are sampled. These results call into question the validity of RI_{min} (and RI_{max}) in the assessment of true vascular resistance. If any resistance index is to be used in characterizing breast tumors, our analysis indicates that only the RI_{mean} should be considered.

References

1 Burns PN, Halliwell M, Wells PNT, Webb AJ: Ultrasonic doppler studies of the breast. Ultrasound in Med. Biol. 8 (1982) 127–143

2 Folkman J: How is blood vessel growth regulated in normal and neoplastic tissue? Cancer Res. 46 (1986) 467–473

3 Less JR, Skalak TC, Sevick EM, Jain RK: Microvascular architecture in a mammary carcinoma: a branching patterns and vessel dimensions. Cancer Res. 51 (1991) 265–273

4 Madjar H, Prömpeler H, Wolfahrt R, Bauknecht T, Pfleiderer A: Farbdopplerflußdaten von Mammatumoren. Ultraschall Med. 15 (1994) 69–76

5 Minasian H, Bamber JC: A preliminary assessment of an ultrasonic doppler method for the study of blood flow in human breast cancer. Ultrasound Med. Biol. 8 (1982) 357–364

6 Villena-Heinsen C, Mink D, Ertan AK, Holländer M, Schmidt W: Bewertung der Aussagekraft der Farb- und Spektral-Doppler-Sonographie bei Mammatumoren. In Schmidt W (ed.): Jahrbuch der Gynäkologie und Geburtshilfe (1995/1996). Biermann, Zülpich 1996, 121–136

7 Villena-Heinsen C, König J, von Tongelen B et al.: Validity of the minimal Resistance Index for discrimination between benign and malignant Breast Tumours. Eur. J. Ultrasound 7 (1998) 189–193

39 Effect of Menopausal Status and Hormone Replacement on Therapy Resistance Indices and Blood Flow Velocities in Breast Tumors

C. Villena-Heinsen, I. Tossounidis, and W. Schmidt

Menopausal Status and Benign–Malignant Tumor Discrimination

Patients in Doppler studies. In the relatively few Doppler ultrasound studies that have been published on breast tumors, the focus has been on testing the ability of the various techniques (CW Doppler, duplex scanning, color Doppler) to discriminate between benign and malignant lesions. For the most part, these studies have disregarded the potential effects of menopausal status and hormone replacement therapy on blood flow parameters. One group of authors[1] mentioned menopausal status but did not make a systematic comparison of premenopausal and postmenopausal women. Another group that made a side-to-side comparison between the tumor and contralateral breast found that the average resistance indices and S/D ratio in premenopausal women were lower than in postmenopausal women[2]. Because these differences did not turn out to be statistically significant, no further importance was attached to menopausal status. Subsequent studies on the discriminating potential of Doppler ultrasound did not take into account menopausal status or hormone replacement therapy.

Divergent results. Generally, then, Patient groups of different ages and menopausal status have been compared with regard to benign–malignant tumor discrimination. This helps to explain why such divergent results have been reported. As our own study results indicate[3], patients with malignant tumors tend to be in an older, predominantly postmenopausal age group, whereas patients with benign tumors are mostly younger, premenopausal women. Additionally, a significant effect of hormone replacement is observed in postmenopausal women.

Because neither menopausal status nor hormone replacement was taken into account in previously published studies, we must question the validity of comparing different patient groups with benign and malignant tumors, as well as the validity of the results.

In this chapter we analyze the effects of menopausal status and hormone replacement on the various flow resistance indices and blood flow velocities that are measured in healthy breast tissue and in breasts with benign and malignant tumors.

Authors' Studies

Patients and Methods

Study design. A total of 114 women were examined. B-mode ultrasound with the assessment of tumor size was followed by the color Doppler imaging of as many tumor vessels as possible and Doppler spectral sampling. Next, all of the healthy breast tissue was examined. For this purpose each breast was divided into four quadrants. After the vessels were defined by color-flow imaging, Doppler velocity spectra were recorded from a maximum of two vessels in each quadrant.

Parameters and velocities. The 114 patients were subdivided into three groups: premenopausal (PRE), postmenopausal (POST), and postmenopausal with hormone replacement therapy (POST HRT). The following parameters were calculated for the tumor area and for each of the eight quadrants in the healthy breast tissue: RI_{min}, RI_{mean}, RI_{max}, PI_{mean}, mean S/D ratio, and the following velocities:

- Maximum peak systolic velocity (peaksys V_{max})
- Mean peak systolic velocity (peaksys V_{mean})
- Minimum peak systolic velocity (peaksys V_{min})
- Mean average velocity over a whole cycle (average V_{mean})
- Mean end-diastolic velocity (enddias V_{mean})

The velocity measurements were not corrected for insonation angle. This step was omitted to save time, as most of the vessels were very small and would have been difficult to define in longitudinal section.

For each parameter, a mean value was taken from the corresponding eight values for the quadrants to represent the healthy breast tissue. The Mann–Whitney *U* test was used for statistical analysis.

Patient ages. The average age of the patient population as a whole was 55 years (21–87). Histological evaluation showed that 63 of the patients had a carcinoma and 51 had a benign lesion. The average age of the carcinoma patients was 60 years (33–87). The average age of the benign cases was 49 years (21–84). The age difference between these two groups was statistically significant ($p = 0.0001$).

Menopausal status and hormone replacement therapy. Thirty-five of the patients were premenopausal, 24 were postmenopausal on hormone replacement therapy (HRT), and 55 were postmenopausal without HRT. Both groups also differed significantly in their menopausal status ($p = 0.0001$). In the group with benign tumors, 49% of the women were pre-

Table 39.**1** Comparison of the benign and malignant tumor groups according to menopausal status (PRE = premenopausal; POST HRT = postmenopausal on hormone replacement therapy; POST = postmenopausal without hormone replacement therapy)

Menopausal status	Number	Benign lesions (%)	Menopausal status	Number	Malignant tumors (%)
PRE	25	49.0	PRE	10	15.9
POST HRT	11	21.6	POST HRT	13	20.6
POST	15	29.4	POST	40	63.5
Total	51	100.0	Total	63	100.0

menopausal ($n = 25$), 21.6% were postmenopausal with HRT ($n = 11$), and 29.4% were postmenopausal without HRT ($n = 15$). The group with malignant tumors was structured as follows: 15.9% premenopausal ($n = 10$), 20.6% postmenopausal with HRT ($n = 13$), and 63.5% postmenopausal ($n = 40$). Table 39.**1** shows a breakdown of the benign and malignant groups according to menopausal status.

Results

Effect of Menopausal Status on Resistance Indices in Healthy Breast Parenchyma

Patient groups. Healthy breast parenchyma was represented by the mean values of the minimum, mean, and maximum RI of the eight quadrants. Since we were dealing with healthy tissue, the effect of menopausal status was investigated regardless of whether the particular case was benign or malignant. As Table 39.**2** shows, the patients were divided into three groups: premenopausal, postmenopausal with HRT, and postmenopausal. The minimum, mean, and maximum RI were calculated for each patient group along with the standard deviation and range of values. The three flow-resistance parameters differed from one another but showed a congruent relationship.

RI_{min}. As an example, we will review our analysis of RI_{min}—the most widely used parameter for determining flow resistance. The mean values of RI_{min} were 0.61 in the premenopausal patients, 0.64 in the postmenopausal patients on HRT, and 0.68 in the postmenopausal patients. Both postmenopausal groups showed a higher RI_{min} than the premenopausal group, but the highest mean value within the postmenopausal group was found in the women who were not receiving HRT.

Pooled groups. Table 39.**3** shows the statistically significant differences between the three groups with regard to the three flow-resistance parameters that were analyzed. It can be seen that both the premenopausal patients and the postmenopausal patients on HRT differed significantly from the postmenopausal patients without HRT. Moreover, there were no significant differences between the premenopausal patients and the postmenopausal patients on HRT. These results demonstrate that premenopausal patients and postmenopausal patients on HRT have very similar resistance indices. We therefore pooled both groups and compared them as a unit with the group of postmenopausal patients without HRT in further analyses.

Table 39.**2** Mean values, standard deviations, and ranges for the parameters RI_{min}, RI_{mean}, and RI_{max} in the healthy breast tissue of the patient groups: premenopausal (PRE), postmenopausal on hormone replacement therapy (POST HRT), and postmenopausal without hormone replacement therapy (POST)

Menopausal status	RI_{min}	RI_{mean}	RI_{max}
PRE ($n = 34$)	0.61 ± 0.07 0.50–0.78	0.64 ± 0.06 0.53–0.78	0.66 ± 0.06 0.53–0.78
POST HRT ($n = 24$)	0.64 ± 0.07 0.50–0.75	0.66 ± 0.06 0.54–0.75	0.68 ± 0.06 0.54–0.77
POST ($n = 54$)	0.68 ± 0.07 0.55–0.85	0.71 ± 0.06 0.59–0.85	0.73 ± 0.06 0.59–0.85

Table 39.**3** Statistical comparison based on the mean values listed in Table 39.**2** (PRE = premenopausal; POST HRT = postmenopausal on hormone replacement therapy; POST = postmenopausal without hormone replacement therapy)

Menopausal status	RI_{min}	RI_{mean}	RI_{max}
PRE vs. POST HRT	$p = 0.12$	$p = 0.26$	$p = 0.30$
PRE vs. POST	$p < 0.0001$	$p = 0.0004$	$p = 0.0001$
POST HRT vs. POST	$p = 0.01$	$p = 0.07$	$p = 0.004$

Effect of Menopausal Status on Resistance Indices in the Tumor Area

Tumor-specific effect. The same type of analysis was done for the tumor area as for the healthy breast tissue. The results are shown in Tables 39.**4** and 39.**5**. They basically confirm all of the previous findings, except that the statistical comparison between the postmenopausal patients with and without HRT shows only minor differences compared with the results in healthy breast tissue. All the resistance indices in the postmenopausal patients are higher than in the postmenopausal patients on HRT. They show different degrees of statistical significance, however: RI_{max} is substantially higher while RI_{mean} is slightly higher and RI_{min} is not significant. This difference compared with healthy breast parenchyma could be related to a tumor-specific effect in postmenopausal women (with or without HRT).

Regardless of whether the tumors were benign or malignant, these results prove the statistically significant effects of menopausal status and hormone replacement on blood flow resistance.

Table 39.**4** Mean values, standard deviations, and ranges for the blood flow parameters RI_{min}, RI_{mean}, and RI_{max} measured in the tumor area of the patient groups: premenopausal (PRE), postmenopausal on hormone replacement therapy (POST HRT), and postmenopausal without hormone replacement therapy (POST)

Menopausal status	RI_{min}	RI_{mean}	RI_{max}
PRE (*n* = 34)	0.58 ± 0.10 0.37 – 0.78	0.65 ± 0.09 0.52 – 0.86	0.71 ± 0.10 0.45 – 0.95
POST HRT (*n* = 25)	0.62 ± 0.13 0.37 – 0.89	0.68 ± 0.10 0.43 – 0.89	0.72 ± 0.10 0.55 – 0.90
POST (*n* = 55)	0.65 ± 0.10 0.42 – 0.81	0.72 ± 0.07 0.50 – 0.85	0.79 ± 0.08 0.50 – 0.96

Table 39.**5** Mean-value comparison of the groups in Table 39.**4** (PRE = premenopausal; POST HRT = postmenopausal on hormone replacement therapy; POST = postmenopausal without hormone replacement therapy)

Menopausal status	RI_{min}	RI_{mean}	RI_{max}
PRE vs. POST HRT	$p = 0.28$	$p = 0.26$	$p = 0.79$
PRE vs. POST	$p = 0.0055$	$p = 0.0004$	$p = 0.0002$
POST HRT vs. POST	$p = 0.25$	$p = 0.07$	$p = 0.004$

Table 39.**6** Comparison of the patient groups: premenopausal (PRE + POST HRT) and postmenopausal (POST) based on the mean values of the blood flow parameters, calculated for healthy breast parenchyma (peaksys V_{max} = maximum peak systolic velocity, average V_{mean} = average mean velocity during one cycle, enddias V_{mean} = mean end-diastolic velocity)

Parameters measured in healthy breast parenchyma	PRE + POST HRT (*n* = 58)	POST (*n* = 54)	*p*-Value
RI_{min}	0.62 ± 0.07 0.50 – 0.78	0.68 ± 0.07 0.55 – 0.85	< 0.0001
RI_{mean}	0.65 ± 0.06 0.53 – 0.78	0.71 ± 0.06 0.59 – 0.85	< 0.0001
RI_{max}	0.67 ± 0.06 0.53 – 0.78	0.73 ± 0.06 0.59 – 0.85	< 0.0001
PI_{mean}	1.16 ± 0.24 0.75 – 1.76	1.42 ± 0.30 0.92 – 2.05	< 0.0001
S/D_{mean}	3.11 ± 0.68 2.14 – 5.40	3.83 ± 0.96 2.59 – 6.69	< 0.0001
peaksys V_{max}	0.12 ± 0.05 0.05 – 0.28	0.12 ± 0.04 0.06 – 0.28	0.54
peaksys V_{mean}	0.10 ± 0.04 0.05 – 0.24	0.11 ± 0.03 0.06 – 0.23	0.23
peaksys V_{min}	0.09 ± 0.04 0.04 – 0.19	0.09 ± 0.03 0.06 – 0.18	0.20
average V_{mean}	0.06 ± 0.03 0.02 – 0.15	0.06 ± 0.02 0.03 – 0.14	0.49
enddias V_{mean}	0.04 ± 0.02 0.01 – 0.10	0.03 ± 0.01 0.01 – 0.09	0.053

Blood Flow Parameters in the Healthy Tissue of Premenopausal and Postmenopausal Patients

As Table 39.**6** indicates, the group of postmenopausal patients shows significantly higher mean values for all resistance indices compared with the hormonally active patients (premenopausal and postmenopausal on HRT). The characteristic resistance indices in the healthy breast tissue of the different groups are shown in Figs. 39.**1**–39.**3**. A similar analysis of the different blood flow velocities shows almost equal values in both groups.

Blood Flow Parameters in the Tumor Area of Premenopausal and Postmenopausal Patients

The above results are also confirmed in the tumor area. As we see in Table 39.**7**, postmenopausal patients have significantly higher resistance indices than premenopausal patients (Figs. 39.**4**, 39.**5**). In the velocity measurements, which were performed without angle correction, hormone-related differences are definitely present but are neutralized by the broad overlaps in the ranges of values.

Discussion

Allowance for menopausal status. Our analysis demonstrates the statistically significant effect of menopausal status on blood flow parameters. Any attempt to improve benign–malignant tumor discrimination by an analysis of blood flow parameters should take into account menopausal status and hor-

Table 39.**7** Same comparison as in Table 39.**6**, calculated for the tumor area

Parameters measured in tumor area	PRE + POST HRT (*n* = 59)	POST (*n* = 55)	*p*-Wert
RI_{min}	0.60 ± 0.11 0.37 – 0.89	0.65 ± 0.10 0.42 – 0.81	0.012
RI_{mean}	0.66 ± 0.10 0.43 – 0.89	0.72 ± 0.07 0.50 – 0.85	0.0007
RI_{max}	0.72 ± 0.10 0.45 – 0.95	0.79 ± 0.08 0.50 – 0.96	< 0.0001
PI_{mean}	1.22 ± 0.43 0.54 – 3.08	1.47 ± 0.34 0.69 – 2.27	< 0.0001
S/D_{mean}	3.42 ± 1.53 1.8 – 10.94	4.18 ± 1.65 2.00 – 11.33	0.0002
peaksys V_{max}	0.17 ± 0.13 0.03 – 0.65	0.18 ± 0.12 0.04 – 0.54	0.31
peaksys V_{mean}	0.12 ± 0.07 0.03 – 0.34	0.13 ± 0.08 0.04 – 0.35	0.51
peaksys V_{min}	0.08 ± 0.04 0.03 – 0.19	0.09 ± 0.06 0.04 – 0.28	0.51
average V_{mean}	0.07 ± 0.04 0.01 – 0.21	0.07 ± 0.04 0.02 – 0.17	0.74
enddias V_{mean}	0.04 ± 0.03 0.01 – 0.15	0.04 ± 0.04 0.01 – 0.33	0.49

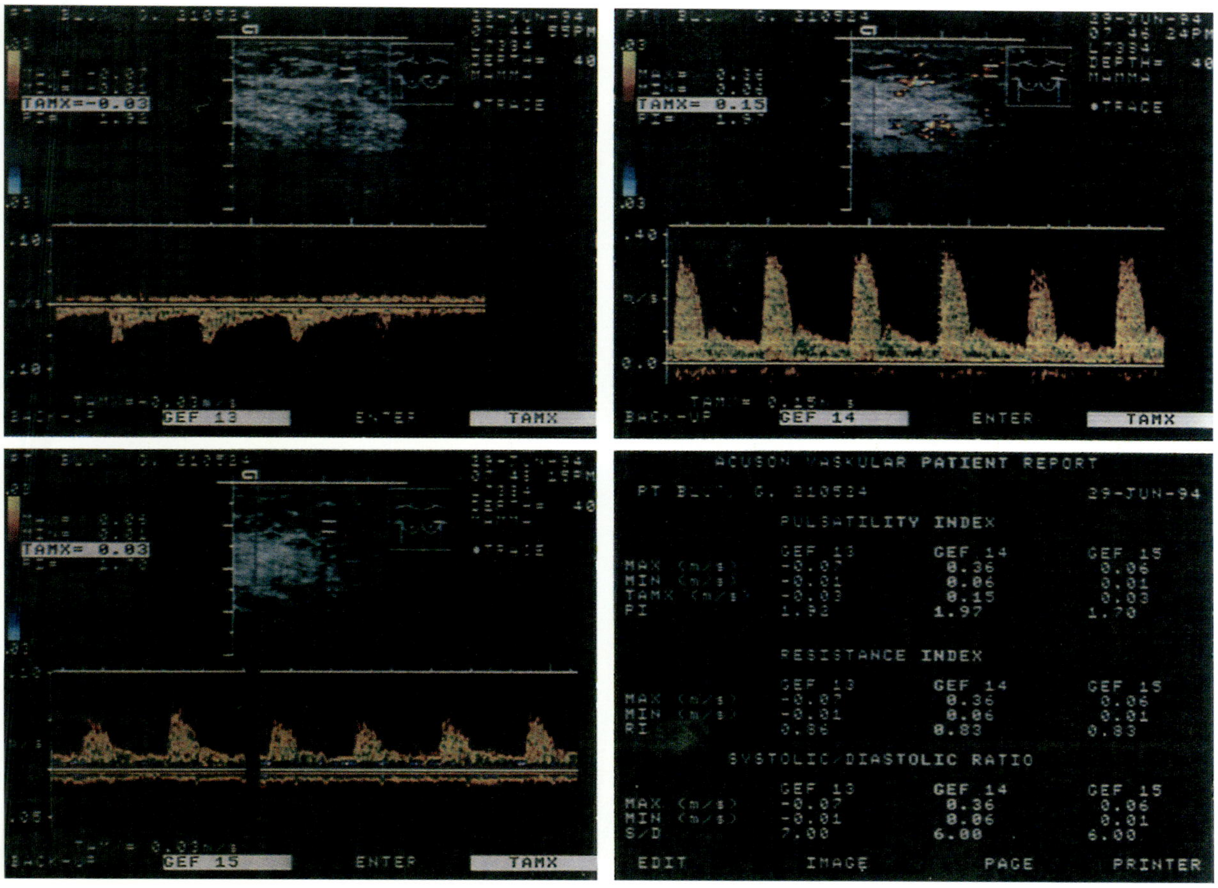

Fig. 39.**1** Very high RI values (0.83–0.86) are found in the healthy breast tissue of a postmenopausal patient not receiving HRT.

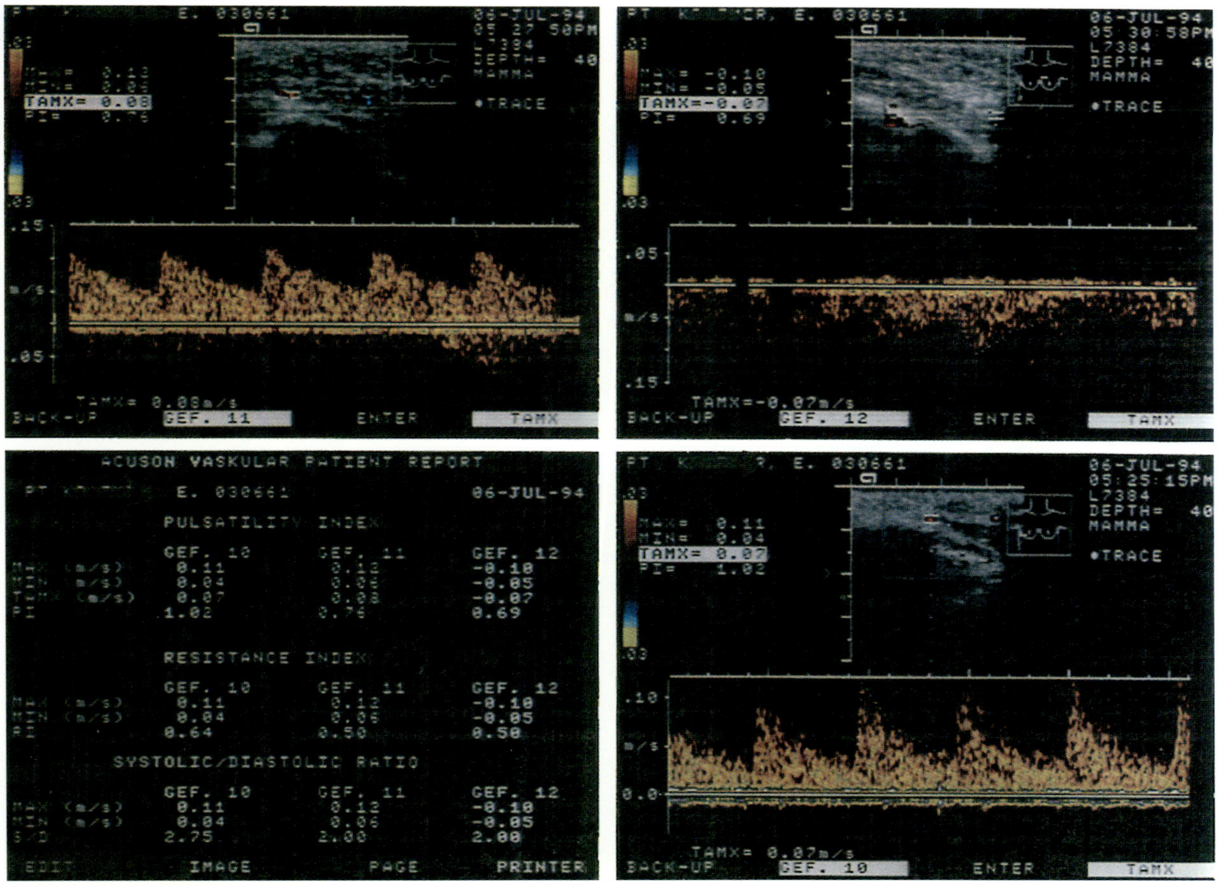

Fig. 39.**2** Low RI values (0.50–0.64) in a pre-menopausal patient.

Gynecological Ultrasound

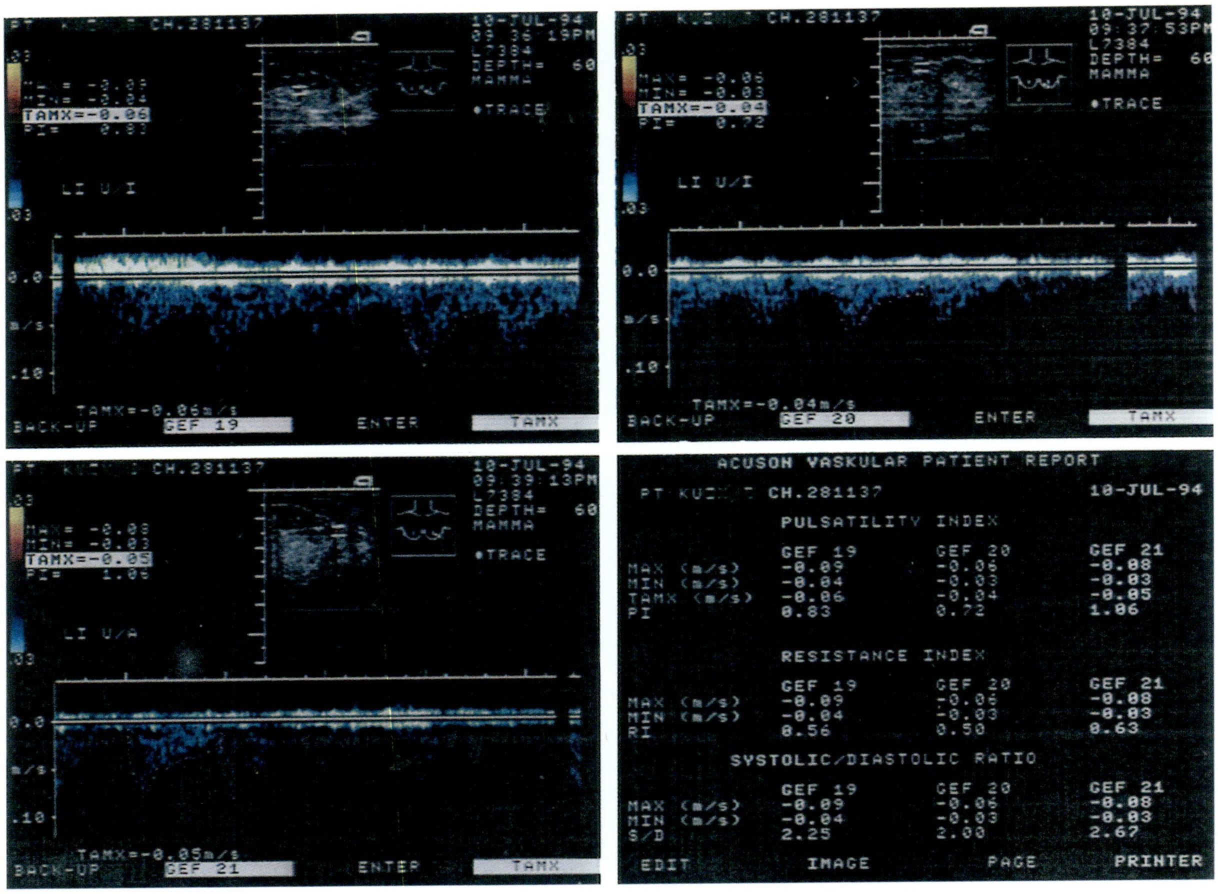

Fig. 39.**3** Low RI values (0.50–0.63) in a postmenopausal patient on HRT. These values occupy the same range as in premenopausal women.

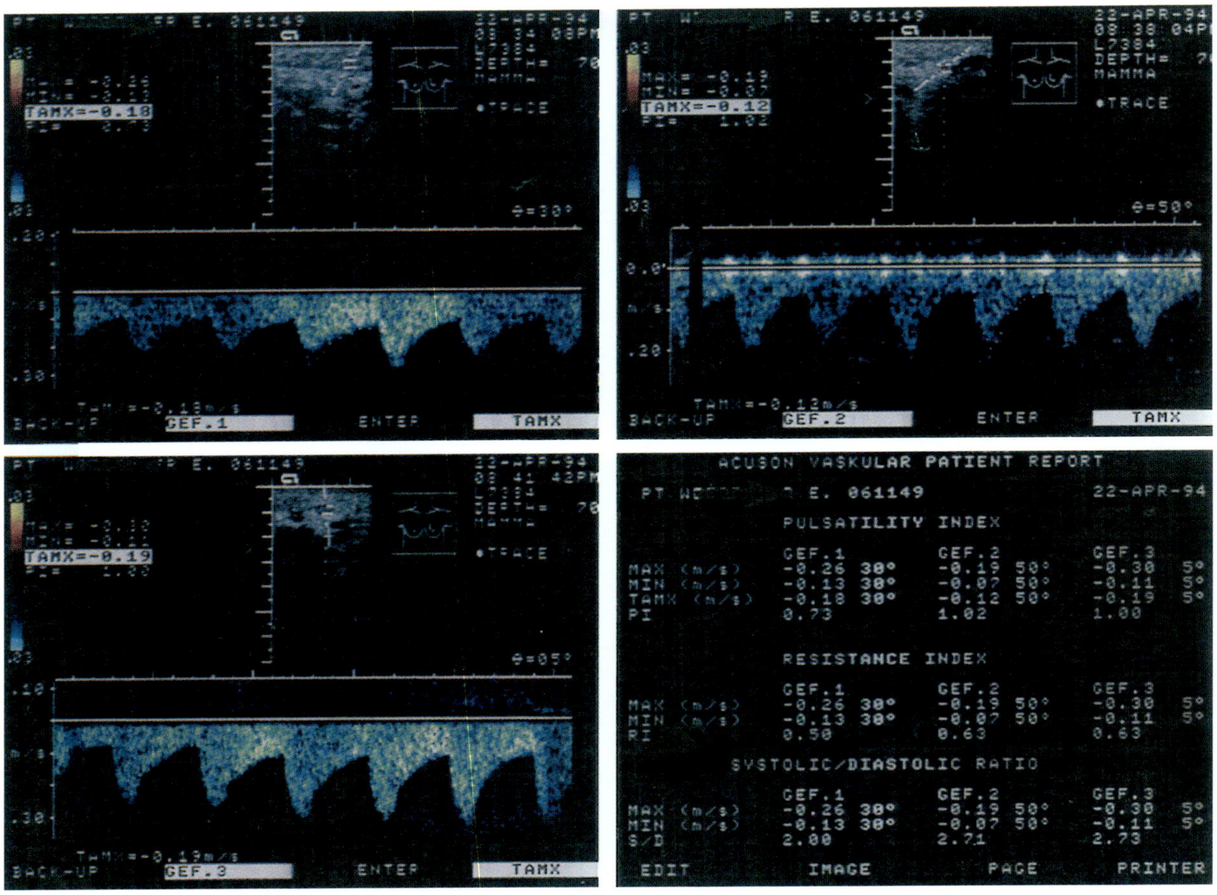

Fig. 39.**4** Low RI in a premenopausal patient with a breast carcinoma. The RI values range from 0.50 to 0.63.

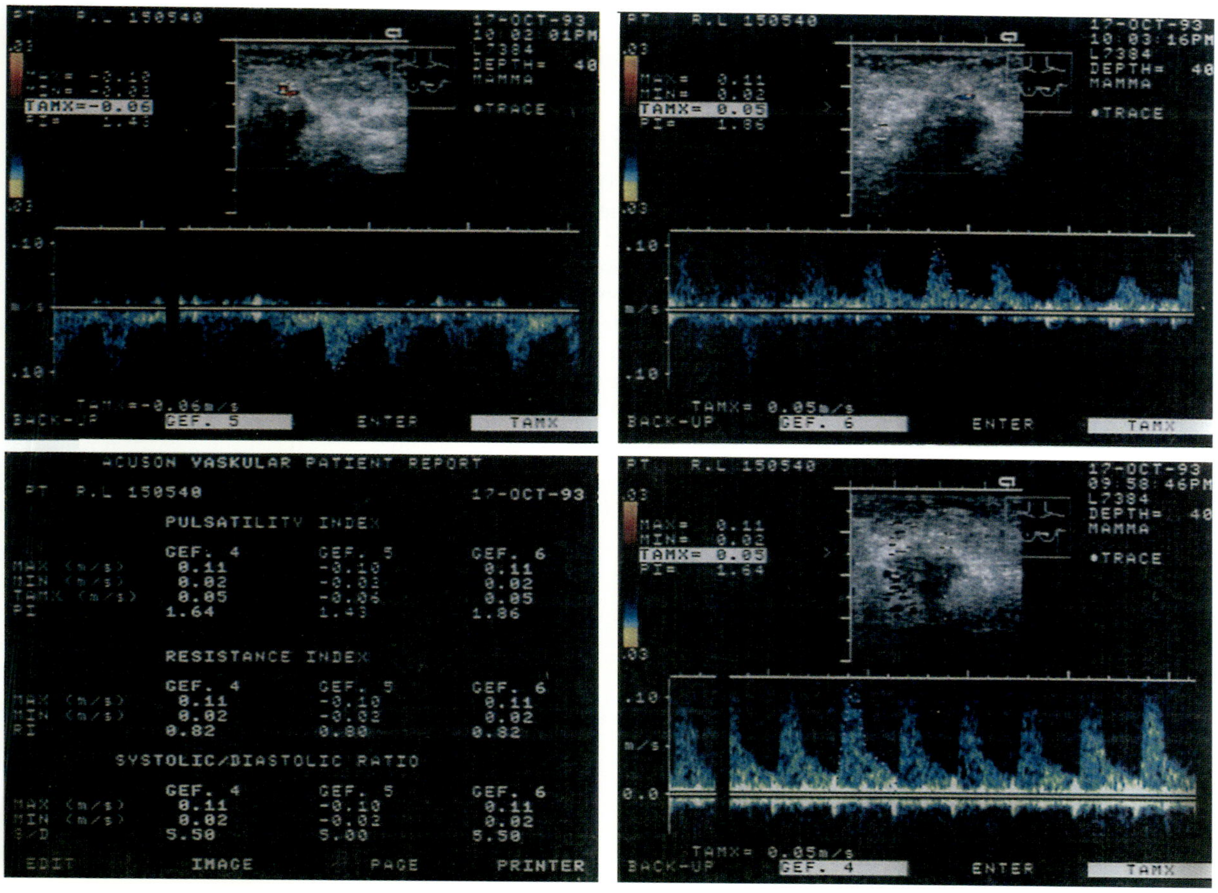

Fig. 39.**5** High RI values (0.80–0.82) are measured in the tumor area of a postmenopausal patient with breast cancer who was not on HRT.

mone replacement therapy. As Table 39.**1** indicates, there were twice as many postmenopausal women in the malignant tumor group as in the group with benign lesions. This fact alone could produce a statistically significant increase in impedance to flow, calling into question the discriminating potential of a blood flow parameter. The discovery that the composition of the groups being compared critically affects the blood flow parameters, could presumably account for many of the divergent and sometimes contradictory results that have been reported.

Significance of hormone replacement therapy. Marked differences in mean values are found between premenopausal patients, postmenopausal patients on HRT, and postmenopausal patients without HRT, both in healthy breast tissue and in breast tumors. It is clear that premenopausal patients do not differ significantly from postmenopausal patients on HRT, but both of these groups differ significantly from postmenopausal patients not receiving HRT. The differences are minor only in the tumor area. The reasons for this may relate to the number of patients or to a possible tumor-specific effect, which would be relevant in postmenopausal women. When the premenopausal patients and postmenopausal patients on HRT are placed in one group and compared with the postmenopausal patients not receiving HRT, the latter group is found to have significantly higher mean values for all resistance indices, as one would expect.

Angle correction in velocity measurements. This clear, systematic dependence is not seen in parameters that are based on the absolute measurement of blood flow velocities. These results show that absolute flow velocities can be used for intergroup comparisons only if they have been corrected for the insonation angle. As noted earlier, this was not done in the present study because of time constraints. Absolute velocity measurements that are not angle-corrected show a large scatter of values. The areas of overlap between different groups—in this case premenopausal and postmenopausal women—are very broad and neutralize the menopause-related difference. These results plainly show that absolute flow velocities without angle correction are not suitable for the comparison of two groups.

Exactly the same principles are found in the tumor area as in healthy breast parenchyma, underscoring the validity of the findings.

Summary

In summary, the effect of menopausal status can be objectively documented with Doppler parameters both in healthy breast tissue and in breast tumors. Postmenopausal women must be viewed differently depending on whether or not they are receiving HRT. Postmenopausal patients on HRT and premenopausal patients comprise a physiological unit. As a basic principle, only angle-independent parameters should be analyzed when two groups are compared. Menopausal status and HRT have a statistically significant effect on Doppler parameters. Menopausal status should be taken into account in studies dealing with the benign–malignant discrimination of breast tumors.

References

1 Minasian H, Bamber JC: A preliminary assessment of an ultrasonic doppler method for the study of blood flow in human breast cancer. Ultrasound Med. Biol. 8 (1982) 357–364

2 Sohn Ch, Stolz W, Grischke EM, Wallwiener D, Bastert G, von Fournier D: Die dopplersonographische Untersuchung von Mammatumoren mithilfe der Farbdopplersonographie, der Duplex-Sonographie und des CW-Dopplers. Zentralbl. Gynäkol. 114 (1992) 249–253

3 Villena-Heinsen C, Ertan AK, Tossounidis I, Holländer M, König J, Schmidt W: Diagnostische Aussagekraft der Farbdoppler-Sonographie bei Mammatumoren. Geburtshilfe. Frauenheilkd. 55 (1995) 541–547

40 Benign–Malignant Tumor Discrimination and Prognostic Evaluation of Breast Tumors with Color Doppler Sonography

C. Villena-Heinsen, A. K. Ertan, D. Mink, and W. Schmidt

Applications of Color Doppler Sonography in Breast Cancer

Technological evolution of color Doppler sonography. In most studies published on color Doppler sonography in patients with breast tumors, the goal has been to use this modality to help discriminate between benign and malignant lesions. Very few publications have dealt with the use of color Doppler in making a prognosis or evaluating response to treatment. Palpation, mammography, and sonography have become the pillars of breast diagnosis. There is still a need, however, to improve sensitivity and specificity through the use of new methods. Today, color Doppler sonography has reached a relatively high level of sophistication that offers decisive advantages. It can easily be incorporated into established preoperative diagnostic protocols. Tumors can be characterized and measured with conventional B-mode imaging, and color Doppler can be added both, to provide a qualitative color display of blood flow and to allow spectra to be sampled from selected vessels. This provides an opportunity for objective flowmetry based on the measurement of blood flow velocities and the calculation of resistance indices.

Blood flow parameters. In the past, resistance indices have been the most commonly used parameters for the benign–malignant discrimination of tumors. But as the preceding chapters have made clear, two important aspects must be considered when resistance indices are used: first, the blood flow parameter used for comparison should be a mean value and not an extreme value (e.g., the maximum or minimum flow resistance) within a tumor; and second, it is important to consider the menopausal status and the use or nonuse of hormone replacement therapy by the patients. The potential of color Doppler sonography in the benign–malignant differentiation and prognostic assessment of breast tumors should be evaluated exclusively by an analysis that takes these criteria into account.

Specificity of benign–malignant discrimination. The color Doppler systems available today are not sensitive enough to detect lesions by an abnormal flow pattern that are not detectable by conventional imaging. The goal of color Doppler imaging, rather, is to improve the specificity of benign–malignant discrimination.

Prognostic evaluation. There is also a need for new parameters in prognostic evaluation, for despite the availability of various prognostic indicators, the future course of a disease after primary treatment remains uncertain in many cases. The relative uncertainty of prognostic assessment is illustrated by the fact that approximately 30% of patients with a prognostically favorable, node-negative breast cancer will experience a progression of disease within 10 years after primary treatment [5,11]. Blood flow can provide information on the metabolism and proliferative behavior of tumors. Weidner et al.[17] documented a positive correlation between the density of neovascularity in breast carcinomas and the risk of distant metastases.

To date there have been only isolated reports on the evaluation of tumor prognosis with Doppler ultrasound[1,8,13]. Madjar et al.[8] found that heavily vascularized tumors had a strong propensity for recurrence, metastasis, and early mortality. Tumor blood flow also correlated with lymph-node and negative hormone-receptor status, but it did not correlate with tumor size or histopathological grade. So far only Cosgrove et al.[3] and Delorme et al.[4] have reported on the vascularity assessment of breast tumors with color Doppler ultrasound. Cosgrove et al.[3] created a special semiquantitative scoring system to assess tumor vascularity and found no correlation between color Doppler scores and conventional prognostic indicators (lymph node status, survival). Delorme et al.[4] found only a weak correlation between the maximum systolic flow velocity in the tumor-feeding vessels and the tumor volume.

Authors' Studies

Methods

The following analyses were performed to evaluate the benign–malignant discrimination and prognostic assessment of breast tumors by color Doppler sonography. The patients consisted of 114 women with a sonographically detectable focal breast lesion who underwent a preoperative Doppler ultrasound examination.

RI_{mean}. We defined and sampled as many blood flow signals from the tumor area as possible. The resistance index RI was calculated from each of the recorded waveforms. All of the RI values were used to calculated an average RI value (RI_{mean}) for the tumor area. Next the remaining healthy breast tissue, divided into four quadrants per breast, was examined with color Doppler ultrasound. RI_{mean} was computed for each healthy breast quadrant and was used as a basis for statistical comparison. Healthy breast tissue was represented by averaging the eight values for the quadrants. Each examination followed a standard protocol and took from 40 to 120 minutes to complete (average of 90 minutes). Premenopausal patients and postmenopausal patients who were receiving hormone replacement therapy (HRT) were grouped together as a physiological unit and compared with postmenopausal patients who were not receiving HRT.

We analyzed the effects of tumor behavior and menopausal status on the RI_{mean}, the relationship between tumor area and healthy tissue, and the relationship between RI_{mean} in the tumor area and established prognostic indicators.

Equipment. The examinations were performed with an Acuson model 128 XP10 system using a 7 MHz linear transducer and a 5 MHz color Doppler transducer. The standard settings were adjusted for small volumes and slow flows to achieve a maximum yield of flow signals.

Results

Flow Detection

The technology used in this study proved to be very sensitive. Blood vessels could be visualized and sampled with color Doppler in all of the 114 patients examined. There were only three cases (0.17%) in which diastolic flow could not be measured. These waveforms were disregarded in our analysis. Between one and nine vascular segments (3.3 ± 1.8 SD) could be defined in all 114 tumor areas. An average of 1.5 vascular segments were examined in each of the eight quadrants. There were 16 cases in which flow detection was unsuccessful in one quadrant or another; a resistance index was not calculated for this small percentage (1.8%) of the quadrants examined.

Histology

Histological evaluation of the 114 specimens revealed 63 carcinomas and 51 benign lesions. The carcinomas consisted of 57 invasive ductal carcinomas, four invasive lobular carcinomas, and two medullary carcinomas. Among the benign lesions were 25 cases with fibrocystic change, 14 with fibroadenomas, three with chronic mastitis, and nine with other lesions (lipomas, papillomas, fibroadenolipoma, fat necrosis, fibroma).

Patients

Age. The average patient age was 55 years (21–87 years). The average age of patients with malignant tumors was 60 years (33–87 years), and that of patients with benign lesions was 49 years (21–84 years). The groups differed significantly from each other with regard to age ($p = 0.0001$).

Menopausal status. A total of 35 patients were premenopausal, 24 were postmenopausal and receiving HRT, and 55 were postmenopausal and not receiving HRT. The groups with malignant and benign lesions differed in their menopausal status. The group with benign tumors were 49% premenopausal ($n = 25$), 22% postmenopausal with HRT ($n = 11$), and 29% postmenopausal without HRT ($n = 15$). The distribution was markedly different in the malignant-tumor group, in which 16% of the women were premenopausal ($n = 10$), 21% were postmenopausal with HRT ($n = 13$), and 63% were postmenopausal without HRT ($n = 40$).

RI_{mean} in the Tumor Area for Malignant and Benign Tumors

When the patients were analyzed by tumor behavior without regard to menopausal status, the RI_{mean} in malignant tumors (0.71 ± 0.09 SD) was significantly higher than in benign tumors (0.66 ± 0.09 SD, $p = 0.003$). Among the premenopausal women, the group with 36 benign tumors had an RI_{mean} of 0.64 ± 0.09 SD and the group with 23 malignant tumors had an RI_{mean} of 0.69 ± 0.10 SD. This difference was not statistically significant. The following mean values were calculated for the postmenopausal patients: 0.71 ± 0.06 SD for the 15 benign cases and 0.72 ± 0.08 SD for the 40 malignant cases. Again, this difference was not statistically significant. Figures 40.**1** and 40.**2** illustrate typical cases with a high RI in a carcinoma and a low RI in a fibroadenoma. Figures 40.**3** and 40.**4** show cases in which exactly the opposite was observed.

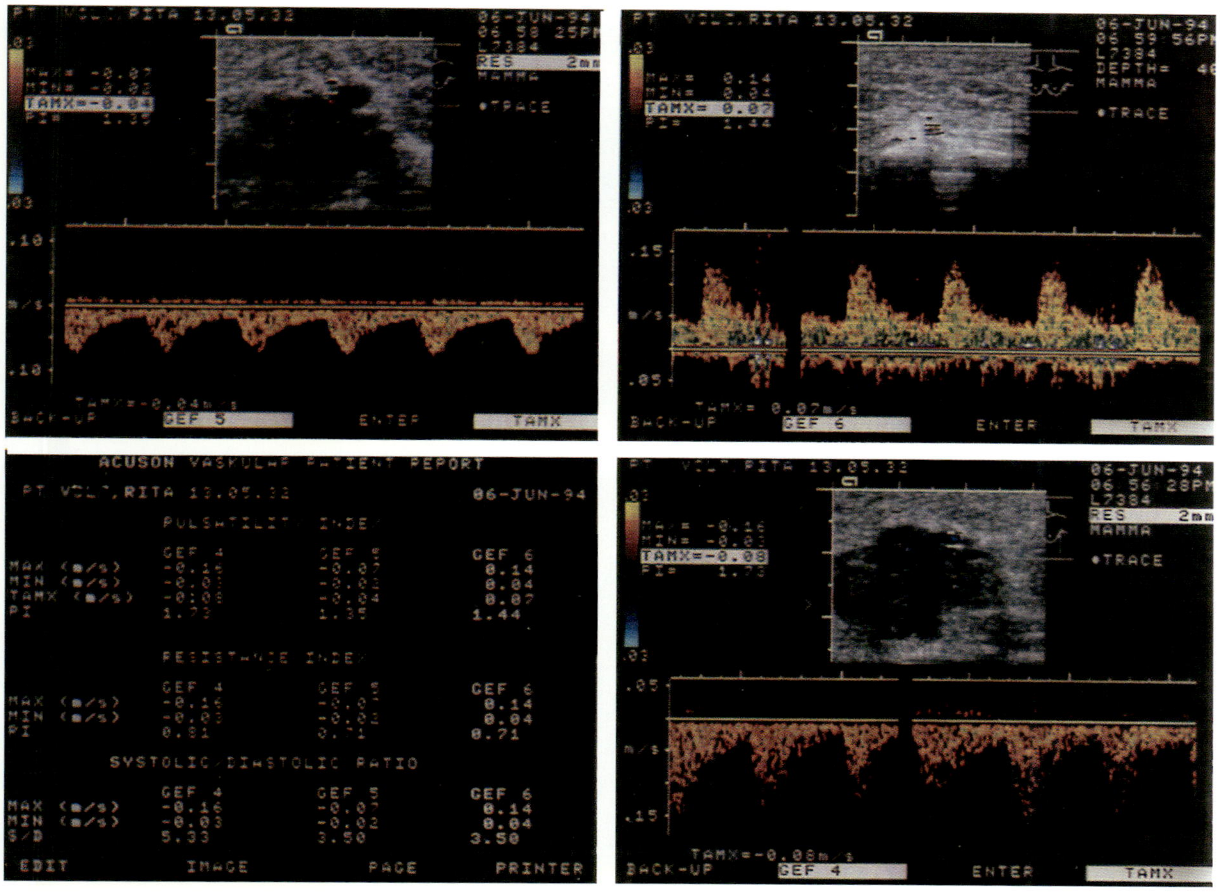

Fig. 40.**1** Breast carcinoma with a high RI (0.81 in vessel 4, 0.71 in vessel 5) in a postmenopausal patient.

a

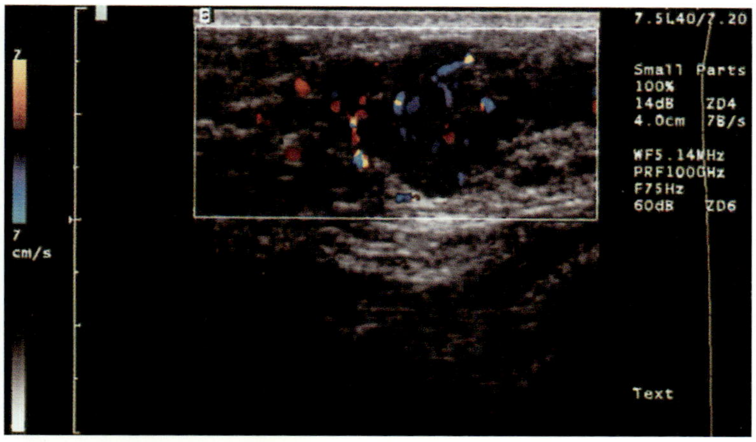

b

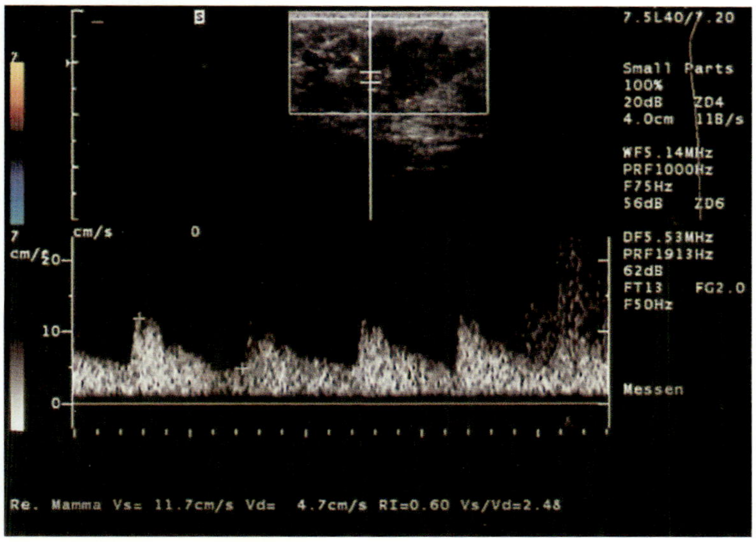

RI_{mean} in the Tumor Area Compared with the Mean Value for All Quadrants with Healthy Breast Tissue

The premenopausal cases with benign lesions had an RI_{mean} of 0.64 ± 0.09 SD in the tumor area. The mean value for all eight quadrants was 0.65 ± 0.07 SD ($p = 0.29$). The premenopausal cases with malignant tumors had an RI_{mean} of 0.69 ± 0.10 SD in the tumor area and an RI_{mean} of 0.65 ± 0.07 SD averaged over the eight quadrants. This difference was statistically significant ($p = 0.02$).

In the postmenopausal cases, the RI_{mean} values in the tumor area of benign and malignant cases were generally higher than the mean values in healthy breast tissue, but the difference was not statistically significant (benign tumors: 0.71 ± 0.06 SD vs. 0.69 ± 0.05 SD; malignant tumors: 0.72 ± 0.08 SD vs. 0.71 ± 0.07 SD). The box plot in Fig. 40.**5** shows the median values for the RI_{mean} in the tumor area and in healthy tissue for all the subgroups. Statistical p-values are also shown to indicate the differences between groups, broken down by menopausal status.

◁ Fig. 40.**2** Well-vascularised fibroadenoma in a premenopausal patient.
a Color Doppler image.
b Spectrum sampled from a vessel within the fibroadenoma, indicating a low RI of 0.60.

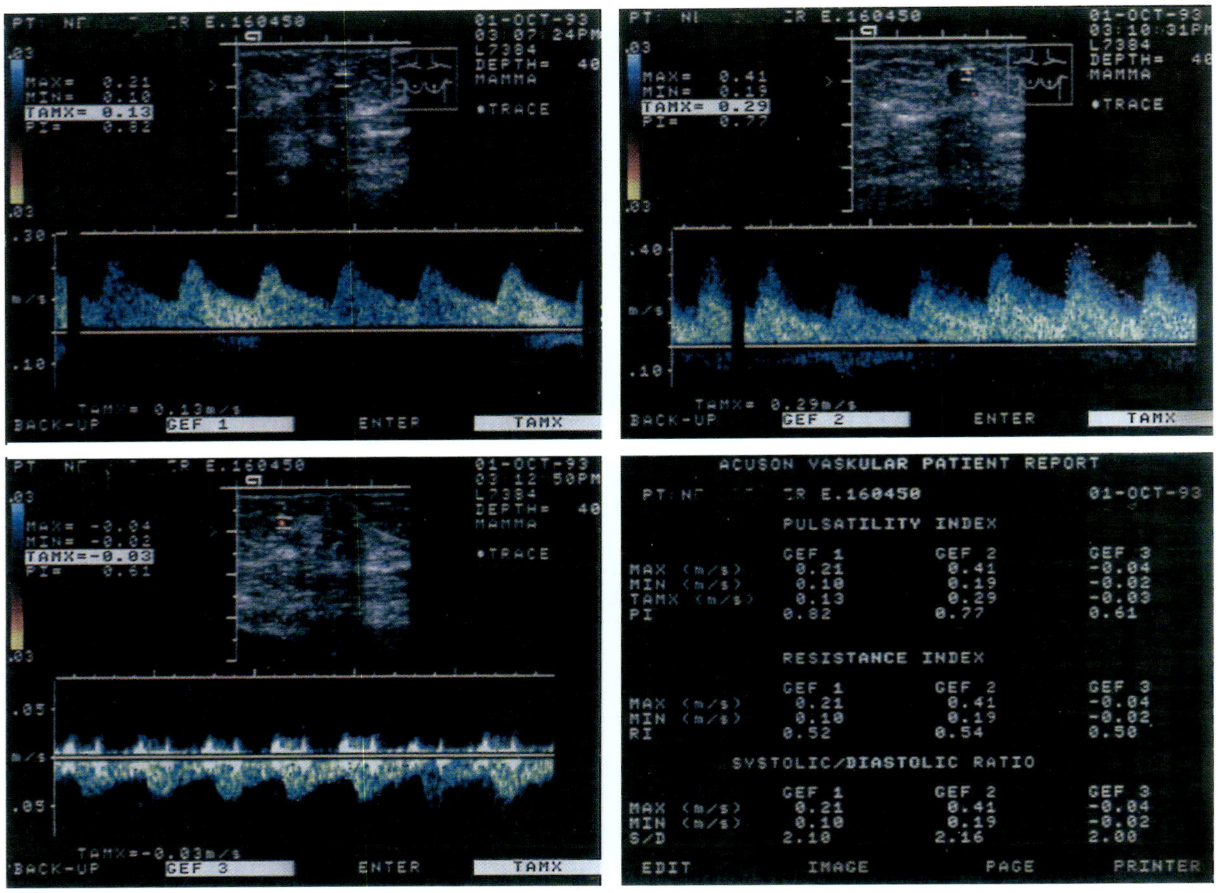

Fig. 40.**3** Breast carcinoma in a premenopausal patient shows a low RI ranging from 0.50 to 0.54.

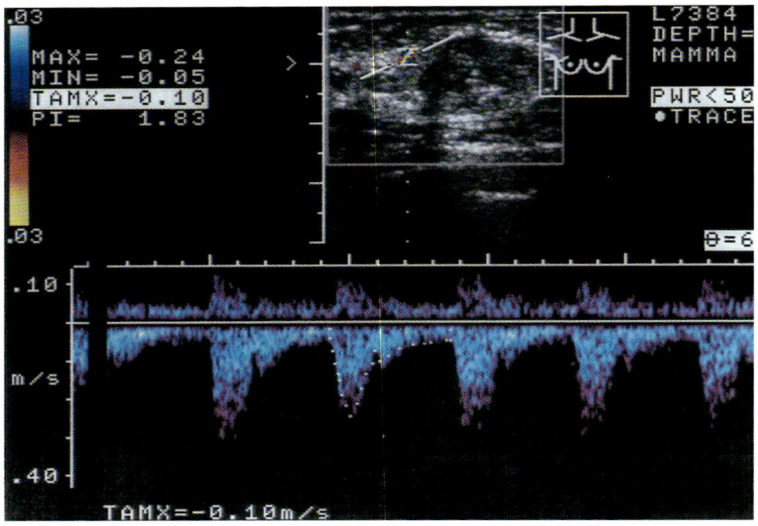

Fig. 40.**4** Fibroadenoma in a premenopausal patient shows a high RI of 0.79.

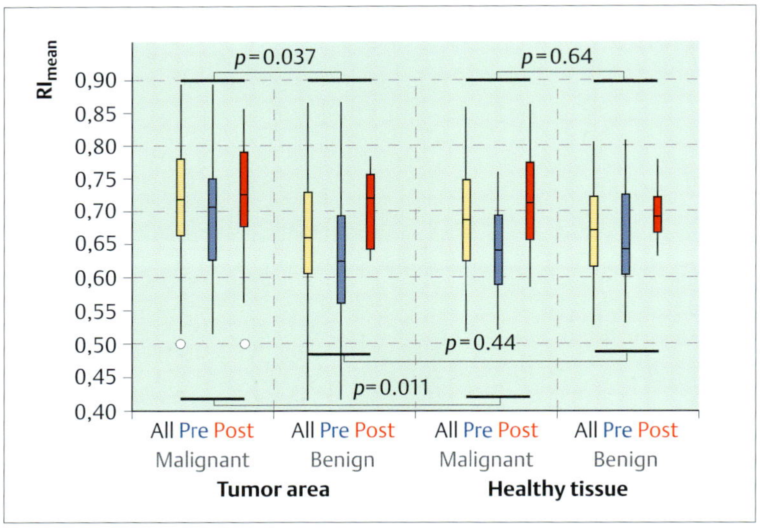

Fig. 40.**5** Box plot diagram of the RI_{mean} values in the patient groups. The values were plotted for the tumor area and "healthy" tissue (mean value for all quadrants), tumor behavior (malignant/benign), and menopausal status (premenopausal/postmenopausal). The boxes cover the values between the 25th and 75th percentiles. The linear extensions from the rectangles indicate the scatter of values. The median values are indicated by a horizontal line. Two "stray points" are also shown. The *p*-values indicate the differences between groups, broken down by menopausal status.

Correlation between the Mean Value for All "Healthy" Quadrants and Patient Age

No correlation was found in the premenopausal patients. The Spearman correlation coefficient was $r = 0.08$ ($p = 0.56$).

In the postmenopausal group, however, a positive correlation was found between the mean value for the healthy breast tissue and patient age. The Spearman correlation coefficient was $r = 0.58$ ($p < 0.0001$). These relationships are shown in Fig. 40.**6**.

Correlation between Tumor Area RI and Menopausal Status

When the cases were analyzed by menopausal status, no statistically significant correlations were found in the 59 premenopausal patients ($r = 0.16$, $p = 0.24$) or the 55 postmenopausal patients ($r = 0.22$, $p = 0.10$). The age structure was as follows: premenopausal patients with benign lesions, 44.3 years ± 15.2 SD with a range of 21–84 years; premenopausal patients with malignant tumors, 50.6 years ± 7.7 SD (33–68 years); postmenopausal patients with benign lesions, 61.4 years ± 7.5 SD (47–70 years); postmenopausal patients with malignant tumors, 65.9 years ± 10.2 SD (43–87 years).

Dependence of RI_{mean} in the Tumor Area on Various Prognostic Factors in 58 Malignant Tumors

Tumor stage and lymph node status. Two groups were formed for comparing the RI_{mean} with the histopathological tumor stage (pT). Comparison of pT1 with pT2, pT3, and pT4 showed no statistically significant difference ($p = 0.10$). The RI_{mean} in the 28 patients with a pT1 tumor was 0.69 ± 0.10 SD (0.50–0.89), compared with 0.73 ± 0.07 SD (0.57–0.85) in the 34 patients with a stage pT2, pT3, or pT4 tumor.

Thirty-four patients had negative axillary lymph nodes and 24 had positive nodes. The RI_{mean} in the tumor area showed no statistically significant difference between these two groups ($p = 0.77$).

Hormone receptor status and tumor grade. Hormone receptor status was classified as positive if the estrogen-receptor or progesterone-receptor content was greater than 20 fmol/mg protein. Forty-five cases had a positive hormone receptor status and 14 had a negative receptor status. The RI_{mean} in the tumor area was not significantly different between the two groups ($p = 0.40$).

For the analysis of histological tumor grade, grades I and II were pooled ($n = 39$) and compared with the grade III group ($n = 19$). Again, no significant difference in RI_{mean} was found ($p = 0.37$).

Tumor diameter and number of positive axillary lymph nodes. A Spearman correlation coefficient of $r = 0.21$ ($p = 0.12$) was found between the maximum diameter of the 58 carcinomas and the RI_{mean} in the tumor area.

The correlation between maximum tumor diameter and the number of positive axillary lymph nodes was $r = 0.06$ ($p = 0.67$). The percentage of immunohistochemically stained Ki-67-positive nuclei ($n = 34$) showed a correlation of $r = -0.23$ ($p = 0.19$)

S-phase fraction. Thirty patients were tested for S-phase fraction (expressed in %). The mean S-phase fraction was 8.85% ± 5.4 SD (minimum 3.2%, maximum 27.5%). A highly significant inverse correlation was found between the RI_{mean} in the tumor area and the S-phase fraction: $r = -0.53$ ($p = 0.003$) (Fig. 40.**7**). This correlation is documented in Figs. 40.**8** and 40.**9**.

The above correlations between RI_{mean} in the tumor area and prognostic indicators are reviewed in Table 40.**1**.

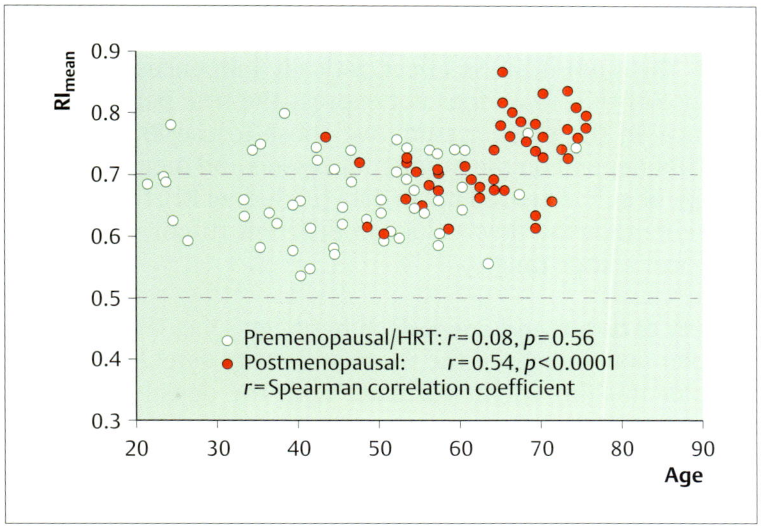

Fig. 40.**6** Relationship between the mean value of RI_{mean} in healthy breast tissue and patient age, broken down by menopausal status.

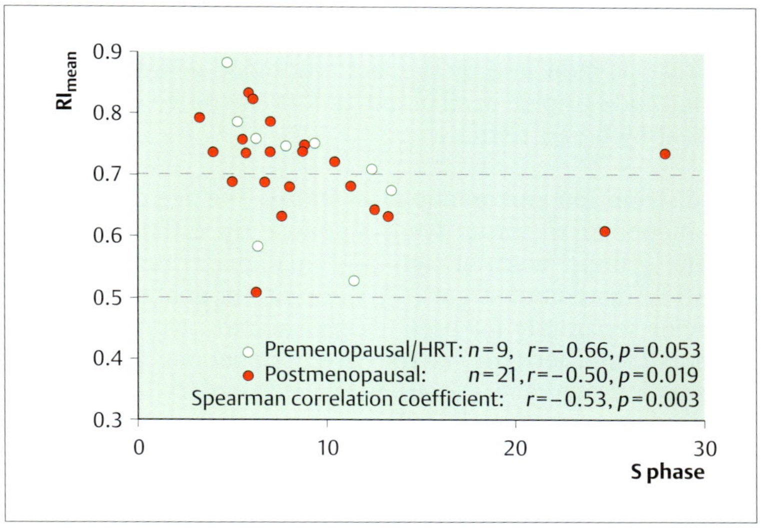

Fig. 40.**7** Distribution of RI_{mean} in the tumor area and correlation with the S-phase fraction.

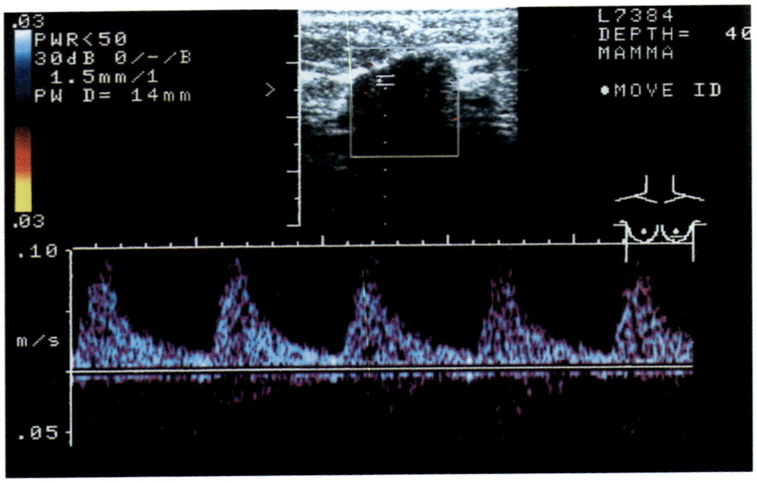

Fig. 40.8 Breast carcinoma 17 mm in diameter in a postmenopausal woman. A total of two vessels were sampled. The RI_{mean} was high, at 0.80, and the S-phase fraction was low, at 3%.

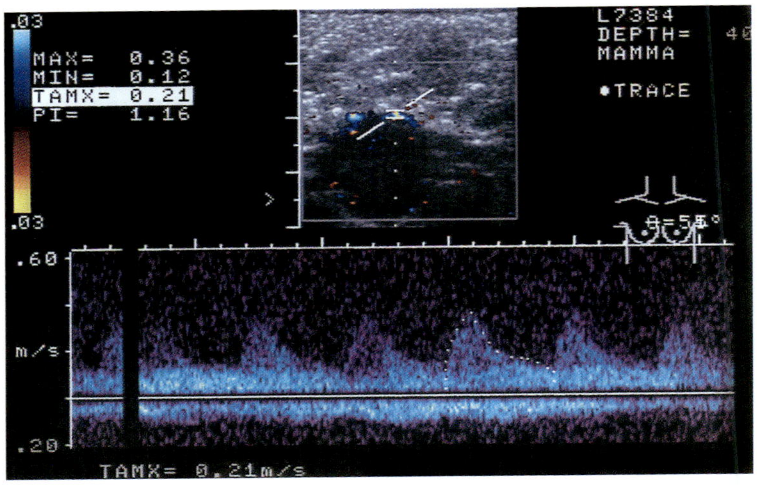

Fig. 40.9 Breast carcinoma 18 mm in diameter in a postmenopausal woman. A total of five vessels were sampled. The RI_{mean} was 0.60 (relatively low), and the S-phase fraction was 24% (extremely high).

Table 40.1 Correlation between RI_{mean} in the carcinoma and prognostic indicators. A highly significant inverse correlation exists with the S-phase fraction

Prognostic indicators	*p*-Value
pT stage (pT1 vs. pT2, 3, 4)	0.10
Maximum tumor diameter	0.12
Lymph node status (N– vs. N+)	0.77
Lymph nodes involved (number)	0.67
Hormone receptor status (+ vs. –)	0.40
Grade (G1 and G2 vs. G3)	0.37
Ki 67	0.19
Ploidy	0.15
S-phase fraction	0.003

Discussion

The Doppler technology used in the present study is extremely sensitive, as indicated by the fact that blood vessels could be defined and sampled in all of the tumors examined and in all of the healthy breast tissue.

Age distribution and menopausal status. The age distribution and menopausal status were significantly different in the patient groups with benign and malignant tumors ($p < 0.0001$). This demonstrates the necessity of analyzing tumor behavior separately for premenopausal and postmenopausal patients. Another key factor is postmenopausal HRT, as patients receiving this therapy have Doppler findings like those in premenopausal patients rather than in other postmenopausal women[12, 15].

Vascular resistance rises substantially in response to the hormone withdrawal that occurs at menopause and, to a lesser degree, in response to the spread of a malignant tumor. In this respect, malignant breast tumors behave differently[9, 12, 14] from ovarian and trophoblastic tumors, for example, which are associated with a decrease in vascular resistance[6].

An intraindividual comparison of vascular resistance between the tumor area (0.69 ± 0.10 SD) and healthy breast tissue (0.65 ± 0.07 SD) shows a statistically significant difference ($p = 0.02$) only in premenopausal patients with a malignant tumor. But while the mean values are significantly different, there is a broad overlap in the total range of values, making the values difficult to evaluate in any given case and calling into question the clinical usefulness of this examination method. The vascular resistance in healthy breast tissue rises with aging, but only in patients not receiving HRT. The tumor area (regardless of whether the lesion is benign or malignant) is subject to different principles.

RI_{min} and RI_{mean}. Our previous study results cast doubt on the value of RI_{min} in the benign–malignant discrimination of breast lesions, because this parameter depends significantly on the number of vascular segments that are defined and sampled[14, 16].

The likelihood of determining the lowest possible value for RI_{min} increases with the number of RI values that are calculated. This minimization effect critically influences the result. Because of this, RI_{min} does not express the true peripheral resistance. In the present study we chose to use RI_{mean}, which does not depend on the number of vessels and measurements. Madjar et al.[10] also used mean data (including RI), their rationale being that the flow values in different tumor vessels can vary over a large range.

Density of tumor angiogenesis. It has been speculated that the Doppler flow indices or the vascular density detected by color Doppler imaging might correlate with the density of tumor vessels determined microscopically. So far, however, there is no proof that current ultrasound technology can detect the capillaries that are formed by tumor angiogenesis[12, 14]. It has been shown that color Doppler sonography can usually detect portions of the vascular network that lead to or away from the tumor and can occasionally detect intratumoral vessels[7, 12, 14].

The most that can be reliably accomplished with color Doppler sonography is an indirect assessment based on the blood flow patterns that are associated with tumor angiogenesis. Lagalla et al.[7] correlated the detection or nondetection of color Doppler signals in 22 breast cancer cases with the angiogenic potential of the lesion as determined in histological section using a scoring system (MAGS = microscopic angiogenesis grading system) based on the vessel count, hyperplasia and mitosis rate of endothelial cells. Building on the discoveries of Weidner et al.[17], who showed that the density of tumor angiogenesis in 49 breast cancer patients correlated with an increased rate of distant metastases, Lagalla et al.[7] expected to find a correlation between angiogenic potential and signal detection or signal intensity. However, the highest angiogenic potential (score > 30) was observed in the four tumors that did not have detectable color Doppler signals, while a score < 30 was found in 17 of 18 tumors with positive flow detection.

Caliber of tumor supply vessels. Microscopically, the detection of flow signals correlates with blood vessels > 1 mm in diameter. This means that color flow detection correlates with the caliber of the tumor supply vessels and not with the density of tumor angiogenesis. Moreover, the absence of flow signals is not a reliable criterion for excluding a malignant tumor[7].

Asymmetry between the tumor and contralateral breast. Several groups of authors[2,9] have noted the importance of asymmetry between the tumor and the contralateral "healthy" breast. Our comparison of the tumor area with healthy breast parenchyma showed a statistically significant difference only in premenopausal patients with malignant tumors ($p = 0.02$). Contrary to expectations, the peripheral resistance in the tumor area was higher than in the "healthy" breast tissue. We consider that this further supports the notion that detection of the capillary network (a low-resistance system) is beyond the capabilities of present-day Doppler ultrasound.

The observation that only postmenopausal patients showed a significant rise of RI_{mean} in the healthy breast tissue with aging underscores the importance of this influence on physiological blood flow. The situation in the tumor area is somewhat different, suggesting that additional factors in that region, such as increased diastolic blood flow, may weaken this physiological effect.

S-phase fraction. Among all the prognostic indicators studied, the S-phase fraction determined by flow cytometry showed a definite correlation with RI_{mean} ($r = -0.53$, $p = 0.003$). This observation suggests that tumor cell kinetics are influenced by tumor blood flow or that spectral Doppler sonography is already sensitive enough to allow a prognostic evaluation.

Conclusion

The present study results indicate that the RI_{mean} reflects the actual peripheral vascular resistance of tumor blood flow. Benign–malignant discrimination cannot be accomplished with this parameter alone, however. Statistically significant differences between the tumor area and "healthy" breast tissue were found only in premenopausal patients with malignant tumors ($p = 0.02$). Because menopausal status significantly affects blood flow, an age dependence of RI_{mean} is observed in "healthy" breast tissue. Of all the prognostic factors investigated, the S-phase fraction, which describes cellular kinetics, showed an inverse correlation with the RI_{mean} in the tumor area. High cellular kinetics are associated with a marked increase in diastolic blood flow and with a low RI in the detectable tumor vessels. This suggests that spectral Doppler sonography may be of some value in evaluating the prognosis of a malignant tumor.

References

1 Belcaro G, Laurora G, Ricci A, Cianchetti E, Legnini M, Napolitano AM: Evaluation of flow in nodular tumours of the breast by Doppler and duplex scanning. Acta Chir. Belg. 88 (1988) 323–327

2 Burns PN, Halliwell M, Wells PNT, Webb AJ: Ultrasonic Doppler studies of the breast. Ultrasound Med. Biol. 8 (1982) 127–143

3 Cosgrove DO, Kedar RP, Bamber JC et al.: Breast Diseases: Color Doppler US in Differential Diagnosis. Radiology 189 (1993) 99–104

4 Delorme S, Anton HW, Knopp MV et al.: Breast Cancer: Assessment of vascularity by color Doppler. Eur. Radiol. 3 (1993) 253–257

5 Early Breast Cancer Trialists Collaborative Group: Systemic treatment of early breast cancer by hormonal, cytotoxic oder immune therapy. The Lancet 339 (1992) 1–15, 72–85

6 Kurjak A, Shalan H, Kupesic S et al.: Transvaginal color Doppler sonography in the assessment of pelvic tumor vascularity. Ultrasound Obstet. Gynecol. 3 (1993) 137–154

7 Lagalla R, Caruso G, Marasa L, D'Angelo I, Cardinale AE: Capacità angiogenetica delle neoplasie mammarie e correlazione con le semeiotica color Doppler. Radiol. Med. 88 (1994) 392–395

8 Madjar H, Giese E, Schillinger H: Durchblutungsmessungen an Mammatumoren. Vergleich mit Prognosefaktoren. Arch. Gynecol. Obstet. 245 (1989) 697–698

9 Madjar H: Breast examinations with continuous wave and color Doppler. Ultrasound Obstet. Gynecol. 2 (1992) 215–220

10 Madjar H, Prömpeler H, Wolfahrt R, Bauknecht T, Pfleiderer A: Farbdopplerflußdaten von Mammatumoren. Ultraschall Med. 15 (1994) 69–76

11 Villena-Heinsen C, Mink D, Kreienberg R, Schmidt W: Die Problematik der adjuvanten Therapie bei nodalnegativen Mammakarzinompatientinnen. Akt. Onkol. 66 (1992) 33–44

12 Villena-Heinsen C, Ertan AK, Tossounidis I, Holländer M, König J, Schmidt W: Diagnostische Aussagekraft der Farb-Doppler-Sonographie bei Mammatumoren. Geburtsh. u. Frauenheilk. 55 (1995) 541–547

13 Villena-Heinsen C, Mink D, Tossounidis I et al.: Ist eine Prognoseeinschätzung beim Mammakarzinom mittels farbkodierter und Spektral-Doppler-Sonographie möglich? Tumordiagn. u. Ther. 16 (1995) 187–193
14 Villena-Heinsen C, Mink D, Ertan AK, Holländer M, Schmidt W: Bewertung der Aussagekraft der Farb- und Spektraldopplersonographie bei Brusttumoren. In Schmidt W (ed.): Jahrbuch der Gynäkologie und Geburtshilfe (1995/96). Biermann, Zülpich 1996, 121–136
15 Villena-Heinsen C, Alexander C, Tossounidis I et al.: Influence of Menopausal State on Colour Doppler Flow Parameters of Breast Tumours and healthy mammary Tissue. Eur. J. Ultrasound 6 (1997) 49–52
16 Villena-Heinsen C, König J, von Tongelen B et al.: Validity of the minimal Resistance Index for discrimination between benign and malignant Breast Tumours. Eur. J. Ultrasound 7 (1998) 189–193
17 Weidner NR, Semple IP, Welch WR: Tumor angiogenesis and metastasis: Correlation in invasive breast carcinoma. New Engl. J. Med. 324 (1991) 1–8

Index

G

H

Q

R

S